Current Cancer Research

Series Editor
Wafik El-Deiry

More information about this series at http://www.springer.com/series/7892

Barbara Burtness • Erica A. Golemis
Editors

Molecular Determinants of Head and Neck Cancer

Second Edition

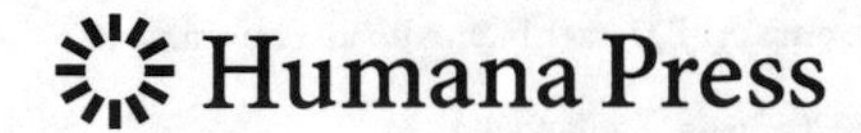
Humana Press

Editors
Barbara Burtness
Department of Internal Medicine
Yale University School of Medicine
New Haven, CT, USA

Erica A. Golemis
Molecular Therapeutics Program
Fox Chase Cancer Center
Philadelphia, PA, USA

ISSN 2199-2584 ISSN 2199-2592 (electronic)
Current Cancer Research
ISBN 978-3-030-08779-1 ISBN 978-3-319-78762-6 (eBook)
https://doi.org/10.1007/978-3-319-78762-6

Printed on acid-free paper

This Humana Press imprint is published by the registered company Springer International Publishing AG part of Springer Nature.
The registered company address is: Gewerbestrasse 11, 6330 Cham, Switzerland

Contents

Chapter 1
Overview: The Pathobiology of Head and Neck Cancer

Barbara Burtness and Erica A. Golemis

Squamous cell cancers arising in the head and neck, from the nasopharynx to the subglottic larynx, are frequently devastating cancers that afflict patients around the world. Early stage cancers are readily cured with surgery or radiation. In contrast, locally advanced cancers require morbid multimodality therapy and nonetheless have high recurrence rates, while metastatic disease has not been curable with cytotoxic chemotherapy. The emergence of more treatment-responsive human papillomavirus (HPV)-driven cancers and the advent of immune checkpoint blockade mean that the outlook for patients with head and neck cancer has improved dramatically since the first edition of this book was published in 2014. Our understanding of the biology of this cancer has deepened considerably in the past 4 years, yet undruggable targets due to the predominance of tumor suppressor gene mutation and other noncatalytic abnormalities continue to present barriers to molecular and personalized therapy and to cure in HPV-negative disease.

The second edition of *Molecular Determinants of Head and Neck Cancer* addresses this difficult disease with a focus on the molecular processes in the carcinogenesis and progression of these cancers which will inform the search for therapeutic targets to enable the prevention and improve the cure of head and neck cancer. With the current volume, we introduce the etiology and subclasses of head and neck squamous cell carcinomas (HNSCC), in the context of how these differences affect prognosis. Second, we summarize the current state of understanding of the genetic, epigenetic, protein expression, and immune environmental changes associated with SCCHN. Thirdly, we situate novel targets in the context of these

B. Burtness (✉)
Department of Internal Medicine, Yale University School of Medicine, New Haven, CT, USA
e-mail: barbara.burtness@yale.edu

E. A. Golemis
Molecular Therapeutics Program, Fox Chase Cancer Center, Philadelphia, PA, USA
e-mail: erica.golemis@fccc.edu

B. Burtness, E. A. Golemis (eds.), *Molecular Determinants of Head and Neck Cancer*, Current Cancer Research, https://doi.org/10.1007/978-3-319-78762-6_1

insights into SCCHN, seeking to provide a template for development of novel treatment strategies.

We begin by introducing the receptor tyrosine kinases (RTKs) and signaling pathways which are central to the biology of HNSCC. Chapter 2 describes the central role of ERBB/HER family proteins in the biology of head and neck cancer and reviews the data regarding inhibition of HER family signaling, given that EGFR remains the sole validated molecular target in HNSCC [1, 2]. Upregulation of RTKs with partially redundant function may provide resistance to cetuximab and more recently developed EGFR-targeting therapies. In particular, abnormal expression and activation of RTKs such as c-MET [3] have emerged as relevant to the pathology of HNSCC and may prove to be an important therapeutic target in this disease (Chap. 3).

Tumor invasion depends in part on epithelial-mesenchymal transition (EMT), which occurs in response to activation of transforming growth factor β (TGFβ) [4], a transmembrane serine-threonine kinase, and its canonical and noncanonical effectors. This is reviewed in Chap. 4. A common feature of RTKs is their activation of downstream effector pathways that support tumor growth, survival, and resistance to therapy. In the case of SCCHN, some of the most important of the effectors are themselves mutated or otherwise constitutively activated. Chapter 5 details mutational and indirect activation of the PTEN-PI3K-AKT-mTOR [5] signaling axis in HNSCC with efforts to target this pathway; while Chap. 6 discusses the role of constitutive JAK/STAT signaling [6] observed in a subset of HNSCC, with the challenges in targeting these noncatalytic signaling proteins. Conversely, Chap. 7 addresses the multiple defects in cell cycle regulation that occur in HNSCC and offer another potential source of targetable vulnerability.

One of the challenges in molecularly directed therapy of HNSCC has been the predominance of tumor suppressor mutations in this disease. Notch signaling is implicated in multiple cellular functions associated both with cancer and with tumor suppression. Notch mutation is a frequent event in HNSCC [7]; however, interestingly, the majority of these mutations are inactivating, indicating that the tumor suppressive function dominates in this epithelial tissue; this is reviewed in Chap. 8. The tumor suppressor *TP53* has long been understood to be the most commonly mutated gene in HPV-negative HNSCC and a major contributor to therapeutic resistance [8, 9]. The biology of this tumor suppressor, as well as new strategies to exploit loss of p53 function with synthetic lethal strategies, are described in Chap. 9.

The entire field of cancer biology is being transformed by the application of powerful new technologies that are elucidating the genome and epigenome. Apolipoprotein B mRNA editing enzyme, catalytic polypeptide-like (APOBEC) proteins are gene editing enzymes which are upregulated in response to viral infection as well as to replication stress and result in mutations and characteristic patterns of mutation. APOBEC family members are active in both HPV-related and HPV-negative HNSCC [10, 11], and both the protein and the mutational burden it elicits have therapeutic implications, discussed in Chap. 10. The theme of insights into the head and neck cancer genome, and the nature of intrinsic HNSCC subsets, is continued in Chap. 11, and the epigenome, with the impact of gene expression, methylation, and histone structure, is considered in Chap. 12.

Since the first edition of this book, the role of inflammation as a factor conditioning growth of tumor cells and the growth-promoting aspects of the tumor microenvironment has taken on much greater prominence. Chapter 13 addresses the process of inflammation as a contributor to tumor aggressiveness, based on activities on tumor cells and immune cells in the tumor microenvironment, and addresses the potential of the inflammatory process as a source of new targets for therapy. In the clinic, the advent of agents to block the immune checkpoint has led to new therapies and a flood of clinical trials. The complex immune microenvironment of HNSCC and the multiplicity of investigational immunotherapy approaches to reverse immune tolerance [12] are reviewed in Chap. 14. Neovascularization and hypoxia have been associated with treatment resistance in HNSCC [13]. Hypoxia-inducible factor is reviewed in Chap. 15, and vascular-endothelial growth factor (VEGF) and the agents which target this angiogenic factor and regulate tumor vascularization [14] are reviewed in Chap. 16.

The theme of the tumor environment and epithelial-mesenchymal transition is continued with chapters discussing integrin-mediated signaling, which mediates interactions between the tumor and the extracellular matrix (ECM), and the stem cell niche. Focal adhesion kinase (FAK) is a key mediator of integrin signaling, mediating tumor-ECM communications in HNC in a manner that affects treatment response [15]: biology of FAK and integrin and the clinical prospects of their inhibitors are the subject of Chap. 17. The Wnt/β-catenin signaling pathway [16] provides an independent input into cell differentiation status, affecting EMT, cancer stem cells, and therapeutic resistance. A growing body of evidence supports the common deregulation of expression of Wnt signaling proteins in HNSCC [16], with early efforts to evaluate therapeutic agents targeting some signaling intermediates. Wnt signaling is presented in Chap. 18 and hyaluronan-mediated activation of head and neck cancer stem cells in Chap. 19.

Historically, habitual exposures such as tobacco, alcohol, and mate have contributed to onset of SCCHN. However, a rising proportion of oropharynx cancers arise from transforming HPV infection [17]. Chapter 20 presents the epidemiology of the various types of HPV that contribute to SCCHN pathogenesis and assesses the potential for targeting viral oncoproteins. For HPV-associated SCCHNs, it will be necessary to identify biomarkers to distinguish between patients with near certainty of cure, and those – perhaps most commonly smokers – with HPV-associated cancer but a higher risk of recurrence [18]. Reduced treatment intensity and concomitant reduction of treatment-related morbidity may be achievable for the former; diverse approaches to treatment-deintensification are the focus of Chap. 21.

The 4 years since the first edition of this book have been marked by increased confidence that treatment for some subsets of HPV-driven HNSCC can be scaled back, by a revolution in our ability to manipulate the immune response to cancer, and by tantalizing clues that the PI3K and angiogenic pathways may also constitute valuable targets in HNSCC. However, HPV-negative HNSCC has not been readily amenable to targeted therapy, and even immunotherapy has had more modest effects in this cancer than in other solid tumors. Patients with HNSCC need treatments that exploit our advancing understanding of the biology of this malignancy. As the chapters collected here make clear, the advances in understanding this cancer bring us progressively closer to improved therapies for many subsets of HNSCC. Going forward, rapid translation of these findings to clinical trials will be essential to extend these insights to the cure of human head and neck cancer.

References

1. Burtness B, Goldwasser MA, Flood W, Mattar B, Forastiere AA. Phase III randomized trial of cisplatin plus placebo compared with cisplatin plus cetuximab in metastatic/recurrent head and neck cancer: an Eastern Cooperative Oncology Group study. J Clin Oncol. 2005;23:8646–54.
2. Vermorken JB, Mesia R, Rivera F, Remenar E, Kawecki A, Rottey S, Erfan J, Zabolotnyy D, Kienzer HR, Cupissol D, Peyrade F, Benasso M, Vynnychenko I, De Raucourt D, Bokemeyer C, Schueler A, Amellal N, Hitt R. Platinum-based chemotherapy plus cetuximab in head and neck cancer. N Engl J Med. 2008;359:1116–27.
3. Seiwert TY, Jagadeeswaran R, Faoro L, Janamanchi V, Nallasura V, El Dinali M, Yala S, Kanteti R, Cohen EE, Lingen MW, Martin L, Krishnaswamy S, Klein-Szanto A, Christensen JG, Vokes EE, Salgia R. The MET receptor tyrosine kinase is a potential novel therapeutic target for head and neck squamous cell carcinoma. Cancer Res. 2009;69:3021–31.
4. White RA, Malkoski SP, Wang XJ. TGFbeta signaling in head and neck squamous cell carcinoma. Oncogene. 2010;29:5437–46.
5. Bauman JE, Arias-Pulido H, Lee SJ, Fekrazad MH, Ozawa H, Fertig E, Howard J, Bishop J, Wang H, Olson GT, Spafford MJ, Jones DV, Chung CH. A phase II study of temsirolimus and erlotinib in patients with recurrent and/or metastatic, platinum-refractory head and neck squamous cell carcinoma. Oral Oncol. 2013;49:461–7.
6. Leeman RJ, Lui VW, Grandis JR. STAT3 as a therapeutic target in head and neck cancer. Expert Opin Biol Ther. 2006;6:231–41.
7. Rothenberg SM, Ellisen LW. The molecular pathogenesis of head and neck squamous cell carcinoma. J Clin Invest. 2012;122:1951–7.
8. Agrawal N, Frederick MJ, Pickering CR, Bettegowda C, Chang K, Li RJ, Fakhry C, Xie TX, Zhang J, Wang J, Zhang N, El-Naggar AK, Jasser SA, Weinstein JN, Trevino L, Drummond JA, Muzny DM, Wu Y, Wood LD, Hruban RH, Westra WH, Koch WM, Califano JA, Gibbs RA, Sidransky D, Vogelstein B, Velculescu VE, Papadopoulos N, Wheeler DA, Kinzler KW, Myers JN. Exome sequencing of head and neck squamous cell carcinoma reveals inactivating mutations in NOTCH1. Science. 2011;333:1154–7.
9. Skinner HD, Sandulache VC, Ow TJ, Meyn RE, Yordy JS, Beadle BM, Fitzgerald AL, Giri U, Ang KK, Myers JN. TP53 disruptive mutations lead to head and neck cancer treatment failure through inhibition of radiation-induced senescence. Clin Cancer Res. 2012;18:290–300.
10. Faden DL, Thomas S, Cantalupo PG, Agrawal N, Myers J, DeRisi J. Multi-modality analysis supports APOBEC as a major source of mutations in head and neck squamous cell carcinoma. Oral Oncol. 2017;74:8–14.
11. Henderson S, Chakravarthy A, Su X, Boshoff C, Fenton TR. APOBEC-mediated cytosine deamination links PIK3CA helical domain mutations to human papillomavirus-driven tumor development. Cell Rep. 2014;7:1833–41.
12. Ling DC, Bakkenist CJ, Ferris RL, Clump DA. Role of immunotherapy in head and neck cancer. Semin Radiat Oncol. 2018;28:12–6.

13. Rischin D, Fisher R, Peters L, Corry J, Hicks R. Hypoxia in head and neck cancer: studies with hypoxic positron emission tomography imaging and hypoxic cytotoxins. Int J Radiat Oncol Biol Phys. 2007;69:S61–3.
14. Vassilakopoulou M, Psyrri A, Argiris A. Targeting angiogenesis in head and neck cancer. Oral Oncol. 2015;51:409–15.
15. Skinner HD, Giri U, Yang L, Woo SH, Story MD, Pickering CR, Byers LA, Williams MD, El-Naggar A, Wang J, Diao L, Shen L, Fan YH, Molkentine DP, Beadle BM, Meyn RE, Myers JN, Heymach JV. Proteomic profiling identifies PTK2/FAK as a driver of radioresistance in HPV-negative head and neck cancer. Clin Cancer Res. 2016;22:4643–50.
16. Rhee CS, Sen M, Lu D, Wu C, Leoni L, Rubin J, Corr M, Carson DA. Wnt and frizzled receptors as potential targets for immunotherapy in head and neck squamous cell carcinomas. Oncogene. 2002;21:6598–605.
17. Chaturvedi AK, Engels EA, Pfeiffer RM, Hernandez BY, Xiao W, Kim E, Jiang B, Goodman MT, Sibug-Saber M, Cozen W, Liu L, Lynch CF, Wentzensen N, Jordan RC, Altekruse S, Anderson WF, Rosenberg PS, Gillison ML. Human papillomavirus and rising oropharyngeal cancer incidence in the United States. J Clin Oncol. 2011;29:4294–301.
18. Gillison ML, Zhang Q, Jordan R, Xiao W, Westra WH, Trotti A, Spencer S, Harris J, Chung CH, Ang KK. Tobacco smoking and increased risk of death and progression for patients with p16-positive and p16-negative oropharyngeal cancer. J Clin Oncol. 2012;30:2102–11.

Chapter 2
Targeting the ErbB Family in Head and Neck Cancer

Anna Kiseleva, Tim N. Beck, Ilya G. Serebriiskii, Hanqing Liu, Barbara Burtness, and Erica A. Golemis

Abstract Members of the ErbB receptor tyrosine kinase family (EGFR, HER2, HER3, and HER4), which regulate cell differentiation, proliferation, and survival, are commonly overexpressed and hyperactivated in squamous cell carcinoma of the head and neck (SCCHN). This abnormal expression and activity triggers multiple effector cascades that promote cancer growth, involving signaling through Ras-Raf-ERK1/2, PI3K/AKT/mTOR, JAK1/STAT3, PLC/PKC, and others. Targeting of EGFR remains one of the most common therapies for patients with SCCHN, with newer therapies also targeting additional ErbB family members and ErbB effectors, and exploring combinatorial approaches. In this chapter, we will describe the biology of ErbB family receptors in normal cells and in SCCHN, current and novel therapeutic approaches, and mechanisms underlying resistance to anti-EGFR therapy.

Keywords Head and neck cancer · ErbB family · EGFR. · EGFR-targeted therapy · Anti-EGFR therapy resistance

---

A. Kiseleva
Department of Biochemistry and Biotechnology, Kazan Federal University, Kazan, Tatarstan, Russia

Molecular Therapeutics Program, Fox Chase Cancer Center, Philadelphia, PA, USA

T. N. Beck
Molecular Therapeutics Program, Fox Chase Cancer Center, Philadelphia, PA, USA

Program in Molecular and Cell Biology and Genetics, Drexel University College of Medicine, Philadelphia, PA, USA

I. G. Serebriiskii · E. A. Golemis (✉)
Molecular Therapeutics Program, Fox Chase Cancer Center, Philadelphia, PA, USA
e-mail: Erica.Golemis@fccc.edu

H. Liu
School of Pharmacy, Jiangsu University, Jiangsu Sheng, China

B. Burtness
Department of Internal Medicine, Yale University School of Medicine, New Haven, CT, USA

B. Burtness, E. A. Golemis (eds.), *Molecular Determinants of Head and Neck Cancer*, Current Cancer Research, https://doi.org/10.1007/978-3-319-78762-6_2

2.1 Introduction

In the past decade, protein-targeted inhibitors have become valuable tools in the treatment of squamous cell carcinoma of the head and neck (SCCHN). The epidermal growth factor receptor (EGFR), also known as avian erythroblastic leukemia viral (v-erb-b) oncogene homolog 1 (ErbB1) or human epidermal receptor (HER1), was one of the first targetable proteins identified as relevant to SCCHN [1, 2]. Subsequent broadened research and development efforts have expanded the armament to encompass agents that also inhibit EGFR family members ErbB2/HER2 (sometimes designated neu), ErbB3/HER3, and ErbB4, as well as key downstream effectors including RAF and phosphoinositol-3-kinase (PI3K).

The ErbB family was first identified as cancer-relevant in the 1980s when an aberrant form of the human epidermal growth factor (EGF) receptor was found to be encoded by the avian erythroblastosis tumor virus [3]. The four members of the ErbB family are structurally related with each containing a large extracellular N-terminal region, a single hydrophobic transmembrane-spanning domain, an intracellular juxtamembrane region, a tyrosine kinase domain, and C-terminal region [4–6]. ErbB3 differs from the other family members in having a kinase domain that was long thought to be a pseudokinase although it has now been shown to have weak autophosphorylation capacity [7] and, through heterodimeric interactions, to serve as an activator of the EGFR kinase domain [8]. Importantly, members of the ErbB family function as homodimers and heterodimers [9, 10]. In normally growing cells, dimer formation and signaling typically involve activating interactions between the extracellular N-terminal domain and small ligands (discussed in Sect. 2.2). Interactions between dimer subunits induce essential phosphorylations within the ErbB C-terminal regions that provide binding sites for partners that interact with effector proteins to initiate downstream signaling cascades and initiate feedback signaling that ultimately restricts ErbB signaling activity. Key effectors that are activated as a result of these phosphorylations include PI3K, PLCγ, GRB2, c-SRC, and JAK. Cyclic, transient activation of ErbB family signaling in normal cells is regulated by a number of factors, including ligand availability, cytoplasmic phosphatases, and the endocytic/degradation machinery (Sect. 2.2).

In human SCCHN, activation of EGFR and its family members occurs by several distinct mechanisms (discussed at length in Sect. 2.3). Elevated expression of EGFR was originally described as characterizing 80–90% of SCCHN [11, 12], and several studies have indicated that overexpression of EGFR correlates with resistance to therapy and reduction of overall survival (OS) [13–15]. However, a meta-analysis evaluating EGFR prognostic value has demonstrated that EGFR overexpression correlates with OS, but not disease-free survival (DFS) [16], and additional recent studies suggest a more complicated relationship between overexpression and survival (Sect. 2.3.1). Although EGFR is by far the most commonly overexpressed ErbB family member in SCCHN, the three other members are also overexpressed in a significant number of cases (ErbB2/HER2, 3–29%; ErbB3/HER3, 21%; and ErbB4/HER4, 26%; [17]). Moreover, ligands contributing to the activation of ErbB proteins are overexpressed in some SCCHN tumors (Sect. 2.3) [18]. In addition, mutational activation of some critical effectors, such as PI3K, defines a subset of

SCCHN [19]. Finally, the past decade has been marked by a growing appreciation of differences in biology and prognosis associated with the presence or absence of human papillomavirus (HPV) as an oncogenic driver in SCCHN [20, 21], and some evidence suggests that HPV status may influence expression and activity of the ErbB proteins [22, 23].

Activation of the ErbB family of transmembrane receptor tyrosine kinases (RTKs) and their downstream effectors is typically associated with rapid cellular growth, as well as activation of the DNA repair machinery induced by DNA-damaging therapies commonly used in treatment of SCCHN, contributing to resistance to cytotoxic therapies such as cisplatin or radiation [14]. Based on abundant evidence, therapeutics targeting the ErbB family and its effectors have appeared to be particularly appropriate for the treatment of SCCHN. Two complementary therapeutic strategies have been developed to target EGFR and its family members. A first strategy involves targeting the extracellular domain of the receptor with monoclonal antibodies, such as cetuximab, panitumumab, zalutumumab, and others that interfere with the processes of dimerization and activation of the intracellular kinase domains [24, 25] (Sect. 2.4.1). A second strategy targets the intracellular domain of the receptor with low-molecular-weight tyrosine kinase inhibitors (TKIs; e.g., gefitinib and erlotinib; see Sect. 2.4.2) [26]. More recently, therapeutic strategies have expanded to include the use of drugs or drug combinations that target multiple ErbB family members or that combine ErbB-targeting drugs with those targeting critical downstream effectors, such as PI3K or MEK1 (Sect. 2.3.3.2) [27]. The nature of EGFR/ErbB signaling and therapeutic strategies to manage tumors with EGFR/ErbB involvement are addressed in detail in the remainder of this chapter.

2.2 Regulation of EGFR and the ErbB Family in Normal Cells

2.2.1 Ligand Binding and Dimerization: Activation of ErbB Proteins in Normal Cells

The extracellular regions of ErbB family members contain two homologous ligand-binding domains (domains I and III) and two cysteine-rich domains (domains II and IV). The ligands required for dimerization and activation of EGFR, ErbB3, and ErbB4 can be separated into five groups: (1) EGFR-specific ligands such as EGF, amphiregulin (AR), epigen (EPN), and transforming growth factor alpha (TGFα); (2) the ErbB3-specific ligands neuregulin1α (NRG1α), NRG2α, and NRG6; (3) NRG3, NRG4, and NRG5 that specifically bind ErbB4; (4) the bispecific ligands betacellulin (BTC), epiregulin (EPR), and heparin-binding EGF-like growth factor (HBEGF), which bind EGFR and ErbB4, and NRG1β which binds ErbB3 and ErbB4; and (5) NRG2β, which is a pan-ErbB ligand and binds to EGFR, ErbB3, and ErbB4 [28] (Fig. 2.4). Uniquely, ErbB2 does not depend on ligands for dimerization or activation. Instead, domains I and III interact directly in a configuration that renders the ligand-binding site inaccessible [29–31]. To date, no high-affinity soluble ligand has been identified for ErbB2 [29, 32]. It is possible that assignment of ligand specificity is not exact; for example, a recent study has demonstrated that stimulation of ErbB4 with NRG1 activates the transcriptional activator YAP, promoting YAP-dependent cell migration [33].

ErbB proteins can homodimerize or heterodimerize [34]. EGFR-EGFR and ErbB4-ErbB4 homodimers and EGFR-ErbB2, EGFR-ErbB3, ErbB2-ErbB3, and ErbB2-ErbB4 heterodimers are abundant in SCCHN tumors and cell lines [17, 35, 36]. There is also evidence that activation of the catalytic domain of EGFR through homo- or heterodimerization occurs due to its increased accumulation at the plasma membrane and can be enhanced by a common mutation in a leucine (*L834R*), which suppresses local disorder [37] and is associated with drug resistance in some tumor types [38]. Some similarities of the EGFR kinase domain with Src and cyclin-dependent kinase (CDK) domains have been observed that support an alternative mechanism for dimerization, in which one EGFR kinase domain interacts asymmetrically with the second domain in a dimer pair, as a cyclin activates a CDK [9].

The configuration changes associated with dimerization lead to transient kinase activation in normal cells. These become constitutive in cancers, in the setting of kinase overexpression. The actual activation process involves an asymmetric interaction between intracellular kinase domains that results in auto- or transphosphorylation of ErbB family members [8, 39, 40]. As ErbB2 is not ligand-responsive, phosphorylation of this kinase can be activated through homodimerization [41, 42] or heterodimerization, frequently with ErbB3 [7, 8]. EGFR and ErbB4 can function independently of other ErbB receptors and autophosphorylate C-terminal tails after binding to activating ligand. These phosphorylations provide binding sites for proteins that transduce activating signals downstream (e.g., STAT5b, GRB2, SHC, GAB1/PI3K(p85), PLCγ), which induce signaling relevant to proliferation, apoptosis resistance, and DNA synthesis [43].

2.2.2 ErbB Trafficking and Other Mechanisms to Limit EGFR Function in Normal Cells

As with most RTKs, duration of ErbB activation is limited by countervailing regulatory processes. Some of the phosphorylations on the C-terminal domains of the ErbB proteins provide binding sites that allow feedback signaling that downregulates the activated ErbB protein through dephosphorylation, ubiquitination, and/or internalization (e.g., SHP1, CBL, CRK). More than one pathway for internalization has been described. In the most studied pathway, binding of the E3 ubiquitin ligase Cbl to phosphorylated Y-1045 of activated EGFR at the plasma membrane triggers clathrin-mediated endocytosis [44]. Multiple additional activation-associated phosphorylations conferred by calmodulin kinase II and p38 enhance the interaction of Cbl with activated EGFR [45, 46].

Subsequently, during EGF-mediated endocytosis, EGFR is either recycled to the plasma membrane or alternatively processed through the late endosome and multivesicular body for proteolytic degradation in the lysosome [47]. An alternative non-clathrin-based endocytotic process has also been described: in this case, the majority of EGFR is targeted for lysosomal destruction [48, 49]. Additional interactions involving the molecular motor dynamin 2 (DYN2) and a scaffolding protein, CIN85, support targeting of EGFR to the lysosome rather than for recycling [50]. As discussed below, reduced phosphorylation of EGFR that limits interaction with Cbl and other internalization proteins often accompanies therapeutic resistance to EGFR inhibitors (EGFRIs). EGFR may also undergo ligand-independent internalization through p38MAPK- and clathrin-mediated, or Src- and caveolin-mediated mechanisms, upon conditions of cell stress [51]. Dutta et al. have recently reported that neuropilin-2 (NRP2) plays an important role as an endocytotic regulator for EGFR, with depletion of NRP2 disrupting normal regulation of endocytic transport of EGFR from the cell surface and leading to the accumulation of active EGFR in endocytic vesicles and abnormal ERK activation [52].

Extending understanding of traffic controls, compartmentalization of the EGFR partner ErbB2 is controlled by a member of the Anks1a adaptor protein family in the endoplasmic reticulum (ER). Within the ER, an ErbB2 complex with another RTK, ephrin A2 (EphA2), allows binding to Anks1a, which in turn regulates EphA2/ErbB2 complex exit. This process is positively associated with tumorigenesis [53]. Once activated, ErbB2 remains at the cell surface, potentially due to an interaction with HSP90 or the plasma membrane calcium ATPase2 (PMCA2) [54, 55]. In breast cancer, inhibition of PMCA2 disrupts binding between ErbB2 and HSP90, leading to ErbB2 internalization and degradation. In MMTV-Neu mice, PMCA2 knockout effectively inhibits tumor formation, suggesting that interaction of ErbB2 and PMCA2 is a potential therapeutic target for this and other cancer types [54] (Fig. 2.1).

2.3 Causes and Consequences of Altered EGFR/ErbB Function in SCCHN

2.3.1 Overexpression of EGFR and Its Ligands

The degree to which EGFR is overexpressed in SCCHN has been reported differently by different groups, reflecting varying approaches used to measure DNA amplification, mRNA overexpression, and protein overexpression, the use of different cutoff values, and (potentially) differences in EGFR expression based on SCCHN sub-site (e.g., oral cavity versus laryngeal). EGFR overexpression in SCCHN is often caused by an increase in the number of gene copies [56] but can also occur at the mRNA or protein level. The original studies reporting overexpression of EGFR in 80–90% of SCCHN [11, 12] were based on analysis of mRNA expression in a limited set of 24 tumor specimens and 10 SCCHN cell lines versus histologically normal mucosa specimens. Follow-up work by Grandis and Tweard found that median EGFR mean optical density (based on IHC analysis using preparations of the EGFR-overexpressing A431 cell line as a positive control) was 54%

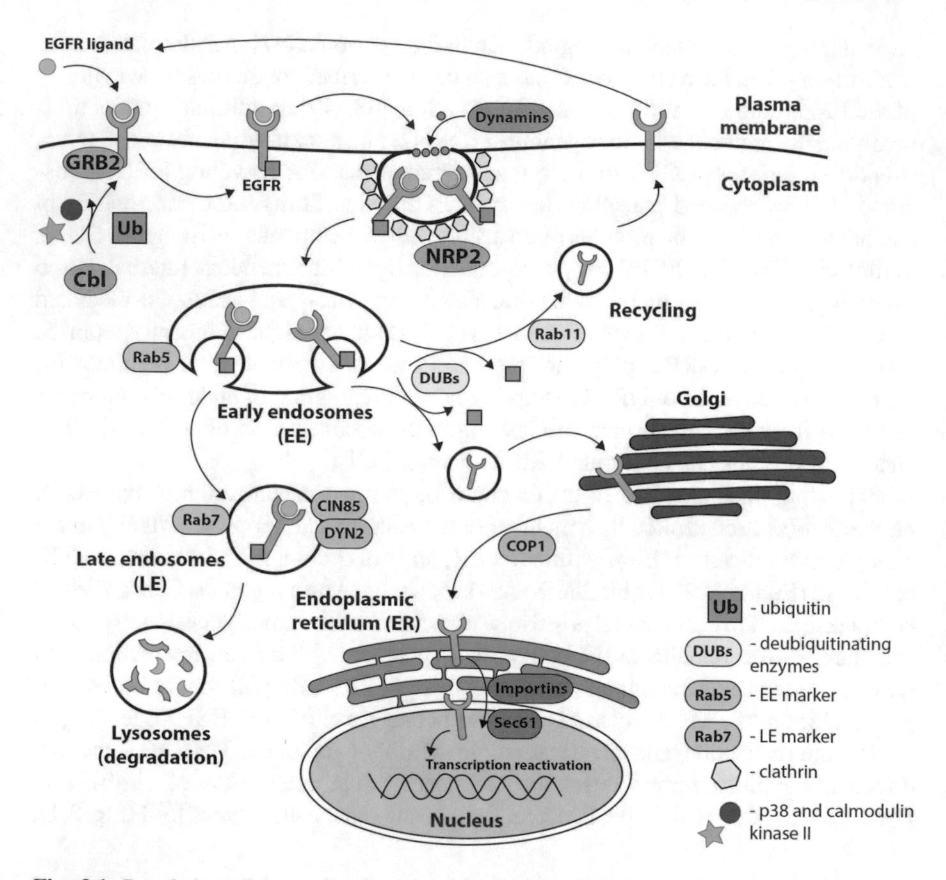

Fig. 2.1 Regulation of internalization and degradation of ErbB proteins. Here we described clathrin-mediated endocytosis of EGFR with further transportation to late endosomes and degradation, recycling, or migration to the nucleus

in a group of 91 patients with SCCHN [57]. In another study, EGFR protein levels in 140 primary laryngeal squamous cell carcinomas were determined using a radioligand receptor assay. The authors established that a cutoff value of 20 fmol mg^{-1} is an effective prognostic marker, and based on this classification, 28 of 140 patients (20%) had elevated levels of EGFR and lower 5-year survival (25%) in comparison with patients with EGFR levels <20 fmol mg^{-1} (81% 5-year survival) [58]. Poor prognosis was also associated with an increased copy number of *EGFR*. *EGFR* gene copy numbers were analyzed in 134 SCCHN tumors using quantitative PCR, with this study finding aberrant *EGFR* copy numbers in 24% of tumors, with 17% of tumors having increased copy numbers [59].

Ongkeko et al. used IHC analysis to show that EGFR was highly expressed (38%–43%) in 21 pharyngeal, 16 laryngeal, and 1 floor of mouth carcinoma compared with benign samples [60], based on qualitative rankings from 3 independent pathologists. Using immunohistochemistry (IHC), Bei et al. found EGFR to be overexpressed in 47% of cases in a group of 38 SCCHN tumor samples in comparison with 24 adjacent normal mucosa specimens [61]. Bernardes et al. analyzed

EGFR in a subset of 52 patients with oral squamous cell carcinoma (OSCC) using 3 different methods: IHC, fluorescent in situ hybridization (FISH), and chromogenic in situ hybridization (CISH). This study showed that EGFR overexpression rates were 53.8% (28/52) by IHC, 5.8% (3/52) by CISH, and 15.4% (8/52) by FISH [62]. Pectasides and colleagues found that increased gene copy number did not directly correlate with protein expression of EGFR, and elevated protein levels of EGFR determined by IHC better correlated with the poor clinical outcome than did *EGFR* copy number determined by FISH [63]. Ang et al. demonstrated that EGFR expression varied widely in a group of 155 patients (based on automatic IHC analysis) and that higher EGFR expression in SCCHN samples correlated with reduced OS and DFS (based on the mean optical density data) [15].

The increasing availability of systematic genomic profiling provides additional data points [64], but does not resolve the issue of how to best define EGFR overexpression values. Among the 357 SCCHN specimens analyzed by The Cancer Genome Atlas (TCGA) Consortium for which genomic data are available, based on default TCGA analysis settings, *EGFR* amplification occurs in 4% of tumors. Among 520 SCCHN specimens with mRNA expression data collected by the TCGA, upregulation at the mRNA level is seen in 15% of cases. Based on these expression data, upregulation results in shorter overall and disease-free survival (z-score >2.5, OS, 28.32 months, versus 57.42 months; DFS, 30.16 months, versus 71.22 months) (Fig. 2.2) [65, 66]. In contrast, a recent study reporting genetic and molecular profiling by Caris Life Sciences of 123 and 236 patients with advanced SCCHN demonstrated that EGFR was overexpressed in 90% of cases by IHC but only 21% by ISH, respectively [67].

Increased EGFR expression is not only found in SCCHN tumor samples but has also been observed in "healthy" mucosa samples of patients with SCCHN [68] and likely reflects a premalignant event in the tissue adjacent to an incipient SCCHN [11, 12]. Hence, elevated EGFR expression is a potential biomarker for early stages of malignant transformation in addition to being a therapeutic target [69]. Similarly, a study of 155 patients found that EGFR expression did not correlate with disease stage at presentation, or other known clinical prognostic variables, in stage II–IV carcinomas of the oral cavity, oropharynx and supraglottic larynx, tongue base, and hypopharynx, although EGFR expression was an independent prognostic indicator of 5-year OS (40% for EGFR negative and 20% for EGFR positive; $p = 0.0006$) as well as disease-free survival (DFS) (25% for EGFR negative and 10% for EGFR positive; $p = 0.0016$) [15]. These findings agreed with an earlier study, based on 140 primary laryngeal squamous cell carcinomas, where the 5-year survival rate was 81% for patients with tumor cells defined as EGFR negative based on biochemical assessment of EGF-binding capacity of membrane fractions prepared from tumors, compared to only 25% for patients with EGFR-positive tumors [58]. This study also reported 5-year relapse-free survival (RFS) of 77% for patients with EGFR-negative tumors, compared to 24% for patients with EGFR-positive tumors [58]. Chang et al. have shown that high EGFR expression also correlates with treatment failure in early glottic cancer treated with radiation alone and that EGFR expression is higher in the tumors of patients with recurrent disease than in controls [70]. In one study, using immunohistochemical (IHC) analysis, EGFR distribution within the tumor

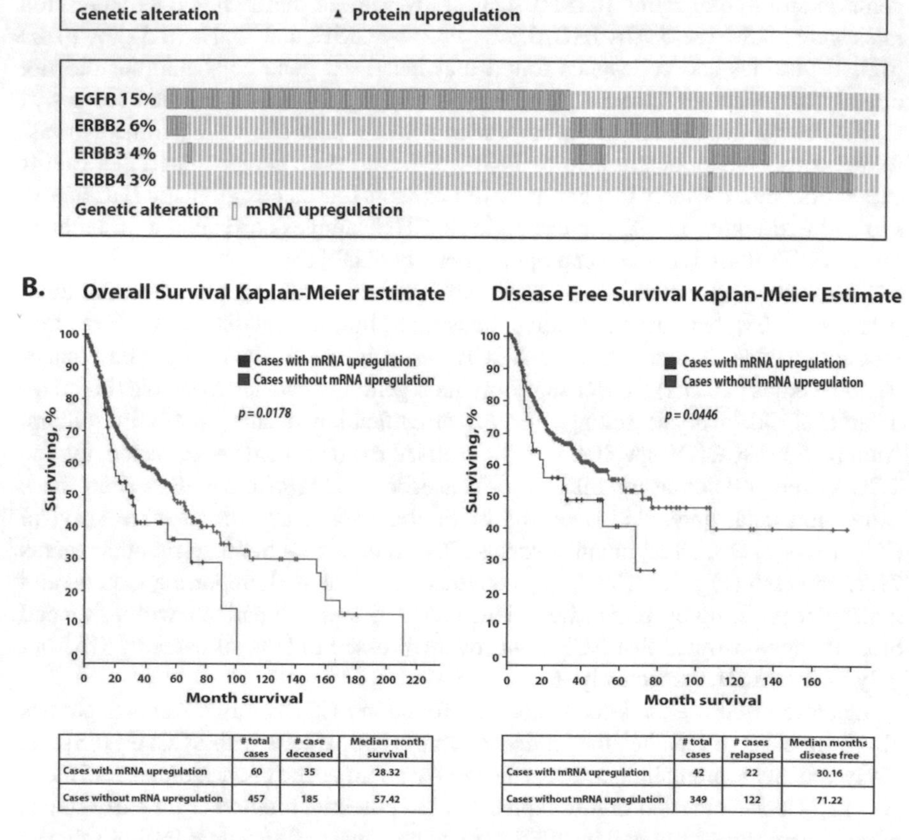

	# total cases	# case deceased	Median month survival
Cases with mRNA upregulation	60	35	28.32
Cases without mRNA upregulation	457	185	57.42

	# total cases	# cases relapsed	Median months disease free
Cases with mRNA upregulation	42	22	30.16
Cases without mRNA upregulation	349	122	71.22

Fig. 2.2 TCGA data for EGFR in SCCHN cancer specimens. (**a**) A representative map of data illustrating TCGA data for EGFR upregulation at the protein level (*top* panel) and mRNA levels for EGFR and other ErbB family members in SCCHN specimens.(**b**) Kaplan-Meier survival curves comparing the SCCHN patients with (*red*, z-score > 2.5) and without (*blue*, z-score < 2.5) EGFR mRNA level upregulation. Overall survival (OS), p-value = 0.0178; progression-free survival (PFS), p-value = 0.0446

tissue was related to patient survival, with heterogeneous distribution of EGFR in tumors significantly associated with poorer OS and DFS, in comparison with homogeneous distribution [71].

Additional members of the ErbB receptor family have been detected in SCCHN at increased expression levels [72, 73], although conflicting reports regarding ErbB3 and ErbB4 expression levels have been published [61, 72, 74]. In the TCGA, protein overexpression was observed for ERBB2 in 2.2% of tumors and for ERBB3 in 5% of tumors: this represents too few cases to perform meaningful analyses and determine potential correlation with survival.

Changes in EGFR and ErbB family internalization and degradation mechanisms have also been associated with SCCHN. Changes in these mechanisms associated with cancer lead to membrane accumulation of ErbB, further contributing to the abnormal activation of EGFR/ErbB signaling, which potentially promotes tumor formation and progression [75]. In this regard, a recent study revealed the relationship between the lysosomal enzyme cathepsin S (CTSS) and EGFR signaling regulation, with increased expression of CTSS detected in a number of types of cancer. Inhibition of CTSS limited EGFR degradation and caused EGFR accumulation in the late endosomal and in the perinuclear region, leading to formation of spatial compartments with extended EGFR, STAT3, and AKT signaling. Combined treatment with the EGFR inhibitor gefitinib and the CTSS inhibitor 6r also significantly increased cellular apoptosis [76]. Downregulation of c-CBL, which mediates internalization and degradation of EGFR, has been identified in a significant subset of SCCHN tumors [77]. Reciprocally, upregulation of the HECT-class ubiquitin ligase SMURF2, which ubiquitinates EGFR in a manner that protects it from c-CBL-dependent degradation, has also been suggested to be important in SCCHN [78].

Upstream of EGFR, overexpressions of ligands such as TGFα have been linked to a poor prognosis [68, 79] and have been associated with malignant tumor development at a number of sites in transgenic mice [80–82]. Additionally, expression of TGFα [17], AR [83, 84], and HB-EGF [85] has been shown to enhance oncogene-induced carcinogenesis and affect the response of tumor cells to EGFR inhibition [86–89], with some evidence suggesting other ligands are likely to also be of importance [90]. Elevated expression of mRNAs for EGFR ligands including AREG (amphiregulin), EGF, HB-EGF, and betacellulin (BTC) was associated with reduced patient survival [91]. Some proteins, such as the CBL-interacting protein of 85 kDa (CIN85), which regulates EGFR internalization, have been shown to be overexpressed in some advanced SCCHN and to increase TGFα-dependent signaling in SCCHN tumors [92].

As Brand et al. have reviewed in detail [93], epithelial cancers such as SCCHN are, surprisingly, characterized by a high frequency of nuclear EGFR localization. Mechanistically, to enter the nucleus, EGFR is passaged from clathrin-coated pits to the Golgi and subsequently via retrograde transport in COPI vesicles to the endoplasmic reticulum (ER) [94], after which the Sec61 translocon moves EGFR from the inner nuclear membrane to the nucleus [95, 96] (Fig. 2.1). Nuclear EGFR acts as a transcription coactivator for many genes associated with cell proliferation, including BCRP, Aurora-A, cyclin D, Myc, c-Myb, Cox-2, and iNOS, and also binds and supports activity of PCNA and DNAPK to enhance DNA synthesis and repair [97]. Increased expression of nuclear EGFR has been associated with a higher incidence of local recurrence and inferior DFS in oropharyngeal squamous cell carcinoma [98, 99]. Nuclear EGFR expression levels retained their prognostic significance in multivariate analysis adjusting for well-characterized prognostic variables [99]. Saloura et al. have reported that posttranslational methylation of the tyrosine kinase domain of EGFR by methyltransferase WHSC1L1 increased activation of the ERK pathway in the absence of EGF stimulation. Interestingly, this methylation appeared to be important for nuclear EGFR, promoting its interaction with PCNA (proliferating cell nuclear antigen) in SCCHN cells and enhancing DNA synthesis and cell cycle progression [100]. At present, it is not clear whether this localization is unique to cancer cells or instead represents an extreme case of a signaling process that also exists in normal cells: in general, this phenomenon requires further study.

2.3.2 Alternative Forms of EGFR and Its Effectors Affecting Signaling Activity in SCCHN

It has been suggested that expression of truncated and activated EGFR is associated with advanced tumor and nodal stage [101]. In studies of SCCHN tumors, Hama et al. detected only 5 different *EGFR* mutations in 6 out of 82 patients [102]. Additional ErbB family members were not identified as commonly mutated in either of these studies. A meta-analysis of multiple studies, including 4122 patients with SCCHN, suggested a 2.8% frequency of mutations affecting the tyrosine kinase domain [102, 118, 119]. Another two studies identified *EGFR* mutations in only 3 of 127 patients (2.4%) and 17 of 110 (16%), respectively [103, 104]. A fourth study found an in-frame deletion mutation in exon 19 of *EGFR* (E746_A750del) in 3 of 41 larynx, tongue, and tonsil tumor samples [105].

One *EGFR* mutation of note reported in SCCHN is EGFR variant III (EGFRvIII), which results in a truncation of the ligand-binding domain that results in ligand-independent, constitutive signaling, greatly potentiating tumorigenicity. EGFRvIII is the most common form of mutant *EGFR* and has been described in several types of cancer [106–111], including SCCHN [102, 112, 113]. However, the reported frequency of EGFRvIII in head and neck cancer is highly inconsistent. The presence of EGFRvIII in SCCHN ranged from none [102] to 15% [114] to 42% [113] and may vary by specific SCCHN subsite [115]. Sok et al. reported that EGFRvIII-transfected SCCHN cells showed increased proliferation in vitro and increased tumor volumes in vivo compared with vector-transfected cells. Furthermore, EGFRvIII-transfected SCCHN cells showed decreased apoptosis in response to cisplatin and decreased growth inhibition following treatment with cetuximab compared with vector-transfected control cells. However, it was not established if the transfected cells expressed EGFRvIII at levels similar to those observed in actual patient samples, given conflicting results in different studies [113, 116]. The significance of this variant remains unclear.

Stransky et al. performed whole exome sequencing on tumor samples from 92 patients with SCCHN and validated known relevant mutations in *TP53*, *CDKN2A*, *PTEN*, *PIK3CA*, and *HRAS* [117]. Agrawal et al. used the same methods to study 32 primary tumors, and 6 of the genes that were mutated in multiple tumors were reassessed in up to 88 additional SCCHN samples. This study identified mutations in *FBXW7* and *NOTCH1* in addition to previously identified genes [118].

Li et al. compared the genomic data of 39 SCCHN cell lines with genomic findings from 106 SCCHN tumors. Their results indicated that eight genes (*PIK3CA, EGFR, CCND2, KDM5A, ERBB2, PMS1, FGFR1*, and *WHSCIL1*) are amplified and five genes (*CDKN2A, SMAD4, NOTCH2, NRAS*, and *TRIM33*) are deleted in both SCCHN cell lines and tumors. Among the mutated genes relevant to the ErbB pathway, activating mutations of the catalytic subunit of *PI3K (PI3KCA)* were shared both in cell lines and in tumors – a result confirmed by a number of other studies [117–120] – and, importantly, based on the pharmacologic profiling results of eight anticancer agents, these mutations influence drug resistance [121].

2.3.3 Consequences of EGF/ErbB Activation

Dimerization of the ErbB RTKs can result in the constitutive activation of a number of intracellular signaling pathways, each of which contributes to the oncogenic activity of this kinase family in SCCHN. Some of the better-studied and physiologically significant effector pathways are represented in Fig. 2.3 and discussed below.

2.3.3.1 Ras/Raf/MAPK

Increased activity of the Ras/Raf/MAPK pathway initiated by EGFR signaling is strongly linked to tumorigenesis in SCCHN [94]. Following EGFR autophosphorylation, mainly on residues Y1068 and Y1086, the growth factor receptor/bound protein 2 (GRB2) adaptor protein is either directly recruited through binding of its Src homology 2 (SH2) domain to the phosphotyrosine residues of the activated receptor or, alternatively, GRB2 is indirectly recruited to active EGFR by interaction with the Src homolog and collagen homolog (SHC) adaptor protein, which directly binds tyrosine-phosphorylated sites on EGFR, itself is tyrosine phosphorylated, and then binds GRB2 [95]. EGFR-bound GRB2 subsequently recruits and activates guanine nucleotide exchange factor Son of Sevenless (SOS). Activated SOS increases the pool of active, GTP-bound Ras, inducing a kinase cascade involving c-Raf, MEK1/2, and ERK1/2 (Fig. 2.3). Phosphorylated ERK1/2 translocates into the nucleus and activates transcription factors that induce transcription of many genes promoting cell growth and survival; a residual pool of active cytoplasmic ERK1/2 also phosphorylates cytoskeletal proteins such as actin, which promotes cell motility, and regulators of cell division and cytokinesis, vesicle and organelle movement, and mitochondrial targets such as Bcl2 that render cells resistant to apoptosis (Fig. 2.3) [95, 96].

2.3.3.2 PI3K/Akt/mTOR

Dimerization of EGFR or ErbB2 with ErbB3 is strongly associated with PI3K activation, because of the high prevalence of PI3K-activating docking sites on ErbB3 [97]. PI3K proteins are composed of a catalytic p110 and a regulatory p85 subunit. The p110 subunits catalyze the phosphorylation of phosphatidylinositol 4,5-diphosphate (PIP2) to the second-messenger phosphatidylinositol 3,4,5-triphosphate (PIP3), which in turn phosphorylates and activates the protein serine/threonine kinase AKT (also known as protein kinase B), inducing protein synthesis and cell growth through activation of the mTOR effector pathway and limiting the apoptotic machinery [98]. AKT activation may also be induced by the binding of serine protease inhibitor Kazal-type 6 (SPINK6) to the EGFR

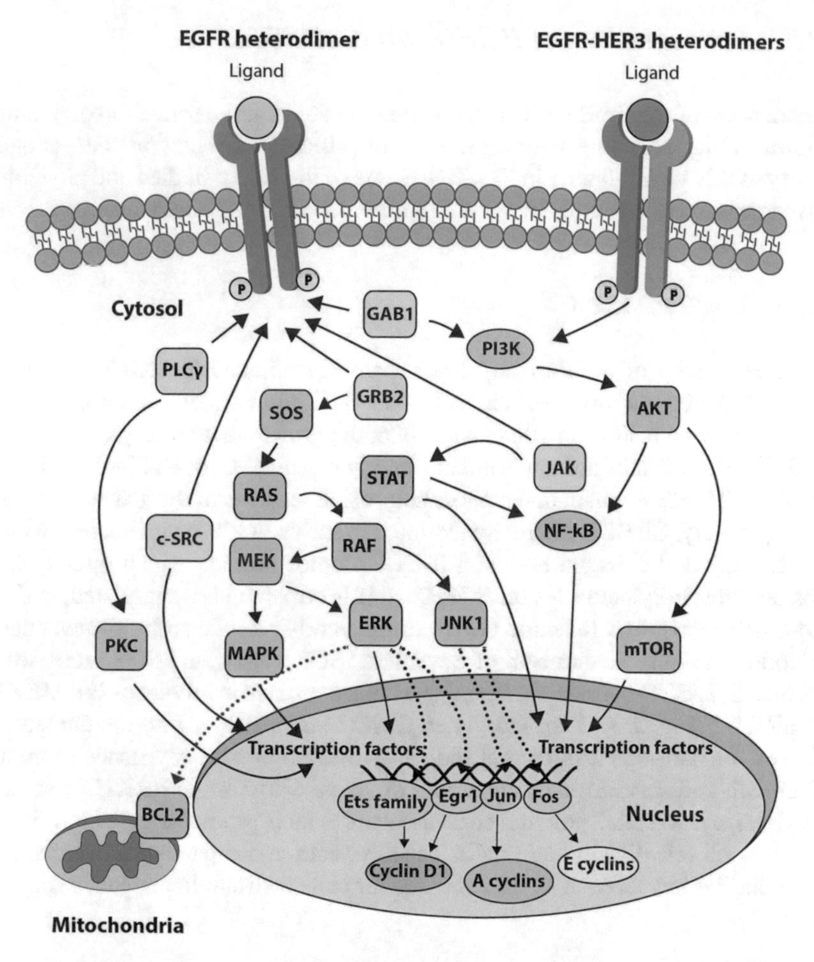

Fig. 2.3 Signaling pathways downstream of EGFR and other ErbB proteins that have been linked to tumorigenesis of SCCHN and/or resistance to ErbB-targeting inhibitors. *Green* boxes indicate targets which bind directly to the EGFR phosphorylation sites. See text for details

extracellular domain, which has been shown to occur in and promote metastasis of nasopharyngeal carcinoma cells [99]. In another study, PIK3CA overexpression in the mouse oral epithelium leads to increased tumor invasiveness and metastasis by inducing epithelial-to-mesenchymal transition (EMT). This study of PIK3CA-driven SCCHN emphasized the importance of 3-phosphoinositide-dependent protein kinase (PDK1) rather than AKT as a key effector [122]. Bozec and colleagues reported that the mTOR inhibitor temsirolimus in combination with cetuximab has a synergistic effect in NOD scid gamma (NSG) mice injected with SCCHN cells into the mouth floor. The combination of these two drugs significantly reduced tumor growth by inhibition of both the MAPK and the PI3K/AKT/mTOR pathway [123].

2.3.3.3 STAT

The signal transducers and activators of transcription (STAT) proteins were originally identified as downstream effectors of non-tyrosine kinase cytokine receptors, such as IL-6, IL-22, IFN-α/β, and IFN-λ. However, STATs can also be directly activated by EGFR, or by EGFR effectors such as c-Src [124], and constitutive activation of STATs has been reported in SCCHN [125]. Activated STATs migrate from cytoplasm to nucleus and upregulate the expression of many proteins associated with tumorigenesis, including the prosurvival factor NF-κB [126]. STAT family activation contributes to cancer cell survival and protects cells from apoptosis, which makes it a potentially useful therapeutic target [127], although there is some debate, as one study of a group of 102 SCCHN patients found nuclear STAT3 localization was associated with improved survival [128]. Additionally, Wheeler et al. reported that STAT3 activation can promote invasion of head and neck cancer cells bearing EGFRvIII and contributes to cetuximab resistance [129]. In vitro studies have shown that simultaneous inhibition of JAK1-STAT3 with JAK1i inhibitor and EGFR (cetuximab) in combination with radiation has a synergistic effect and leads to radiosensitization of human head and neck cancer cells and apoptosis [130]. Pre-irradiation inhibition of STAT5, STAT6, and MEK1/2 by 573,108, leflunomide, and U0126, respectively, in a panel of SCCHN cells, led to decreased survival following irradiation [131].

2.3.3.4 PLC/PKC

PLC is recruited by phosphorylated EGFR and subsequently activated. Primary tumors express elevated levels of total and phosphorylated PLCγ (one of six isotypes: β, γ, δ, ε, ζ, and η; [132]), and EGFR-stimulated activation of PLCγ promotes invasion of SCCHN [133]. PLCγ inhibition decreases the invasive potential of prostate, breast, and head and neck carcinoma cells [134, 135]. Once activated, PLC hydrolyzes PIP2 to diacylglycerol (DAG) and inositol 1,4,5-triphosphate (IP3). DAG in turn activates members of the PKC family, which is composed of 12 different isoforms in mammals [18]. Protein kinase Cε [136] has been proposed as a promising prognostic factor for relapse and OS of SCCHN [137]. PKCζ is highly expressed in SCCHN tumors and mediates EGF-induced growth of SCCHN tumor cells by regulating MAPK [138]. A recent study demonstrated that resistance to PIK3CA inhibitors occurs due to the induction of the RTK AXL, which interacts with EGFR and activates PKC and mTOR, leading to cancer cell survival [139].

2.3.3.5 Src

Activation of members of the Src kinase family (Blk, Fgr, Fyn, Hck, Lck, Lyn, Src, Yes, and Yrk; [140]) by EGFR and ErbB2 positively regulates cell proliferation, migration, adhesion, and tumor angiogenesis, with activation seen in many cancer types, including SCCHN [141–143]. In SCCHN, Src contributes to EGFR-dependent activation of STAT3 and STAT5, which, as mentioned above, are important for tumor growth [144]. Reciprocally, Src helps to activate EGFR by participating in G protein-coupled receptor-initiated TGFα release [145]. Changes in the interaction between Src and EGFR have been suggested to be involved in resistance to EGFR-targeting antibodies such as cetuximab by increasing translocation of EGFR to the nucleus (Sect. 2.2.2) [143, 146]. Src additionally interacts with other RTKs that are upregulated during acquisition of resistance to EGFRIs, such as IGF-1R (insulin-like growth factor-1 receptor) and others [147]. In gastric cancer, c-SRC-mediated activation of EGFR was shown to be induced by the receptor activator of NF-κB ligand (RANKL)/RANK pathway, promoting resistance to cetuximab [148]: whether this mechanism is relevant to SCCHN remains to be determined, although the fact that RANKL function has recently been found to be important for SCCHN progression is suggestive [149].

2.3.3.6 Nuclear Factor-κB (NF-κB)

High expression and constitutive activation of the transcription factor nuclear factor-κB (NF-κB) has been directly linked to tumorigenesis, metastasis, and chemoresistance in many cancers including SCCHN [150–154], with particularly high levels of NF-κB in highly metastatic cells [150, 155]. NF-κB induction of matrix metalloproteinases MMP-1, MMP-2, MMP-9, and MMP-14, fibronectin, ß1 integrin, and vascular endothelial growth factor C is strongly associated with tumor progression and metastasis [155]. In SCCHN, NF-κB activation has been described as both independent of and dependent on EGFR signaling [156–158]. In the EGFR-dependent activation of NF-κB, phosphorylated EGFR activates PI3K, ERK1/2, and STAT3, all of which are associated with increased NF-κB activity (Fig. 2.3) [156]. Depletion of NF-κB pathway components or pharmacological inhibition of NF-κB significantly increased cell death induced by erlotinib in EGFR-mutant lung cancer cells [159], supporting relevance of this signaling axis [160].

2.3.3.7 Cyclin D1

Cyclically induced expression of the cell cycle regulatory protein cyclin D1 (CCND1) promotes the key G1-to-S phase transition through formation of complexes with CDK4 and CDK6. The CCND1-CDK4/6 complex phosphorylates retinoblastoma tumor suppressor protein (Rb), inhibiting its activity and allowing changes in gene expression that promote S phase progression [161]. Analysis of SCCHN cell lines found common *CCND1* amplification and/or overexpression, which was associated with resistance to gefitinib [162]. In a group of 103 SCCHN patients' tumor samples, *CCND1* amplification was observed in 30% (31/103) of patients and had a statistically significant association with recurrence, distant metastasis, and survival at 36 months [163]. EGFR expression coupled with CCND1 overexpression was found to be associated with sensitivity to combination therapy with CDK4/6 and ERBB-targeting inhibitors (palbociclib and lapatinib) in HPV(-) SCCHN cell models, while integrated analysis of CDK4/6, CCND1, and EGFR expression refined ability to predict SCCHN prognosis [164].

2.4 EGFR and Targeted Inhibitors

For patients with locally advanced SCCHN, randomized controlled trials have shown that the addition of chemotherapy to radiotherapy improves 3-year overall survival (51% for chemoradiotherapy compared to 31% for radiation alone) [165] and disease-free survival (37% for radiation plus chemoradiotherapy compared to 23% for radiation alone) [166], albeit at the cost of increased toxicity [167]. Targeting EGFR is now a well-established therapeutic strategy for SCCHN treatment [168], as EGFR inhibition seems to prevent activation of DNA repair mechanisms that enable cancer cells to survive radiation- or chemotherapy-induced DNA damage [169, 170]. However, the monotherapy response rate to cetuximab is only 10% in patients with platinum-refractory SCCHN [171]. It is unknown if this observation is due to cell-mediated immunity; data from R0522 suggests not, as cetuximab benefit did not correlate with FCγ subtype or markers of inflammation [172]. Alternatively, it may reflect the small proportion of cancers that are truly EGFR dependent. Numerous data suggest that high levels of EGFR may accelerate repopulation, a condition of enhanced cellular proliferation after exposure to ionizing radiation, contributing to the radioresistance associated with head and neck cancers [173, 174].

Despite potential secondary mechanisms, preclinical and clinical data support the premise that the inhibition of EGFR activity increases radio- and chemosensitivity of SSCHN tumors [173–175]. However, conflicting data regarding the sensitizing potential of EGFR inhibitors (EGFRIs) exist: for example, the RTOG 0522 phase III trial showed that the addition of the EGFRI cetuximab (Sect. 2.4.1.1) does not improve PFS rates to chemoradiation in patients with stage III and IV SCCHN, potentially due to overlapping radiosensitization properties of cisplatin and cetuximab treatment [176]. Newer agents that target EGFR remain under investigation (e.g., NCT02555644, NCT01427478, NCT00588770, and others: see also Table 2.1). Further, the side effect profiles of EGFRIs have been generally favorable compared to standard chemotherapeutics [177–179].

2.4.1 Monoclonal Antibodies Targeting EGFR and Other ErbB Proteins

Cetuximab, a monoclonal antibody targeting EGFR, plays a significant role in the treatment of SCCHN. Cetuximab, the pioneer for antibody-based anti-ErbB therapy in SCCHN, was approved for treatment of locally or regionally advanced SCCHN in 2006 [175] and for metastatic SCCHN in 2011 [180]. While cetuximab represents the EGFR inhibitor with the most clinical data and the most significant results in the treatment of SCCHN, multiple additional antibodies targeting ErbB receptors are currently being investigated in clinical trials. Promising results of EGFR-targeting antibodies used in mice were first published in 1984 [181], over two decades before cetuximab was approved for clinical use. This section introduces several relevant antibodies and covers their current status as it pertains to SCCHN, with some additional information found in Table 2.1.

2.4.1.1 Cetuximab

Cetuximab is a chimeric monoclonal antibody that inhibits EGFR by binding to its extracellular domain (Fig. 2.4). Cetuximab binds to EGFR with a higher affinity than its natural ligands EGF and TGFα [182–184]. Once bound to the EGFR extracellular domain, cetuximab occludes the ligand-binding site, thus inhibiting ligand-dependent EGFR signaling [185]. Depletion of the targeted receptors from the cell surface via downregulation is a second mechanism of effective EGFR inhibition [186]. Additionally, binding of cetuximab to EGFR enhances antibody-dependent, cell-mediated cytotoxicity via natural killer cells and macrophages in model systems [187, 188]. Cetuximab has been approved for three indications in patients with SCCHN. These are patients with locally or regionally advanced SCCHN (cetuximab in combination with radiation therapy; [175]), patients with recurrent or metastatic platinum-refractory SCCHN (cetuximab monotherapy; [189]), and patients with recurrent locoregional and/or metastatic SCCHN not refractory to platinum-based therapies (cetuximab in combination with platinum chemotherapy and 5-fluorouracil as first-line therapy; [180]).

Phase I studies of cetuximab defined the dose and schedule required to maintain biologically active and tolerable levels [190, 191]. Whether used in combination with chemotherapy or radiotherapy, or as monotherapy, cetuximab was found to have nonlinear saturation kinetics. Median serum cetuximab terminal half-life ranged from 14 to 97 h with doses from 5 to 300 mg/m2. Skin reactions increased significantly at doses of 500 mg/m^2 or higher. Given the results of these phase I trials, the recommended cetuximab regimen was established as an initial loading dose of 400 mg/m^2 followed by weekly doses of 250 mg/m^2 [190, 191].

The landmark phase III study of cetuximab added to radiation published by Bonner and colleagues led to the approval of cetuximab for the treatment of patients with locally or regionally advanced SCCHN [175]. Four hundred and twenty-four

Table 2.1 Monoclonal antibody and small-molecule inhibitors of EGFR

Monoclonal antibodies

Name	*Target*	*Stage of development/ trials*	*Administration*	*Comments*
Cetuximab	EGFR; extracellular; domain III	Approved for use in SCCHN/ EXTREME	IV, weekly	First targeted therapy for SCCHN [175, 180]. In combination with radiotherapy or cisplatin significantly increases progression-free survival [175, 192, 195]. Mediates antibody-dependent cellular cytotoxicity (ADCC) [198]
Panitumumab	EGFR; extracellular; domain III	Phase III/ PRISM, PARTNER, SPECTRUM (SCCHN)	IV, every 3 weeks	Potentially less immunogenic than chimeric mAbs; low rate of infusion-related hypersensitivity reaction [199]. In a combination with cisplatin is tolerable and demonstrates improved clinical outcome for high-risk, resected, HPV-negative SCCHN patients [357]. A phase I study of panitumumab, carboplatin, paclitaxel, and radiation for locally advanced disease has indicated that this combination is feasible with tolerable toxicity, and 69% of patients had a complete response and 34% had a partial response [201]
Zalutumumab	EGFR; extracellular; domain III	Phase III, DAHANCA 19 (SCCHN)	IV, every 2 weeks	Particularly effective induction of ADCC. In 286 patients with metastatic/recurrent SCCHN after failure of platinum-based therapy, zalutumumab plus best supportive care was compared with best supportive care plus methotrexate. Zalutumumab did not increase OS, although PFS was extended [358]
Nimotuzumab	EGFR; extracellular; domain III	Phase I/II (SCCHN)	IV, weekly	Binds with less affinity than cetuximab; mild to no skin toxicity. In a double-blind trial, patients with unresectable locoregional SCCHN were assigned to receive first-line therapy with nimotuzumab plus radiotherapy versus placebo plus radiotherapy. Complete response rates were significantly better in the nimotuzumab group with 59.5% for patients receiving nimotuzumab and radiotherapy versus 34.2% of patients receiving radiotherapy alone [359]. Nimotuzumab was found to be safe and well tolerated in a group of 92 treatment-naïve SCCHN patients and led to a benefit to long-term survival in combination with chemoradiotherapy (CRT) [360]

(continued)

Table 2.1 (continued)

Matuzumab	EGFR; extracellular domain III	No active trials	N/A	Binds at a completely different epitope than cetuximab [208]. Matuzumab has been evaluated in a phase I dose escalation study focused on patients with advanced EGFR-positive esophagogastric cancer. Matuzumab in combination with epirubicin, cisplatin, and capecitabine (ECX) was well tolerated. furthermore, in skin biopsies, decreased phosphorylation of EGFR and MAPK was detected [361]. Surprisingly, the phase II study of matuzumab in combination with ECX did not increase response or survival in patients with metastatic esophagogastric cancer [362]
XGFR*	EGFR; IGF-1R	No active trials	N/A	Binds EGFR and insulin growth factor receptor 1 (IGF-1R), which was shown to overcome resistance to EGFR inhibitors [210]. XGFR* is more selective for IGF-1R binding due to affinity maturation and stability to oxidative and thermal stress. In addition, the XGFR* induces ADCC. Showed a significant effect on different cancer types, including lung and pancreatic, in vivo and in vitro [211]
Tyrosine kinase inhibitors				
Gefitinib	ATP binding site; intracellular; TK domain of EGFR; reversible binding	Phase III	PO, daily	First TKI to reach a phase III investigation; trial failures led to its withdrawal from clinical investigation in SCCHN
Erlotinib	ATP binding site; intracellular; TK domain of EGFR; reversible binding	Phase II and phase III studies	PO, daily	Investigated as first-line treatment with radiotherapy or chemoradiotherapy in locally advanced SCCHN. Erlotinib has been evaluated in multiple phase II and phase III studies including studies focused on erlotinib combined with cetuximab, carboplatin, and paclitaxel (NCT01316757), docetaxel and radiation therapy (NCT00720304), docetaxel and cisplatin or carboplatin (NCT01064479), and cisplatin (NCT00410826) [168]. In patients with locally advanced SCCHN, the addition of erlotinib to standard cisplatin-radiotherapy regimens has not improved to therapeutic efficacy (CRR or PFS) [363]. However, a phase II clinical study of combination of erlotinib and docetaxel with intensity-modulated radiotherapy (IMRT) for locally advanced SCCHN suggested the inclusion of erlotinib led to a significantly better outcome [364]

(second generation)	site; intracellular; TK domain of EGFR and ErbB2/HER2; reversible binding		daily	dual-action small-molecule drug as an adjuvant to postoperative chemoradiation in SCCHN patients (e.g., NCT00387127), although addition of lapatinib to standard therapy RT/CDDP was found not to extend DFS [250, 365]. In general developments with this reagent, a novel type of lapatinib delivery, using hybrid nanoparticles, demonstrated a great potential in enhanced therapeutic effect in breast tumors and could be potentially applied for the treatment of SCCHN [366]
Afatinib (second generation)	ATP binding site; intracellular; TK domain of EGFR, ErbB2, ErbB4; irreversible binding	Phase III	PO, daily	Multi-specificity, comparable outcome to cetuximab monotherapy in a randomized phase II study. Phase III clinical trial has demonstrated that afatinib treatment significantly improved PFS and had a manageable safety profile in a group of 322 SCCHN patients in comparison with methotrexate (NCT01345682) [367]. In another ongoing phase III study, a group of patients with SCCHN after postoperative radiochemotherapy was randomized to afatinib or placebo for 18 months in a phase III trial designed to detect an improvement in PFS (NCT01427478)
Dacomitinib (second generation)	ATP binding site; intracellular; TK domain of EGFR, ErbB2, ErbB4; irreversible binding	Phase I and phase II	PO, daily	Has an activity against wild-type and mutant receptors, including EGFRvIII. Compared to phase II studies involving gefitinib or erlotinib, dacomitinib produces favorable outcomes in terms of disease control and survival [368]. In another clinical trial, dacomitinib demonstrated clinical efficacy with manageable toxicity in R/M-SCCHN patients who had failed prior treatment with platinum agents [369]

(continued)

Table 2.1 (continued)

CUDC-101	ATP binding site; intracellular; TK domain of EGFR and ErbB2; also histone deacetylase (HDAC)	Phase I	IV, 3 times per week	Structurally contains elements of TKIs (e.g., erlotinib) and elements of HDAC inhibitors (e.g., vorinostat). In a phase I clinical trial, CUDC-101 in combination with cisplatin and radiation was well tolerated. further patients' biopsies analysis demonstrated an inhibition of EGFR [264]
TX1-121-1	Binds the unique ErbB3 ATP-binding site (Cys721)	No active trials	N/A	Showed no significant effect against ErbB3-dependent signaling and growth in vitro. Subsequent conjugation of TX1-85-1 with adamantane (called TX2-121-1) caused partial proteolytic degradation of ErbB3 and inhibited ErbB3/ErbB2 and ErbB3/c-met heterodimerization in vitro [265]

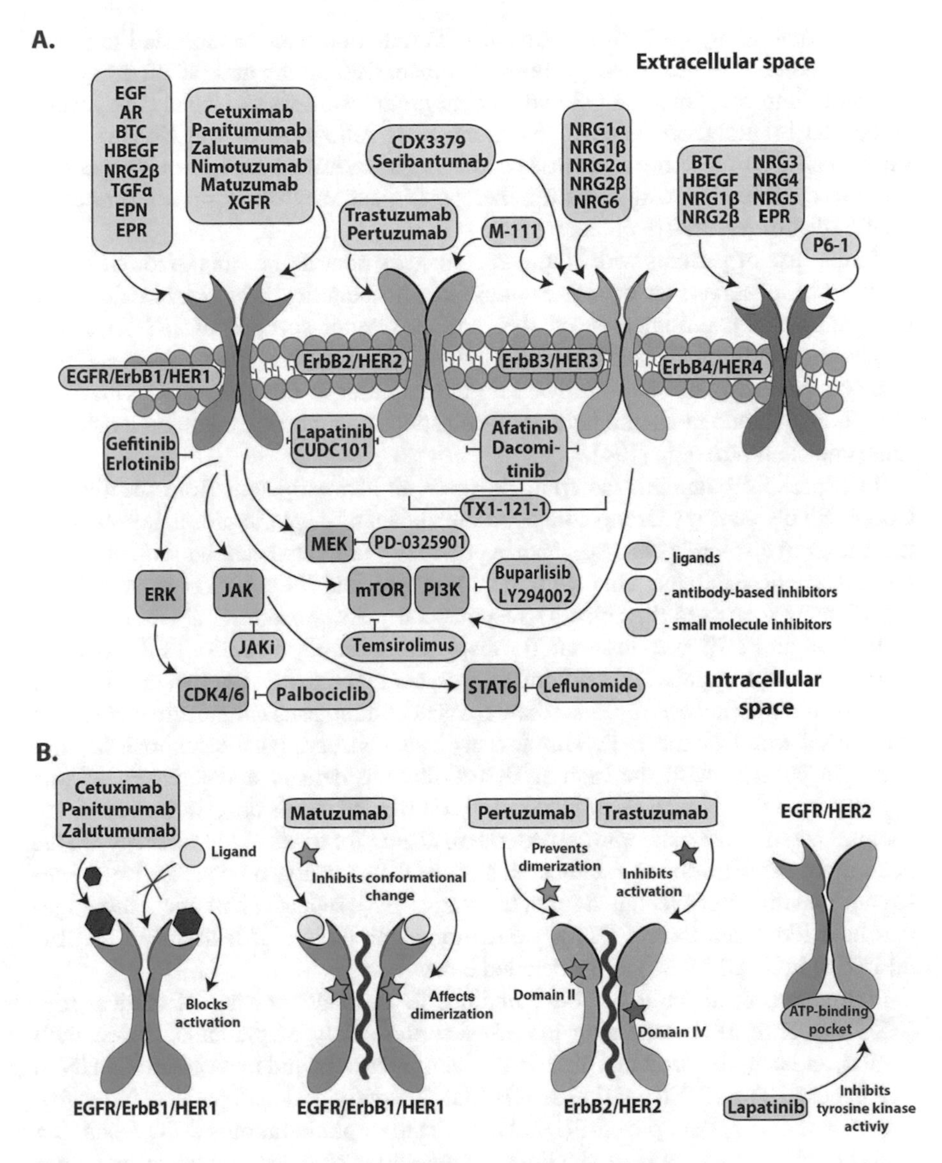

Fig. 2.4 Activators and inhibitors of the ErbB proteins and critical signaling effectors. (**a**) *Blue* boxes indicate activating ligands for the indicated ErbB dimers. *Pink* boxes indicate monoclonal antibodies targeting ErbB family proteins. *Green* boxes indicate small-molecule inhibitors of ErbB proteins or specific signaling effectors. (**b**) Mechanism of action for selected ErbB family inhibitors

patients undergoing definitive treatment with radiation were randomized to radiotherapy alone or to radiotherapy plus cetuximab. Cetuximab and radiotherapy significantly improved median OS and median progression-free survival (PFS) when compared to radiation alone [175]. Long-term follow-up of the Bonner study showed an absolute survival increase of 9% in the treatment group receiving cetuximab in combination with radiation therapy (5-year survival of 45.6% for cetuximab/radiation vs. 36.4% radiation alone) [192].

In the case of patients with platinum-refractory recurrent or metastatic SCCHN, Trigo et al. observed an overall response rate to cetuximab in combination with platinum-based chemotherapy of 10% and an overall survival of 183 days in a single-arm study of 96 patients with platinum-refractory recurrent disease [193]. The observed response and survival rate were similar to rates expected with platinum therapy alone in chemotherapy-naïve patients, supporting combination cetuximab and chemotherapy [194].

In a phase III randomized trial, Burtness and investigators from the Eastern Cooperative Oncology Group compared the impact of cisplatin plus a placebo with the impact of cisplatin plus cetuximab in previously untreated patients with recurrent and metastatic SCCHN. This study demonstrated that the addition of cetuximab significantly increased the objective response rate (26% response rate for cisplatin/cetuximab and 10% response rate for cisplatin/placebo; $p = 0.03$). PFS increased from 2.7 to 4.2 months; this difference was not statistically significant, in a trial which was underpowered because the statistical assumptions underestimated PFS in the control arm. Unexpectedly, Burtness et al. also observed that cetuximab was not active in patients with the highest EGFR staining density and intensity. It was hypothesized that several factors may have contributed to this observation: the small sample size ($n = 123$), suboptimal cetuximab dosing for cases of high-density EGFR occurrence, stochastic interactions at high EGFR density, or constitutive downstream signaling not accounted for in the study [195]. Indeed, subsequent investigators have identified loss of PTEN expression as predictive of resistance to EGFR inhibition in SCCHN, as will be detailed below.

Vermorken et al. built on these findings to conduct a phase III clinical trial (EXTREME trial) investigating the efficacy and safety of platinum, fluorouracil (5-FU), and cetuximab as first-line treatment of recurrent and metastatic SCCHN in 442 patients. The EXTREME phase III trial randomly assigned patients to receive cisplatin or carboplatin plus 5-FU and cetuximab or platinum plus 5-FU alone. Six cycles of chemotherapy were the limit for both arms of the study; however, cetuximab was continued until disease progression or prohibitive toxicity. The addition of cetuximab to chemotherapy significantly increased median OS (10.1 months in the cetuximab group and 7.4 months in the chemotherapy-alone group; $p = 0.04$) and PFS (5.6 months in the cetuximab group to 3.3 months in the chemotherapy-alone group; $p < 0.001$) when compared to standard chemotherapy alone [180]. Importantly, additional analysis of the EXTREME data provided further evidence that, in the case of SCCHN, EGFR expression level is not a clinically useful predictive biomarker [196]. Lei et al. demonstrated that resistance to cetuximab occurs in part through autophagy activation involving NLRX1-TUFM protein complex.

Analysis of patients' tumor specimens also confirmed that cetuximab therapy leads to the increased expression of autophagy SQSTM1/p62 protein, and data analysis from clinical trials showed a positive correlation between cetuximab-induced autophagy and poor prognosis [197].

2.4.1.2 Additional EGFR-Targeting Antibodies

Panitumumab is a fully humanized immunoglobulin IgG2 monoclonal antibody that, like cetuximab, binds to EGFR domain III and, in the process, inhibits EGF and TGFα binding [185] (Fig. 2.4). In contrast to cetuximab [198], panitumumab does not mediate antibody-dependent cellular cytotoxicity and has been shown to have a very low rate of infusion-related hypersensitivity reaction [199]. Another study found that panitumumab effectively inhibits EGFR signaling, but this antibody had a reduced ability to enhance dendritic cell maturation, in comparison with cetuximab, a finding of uncertain clinical significance [200] [201]. The SPECTRUM trial (phase III; NCT00460265) compared cisplatin/5-FU plus panitumumab to cisplatin/5-FU alone in patients with metastatic/recurrent SCCHN. The addition of panitumumab to chemotherapy did not significantly improve median OS versus chemotherapy alone but did improve median PFS (5.8 vs. 4.6 months) [184, 202]. Another study comparing PFS in patients with locally advanced (LA) SCCHN treated with standard-fractionation RT plus high-dose cisplatin versus accelerated-fractionation RT plus panitumumab demonstrated no difference between these two treatment types [203], providing further evidence that panitumumab has activity in SCCHN.

Zalutumumab is a human IgG1 high-affinity antibody also targeting EGFR domain III, and, just like panitumumab and cetuximab, zalutumumab is thought to block ligand binding, but with exceptional tumor specificity at lower doses [185] (Fig. 2.4). A phase III trial in metastatic or recurrent SCCHN following platinum failure compared zalutumumab to best supportive care, defined to include methotrexate monotherapy. Median overall survival was modestly increased by zalutumumab (hazard ratio [HR] for death, stratified by performance status, was 0.77; unadjusted $p = 0.0648$). Progression-free survival was significantly longer in the zalutumumab group (HR for progression or death was 0.63, 95% CI 0.47–0.84; $p = 0.0012$). A phase III trial is currently underway (DAHANCA 19; NCT00496652) to determine if the addition of zalutumumab to radiotherapy improves locoregional control. Preliminary results from this trial did not demonstrate beneficial effect from addition of zalutumumab [204].

Nimotuzumab has been approved for SCCHN in several countries, not including the USA. Nimotuzumab is a humanized murine IgG1 monoclonal antibody that also blocks interaction between ligand and receptor by binding to EGFR domain III, but with lesser affinity than some of the other antibodies [205]. The therapeutic implications of this reduced affinity are unclear, but nimotuzumab has been shown to have mild to absent skin toxicity, eliminating a clinically important adverse effect commonly associated with cetuximab [206]. An early pharmacodynamic study showed nimotuzumab plus radiotherapy was tolerated with no evidence of skin rash in patients with unresectable SCCHN [207]. Hence, nimotuzumab may offer an EGFR-targeted therapy with a favorable side effect profile. The possibility of combining high- and low-affinity antibodies to optimize antibody penetration in larger tumors has not been explored clinically.

Matuzumab, another humanized mouse monoclonal antibody, also binds to EGFR domain III, but at a completely different epitope than the previously mentioned antibodies (Fig. 2.4). This was confirmed by experiments in which cetuximab and matuzumab were observed to simultaneously bind to EGFR [208]. When bound to EGFR, matuzumab was determined to predominantly prevent domain II from assuming the configuration, in relation to domain III, necessary for high-affinity ligand binding [185], interrupting EGFR signaling. Matuzumab has not been tested in SCCHN.

XGFR*. In 2014, Schanzer et al. developed a novel one-arm single-chain Fab heterodimeric bispecific IgG (OAscFab-IgG; XGFR*) antibody, which targets both the insulin-like growth factor receptor type I (IGF-1R) and EGFR; importantly, this antibody has only one binding site for each target antigen [209]. It was previously shown that signaling through IGF-1R can overcome resistance to EGFR inhibitors, and EGFR-dependent signaling can confer resistance to IGF-1R inhibitors [210]. In addition, the XGFR* antibody has an afucosylated Fc portion and induces antibody-dependent cell-mediated cytotoxicity (ADCC). Thus, inhibition of both these receptors by XGFR* antibody had a significant effect on different cancer types, including lung and pancreatic, in vivo and in vitro [211]. Clinical evaluation of this agent is required.

2.4.1.3 Monoclonal Antibodies Targeting Other ErbB Proteins

Given the heterodimerization of EGFR with other ErbB family proteins, and the fact that overexpression of some of these proteins can compensate for EGFR inhibition during development of therapeutic resistance, a natural development has been to explore inhibition of additional EGFR family members in SCCHN [212–214].

Pertuzumab binds the ErbB2 dimerization domain and blocks its interaction with all four ErbB family members [215] (Fig. 2.4). Erjala et al. observed that increased expression levels of phosphorylated ErbB2 and total ErbB3 were associated with SCCHN cell line resistance to gefitinib [216]. Confirming the importance of ErbB2 in resistance, when gefitinib was combined with pertuzumab, significant growth inhibition of relatively gefitinib-resistant SCCHN cell lines was observed. Phosphorylated ErbB2 and total ErbB3 were not predictive of resistance to cetuximab [216].

Seribantumab (SAR256212, MM-121), targeting ErbB3, has been shown to be effective by inhibiting ligand-induced ErbB3 signaling [50, 217]. In another study, seribantumab demonstrated antitumor activity in breast cancer cell models [218]. More recently, in a randomized phase II trial of seribantumab in combination with weekly paclitaxel compared with paclitaxel alone, the combination showed no effect in PFS among 140 patients with ovarian cancer [219].

MM-111 is a bispecific single-chain antibody that simultaneously targets ErbB2 and ErbB3. An antitumor effect of MM-111 was shown in several in vivo cancer models [220].

P6-1. Recently, a new monoclonal antibody that specifically targets neuregulin-1-induced ErbB4 activation was developed. P6-1 was tested on breast cancer cell lines and showed moderate anticancer activity [221].

Trastuzumab. The ErbB2/HER2-targeting antibody trastuzumab is an invaluable drug for breast cancer and other epithelial tumors [222–225]. In vitro studies have shown that trastuzumab enhances the efficacy of gefitinib [226] and cetuximab [227] in SCCHN cells. Surprisingly, analysis of the mRNA expression of EGFR and ErbB2 indicated lack of correlation with efficacy of the combination therapy [227]. Moreover, an independent study found that a subset of non-ErbB2-amplified SCCHN cells was nevertheless extremely responsive to the small molecule multi-ErbB inhibitor lapatinib, based on activation of a neuregulin-ErbB3 loop [228]. A number of studies have been investigating factors contributing to trastuzumab. Kulkarni and colleagues found that the Ca^{2+}-activated Cl^- channel TMEM16A significantly contributes to tumor growth in SCCHN and some other tumors, with levels of TMEM1A increased in trastuzumab-resistant cells. They also found that concurrent treatment of these cells with cetuximab and TMEM16A inhibitor led to cell death, demonstrating a novel role of TMEM16A in regulation of the EGFR and HER2 pathways [229].

Duligotuzumab (MEHD7945A) blocks ligand binding to EGFR and HER3 and may contribute to antibody-dependent cell-mediated cytotoxicity (ADCC) in cells overexpressing these proteins. A randomized phase II study (MEHGAN, NCT01577173) evaluated drug efficacy in patients with recurrent/metastatic (R/M) SCCHN, treated with duligotuzumab or cetuximab. In this study, duligotuzumab demonstrated a higher efficacy than cetuximab only in tumors with greater expression of NRG1, a ligand for ERBB3 and ERBB4 [230]. Duligotuzumab in combination with cisplatin/5-fluorouracil or carboplatin/paclitaxel demonstrated a promising effect in patients with recurrent/metastatic SCCHN [231]. An independent study performed in a group of SCCHN and colorectal cancer patients demonstrated similar results, with duligotuzumab as a single agent possessing pronounced antitumor activity [232].

CDX3379 (formerly KTN3379) locks HER3 in an inactive confirmation by docking to its extracellular domain, effectively promoting inhibition of both ligand-dependent or ligand-independent activation of HER3 [233]. CDX3379 was found to be efficient against HPV-positive SCCHN cell lines and PDXs [234]. CDX3379 is currently being evaluated for safety in a phase I clinical trial (NCT02014909).

2.4.2 Tyrosine Kinase Inhibitors (TKIs) Targeting EGFR and Other ErbB Proteins

TKIs block EGFR activation by inhibiting the cytoplasmic tyrosine kinase domain and have proven valuable agents in a number of cancer types. First-generation TKIs for EGFR, including gefitinib and erlotinib, reversibly bind the ATP-binding pocket of the kinase domain and are EGFR specific. Second-generation TKIs relevant to SCCHN, including lapatinib, afatinib, and dacomitinib, target multiple ErbB members (in the case of afatinib and dacomitinib, irreversibly) (Table 2.1; Fig. 2.4) [168, 182].

Gefitinib, an orally administered, small-molecule, reversible EGFR TKI, was the first TKI to reach phase III trials; however, overall results have dampened expectations. Early studies suggested a clinical benefit of gefitinib similar to cetuximab [235, 236]; unfortunately, more recent results do not indicate a significant role for gefitinib in the management of SCCHN [168, 237, 238]. In treatment-refractory SCCHN, gefitinib did not improve OS when compared to methotrexate [239]. These findings were consistent when gefitinib was administered orally at 250 or 500 mg daily, despite the fact that single-arm studies had demonstrated favorable response rates when gefitinib was administered at 500 mg/day [235, 239]. The maximum tolerated dose of gefitinib has been established to be 600 mg/day, with dose-limiting toxicity observed at 1000 mg/day [240]. In a phase II study of 44 patients with SCCHN, Perez et al. investigated if doses higher than 500 mg/day would produce increased skin toxicity, which is thought to be associated with improved response to EGFRIs. Patients treated with 750 mg/day had increased incidence of skin toxicity compared with patients receiving 500 mg/day (58 and 9% grade 2 skin toxicity for 750 and 500 mg/day, respectively); however, the higher dose of gefitinib failed to significantly improve outcome [241]. Gefitinib (250 mg/day) in combination with docetaxel (docetaxel/gefitinib vs. docetaxel alone) was evaluated in a phase III trial of patients with metastatic or locally recurrent SCCHN. In this study, Argiris et al. were unable to demonstrate a statistically significant survival benefit for patients receiving the docetaxel/gefitinib combination. However, subgroup analysis demonstrated that for patients younger than 65 years of age, the addition of gefitinib to docetaxel did increase survival significantly (median survival of 7.6 vs. 5.2 months; $p = 0.04$) [242]. In the case of NSCLC, non-smoking patients had a significantly better response to gefitinib treatment and demonstrated prolonged PFS in comparison with smoking patients [243]; this may be related to the fact that smoking induces detoxification enzymes such as cytochromes (CYP2D6, CYP1A2) that contribute to the metabolism of gefitinib and erlotinib [244], limiting their efficacy. It is important to note that gefitinib therapy demonstrates high interpatient variability in plasma levels even in healthy male volunteers, varying up to 15-fold between individuals [245], suggesting the need for further studies to allow more accurate individual dosing with gefitinib and other 4-anilinoquinazolines (including erlotinib and lapatinib).

Erlotinib, the second first-generation TKI, like gefitinib, is an orally administered, small-molecule, reversible TKI. In 115 patients with refractory SCCHN, erlotinib led to disease stabilization in 38.3% of patients for a median duration of 16.1 weeks. The median PFS was 9.6 weeks and the median OS was 6.0 months [246]. Smoking status of lung cancer patients was also found to be an important determinant of the pharmacokinetics of erlotinib during therapy, leading to a 2.8-fold increased C_{max} and decreased metabolic clearance of erlotinib in non-smokers to compare with smoking patients [247].

Lapatinib has dual specificity, targeting EGFR and ErbB2 [248] (Fig. 2.4). Surprisingly, in a phase II study focused on recurrent/metastatic SCCHN, lapatinib, as monotherapy, failed to lead to objective responses. On-treatment biopsies were performed: although ErbB2 levels were significantly decreased, EGFR phosphorylation remained unaffected in these specimens [249]. Addition of lapatinib to standard therapy radiation and cisplatin": standard therapy with radiation and cisplatin was found not to extend DFS [250]. Another clinical trial focusing on the combination of lapatinib with capecitabine in patients with metastatic SCCHN (NCT01044433) is still in process. Additional phase II trials are investigating the effect of lapatinib in combination with definitive chemoradiation followed by 1 year of lapatinib maintenance for locally advanced (LA) SCCHN (NCT00387127) and definitive radiation for the patients who had low tolerance to CRT (NCT00490061) [251].

Afatinib, also a multi-specific TKI, irreversibly targets three of the four ErbB family members: EGFR, ErbB2, and ErbB4 [252]. In a comparison to cetuximab, afatinib showed similar antitumor activity in patients with recurrent or metastatic SCCHN after failing platinum therapy. Median PFS was 15.9 weeks with afatinib and 15.1 weeks with cetuximab [253]. In another clinical trial, 2 weeks of pretreatment of newly diagnosed SCCHN patients with afatinib led to a better metabolic FDG-PET profile than seen in patients receiving no treatment (NCT01538381) [254]. Based on the results from a group of R/M SCCHN patients, Cohen et al. established potential biomarkers of afatinib clinical outcome. Prolonged PFS after afatinib treatment was associated with amplified *EGFR*, negative p16 status, high expression of PTEN, and low expression of HER3 [255]. According to the most recent data, afatinib treatment of 411 SCCHN patients with complete response to chemoradiation did not demonstrate any improvement in disease-free survival to compare with placebo group (NCT01345669) [256].

Dacomitinib, like afatinib, is an irreversible inhibitor of EGFR, ErB2, and ErbB4 [257]. Dacomitinib is a potent inhibitor of wild-type EGFR as well as EGFR with activating mutations. Furthermore, dacomitinib appears to be active against the *T790 M* secondary *EGFR* mutation, which generally renders cancer cells resistant to erlotinib and gefitinib, in NSCLC [258]. Dacomitinib reduced the viability of SCCHN cells and in combination with ionizing radiation (IR) more effectively delayed tumor growth in vivo in a dose-dependent manner [259]. A phase I clinical trial investigating the efficacy of dacomitinib plus chemotherapy in treating SCCHN was completed (NCT01737008), but the results are not yet published. In another study (NCT01484847), dacomitinib was administered via gastrostomy feeding tube (GT) to evaluate the efficacy of this method. The results have demonstrated that the pharmacokinetics of dacomitinib administered by GT are significantly decreased to compare with the oral administration [260]. In a completed phase II trial (NCT00768664) of 69 patients with recurrent/metastatic SCCHN treated with dacomitinib, pretreatment tumor and normal tissue specimens were analyzed. No biomarkers of dacomitinib treatment efficacy were identified. Nevertheless, dacomitinib treatment was associated with increased OS in HPV-driven cancers [261].

CUDC-101 is a multi-targeted hybrid anticancer drug candidate with a complex mode of action. CUDC101 effectively inhibits EGFR, HER2, and histone deacetylase (HDAC) and has shown impressive activity in in vitro as well as in vivo cancer models [262, 263]. This hybrid inhibitor has been investigated in combination with cisplatin and radiation therapy in patients with locally advanced head and neck cancer as part of a phase I drug escalation trial (NCT01384799). The results of this study established the maximum tolerated dose (MTD) of CUDC-101 in combination with cisplatin and radiation. Further analysis of tumor biopsies demonstrated inhibition of EGFR [264].

TX1-121-1. ErbB3 plays an important role in the EGFR signaling through activation of EGF receptor [9]. The small-molecule agent TX1-85-1 covalently binds the unique ErbB3 ATP-binding site (Cys721). However, this compound showed no activity against ErbB3-dependent signaling and growth [265].

2.4.3 Targeting Critical EGFR/ErbB Effectors

A classic means by which cancer cells overcome resistance to inhibition of upstream activating oncogenes is to upregulate or activate one or more essential effectors operating downstream. A reactive therapeutic approach is to inhibit these downstream effectors, sometimes in combination with agents targeting the upstream oncogenic driver. In SCCHN, resistance to inhibitors of ErbB proteins has been associated with upregulation or activation of a number of downstream or lateral signaling effectors, including PI3K, MEK, ALK, and others. Some recent studies have evaluated inhibition of these targets in SCCHN.

Using xenograft experiments, Mizrachi et al. found that the inhibitor BYL719, which targets PI3Kα (p110α subunit of PI3K), is effective against SCCHN. Furthermore, they demonstrated that encapsulation of BYL719 into P-selectin-targeted nanoparticles leads to an accumulation of BYL719 in tumors, which promotes more effective inhibition of tumor growth and ameliorates side effects associated with BYL719 treatment [266]. Several isoform-specific PI3K inhibitors are under development and in clinical trials [267]. In preclinical studies, the specific PI3KCA inhibitor GDC-0032 was effective in controlling SCCHN in vitro and in vivo [267]. Co-targeting of EGFR and PI3K with erlotinib and BKM20, respectively, had synergistic antitumor effects and apoptosis induction in a panel of SCCHN cell lines and xenograft models of SCCHN.

There is some evidence that, in several types of cancer, including breast, non-small cell lung, and glioblastoma, inhibition of PI3K or mTOR can abrogate cell resistance to anticancer therapy [268]. Apitolisib (GDC-0980) is a dual inhibitor of class I PI3Ks and mTOR kinases. A phase I clinical trial of apitolisib indicated durable antitumor activity in a group of patients with solid tumors, including SCCHN [269]. This may predominantly reflect inhibitory activity against PI3K, as treatment of recurrent/metastatic SCCHN patients with the mTOR inhibitor everolimus showed no response to the therapy [270] [271]. However, other studies have concluded that concurrent targeting of EGFR and PI3K is synergistic specifically because of inhibition of both axes of the AKT-mTOR pathway, coupled with translational regulation of anti-apoptotic Bcl-2 proteins [271].

Activation of ErbB kinases signals through RAS and RAF to activate downstream MEK/ERK kinases; RAS, PI3K, and other ErbB effectors activate AKT. Cetuximab inhibits ERK and AKT phosphorylation in cetuximab-sensitive SCCHN cells, whereas the level of AKT phosphorylation is unmodified in cetuximab-resistant cells [272]. Mohan et al. demonstrated that inhibition of MEK by PD-0325901 overcame resistance to the PI3K/mTOR inhibitor PF-5212384 and had a potential antitumor effect in SCCHN [273]. Increased activity of the anaplastic lymphoma kinase (ALK) was observed in late-stage human OSCC tumors and invasive OSCC cell lines. Both in vitro and in xenografts, concurrent inhibition of ALK (using the TAE684 inhibitor) and EGFR (with gefitinib) significantly reduced OSCC cell proliferation and tumor volume in comparison with ALK inhibition alone. Dual inhibition was associated with complete abolishment of AKT activation, whereas separate inhibition of EGFR or ALK only reduced it [274].

A study of esophageal squamous cell carcinoma showed that tumor cells treated with afatinib and erlotinib become rapidly resistant due to reactivation of the MEK/ERK pathway. This can be delayed by initial dual treatment with the MEK inhibitor trametinib used in combination with an EGFR inhibitor, which decreased tumor cell proliferation and survival [275]. Such strategies may be useful for SCCHN. The same study also noted the potential utility of combining CDK4/6 inhibitors with EGFR inhibitors in *EGFR*-amplified tumors [275]. At present, the CDK4/6 inhibitor palbociclib is in a phase I clinical trial in a complex with cetuximab for SCCHN patients. Preliminary results have demonstrated the safety of palbociclib with cetuximab in patients with recurrent/metastatic SCCHN. Encouragingly, tumor responses were observed, even in cetuximab- or platinum-resistant disease [276].

Checkpoint kinases 1 and 2 (Chk1/2) are critical regulators of the DNA damage response and important for regulation of cell cycle arrest in S phase to allow repair. In preclinical studies, simultaneous treatment with the Chk1/2 inhibitor prexasertib, cetuximab, and irradiation significantly decreased cell proliferation and survival of SCCHN cells both in vitro and in vivo. A clinical trial to test this treatment for patients with SCCHN is ongoing (NCT02555644) [277]. Huang et al. have performed whole-genome sequencing of cisplatin-resistant SCCHN tumors to find potential predictive biomarkers of dacomitinib resistance and identified a "platinum" mutational signature, involving a number of genes, including *REV3L*, which encodes the catalytic subunit of DNA polymerase ζ. Further investigations have demonstrated that depletion of REV3L dramatically enhanced the sensitivity of SCCHN cells to the ErbB2 inhibitor dacomitinib through translesion synthesis and homologous recombination [278].

FGFR1 was identified as a prognostic marker of SCCHN, and shown to be highly expressed in 82% (36/44) of HPV(+) and 75% (294/392) of HPV(-) SCCHN samples, and associated with poor OS and DFS in samples with HPV(-) status. Mechanisms of resistance to the FGFR inhibitor AZD4547 occur due to compensatory EGFR signaling [279], suggesting potential synergy in combining EGFR- and FGFR-targeting agents.

2.4.4 EGFR/ErbB2 and Immunotherapy/Immune Response

The past 5 years have seen the emergence of several distinct classes of immunotherapies as potent anticancer treatments. Some immunotherapies can be used to potentiate natural killer (NK) cell-mediated antibody-dependent cellular cytotoxicity (ADCC) against antibody-coated tumor cells, enhancing the antitumor activity of monoclonal antibodies [280]. Of specific relevance to EGFR inhibition, therapy with cetuximab mediates NK cell/dendritic cell (DC) cross talk by cross-linking FcγRIIIa. NK cells activated by cetuximab can upregulate the costimulatory receptor CD137 (4-1BB), which enhances NK cell activity upon treatment with urelumab (CB137 agonist). In a phase Ib trial, concurrent treatment of SCCHN patients with cetuximab and urelumab showed modulation of immune biomarkers and better survival of activated NK cells taking part in antitumor cell immunity [281]. Another group showed that NK cell-mediated ADCC increases upon stimulation with IL-12 in the presence of cetuximab-treated SCCHN cell lines. Combination IL-12 and cetuximab also significantly reduced tumor volume in a group of mice injected with a squamous cell carcinoma of tongue cell line (Cal-27). These results suggest that concurrent treatment with cytokines and cetuximab might have a beneficial effect in SCCHN therapy [282]. Kumai et al. have exploited the fact that HER3 expression is significantly increased upon EGFR inhibition therapy in SCCHN. They have used a HER3 peptide analog as a helper epitope for antigen presentation and found this induces cytolytic activity of CD4 T cells against tumor cells in vitro [283].

Conversely, cetuximab treatment has been linked to suppression of ADCC by increasing activity of regulatory T cells (Tregs), expressing CTLA-4, CD39, and TGFβ. Levels of Tregs correlated with poor clinical outcome in patient cohorts treated with cetuximab. Inhibition of CTLA-4(+) Treg by the CTLA-4-targeting antibody ipilimumab restored NK cell-mediated ADCC, resulting in better response to cetuximab treatment in vitro [284]. Finally, programmed death 1 (PD-1)/programmed death ligand 1 (PD-L1) immune checkpoint inhibitors are emerging as potent therapeutic agents. PD-1/PD-L1 limit T lymphocyte activation and promote immune resistance in SCCHN [285, 286]. In a cohort of 134 SCCHN specimens, PD-L1 levels were elevated and notably higher in HPV+ samples. This elevation was also positively correlated with EGFR and JAK2 levels; moreover, EGFR was identified as a mediator for activation of PD-L1 in a JAK2- and/or STAT1-dependent manner. In vitro studies using coculturing of tumor and NK cells showed that JAK2 inhibition using BMS-911543 in combination with cetuximab resulted in prevention of PD-L1 upregulation and increase of immunogenicity in tumor cells [287]. The rapid growth of the PD-1/PD-L1 inhibitor armamentarium and the large number of agents moving to clinical trial suggest that this field will expand enormously in the next several years.

2.5 Mechanisms of Resistance to Anti-EGFR Therapies

Patients initially responsive to anti-EGFR therapy often develop resistance during the course of treatment [18]. A number of specific factors associated with resistance have been identified [288]. These include altered ubiquitination and trafficking (Sect. 2.5.1) [289–291], overexpression and amplification of ErbB2 (Sect. 2.5.2) [216], altered expression levels of VEGF (Sect. 2.5.3) [292], altered expression levels of STAT3 (Sect. 2.5.4) [293], *KRAS* mutations (Sect. 2.5.5) [294, 295], changes in the tumor microenvironment (Sect. 2.5.6) [296], epigenetic compensation (Sect. 2.5.7) [297], and several other factors (Sect. 2.5.8).

2.5.1 EGFR-Intrinsic Resistance

Altered EGFR ubiquitination represents a mechanism of acquired resistance to cetuximab [289–291]. In vitro resistance to cetuximab was established by exposing cells to subeffective doses of cetuximab [291] or by prolonged exposure to escalating doses [290]. Lu et al. found that the cetuximab-resistant colorectal cancer cell line DiFi5 (rendered resistant through prolonged exposure to cetuximab) had markedly lower levels of EGFR. However, DiFi5 cells had enhanced associations between EGFR and the E3 ubiquitin ligase Cbl, as well as increased levels of ubiquitinated EGFR. DiFi5 also had significantly higher levels of active, Y16-phosphorylated Src, both at baseline and post-EGF stimulation, with inhibition of Src with the nonselective kinase inhibitor PP2 reversing cetuximab resistance. In addition, DiFi5 cells responded to EGF stimulation with more robust phosphorylation of EGFR at Y845 and strong phosphorylation of AKT and other extracellular EGFR signal-regulated kinases. These observations suggest that colorectal cancer cells may develop resistance to cetuximab by reducing EGFR levels via increased ubiquitination and degradation and via increased Src kinase-mediated cell signaling to bypass dependency on EGFR for cell growth and survival [291].

On the other hand, Wheeler et al. reported increased EGFR expression levels associated with deregulation of EGFR internalization and degradation in several resistant clones of NSCLC cell lines [289]. Loss of c-Cbl association with EGFR was reported to significantly lessen EGFR ubiquitination after EGF stimulation in the cetuximab-resistant cells compared to the nonresistant parent cells. These findings suggest that acquired resistance to cetuximab is accompanied by deregulation of EGFR internalization/degradation and subsequent EGFR-dependent activation of ErbB3 [289]. Further supporting the role of decreased EGFR ubiquitination in treatment resistance, Ahsan et al. found that cisplatin-resistant head and neck cancer cell lines undergo minimal EGFR phosphorylation at the Y1045 site and minimal ubiquitination [298].

Genetic variance in patient populations can affect response to EGFR inhibitors. For example, in >40% of patients with SCCHN, a single nucleotide variant polymorphism of *EGFR* (*EGFR R521K*) is present [299, 300]. This polymorphism has been linked to primary resistance to cetuximab in SCCHN in vitro and in vivo models and suggested as a potential marker for response to therapy. Cetuximab has a lower affinity to EGFR bearing a lysine residue at aa 521, affecting some patient populations. However, a next-generation EGFR antibody, GT-MAB 5.2-GEX with an Fc optimized by glycosylation [301, 302] to bind with higher affinity to the FcγRIIIa on immune effector cells and promote enhanced ADCC activity, targets EGFRK521 more efficiently [303].

2.5.2 Elevated Expression of ErbB Family Members

Ritter et al. demonstrated elevated levels of phosphorylated EGFR, EGFR/ErbB2 heterodimers, TGFα, hairpin-binding EGF, and heregulin RNA in trastuzumab-resistant human breast cancer cells. These findings suggest that enhanced EGFR-mediated activation of ErbB2 may be a potential mechanism of acquired resistance to trastuzumab [304]. A study by Yonesaka et al. (2011) identified a new mechanism of de novo and acquired resistance to cetuximab via increased signaling through ErbB2. Yonesaka et al. have shown that amplification of ErbB2 or upregulation of heregulin (ErbB3/ErbB4 ligand) is present in cetuximab-resistant colorectal cancer patients. This study suggests that ErbB2 inhibitors, in combination with cetuximab, represent a rational therapeutic strategy that should be assessed in patients with cetuximab-resistant SCCHN [212]. The same was demonstrated for ErbB3. In vitro experiments with SCCHN cell lines showed that treatment with cetuximab caused HER3 activation and HER2/HER3 dimerization. However, combined treatment with cetuximab and MM-121/seribantumab significantly decreased cell growth and downregulated the PI3K/AKT and ERK pathways, in comparison with each antibody alone. A similar effect was found in cetuximab-resistant xenografts and PDX models [305]. It is possible that such synergies may be achievable using a single therapeutic agent. A recent study in non-small cell lung cancer, using the novel Pan-HER inhibitor from Symphogen A/S (Ballerup, Denmark), which is an antibody mixture targeting EGFR, HER2, and HER3, effectively overcame resistance to cetuximab [214].

2.5.3 VEGF Expression

Enhanced angiogenesis is a fundamental step in the transition of tumors from a dormant state to a malignant one and correlates with tumor progression and metastasis [306]. Angiogenesis is elevated in various human tumors, including SCCHN, and VEGF has been demonstrated to be a major angiogenic factor [307]. Preclinical and early clinical data imply a central role of angiogenesis in SCCHN: up to 90% of SCCHNs express vascular endothelial growth factor (VEGF) and the respective receptors (VEGFRs) [308].

Multiple studies support the prognostic implications of angiogenic markers in SCCHN and functional connections between angiogenic and EGFR signaling [309]. One recent study has identified angiopoietin-like 4 (ANGPTL4) as a regulator of EGF-induced cancer metastasis. The authors suggested that expression and autocrine production of ANGPTL4 mediated by EGF promotes SCCHN metastasis through the expression of metalloprotease-1 (MMP-1) [310]. Another study used immunofluorescence and single-cell segmentation to analyze expression and localization of EGFR in relation to the endogenous hypoxia marker CA IX and the intratumoral diffusion distance of EGFR+ cells from microvessels. Analysis was done using 58 human SCCHNs and a set of normal versus cancer-adjacent tissues [311]. This study found that the resistance to cetuximab was associated with downregulation of membrane EGFR expression in the hypoxic tissue [311]. Bossi et al. analyzed a group of recurrent-metastatic SCCHN patients pretreated with cetuximab and platinum chemotherapy, using whole transcriptome screening to find potential unique signatures of resistance. They found a connection between prolonged PFS and enriched expression of genes associated with active EGFR pathway signaling and hypoxic differentiation. In contrast, short PFS was associated with elevated RAS (but not EGFR) signaling, implying compensatory activation of downstream signaling [312].

EGFR activation and the overexpression of the three major ErbB-associated ligands trigger upregulation of multiple VEGF members and may induce resistance to anti-EGFR agents in vitro [292]. Riedel et al. showed that an EGFR antisense oligonucleotide treatment resulted in a significant reduction of VEGF protein expression, and addition of conditioned medium from EGFR antisense-treated tumor cells resulted in decreased endothelial cell migration [313]. The combination of bevacizumab (a humanized monoclonal IgG1 antibody targeting VEGF) with erlotinib was well-tolerated and had a response rate of 15% [314], which, in a cross-trial comparison, was higher than the response rate for erlotinib alone (5%) [315] or the VEGFR inhibitors SU5416 alone (5%) [316] or sorafenib (an inhibitor of VEGFR, PDGFR, Raf kinase, and others) alone (3–4%) [317]. A phase II clinical trial of sorafenib in combination with cetuximab demonstrated no clinical benefit in a group of patients with recurrent and/or metastatic (R/M) SCCHN [318]. Argiris et al. demonstrated that the combination of bevacizumab and cetuximab enhanced growth inhibition both in vivo and in vitro in preclinical models and resulted in a SCCHN disease control rate of 46% [319]. However, a phase II trial showed that radiotherapy and cetuximab with the addition of bevacizumab led to no improvement in treatment efficacy compared with radiation/cetuximab alone, showing no clinical benefits of dual targeting of VEGF and EGFR for a group of previously untreated stage III–IVB SCCHN patients [320]. Chemotherapy with or without bevacizumab is being investigated in a phase III trial with patients with recurrent or metastatic head and neck cancer (NCT00588770).

A novel fully humanized dual-targeting IgG (DT-IgG) antibody that simultaneously targets VEGF and EGFR has been designed to optimize tumor targeting and maximize potential clinical benefits [321]. Hurwitz et al. tested DT-IgG on SCCHN, lung adenocarcinoma, and colon cancer xenograft models and discovered that DT-IgG had a lower in vivo IC50 than bevacizumab (VEGF-targeting antibody) and cetuximab; however, a higher dose of DT-IgG was needed to produce efficacy similar to that observed with combined bevacizumab and cetuximab treatment [321].

Zhang et al. showed in SCCHN in vitro studies that DT-IgG neutralizes VEGF as effectively as bevacizumab and inhibits EGFR activation and cell proliferation as effectively as cetuximab [322]. One obvious benefit of DT-IgG therapy would be avoidance of dosing complications associated with drug combinations [321, 322]. Lecaros et al., using human SCCHN xenograft models, have found that delivery of siRNA against VEGF-A using lipid-calcium-phosphate nanoparticles (LCP NPs) in combination with photodynamic therapy promotes apoptosis and controls tumor growth [323]; this may be useful in conjunction with EGFR-targeting therapies.

2.5.4 STAT3 Expression

Sen et al. found that increased STAT3 may contribute to cetuximab resistance in SCCHN [293]. STAT3 inhibition in cetuximab-resistant SCCHN cells using a STAT3 decoy oligonucleotide to inhibit STAT3-mediated transcription reduced cellular viability and the expression of STAT3 target genes. STAT3 decoy treatment also successfully decreased tumor growth in vivo [293]. In SCCHN cells, activation of the JAK2/STAT3 pathway was associated with EMT and metastasis through induction of the chemokine CCL19 and its receptor CCR7, which was previously found to play an important role in chemotaxis and migration of immune cells, such as leukocytes. In a panel of 78 human SCCHN specimens, phosphorylation of CCR7 and STAT3 positively correlated with lymph node metastasis [324].

2.5.5 KRAS *and* PI3K *Mutation*

KRAS mutations are fairly rare in SCCHN compared to other types of cancer [325]. Mutational activation of KRAS only occurred in 2.6% of 115 clinical specimens of SCCHN, although copy number amplification of *KRAS* was found in 10 samples (8.7%) in the same study [326]. Chau et al., using next-generation sequencing (NGS) in a group of 213 SCCHN patients, demonstrated that oncogenic *RAS* mutations are associated with poorer PFS. In another study, *KRAS* mutations were found in 4 out of 29 patients with SCCHN, and the presence of the *G12 V KRAS* mutation was associated with an absence of response to cetuximab and radiotherapy [327]. Monitoring of *RAS* mutations as well as mutations in the cetuximab-interacting ectodomain of the EGFR may be a good predictive marker for cetuximab resistance [328, 329]. *KRAS* mutation has been associated with HPV status. In a cohort of 179 SCCHN, *KRAS* mutations occurred more frequently in HPV(+) tumors (6%) versus 1% for HPV(-) tumors [330]. Finally, as noted above, *PIK3CA* is one of the most frequently amplified or mutated oncogenes in SCCHN tumors [67, 119, 121], with mutation associated with activation of the AKT signaling pathway [331]. Additionally, phosphatase and tensin homolog (PTEN) acts as a tumor suppressor and an important negative regulator of the PI3K/AKT pathway. Low expression of PTEN (44.4% with low vs. 71.7% with high) and pAKT (75.2% with low vs. 52.5% with high) were shown to be important for prolonged 2-year overall survival in a group of 49 patients with SCCHN previously treated with cetuximab-based induction chemotherapy (a combination of cetuximab, docetaxel, and cisplatin) [332]. Burtness et al. demonstrated that loss of PTEN (PTEN null) was observed in 23 of 67 (34%) SCCHN patient samples analyzed by AQUA and that treatment of these patients with cetuximab or placebo leads to better PFS in comparison with patients with tumors expressing PTEN (4.2 months for cetuximab vs. 2.9 months for placebo for patients expressing PTEN and 4.6 months for cetuximab vs. 3.5 months for placebo for PTEN null) [333].

2.5.6 Microenvironment

A growing body of evidence suggests that components of the tumor microenvironment may also contribute to tumorigenesis in cancers of epithelial origin and may modulate the treatment sensitivity of tumor cells [334]. Johansson et al. reported that cancer-associated fibroblasts (CAFs) offer protection from cetuximab treatment and negate cetuximab-induced growth inhibition [296]. They further described that SCCHN cell lines cocultured with CAFs from patients with SCCHN result in elevated expression of matrix metalloproteinase-1 (MMP-1) in both the tumor cells and the CAFs. MMP inhibitors can partly abolish CAF-induced resistance; however, siRNA knockdown of MMP-1 in CAFs did not abolish resistance, suggesting that other MMP family members may be involved (Johansson et al. 2012). The mechanism of MMP-associated cetuximab resistance is not clear, and further investigation is warranted. Quantitative RT-PCR analysis of 31 tumor-adjacent tissues, 94 SCCHN tumors, and 10 tonsillectomy noncancerous specimens has revealed that expression of metalloproteins (MT) MT1A and MT2A is significantly higher in the tumors and correlates with a higher tumor grade. Expression of MT was also observed in tonsillectomy samples, gradually increasing in adjacent tissues and tumors, possibly in response to the oxidative stress. The authors suggested that MT accumulation in adjacent tissue occurs as a response to the tumor cells [335].

2.5.7 Epigenetic Changes

Emerging evidence has indicated connections between epigenetic changes, such as DNA methylation at CpG islands, and development of resistance to multiple cancer therapeutics [336–339]. Ogawa et al. tested a panel of 56 genes (including death-associated protein kinase (*DAPK*), *MGMT*, and *SRBC*, commonly known to be regulated through promoter methylation, using array-based methylation analysis of two parental NSCLC and SCCHN cell lines and progeny rendered resistant to either erlotinib or cetuximab. The study found that DAPK was hypermethylated in NSCLC and SCCHN drug-resistant cells. Subsequent demethylation of DAPK in the resistant NSCLC cells restored sensitivity to both erlotinib and cetuximab. siRNA-mediated knockdown of DAPK validated the array-based findings by inducing erlotinib and cetuximab resistance in cells normally sensitive to either agent [297].

2.5.8 Other Factors

Transforming growth factor beta (TGFβ) has recently been shown to be a key molecular determinant of de novo and acquired resistance of cancers to EGFR-targeted antibodies [340]. Bedi et al. found that treatment of mice bearing xenografts of human SCCHN cells with cetuximab resulted in emergence of resistant tumor cells that expressed relatively higher levels of TGFβ compared to the control group. Also, treatment with cetuximab alone induced an apparent natural selection of TGFβ-overexpressing tumor cells in nonregressing tumors. Combinatorial treatment with cetuximab and a TGFβ-blocking antibody prevented the emergence of resistant tumor cells and induced complete tumor regression [340].

Coexpression of elevated levels of Aurora-A and EGFR was an adverse prognostic factor with poor disease-free and overall survival in a cohort of 180 patients [341]. In vitro studies showed that simultaneous targeting of Aurora kinase and EGFR using cetuximab and a pan-Aurora kinase inhibitor (R763) was more effective than mono-EGFR or mono-Aurora kinase inhibition. Interestingly, growth inhibitory effects were noticeable with the addition of R763 to cell lines with no or very moderate response to mono-EGFR-targeted treatment and/or with very low EGFR expression [341]. Independent studies have shown efficacy of combination of a specific Aurora-A inhibitor with erlotinib and cetuximab in an EGFR-dependent cancer cell line [342]. These findings suggest that Aurora kinase inhibitors may help overcome cetuximab resistance in the treatment of SCCHN; however, more work is needed.

Recent studies have revealed a connection between a fibroblast growth factor receptor 3-transforming acidic coiled-coil-containing protein 3 (FGFR3-TACC3) fusion protein and drug resistance. This fusion gene has been detected in nasopharyngeal carcinoma, SCCHN, esophageal squamous cell carcinoma, and lung cancer. In vitro studies have identified that FGFR3-TACC3 fusion gene promotes survival of cancer cells [343]. In a murine SCCHN xenograft model, simultaneous targeting of both EGFR and ERBB3 leads to decreased tumor progression. However, it results in the outgrowth of resistant cells. Daly et al. revealed that in these tumors levels of FGFR3-TACC3 fusion protein are significantly elevated. Further investigation demonstrated that FGFR3-TACC3 overexpression promotes cancer cell resistance through the blockade of EGFR/RAS/ERK signaling, but not through ERBB3/PI3K/AKT signaling [344].

2.6 Toxicity and Tolerance

EGFR is widely expressed at the basal level of the epidermis [293, 294], and EGFR-directed antibodies have predictable dermatologic side effects including follicular eruption, scaling, paronychia, and skin fissures. Although a significant difference in high-grade in field skin toxicity was not reported by Bonner et al. [175], subsequent studies have demonstrated increased high-grade radiation dermatitis and mucositis [176]. Initial management of skin toxicity is with topical steroids; antibiotics may be indicated in the case of superinfection, and systemic steroids are reserved for desquamation or very high proportion of body surface involved [345]. Skin toxicity is associated with benefit from cetuximab in both SCCHN and colon cancer [193, 295]. Interestingly, there is some evidence that intrinsic germline genetic variation of *EGFR* (rs2227983), *KRAS* (rs61764370), and *FCGR2A* (rs180127) can be used as predictive markers of reduced skin toxicity in SCCHN patients treated with a cetuximab-based therapy [346]. Allergic and anaphylactoid reactions can be observed in patients treated with cetuximab and, less often, with panitumumab; however, both of these monoclonal antibodies have fewer nonspecific and hematopoietic side effects compared to other chemotherapeutics [347, 348]. Electrolyte abnormities, specifically hypomagnesemia, are also commonly observed and should be monitored during treatment with EGFRIs [195].

2.7 Conclusions and New Frontiers in Drug Discovery

As study of EGFR and the ErbB family advances, several themes emerge for future investigation.

First, major, ongoing investments in personalized medicine, involving in part studies of exceptional responders for targeted therapy, and growth of large databases integrating sequence and clinical data, will help provide a panel of predictive response biomarkers. However, based on data summarized above, interpreting these data will be a challenge. It is likely that there will be multiple potential mechanisms for resistance, each operative in small subsets of patients: further, given that targeted therapies are typically administered in combinations, and many combinations are under evaluation, it will be challenging to develop statistically robust datasets to predict response to drug combinations. Systems biology may be able to help guide the integration of this information [349]. In complementary work, high-throughput screening may identify new targets relevant to specific subtypes of SCCHN. In another example, recent studies applying whole exome sequencing to nasopharyngeal cancers identified specific mutational signatures in this disease subclass, some of which are potentially druggable and may provide patient-tailored options [350] that can augment classic EGFR-ErbB-targeted therapeutics.

Second, existing EGFR-/ErbB-targeting agents may find new uses in the fields of SCCHN prevention, while efforts continue to develop new therapeutic strategies for EGFR and ErbB. In some intriguing recent work, and echoing the fact that some studies have found elevated EGFR in noncancerous premalignant tissue [11, 12, 351], cetuximab treatment of patients with high-risk premalignancy of the upper aerodigestive tract was demonstrated as a potentially effective treatment of moderate to severe dysplasia in a subset of patients [352]. However, a conceptually related study using erlotinib was not effective [351], and more work is required. In terms of developing new approaches to targeting EGFR, a number of studies support the idea that EGFR forms higher-order oligomers, such as tetramers, during signaling [353–355]. Ramirez et al. used a virtual docking model of extracellular tetrameric EGFR configuration coupled with functional screening to identify compounds that affect internalization of adaptor responsive to EGFR activation Grb2. This work indicates it may be possible to develop new classes of EGFR-targeting agents, based on targeting epitopes formed by functional, higher-order oligomers [356].

Third, the rapid rise of immunotherapies, with results from major trials expected in the next several years, is likely to transform the use of EGFR-/ErbB-targeting therapies. As effective immunotherapies have only recently been developed, much effort will be needed to understand how these agents interact with EGFR- or ErbB-targeting agents and cytotoxic therapies, develop appropriate dosing regimens that maximize effectiveness and minimize toxicities, establish genomic signatures and clinical profiles that are best suited for their use, and many other considerations. It is to be hoped that successful integration of these new approaches with EGFR-targeting agents will ultimately result in better patient outcomes.

Acknowledgments The authors were supported by R21CA191425 and R01DK108195 (to EAG), Ruth L. Kirschstein National Research Service Award F30 Fellowship (F30 CA180607 to TNB), and NIH core grant CA06927 (to Fox Chase Cancer Center).

References

1. Cowley GP, Smith JA, Gusterson BA. Increased EGF receptors on human squamous carcinoma cell lines. Br J Cancer. 1986;53(2):223–9.
2. Stanton P, et al. Epidermal growth factor receptor expression by human squamous cell carcinomas of the head and neck, cell lines and xenografts. Br J Cancer. 1994;70(3):427–33.
3. Bassiri M, Privalsky ML. Mutagenesis of the avian erythroblastosis virus erbB coding region: an intact extracellular domain is not required for oncogenic transformation. J Virol. 1986;59(2):525–30.

4. Bouyain S, et al. The extracellular region of ErbB4 adopts a tethered conformation in the absence of ligand. Proc Natl Acad Sci. 2005;102(42):15024–9.
5. Cho HS. Structure of the extracellular region of HER3 reveals an interdomain tether. Science. 2002;297(5585):1330–3.
6. Franklin MC, et al, Insights into ErbB signaling from the structure of the ErbB2-pertuzumab complex. 2004, Protein Data Bank, Rutgers University.
7. Shi F, et al. ErbB3/HER3 intracellular domain is competent to bind ATP and catalyze autophosphorylation. Proc Natl Acad Sci. 2010;107(17):7692–7.
8. Jura N, et al. Catalytic control in the EGF receptor and its connection to general kinase regulatory mechanisms. Mol Cell. 2011;42(1):9–22.
9. Zhang X, et al. An allosteric mechanism for activation of the kinase domain of epidermal growth factor receptor. Cell. 2006;125(6):1137–49.
10. Jura N, et al. Structural analysis of the catalytically inactive kinase domain of the human EGF receptor 3. Proc Natl Acad Sci U S A. 2009;106(51):21608–13.
11. Grandis JR, Tweardy DJ. TGF-α and EGFR in head and neck cancer. J Cell Biochem. 1993;53(S17F):188–91.
12. Grandis JR, Tweardy DJ. Elevated levels of transforming growth factor alpha and epidermal growth factor receptor messenger RNA are early markers of carcinogenesis in head and neck cancer. Cancer Res. 1993;53(15):3579–84.
13. Hitt R, et al. Prognostic value of the epidermal growth factor receptor (EGRF) and p53 in advanced head and neck squamous cell carcinoma patients treated with induction chemotherapy. Eur J Cancer. 2005;41(3):453–60.
14. Liccardi G, Hartley JA, Hochhauser D. EGFR nuclear translocation modulates DNA repair following cisplatin and ionizing radiation treatment. Cancer Res. 2011;71(3):1103–14.
15. Ang KK, et al. Impact of epidermal growth factor receptor expression on survival and pattern of relapse in patients with advanced head and neck carcinoma. Cancer Res. 2002;62(24):7350–6.
16. Keren S, et al. Role of EGFR as a prognostic factor for survival in head and neck cancer: a meta-analysis. Tumour Biol. 2014;35(3):2285–95.
17. Kalyankrishna S, Grandis JR. Epidermal growth factor receptor biology in head and neck cancer. J Clin Oncol. 2006;24(17):2666–72.
18. Mehra R, et al. Protein-intrinsic and signaling network-based sources of resistance to EGFR- and ErbB family-targeted therapies in head and neck cancer. Drug Resist Updat. 2011;14(6):260–79.
19. Lui VW, et al. Frequent mutation of the PI3K pathway in head and neck cancer defines predictive biomarkers. Cancer Discov. 2013;3(7):761–9.
20. O'Rorke MA, et al. Human papillomavirus related head and neck cancer survival: a systematic review and meta-analysis. Oral Oncol. 2012;48(12):1191–201.
21. Husain H, et al. Nuclear epidermal growth factor receptor and p16 expression in head and neck squamous cell carcinoma. Laryngoscope. 2012;122(12):2762–8.
22. Kim SH, et al. Human papillomavirus 16 E5 up-regulates the expression of vascular endothelial growth factor through the activation of epidermal growth factor receptor, MEK/ ERK1,2 and PI3K/Akt. Cell Mol Life Sci. 2006;63(7–8):930–8.
23. Straight SW, Herman B, McCance DJ. The E5 oncoprotein of human papillomavirus type 16 inhibits the acidification of endosomes in human keratinocytes. J Virol. 1995;69(5):3185–92.
24. Weiner LM, Surana R, Wang S. Monoclonal antibodies: versatile platforms for cancer immunotherapy. Nat Rev Immunol. 2010;10(5):317–27.
25. Rabinowits G, Haddad RI. Overcoming resistance to EGFR inhibitor in head and neck cancer: a review of the literature. Oral Oncol. 2012;48(11):1085–9.
26. Zhang J, Yang PL, Gray NS. Targeting cancer with small molecule kinase inhibitors. Nat Rev Cancer. 2009;9(1):28–39.
27. Mancini M, et al. Combining three antibodies nullifies feedback-mediated resistance to erlotinib in lung cancer. Sci Signal. 2015;8(379):ra53.

28. Wilson KJ, et al. Functional selectivity of EGF family peptide growth factors: implications for cancer. Pharmacol Ther. 2009;122(1):1–8.
29. Burgess AW, et al. An open-and-shut case? Recent insights into the activation of EGF/ErbB receptors. Mol Cell. 2003;12(3):541–52.
30. Ferguson KM, et al. EGF activates its receptor by removing interactions that autoinhibit ectodomain dimerization. Mol Cell. 2003;11(2):507–17.
31. Cho H-S, et al. Structure of the extracellular region of HER2 alone and in complex with the Herceptin fab. Nature. 2003;421(6924):756–60.
32. Garrett TPJ, et al. Crystal structure of a truncated epidermal growth factor receptor extracellular domain bound to transforming growth factor α. Cell. 2002;110(6):763–73.
33. Haskins JW, Nguyen DX, Stern DF. Neuregulin 1-activated ERBB4 interacts with YAP to induce hippo pathway target genes and promote cell migration. Sci Signal. 2014;7(355):ra116.
34. Schlessinger J. Ligand-induced, receptor-mediated dimerization and activation of EGF receptor. Cell. 2002;110(6):669–72.
35. Rogers SJ, et al. Biological significance of c-erbB family oncogenes in head and neck cancer. Cancer Metastasis Rev. 2005;24(1):47–69.
36. Bragin PE, et al. HER2 transmembrane domain dimerization coupled with self-association of membrane-embedded cytoplasmic juxtamembrane regions. J Mol Biol. 2016;428(1):52–61.
37. Shan Y, et al. Oncogenic mutations counteract intrinsic disorder in the EGFR kinase and promote receptor dimerization. Cell. 2012;149(4):860–70.
38. Red Brewer M, et al. Mechanism for activation of mutated epidermal growth factor receptors in lung cancer. Proc Natl Acad Sci U S A. 2013;110(38):E3595–604.
39. Endres NF, et al. Regulation of the catalytic activity of the EGF receptor. Curr Opin Struct Biol. 2011;21(6):777–84.
40. Alvarado D, Klein DE, Lemmon MA. Structural basis for negative cooperativity in growth factor binding to an EGF receptor. Cell. 2010;142(4):568–79.
41. Ghosh R, et al. Trastuzumab has preferential activity against breast cancers driven by HER2 homodimers. Cancer Res. 2011;71(5):1871–82.
42. Landgraf R. HER2 therapy. HER2 (ERBB2): functional diversity from structurally conserved building blocks. Breast Cancer Res. 2007;9(1):202.
43. Wieduwilt MJ, Moasser MM. The epidermal growth factor receptor family: biology driving targeted therapeutics. Cell Mol Life Sci. 2008;65(10):1566–84.
44. Oksvold MP, et al. Serine mutations that abrogate ligand-induced ubiquitination and internalization of the EGF receptor do not affect c-Cbl association with the receptor. Oncogene. 2003;22(52):8509–18.
45. Oksvold MP, et al. UV-radiation-induced internalization of the epidermal growth factor receptor requires distinct serine and tyrosine residues in the cytoplasmic carboxy-terminal domain. Radiat Res. 2004;161(6):685–91.
46. Tong J, et al. Epidermal growth factor receptor phosphorylation sites Ser991 and Tyr998 are implicated in the regulation of receptor endocytosis and phosphorylations at Ser1039 and Thr1041. Mol Cell Proteomics. 2009;8(9):2131–44.
47. Sousa LP, et al. Suppression of EGFR endocytosis by dynamin depletion reveals that EGFR signaling occurs primarily at the plasma membrane. Proc Natl Acad Sci U S A. 2012;109(12):4419–24.
48. Orth JD, et al. A novel endocytic mechanism of epidermal growth factor receptor sequestration and internalization. Cancer Res. 2006;66(7):3603–10.
49. Sigismund S, et al. Clathrin-mediated internalization is essential for sustained EGFR signaling but dispensable for degradation. Dev Cell. 2008;15(2):209–19.
50. Schoeberl B, et al. An ErbB3 antibody, MM-121, is active in cancers with ligand-dependent activation. Cancer Res. 2010;70(6):2485–94.
51. Tan X, et al. Stress-induced EGFR trafficking: mechanisms, functions, and therapeutic implications. Trends Cell Biol. 2016;26(5):352–66.

52. Dutta S, et al. Neuropilin-2 regulates endosome maturation and EGFR trafficking to support Cancer cell pathobiology. Cancer Res. 2016;76(2):418–28.
53. Lee H, et al. Anks1a regulates COPII-mediated anterograde transport of receptor tyrosine kinases critical for tumorigenesis. Nat Commun. 2016;7:12799.
54. Jeong J, et al. PMCA2 regulates HER2 protein kinase localization and signaling and promotes HER2-mediated breast cancer. Proc Natl Acad Sci U S A. 2016;113(3):E282–90.
55. Lang SA, et al. Inhibition of heat shock protein 90 impairs epidermal growth factor-mediated signaling in gastric cancer cells and reduces tumor growth and vascularization in vivo. Mol Cancer Ther. 2007;6(3):1123–32.
56. Chung CH, et al. Increased epidermal growth factor receptor gene copy number is associated with poor prognosis in head and neck squamous cell carcinomas. J Clin Oncol. 2006;24(25):4170–6.
57. Grandis JR, et al. Levels of TGF- and EGFR protein in head and neck squamous cell carcinoma and patient survival. J Natl Cancer Inst. 1998;90(11):824–32.
58. Maurizi M, et al. Prognostic significance of epidermal growth factor receptor in laryngeal squamous cell carcinoma. Br J Cancer. 1996;74(8):1253–7.
59. Temam S, et al. Epidermal growth factor receptor copy number alterations correlate with poor clinical outcome in patients with head and neck squamous cancer. J Clin Oncol. 2007;25(16):2164–70.
60. Ongkeko WM, et al. Expression of protein tyrosine kinases in head and neck squamous cell carcinomas. Am J Clin Pathol. 2005;124(1):71–6.
61. Bei R, et al. Frequent overexpression of multiple ErbB receptors by head and neck squamous cell carcinoma contrasts with rare antibody immunity in patients. J Pathol. 2004;204(3):317–25.
62. Bernardes VF, et al. EGFR status in oral squamous cell carcinoma: comparing immunohistochemistry, FISH and CISH detection in a case series study. BMJ Open. 2013;3(1):e002077.
63. Pectasides E, et al. Comparative prognostic value of epidermal growth factor quantitative protein expression compared with FISH for head and neck squamous cell carcinoma. Clin Cancer Res. 2011;17(9):2947–54.
64. Beck TN, Golemis EA. Genomic insights into head and neck cancer. Cancers of the Head & Neck. 2016;1(1):1.
65. Gao J, et al. Integrative analysis of complex cancer genomics and clinical profiles using the cBioPortal. Sci Signal. 2013;6(269):pl1.
66. Cerami E, et al. The cBio cancer genomics portal: an open platform for exploring multidimensional cancer genomics data. Cancer Discov. 2012;2(5):401–4.
67. Feldman R, et al. Molecular profiling of head and neck squamous cell carcinoma. Head Neck. 2015;38(S1):E1625–38.
68. Rubin Grandis J, et al. Quantitative immunohistochemical analysis of transforming growth factor-alpha and epidermal growth factor receptor in patients with squamous cell carcinoma of the head and neck. Cancer. 1996;78(6):1284–92.
69. Shin DM, et al. Dysregulation of epidermal growth factor receptor expression in premalignant lesions during head and neck tumorigenesis. Cancer Res. 1994;54(12):3153–9.
70. Chang AR, et al. Expression of epidermal growth factor receptor and cyclin D1 in pretreatment biopsies as a predictive factor of radiotherapy efficacy in early glottic cancer. Head Neck. 2008;30(7):852–7.
71. Alterio D, et al. Role of EGFR as prognostic factor in head and neck cancer patients treated with surgery and postoperative radiotherapy: proposal of a new approach behind the EGFR overexpression. Med Oncol. 2017;34(6):107.
72. Morgan S, Grandis JR. ErbB receptors in the biology and pathology of the aerodigestive tract. Exp Cell Res. 2009;315(4):572–82.
73. Bei R, et al. Co-localization of multiple ErbB receptors in stratified epithelium of oral squamous cell carcinoma. J Pathol. 2001;195(3):343–8.

74. Ekberg T, et al. Expression of EGFR, HER2, HER3, and HER4 in metastatic squamous cell carcinomas of the oral cavity and base of tongue. Int J Oncol. 2005;26(5):1177–85.
75. Roepstorff K, et al. Endocytic downregulation of ErbB receptors: mechanisms and relevance in cancer. Histochem Cell Biol. 2008;129(5):563–78.
76. Huang C-C, et al. Cathepsin S attenuates endosomal EGFR signalling: a mechanical rationale for the combination of cathepsin S and EGFR tyrosine kinase inhibitors. Sci Rep. 2016;6:29256.
77. Rolle CE, et al. Expression and mutational analysis of c-CBL and its relationship to the MET receptor in head and neck squamous cell carcinoma. Oncotarget. 2017 Mar 21;8(12):18726–18734.
78. Ray D, et al. Regulation of EGFR protein stability by the HECT-type ubiquitin ligase SMURF2. Neoplasia. 2011;13(7):570–8.
79. Rajput A, et al. A novel mechanism of resistance to epidermal growth factor receptor antagonism in vivo. Cancer Res. 2007;67(2):665–73.
80. Jhappan C, et al. TGFα overexpression in transgenic mice induces liver neoplasia and abnormal development of the mammary gland and pancreas. Cell. 1990;61(6):1137–46.
81. Matsui Y, et al. Development of mammary hyperplasia and neoplasia in MMTV-TGFα transgenic mice. Cell. 1990;61(6):1147–55.
82. Sandgren EP, et al. Overexpression of TGFα in transgenic mice: induction of epithelial hyperplasia, pancreatic metaplasia, and carcinoma of the breast. Cell. 1990;61(6):1121–35.
83. Tinhofer I, et al. Expression of Amphiregulin and EGFRvIII affect outcome of patients with squamous cell carcinoma of the head and neck receiving Cetuximab-docetaxel treatment. Clin Cancer Res. 2011;17(15):5197–204.
84. Yonesaka K, et al. Autocrine production of Amphiregulin predicts sensitivity to both Gefitinib and Cetuximab in EGFR wild-type cancers. Clin Cancer Res. 2008;14(21):6963–73.
85. Hatakeyama H, et al. Regulation of heparin-binding EGF-like growth factor by miR-212 and acquired Cetuximab-resistance in head and neck squamous cell carcinoma. PLoS One. 2010;5(9):e12702.
86. Murakami H, et al. Transgenic mouse model for synergistic effects of nuclear oncogenes and growth factors in tumorigenesis: interaction of c-myc and transforming growth factor alpha in hepatic oncogenesis. Cancer Res. 1993;53(8):1719–23.
87. Sandgren EP, et al. Transforming growth factor alpha dramatically enhances oncogene-induced carcinogenesis in transgenic mouse pancreas and liver. Mol Cell Biol. 1993;13(1):320–30.
88. Sandgren EP, et al. Inhibition of mammary gland involution is associated with transforming growth factor alpha but not c-myc-induced tumorigenesis in transgenic mice. Cancer Res. 1995;55(17):3915–27.
89. Oshima G, et al. Autocrine epidermal growth factor receptor ligand production and cetuximab response in head and neck squamous cell carcinoma cell lines. J Cancer Res Clin Oncol. 2012;138(3):491–9.
90. O-Charoenrat P, Rhys-Evans P, Eccles S. Expression and regulation of c-erbB ligands in human head and neck squamous carcinoma cells. Int J Cancer. 2000;88(5):759–65.
91. Gao J, Ulekleiv CH, Halstensen TS. Epidermal growth factor (EGF) receptor-ligand based molecular staging predicts prognosis in head and neck squamous cell carcinoma partly due to deregulated EGF-induced amphiregulin expression. J Exp Clin Cancer Res. 2016;35(1):151.
92. Wakasaki T, et al. A critical role of c-Cbl-interacting protein of 85 kDa in the development and progression of head and neck squamous cell carcinomas through the ras-ERK pathway. Neoplasia. 2010;12(10):789–96.
93. Brand TM, et al. The nuclear epidermal growth factor receptor signaling network and its role in cancer. Discov Med. 2011;12(66):419–32.
94. Matta A, Ralhan R. Overview of current and future biologically based targeted therapies in head and neck squamous cell carcinoma. Head Neck Oncol. 2009;1(1):6.
95. Roskoski R. ERK1/2 MAP kinases: structure, function, and regulation. Pharmacol Res. 2012;66(2):105–43.

96. Wortzel I, Seger R. The ERK Cascade: distinct functions within various subcellular organelles. Genes Cancer. 2011;2(3):195–209.
97. Lurje G, Lenz H-J. EGFR signaling and drug discovery. Oncology. 2009;77(6):400–10.
98. Faivre S, Kroemer G, Raymond E. Current development of mTOR inhibitors as anticancer agents. Nat Rev Drug Discov. 2006;5(8):671–88.
99. Zheng LS, et al. SPINK6 promotes metastasis of nasopharyngeal carcinoma via binding and activation of epithelial growth factor receptor. Cancer Res. 2017;77(2):579–89.
100. Saloura V, et al. WHSC1L1-mediated EGFR mono-methylation enhances the cytoplasmic and nuclear oncogenic activity of EGFR in head and neck cancer. Sci Rep. 2017;7:40664.
101. Keller J, et al. Combination of phosphorylated and truncated EGFR correlates with higher tumor and nodal stage in head and neck cancer. Cancer Investig. 2010;28(10):1054–62.
102. Hama T, et al. Prognostic significance of epidermal growth factor receptor phosphorylation and mutation in head and neck squamous cell carcinoma. Oncologist. 2009;14(9):900–8.
103. Schwentner I, et al. Identification of the rare EGFR mutation p.G796S as somatic and germline mutation in white patients with squamous cell carcinoma of the head and neck. Head Neck. 2008;30(8):1040–4.
104. Na II, et al. EGFR mutations and human papillomavirus in squamous cell carcinoma of tongue and tonsil. Eur J Cancer. 2007;43(3):520–6.
105. Lee JW. Somatic mutations of EGFR gene in squamous cell carcinoma of the head and neck. Clin Cancer Res. 2005;11(8):2879–82.
106. Diedrich U, et al. Distribution of epidermal growth factor receptor gene amplification in brain tumours and correlation to prognosis. J Neurol. 1995;242(10):683–8.
107. Garcia de Palazzo IE, et al. Expression of mutated epidermal growth factor receptor by non-small cell lung carcinomas. Cancer Res. 1993;53(14):3217–20.
108. Ge H, Gong X, Tang CK. Evidence of high incidence of EGFRvIII expression and coexpression with EGFR in human invasive breast cancer by laser capture microdissection and immunohistochemical analysis. Int J Cancer. 2002;98(3):357–61.
109. Moscatello DK, et al. Frequent expression of a mutant epidermal growth factor receptor in multiple human tumors. Cancer Res. 1995;55(23):5536–9.
110. Nishikawa R, et al. A mutant epidermal growth factor receptor common in human glioma confers enhanced tumorigenicity. Proc Natl Acad Sci. 1994;91(16):7727–31.
111. Okamoto I, et al. Expression of constitutively activated EGFRvlll in non-small cell lung cancer. Cancer Sci. 2003;94(1):50–6.
112. Chau NG, et al. The association between EGFR variant III, HPV, p16, c-MET, EGFR gene copy number and response to EGFR inhibitors in patients with recurrent or metastatic squamous cell carcinoma of the head and neck. Head Neck Oncol. 2011;3(1):11.
113. Sok JC. Mutant epidermal growth factor receptor (EGFRvIII) contributes to head and neck Cancer growth and resistance to EGFR targeting. Clin Cancer Res. 2006;12(17):5064–73.
114. Yang B, et al. Expression of epidermal growth factor receptor variant III in laryngeal carcinoma tissues. Auris Nasus Larynx. 2009;36(6):682–7.
115. McIntyre JB, et al. Specific and sensitive hydrolysis probe-based real-time PCR detection of epidermal growth factor receptor variant III in oral squamous cell carcinoma. PLoS One. 2012;7(2):e31723.
116. Melchers LJ, et al. Head and neck squamous cell carcinomas do not express EGFRvIII. Int J Radiat Oncol Biol Phys. 2014;90(2):454–62.
117. Stransky N, et al. The mutational landscape of head and neck squamous cell carcinoma. Science. 2011;333(6046):1157–60.
118. Agrawal N, et al. Exome sequencing of head and neck squamous cell carcinoma reveals inactivating mutations in NOTCH1. Science. 2011;333(6046):1154–7.
119. Chau NG, et al. Incorporation of next-generation sequencing into routine clinical care to direct treatment of head and neck squamous cell carcinoma. Clin Cancer Res. 2016;22(12):2939–49.
120. The Cancer Genome Atlas, N. Comprehensive genomic characterization of head and neck squamous cell carcinomas. Nature. 2015;517(7536):576–82.

121. Li H, et al. Genomic analysis of head and neck squamous cell carcinoma cell lines and human tumors: a rational approach to preclinical model selection. Mol Cancer Res. 2014;12(4):571–82.
122. Du L, et al. Overexpression of PIK3CA in murine head and neck epithelium drives tumor invasion and metastasis through PDK1 and enhanced TGFbeta signaling. Oncogene. 2016;35(35):4641–52.
123. Bozec A, et al. Combination of mTOR and EGFR targeting in an orthotopic xenograft model of head and neck cancer. Laryngoscope. 2016;126(4):E156–63.
124. Lai SY, Johnson FM. Defining the role of the JAK-STAT pathway in head and neck and thoracic malignancies: implications for future therapeutic approaches. Drug Resist Updat. 2010;13(3):67–78.
125. Quesnelle KM, Boehm AL, Grandis JR. STAT-mediated EGFR signaling in cancer. J Cell Biochem. 2007;102(2):311–9.
126. Yu H, Pardoll D, Jove R. STATs in cancer inflammation and immunity: a leading role for STAT3. Nat Rev Cancer. 2009;9(11):798–809.
127. Mali SB. Review of STAT3 (signal transducers and activators of transcription) in head and neck cancer. Oral Oncol. 2015;51(6):565–9.
128. Pectasides E, et al. Nuclear localization of signal transducer and activator of transcription 3 in head and neck squamous cell carcinoma is associated with a better prognosis. Clin Cancer Res. 2010;16(8):2427–34.
129. Wheeler SE, et al. Epidermal growth factor receptor variant III mediates head and neck cancer cell invasion via STAT3 activation. Oncogene. 2010;29(37):5135–45.
130. Bonner JA, et al. Enhancement of cetuximab-induced radiosensitization by JAK-1 inhibition. BMC Cancer. 2015;15:673.
131. Stegeman H, et al. Combining radiotherapy with MEK1/2, STAT5 or STAT6 inhibition reduces survival of head and neck cancer lines. Mol Cancer. 2013;12(1):133.
132. Hwang J-I, et al. Molecular cloning and characterization of a novel phospholipase C, PLC-η. Biochem J. 2005;389(1):181–6.
133. Thomas SM, et al. Epidermal growth factor receptor-stimulated activation of phospholipase Cgamma-1 promotes invasion of head and neck squamous cell carcinoma. Cancer Res. 2003;63(17):5629–35.
134. Kassis J, et al. A role for phospholipase C-gamma-mediated signaling in tumor cell invasion. Clin Cancer Res. 1999;5(8):2251–60.
135. Nozawa H, et al. Combined inhibition of PLC-1 and c-Src abrogates epidermal growth factor receptor-mediated head and neck squamous cell carcinoma invasion. Clin Cancer Res. 2008;14(13):4336–44.
136. Rosse C, et al. PKC and the control of localized signal dynamics. Nat Rev Mol Cell Biol. 2010;11(2):103–12.
137. Martínez-Gimeno C, et al. Alterations in levels of different protein kinase C isotypes and their influence on behavior of squamous cell carcinoma of the oral cavity: εPKC, a novel prognostic factor for relapse and survival. Head Neck. 1995;17(6):516–25.
138. Cohen EEW. Protein kinase C mediates epidermal growth factor-induced growth of head and neck tumor cells by regulating mitogen-activated protein kinase. Cancer Res. 2006;66(12):6296–303.
139. Elkabets M, et al. AXL mediates resistance to PI3Kalpha inhibition by activating the EGFR/PKC/mTOR axis in head and neck and esophageal squamous cell carcinomas. Cancer Cell. 2015;27(4):533–46.
140. Parsons SJ, Parsons JT. Src family kinases, key regulators of signal transduction. Oncogene. 2004;23(48):7906–9.
141. Koppikar P, et al. Combined inhibition of c-Src and epidermal growth factor receptor abrogates growth and invasion of head and neck squamous cell carcinoma. Clin Cancer Res. 2008;14(13):4284–91.

142. Egloff AM, Grandis JR. Improving response rates to EGFR-targeted therapies for head and neck squamous cell carcinoma: candidate predictive biomarkers and combination treatment with Src inhibitors. J Oncol. 2009;2009:1–12.
143. Wheeler DL, Iida M, Dunn EF. The role of Src in solid tumors. Oncologist. 2009;14(7):667–78.
144. Xi S, et al. Src kinases mediate STAT growth pathways in squamous cell carcinoma of the head and neck. J Biol Chem. 2003;278(34):31574–83.
145. Zhang Q. Src family kinases mediate epidermal growth factor receptor ligand cleavage, proliferation, and invasion of head and neck Cancer cells. Cancer Res. 2004;64(17):6166–73.
146. Li C, et al. Nuclear EGFR contributes to acquired resistance to cetuximab. Oncogene. 2009;28(43):3801–13.
147. Yeatman TJ. A renaissance for SRC. Nat Rev Cancer. 2004;4(6):470–80.
148. Zhang X, et al. RANKL/RANK pathway abrogates cetuximab sensitivity in gastric cancer cells via activation of EGFR and c-Src. OncoTargets and Therapy. 2017;10:73–83.
149. Yamada T, et al. Receptor activator of NF-kappaB ligand induces cell adhesion and integrin alpha2 expression via NF-kappaB in head and neck cancers. Sci Rep. 2016;6:23545.
150. Yan M, et al. Correlation of NF-kappaB signal pathway with tumor metastasis of human head and neck squamous cell carcinoma. BMC Cancer. 2010;10:437.
151. Ferris RL, Grandis JR. NF-B gene signatures and p53 mutations in head and neck squamous cell carcinoma. Clin Cancer Res. 2007;13(19):5663–4.
152. Nakayama H, et al. High expression levels of nuclear factor κB, IκB kinase α and Akt kinase in squamous cell carcinoma of the oral cavity. Cancer. 2001;92(12):3037–44.
153. Yan B, et al. Genome-wide identification of novel expression signatures reveal distinct patterns and prevalence of binding motifs for p53, nuclear factor-κB and other signal transcription factors in head and neck squamous cell carcinoma. Genome Biol. 2007;8(5):R78.
154. Arun P, et al. Nuclear NF-κB p65 phosphorylation at serine 276 by protein kinase A contributes to the malignant phenotype of head and neck cancer. Clin Cancer Res. 2009;15(19):5974–84.
155. Tanaka T, et al. Selective inhibition of nuclear factor-κB by nuclear factor-κB essential modulator-binding domain peptide suppresses the metastasis of highly metastatic oral squamous cell carcinoma. Cancer Sci. 2012;103(3):455–63.
156. Aravindan N, et al. Irreversible EGFR inhibitor EKB-569 targets low-LET γ-radiation-triggered Rel orchestration and potentiates cell death in squamous cell carcinoma. PLoS One. 2011;6(12):e29705.
157. Wilken R, et al. Curcumin: a review of anti-cancer properties and therapeutic activity in head and neck squamous cell carcinoma. Mol Cancer. 2011;10(1):12.
158. Lee TL, et al. A signal network involving coactivated NF-κB and STAT3 and altered p53 modulates BAX/BCL-XL expression and promotes cell survival of head and neck squamous cell carcinomas. Int J Cancer. 2008;122(9):1987–98.
159. Bivona TG, et al. FAS and NF-κB signalling modulate dependence of lung cancers on mutant EGFR. Nature. 2011;471(7339):523–6.
160. Li Z, et al. A positive feedback loop involving EGFR/Akt/mTORC1 and IKK/NF-kB regulates head and neck squamous cell carcinoma proliferation. Oncotarget. 2016;7(22):31892–906.
161. Kang H, Kiess A, Chung CH. Emerging biomarkers in head and neck cancer in the era of genomics. Nat Rev Clin Oncol. 2015;12(1):11–26.
162. Kalish LH, et al. Deregulated cyclin D1 expression is associated with decreased efficacy of the selective epidermal growth factor receptor tyrosine kinase inhibitor gefitinib in head and neck squamous cell carcinoma cell lines. Clin Cancer Res. 2004;10(22):7764–74.
163. Namazie A, et al. Cyclin D1 amplification and p16(MTS1/CDK4I) deletion correlate with poor prognosis in head and neck tumors. Laryngoscope. 2002;112(3):472–81.
164. Beck TN, et al. EGFR and RB1 as dual biomarkers in HPV-negative head and neck Cancer. Mol Cancer Ther. 2016;15(10):2486.
165. Calais G, et al. Randomized trial of radiation therapy versus concomitant chemotherapy and radiation therapy for advanced-stage oropharynx carcinoma. J Natl Cancer Inst. 1999;91(24):2081–6.

166. Adelstein DJ, et al. An intergroup phase III comparison of standard radiation therapy and two schedules of concurrent Chemoradiotherapy in patients with Unresectable squamous cell head and neck Cancer. J Clin Oncol. 2003;21(1):92–8.
167. Trotti A, et al. TAME: development of a new method for summarising adverse events of cancer treatment by the radiation therapy oncology group. Lancet Oncol. 2007;8(7):613–24.
168. Agulnik M. New approaches to EGFR inhibition for locally advanced or metastatic squamous cell carcinoma of the head and neck (SCCHN). Med Oncol. 2012;29(4):2481–91.
169. Dittmann K, et al. Radiation-induced epidermal growth factor receptor nuclear import is linked to activation of DNA-dependent protein kinase. J Biol Chem. 2005;280(35):31182–9.
170. Mendelsohn J, Baselga J. Status of epidermal growth factor receptor antagonists in the biology and treatment of Cancer. J Clin Oncol. 2003;21(14):2787–99.
171. Vermorken JB, et al. Open-label, uncontrolled, multicenter phase II study to evaluate the efficacy and toxicity of Cetuximab as a single agent in patients with recurrent and/or metastatic squamous cell carcinoma of the head and neck who failed to respond to platinum-based therapy. J Clin Oncol. 2007;25(16):2171–7.
172. Ferris RL, et al. Correlation of fc gamma receptor (fcγR) IIa and IIIa polymorphisms with clinical outcome in patients treated with Cetuximab-based Chemoradiation in the RTOG 0522 trial. Int J Radiat Oncol Biol Phys. 2014;88(2):467.
173. Dent P, et al. Radiation-induced release of transforming growth factor alpha activates the epidermal growth factor receptor and mitogen-activated protein kinase pathway in carcinoma cells, leading to increased proliferation and protection from radiation-induced cell death. Mol Biol Cell. 1999;10(8):2493–506.
174. Schmidt-Ullrich RK, et al. Altered expression of epidermal growth factor receptor and estrogen receptor in MCF-7 cells after single and repeated radiation exposures. Int J Radiat Oncol Biol Phys. 1994;29(4):813–9.
175. Bonner JA, et al. Radiotherapy plus cetuximab for squamous-cell carcinoma of the head and neck. N Engl J Med. 2006;354(6):567–78.
176. Ang KK, et al. Randomized phase III trial of concurrent accelerated radiation plus cisplatin with or without cetuximab for stage III to IV head and neck carcinoma: RTOG 0522. J Clin Oncol. 2014;32(27):2940–50.
177. Fakih M, Vincent M. Adverse events associated with anti-EGFR therapies for the treatment of metastatic colorectal cancer. Curr Oncol. 2010;17(Suppl 1):S18–30.
178. Moon C, Chae YK, Lee J. Targeting epidermal growth factor receptor in head and neck cancer: lessons learned from cetuximab. Exp Biol Med. 2010;235(8):907–20.
179. Sipples R. Common side effects of anti-EGFR therapy: Acneform rash. Semin Oncol Nurs. 2006;22:28–34.
180. Vermorken JB, et al. Platinum-based chemotherapy plus Cetuximab in head and neck Cancer. N Engl J Med. 2008;359(11):1116–27.
181. Masui H, et al. Growth inhibition of human tumor cells in athymic mice by anti-epidermal growth factor receptor monoclonal antibodies. Cancer Res. 1984;44(3):1002–7.
182. Wheeler DL, Dunn EF, Harari PM. Understanding resistance to EGFR inhibitors—impact on future treatment strategies. Nat Rev Clin Oncol. 2010;7(9):493–507.
183. Mendelsohn J. Blockade of receptors for growth factors: an anticancer therapy—the fourth annual Joseph H Burchenal American Association of Cancer Research Clinical Research Award Lecture. Clin Cancer Res. 2000;6(3):747–53.
184. Specenier P, Vermorken JB. Biologic therapy in head and neck Cancer: a road with hurdles. ISRN Oncology. 2012;2012:1–15.
185. Schmitz KR, Ferguson KM. Interaction of antibodies with ErbB receptor extracellular regions. Exp Cell Res. 2009;315(4):659–70.
186. Marshall J. Clinical implications of the mechanism of epidermal growth factor receptor inhibitors. Cancer. 2006;107(6):1207–18.
187. Kurai J, et al. Antibody-dependent cellular cytotoxicity mediated by Cetuximab against lung Cancer cell lines. Clin Cancer Res. 2007;13(5):1552–61.

188. Mehra R, Cohen RB, Burtness BA. The role of cetuximab for the treatment of squamous cell carcinoma of the head and neck. Clin Adv Hematol Oncol. 2008;6(10):742–50.
189. Trigo J, et al. Cetuximab monotherapy is active in patients (pts) with platinum-refractory recurrent/metastatic squamous cell carcinoma of the head and neck (SCCHN): results of a phase II study. J Clin Oncol. 2004;22(14_suppl): 5502–5502.
190. Baselga J, et al. Phase I studies of anti-epidermal growth factor receptor chimeric antibody C225 alone and in combination with cisplatin. J Clin Oncol. 2000;18(4):904–14.
191. Robert F, et al. Phase I study of anti-epidermal growth factor receptor antibody Cetuximab in combination with radiation therapy in patients with advanced head and neck Cancer. J Clin Oncol. 2001;19(13):3234–43.
192. Bonner JA, et al. Radiotherapy plus cetuximab for locoregionally advanced head and neck cancer: 5-year survival data from a phase 3 randomised trial, and relation between cetuximab-induced rash and survival. Lancet Oncol. 2010;11(1):21–8.
193. Baselga J, et al. Phase II multicenter study of the antiepidermal growth factor receptor monoclonal antibody cetuximab in combination with platinum-based chemotherapy in patients with platinum-refractory metastatic and/or recurrent squamous cell carcinoma of the head and neck. J Clin Oncol. 2005;23(24):5568–77.
194. Merlano M, Occelli M. Review of cetuximab in the treatment of squamous cell carcinoma of the head and neck. Ther Clin Risk Manag. 2007;3(5):871–6.
195. Burtness B, et al. Phase III randomized trial of cisplatin plus placebo compared with cisplatin plus Cetuximab in metastatic/recurrent head and neck Cancer: an eastern cooperative oncology group study. J Clin Oncol. 2005;23(34):8646–54.
196. Licitra L, et al. Predictive value of epidermal growth factor receptor expression for first-line chemotherapy plus cetuximab in patients with head and neck and colorectal cancer: analysis of data from the EXTREME and CRYSTAL studies. Eur J Cancer. 2013;49(6):1161–8.
197. Lei Y, et al. EGFR-targeted mAb therapy modulates autophagy in head and neck squamous cell carcinoma through NLRX1-TUFM protein complex. Oncogene. 2016;35(36):4698–707.
198. Srivastava RM, et al. Cetuximab-activated natural killer and dendritic cells collaborate to trigger tumor antigen-specific T-cell immunity in head and neck cancer patients. Clin Cancer Res. 2013;19(7):1858–72.
199. Nielsen DL, Pfeiffer P Jensen BV, Six cases of treatment with panitumumab in patients with severe hypersensitivity reactions to cetuximab. Ann Oncol. 2009;20(4):798–798.
200. Trivedi S, et al. Anti-EGFR targeted monoclonal antibody isotype influences antitumor cellular immunity in head and neck Cancer patients. Clin Cancer Res. 2016;22(21):5229–37.
201. Wirth LJ, et al. Phase I dose-finding study of paclitaxel with panitumumab, carboplatin and intensity-modulated radiotherapy in patients with locally advanced squamous cell cancer of the head and neck. Ann Oncol. 2009;21(2):342–7.
202. Vermorken JB, et al. Cisplatin and fluorouracil with or without panitumumab in patients with recurrent or metastatic squamous-cell carcinoma of the head and neck (SPECTRUM): an open-label phase 3 randomised trial. Lancet Oncol. 2013;14(8):697–710.
203. Siu LL, et al. Effect of standard radiotherapy with cisplatin vs accelerated radiotherapy with panitumumab in locoregionally advanced squamous cell head and neck carcinoma: a randomized clinical trial. JAMA Oncol. 2017;3(2):220–6.
204. Eriksen JG, et al. OC-009: update of the randomised phase III trial DAHANCA 19: primary C-RT or RT and zalutumumab for squamous cell carcinomas of head and neck. Radiother Oncol. 2015;114:10.
205. Mateo C, et al. Humanization of a mouse monoclonal antibody that blocks the epidermal growth factor receptor: recovery of antagonistic activity. Immunotechnology. 1997;3(1):71–81.
206. Rivera F, et al. Current situation of Panitumumab, Matuzumab, Nimotuzumab and Zalutumumab. Acta Oncol. 2008;47(1):9–19.
207. Rojo F, et al. Pharmacodynamic trial of Nimotuzumab in Unresectable squamous cell carcinoma of the head and neck: a SENDO foundation study. Clin Cancer Res. 2010;16(8):2474–82.

208. Schmiedel J, et al. Matuzumab binding to EGFR prevents the conformational rearrangement required for dimerization. Cancer Cell. 2008;13(4):365–73.
209. Schanzer JM, et al. A novel glycoengineered bispecific antibody format for targeted inhibition of epidermal growth factor receptor (EGFR) and insulin-like growth factor receptor type I (IGF-1R) demonstrating unique molecular properties. J Biol Chem. 2014;289(27):18693–706.
210. Jones HE, et al. Insulin-like growth factor-I receptor signalling and acquired resistance to gefitinib (ZD1839; Iressa) in human breast and prostate cancer cells. Endocr Relat Cancer. 2004;11(4):793–814.
211. Schanzer JM, et al. XGFR*, a novel affinity-matured bispecific antibody targeting IGF-1R and EGFR with combined signaling inhibition and enhanced immune activation for the treatment of pancreatic cancer. MAbs. 2016;8(4):811–27.
212. Yonesaka K, et al. Activation of ERBB2 signaling causes resistance to the EGFR-directed therapeutic antibody Cetuximab. Sci Transl Med. 2011;3(99):99ra86–99ra86.
213. Gutiérrez VF, et al. Genetic profile of second primary tumors and recurrences in head and neck squamous cell carcinomas. Head Neck. 2011;34(6):830–9.
214. Iida M, et al. Targeting the HER family with pan-HER effectively overcomes resistance to Cetuximab. Mol Cancer Ther. 2016;15(9):2175–86.
215. Adams CW, et al. Humanization of a recombinant monoclonal antibody to produce a therapeutic HER dimerization inhibitor, pertuzumab. Cancer Immunol Immunother. 2005;55(6):717–27.
216. Erjala K. Signaling via ErbB2 and ErbB3 associates with resistance and epidermal growth factor receptor (EGFR) amplification with sensitivity to EGFR inhibitor Gefitinib in head and neck squamous cell carcinoma cells. Clin Cancer Res. 2006;12(13):4103–11.
217. Schoeberl B, et al. Therapeutically targeting ErbB3: a key node in ligand-induced activation of the ErbB receptor-PI3K Axis. Sci Signal. 2009;2(77):ra31–ra31.
218. Huang J, et al. The anti-erbB3 antibody MM-121/SAR256212 in combination with trastuzumab exerts potent antitumor activity against trastuzumab-resistant breast cancer cells. Mol Cancer. 2013;12(1):134.
219. Liu JF, et al. Randomized phase II trial of Seribantumab in combination with paclitaxel in patients with advanced platinum-resistant or -refractory ovarian Cancer. J Clin Oncol. 2016;34(36):4345–53.
220. McDonagh CF, et al. Antitumor activity of a novel bispecific antibody that targets the ErbB2/ErbB3 oncogenic unit and inhibits Heregulin-induced activation of ErbB3. Mol Cancer Ther. 2012;11(3):582–93.
221. Okazaki S, et al. Development of an ErbB4 monoclonal antibody that blocks neuregulin-1-induced ErbB4 activation in cancer cells. Biochem Biophys Res Commun. 2016;470(1):239–44.
222. Lange T, et al. Trastuzumab has anti-metastatic and anti-angiogenic activity in a spontaneous metastasis xenograft model of esophageal adenocarcinoma. Cancer Lett. 2011;308(1):54–61.
223. Norman G, et al. Trastuzumab for the treatment of HER2-positive metastatic adenocarcinoma of the stomach or gastro-oesophageal junction. Health Technol Assess. 2011;15(Suppl_1):33–42.
224. Slamon DJ, et al. Use of chemotherapy plus a monoclonal antibody against HER2 for metastatic breast Cancer that overexpresses HER2. N Engl J Med. 2001;344(11):783–92.
225. Bang Y-J, et al. Trastuzumab in combination with chemotherapy versus chemotherapy alone for treatment of HER2-positive advanced gastric or gastro-oesophageal junction cancer (ToGA): a phase 3, open-label, randomised controlled trial. Lancet. 2010;376(9742):687–97.
226. Kondo N, et al. Antitumor effect of gefitinib on head and neck squamous cell carcinoma enhanced by trastuzumab. Oncol Rep. 2008;20(2):373–8.
227. Kondo N, et al. Combined molecular targeted drug therapy for EGFR and HER-2 in head and neck squamous cell carcinoma cell lines. Int J Oncol. 2012;40(6):1805–12.
228. Wilson TR, et al. Neuregulin-1-mediated autocrine signaling underlies sensitivity to HER2 kinase inhibitors in a subset of human cancers. Cancer Cell. 2011;20(2):158–72.

229. Kulkarni S, et al. TMEM16A/ANO1 suppression improves response to antibody mediated targeted therapy of EGFR and HER2/ERBB2. Genes Chromosom Cancer. 2017;56:460–71.
230. Fayette J, et al. Randomized phase II study of Duligotuzumab (MEHD7945A) vs. Cetuximab in squamous cell carcinoma of the head and neck (MEHGAN study). Front Oncol. 2016;6:232.
231. Jimeno A, et al. Phase Ib study of duligotuzumab (MEHD7945A) plus cisplatin/5-fluorouracil or carboplatin/paclitaxel for first-line treatment of recurrent/metastatic squamous cell carcinoma of the head and neck. Cancer. 2016;122(24):3803–11.
232. Juric D, et al. Safety and pharmacokinetics/pharmacodynamics of the first-in-class dual action HER3/EGFR antibody MEHD7945A in locally advanced or metastatic epithelial tumors. Clin Cancer Res. 2015;21(11):2462–70.
233. Lee S, et al. Inhibition of ErbB3 by a monoclonal antibody that locks the extracellular domain in an inactive configuration. Proc Natl Acad Sci U S A. 2015;112(43):13225–30.
234. Brand TM, et al. Human papillomavirus regulates HER3 expression in head and neck Cancer: implications for targeted HER3 therapy in HPV+ patients. Clin Cancer Res. 2017;23(12):3072–83.
235. Cohen EEW, et al. Phase II trial of ZD1839 in recurrent or metastatic squamous cell carcinoma of the head and neck. J Clin Oncol. 2003;21(10):1980–7.
236. Cohen EEW. Phase II trial of Gefitinib 250 mg daily in patients with recurrent and/or metastatic squamous cell carcinoma of the head and neck. Clin Cancer Res. 2005;11(23):8418–24.
237. Caponigro F, et al. A phase I/II trial of gefitinib and radiotherapy in patients with locally advanced inoperable squamous cell carcinoma of the head and neck. Anti-Cancer Drugs. 2008;19(7):739–44.
238. Hainsworth JD, et al. Neoadjuvant chemotherapy/gefitinib followed by concurrent chemotherapy/radiation therapy/gefitinib for patients with locally advanced squamous carcinoma of the head and neck. Cancer. 2009;115(10):2138–46.
239. Stewart JSW, et al. Phase III study of Gefitinib compared with intravenous methotrexate for recurrent squamous cell carcinoma of the head and neck. J Clin Oncol. 2009;27(11):1864–71.
240. Baselga J, et al. Phase I safety, pharmacokinetic, and pharmacodynamic trial of ZD1839, a selective oral epidermal growth factor receptor tyrosine kinase inhibitor, in patients with five selected solid tumor types. J Clin Oncol. 2002;20(21):4292–302.
241. Perez CA, et al. Phase II study of gefitinib adaptive dose escalation to skin toxicity in recurrent or metastatic squamous cell carcinoma of the head and neck. Oral Oncol. 2012;48(9):887–92.
242. Argiris A, et al. Phase III randomized, placebo-controlled trial of docetaxel with or without Gefitinib in recurrent or metastatic head and neck Cancer: an eastern cooperative oncology group trial. J Clin Oncol. 2013;31(11):1405–14.
243. Chen L, et al. Predictive factors associated with gefitinib response in patients with advanced non-small-cell lung cancer (NSCLC). Chin J Cancer Res. 2014;26(4):466–70.
244. Scheffler M, et al. Clinical pharmacokinetics of tyrosine kinase inhibitors: focus on 4-anilinoquinazolines. Clin Pharmacokinet. 2011;50(6):371–403.
245. Swaisland HC, et al. Single-dose clinical pharmacokinetic studies of gefitinib. Clin Pharmacokinet. 2005;44(11):1165–77.
246. Siu LL, et al. Phase I/II trial of Erlotinib and cisplatin in patients with recurrent or metastatic squamous cell carcinoma of the head and neck: a Princess Margaret Hospital phase II consortium and National Cancer Institute of Canada clinical trials group study. J Clin Oncol. 2007;25(16):2178–83.
247. Hamilton M, et al. Effects of smoking on the pharmacokinetics of erlotinib. Clin Cancer Res. 2006;12(7 Pt 1):2166–71.
248. Xia W, et al. Anti-tumor activity of GW572016: a dual tyrosine kinase inhibitor blocks EGF activation of EGFR/erbB2 and downstream Erk1/2 and AKT pathways. Oncogene. 2002;21(41):6255–63.
249. de Souza JA, et al. A phase II study of Lapatinib in recurrent/metastatic squamous cell carcinoma of the head and neck. Clin Cancer Res. 2012;18(8):2336–43.
250. Harrington K, et al. Postoperative adjuvant Lapatinib and concurrent Chemoradiotherapy followed by maintenance Lapatinib monotherapy in high-risk patients with resected squamous

cell carcinoma of the head and neck: a phase III, randomized, double-blind, placebo-controlled study. J Clin Oncol. 2015;33(35):4202–9.
251. Sacco AG, Worden FP. Molecularly targeted therapy for the treatment of head and neck cancer: a review of the ErbB family inhibitors. OncoTargets and therapy. 2016;9:1927.
252. Solca F, et al. Target binding properties and cellular activity of Afatinib (BIBW 2992), an irreversible ErbB family blocker. J Pharmacol Exp Ther. 2012;343(2):342–50.
253. Markovic A, Chung CH. Current role of EGF receptor monoclonal antibodies and tyrosine kinase inhibitors in the management of head and neck squamous cell carcinoma. Expert Rev Anticancer Ther. 2012;12(9):1149–59.
254. Machiels J-PH, et al. Activity of afatinib administered in a window pre-operative study in squamous cell carcinoma of the head and neck (SCCHN) : EORTC-90111. J Clin Oncol. 2016;34(15_suppl):6049–6049.
255. Cohen EEW, et al. Biomarker analysis in recurrent and/or metastatic head and neck squamous cell carcinoma (R/M HNSCC) patients (pts) treated with second-line afatinib versus methotrexate (MTX): LUX-Head & Neck 1 (LUX-H&N1). J Clin Oncol. 2015;33(15_suppl):6023–6023.
256. Burtness B, et al. LUX-head and neck 2: randomized, double-blind, placebo-controlled, phase III trial of afatinib as adjuvant therapy after chemoradiation (CRT) in primary unresected, high/intermediate-risk, squamous cell cancer of the head and neck (HNSCC) patients (pts). J Clin Oncol. 2017;35(15_suppl):6001–6001
257. Kalous O, et al. Dacomitinib (PF-00299804), an irreversible pan-HER inhibitor, inhibits proliferation of HER2-amplified breast Cancer cell lines resistant to Trastuzumab and Lapatinib. Mol Cancer Ther. 2012;11(9):1978–87.
258. Engelman JA, et al. PF00299804, an irreversible pan-ERBB inhibitor, is effective in lung Cancer models with EGFR and ERBB2 mutations that are resistant to Gefitinib. Cancer Res. 2007;67(24):11924–32.
259. Williams JP, et al. Pre-clinical characterization of Dacomitinib (PF-00299804), an irreversible pan-ErbB inhibitor, combined with ionizing radiation for head and neck squamous cell carcinoma. PLoS One. 2014;9(5):e98557.
260. Chiu JW, et al. Pharmacokinetic assessment of dacomitinib (pan-HER tyrosine kinase inhibitor) in patients with locally advanced head and neck squamous cell carcinoma (LA SCCHN) following administration through a gastrostomy feeding tube (GT). Investig New Drugs. 2015;33(4):895–900.
261. Audet M-L, et al. Evaluation of potential predictive markers of efficacy of dacomitinib in patients (pts) with recurrent/metastatic SCCHN from a phase II trial. J Clin Oncol. 2013;31(15_suppl):6041–6041.
262. Cai X, et al. Discovery of 7-(4-(3-Ethynylphenylamino)-7-methoxyquinazolin-6-yloxy)-N-hydroxyheptanamide (CUDC-101) as a potent multi-acting HDAC, EGFR, and HER2 inhibitor for the treatment of Cancer. J Med Chem. 2010;53(5):2000–9.
263. Lai CJ, et al. CUDC-101, a multitargeted inhibitor of histone deacetylase, epidermal growth factor receptor, and human epidermal growth factor receptor 2, exerts potent anticancer activity. Cancer Res. 2010;70(9):3647–56.
264. Galloway TJ, et al. A phase I study of CUDC-101, a multitarget inhibitor of HDACs, EGFR, and HER2, in combination with Chemoradiation in patients with head and neck squamous cell carcinoma. Clin Cancer Res. 2015;21(7):1566–73.
265. Xie T, et al. Pharmacological targeting of the pseudokinase Her3. Nat Chem Biol. 2014;10(12):1006–12.
266. Mizrachi A, et al. Tumour-specific PI3K inhibition via nanoparticle-targeted delivery in head and neck squamous cell carcinoma. Nat Commun. 2017;8:14292.
267. Zumsteg ZS, et al. Taselisib (GDC-0032), a potent beta-sparing small molecule inhibitor of PI3K, Radiosensitizes head and neck squamous carcinomas containing activating PIK3CA alterations. Clin Cancer Res. 2016;22(8):2009–19.
268. Burris HA, 3rd, Overcoming acquired resistance to anticancer therapy: focus on the PI3K/AKT/mTOR pathway. Cancer Chemother Pharmacol. 2013;71(4):829–842.

269. Dolly SO, et al. Phase I study of Apitolisib (GDC-0980), dual Phosphatidylinositol-3-kinase and mammalian target of rapamycin kinase inhibitor, in patients with advanced solid tumors. Clin Cancer Res. 2016;22(12):2874–84.
270. Geiger JL, et al. Phase II trial of everolimus in patients with previously treated recurrent or metastatic head and neck squamous cell carcinoma. Head Neck. 2016;38(12):1759–64.
271. Anisuzzaman AS, et al. In vitro and in vivo synergistic antitumor activity of the combination of BKM120 and erlotinib in head and neck cancer: mechanism of apoptosis and resistance. Mol Cancer Ther. 2017;16:729–38.
272. Rebucci M, et al. Mechanisms underlying resistance to cetuximab in the HNSCC cell line: role of AKT inhibition in bypassing this resistance. Int J Oncol. 2011;38(1):189–200.
273. Mohan S, et al. MEK inhibitor PD-0325901 overcomes resistance to PI3K/mTOR inhibitor PF-5212384 and potentiates antitumor effects in human head and neck squamous cell carcinoma. Clin Cancer Res. 2015;21(17):3946–56.
274. Gonzales CB, et al. Co-targeting ALK and EGFR parallel signaling in oral squamous cell carcinoma. Oral Oncol. 2016;59:12–9.
275. Zhou J, et al. CDK4/6 or MAPK blockade enhances efficacy of EGFR inhibition in oesophageal squamous cell carcinoma. Nat Commun. 2017;8:13897.
276. Michel L, et al. Phase I trial of palbociclib, a selective cyclin dependent kinase 4/6 inhibitor, in combination with cetuximab in patients with recurrent/metastatic head and neck squamous cell carcinoma. Oral Oncol. 2016;58:41–8.
277. Zeng L, et al. Combining Chk1/2 inhibition with cetuximab and radiation enhances in vitro and in vivo cytotoxicity in head and neck squamous cell carcinoma. Mol Cancer Ther. 2017;16(4):591–600.
278. Huang KK, et al. Exome sequencing reveals recurrent REV3L mutations in cisplatin-resistant squamous cell carcinoma of head and neck. Sci Rep. 2016;6:19552.
279. Koole K, et al. FGFR1 is a potential prognostic biomarker and therapeutic target in head and neck squamous cell carcinoma. Clin Cancer Res. 2016;22(15):3884–93.
280. Muntasell A, et al. Targeting NK-cell checkpoints for cancer immunotherapy. Curr Opin Immunol. 2017;45:73–81.
281. Srivastava RM, et al. CD137 stimulation enhances Cetuximab-induced natural killer: dendritic cell priming of antitumor T-cell immunity in patients with head and neck Cancer. Clin Cancer Res. 2017;23(3):707–16.
282. Luedke E, et al. Cetuximab therapy in head and neck cancer: immune modulation with interleukin-12 and other natural killer cell-activating cytokines. Surgery. 2012;152(3):431–40.
283. Kumai T, et al. Targeting HER-3 to elicit antitumor helper T cells against head and neck squamous cell carcinoma. Sci Rep. 2015;5:16280.
284. Jie HB, et al. CTLA-4(+) regulatory T cells increased in Cetuximab-treated head and neck Cancer patients suppress NK cell cytotoxicity and correlate with poor prognosis. Cancer Res. 2015;75(11):2200–10.
285. Pardoll DM. The blockade of immune checkpoints in cancer immunotherapy. Nat Rev Cancer. 2012;12(4):252–64.
286. Lyford-Pike S, et al. Evidence for a role of the PD-1:PD-L1 pathway in immune resistance of HPV-associated head and neck squamous cell carcinoma. Cancer Res. 2013;73(6):1733–41.
287. Concha-Benavente F, et al. Identification of the cell-intrinsic and -extrinsic pathways downstream of EGFR and IFNgamma that induce PD-L1 expression in head and neck Cancer. Cancer Res. 2016;76(5):1031–43.
288. Boeckx C, et al. Anti-epidermal growth factor receptor therapy in head and neck squamous cell carcinoma: focus on potential molecular mechanisms of drug resistance. Oncologist. 2013;18(7):850–64.
289. Wheeler DL, et al. Mechanisms of acquired resistance to cetuximab: role of HER (ErbB) family members. Oncogene. 2008;27(28):3944–56.
290. Wheeler DL, et al. Epidermal growth factor receptor cooperates with Src family kinases in acquired resistance to cetuximab. Cancer Biol Ther. 2009;8(8):696–703.

291. Lu Y, et al. Epidermal growth factor receptor (EGFR) ubiquitination as a mechanism of acquired resistance escaping treatment by the anti-EGFR monoclonal antibody Cetuximab. Cancer Res. 2007;67(17):8240–7.
292. P O-charoenrat, et al. Vascular endothelial growth factor family members are differentially regulated by c-erbB signaling in head and neck squamous carcinoma cells. Clin Exp Metastasis. 2000;18(2):155–61.
293. Sen M, et al. Targeting Stat3 abrogates EGFR inhibitor resistance in Cancer. Clin Cancer Res. 2012;18(18):4986–96.
294. Misale S, et al. Emergence of KRAS mutations and acquired resistance to anti-EGFR therapy in colorectal cancer. Nature. 2012;486:532–6.
295. Diaz LA Jr, et al. The molecular evolution of acquired resistance to targeted EGFR blockade in colorectal cancers. Nature. 2012;486(7404):537–40.
296. Johansson AC, et al. Cancer-associated fibroblasts induce matrix metalloproteinase-mediated Cetuximab resistance in head and neck squamous cell carcinoma cells. Mol Cancer Res. 2012;10(9):1158–68.
297. Ogawa T, et al. Methylation of death-associated protein kinase is associated with cetuximab and erlotinib resistance. Cell Cycle. 2012;11(8):1656–63.
298. Ahsan A, et al. Role of epidermal growth factor receptor degradation in cisplatin-induced cytotoxicity in head and neck cancer. Cancer Res. 2010;70(7):2862–9.
299. Stoehlmacher-Williams J, et al. Polymorphisms of the epidermal growth factor receptor (EGFR) and survival in patients with advanced cancer of the head and neck (HNSCC). Anticancer Res. 2012;32(2):421–5.
300. Wang Y, et al. A meta-analysis on the relations between EGFR R521K polymorphism and risk of Cancer. Int J Genomics. 2014;2014:1–7.
301. Shinkawa T, et al. The absence of Fucose but not the presence of galactose or bisecting N-Acetylglucosamine of human IgG1 complex-type oligosaccharides shows the critical role of enhancing antibody-dependent cellular cytotoxicity. J Biol Chem. 2002;278(5):3466–73.
302. Suzuki E, et al. A Nonfucosylated anti-HER2 antibody augments antibody-dependent cellular cytotoxicity in breast Cancer patients. Clin Cancer Res. 2007;13(6):1875–82.
303. Braig F, et al. Cetuximab resistance in head and neck cancer is mediated by EGFR-K521Polymorphism. Cancer Res. 2016;77(5):1188–99.
304. Ritter CA, et al. Human breast cancer cells selected for resistance to Trastuzumab in vivo overexpress epidermal growth factor receptor and ErbB ligands and remain dependent on the ErbB receptor network. Clin Cancer Res. 2007;13(16):4909–19.
305. Wang D, et al. HER3 targeting sensitizes HNSCC to Cetuximab by reducing HER3 activity and HER2/HER3 dimerization: evidence from cell line and patient-derived xenograft models. Clin Cancer Res. 2016;23(3):677–86.
306. Cabebe E, Wakelee H. Role of anti-angiogenesis agents in treating NSCLC: focus on bevacizumab and VEGFR tyrosine kinase inhibitors. Curr Treat Options in Oncol. 2007;8(1):15–27.
307. Ellis LM, Hicklin DJ. VEGF-targeted therapy: mechanisms of anti-tumour activity. Nat Rev Cancer. 2008;8(8):579–91.
308. Ratushny V, et al. Targeting EGFR resistance networks in head and neck cancer. Cell Signal. 2009;21(8):1255–68.
309. Seiwert TY, Cohen EEW. Targeting angiogenesis in head and neck cancer. Semin Oncol. 2008;35(3):274–85.
310. Liao YH, et al. Epidermal growth factor-induced ANGPTL4 enhances anoikis resistance and tumour metastasis in head and neck squamous cell carcinoma. Oncogene. 2017;36(16):2228–42.
311. Mayer A, et al. Downregulation of EGFR in hypoxic, diffusion-limited areas of squamous cell carcinomas of the head and neck. Br J Cancer. 2016;115(11):1351–8.
312. Bossi P, et al. Functional genomics uncover the biology behind the responsiveness of head and neck squamous cell Cancer patients to Cetuximab. Clin Cancer Res. 2016;22(15):3961–70.
313. Riedel F, et al. EGFR antisense treatment of human HNSCC cell lines down-regulates VEGF expression and endothelial cell migration. Int J Oncol. 2002;21(1):11–6.

314. Cohen EEW, et al. Erlotinib and bevacizumab in patients with recurrent or metastatic squamous-cell carcinoma of the head and neck: a phase I/II study. Lancet Oncol. 2009;10(3):247–57.
315. Soulieres D, et al. Multicenter phase II study of Erlotinib, an oral epidermal growth factor receptor tyrosine kinase inhibitor, in patients with recurrent or metastatic squamous cell Cancer of the head and neck. J Clin Oncol. 2004;22(1):77–85.
316. Fury MG, et al. A phase II study of SU5416 in patients with advanced or recurrent head and neck cancers. Investig New Drugs. 2006;25(2):165–72.
317. Elser C, et al. Phase II trial of Sorafenib in patients with recurrent or metastatic squamous cell carcinoma of the head and neck or nasopharyngeal carcinoma. J Clin Oncol. 2007;25(24):3766–73.
318. Gilbert J, et al. A randomized phase II efficacy and correlative studies of cetuximab with or without sorafenib in recurrent and/or metastatic head and neck squamous cell carcinoma. Oral Oncol. 2015;51(4):376–82.
319. Argiris A, et al. Cetuximab and bevacizumab: preclinical data and phase II trial in recurrent or metastatic squamous cell carcinoma of the head and neck. Ann Oncol. 2012;24(1):220–5.
320. Argiris A, et al. Phase II randomized trial of radiation therapy, cetuximab, and pemetrexed with or without bevacizumab in patients with locally advanced head and neck cancer. Ann Oncol. 2016;27(8):1594–600.
321. Hurwitz SJ, et al. Pharmacodynamics of DT-IgG, a dual-targeting antibody against VEGF-EGFR, in tumor xenografted mice. Cancer Chemother Pharmacol. 2011;69(3):577–90.
322. Zhang H, et al. A dual-targeting antibody against EGFR-VEGF for lung and head and neck cancer treatment. Int J Cancer. 2011;131(4):956–69.
323. Lecaros RLG, et al. Nanoparticle delivered VEGF-A siRNA enhances photodynamic therapy for head and neck cancer treatment. Mol Ther. 2016;24(1):106–16.
324. Liu F-Y, et al. CCR7 regulates cell migration and invasion through JAK2/STAT3 in metastatic squamous cell carcinoma of the head and neck. Biomed Res Int. 2014;2014:1–11.
325. Karnoub AE, Weinberg RA. Ras oncogenes: split personalities. Nat Rev Mol Cell Biol. 2008;9(7):517–31.
326. Suda T, et al. Copy number amplification of the PIK3CA gene is associated with poor prognosis in non-lymph node metastatic head and neck squamous cell carcinoma. BMC Cancer. 2012;12:416.
327. Smilek P, et al. Epidermal growth factor receptor (EGFR) expression and mutations in the EGFR signaling pathway in correlation with anti-EGFR therapy in head and neck squamous cell carcinomas. Neoplasma. 2012;59(05):508–15.
328. Braig F, et al. Liquid biopsy monitoring uncovers acquired RAS-mediated resistance to cetuximab in a substantial proportion of patients with head and neck squamous cell carcinoma. Oncotarget. 2016;7(28):42988–95.
329. Rampias T, et al. RAS/PI3K crosstalk and cetuximab resistance in head and neck squamous cell carcinoma. Clin Cancer Res. 2014;20(11):2933–46.
330. Tinhofer I, et al. Targeted next-generation sequencing of locally advanced squamous cell carcinomas of the head and neck reveals druggable targets for improving adjuvant chemoradiation. Eur J Cancer. 2016;57:78–86.
331. Pedrero JM, et al. Frequent genetic and biochemical alterations of the PI 3-K/AKT/PTEN pathway in head and neck squamous cell carcinoma. Int J Cancer. 2005;114(2):242–8.
332. Lyu J, et al. Predictive value of pAKT/PTEN expression in oral squamous cell carcinoma treated with cetuximab-based chemotherapy. Oral Surg Oral Med Oral Pathol Oral Radiol. 2016;121(1):67–72.
333. Burtness B, et al. Activity of cetuximab (C) in head and neck squamous cell carcinoma (HNSCC) patients (pts) with PTEN loss or PIK3CA mutation treated on E5397, a phase III trial of cisplatin (CDDP) with placebo (P) or C. J Clin Oncol. 2013;31(15_suppl):6028–6028
334. McAllister SS, Weinberg RA. Tumor-host interactions: a far-reaching relationship. J Clin Oncol. 2010;28(26):4022–8.

335. Raudenska M, et al. Prognostic significance of the tumour-adjacent tissue in head and neck cancers. Tumour Biol. 2015;36(12):9929–39.
336. Baylin SB. Aberrant patterns of DNA methylation, chromatin formation and gene expression in cancer. Hum Mol Genet. 2001;10(7):687–92.
337. Chang X, et al. Identification of Hypermethylated genes associated with cisplatin resistance in human cancers. Cancer Res. 2010;70(7):2870–9.
338. Gifford G. The acquisition of hMLH1 methylation in plasma DNA after chemotherapy predicts poor survival for ovarian cancer patients. Clin Cancer Res. 2004;10(13):4420–6.
339. Segura-Pacheco B, et al. Global DNA hypermethylation-associated cancer chemotherapy resistance and its reversion with the demethylating agent hydralazine. J Transl Med. 2006;4:32.
340. Bedi A, et al. Inhibition of TGF-enhances the in vivo antitumor efficacy of EGF receptor-targeted therapy. Mol Cancer Ther. 2012;11(11):2429–39.
341. Hoellein A, et al. Aurora kinase inhibition overcomes Cetuximab resistance in squamous cell Cancer of the head and neck. Oncotarget. 2011;2(8):599–609.
342. Astsaturov I, et al. Synthetic lethal screen of an EGFR-centered network to improve targeted therapies. Sci Signal. 2010;3(140):ra67-ra67.
343. Yuan L, et al. Recurrent FGFR3-TACC3 fusion gene in nasopharyngeal carcinoma. Cancer Biol Ther. 2014;15(12):1613–21.
344. Daly C, et al. FGFR3-TACC3 fusion proteins act as naturally occurring drivers of tumor resistance by functionally substituting for EGFR/ERK signaling. Oncogene. 2017;36(4):471–81.
345. Burtness B, et al. NCCN task force report: management of dermatologic and other toxicities associated with EGFR inhibition in patients with cancer. J Natl Compr Cancer Netw. 2009;7(Suppl 1):S5–21. quiz S22-4
346. Fernandez-Mateos J, et al. Epidermal growth factor receptor (EGFR) pathway polymorphisms as predictive markers of cetuximab toxicity in locally advanced head and neck squamous cell carcinoma (HNSCC) in a Spanish population. Oral Oncol. 2016;63:38–43.
347. O'Neil BH, et al. High incidence of Cetuximab-related infusion reactions in Tennessee and North Carolina and the association with atopic history. J Clin Oncol. 2007;25(24):3644–8.
348. Baas JM, et al. Recommendations on management of EGFR inhibitor-induced skin toxicity: a systematic review. Cancer Treat Rev. 2012;38(5):505–14.
349. Kohl P, et al. Systems biology: an approach. Clinical Pharmacology & Therapeutics. 2010;88(1):25–33.
350. Chow YP, et al. Exome sequencing identifies potentially Druggable mutations in nasopharyngeal carcinoma. Sci Rep. 2017;7:42980.
351. William WN, et al. Erlotinib and the risk of oral cancer: the erlotinib prevention of oral cancer (epoc) randomized clinical trial. JAMA Oncol. 2016;2(2):209–16.
352. Khan Z, et al. Cetuximab activity in dysplastic lesions of the upper aerodigestive tract. Oral Oncol. 2016;53:60–6.
353. Ariotti N, et al. Epidermal growth factor receptor activation remodels the plasma membrane lipid environment to induce nanocluster formation. Mol Cell Biol. 2010;30(15):3795–804.
354. Clayton AH, et al. Ligand-induced dimer-tetramer transition during the activation of the cell surface epidermal growth factor receptor—a multidimensional microscopy analysis. J Biol Chem. 2005;280(34):30392–9.
355. Needham SR, et al. EGFR oligomerization organizes kinase-active dimers into competent signalling platforms. Nat Commun. 2016;7:13307.
356. Ramirez UD, et al. Compounds identified by virtual docking to a tetrameric EGFR extracellular domain can modulate Grb2 internalization. BMC Cancer. 2015;15:436.
357. Ferris RL, et al. Phase II trial of post-operative radiotherapy with concurrent cisplatin plus panitumumab in patients with high-risk, resected head and neck cancer. Ann Oncol. 2016;27(12):2257–62.
358. Machiels J-P, et al. Zalutumumab plus best supportive care versus best supportive care alone in patients with recurrent or metastatic squamous-cell carcinoma of the head and neck after

failure of platinum-based chemotherapy: an open-label, randomised phase 3 trial. Lancet Oncol. 2011;12(4):333–43.
359. Rodriguez MO, et al. Nimotuzumab plus radiotherapy for unresectable squamous-cell carcinoma of the head and neck. Cancer Biol Ther. 2010;9(5):343–9.
360. Reddy BK, et al. Nimotuzumab provides survival benefit to patients with inoperable advanced squamous cell carcinoma of the head and neck: a randomized, open-label, phase IIb, 5-year study in Indian patients. Oral Oncol. 2014;50(5):498–505.
361. Rao S, et al. Phase I study of epirubicin, cisplatin and capecitabine plus matuzumab in previously untreated patients with advanced oesophagogastric cancer. Br J Cancer. 2008;99(6):868–74.
362. Rao S, et al. Matuzumab plus epirubicin, cisplatin and capecitabine (ECX) compared with epirubicin, cisplatin and capecitabine alone as first-line treatment in patients with advanced oesophago-gastric cancer: a randomised, multicentre open-label phase II study. Ann Oncol. 2010;21(11):2213–9.
363. Martins RG, et al. Cisplatin and radiotherapy with or without Erlotinib in locally advanced squamous cell carcinoma of the head and neck: a randomized phase II trial. J Clin Oncol. 2013;31(11):1415–21.
364. Yao M, et al. Phase II study of erlotinib and docetaxel with concurrent intensity-modulated radiotherapy in locally advanced head and neck squamous cell carcinoma. Head Neck. 2016;38(Suppl 1):E1770–6.
365. Harrington KJ, et al. Final analysis: a randomized, blinded, placebo (P)-controlled phase III study of adjuvant postoperative lapatinib (L) with concurrent chemotherapy and radiation therapy (CH-RT) in high-risk patients with squamous cell carcinoma of the head and neck (SCCHN). J Clin Oncol 2014;32(15_suppl):6005–6005
366. Huo ZJ, et al. Novel nanosystem to enhance the antitumor activity of lapatinib in breast cancer treatment: therapeutic efficacy evaluation. Cancer Sci. 2015;106(10):1429–37.
367. Machiels JP, et al. Afatinib versus methotrexate as second-line treatment in patients with recurrent or metastatic squamous-cell carcinoma of the head and neck progressing on or after platinum-based therapy (LUX-Head & Neck 1): an open-label, randomised phase 3 trial. Lancet Oncol. 2015;16(5):583–94.
368. Abdul Razak AR, et al. A phase II trial of dacomitinib, an oral pan-human EGF receptor (HER) inhibitor, as first-line treatment in recurrent and/or metastatic squamous-cell carcinoma of the head and neck. Ann Oncol. 2012;24(3):761–9.
369. Kim HS, et al. Phase II clinical and exploratory biomarker study of dacomitinib in patients with recurrent and/or metastatic squamous cell carcinoma of head and neck. Clin Cancer Res. 2015;21(3):544–52.

Chapter 3
c-MET in Head and Neck Squamous Cell Carcinoma

John Kaczmar and Tim N. Beck

Abstract The kinase receptor c-MET (MET proto-oncogene, receptor tyrosine kinase; also known as the hepatocyte growth factor receptor) and its ligand hepatocyte growth factor/scatter factor (HGF/SF) are two promising, potentially therapeutically exploitable targets in head and neck squamous cell carcinoma (HNSCC). c-MET is commonly overexpressed in head and neck cancer cells compared to normal epithelial cells and HGF/SF is often detected at high expression levels in tumor-adjacent mesenchymal cells, inducing paracrine activation of c-MET to support tumor growth and proliferation. Blocking this paracrine activity has been shown to reduce the proliferative capacity of HNSCC cells. Importantly, c-MET signaling outputs intersect with those of multiple other signaling pathways that drive or otherwise contribute to HNSCC cell survival and spread, including EGFR, HER2, SRC, STAT3, PI3K, RAS, GRB2, and others. In this review, we emphasize the roles of c-MET and HGF in HNSCC as well as the potential for therapeutic targeting of this signaling axis.

Keywords c-MET · HGFR · HGF/SF · Head and neck cancer · Squamous cell cancer · Receptor tyrosine kinases · Targeted therapy

J. Kaczmar
Hollings Cancer Center, Division of Hematology and Oncology, Department of Medicine, Medical University of South Carolina, Charleston, SC, USA

T. N. Beck (✉)
Molecular Therapeutics Program, Fox Chase Cancer Center, Philadelphia, PA, USA

Program in Molecular and Cell Biology and Genetics, Drexel University College of Medicine, Philadelphia, PA, USA
e-mail: Tim.Beck@fccc.edu

B. Burtness, E. A. Golemis (eds.), *Molecular Determinants of Head and Neck Cancer*, Current Cancer Research, https://doi.org/10.1007/978-3-319-78762-6_3

3.1 Introduction

Head and neck squamous cell carcinoma (HNSCC) remains a formidable treatment challenge, particularly in the setting of advanced locoregional and metastatic disease [1–3]. Despite significant scientific and medical advances over the past decades, the 5-year survival for HNSCC remains unacceptably low at 63.3% and 38% for advanced locoregional and metastatic disease, respectively [4]. Recent advances in the application of immunotherapies may improve these numbers modestly, but immunotherapy response rates currently remain below 20% [5, 6]. The advent of genomic and transcriptomic sequencing and proteomic studies have provided encouraging insights into the molecular landscape of HNSCC and have contributed to the identification and expanded study of a number of promising targets [1, 7–9]. Epidermal growth factor receptor (EGFR; discussed in Chap. 2), the first tyrosine kinase receptor successfully targeted in HNSCC, has been validated as an important molecular entity whose inhibition provides independent treatment efficacy [10] and synergizes with radiation [11] and chemotherapy [12]. The success of EGFR inhibition has prompted continued interest in identifying additional pathways involved in the proliferation or regulation of HNSCC mediated by receptor tyrosine kinases amenable to molecular targeted therapy. One such pathway is regulated by the hepatocyte growth factor receptor c-MET (also known as hepatocyte growth factor receptor, HGFR) and its ligand hepatocyte growth factor/scatter factor (HGF/SF), the focus of this chapter.

A number of studies have identified c-MET and its ligand as important in the pathogenesis of different subpopulations of head and neck cancers [13, 14]. Human papillomavirus (HPV) status – infection with this virus is an important cause of HNSCC (discussed in Chaps. 20 and 21) – broadly defines two subpopulations of HNSCC, HPV-positive and HPV-negative HNSCC [15]. The relationship between HPV and c-MET has not been wholly elucidated and available studies are incongruous, with some work suggesting that tumors with high c-MET expression are more likely to be HPV-negative, whereas other work suggests the opposite [16–20]. In general, elevated expression of c-MET in HNSCC has been associated with significantly worse prognosis compared to cases with low levels of the receptor [14, 21, 22]. c-MET is an important mediator of critical cellular processes involved in cancer cell survival, proliferation, and motility [23–26]. Signal transduction frequently involves crosstalk and cross-regulation of multiple pathways or signaling networks [27–30], and c-MET-associated signaling in HNSCC co-activates multiple signaling pathways to precisely regulate cancer cell behavior and therapy resistance [2, 31, 32].

To date, extensive preclinical studies have highlighted the potential of targeting c-MET in HNSCC [22, 33–35]. Inhibition of c-MET, similar to the inhibition of EGFR, may be most viable in combination with other therapies and treatment modalities. Evidence from HNSCC cell line models tested in vitro and in vivo indicates that c-MET inhibition can improve the efficacy of other therapeutics, including that of EGFR inhibitors and of standard chemotherapeutics [35]. In contrast, clinical evaluation of c-MET inhibitors has been slow, and only limited results are available, highlighting the need for additional large clinical trials and research into the underlying molecular mechanisms driving susceptibility and resistance [36].

3.2 c-MET Structure and Normal Function

c-MET is a receptor tyrosine kinase (RTK) phylogenetically and structurally related to semaphorins (a family of at least 20 known membrane-bound and secreted proteins; [37]) and plexins [37–41]. This RTK was first discovered as the product of the oncogene translocated promoter region (TPR)-MET during the study of human osteogenic sarcoma cell lines treated with *N*-methyl-*N'*-nitro-*N*-nitrosoguanidine, a potent carcinogen [42, 43]. Translocations involve rearrangement of chromosomal regions [44], which in the case of TPR-MET involved fusion of the TPR region from chromosome 1q25 with c-MET found on chromosome 7q31 [45].

Subsequent studies established that the non-fusion c-MET gene normally encodes a single precursor amino acid chain with two distinct transcript variants: isoform a, which codes for 1408 amino acids, and isoform b, which codes for 1390 amino acids. No functional differences have been described for the two isoforms and both contain the same N- and C-termini [46–48]. Functional c-MET signaling is present in both normal and cancer cells and is initiated by binding of the HGF/SF ligand, leading to activation of downstream signaling proteins including GAB1, PI3K, AKT, GRB2, RAS, RAF, and ERK/MAPK (Fig. 3.1; [23, 24, 49–52]).

3.2.1 c-MET General Structure

Posttranslational modification is a critical early step in the processing of c-MET. Once the single amino acid chain has been synthesized, it is cleaved within the endoplasmic reticulum by the endoprotease furin, to give rise to an α-subunit, which exists in the extracellular space, and a much larger membrane-spanning β-subunit [51–55]. The two subunits dimerize via a disulfide bond. This heterodimer has a molecular weight of 190 kDa. The 145 kDa β-subunit includes a single-pass transmembrane region in addition to three cytoplasmic domains and three extracellular domains (Fig. 3.1). The most membrane-distal domain of the extracellular component of c-MET includes the 50 kDa α-subunit, tethered by the disulfide bond [56, 57]. Collectively, the extracellular region of the α-β-subunit heterodimer forms the "SEMA domain" (Fig. 3.1), which has significant homology with recepteur d'origine nantais (RON), semaphorins, and plexins [58]. The SEMA domain is necessary for binding to the HGF/SF ligand, a molecular entity also composed of an α- and a β-subunit (discussed in greater detail in the next section).

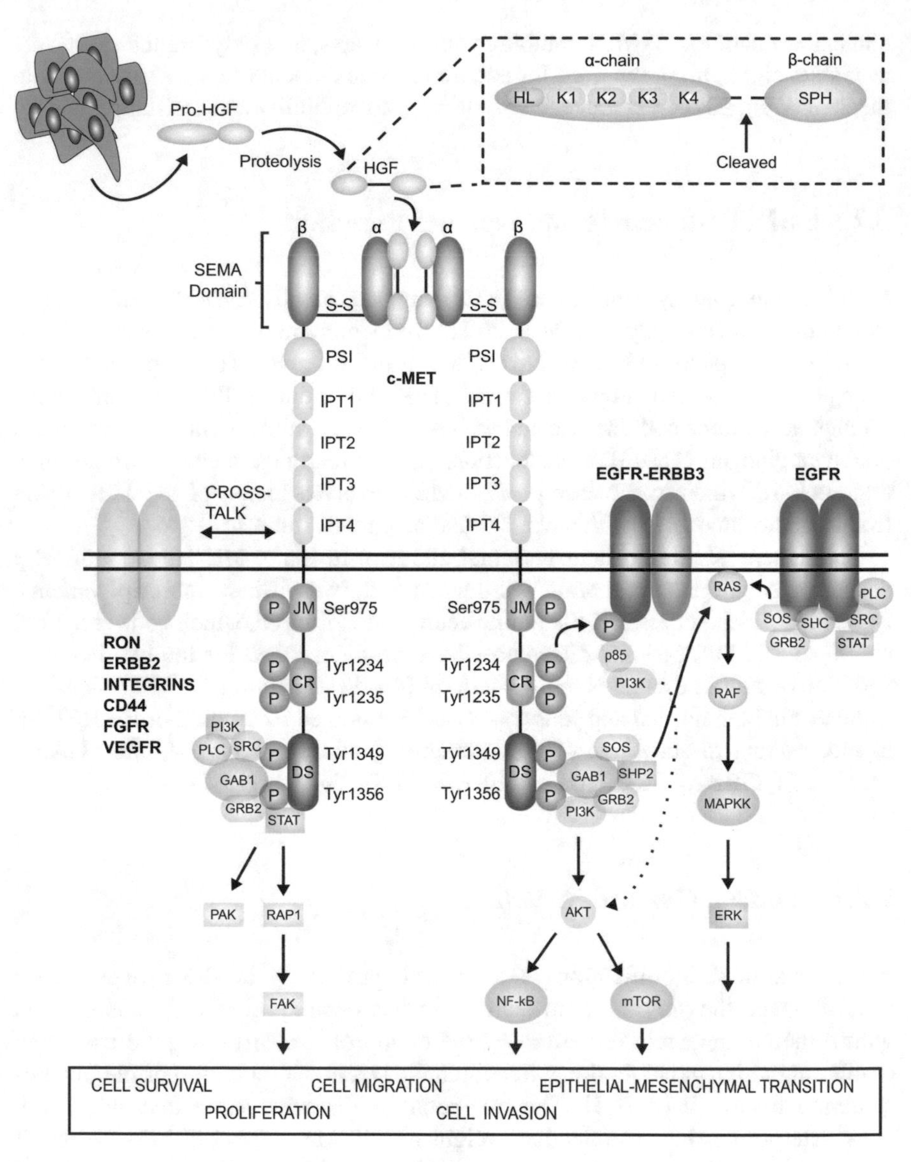

Fig. 3.1 HGF/c-MET structure, signaling, and cross talk. c-MET is shown in its homodimeric form. Docking site (DS); catalytic region (CR); juxtamembrane sequence (JM); immunoglobulin-like fold shared by plexins and transcriptional factors (IPT1–4); plexins, semaphorins, and integrins (PSI; domain is present in all three classes); and semaphorin (SEMA) domain

In addition to the SEMA domain, extracellular components of c-MET receptors include four immunoglobulin-like regions in plexins and transcription factors (IPT) domains. The IPT domains, particularly IPT3 and IPT4, are critical for binding of the α-subunit of the HGF ligand. Interestingly, in vitro work has shown that ligand binding to IPT initiates signaling even in the absence of the SEMA domain;

however, SEMA appears to allow the receptor to preferentially bind activated HGF over inactive HGF [59], suggesting that the SEMA domain is an important regulatory element.

The β-chain also contains a transmembrane domain and a cytoplasmic juxtamembrane (JM) domain and a functional kinase domain [53, 60, 61]. As is the case for many receptors, ligand binding induces homodimerization of c-MET, positioning the receptor kinase domain of one receptor near residues Y1234 and Y1235 on the cytoplasmic tail of the second receptor, resulting in autophosphorylation and subsequent recruitment of intracellular signaling proteins (Fig. 3.1; [62]). c-MET does not exclusively homodimerize. It has been experimentally shown that c-MET is capable of forming heterodimers with the RTK RON [63]. RON, like c-MET, has been implicated in HNSCC [64], as well as in several other cancers, and presents another potential therapeutic target [65]. It is not clear if c-MET-RON heterodimers are of relevance in HNSCC. Further study is needed to explore this possibility.

3.2.2 c-MET Ligand: HGF/SF

HGF/SF is the ligand associated with c-MET signaling [51]. Initially, studies suggested the existence of two independent ligands, HGF and SF, which were subsequently reconciled as representing a single protein [52]. In structural terms, HGF/SF is related to the protease zymogen plasminogen, an important component of the coagulation cascade; however, unlike plasminogen, HGF/SF does not have any known enzymatic capacities and simply functions as a MET receptor agonist [24, 52, 66, 67].

HGF/SF is synthesized as an inactive pro-peptide and requires precise proteolytic cleavage of the pro-peptide form of HGF/SF between residues Arg^{494}-Val^{495} [59, 67, 68] for activation. Matriptase (ST14), hepsin, plasma kallikrein, factor XIa, and HGF activator (HGFA) – frequently in the context of tissue damage – have been identified as the dominant proteases involved in this extracellular process. Negative regulation is conferred by HGF activator inhibitors 1 (HAI1) and 2 (HAI2), which block HGF/SF activation by inhibiting the enzymatic activity of the activating proteases [52, 69–71].

The pro-HGF/SF zymogen form can bind c-MET with high affinity through the ligand's α-chain, but this interaction does not initiate receptor signaling [67]. The cleaved, activated form of HGF/SF, like c-MET, is composed of a 68 kDa α-chain and a 34 kDa β-chain, linked by a disulfide bond [54, 55, 72]. The α-chain contains four kringle domains (K1–K4) as well as an N-terminal finger domain. The smaller β-chain only encodes one defined domain, a serine proteinase homology (SPH) domain (Fig. 3.1; [60, 73, 74]).

Activated HGF/SF predominantly binds to the SEMA domain of c-MET but can also bind to IPT domains to initiate signaling [59]. Interaction of HGF/SF with the SEMA domain induces receptor dimerization and phosphorylation of receptors Tyr1230, Tyr1235, and Tyr1243, resulting in subsequent autophosphorylation of two additional tyrosine residues: Tyr 1349 and Tyr 1356 (Fig. 3.1; [23, 62, 75]). Paracrine signaling generally involves secretion of pro-HGF/SF by mesenchymal cells and extracellular activation of the ligand leading to c-MET activation on epithelial cells [76]. c-MET activation has also been shown to be initiated via direct HGF/SF autocrine loops, to support cancer cell invasiveness, metastasis, and proliferation [16, 77–79]. In HNSCC, it has been suggested that paracrine activation of c-MET predominates [33, 80]; additional work may clarify this point in the future.

3.2.3 c-MET Trafficking

Receptor trafficking and endosomal compartmentalization are critical components of many membrane receptors, including c-MET [27, 81, 82]. During synthesis, c-MET is trafficked through the Golgi apparatus to the plasma membrane. At the membrane, an appropriate equilibrium is obtained via steady-state internalization and degradation. The equilibrium kinetics are regulated by the level of available ligand [42, 83–85]. This process provides cells with rapid spatiotemporal control to fine-tune response-specific signaling [86]. The signal transducer and activator of transcription 3 (STAT3; [81], of great relevance in HNSCC [87]) and ERK1/2 [88] signaling effector proteins are in part regulated by c-MET trafficking patterns.

Recent work using cell lines has suggested that c-MET trafficking is dependent on palmitoylation: inhibition of palmitoylation resulted in reduced c-MET expression posttranslationally, with diminished levels of c-MET on the cell surface and increased accumulation in a (likely Golgi) perinuclear compartment [89].

Canonical endocytosis of c-MET is mediated by both clathrin-dependent and clathrin-independent pathways [90]. The endocytotic process of c-MET is not completely understood, but studies have shown that PKC [88], Cbl [91], clathrin [92], dynamin 2 [93], sorting nexin 2 (SNX2) [94], and the Rho, Rac, and Rab GTPases [95] are all involved to varying degrees. Specifics of c-MET trafficking and endosomal compartmentalization in HNSCC have yet to be fully elucidated.

3.2.4 c-MET-Associated Signaling

c-MET signaling promotes cell migration, cellular motility, invasion, survival, and proliferation [96]. Signaling commences with binding of the ligand, HGF/SF, to the extracellular domain of c-MET, initiating tyrosine phosphorylation. Phosphorylated tyrosines 1349 and 1356 of c-MET have been identified as critical binding sites for several effector proteins, including growth factor receptor-bound protein 2 (GRB2), SHC, SRC, SHP2, SHIP1, phosphatidylinositol 3-kinase (PI3K), and Grb2-associated binder 1 (GAB1) (Fig. 3.1; [75, 97]).

GAB1 is particularly important for recruitment and activation of additional signaling proteins, specifically STAT3, GRB2/SOS, PI3K, SRC, and PLCγ, all critical for sustained c-MET signaling. GRB2/SOS activate downstream effectors that include the RAS/RAF/MEK/ERK signaling cascade and RAC/PAK to increase cell motility. c-MET activation also induces reorganization of focal adhesion structures to further promote cell motility, particularly through focal adhesion kinase (p125FAK) and paxillin [32, 98, 99]. c-MET-dependent activation of PI3K-AKT-mTOR promotes cancer cell survival (Fig. 3.1; [24, 51, 52, 100]).

Cytokine release is another important consequence of c-MET activity, particularly release of interleukin-8 (IL-8, [101]). IL-8 is categorized as a pro-inflammatory chemokine and has been shown to contribute to angiogenesis, proliferation, and survival of cancer cells [102] and as relevant in head and neck cancer [103]. In HNSCC, studies have provided evidence that co-expression of IL-8 and HGF is linked to more aggressive disease progression and response to therapy [101, 104].

3.3 c-MET Critical Interactions and Functions

c-MET plays a critical role in signal transduction crosstalk with other effector pathways. These intricate processes, critical for normal cell function and survival, require complex interactions and rapid changes in signal transduction circuits. The concept of adaptive signaling crosstalk is of great relevance in cancer, because it provides an important mechanism for cells to respond to therapeutic assaults [105, 106]. For example, redundancy and signaling crosstalk provide a platform for cancer cells to survive molecular inhibition of important proteins [106–109]. In HNSCC, crosstalk between c-MET and members of the ERBB family, SRC, and regulators of epithelial-mesenchymal transition have been described as particularly relevant.

3.3.1 c-MET and the Epidermal Growth Factor (EGF) Family of Receptors

The most clinically relevant and certainly most studied signaling c-MET crosstalk involves EGFR (HER1/ERBB1). Multiple studies have indicated that increased c-MET signaling activity plays a vital role in resistance to EGFR-targeted drugs (see Sect. 3.5; [32, 34, 35, 94, 110–112]). c-MET and EGFR share many common effector proteins (Fig. 3.1). Convergence of effector proteins enables cancer cells to reestablish signaling activities despite targeted inhibition of EGFR, thereby enhancing therapeutic resistance [106, 113].

In addition to downstream signaling crosstalk, co-immunoprecipitation studies have provided evidence for heterodimeric c-MET-EGFR complex formation, although further work is needed, particularly in HNSCC, to unambiguously demonstrate physical interactions between the two RTKs [32]. Garofalo et al. further demonstrated the close collaboration of EGFR and c-MET in signaling, with experiments showing that the EGFR inhibitor gefitinib downregulated oncogenic miRNAs controlled by EGFR and c-MET (miR-30b, miR-30c, miR-221, and miR-222), thereby promoting apoptosis. Furthermore, Garofalo et al. demonstrated that in multiple gefitinib-resistant cell lines, the miRNAs were not downregulated, likely due to c-MET overexpression [110].

c-MET has also been implicated in signaling crosstalk with additional ERBB family members, specifically HER2 (ERBB2) and HER3 (ERBB3) in some cancers, although these interactions have not yet been observed in HNSCC. In HER2-positive breast cancer, c-MET signaling has been shown to synergize with HER2 signaling, to reduce cell-cell adherence and induce a more aggressive phenotype [114]. In gastric cancer with HER2 amplification, MET activation was strongly associated with response to the small-molecule inhibitor lapatinib [115].

The relationship between c-MET and the ERBB family member ERBB3 has also not yet been evaluated in HNSCC; however, studies in other cancer types have provided important, potentially relevant findings. For example, in lung cancer, amplification of c-MET has been reported to cause resistance to gefitinib by driving ERBB3 activation of the effector protein PI3K. In this setting, ERBB3 activating phosphorylation was only abrogated by simultaneous inhibition of c-MET and ERBB3 [116]. Interestingly, c-MET and ERBB3 can heterodimerize and induce signaling [116]. These findings are particularly intriguing given the recent development of ERBB3/HER3-targeting antibodies, such as MM-121 (seribantumab; [117, 118]), AV-203 [119], CDX3379 [120], and lumretuzumab [121], and the interest of these agents for HNSCC.

Future studies are needed to explore the existence and therapeutic exploitability of c-MET-ERBB signaling in HNSCC. Given that ERBB family members are frequently expressed in HNSCC [122, 123] and that single-agent ERBB targeting yields response rates of 4–13% [10, 124, 125], this strongly suggests exploration of c-MET as a resistance driver. Important preclinical work by Xu et al. and by Seiwert et al. suggests that targeting c-MET may enhance the efficacy of ERBB-targeted therapy in HNSCC [22, 35]. A small retrospective study of samples from 33 recurrent/metastatic patients and 24 non-recurrent/non-metastatic patients indicated that overexpression of c-MET or phosphorylated c-MET correlated with worse outcome, including in cases with a cetuximab-based treatment regimen [126].

3.3.2 c-MET and SRC

Signaling interactions between the cytoplasmic tyrosine kinase SRC and c-MET have also been implicated as important for drug resistance and tumorigenesis in HNSCC. Interactions between SRC and EGFR and SRC and c-MET have been studied extensively [127], and given the interconnectedness of these three proteins (Fig. 3.1), findings by Stabile et al. that SRC activation in HNSCC drives resistance to erlotinib by stimulating c-MET align with previous assumptions [111].

The relevance of c-MET-SRC cross-activation was further demonstrated by the observation that c-MET activity increases HNSCC cell survival following SRC inhibition [128]. Furthermore, SRC kinase has been implicated with increased epithelial-to-mesenchymal transition (EMT; [129]) and more aggressive tumor progression, features particularly relevant to HNSCC, where locoregional invasion is strongly associated with worse outcome [3, 130]. Collectively, these findings suggest that combination therapy featuring a SRC inhibitor, such as dasatinib [131], c-MET-targeted therapy, or both used in combination, may successfully address tumor resistance to EGFR-targeted therapy. However in an unselected population of operable HNSCC, the addition of dasatinib to erlotinib failed to demonstrate increased antitumor activity, suggesting that the addition of SRC inhibition may not be able to overcome EGFR inhibitor resistance or that precise selection of patients based on SRC activity is necessary [132].

3.3.3 c-MET in Epithelial-to-Mesenchymal Transition (EMT)

EMT [133], a critical process during embryogenesis, is subverted in tumorigenesis to support cancer cell invasion, metastasis, and therapy resistance [27, 129, 134, 135]. Studies have shown that c-MET regulates and is critical for proper embryologic development by initiating EMT of myogenic progenitor cells [52, 136]. However, the precise role of c-MET in HNSCC as a regulator of EMT has not been exhaustively studied, with only a few published articles available. For example, cell culture experiments have shown that the addition of HGF to HNSCC cells significantly reduces the number of cell-cell junctions [137], while increasing wound closure, cell proliferation, and invasion [33], all important aspects of EMT [138–140].

The study of EMT in HNSCC is challenging due to a number of factors, including the subsite diversity (e.g., nasopharynx, oropharynx, oral, tongue base, and larynx), the influence of HPV, and the complex mutational landscape frequently shaped by continuous exposure to mutagens (i.e., tobacco, alcohol, betel nuts [12, 111, 141–144]). Nevertheless, analysis of HNSCC primary tumor tissue has found that high levels of activated SRC correlated with reduced expression of E-cadherin (important cell-cell junctional protein [145]) and with more aggressive tumor features, including penetrating invasive fronts and lymph node metastasis [130]. The relevance of E-cadherin was further emphasized in a meta-analysis of 19 studies of HNSCC, where high expression of E-cadherin was found to correlate with improved overall survival and disease-free survival; significant statistical heterogeneity was however noted [146]. The role of c-MET in this regard continues to be elusive.

3.4 Dysregulation of HGF/c-MET in HNSCC

Dysregulation of c-MET signaling has been shown to initiate or support malignant growth, metastasis, and therapy resistance in many cancers, including HNSCC [21, 24, 26, 33, 51, 56, 147]. Processes leading to c-MET dysregulation include its overexpression, amplification, mutation, and microenvironment/ligand-associated triggers. Some research has suggested a role for c-MET in human papillomavirus (HPV)-related HNSCC [44, 148] (see Chaps. 20 and 21), with the HPV oncoprotein E6 [149, 150] upregulating expression of c-MET in part through downregulation of p53 [20].

3.4.1 c-MET Overexpression and Amplification

An increase in copy number of c-MET and robust overexpression of c-MET and HGF have been detected in up to 60–80% of HNSCC [21, 22, 33, 122]. In vitro work first suggested that several HNSCC cell lines had a low level of c-MET copy number increases, based on FISH (fluorescence in situ hybridization) and qPCR (quantitative polymerase chain reaction), with a substantial number of patient-derived tumors harboring much more substantial c-MET copy number increases [22]. Of note, expression levels of c-MET seem to positively correlate with tissue progression from normal to dysplastic to carcinomatous [22, 33, 35]. Large-scale genomic analyses have reported similar findings, albeit with few HNSCC cases presenting with c-MET amplifications (5 of 528; accessed using http://www.cbioportal.org; see [151, 152] for cBioPortal information).

Expression levels of c-MET in HNSCC appear to mediate resistance to the commonly used chemotherapeutic, cisplatin [21]. Akervall et al. used cDNA microarray and RT-PCR to explore expression levels of c-MET in HNSCC cell lines that were either chemosensitive or chemoresistant. There was significantly higher expression of c-MET in chemoresistant versus cisplatin-sensitive cell lines. Importantly, expression analysis of c-MET in 29 patient HNSCC samples by immunohistochemistry yielded similar findings: 5 of 9 (56%) patients with low c-MET responded to induction chemotherapy (cisplatin and 5-fluorouracil), whereas only 4 of 20 (20%) patients with high c-MET-expressing tumors responded [21]. These findings could be of clinical relevance, considering that approximately 10% of HNSCC express high levels of c-MET mRNA or protein (52 of 528; accessed using http://www.cbioportal.org; z-score for mRNA and RPPA (reverse phase protein array) set at 2; see [151, 152] for cBioPortal information). Hence, monitoring c-MET expression may be important for clinical trial stratification and treatment strategies (see Sect. 3.5).

3.4.2 *c-MET Mutation*

Genetic and protein level aberrations associated with c-MET activation, including several single nucleotide polymorphisms (SNPs), have been reported in HNSCC [22–24]. Mutations affecting each of the functional domains of c-MET (Fig. 3.1) have been detected throughout the protein, affecting the tyrosine kinase, JM domain, IPT, and SEMA domains [22, 76, 153–155]. Clinical significance of these mutations is not yet known.

Based on data provided by The Cancer Genome Atlas (TCGA) consortium, the frequency of c-MET mutations in HNSCC is relatively low (4 truncating/missense mutations out of 528 cases; accessed using http://www.cbioportal.org; see [151, 152] for cBioPortal information). The TCGA data also indicate a low prevalence (2 out of 279 samples) of an alternative c-MET transcript with skipped exon 14 [8], an alteration shown to be c-MET activating in lung cancer [156]. Further work is needed to determine if individuals with this specific c-MET mutation benefit disproportionally from treatment with c-MET inhibitors.

Exome sequencing of 149 matched esophageal adenocarcinoma tumor-normal tissue samples, with whole genome sequencing of 15 of the samples, identified c-MET mutations in 2% and gain-of-function events in 6% of tested samples [157]. In these cases, c-MET may be a putative therapeutic target. Based on results from a phase III clinical trial, Argiris et al. have confirmed that c-MET mutations are possible prognostic markers of disease progression and for survival [158].

Gain-of-function mutations and cancer-associated SNPs that affect the SEMA domain or the JM domain have been identified in 12% (8 out of 66) analyzed HNSCC tumors [22, 159]. These specific alterations lead to increased responsiveness to HGF, which in vitro increased soft-agar colony formation [22]. Most importantly, mutated or elevated c-MET, be it in cell lines, xenograft models, or patient tumors, tends to correlate with exquisite sensitivity to c-MET inhibition, including to inhibition by the small-molecule inhibitor crizotinib [22], further discussed in Sect. 3.5.

3.4.3 *Microenvironment and HGF/SF*

HGF is typically secreted by mesenchymal cells in the tumor microenvironment [160] but can also be produced by cancer cells in an autocrine fashion [33]. The role of HGF in HNSCC has been less studied than the role of its corresponding receptor; it is likely that elevated expression of HGF contributes to the pathogenesis of a subpopulation of HNSCC cases. HNSCC tumor-derived fibroblasts produce more HGF than non-tumor-associated fibroblasts, and this may be one of the dominant mechanisms of c-MET activation in HNSCC [33]. Data from the TCGA indicate that HGF is amplified in only 2% (11 out of 510) of HNSCC, with another 1% (6 out of 510) overexpressing HGF mRNA (accessed using http://www.cbioportal.org; see [151,

152] for cBioPortal information). How this corresponds to protein expression, activation, or availability is not known. Activation of c-MET is tightly regulated by control over the interaction between ligand and receptor. Once bound, HGF induces signaling responsible for the upregulation of hundreds of genes, including c-MET itself and the proteases needed for HGF and c-MET processing [21, 24].

3.5 Clinical Applications of HGF/c-MET-Targeted Therapy

Through the use of siRNA, researchers have established the c-MET dependency of HNSCC tumor-derived cell lines [22, 111]. Based on these observations and the current understanding of frequent MET overexpression in more advanced HNSCC, broad interest has grown in the rational targeting of this pathway. Currently, a large number of therapeutic strategies are being utilized against the HGF/c-MET axis (summarized by Sadiq et al. [26]), and a large number of inhibitors have been developed. These include multiple small-molecule inhibitors (tivantinib, crizotinib) as well as antibodies against the receptor (MetMAb [onartuzumab], ABT700) and the ligand (ficlatuzumab, rilotumumab; [26, 63]). There are currently a multitude of clinical trials evaluating c-MET or HGF targeting for several different cancers.

Perhaps the area of greatest interest involving c-MET-targeted therapy centers on the ability of these agents to synergize with and overcome resistance to standard of care therapies. As previously described, increasing c-MET expression reduces sensitivity to cisplatin treatment [21, 161], and c-MET is commonly amplified specifically in the context of EGFR resistance [162]. In addition to resistance to traditional chemotherapy and targeted therapies, research has begun to delineate a role for c-MET in cell repopulation after radiation in HNSCC xenografts [163] and resistance to radiation in HNSCC cell lines [164]. Preliminary research suggests that immunotherapy could be enhanced by increasing immune recognition of the often overexpressed c-MET through combinatorial regimens or peptide vaccine technologies [165]. In all of these contexts, there is a clear clinical rationale to consider the addition of c-MET-targeted therapy to overcome treatment resistance.

Research in this area is perhaps furthest along in investigating strategies to overcome EGFR tyrosine-kinase inhibitor (TKI) resistance. Engleman et al. reported detection of c-MET amplification in 4 of 18 lung cancer specimens that had developed resistance to the EGFR TKIs gefitinib or erlotinib [116]. Elegant work by Turke et al. showed that in NSCLC, resistance to EGFR TKIs can be established through c-MET amplification or autocrine HGF production and that combined EGFR and c-MET inhibition is curative in a xenograft model. In the same study, analysis of clinical data identified c-MET amplification in 4 of 27 EGFR inhibitor-resistant NSCLC tumor specimens [112].

Bean et al. reported that 9 of 43 patients with acquired resistance to erlotinib or gefitinib presented with c-MET amplification; whereas, among 62 untreated patients, only 2 presented with amplified c-MET. Notably, Bean et al. found no correlation between c-MET amplification and EGFR mutation status [166]. These reports provide a strong rationale for combined c-MET and EGFR inhibition in specific cases; one of the obvious challenges will be identifying the appropriate patient populations who might benefit from such synergistic treatments.

Most investigation regarding c-MET and EGFR has thus far been conducted in lung cancer; nevertheless, marked synergy combining c-MET and EGFR inhibition has been observed in HNSCC [22, 35]. For HNSCC, a randomized phase II trial [NCT01696955] to evaluate the response of patients with recurrent, metastatic, or inoperable HNSCC to the EGFR-targeting antibody cetuximab (see Chap. 2) with or without tivantinib (Sect. 3.5.2) demonstrated that the addition of tivantinib to cetuximab was tolerable. The study failed to show improvement in response rate or overall survival (OS) [167]. Notably, treated patients were not selected based on c-MET expression status, a potential weakness of the trial design [168]. Several trials featuring tivantinib in combination with EGFR-directed agents performed in the NSCLC domain demonstrate the safety of these combinations but only modest, if any, clinical activity [169].

Given the feasibility of combination treatment with an HGF/c-MET inhibitor, the primary effort must now turn to defining whether there are specific biomarkers applicable for the selection of patients most likely to respond [168]. Again, research in NSCLC provides context. A recent phase III study [NCT01244191] of tivantinib in combination with erlotinib versus erlotinib alone demonstrated significant improvement in progression-free survival (PFS) but failed to demonstrate improved OS [170]. While the trial enrolled an unselected population in regard to c-MET status, the study did perform tumor testing for both c-MET expression and amplification. An exploratory subgroup analysis suggested that patients whose tumors had high c-MET expression might have improved OS. In contrast, a recent randomized study investigating the addition of onartuzumab to first-line chemotherapy in HER2-negative advanced gastric cancer failed to demonstrate improvement in OS or PFS in patients with c-MET high tumors. These divergent findings highlight the need for prospective HNSCC trials featuring agents targeting HGF/c-MET that stratify patients based on HGF/c-MET status.

Ongoing trials in HNSCC are doing just that, building on the concept of selecting specific subpopulations for treatment. For example, in clinical trial NCT02205398, a study of safety and efficacy of the c-MET kinase inhibitor capmatinib (INC280; [171]) and cetuximab, only adults with metastatic colorectal cancer or HNSCC and ≥50% c-MET positivity of tumor cells were eligible. This shift is emblematic of the concept of biomarker-driven clinical trial designs [168]. This approach seems critical for the success of HGF/c-MET inhibitors.

3.5.1 Foretinib

Foretinib is an orally available inhibitor of MET, RON, AXL, Tie-1, KIT, PDGFR, and VEGFR. Importantly, foretinib inhibits HGF-induced MET phosphorylation and extracellular VEGF-induced kinase phosphorylation events (Fig. 3.2; [172]). In

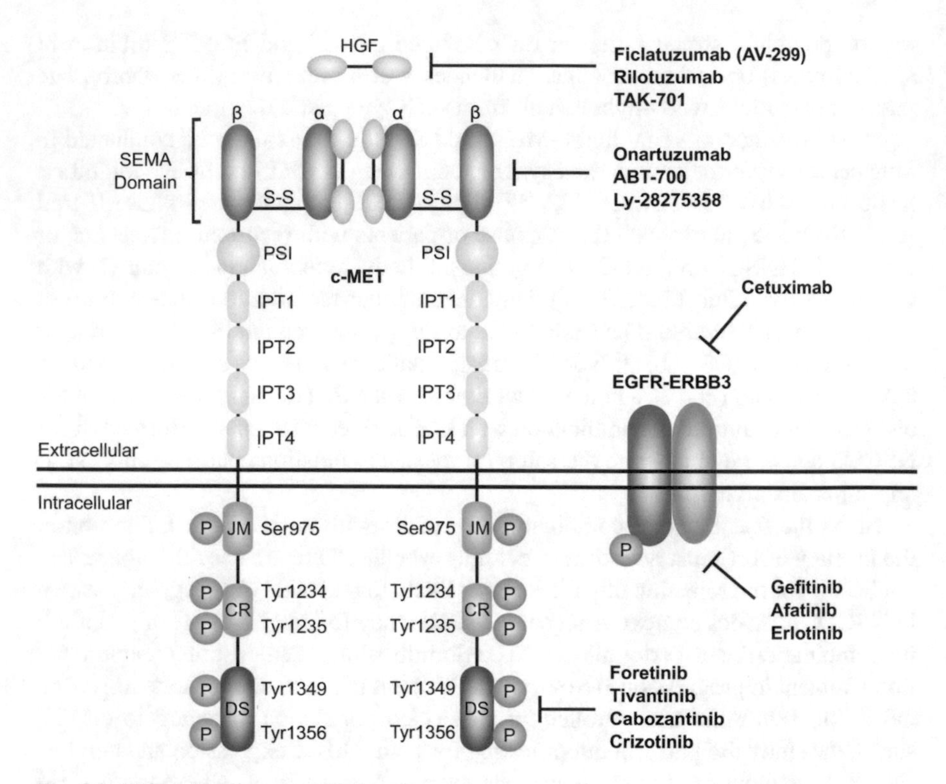

Fig. 3.2 HGF/c-MET-targeted inhibitors

HNSCC, foretinib has been studied in a phase II single-agent clinical trial [36]. Foretinib was administered orally at 240 mg for 5 consecutive days of a 14-day cycle (a maximum of 30 cycles was administered). Among the 14 patients treated, no objective responses were observed, but 43% (6 of 14) of patients had tumor shrinkage of 5–21%, and the treatment was generally well tolerated [36]. Despite these data, the development of foretinib has been discontinued, and no registered clinical trials currently feature this inhibitor.

3.5.2 Tivantinib

A more potent c-MET inhibitor than foretinib, tivantinib, is currently being explored in numerous cancer types including HNSCC. Most recently, a randomized study of tivantinib versus tivantinib plus cetuximab [NCT01696955] failed to demonstrate improvement in response rate or OS [167].The study was built on the observed synergy between c-MET and EGFR inhibition [22, 173]. Tivantinib, a staurosporine derivative that targets dephosphorylated c-MET, has shown promising activity in several phase I and phase II trials and was recently evaluated in a large phase III trial

[NCT01244191] in combination with erlotinib for the treatment of NSCLC [174, 175]. As highlighted earlier, the study demonstrated improved PFS, but not OS in a c-MET unselected population. Subgroup analysis suggested a trend to median OS benefit (HR 0.70 95% CI, 0.49–1.01) for patients with c-MET high tumors (9.3 vs. 5.9 months; [170]). Tivantinib is also currently under study in c-MET high hepatocellular carcinoma [NCT01755767]. Numerous trials involving other malignancies are also ongoing.

3.5.3 Additional Small-Molecule Inhibitors

Cabozantinib is a promising multitargeted TKI with activity against c-MET, VEGFR2, and RET that has been approved for use against advanced medullary thyroid cancer ([NCT00704730]; [176]) and in metastatic renal cell carcinoma [12]. Additionally, cabozantinib was assessed in combination with erlotinib in a phase Ib/II study in patients with NSCLC: 17% of patients achieved 30% or greater reduction in tumor burden [177]. To our knowledge, this inhibitor has not yet been studied in HNSCC.

Crizotinib, another small-molecule inhibitor, has recently emerged as a promising treatment option for ALK-positive NSCLC [178]; in addition to targeting ALK, crizotinib also inhibits c-MET [24]. Ou et al. have reported the observation that an NSCLC patient with de novo c-MET amplification but no ALK fusion protein had a sustained response to crizotinib [179]. Additionally, it has been shown that crizotinib appeared to have efficacy against cancer stem-like cells and synergized with docetaxel or cisplatin to improve tumor inhibitory effect in a mouse xenograft model of HNSCC [180]. These findings suggest a role for c-MET inhibition in combination with conventional chemotherapy in HNSCC. There are several additional small-molecule c-MET inhibitors (golvatinib, AMG-208, INC-280, and PF-04217903 among others) that may be worth consideration for the treatment of HNSCC [24, 177].

3.5.4 Antibodies Targeting c-MET

Onartuzumab (MetMAb) is a recombinant monoclonal antibody that binds the SEMA domain of c-MET (Fig. 3.2). By binding to c-MET, onartuzumab blocks HGF from binding and activating c-MET. There is also evidence that HGF may serve as a reliable biomarker for onartuzumab target engagement [181]. In a phase II study of second-/third-line treatment in NSCLC with onartuzumab as monotherapy or in combination with erlotinib, patients with c-MET immunohistochemistry-positive (IHC+) tumors significantly benefited from onartuzumab plus erlotinib treatment compared to placebo plus erlotinib (c-MET IHC+, OS of 4.6 months versus and 12.6 months; p = 0.002; c-MET IHC-, OS of 9.2 months versus of

5.5 months; $p = 0.021$; [182]). Other subgroups did not benefit with onartuzumab. This strongly suggested that a population of higher c-MET-expressing patients derived disproportionate benefit from the combined therapy [182, 183].

Unfortunately, these very exciting findings were not replicated in a phase III study of patients with c-MET-positive NSCLC treated with onartuzumab/erlotinib [NCT01456325]; the study was terminated early due to futility with regard to its primary OS outcome goal [184]. Moreover, onartuzumab has been studied in metastatic colorectal cancer and did not improve OS, nor did IHC-verified c-MET expression prove to be a predictive biomarker [185].

Another c-MET-targeting antibody that is being actively studied is emibetuzumab (LY-2875358), a humanized IgG4 monoclonal antibody. Emibetuzumab binds to c-MET and thus inhibits binding of HGF/SF; additionally, the antibody induces c-MET internalization and degradation; therefore, emibetuzumab can disrupt ligand-dependent and ligand-independent activation of c-MET [186]. The agent has proven safe and tolerable as monotherapy [187] and in combination with EGFR TKIs in phase I studies, achieving a partial response in one patient with MET-high hepatocellular carcinoma [188]. A recent phase II trial in which patients with erlotinib-refractory NSCLC received emibetuzumab alone or in combination with erlotinib demonstrated a disappointing objective response rate of 3.8% in patients classified as MET DX-high (≥60% cells with ≥2+ IHC staining; [189]).

3.5.5 Antibodies Targeting HGF/SF

An alternative therapeutic approach to inhibiting HGF/c-MET signaling is by targeting HGF with antibodies (Fig. 3.2). Ficlatuzumab (AV-299) is a humanized IgG1 antibody that specifically binds HGF. In a phase II study designed to compare gefitinib as a single agent versus gefitinib and ficlatuzumab in patients with adenocarcinoma of the lung, the drug combination did not significantly improve response rate or PFS [190]. However, there is interest in this antibody for HNSCC, as preclinical data demonstrated the agent is capable of decreasing proliferation, migration, and invasion of HNSCC cell lines [191]. A phase I study investigating ficlatuzumab in combination with cetuximab is ongoing [NCT02277197].

Rilotumumab is a fully human monoclonal antibody that targets HGF to reduced HGF/c-MET signaling. In a phase II study, the efficacy of rilotumumab plus mitoxantrone and prednisone was tested in castration-resistant prostate cancer. The addition of rilotumumab did not have any beneficial effect [192]. In general, the antibody has not demonstrated significant clinical activity to date, perhaps explained by the recent observation that rilotumumab only serves as a partial c-MET antagonist, unable to reduce c-MET signaling in the presence of EGFR activation [193].

Evidence for the value of combined targeting of HGF and EGFR has been provided by a study of Tak-701, another humanized monoclonal antibody that targets HGF. Combining this agent with gefitinib has been shown capable of successfully restoring sensitivity to gefitinib in gefitinib-resistant human lung cancer cells [194].

3.6 Conclusions and Future Directions

Cancer genomic studies have provided sufficient data to propose that "driver" genes can be assigned to at least twelve different signaling pathways [142]. This realization has created a major impetus to the growth of genomic evaluation of individual patients to usher in the age of precision medicine/oncology [195, 196]. However, it has also become evident that to more effectively treat cancer, multiple pathways must be targeted concurrently, a strategy that potentially reduces opportunities for the emergence of tumor resistance. Given the frequent overexpression of c-MET in advanced HNSCC, the potential of c-MET-targeted therapy, particularly in combination with other inhibitors targeting interlinked pathways (e.g., cetuximab), has proven feasible and tolerable, but unfortunately, to date, meaningful clinical activity for these approaches has not been demonstrated.

Precision medicine requires discovery and exploitation of good predictive biomarkers. While data suggest that elevated c-MET expression is prognostic of worse outcomes in HNSCC and other malignancies, in many cases, it does not appear predictive of response to therapy. The clinical trial data reviewed here suggest that c-MET expression status alone is not sufficient to guide patient selection for treatment with c-MET inhibitors. The powerful ability of these agents to arrest HNSCC tumor growth in vitro and in xenograft models continues to suggest the possibility that in the right context, these agents could make a contribution to management of HNSCC. Trials do show that there are some patients who benefited from c-MET inhibition in one form or another, but more work is needed to identify likely responders.

An additional factor that must be considered in any discussion of HNSCC is the bifurcation between HPV-initiated disease and HPV-negative disease. It currently remains unknown whether HPV status impacts efficacy of HGF/c-MET-targeted agents. Preliminary evidence suggests that HGF/c-MET mutational or amplifying events are less common in HPV(+) tumors, based on early reports by The Cancer Genome Atlas Project (TCGA; [159]). Still, validation will be important, including evaluation of tumors and tumor stroma from HPV(+) HNSCC. Where possible, future studies should report outcomes stratified by HPV status as differential response to c-MET-directed therapy could be of clinical interest.

Importantly, the very recent advent of immunotherapy as an approved and efficacious treatment option in HNSCC may provide an additional role for targeting of the c-MET axis to either increase antigen presentation or improve immune recognition of malignant cells. Studies in this area are in their relative infancy, and time will tell whether there is synergy to be found in this space.

The HGF/c-MET axis has been studied for a considerable period and is altered and of biological importance in multiple cancer types. If there is to be clinical success in targeting the HGF/c-MET axis, patient selection must be refined using novel predictive biomarkers. It is expected, based on knowledge from other tumor types, that the greatest response will occur in a subset of patients that have an activated c-MET phenotype, which presumably is not fully captured by any single assay modality. To better hone the ability to predict therapeutic benefit of c-MET inhibition, it may be necessary for future studies to perform detailed genomic, transcriptomic, epigenomic, and proteomic analysis of matched tumor, tumor stroma, and normal tissue from patients who responded and failed to respond to c-MET-directed therapy. The analysis of circulating tumor cells and genetic material pre- and post-therapy may provide additional relevant information. Analysis of tissue for HGF/c-MET expression, mutations/SNPs, amplification, immune system activity, and relationships with downstream signaling pathways with known crosstalk might help determine mechanisms of resistance to c-MET axis inhibition. Furthering our understanding of the drivers of HNSCC growth and treatment resistance will potentially enable targeted combinational strategies that support durable treatment responses and lay the groundwork for future treatment paradigms.

References

1. Haddad RI, Shin DM. Recent advances in head and neck cancer. N Engl J Med. 2008;359(11):1143–54.
2. Leemans CR, Braakhuis BJ, Brakenhoff RH. The molecular biology of head and neck cancer. Nat Rev Cancer. 2011;11(1):9–22.
3. Pfister DG, et al. Head and neck cancers. J Natl Compr Cancer Netw. 2011;9(6):596–649.
4. Howlader N, et al. SEER cancer statistics review, 1975–2013. April 2016. Available from: http://seer.cancer.gov/csr/1975_2013/.
5. Ferris RL, et al. Nivolumab for recurrent squamous-cell carcinoma of the head and neck. N Engl J Med. 2016;375(19):1856–67.
6. Chow LQM, et al. Antitumor activity of pembrolizumab in biomarker-unselected patients with recurrent and/or metastatic head and neck squamous cell carcinoma: results from the phase Ib KEYNOTE-012 expansion cohort. J Clin Oncol. 2016;34(32):3838–45.
7. Beck TN, Golemis EA. Genomic insights into head and neck cancer. Cancers Head Neck. 2016;1(1):1.
8. Network CGA. Comprehensive genomic characterization of head and neck squamous cell carcinomas. Nature. 2015;517(7536):576–82.
9. The Cancer Genome Atlas Research Network. Integrated genomic characterization of oesophageal carcinoma. Nature. 2017;541(7636):169–75.
10. Vermorken JB, et al. Open-label, uncontrolled, multicenter phase II study to evaluate the efficacy and toxicity of cetuximab as a single agent in patients with recurrent and/or metastatic squamous cell carcinoma of the head and neck who failed to respond to platinum-based therapy. J Clin Oncol. 2007;25(16):2171–7.

11. Bonner JA, et al. Radiotherapy plus cetuximab for squamous-cell carcinoma of the head and neck. N Engl J Med. 2006;354(6):567–78.
12. Maxwell JH, Grandis JR, Ferris RL. HPV-associated head and neck cancer: unique features of epidemiology and clinical management. Annu Rev Med. 2016;67:91–101.
13. De Herdt MJ, de Jong RJB. HGF and c-MET as potential orchestrators of invasive growth in head and neck squamous cell carcinoma. Front Biosci (Landmark). 2008;13:2516–26.
14. Lo Muzio L, et al. Effect of c-Met expression on survival in head and neck squamous cell carcinoma. Tumor Biol. 2006;27(3):115–21.
15. Vokes EE, Agrawal N, Seiwert TY. HPV-associated head and neck cancer. J Natl Cancer Inst. 2015;107(12):djv344.
16. Szturz P, et al. Understanding c-MET signalling in squamous cell carcinoma of the head & neck. Crit Rev Oncol Hematol. 2017;111:39–51.
17. Choe JY, et al. Expression of c-Met is different along the location and associated with lymph node metastasis of head and neck carcinoma. Korean J Pathol. 2012;46(6):515–22.
18. Kwon MJ, et al. Frequent hepatocyte growth factor overexpression and low frequency of c-Met gene amplification in human papillomavirus-negative tonsillar squamous cell carcinoma and their prognostic significances. Hum Pathol. 2014;45(7):1327–38.
19. Baschnagel AM, et al. c-Met expression is a marker of poor prognosis in patients with locally advanced head and neck squamous cell carcinoma treated with chemoradiation. Int J Radiat Oncol Biol Phys. 2014;88(3):701–7.
20. Qian GQ, et al. Human papillomavirus oncoprotein E6 upregulates c-Met through p53 downregulation. Eur J Cancer. 2016;65:21–32.
21. Akervall J, et al. Genetic and expression profiles of squamous cell carcinoma of the head and neck correlate with cisplatin sensitivity and resistance in cell lines and patients. Clin Cancer Res. 2004;10(24):8204–13.
22. Seiwert TY, et al. The MET receptor tyrosine kinase is a potential novel therapeutic target for head and neck squamous cell carcinoma. Cancer Res. 2009;69(7):3021–31.
23. Birchmeier C, et al. Met, metastasis, motility and more. Nat Rev Mol Cell Biol. 2003;4(12):915–25.
24. Blumenschein GR Jr, Mills GB, Gonzalez-Angulo AM. Targeting the hepatocyte growth factor-cMET axis in cancer therapy. J Clin Oncol. 2012;30(26):3287–96.
25. Boccaccio C, Comoglio PM. Invasive growth: a MET-driven genetic programme for cancer and stem cells. Nat Rev Cancer. 2006;6(8):637–45.
26. Sadiq AA, Salgia R. Inhibition of MET receptor tyrosine kinase and its ligand hepatocyte growth factor. J Thorac Oncol. 2012;7(16 suppl 5):S372–4.
27. Beck TN, et al. Anti-Mullerian hormone signaling regulates epithelial plasticity and chemoresistance in lung cancer. Cell Rep. 2016;16(3):657–71.
28. Rolland T, et al. A proteome-scale map of the human interactome network. Cell. 2014;159(5):1212–26.
29. Pawson T. Protein modules and signalling networks. Nature. 1995;373(6515):573–80.
30. Chen JC, et al. Identification of causal genetic drivers of human disease through systems-level analysis of regulatory networks. Cell. 2014;159(2):402–14.
31. Molinolo AA, et al. Dysregulated molecular networks in head and neck carcinogenesis. Oral Oncol. 2009;45(4–5):324–34.
32. Lai AZ, Abella JV, Park M. Crosstalk in Met receptor oncogenesis. Trends Cell Biol. 2009;19(10):542–51.
33. Knowles LM, et al. HGF and c-Met participate in paracrine tumorigenic pathways in head and neck squamous cell cancer. Clin Cancer Res. 2009;15(11):3740–50.
34. Singleton KR, et al. A receptor tyrosine kinase network composed of fibroblast growth factor receptors, epidermal growth factor receptor, v-erb-b2 erythroblastic leukemia viral oncogene homolog 2, and hepatocyte growth factor receptor drives growth and survival of head and neck squamous carcinoma cell lines. Mol Pharmacol. 2013;83(4):882–93.

35. Xu H, et al. Dual blockade of EGFR and c-Met abrogates redundant signaling and proliferation in head and neck carcinoma cells. Clin Cancer Res. 2011;17(13):4425–38.
36. Seiwert T, et al. Phase II trial of single-agent foretinib (GSK1363089) in patients with recurrent or metastatic squamous cell carcinoma of the head and neck. Investig New Drugs. 2013;31(2):417–24.
37. Trusolino L, Comoglio PM. Scatter-factor and semaphorin receptors: cell signalling for invasive growth. Nat Rev Cancer. 2002;2(4):289–300.
38. Committee SN. Unified nomenclature for the semaphorins/collapsins. Cell. 1999;97(5):551–2.
39. Tamagnone L, Comoglio PM. Control of invasive growth by hepatocyte growth factor (HGF) and related scatter factors. Cytokine Growth Factor Rev. 1997;8(2):129–42.
40. Winberg ML, et al. Plexin A is a neuronal semaphorin receptor that controls axon guidance. Cell. 1998;95(7):903–16.
41. Giordano S, et al. The semaphorin 4D receptor controls invasive growth by coupling with Met. Nat Cell Biol. 2002;4(9):720–4.
42. Park M, et al. Mechanism of met oncogene activation. Cell. 1986;45(6):895–904.
43. Cooper CS, et al. Molecular cloning of a new transforming gene from a chemically transformed human cell line. Nature. 1984;311(5981):29–33.
44. Beck TN, Gabitova L, Serebriiskii IG. Targeted therapy: genomic approaches. In: Reviews in cell biology and molecular medicine. Weinheim, Germany: Wiley-VCH Verlag GmbH & Co. KGaA; 2015.
45. Peschard P, Park M. From Tpr-Met to Met, tumorigenesis and tubes. Oncogene. 2007;26(9):1276–85.
46. Park M, et al. Sequence of MET protooncogene cDNA has features characteristic of the tyrosine kinase family of growth-factor receptors. Proc Natl Acad Sci U S A. 1987;84(18):6379–83.
47. Bottaro DP, et al. Identification of the hepatocyte growth-factor receptor as the c-met protooncogene product. Science. 1991;251(4995):802–4.
48. Rodrigues GA, Naujokas MA, Park M. Alternative splicing generates isoforms of the met receptor tyrosine kinase which undergo differential processing. Mol Cell Biol. 1991;11(6):2962–70.
49. Kim ES, Salgia R. MET pathway as a therapeutic target. J Thorac Oncol. 2009;4(4):444–7.
50. Organ SL, Tsao MS. An overview of the c-MET signaling pathway. Ther Adv Med Oncol. 2011;3(1 Suppl):S7–S19.
51. Trusolino L, Bertotti A, Comoglio PM. MET signalling: principles and functions in development, organ regeneration and cancer. Nat Rev Mol Cell Biol. 2010;11(12):834–48.
52. Gherardi E, et al. Targeting MET in cancer: rationale and progress. Nat Rev Cancer. 2012;12(2):89–103.
53. Giordano S, et al. Tyrosine kinase receptor indistinguishable from the c-met protein. Nature. 1989;339(6220):155–6.
54. Gherardi E, et al. Structural basis of hepatocyte growth factor/scatter factor and MET signalling. Proc Natl Acad Sci U S A. 2006;103(11):4046–51.
55. Komada M, et al. Proteolytic processing of the hepatocyte growth factor/scatter factor receptor by furin. FEBS Lett. 1993;328(1–2):25–9.
56. Chen Z. Aberrant activation of HGF/c-MET signaling and targeted therapy in squamous cancer. In: Glick AB, Van Waes C, editors. Signaling pathways in squamous cancer. New York: Springer; 2011. p. 462.
57. Prat M, et al. C-terminal truncated forms of Met, the hepatocyte growth factor receptor. Mol Cell Biol. 1991;11(12):5954–62.
58. Gherardi E, et al. The sema domain. Curr Opin Struct Biol. 2004;14(6):669–78.
59. Basilico C, et al. A high affinity hepatocyte growth factor-binding site in the immunoglobulin-like region of Met. J Biol Chem. 2008;283(30):21267–77.
60. Stamos J, et al. Crystal structure of the HGF beta-chain in complex with the Sema domain of the Met receptor. EMBO J. 2004;23(12):2325–35.
61. Lemmon MA, Schlessinger J. Cell signaling by receptor tyrosine kinases. Cell. 2010;141(7):1117–34.

62. Ponzetto C, et al. A multifunctional docking site mediates signaling and transformation by the hepatocyte growth factor/scatter factor receptor family. Cell. 1994;77(2):261–71.
63. Sadiq AA, Salgia R. MET as a possible target for non-small-cell lung cancer. J Clin Oncol. 2013;31(8):1089–96.
64. Yoon TM, et al. Expression of the receptor tyrosine kinase recepteur d'origine nantais and its association with tumor progression in hypopharyngeal cancer. Head Neck. 2013;35(8):1106–13.
65. Yao HP, et al. The monoclonal antibody Zt/f2 targeting RON receptor tyrosine kinase as potential therapeutics against tumor growth-mediated by colon cancer cells. Mol Cancer. 2011;10:82.
66. Donate LE, et al. Molecular evolution and domain-structure of plasminogen-related growth-factors (Hgf/Sf and Hgf1/Msp). Protein Sci. 1994;3(12):2378–94.
67. Landgraf KE, et al. An allosteric switch for pro-HGF/Met signaling using zymogen activator peptides. Nat Chem Biol. 2014;10(7):567–73.
68. Naldini L, et al. Extracellular proteolytic cleavage by urokinase is required for activation of hepatocyte growth-factor scatter factor. EMBO J. 1992;11(13):4825–33.
69. Kojima K, et al. Roles of functional and structural domains of hepatocyte growth factor activator inhibitor type 1 in the inhibition of matriptase. J Biol Chem. 2008;283(5):2478–87.
70. Szabo R, et al. Matriptase inhibition by hepatocyte growth factor activator inhibitor-1 is essential for placental development. Oncogene. 2007;26(11):1546–56.
71. Kawaguchi T, et al. Purification and cloning of hepatocyte growth factor activator inhibitor type 2, a Kunitz-type serine protease inhibitor. J Biol Chem. 1997;272(44):27558–64.
72. Ultsch M, et al. Crystal structure of the NK1 fragment of human hepatocyte growth factor at 2.0 A resolution. Structure. 1998;6(11):1383–93.
73. Kirchhofer D, et al. Structural and functional basis of the serine protease-like hepatocyte growth factor beta-chain in Met binding and signaling. J Biol Chem. 2004;279(38):39915–24.
74. Chirgadze DY, et al. Crystal structure of the NK1 fragment of HGF/SF suggests a novel mode for growth factor dimerization and receptor binding. Nat Struct Biol. 1999;6(1):72–9.
75. Xu Y, et al. Receptor-type protein tyrosine phosphatase beta (RPTP-beta) directly dephosphorylates and regulates hepatocyte growth factor receptor (HGFR/Met) function. J Biol Chem. 2011;286(18):15980–8.
76. Sattler M, et al. The role of the c-Met pathway in lung cancer and the potential for targeted therapy. Ther Adv Med Oncol. 2011;3(4):171–84.
77. Rong S, et al. Invasiveness and metastasis of NIH 3T3 cells induced by Met-hepatocyte growth factor/scatter factor autocrine stimulation. Proc Natl Acad Sci U S A. 1994;91(11):4731–5.
78. Otsuka T, et al. c-Met autocrine activation induces development of malignant melanoma and acquisition of the metastatic phenotype. Cancer Res. 1998;58(22):5157–67.
79. Ferrucci A, et al. A HGF/cMET autocrine loop is operative in multiple myeloma bone marrow endothelial cells and may represent a novel therapeutic target. Clin Cancer Res. 2014;20(22):5796–807.
80. Rothenberger NJ, Stabile LP. Hepatocyte growth factor/c-Met signaling in head and neck cancer and implications for treatment. Cancers (Basel). 2017;9(4):pii: E39.
81. Kermorgant S, Parker PJ. Receptor trafficking controls weak signal delivery: a strategy used by c-Met for STAT3 nuclear accumulation. J Cell Biol. 2008;182(5):855–63.
82. Parachoniak CA, Park M. Dynamics of receptor trafficking in tumorigenicity. Trends Cell Biol. 2012;22(5):231–40.
83. Abella JV, et al. Dorsal ruffle microdomains potentiate Met receptor tyrosine kinase signaling and down-regulation. J Biol Chem. 2010;285(32):24956–67.
84. Parachoniak CA, et al. GGA3 functions as a switch to promote Met receptor recycling, essential for sustained ERK and cell migration. Dev Cell. 2011;20(6):751–63.
85. Lefebvre J, et al. Met degradation: more than one stone to shoot a receptor down. FASEB J. 2012;26(4):1387–99.
86. Joffre C, et al. A direct role for Met endocytosis in tumorigenesis. Nat Cell Biol. 2011;13(7):827–U227.

87. Geiger JL, Grandis JR, Bauman JE. The STAT3 pathway as a therapeutic target in head and neck cancer: barriers and innovations. Oral Oncol. 2016;56:84–92.
88. Kermorgant S, Zicha D, Parker PJ. PKC controls HGF-dependent c-Met traffic, signalling and cell migration. EMBO J. 2004;23(19):3721–34.
89. Coleman DT, et al. Palmitoylation regulates the intracellular trafficking and stability of c-Met. Oncotarget. 2016;7(22):32664–77.
90. Cho KW, et al. Identification of a pivotal endocytosis motif in c-Met and selective modulation of HGF-dependent aggressiveness of cancer using the 16-mer endocytic peptide. Oncogene. 2013;32(8):1018–29.
91. Petrelli A, et al. The endophilin-CIN85-Cbl complex mediates ligand-dependent downregulation of c-Met. Nature. 2002;416(6877):187–90.
92. Li N, et al. Specific Grb2-mediated interactions regulate clathrin-dependent endocytosis of the cMet-tyrosine kinase. J Biol Chem. 2007;282(23):16764–75.
93. Singleton PA, et al. CD44 regulates hepatocyte growth factor-mediated vascular integrity: role of c-Met, Tiam1/Rac1, dynamin 2, and cortactin. J Biol Chem. 2007;282(42):30643–57.
94. Ogi S, et al. Sorting nexin 2-mediated membrane trafficking of c-Met contributes to sensitivity of molecular-targeted drugs. Cancer Sci. 2013;104(5):573–83.
95. Kamei T, et al. Coendocytosis of cadherin and c-Met coupled to disruption of cell-cell adhesion in MDCK cells - regulation by Rho, Rac and Rab small G proteins. Oncogene. 1999;18(48):6776–84.
96. Hartmann S, Bhola NE, Grandis JR. HGF/Met signaling in head and neck cancer: impact on the tumor microenvironment. Clin Cancer Res. 2016;22(16):4005–13.
97. Weidner KM, et al. Interaction between Gab1 and the c-Met receptor tyrosine kinase is responsible for epithelial morphogenesis. Nature. 1996;384(6605):173–6.
98. Grotegut S, et al. Hepatocyte growth factor induces cell scattering through MAPK/Egr-1-mediated upregulation of Snail. EMBO J. 2006;25(15):3534–45.
99. Nikonova AS, et al. CAS proteins in health and disease: an update. IUBMB Life. 2014;66(6):387–95.
100. Comoglio PM, Giordano S, Trusolino L. Drug development of MET inhibitors: targeting oncogene addiction and expedience. Nat Rev Drug Discov. 2008;7(6):504–16.
101. Dong G, et al. Hepatocyte growth factor/scatter factor-induced activation of MEK and PI3K signal pathways contributes to expression of proangiogenic cytokines interleukin-8 and vascular endothelial growth factor in head and neck squamous cell carcinoma. Cancer Res. 2001;61(15):5911–8.
102. Waugh DJ, Wilson C. The interleukin-8 pathway in cancer. Clin Cancer Res. 2008;14(21):6735–41.
103. Cohen RF, et al. Interleukin-8 expression by head and neck squamous cell carcinoma. Arch Otolaryngol Head Neck Surg. 1995;121(2):202–9.
104. Le QT, et al. Prognostic and predictive significance of plasma HGF and IL-8 in a phase III trial of chemoradiation with or without tirapazamine in locoregionally advanced head and neck cancer. Clin Cancer Res. 2012;18(6):1798–807.
105. Duncan JS, et al. Dynamic reprogramming of the kinome in response to targeted MEK inhibition in triple-negative breast cancer. Cell. 2012;149(2):307–21.
106. Guo A, et al. Signaling networks assembled by oncogenic EGFR and c-Met. Proc Natl Acad Sci U S A. 2008;105(2):692–7.
107. Beck TN, et al. EGFR and RB1 as dual biomarkers in HPV-negative head and neck cancer. Mol Cancer Ther. 2016;15(10):2486–97.
108. Logue JS, Morrison DK. Complexity in the signaling network: insights from the use of targeted inhibitors in cancer therapy. Genes Dev. 2012;26(7):641–50.
109. Sun C, Bernards R. Feedback and redundancy in receptor tyrosine kinase signaling: relevance to cancer therapies. Trends Biochem Sci. 2014;39(10):465–74.
110. Garofalo M, et al. EGFR and MET receptor tyrosine kinase-altered microRNA expression induces tumorigenesis and gefitinib resistance in lung cancers. Nat Med. 2012;18(1):74–82.

111. Stabile LP, et al. c-Src activation mediates erlotinib resistance in head and neck cancer by stimulating c-Met. Clin Cancer Res. 2013;19(2):380–92.
112. Turke AB, et al. Preexistence and clonal selection of MET amplification in EGFR mutant NSCLC. Cancer Cell. 2010;17(1):77–88.
113. Tang Z, et al. Dual MET-EGFR combinatorial inhibition against T790M-EGFR-mediated erlotinib-resistant lung cancer. Br J Cancer. 2008;99(6):911–22.
114. Khoury H, et al. HGF converts ErbB2/Neu epithelial morphogenesis to cell invasion. Mol Biol Cell. 2005;16(2):550–61.
115. Chen CT, et al. MET activation mediates resistance to lapatinib inhibition of HER2-amplified gastric cancer cells. Mol Cancer Ther. 2012;11(3):660–9.
116. Engelman JA, et al. MET amplification leads to gefitinib resistance in lung cancer by activating ERBB3 signaling. Science. 2007;316(5827):1039–43.
117. Schoeberl B, et al. An ErbB3 antibody, MM-121, is active in cancers with ligand-dependent activation. Cancer Res. 2010;70(6):2485–94.
118. Liu JF, et al. Randomized phase II trial of seribantumab in combination with paclitaxel in patients with advanced platinum-resistant or -refractory ovarian cancer. J Clin Oncol. 2016;34(36):4345–53.
119. Meetze K, et al. Neuregulin 1 expression is a predictive biomarker for response to AV-203, an ERBB3 inhibitory antibody, in human tumor models. Clin Cancer Res. 2015;21(5):1106–14.
120. Lee S, et al. Inhibition of ErbB3 by a monoclonal antibody that locks the extracellular domain in an inactive configuration. Proc Natl Acad Sci U S A. 2015;112(43):13225–30.
121. Meulendijks D, et al. First-in-human phase I study of lumretuzumab, a glycoengineered humanized anti-HER3 monoclonal antibody, in patients with metastatic or advanced HER3-positive solid tumors. Clin Cancer Res. 2016;22(4):877–85.
122. Mehra R, et al. Protein-intrinsic and signaling network-based sources of resistance to EGFR- and ErbB family-targeted therapies in head and neck cancer. Drug Resist Updat. 2011;14(6):260–79.
123. Price KA, Cohen EE. Mechanisms of and therapeutic approaches for overcoming resistance to epidermal growth factor receptor (EGFR)-targeted therapy in squamous cell carcinoma of the head and neck (SCCHN). Oral Oncol. 2015;51(5):399–408.
124. Soulieres D, et al. Multicenter phase II study of erlotinib, an oral epidermal growth factor receptor tyrosine kinase inhibitor, in patients with recurrent or metastatic squamous cell cancer of the head and neck. J Clin Oncol. 2004;22(1):77–85.
125. Machiels J-PH, et al. Afatinib versus methotrexate as second-line treatment in patients with recurrent or metastatic squamous-cell carcinoma of the head and neck progressing on or after platinum-based therapy (LUX-Head & Neck 1): an open-label, randomised phase 3 trial. Lancet Oncol. 2015;16(5):583–94.
126. Madoz-Gurpide J, et al. Activation of MET pathway predicts poor outcome to cetuximab in patients with recurrent or metastatic head and neck cancer. J Transl Med. 2015;13:282.
127. Zhang S, Yu D. Targeting Src family kinases in anti-cancer therapies: turning promise into triumph. Trends Pharmacol Sci. 2012;33(3):122–8.
128. Sen B, et al. Distinct interactions between c-Src and c-Met in mediating resistance to c-Src inhibition in head and neck cancer. Clin Cancer Res. 2011;17(3):514–24.
129. Thiery JP, et al. Epithelial-mesenchymal transitions in development and disease. Cell. 2009;139(5):871–90.
130. Mandal M, et al. Epithelial to mesenchymal transition in head and neck squamous carcinoma: association of Src activation with E-cadherin down-regulation, vimentin expression, and aggressive tumor features. Cancer. 2008;112(9):2088–100.
131. Shah NP, et al. Overriding imatinib resistance with a novel ABL kinase inhibitor. Science. 2004;305(5682):399–401.
132. Bauman JE, et al. Randomized, placebo-controlled window trial of EGFR, Src, or combined blockade in head and neck cancer. JCI Insight. 2017;2(6):e90449.

133. Hay ED. An overview of epithelio-mesenchymal transformation. Cells Tissues Organs. 1995;154(1):8–20.
134. Fischer KR, et al. Epithelial-to-mesenchymal transition is not required for lung metastasis but contributes to chemoresistance. Nature. 2015;527(7579):472–6.
135. Zheng X, et al. Epithelial-to-mesenchymal transition is dispensable for metastasis but induces chemoresistance in pancreatic cancer. Nature. 2015;527(7579):525–30.
136. Bladt F, et al. Essential role for the c-Met receptor in the migration of myogenic precursor cells into the limb bud. Nature. 1995;376(6543):768–71.
137. Dawson JC, et al. Mtss1 promotes cell-cell junction assembly and stability through the small GTPase Rac1. PLoS One. 2012;7(3):e31141.
138. Lamouille S, Xu J, Derynck R. Molecular mechanisms of epithelial–mesenchymal transition. Nat Rev Mol Cell Biol. 2014;15(3):178–96.
139. Nieto MA, et al. Emt: 2016. Cell. 2016;166(1):21–45.
140. Nieto MA. Epithelial plasticity: a common theme in embryonic and cancer cells. Science. 2013;342(6159):1234850.
141. Smith A, Teknos TN, Pan Q. Epithelial to mesenchymal transition in head and neck squamous cell carcinoma. Oral Oncol. 2013;49(4):287–92.
142. Vogelstein B, et al. Cancer genome landscapes. Science. 2013;339(6127):1546–58.
143. Kandoth C, et al. Mutational landscape and significance across 12 major cancer types. Nature. 2013;502(7471):333–9.
144. McGranahan N, et al. Clonal status of actionable driver events and the timing of mutational processes in cancer evolution. Sci Transl Med. 2015;7(283):283ra54.
145. van Roy F, Berx G. The cell-cell adhesion molecule E-cadherin. Cell Mol Life Sci. 2008;65(23):3756–88.
146. Ren X, et al. E-cadherin expression and prognosis of head and neck squamous cell carcinoma: evidence from 19 published investigations. Onco Targets Ther. 2016;9:2447–53.
147. Cecchi F, Rabe DC, Bottaro DP. Targeting the HGF/Met signalling pathway in cancer. Eur J Cancer. 2010;46(7):1260–70.
148. Bhatia A, Burtness B. Human papillomavirus-associated oropharyngeal cancer: defining risk groups and clinical trials. J Clin Oncol. 2015;33(29):3243–50.
149. Beck TN, et al. Head and neck squamous cell carcinoma: ambiguous human papillomavirus status, elevated p16, and deleted retinoblastoma 1. Head Neck. 2017;39(3):E34–9.
150. Moody CA, Laimins LA. Human papillomavirus oncoproteins: pathways to transformation. Nat Rev Cancer. 2010;10(8):550–60.
151. Gao JJ, et al. Integrative analysis of complex cancer genomics and clinical profiles using the cBioPortal. Sci Signal. 2013;6(269):pl1.
152. Cerami E, et al. The cBio cancer genomics portal: an open platform for exploring multidimensional cancer genomics data (vol 2, pg 401, 2012). Cancer Discov. 2012;2(10):960.
153. Tyner JW, et al. MET receptor sequence variants R970C and T992I lack transforming capacity. Cancer Res. 2010;70(15):6233–7.
154. Tengs T, et al. A transforming MET mutation discovered in non-small cell lung cancer using microarray-based resequencing. Cancer Lett. 2006;239(2):227–33.
155. Lengyel E, Sawada K, Salgia R. Tyrosine kinase mutations in human cancer. Curr Mol Med. 2007;7(1):77–84.
156. Kong-Beltran M, et al. Somatic mutations lead to an oncogenic deletion of met in lung cancer. Cancer Res. 2006;66(1):283–9.
157. Dulak AM, et al. Exome and whole-genome sequencing of esophageal adenocarcinoma identifies recurrent driver events and mutational complexity. Nat Genet. 2013;45(5):478–U37.
158. Argiris A, et al. Phase III randomized, placebo-controlled trial of docetaxel with or without gefitinib in recurrent or metastatic head and neck cancer: an eastern cooperative oncology group trial. J Clin Oncol. 2013;31(11):1405–14.
159. Hayes DN, Grandis J, El-Naggar AK, et al. Comprehensive genomic characterization of squamous cell carcinoma of the head and neck in the Cancer Genome Atlas. In: AACR annual meeting 2013. Washington, DC: AACR; 2013.

160. Sonnenberg E, et al. Scatter factor/hepatocyte growth factor and its receptor, the c-met tyrosine kinase, can mediate a signal exchange between mesenchyme and epithelia during mouse development. J Cell Biol. 1993;123(1):223–35.
161. Sun S, Wang Z. Head neck squamous cell carcinoma c-Met(+) cells display cancer stem cell properties and are responsible for cisplatin-resistance and metastasis. Int J Cancer. 2011;129(10):2337–48.
162. Sierra J. c-MET as a potential therapeutic target and biomarker in cancer. Ther Adv Med Oncol. 2011;3(S1):S21–35.
163. Wilson GD, et al. Cancer stem cell signaling during repopulation in head and neck cancer. Stem Cells Int. 2016;2016:1894782.
164. Ettl T, et al. AKT and MET signalling mediates antiapoptotic radioresistance in head neck cancer cell lines. Oral Oncol. 2015;51(2):158–63.
165. Kumai T, et al. EGFR inhibitors augment antitumour helper T-cell responses of HER family-specific immunotherapy. Br J Cancer. 2013;109(8):2155–66.
166. Bean J, et al. MET amplification occurs with or without T790M mutations in EGFR mutant lung tumors with acquired resistance to gefitinib or erlotinib. Proc Natl Acad Sci U S A. 2007;104(52):20932–7.
167. Vokes EE, et al. A randomized phase II trial of the MET inhibitor tivantinib + cetuximab versus cetuximab alone in patients with recurrent/metastatic head and neck cancer. J Clin Oncol. 2015;33(15_suppl):6060.
168. Biankin AV, Piantadosi S, Hollingsworth SJ. Patient-centric trials for therapeutic development in precision oncology. Nature. 2015;526(7573):361–70.
169. Szturz P, Raymond E, Faivre S. c-MET-mediated resistance to EGFR inhibitors in head and neck cancer: how to move from bench to bedside. Oral Oncol. 2016;59:E12–4.
170. Scagliotti G, et al. Phase III multinational, randomized, double-blind, placebo-controlled study of tivantinib (ARQ 197) plus erlotinib versus erlotinib alone in previously treated patients with locally advanced or metastatic nonsquamous non-small-cell lung cancer. J Clin Oncol. 2015;33(24):2667–74.
171. Liu X, et al. A novel kinase inhibitor, INCB28060, blocks c-MET-dependent signaling, neoplastic activities, and cross-talk with EGFR and HER-3. Clin Cancer Res. 2011;17(22):7127–38.
172. Dufies M, et al. Mechanism of action of the multikinase inhibitor Foretinib. Cell Cycle. 2011;10(23):4138–48.
173. Xu L, et al. Combined EGFR/MET or EGFR/HSP90 inhibition is effective in the treatment of lung cancers codriven by mutant EGFR containing T790M and MET. Cancer Res. 2012;72(13):3302–11.
174. Michieli P, Di Nicolantonio F. Targeted therapies: Tivantinib-a cytotoxic drug in MET inhibitor's clothes? Nat Rev Clin Oncol. 2013;10(7):372–4.
175. Scagliotti GV, et al. Rationale and design of MARQUEE: a phase III, randomized, double-blind study of tivantinib plus erlotinib versus placebo plus erlotinib in previously treated patients with locally advanced or metastatic, nonsquamous, non-small-cell lung cancer. Clin Lung Cancer. 2012;13(5):391–5.
176. Schlumberger M, et al. Final overall survival analysis of EXAM, an international, double-blind, randomized, placebo-controlled phase III trial of cabozantinib (Cabo) in medullary thyroid carcinoma (MTC) patients with documented RECIST progression at baseline. J Clin Oncol. 2015;33(suppl 15):abstr 6012.
177. Scagliotti GV, Novello S, von Pawel J. The emerging role of MET/HGF inhibitors in oncology. Cancer Treat Rev. 2013;39(7):793–801.
178. Shaw AT, et al. Crizotinib versus chemotherapy in advanced ALK-positive lung cancer. N Engl J Med. 2013;368(25):2385–94.
179. Ou SH, et al. Activity of crizotinib (PF02341066), a dual mesenchymal-epithelial transition (MET) and anaplastic lymphoma kinase (ALK) inhibitor, in a non-small cell lung cancer patient with de novo MET amplification. J Thorac Oncol. 2011;6(5):942–6.

180. Sun S, et al. Targeting the c-Met/FZD8 signaling axis eliminates patient-derived cancer stem-like cells in head and neck squamous carcinomas. Cancer Res. 2014;74(24):7546–59.
181. Penuel J, Smith JC, Shen SQ. Integer programming models and algorithms for the graph decontamination problem with mobile agents. Networks. 2013;61(1):1–19.
182. Spigel D, et al. Final efficacy results from OAM4558g, a randomized phase II study evaluating MetMAb or placebo in combination with erlotinib in advanced NSCLC. J Clin Oncol. 2011;29(suppl 15):abstr 7505.
183. Spigel DR, et al. Treatment rationale study design for the MetLung trial: a randomized, double-blind phase III study of onartuzumab (MetMAb) in combination with erlotinib versus erlotinib alone in patients who have received standard chemotherapy for stage IIIB or IV met-positive non-small-cell lung cancer. Clin Lung Cancer. 2012;13(6):500–4.
184. Spigel DR, et al. Onartuzumab plus erlotinib versus erlotinib in previously treated stage IIIb or IV NSCLC: results from the pivotal phase III randomized, multicenter, placebo-controlled METLung (OAM4971g) global trial. J Clin Oncol. 2014;32(suppl 5):abstr 8000.
185. Bendell JC, et al. A phase II randomized trial (GO27827) of first-line FOLFOX plus bevacizumab with or without the MET inhibitor onartuzumab in patients with metastatic colorectal cancer. Oncologist. 2017;22(3):264–71.
186. Zeng W, et al. Abstract 2734: c-Met antibody LY2875358 (LA480) shows differential antitumor effects in non-small cell lung cancer. Cancer Res. 2012;72(8):abstr 2734.
187. Banck MS, et al. Abstract A55: phase 1 results of emibetuzumab (LY2875358), a bivalent MET antibody, in patients with advanced castration-resistant prostate cancer, and MET positive renal cell carcinoma, non-small cell lung cancer, and hepatocellular carcinoma. Mol Cancer Ther. 2015;14(12 Supplement 2):A55.
188. Yoh K, et al. A phase I dose-escalation study of LY2875358, a bivalent MET antibody, given as monotherapy or in combination with erlotinib or gefitinib in Japanese patients with advanced malignancies. Investig New Drugs. 2016;34(5):584–95.
189. Camidge DR, et al. A randomized, open-label, phase 2 study of emibetuzumab plus erlotinib (LY+E) and emibetuzumab monotherapy (LY) in patients with acquired resistance to erlotinib and MET diagnostic positive (MET Dx+) metastatic NSCLC. J Clin Oncol. 2016;34(suppl 15):abstr 9070.
190. D'Arcangelo M, Cappuzzo F. Focus on the potential role of ficlatuzumab in the treatment of non-small cell lung cancer. Biologics. 2013;7:61–8.
191. Kumar D, et al. Mitigation of tumor-associated fibroblast-facilitated head and neck cancer progression with anti-hepatocyte growth factor antibody ficlatuzumab. JAMA Otolaryngol Head Neck Surg. 2015;141(12):1133–9.
192. Ryan CJ, et al. Targeted MET inhibition in castration-resistant prostate cancer: a randomized phase II study and biomarker analysis with rilotumumab plus mitoxantrone and prednisone. Clin Cancer Res. 2013;19(1):215–24.
193. Greenall SA, Adams TE, Johns TG. Incomplete target neutralization by the anti-cancer antibody rilotumumab. MAbs. 2016;8(2):246–52.
194. Okamoto W, et al. TAK-701, a humanized monoclonal antibody to hepatocyte growth factor, reverses gefitinib resistance induced by tumor-derived HGF in non-small cell lung cancer with an EGFR mutation. Mol Cancer Ther. 2010;9(10):2785–92.
195. Mirnezami R, Nicholson J, Darzi A. Preparing for precision medicine. N Engl J Med. 2012;366(6):489–91.
196. Garraway LA, Verweij J, Ballman KV. Precision oncology: an overview. J Clin Oncol. 2013;31(15):1803–5.

Chapter 4
Transforming Growth Factor Beta (TGF-β) Signaling in Head and Neck Squamous Cell Carcinoma (HNSCC)

Alexander E. Kudinov and Tim N. Beck

Abstract Transforming growth factor receptor beta (TGF-β) signaling is commonly dysregulated in head and neck squamous cell carcinoma (HNSCC). TGF-β signaling influences homeostasis in normal epithelial cells and regulates a critical signaling network during development. In HNSCC, TGF-β signaling frequently promotes cell invasion, metastasis, proliferation, and drug resistance and may present an important therapeutic target. Canonical TGF-β signaling generally involves activation of SMAD effector proteins, most prominently SMAD2 and SMAD3, whereas noncanonical TGF-β signaling requires signal propagators including ERK, AKT, and RAF, also commonly employed by receptor tyrosine kinases (RTKs), thereby providing opportunities for signaling crosstalk. Several members of the TGF-β superfamily are being explored as potential targets to control drug resistance and metastatic spread, both important barriers to cure in HNSCC. In this chapter, the roles of TGF-β in HNSCC are described, with particular focus on molecular signaling, TGF-β's role in controlling gene expression, and relevant therapeutic directions involving TGF-β.

Keywords TGF-β · TGFBRII · Head and neck cancer · Targeted therapy · Metastasis · Proliferation · Epithelial-mesenchymal transition · Cancer stem cells

---

A. E. Kudinov
Molecular Therapeutics Program, Fox Chase Cancer Center, Philadelphia, PA, USA

Department of Internal Medicine, University of New Mexico, Albuquerque, NM, USA

T. N. Beck (✉)
Molecular Therapeutics Program, Fox Chase Cancer Center, Philadelphia, PA, USA

Program in Molecular and Cell Biology and Genetics, Drexel University College of Medicine, Philadelphia, PA, USA
e-mail: Tim.Beck@fccc.edu

B. Burtness, E. A. Golemis (eds.), *Molecular Determinants of Head and Neck Cancer*, Current Cancer Research, https://doi.org/10.1007/978-3-319-78762-6_4

4.1 Introduction

The majority of head and neck squamous cell carcinomas (HNSCCs) are associated with tobacco, alcohol, and, in some populations, betel nuts [1, 2]. A more recently identified risk factor for HNSCC is infection with human papillomavirus (HPV; discussed in Chaps. 20 and 21), most commonly HPV type 16 [3]: HPV-related HNSCCs present at a younger median age and carry a significantly better prognosis compared to HPV-unrelated HNSCC [4–7]. Similar to other squamous cell carcinomas (SCC), HNSCC typically presents with an aggressive growth pattern, high tumor heterogeneity, early metastatic spread to lymph nodes, and poor outcomes once the disease has spread beyond the locoregional milieu [2, 8, 9]. Tumor dispersion, aggressiveness, and drug resistance often occur subsequent to cancer cells undergoing epithelial-mesenchymal transition (EMT), which is regulated by complex signaling changes in response to internal and external stimuli [1, 10–13]. One signaling network critical for EMT, a process also linked to cancer stem cells [14–16], is orchestrated by the transforming growth factor beta (TGF-β; [17]) family of ligands and receptors [18–20].

The TGF-β polypeptide ligand was first purified from murine sarcoma cells in 1982 [21] and was further described by Anzano et al. [17] and other members of the Sporn laboratory [22–25]. These studies provided many foundational observations including a synergistic relationship between TGF-β, TGF-α, and epidermal growth factor (EGF) in terms of promoting colony formation of sarcoma cells [17]. A particularly important aspect of this work was the discovery that bioactive TGF-β arises from a longer precursor polypeptide that is proteolytically modified to produce the shorter, homodimeric ligand [22]; it was also observed that the TGF-β ligand was not only expressed in neoplastic tissue but also in healthy tissue [22, 24, 26].

Critical technical milestones have accelerated the study of TGF-β: the ligand was purified efficiently, and epitope-specific antibodies have been developed [25, 27], and Derynck et al. cloned TGF-β1, after which it was determined that three distinct TGF-β ligands (TGF-β1, TGF-β2, and TGF-β3), encoded by three paralogous genes, exist in humans [22]. These tools collectively provided the foundation for future study of the TGF-β ligands [28] as well as their associated transmembrane receptors [29, 30]. The extended TGF-β superfamily is known to include over 30 ligands with recognizable structural and sequence homology, with at least twelve known corresponding receptors [19, 20, 31].

Breakthroughs in the identification of TGF-β-activated signaling effectors came from genetic analyses of *Drosophila melanogaster* and *Caenorhabditis elegans*. Particularly important was the discovery of SMAD effector proteins (Fig. 4.1), a unifying label in vertebrates that combined the previous names given to the orthologs of these proteins that had been discovered in *C. elegans* and *Drosophila* and had been designated "sma" and "mothers against decapentaplegic (mad)", respectively [33–38].

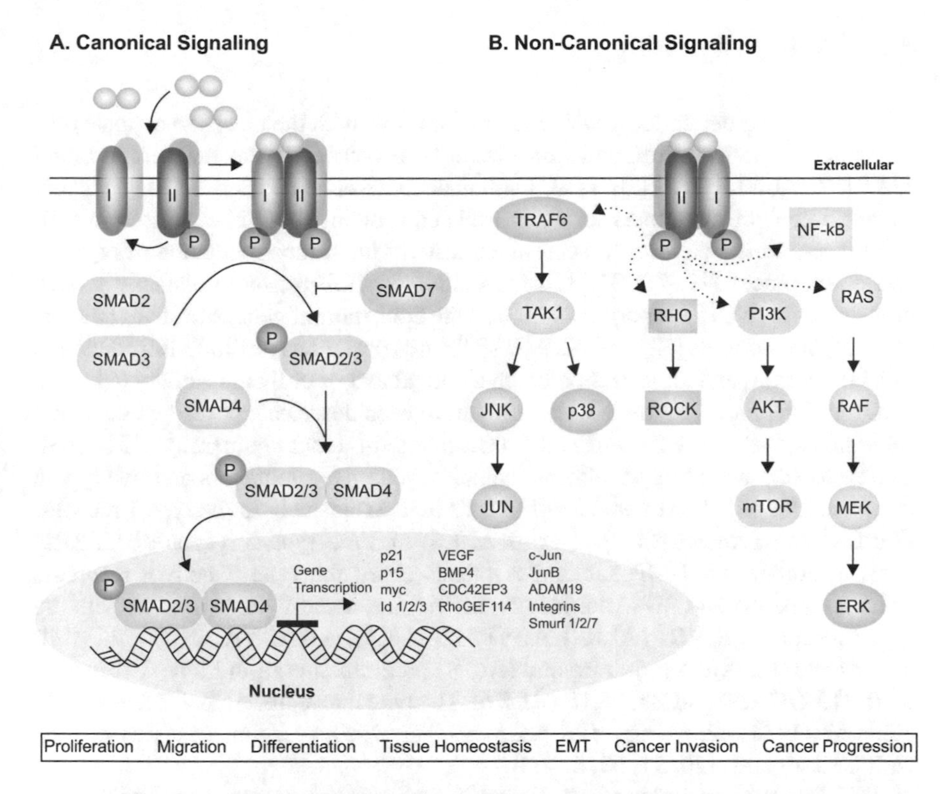

Fig. 4.1 TGF-β signaling. (**a**) Shown is the most common canonical signaling pathway associated with TGF-β signaling. (**b**) Shown are the major effector proteins associated with noncanonical TGF-β signaling. TGF-β transcriptional program is elaborated on in [31, 32]

For researchers interested in tumor invasion and metastasis, the observation that members of the TGF-β superfamily were critical for proper migration of cells during embryogenesis as well as during wound healing was particularly important [39–41]. TGF-β was subsequently found to play a similar role in regulating cell migration and EMT in tumor cells, processes not only associated with cancer cell metastasis and cancer stem cells but also with resistance to chemo- and targeted therapies [15, 16, 42–45]. As of 2017, TGF-β signaling has been implicated in many types of cancer, including lung cancer [42, 46], breast cancer [47], colorectal cancer [48], and head and neck cancer [18, 49]. Early studies also uncovered bifunctional characteristics associated with TGF-β signaling, specifically, context-dependent inhibitory or activating functions [31, 50]. The bifunctional nature of TGF-β signaling has added significant complexity to its study in HNSCC and other cancers.

The remainder of this chapter describes the roles of TGF-β in HNSCC and the development of treatment strategies to target the TGF-β signaling axis.

4.2 TGF-β Signaling

TGF-β signaling networks provide mammalian cells with the means to regulate proliferation, migration, differentiation, tissue homeostasis, tissue regeneration, and EMT [19, 20, 31, 32, 51]. In total, the human genome encodes over 30 genes that encode precursor monomers for TGF-β-related proteins [31, 52] expressed in cell- and tissue-specific patterns to achieve exquisitely fine-tuned regulation of organismal development [19, 20, 31, 32]. In addition to TGF-βs, more distantly related branches of the TGF-β superfamily include the bone morphogenetic proteins (BMPs; [53–55]) and activins [52, 56–58]. TGF-β ligands present as disulfide-linked dimers [59, 60], with ligand dimerization being a critical aspect of ligand activity [61, 62].

Over 30 members of the TGF-β family have been described, covering two ligand subfamilies, the BMP and the TGF-β-activin-nodal subfamily [31, 63]. Corresponding receptors in humans include seven type 1 receptors and five type 2 receptors [31, 52]. The three TGF-β ligands bind exclusively to the type 1 receptor TGFBR1 (also known as TβRI or as ALK5) and the type 2 receptor TGFBR2. Activin, nodal, and BMP subfamily ligands share the type 2 activin receptors (ACVR) type 2A and 2B. Additionally, activin and nodal ligands interact with the type 1 receptors ACVR1 (ALK2), ACVR1B (ALK4), and ACVR1C (ALK7). BMP ligands and anti-Müllerian hormone (AMH) predominantly bind type 1 receptors BMPR1A (ALK3) and BMPR1B (ALK6). The type 2 receptor AMHR2 exclusively binds AMH [31, 42, 64, 65]. BMPR2 is another type 2 receptor mostly associated with BMP ligands [20, 31, 63, 66–68].

TGF-β signaling is based on ligand-initiated signal propagation, originating from a membrane-bound receptor complex (Fig. 4.1). Most signaling by TGF-β members follows a general paradigm: ligand-receptor interaction initiates tetramerization of two type 1 receptors (e.g., TGFBR1) and two type 2 receptors (e.g., TGFBR2; Fig. 4.1; [63]). The two receptor types, type 1 and type 2, are both serine/threonine (Ser/Thr) protein kinases, a unique characteristic among membrane-bound receptors [69]. In humans, no other known surface receptors are classified as Ser/Thr protein kinases [31]. Functionally, a ligand binds to and induces dimerization of two TGFBR2 receptors (note, type 2 receptors are constitutively active; [68]). The type 2 receptor-bound TGF-β ligand then recruits type 1 receptors into the complex at the interface created by the ligand-type 2 receptor interaction, leading to phosphorylation of the type 1 receptors by a type 2 receptor at the juxtamembrane part of the cytoplasmic domain [63, 68, 70]. Phosphorylated, activated type 1 receptors are primed to interact with downstream effectors, the SMAD proteins and other proteins in noncanonical signaling (Fig. 4.1; Sect. 4.2.2; [32, 51]).

TGF-β signaling is highly versatile and can be influenced by contextual cues from the microenvironment, exposure to different treatment modalities, tumor inflammation, cross-regulation involving other members of the TGF-β family, and the mutational landscape of a given tumor [19, 20, 31, 32, 71]. Typically, in tumors, signaling by TGF-β is autocrine and/or paracrine, with ligand and receptors produced by the same tumor cell or by stromal cells or neighboring tumor cells [42, 71–74].

4.2.1 Canonical TGF-β Signaling

The dominant effector proteins of canonical TGF-β signaling are the SMAD proteins. In mammals, eight SMAD proteins are coded for, and five of the mammalian SMADs – SMAD1, SMAD2, SMAD3, SMAD5, and SMAD8 – are direct binding partners and substrates for the TGF-β family receptor kinases. These are commonly referred to as receptor-regulated SMADs (R-SMADs; [75]). The remaining three SMAD proteins include SMAD4, a common protein that interacts with the receptor-regulated SMADs, and two inhibitory SMAD proteins (SMAD6 and SMAD7; [19, 31, 75]). SMAD6 and SMAD7 were first described in endothelial cells [76] and form stable interactions with activated type 1 receptors, thereby effectively blocking phosphorylation of downstream effector SMADs (Fig. 4.1; [77, 78]).

The C-terminal domain, also known as Mad-homology 2 (MH2), is conserved in all eight mammalian SMAD proteins; however, the N-terminal domain (MH1) is only conserved in the R-SMADs and in SMAD4, but not in the two inhibitory SMADs [75]. MH2 includes a Ser-X-Ser motif that is phosphorylated by activated TGF-β receptors. The two MH domains are separated by a less well-conserved serine- and proline-rich linker region, which is an important site for posttranslational regulatory modifications [79, 80] and in the case of SMAD4 includes a nuclear export signal (NES; [75]). MH1 serves as a DNA-binding domain, allowing R-SMADs and SMAD4 to interact with SMAD-binding elements (SBE) in target gene promoters [81].

Canonical TGF-β ligand signaling is most commonly propagated by effector proteins SMAD2, SMAD3, and SMAD4, although additional SMADs (e.g., SMAD1, SMAD5, and SMAD8) have been shown to respond to TGF-β activation in some cases [42, 82, 83]. SMAD1, SMAD5, and SMAD8 are frequently associated with BMP and activin signaling [54].

In the nucleus, SMAD proteins interact with a large number of transcription factors, activators, and repressors, including FoxH, FoxO, p300, E2F4/5, Runx2, and others [75, 80]. Target genes activated or repressed by TGF-β signaling are numerous and include p21, p15, IL-11, c-Myc, VEGF, BMP4, Id1, Id2, Id3, c-JUN, JunB, and many more that promote mesenchymal transition and enhanced migration [32]. Specific events, including tumorigenesis, inflammation, and wound healing, alter the extent of TGF-β signaling by modifying aspects of the network at several different levels: the ligand level, the receptor level, the SMAD protein level, and the DNA-binding level.

4.2.2 Noncanonical TGF-β Signaling

In 1992, Mulder and Morris unexpectedly identified fast (within minutes) activation of p21Ras (an important mediator of cellular differentiation and proliferation [84, 85]) by treatment of intestinal epithelial cells with TGF-β1 or TGF-β2 [86]. This spurred the search for alternative mechanisms of TGF-β signal transduction, as this activation could not be explained by triggering of the canonical SMAD signaling pathway [86]. Additional insight was obtained from research by Yan et al., who showed that myelin basic protein (MBP) kinases – kinases for which MBP is a peptide substrate, which include ERK1, ERK2, other MAPKs, PKA, and PKC [87–89] – serve as intracellular effector proteins following TGF-β activation of p21Ras (Fig. 4.1; [89]).

Further studies confirmed SMAD-independent activation of ERK/MAP kinases by TGF-β in breast cancer as well as in normal epithelial cells [90, 91]. TGF-β activation of ERK/MAP kinases enhanced EMT (Section 3) and increased motility and invasiveness of breast cancer cells [92, 93]. Additionally, the MAPKs JNK and p38 have been identified as molecular transducers of TGF-β signal in SMAD2/3-depleted in vitro cell models [94–97]. Through regulation p21Ras, TGF-β also activates PI3K/AKT in myofibroblasts, in a manner independent of SMAD2 and SMAD3 [98]. Through this crosstalk, TGF-β also supports activity of mTOR, an inducer of protein translation and an important hub for AKT signaling (Fig. 5.1, Chap. 5; [99]). These findings highlight the connection between two major pathways regulating cancer invasion and cancer progression, TGF-β and PI3K/AKT/mTOR [100].

In additional SMAD-independent TGF-β signaling, Yamaguchi et al. showed that the activity of the MAPK TAK1 kinase was stimulated by both TGF-β and BMP [101]. The TGF-β receptor complex, specifically the TGFBR1 component, induces TRAF6 ubiquitylation of TAK1, thereby activating TAK1. The ligand bound TGF-β receptor complex interacts with TRAF6 and induces its polyubiquitination on residue K63, which is required for its activation of JNK and p38 (Fig. 4.1; [102, 103]). Work with siRNA and dominant-negative mutants established an essential role for TRAF6 as a critical activator of TAK1. Deletion of TAK1 has also been reported to impair NF-κB and JNK activation without diminishing expression of SMAD2-associated TGF-β response genes [104].

Rho-like GTPases, specifically RhoA, play an important role in controlling mesenchymal cellular characteristics, including cytoskeletal organization and cell motility [105, 106]. RhoA is rapidly activated by TGF-β (Fig. 4.1). This activation may be SMAD-independent, as in SMAD3-negative mutant epithelial cells, and RhoA activation by TGF-β was still observed [107]. RhoA is an important mediator of TGF-β-induced EMT, as depletion of RhoA or its downstream target RhoA-induced Rho-kinase (ROCK) inhibited TGF-β1-mediated EMT, as indicated by the absence of stress fiber formation and mesenchymal characteristics [107].TGF-β also induces activation of the Rho-like GTPase Cdc42, again in a SMAD-independent manner (Fig. 4.2; [110]). Activation of Cdc42 by TGF-β in turn induces activation of p21-activated serine/threonine kinase (PAK) 2 [111]. Depletion of SMAD2 and/or SMAD3 does not impact PAK2 activation by TGF-β [110]. In dominant-negative PAK2 cells, TGF-β-mediated transformation was significantly reduced [110].

The TGF-β1 receptor TGFBR2 also contributes to SMAD-independent signaling by directly phosphorylating cell polarity regulator partitioning defective 6 (PAR6), initiating loss of tight junctions, and inducing changes in cell polarity [112, 113].

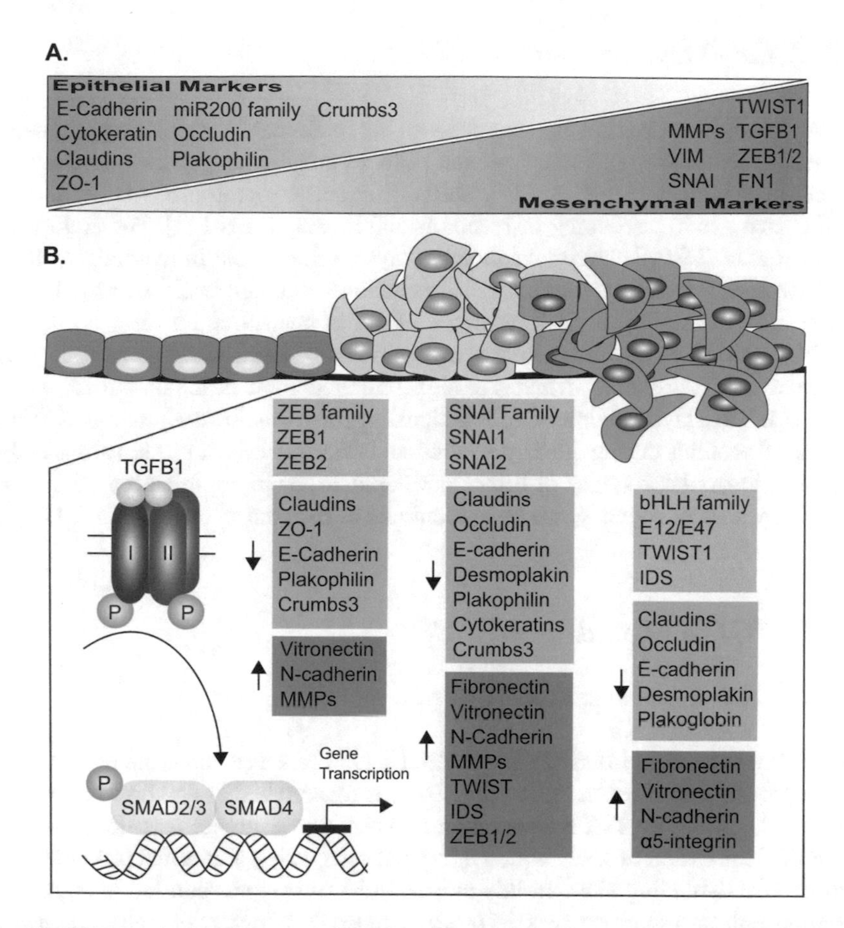

Fig. 4.2 TGF-β and epithelial-mesenchymal transition. (**a**) Dominant markers associated with the epithelial (pink) and mesenchymal (purple) cell state, respectively [31, 43, 100, 105, 108, 109]. (**b**) Three families of transcription factors (in gray) with associated epithelial (pink) and mesenchymal (purple) markers that have been shown to be up-/downregulated by TGF-β activity

Depletion of PAR6 in an in vivo orthotopic model of breast cancer has been shown to increase expression of tight junction proteins and suppress metastasis to the lungs [114]. PAR6 is localized to the tight junctions between cells, as is TGFBR1, allowing complex formation, likely enabled by the similar localization patterns between TGFBR1 and PAR6 even in unstimulated cells [112, 115]. Stimulation of cells with TGF-β1 induces redistribution of TGFBR2 to the tight junctions and induces interactions between TGFBR2 and the PAR6-TGFBR1 complex, resulting in increased phosphorylation of PAR6, which increases PAR6 binding to SMURF1, mediating localized ubiquitination and degradation of RhoA and subsequent disruption of tight cell-cell junctions and EMT (Fig. 4.1; [112]). In this context, RhoA stabilizes cell-cell junctions most likely through the effector protein diaphanous-related formin-1 (mDia1) and degradation of RhoA thus leads to degeneration of the junctions [116]. This contrasts with RhoA activation of ROCK described in the previous paragraph.

4.3 TGF-β Dysfunction in HNSCC

TGF-β can drive malignancy-suppressive or malignancy-promoting signaling, depending on the overall context, which includes epigenetic, genetic, and microenvironmental factors [31, 117, 118], shifts in ligand concentrations [119, 120], as well as changes in trafficking of receptors and ligands [121–123]. For example, in normal tissue, TGF-β activity frequently plays a critical role in maintaining tissue homeostasis by inducing apoptosis, differentiation, and cell cycle arrest [31, 124]; however, in the setting of malignancy, TGF-β has been found to play a predominantly suppressive role in the early stages of carcinogenesis, but as the cancer advances, the TGF-β axis switches to signaling in support of the malignant phenotype [125, 126]. Dysfunction of TGF-β signaling, be it via downregulation of TGF-β associated proteins during the earlier stages of tumorigenesis or by activating TGF-β signaling in the later stages of tumor development, or in combination with mutations in specific oncogenes, has been identified as relevant in HNSCC [20, 127].

4.3.1 TGF-β Ligands in HNSCC

4.3.1.1 TGF-β Ligands: Model Systems

In the tumor microenvironment, the TGFB1 ligand acts as a paracrine modulator of tumor growth and invasion [128, 129]. To investigate whether and how cancer progression is driven by TGFB1 overexpression, a mouse model was designed with inducible expression of *tgfb1* in oral epithelial cells, using a minimal promoter containing GAL4-binding sites. In this model, *tgfb1* overexpression led to significant inflammation and angiogenesis in the oral mucosa, which eventually resulted in escape from the abovementioned TGF-β inhibition capacity during the early stages – most likely driven by the significant and consistent increase in TGFB1 levels above baseline – inducing proliferation and providing evidence in support of the activating contribution of TGF-β in later-stage cancer growth of oral epithelial cell origin [130].

4.3.1.2 TGF-β Ligands: Clinical Specimens

In a close analysis of normal stroma and tumor-associated stroma from twelve patients with HNSCC, Rosenthal and colleagues identified TGFB1 as highly overexpressed in the tumor stromal compartment compared to normal mucosa [131]. Additional work using HNSCC tissue with case-matched adjacent normal tissue and independently acquired normal oropharyngeal tissue from sleep apnea patients showed that HNSCC epithelial cells, in addition to the stromal cells, also have a propensity to overexpress TGFB1 [130].

Analysis of thirty-two specimens from patients with laryngeal squamous cell carcinoma further validated the frequent overexpression of TGFB1 and TGFB2 in tumors [132]. In this study, TGFB1 and TGFB2 mRNA was expressed at significantly higher levels in tumors compared to adjacent nonneoplastic tissue ($p < 0.001$), prompting the authors to suggest TGFB1 and TGFB2 mRNA as possible tumor markers [133].

4.3.1.3 TGF-β Ligands: Cancer Stem Cells

Regulation of TGF-β activity is critical for normal stem cell function during development, and dysregulation thereof can contribute to the expansion of cancer stem cell populations [31, 134]. Analysis of K3 HNSCC cells, isolated from a clinical HNSCC specimen and previously described in terms of cancer stem cell (CSC) characteristics [135], indicated that treatment of these cells with TGFB1 increases expression of the cancer stem cell markers Sox2 and Oct4 and increased the population of aldehyde dehydrogenase positive ($ALDH^+$) cells (7.24% DMSO treated versus 19.28% TGFB1 treated; [133]). Sox2 and Oct4 are two of the four Yamanaka factors (Klf4 and c-Myc being the other two) that, when activated, have been shown to transform fibroblasts into pluripotent stem cells [136]. In studies of breast cancer, $ALDH^+CD44^+CD24^-$ expression identified cell populations enriched for tumor-propagating cells and ALDH expression correlated with poor prognosis [137]. Further studies of cancer stem cell regulation by TGFB1 in head and neck cancer are needed to better understand how this population of cells impacts prognosis and treatment response, particularly because it has been shown that cancer stem cell populations are enriched following chemo- or radiotherapy [138]; furthermore, one of the most important processes associated with cancer stemness is EMT, which is strongly linked to TGF-β activity [15, 19, 31, 108] and discussed extensively in Sect. 4.4.

4.3.2 TGF-β Receptors in HNSCC

4.3.2.1 TGF-β Receptors: Model Systems

Loss of the transforming growth factor receptor type 2 has been shown to be an important promoter of metastatic head and neck cancer in a transgenic mouse model. However, for SCC to occur, alteration of additional genes, specifically KRAS and HRAS, is necessary for the induction of tumors [139]. The model was generated by crossing K5.CrePR1 [140] and *tgfbr2*$^{f/f}$ mice [141]. Exon two of the *tgfbr2* gene was floxed in the *tgfbr2*$^{f/f}$ mice and in the *K5.CrePR1* mice; Cre recombinase was fused to a truncated human progesterone receptor (PR) – which can be activated by the compound RU486 (mifepristone) – driven by the keratin 5 (K5) promoter. The K5 promoter targets gene expression specifically to the epidermis and head and neck epithelium. Thus, application of RU486 to the oral cavities of bigenic K5.CrePR1/ *tgfbr2*$^{f/f}$ mice allowed targeted deletion of *tgfbr2* in the head and neck epithelial cells [139].

Whereas deletion of *tgfbr2* by itself had little effect, deletion of *tgfbr2* combined with mutational activation of KRAS (codon 12 G-to-D mutation [142]) or HRAS caused HNSCC with complete penetrance [139]). This elegant work underscores the potent tumor-suppressive role of TGFBR2 during early stages of cancer formation and emphasizes that tumorigenesis requires alterations in multiple genes [139]. In mice with normal TGFBR2 expression (*tgfbr2*$^{+/+}$), similar KRAS activation caused the development of large, benign papillomas, and HRAS activation did not cause the appearance of any tumors within a 60-week period [139].

In the same model, increased endogenous *tgfb1* levels were detected in *tgfbr2*$^{-/-}$ tissues and tumors, but not in KRAS-driven papillomas [139]. This observation suggests a feedback loop between TGFBR2 and TGFB1 expression; for cells to overexpress a specific ligand to compensate for depressed expression of the corresponding receptor is perhaps an expected response. TGFB1 is a known stimulator of angiogenesis and inflammation [143, 144], and the *tgfbr2*$^{-/-}$ lesions did indeed show increased angiogenesis and increased macrophages and neutrophils. Interestingly, these changes were seen prior to HNSCC formation and thus suggest direct, non-tumor stage-specific effects of TGFB1 on nonepithelial cells [139].

The TGF-β type 1 receptor has also been implicated in the oncogenesis of HNSCC in several studies. Functionally, based on the interactions between TGFBR2 and TGFBR1 (Fig. 4.1), downregulation of TGFBR1 would be predicted to induce phenotypes that parallel those observed for cases with TGFBR2 downregulation. Work on mouse models indicated that deletion of *tgfbr1* by itself led to development of squamous cell carcinoma in 10% of mice (3 of 31) after 1 year [145]. In contrast, conditionally deleted *tgfbr1* and *pten* resulted in tumors of the head and neck in 100% of animals (42 of 42) within 10 weeks, paralleling results for the mouse model of TGFBR2 and RAS described above [139]. Mechanistically, loss of *tgfbr1* and *pten* inhibits apoptosis and increased cell proliferation. Loss of *tgfbr1* also reduced AKT-induced senescence [145].

In another mouse model, conditional deletion of *tgfbr1* resulted in spontaneous development of periorbital (42%) and perianal tumors [146]. Fourteen of the twenty-six tumors were squamous cell carcinoma based on histological analysis [146]. Interestingly, the tumors expressed high levels of IL-13Ra2, which has also been observed in human head and neck squamous cell carcinoma [147, 148]. In human HNSCC, high IL-13Ra2 correlates with advanced disease [147], and it has been proposed that IL-13Ra2 is involved in tumorigenesis by enhancing paracrine effects of TGFBR1 and by allowing tumor cells to escape immunosurveillance [146].

4.3.2.2 TGF-β Receptors: Clinical Specimens

In general, mutations affecting components of the canonical TGF-β signaling network have been detected with varying frequency by different studies [1, 10, 149–152] with TGFBR2 genomic mutations having been extensively validated [10, 153, 154]. Surprisingly, TGFBR2 was the only TGF-β signaling component identified by

The Cancer Genome Atlas (TCGA) as mutated. A TCGA study of 279 cases of HNSCC identified mutated (nonsense or missense) TGFBR2 in 1% of cases, primarily localized to the oral cavity (seven out of 172 oral cancers and one out of thirty-three oropharyngeal cancers [10]). It is possible that with analysis of larger cohorts, mutations will also be detected in additional components of the TGF-β signaling network, many of which have been implicated by other studies [1, 10, 149–152].

mRNA expression of TGFBR1 was significantly reduced in ten out of ten HNSCC cell lines revealed in comparison to untransformed human oral keratinocytes [145]. Importantly, subsequent immunostaining of twenty human HNSCC samples for TGFBR1 found that 50% of samples had undetectable or reduced TGFBR1 [145]. Immunohistochemical staining for TGFBR1 and TGFBR2 in eighty HNSCC tumor specimens showed that expression of TGFBR1 and TGFBR2 was reduced in forty-three and twenty-three specimens, respectively [155]. Reduced expression of either receptor was associated with increased depth of invasion (TGFBR1, $P = 0.0309$; TGFBR2, $P = 0.0059$), increased lymph node metastasis (TGFBR1, $P = 0.0103$; TGFBR2 $P = 0.0401$), and reduced cancer-specific survival (TGFBR1, $P = 0.0324$; TGFBR2 $P = 0.0243$; [155]). Alterations in the expression of the two receptors were significantly correlated ($P = 0.0002$). These findings parallel previous analysis of thirty-eight HNSCC specimens, which showed significant reduction of TGFBR1 mRNA expression in the most invasive and most poorly differentiated areas of tumors [156]. Reduced expression of TGFBR2 or TGFBR1 also correlated with overexpression of TGF-β1 [155]. It is possible that downregulation of the receptors initiates overexpression of TGF-β1, which perhaps supports tumorigenesis in terms of angiogenesis and by creating a pro-tumorigenic microenvironment [157–159].

In a study of twenty-three HNSCC samples using semiquantitative multiplex RT-PCR, TGFBR2 expression was reduced compared to matched normal tissue by 24–74% in 87% of cases [154]. This study also detected six cases (26%) with mutations in the coding region of TGFBR2: all six mutations were specifically in the serine/threonine kinase domain [154]. This work highlights the importance of TGFBR2 in HNSCC, but, considering the work by the TCGA, it is unlikely that 26% is the true incidence rate of TGFBR2 mutations [10, 154]. A better estimate will surely emerge as more HNSCC specimens are sequenced.

In addition to primary HNSCC tumors, head and neck cancer metastases have also been reported to downregulate TGFBR1. Among twenty-three cases, four had mutations of TGFBR1 [160]. One of the four mutations was a somatic intragenic four base-pair deletion predicted to produce a truncated protein. The other three mutation types were classified as a missense and two substitution mutations [160].

Lastly, a large multicenter study discovered that the common polymorphism TGFBR1*6A, proposed to be a tumor susceptibility allele [161], was somatically acquired in roughly 2% of head and neck cancers analyzed (4 of 226; [162]). TGFBR1*6A contains a three alanine deletion within a nine alanine repeat at the 3′-end of exon 1 and codes for a TGFBR1 receptor that has been shown to have less antiproliferative activity compared to other variants [161, 163].

4.3.2.3 TGF-β Receptors: Cancer Stem Cells

In the previously described *tgfbr1/Pten* mouse model, the combined depletion of *tgfbr1* and *Pten* significantly increased mRNA expression of *Nanog* ($p < 0.01$) and *Oct4* ($p < 0.001$), two stem cell transcription factors [136], as well as of CD44 and CD133, two markers for cancer stem cells in HNSCC [145, 164–166]. Mouse xenograft models using primary HNSCC specimens obtained from patients undergoing surgical resection have strongly supported the relevance of CD44 in HNSCC: twenty of thirty-one injections of $CD44^+$ cells formed tumors, whereas, only one of fourty injections of $CD44^-$ cells produced tumors. Furthermore, as few as 5×10^3 $CD44^+Lin^-$ cells reliably produced tumors in this model, in contrast to $CD44^-$ cells, which did not produce any identifiable tumor growth even with injections of up to 5×10^5 cells [166]. This study of *tgfbr1/Pten* knockout mice did not elucidate the molecular mechanism by which the two deleted genes regulate cancer stem cell self-renewal and differentiation, two points of high relevance for future studies to explore.

4.3.3 *TGF-β Effectors in HNSCC*

Mutations and expression changes affecting the canonical SMAD effectors of TGFBR1 and TGFBR2 proteins (Fig. 4.1) are also associated with malignant phenotypes. Several of the SMAD genes are clustered along the same chromosomal region, with SMAD2, SMAD4, and SMAD7 all found at 18q21 [167–169] and SMAD3 and SMAD7 at 15q21–22 [167, 170]. Deletion of chromosome 18q is of known significance in HNSCC and was detected in a genome-wide analysis of copy number aberrations in eighty-nine samples [171] and in an independent study using twenty-one prospectively collected fresh-frozen oral SCC specimens [172]. Studies in the 1990s uncovered that loss of heterogeneity of 18q in HNSCC tumors was detectable with high frequency (twelve of sixteen patients [173]), a finding confirmed by additional studies [174–179].

In a small study of thirteen head and neck cancers, protein expression of SMAD2 was lost in 38% (5 of 13) of cases, and it was noted that in poorly differentiated tumors, phosphorylated (activated) SMAD2 was undetectable. On the mRNA level, SMAD2 remained detectable for all specimens in this study, suggesting that SMAD2 expression was regulated posttranscriptionally in HNSCC [180], unlike pancreatic cancer and colorectal cancer, where SMAD2 tends to be altered on the genetic level [168, 169, 181]. Interestingly, SMAD3 was not lost in any of the thirteen HNSCC specimens examined [180], potentially suggesting similarities to studies in mouse models, where deletion of SMAD4 induced nuclear activation of SMAD3 and TGFB1, leading to inflammation and tumorigenesis [182].

Phosphorylated SMAD2 and SMAD3 interact with the central mediator of TGF-β signaling, SMAD4 (Fig. 4.1). SMAD4 was first described as a tumor suppressor in pancreatic cancer [168] and changes in SMAD4 function have since been linked to a number of additional cancer types, including HNSCC [182]. As is true for TGF-β in general, the impact of SMAD4 alterations is highly context specific. For example, in pancreatic cancer, deletion of SMAD4 without mutations in additional driver mutations is not sufficient to initiate tumor formation; however, mouse models showed that in head and neck cancer, deletion of *smad4* initiates genomic instability and inflammation, resulting in tumor formation and cancer progression [182, 183].

Conditional depletion of *smad4* in the head and neck epithelial tissue of mice led to the development of spontaneous oral tumors in 74% (twenty-six of thirty-five) of tested animals [182]. In the same study, *smad4* knockdown caused a reduction in the expression of FANC/BRCA genes and restoration of normal *smad4* in a SMAD4-deficient HNSCC cell line increased the expression of genes in the FANC/BRCA pathway. These observations propose that genomic instability due to SMAD4 deletion is part of the mechanism underlying tumor development in this model [182].

Unlike TGFBR2, SMAD4 was found to not only be downregulated in malignant HNSCC, it was also downregulated in grossly normal tumor-adjacent mucosa compared to normal non-tumor associated head and neck tissue [182]. The discovery that normal tumor-adjacent tissue is not an equivalent control to normal non-tumor associated head and neck tissue is a likely contributing factor to the variation in reported incidence of SMAD4 expression reduction in HNSCC, which ranges from 12% [184] up to 86% [182, 183].

The finding that 86% (31 of 36) of analyzed HNSCC cases had downregulated SMAD4 was based on quantitative RT-PCR results benchmarked to normal tissue from cancer-free patients with sleep apnea. Furthermore, 67% (twenty-four of thirty-six) of samples from normal tumor-adjacent mucosa had less than 50% the expression level of SMAD4 compared to normal control tissue [182]. These findings suggest that reduction of SMAD4 may play an important role early during tumor development in HNSCC.

4.4 TGF-β Driven EMT in HNSCC

TGF-β is a key regulator of the EMT required for normal cell migration during embryogenesis, and activation of this pathway is often coopted to promote migration and invasion in disease states such as cancer. As with other tumor types, TGF-β signaling is an important regulator of EMT in HNSCC [185–187]). In one well-studied example, in oral squamous cell carcinoma (OSCC), stromally synthesized TGFB1 was shown to promote cancer cell invasion [188]. This was mechanistically explained by reduced epithelial morphology and reduced apicobasal polarization of tumor cells and stimulated invadopodia formation, leading to a more invasive phenotype [189]. Invadopodia are specialized cell protrusions composed of an actin filament backbone that mediate localized secretion of basement membrane-degrading matrix metalloproteases (MMPs) and are thus associated with initiating invasion of tumor-surrounding tissues and cancer cell extravasation, which eventually results in increased metastasis to lymph nodes [188, 190, 191].

In OSCC, another study reported TGFB1 activation of invadopodia to be driven by upregulation of podoplanin (PDPN) [188]. Other studies have shown that TGFB1 enhances OSCC invasiveness via a different mechanism, specifically, a SNAIL-dependent upregulation of two MMPs, MMP2 and MMP9 [192, 193]. SNAIL is a potent regular of EMT and a TGFB1 effector in many cancer types [194–198]. Using in vitro and in vivo models, it has also been shown that TGFB1 induces receptor activator of nuclear factor kappa-B ligand (RANKL) and that TGFB1 synthesis by cancer cells and stroma cells increases phosphorylation of SMAD2 and increases OSCC-induced bone destruction when activated HSC3 cells (human oral squamous cell carcinoma cells) are transplanted onto the skulls of athymic mice [199].

4.5 TGF-β-Targeted Therapies

The dual nature of TGF-β signaling is a serious challenge for successful implementation of TGF-β-targeted drugs in clinical practice. An obvious concern is that blocking the TGF-β pathway may result not only in deceleration of tumor invasion and metastasis in established tumors but may promote progression of early lesions and thus ultimately worsen disease [200]. Nevertheless, careful consideration of TGF-β signaling implies some potential therapeutic opportunities for the treatment of HNSCC (Fig. 4.3; [201, 202]).

Several different therapeutic agents are currently in preclinical and clinical stages of development. These agents can be classified based on the level at which they disrupt TGF-β signaling: at the ligand level (ligand traps, antibodies, antisense oligonucleotides, and cancer vaccines), at the receptor level (small molecule receptor kinase inhibitors and antibodies), or at the intracellular level (peptide aptamers that disrupt SMAD complexes). However, only a limited number of inhibitors are actively being pursued for the treatment of HNSCC.

4.5.1 Targeting the Receptors

Small molecular inhibitors of TGFBR1 stand out as having had the most clinical success to date. Galunisertib (LY2157299) is currently undergoing clinical trials in patients with glioblastoma, pancreatic carcinoma, or hepatocellular carcinoma

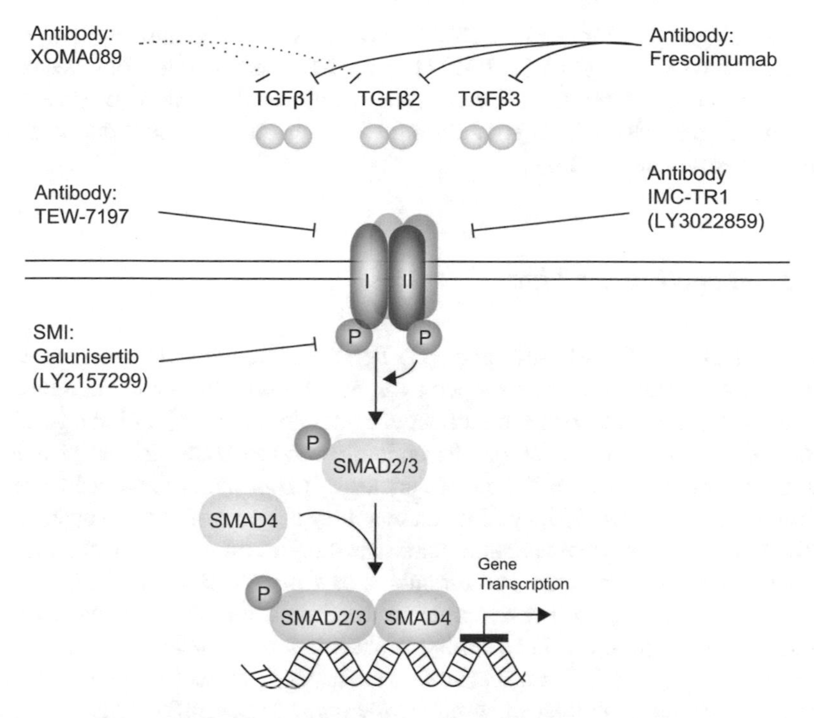

Fig. 4.3 Targeting TGF-β signaling in HNSCC. SMI = small molecule inhibitor

[203]. The inhibitor specifically reduces phosphorylation of SMAD2 and thus abrogates canonical TGF-β signaling (Fig. 4.1). Bhola et al. used preclinical in vivo evaluations to show that galunisertib and TR1 (a monoclonal antibody against TGFBR2) were effective in preventing post-chemotherapy growth of cancer stem cell subpopulations in triple-negative breast cancer [204]. Cardiac toxicity is a significant concern in cases of long-term, continuous use of galunisertib, a side effect also observed in animal studies [203]. However, careful dosing strategies identified a therapeutic window for the drug, which has allowed clinical trials to commence. The currently recommended dosing regimen is based on a 14 days on/14 days off, 28 day cycle [203].

Less advanced inhibitors include the monoclonal antibodies IMC-TR1 (LY3022859; a monoclonal anti-TGFBR2 antibody; [205]) and TEW-7197 (an anti-TGFBR1 antibody; [201]). Clinical and preclinical studies of these therapeutic agents have suggested reasonable safety profiles but often relatively limited therapeutic efficacy. TEW-7191 has been shown in vitro to reduce the growth and viability of multiple myeloma cells by inhibiting SMAD2/3 signaling and subsequent apoptosis [206] and is currently being evaluated in a Phase 1 clinical trial for advanced stage solid tumors [NCT02160106]. Preclinical studies of 50-week-old mice treated with 40 mg/kg of IMC-TR1 three times per week for four weeks indicated significant reduction of fibrosis [205]. Unfortunately, in a recent Phase 1 study, doses for IMC-TR1 beyond 25 mg were considered unsafe due to uncontrolled cytokine release [207].

4.5.2 Targeting the Ligand

The Pan-TGF-β (TGF-β1, -β2, and -β3) ligand-neutralizing human monoclonal antibody fresolimumab was developed as an anti-fibrotic therapeutic and has been studied in Phase 1 trials for the treatment of systemic sclerosis [208] and focal segmental glomerulosclerosis [209]. The antibody showed some limited promise in both trials. Previous work with a mouse analog of fresolimumab had been effective in preserving significant kidney function in a mouse model of chronic nephropathy [210]. Fresolimumab demonstrated some anti-tumor activity in the treatment of advanced melanoma and renal cell carcinoma in a Phase 1 trial but failed in clinical trials due to the unexpected development of reversible keratoacanthomas and early squamous cell carcinomas [211], similar to observations made for the BRAF inhibitor vemurafenib [212, 213]. Further studies are needed to establish if the underlying mechanism causing keratoacanthomas is the same for fresolimumab and vemurafenib; it could possibly be driven by disruption of RAF signaling (Fig. 4.1).

The potential of fresolimumab in HNSCC was explored by Bedi et al., who showed in a xenograft mouse model that cetuximab (EGFR-targeting antibody) resistance is in part due to tumor cell-autonomous expression of TGF-β and subsequent activation of AKT (Fig. 4.1) and that this resistance could be successfully addressed, at least in vivo, with the mouse analog of fresolimumab 1D11 [214]. Optimizing EGFR-targeted therapy in HNSCC has been challenging, and utilizing TGF-β inhibitors in combination with EGFR inhibition is worth further consideration.

Various efforts have been undertaken to overcome the challenge of balancing the tumor suppressive versus the tumor promoting activity of TGF-β targeting agents. For example, the recently developed human antibody XOMA089 was designed to target TGF-β1 and TGF-β2 while sparing TGF-β3 (Fig. 4.3). Available data are extremely limited; however, early preclinical results indicated significant growth arrest of both HNSCC and of breast cancer cells in preclinical models. Possible synergy between XOMA089 and PD-1 inhibitors based on animal data has also been suggested [215]. More data are needed to fully assess the potential of XOMA089.

4.5.3 *TGF-β and Immunotherapy*

TGF-β is a major regulator of the tumor microenvironment and of immune cells [18, 129, 182, 183, 216, 217]; therefore, consideration of TGF-β activity in the context of immunotherapy seems intuitive. It has been proposed to use immunoassays to screen for tumors responsive to TGF-β modulated therapy and to investigate the use of TGF-β docking receptor GARP as a response-predictive biomarker for immunotherapy [215, 216, 218].

Interesting early efforts have provided some evidence of complementary activity between TGF-β inhibition and immunotherapeutic drugs [219–223]. Preclinical melanoma models have shown that combinations of galunisertib and an anti-cytotoxic T-lymphocyte-associated protein 4 (CTLA-4) antibody had significant synergy and prevented in vivo tumor growth and melanoma metastasis [224]. Similar studies are yet to be completed in HNSCC.

4.6 Conclusions and Future Directions

The importance of TGF-β in HNSCC is undeniable but complicated. Of increasing interest is the role this family of receptors and ligands plays in terms of immunoregulatory functions [129, 216, 225]. Considering the breakthrough of immunotherapies across many cancer types [226], including HNSCC ([227–231]; see Chap. 14), the need to better understand TGF-β signaling will continue to be high.

Considering the challenges of targeting TGF-β signaling clinically [232], strategies such as identification of response-predictive biomarkers should be encouraged [233]. Specifically, more investigative efforts should be put into solving some of the technical difficulties of working with phosphorylated proteins, such as phospho-SMAD2 and possibly phospho-SMAD3, to more accurately assess the activity level of canonical TGF-β signaling [234, 235]. Focusing this work on HNSCC has the potential to produce high-impact findings, as the need for improved treatment strategies for this malignancy continues to be significant.

Acknowledgments We thank Drs. Golemis and Burtness for critical reading of our chapter. The authors were supported by the Ruth L. Kirschstein NRSA F30 fellowship (F30 CA180607) from the NIH (to TNB).

References

1. Beck TN, Golemis EA. Genomic insights into head and neck cancer. Cancers Head Neck. 2016;1(1)
2. Argiris A, Karamouzis MV, Raben D, Ferris RL. Head and neck cancer. Lancet. 2008;371:1695–709.

3. Marur S, D'Souza G, Westra WH, Forastiere AA. HPV-associated head and neck cancer: a virus-related cancer epidemic. Lancet Oncol. 2010;11:781–9.
4. Kelly JR, Husain ZA, Burtness B. Treatment de-intensification strategies for head and neck cancer. Eur J Cancer. 2016;68:125–33.
5. Chaturvedi AK, Engels EA, Pfeiffer RM, et al. Human papillomavirus and rising oropharyngeal cancer incidence in the United States. J Clin Oncol. 2011;29:4294–301.
6. Ang KK, Harris J, Wheeler R, et al. Human papillomavirus and survival of patients with oropharyngeal cancer. N Engl J Med. 2010;363:24–35.
7. Bhatia A, Burtness B. Human papillomavirus-associated oropharyngeal cancer: defining risk groups and clinical trials. J Clin Oncol. 2015;33:3243–50.
8. Magrini SM, Buglione M, Corvo R, et al. Cetuximab and radiotherapy versus cisplatin and radiotherapy for locally advanced head and neck cancer: a randomized phase II trial. J Clin Oncol. 2016;34:427–35.
9. Sacco AG, Cohen EE. Current treatment options for recurrent or metastatic head and neck squamous cell carcinoma. J Clin Oncol. 2015;33:3305–13.
10. Cancer Genome Atlas N. Comprehensive genomic characterization of head and neck squamous cell carcinomas. Nature. 2015;517:576–82.
11. Liu H, Beck TN, Golemis EA, Serebriiskii IG. Integrating in silico resources to map a signaling network. Methods Mol Biol. 2014;1101:197–245.
12. Creixell P, Schoof EM, Erler JT, Linding R. Navigating cancer network attractors for tumor-specific therapy. Nat Biotechnol. 2012;30:842–8.
13. Creixell P, Schoof EM, Simpson CD, et al. Kinome-wide decoding of network-attacking mutations rewiring cancer signaling. Cell. 2015;163:202–17.
14. Nassar D, Blanpain C. Cancer stem cells: basic concepts and therapeutic implications. Annu Rev Pathol. 2016;11:47–76.
15. Oshimori N, Oristian D, Fuchs E. TGF-beta promotes heterogeneity and drug resistance in squamous cell carcinoma. Cell. 2015;160:963–76.
16. Mani SA, Guo W, Liao MJ, et al. The epithelial-mesenchymal transition generates cells with properties of stem cells. Cell. 2008;133:704–15.
17. Anzano MA, Roberts AB, Meyers CA, et al. Synergistic interaction of two classes of transforming growth factors from murine sarcoma cells. Cancer Res. 1982;42:4776–8.
18. White RA, Malkoski SP, Wang XJ. TGFbeta signaling in head and neck squamous cell carcinoma. Oncogene. 2010;29:5437–46.
19. Massague J. TGFbeta Cancer Cell. 2008;134:215–30.
20. Levy L, Hill CS. Alterations in components of the TGF-beta superfamily signaling pathways in human cancer. Cytokine Growth Factor Rev. 2006;17:41–58.
21. Roberts AB, Anzano MA, Lamb LC, et al. Isolation from murine sarcoma cells of novel transforming growth factors potentiated by EGF. Nature. 1982;295:417–9.
22. Derynck R, Jarrett JA, Chen EY, et al. Human transforming growth factor-beta complementary DNA sequence and expression in normal and transformed cells. Nature. 1985;316:701–5.
23. Derynck R, Rhee L, Chen EY, Van Tilburg A. Intron-exon structure of the human transforming growth factor-beta precursor gene. Nucleic Acids Res. 1987;15:3188–9.
24. Assoian RK, Komoriya A, Meyers CA, et al. Transforming growth factor-beta in human platelets. Identification of a major storage site, purification, and characterization. J Biol Chem. 1983;258:7155–60.
25. Sporn MB. The early history of TGF-beta, and a brief glimpse of its future. Cytokine Growth Factor Rev. 2006;17:3–7.
26. Roberts AB, Frolik CA, Anzano MA, Sporn MB. Transforming growth factors from neoplastic and nonneoplastic tissues. Fed Proc. 1983;42:2621–6.
27. Flanders KC, Roberts AB, Ling N, et al. Antibodies to peptide determinants in transforming growth factor beta and their applications. Biochemistry. 1988;27:739–46.
28. Weiss A, Attisano L. The TGFbeta superfamily signaling pathway. Wiley Interdiscip Rev Dev Biol. 2013;2:47–63.

29. Tucker RF, Branum EL, Shipley GD, et al. Specific binding to cultured cells of 125I-labeled type beta transforming growth factor from human platelets. Proc Natl Acad Sci U S A. 1984;81:6757–61.
30. Frolik CA, Wakefield LM, Smith DM, Sporn MB. Characterization of a membrane receptor for transforming growth factor-beta in normal rat kidney fibroblasts. J Biol Chem. 1984;259:10995–1000.
31. Massague J. TGFbeta signalling in context. Nat Rev Mol Cell Biol. 2012;13:616–30.
32. Massague J, Gomis RR. The logic of TGFbeta signaling. FEBS Lett. 2006;580:2811–20.
33. Ruberte E, Marty T, Nellen D, et al. An absolute requirement for both the type II and type I receptors, punt and thick veins, for dpp signaling in vivo. Cell. 1995;80:889–97.
34. Raftery LA, Twombly V, Wharton K, Gelbart WM. Genetic screens to identify elements of the decapentaplegic signaling pathway in Drosophila. Genetics. 1995;139:241–54.
35. Sekelsky JJ, Newfeld SJ, Raftery LA, et al. Genetic characterization and cloning of mothers against dpp, a gene required for decapentaplegic function in Drosophila melanogaster. Genetics. 1995;139:1347–58.
36. Estevez M, Attisano L, Wrana JL, et al. The daf-4 gene encodes a bone morphogenetic protein receptor controlling C. elegans dauer larva development. Nature. 1993;365:644–9.
37. Savage C, Das P, Finelli AL, et al. Caenorhabditis elegans genes sma-2, sma-3, and sma-4 define a conserved family of transforming growth factor beta pathway components. Proc Natl Acad Sci U S A. 1996;93:790–4.
38. Derynck R, Gelbart WM, Harland RM, et al. Nomenclature: vertebrate mediators of TGFbeta; family signals. Cell. 87:173.
39. Akhurst RJ, Fee F, Balmain A. Localized production of TGF-beta mRNA in tumour promoter-stimulated mouse epidermis. Nature. 1988;331:363–5.
40. Akhurst RJ, Lehnert SA, Gatherer D, Duffie E. The role of TGF beta in mouse development. Ann N Y Acad Sci. 1990;593:259–71.
41. Sporn MB, Roberts AB, Shull JH, et al. Polypeptide transforming growth factors isolated from bovine sources and used for wound healing in vivo. Science. 1983;219:1329–31.
42. Beck TN, Korobeynikov VA, Kudinov AE, et al. Anti-Mullerian hormone signaling regulates epithelial plasticity and chemoresistance in lung cancer. Cell Rep. 2016;16:657–71.
43. Xu J, Lamouille S, Derynck R. TGF-beta-induced epithelial to mesenchymal transition. Cell Res. 2009;19:156–72.
44. Katsuno Y, Lamouille S, Derynck R. TGF-beta signaling and epithelial-mesenchymal transition in cancer progression. Curr Opin Oncol. 2013;25:76–84.
45. Scheel C, Weinberg RA. Cancer stem cells and epithelial-mesenchymal transition: concepts and molecular links. Semin Cancer Biol. 2012;22:396–403.
46. Kudinov AE, Deneka A, Nikonova AS, et al. Musashi-2 (MSI2) supports TGF-beta signaling and inhibits claudins to promote non-small cell lung cancer (NSCLC) metastasis. Proc Natl Acad Sci U S A. 2016;113:6955–60.
47. Moses H, Barcellos-Hoff MH. TGF-beta biology in mammary development and breast cancer. Cold Spring Harb Perspect Biol. 2011;3:a003277.
48. Calon A, Espinet E, Palomo-Ponce S, et al. Dependency of colorectal cancer on a TGF-beta-driven program in stromal cells for metastasis initiation. Cancer Cell. 2012;22:571–84.
49. Graves CA, Abboodi FF, Tomar S, et al. The translational significance of epithelial-mesenchymal transition in head and neck cancer. Clin Transl Med. 2014;3:60.
50. Roberts AB, Anzano MA, Wakefield LM, et al. Type beta transforming growth factor: a bifunctional regulator of cellular growth. Proc Natl Acad Sci U S A. 1985;82:119–23.
51. Wrana JL, Attisano L, Wieser R, et al. Mechanism of activation of the TGF-beta receptor. Nature. 1994;370:341–7.
52. Moses HL, Roberts AB, Derynck R. The discovery and early days of TGF-beta: a historical perspective. Cold Spring Harb Perspect Biol. 2016;8
53. Bragdon B, Moseychuk O, Saldanha S, et al. Bone morphogenetic proteins: a critical review. Cell Signal. 2011;23:609–20.

54. Wang RN, Green J, Wang Z, et al. Bone morphogenetic protein (BMP) signaling in development and human diseases. Genes Dis. 2014;1:87–105.
55. Salazar VS, Gamer LW, Rosen V. BMP signalling in skeletal development, disease and repair. Nat Rev Endocrinol. 2016;12:203–21.
56. Pickup MW, Owens P, Moses HL. TGF-beta, bone morphogenetic protein, and activin signaling and the tumor microenvironment. Cold Spring Harb Perspect Biol. 2017;9
57. Attisano L, Wrana JL, Cheifetz S, Massague J. Novel activin receptors: distinct genes and alternative mRNA splicing generate a repertoire of serine/threonine kinase receptors. Cell. 1992;68:97–108.
58. Xia Y, Schneyer AL. The biology of activin: recent advances in structure, regulation and function. J Endocrinol. 2009;202:1–12.
59. Daopin S, Li M, Davies DR. Crystal structure of TGF-beta 2 refined at 1.8 A resolution. Proteins. 1993;17:176–92.
60. Daopin S, Davies DR, Schlunegger MP, Grutter MG. Comparison of two crystal structures of TGF-beta2: the accuracy of refined protein structures. Acta Crystallogr D Biol Crystallogr. 1994;50:85–92.
61. Zhang W, Jiang Y, Wang Q, et al. Single-molecule imaging reveals transforming growth factor-beta-induced type II receptor dimerization. Proc Natl Acad Sci U S A. 2009;106:15679–83.
62. Zhang W, Yuan J, Yang Y, et al. Monomeric type I and type III transforming growth factor-beta receptors and their dimerization revealed by single-molecule imaging. Cell Res. 2010;20:1216–23.
63. Shi Y, Massague J. Mechanisms of TGF-beta signaling from cell membrane to the nucleus. Cell. 2003;113:685–700.
64. di Clemente N, Josso N, Gouedard L, Belville C. Components of the anti-Mullerian hormone signaling pathway in gonads. Mol Cell Endocrinol. 2003;211:9–14.
65. Kim JH, MacLaughlin DT, Donahoe PK. Mullerian inhibiting substance/anti-Mullerian hormone: a novel treatment for gynecologic tumors. Obstet Gynecol Sci. 2014;57:343–57.
66. Daly AC, Randall RA, Hill CS. Transforming growth factor beta-induced Smad1/5 phosphorylation in epithelial cells is mediated by novel receptor complexes and is essential for anchorage-independent growth. Mol Cell Biol. 2008;28:6889–902.
67. Beck TN, Georgopoulos R, Shagisultanova EI, et al. EGFR and RB1 as dual biomarkers in HPV-negative head and neck cancer. Mol Cancer Ther. 2016;15:2486–97.
68. Moustakas A, Heldin CH. The regulation of TGFbeta signal transduction. Development. 2009;136:3699–714.
69. Lemmon MA, Schlessinger J. Cell signaling by receptor tyrosine kinases. Cell. 141:1117–34.
70. Heldin CH, Moustakas A. Signaling receptors for TGF-beta family members. Cold Spring Harb Perspect Biol. 2016;8:a022053. https://doi.org/10.1101/cshperspect.a022053.
71. Oshimori N, Fuchs E. Paracrine TGF-beta signaling counterbalances BMP-mediated repression in hair follicle stem cell activation. Cell Stem Cell. 2012;10:63–75.
72. Turley JM, Falk LA, Ruscetti FW, et al. Transforming growth factor beta 1 functions in monocytic differentiation of hematopoietic cells through autocrine and paracrine mechanisms. Cell Growth Differ. 1996;7:1535–44.
73. Vilar JM, Jansen R, Sander C. Signal processing in the TGF-beta superfamily ligand-receptor network. PLoS Comput Biol. 2006;2:e3.
74. Lehnert SA, Akhurst RJ. Embryonic expression pattern of TGF beta type-1 RNA suggests both paracrine and autocrine mechanisms of action. Development. 1988;104:263–73.
75. Massague J, Seoane J, Wotton D. Smad transcription factors. Genes Dev. 2005;19:2783–810.
76. Topper JN, Cai J, Qiu Y, et al. Vascular MADs: two novel MAD-related genes selectively inducible by flow in human vascular endothelium. Proc Natl Acad Sci U S A. 1997;94:9314–9.
77. Imamura T, Takase M, Nishihara A, et al. Smad6 inhibits signalling by the TGF-beta superfamily. Nature. 1997;389:622–6.
78. Nakao A, Afrakhte M, Moren A, et al. Identification of Smad7, a TGFbeta-inducible antagonist of TGF-beta signalling. Nature. 1997;389:631–5.

79. Gaarenstroom T, Hill CS. TGF-beta signaling to chromatin: how Smads regulate transcription during self-renewal and differentiation. Semin Cell Dev Biol. 2014;32:107–18.
80. Ross S, Hill CS. How the Smads regulate transcription. Int J Biochem Cell Biol. 2008;40:383–408.
81. Chen W, Fu X, Sheng Z. Review of current progress in the structure and function of Smad proteins. Chin Med J. 2002;115:446–50.
82. Liu IM, Schilling SH, Knouse KA, et al. TGFbeta-stimulated Smad1/5 phosphorylation requires the ALK5 L45 loop and mediates the pro-migratory TGFbeta switch. EMBO J. 2009;28:88–98.
83. Holtzhausen A, Golzio C, How T, et al. Novel bone morphogenetic protein signaling through Smad2 and Smad3 to regulate cancer progression and development. FASEB J. 2014;28:1248–67.
84. Pronk GJ, Bos JL. The role of p21ras in receptor tyrosine kinase signalling. Biochim Biophys Acta. 1994;1198:131–47.
85. Shih TY, Hattori S, Clanton DJ, et al. Structure and function of p21 ras proteins. Gene Amplif Anal. 1986;4:53–72.
86. Mulder KM, Morris SL. Activation of p21ras by transforming growth factor beta in epithelial cells. J Biol Chem. 1992;267:5029–31.
87. Jacobs D, Glossip D, Xing H, et al. Multiple docking sites on substrate proteins form a modular system that mediates recognition by ERK MAP kinase. Genes Dev. 1999;13:163–75.
88. Vartanian T, Dawson G, Soliven B, et al. Phosphorylation of myelin basic protein in intact oligodendrocytes: inhibition by galactosylsphingosine and cyclic AMP. Glia. 1989;2:370–9.
89. Yan Z, Winawer S, Friedman E. Two different signal transduction pathways can be activated by transforming growth factor beta 1 in epithelial cells. J Biol Chem. 1994;269:13231–7.
90. Frey RS, Mulder KM. TGFbeta regulation of mitogen-activated protein kinases in human breast cancer cells. Cancer Lett. 1997;117:41–50.
91. Hartsough MT, Mulder KM. Transforming growth factor beta activation of p44mapk in proliferating cultures of epithelial cells. J Biol Chem. 1995;270:7117–24.
92. Galliher-Beckley AJ, Schiemann WP. Grb2 binding to Tyr284 in TbetaR-II is essential for mammary tumor growth and metastasis stimulated by TGF-beta. Carcinogenesis. 2008;29:244–51.
93. Northey JJ, Chmielecki J, Ngan E, et al. Signaling through ShcA is required for transforming growth factor beta- and Neu/ErbB-2-induced breast cancer cell motility and invasion. Mol Cell Biol. 2008;28:3162–76.
94. Engel ME, McDonnell MA, Law BK, Moses HL. Interdependent SMAD and JNK signaling in transforming growth factor-beta-mediated transcription. J Biol Chem. 1999;274:37413–20.
95. Hocevar BA, Brown TL, Howe PH. TGF-beta induces fibronectin synthesis through a c-Jun N-terminal kinase-dependent, Smad4-independent pathway. EMBO J. 1999;18:1345–56.
96. Itoh S, Thorikay M, Kowanetz M, et al. Elucidation of Smad requirement in transforming growth factor-beta type I receptor-induced responses. J Biol Chem. 2003;278:3751–61.
97. Yu L, Hebert MC, Zhang YE. TGF-beta receptor-activated p38 MAP kinase mediates Smad-independent TGF-beta responses. EMBO J. 2002;21:3749–59.
98. Wilkes MC, Mitchell H, Penheiter SG, et al. Transforming growth factor-beta activation of phosphatidylinositol 3-kinase is independent of Smad2 and Smad3 and regulates fibroblast responses via p21-activated kinase-2. Cancer Res. 2005;65:10431–40.
99. Song MS, Salmena L, Pandolfi PP. The functions and regulation of the PTEN tumour suppressor. Nat Rev Mol Cell Biol. 2012;13:283–96.
100. Lamouille S, Derynck R. Cell size and invasion in TGF-beta-induced epithelial to mesenchymal transition is regulated by activation of the mTOR pathway. J Cell Biol. 2007;178:437–51.
101. Yamaguchi K, Shirakabe K, Shibuya H, et al. Identification of a member of the MAPKKK family as a potential mediator of TGF-beta signal transduction. Science. 1995;270:2008–11.
102. Sorrentino A, Thakur N, Grimsby S, et al. The type I TGF-beta receptor engages TRAF6 to activate TAK1 in a receptor kinase-independent manner. Nat Cell Biol. 2008;10:1199–207.

103. Yamashita M, Fatyol K, Jin C, et al. TRAF6 mediates Smad-independent activation of JNK and p38 by TGF-beta. Mol Cell. 2008;31:918–24.
104. Shim JH, Xiao C, Paschal AE, et al. TAK1, but not TAB1 or TAB2, plays an essential role in multiple signaling pathways in vivo. Genes Dev. 2005;19:2668–81.
105. Lamouille S, Xu J, Derynck R. Molecular mechanisms of epithelial-mesenchymal transition. Nat Rev Mol Cell Biol. 2014;15:178–96.
106. Zhang YE. Non-Smad pathways in TGF-[beta] signaling. Cell Res. 2009;19:128–39.
107. Bhowmick NA, Ghiassi M, Bakin A, et al. Transforming growth factor-beta 1 mediates epithelial to mesenchymal transdifferentiation through a RhoA-dependent mechanism. Mol Biol Cell. 2001;12:27–36.
108. Thiery JP, Acloque H, Huang RY, Nieto MA. Epithelial-mesenchymal transitions in development and disease. Cell. 2009;139:871–90.
109. Kalluri R, Weinberg RA. The basics of epithelial-mesenchymal transition. J Clin Invest. 2009;119:1420–8.
110. Wilkes MC, Murphy SJ, Garamszegi N, Leof EB. Cell-type-specific activation of PAK2 by transforming growth factor beta independent of Smad2 and Smad3. Mol Cell Biol. 2003;23:8878–89.
111. Jaffer ZM, Chernoff J. p21-activated kinases: three more join the Pak. Int J Biochem Cell Biol. 2002;34:713–7.
112. Ozdamar B, Bose R, Barrios-Rodiles M, et al. Regulation of the polarity protein Par6 by TGFbeta receptors controls epithelial cell plasticity. Science. 2005;307:1603–9.
113. Mu Y, Zang G, Engstrom U, et al. TGF beta-induced phosphorylation of Par6 promotes migration and invasion in prostate cancer cells. Br J Cancer. 2015;112:1223–31.
114. Viloria-Petit AM, David L, Jia JY, et al. A role for the TGFbeta-Par6 polarity pathway in breast cancer progression. Proc Natl Acad Sci U S A. 2009;106:14028–33.
115. Gao L, Joberty G, Macara IG. Assembly of epithelial tight is negatively regulated by junctions Par6. Curr Biol. 2002;12:221–5.
116. Sahai E, Marshall CJ. ROCK and Dia have opposing effects on adherens junctions downstream of Rho. Nat Cell Biol. 2002;4:408–15.
117. Lebrun JJ. The dual role of TGFbeta in human cancer: from tumor suppression to cancer metastasis. ISRN Mol Biol. 2012;2012:381428.
118. Derynck R, Akhurst RJ, Balmain A. TGF-beta signaling in tumor suppression and cancer progression. Nat Genet. 2001;29:117–29.
119. Bier E, De Robertis EM. EMBRYO DEVELOPMENT. BMP gradients: A paradigm for morphogen-mediated developmental patterning. Science. 2015;348:aaa5838.
120. Green JB, New HV, Smith JC. Responses of embryonic Xenopus cells to activin and FGF are separated by multiple dose thresholds and correspond to distinct axes of the mesoderm. Cell. 1992;71:731–9.
121. Akhurst RJ, Padgett RW. Matters of context guide future research in TGFbeta superfamily signaling. Sci Signal. 2015;8:re10.
122. Lu Z, Murray JT, Luo W, et al. Transforming growth factor beta activates Smad2 in the absence of receptor endocytosis. J Biol Chem. 2002;277:29363–8.
123. Pomeraniec L, Hector-Greene M, Ehrlich M, et al. Regulation of TGF-beta receptor hetero-oligomerization and signaling by endoglin. Mol Biol Cell. 2015;26:3117–27.
124. Tian M, Schiemann WP. The TGF-beta paradox in human cancer: an update. Future Oncol. 2009;5:259–71.
125. Cui W, Fowlis DJ, Bryson S, et al. TGF beta 1 inhibits the formation of benign skin tumors, but enhances progression to invasive spindle carcinomas in transgenic mice. Cell. 1996;86:531–42.
126. Moses HL, Yang EY, Pietenpol JA. TGF-beta stimulation and inhibition of cell proliferation: new mechanistic insights. Cell. 1990;63:245–7.
127. Hannigan A, Smith P, Kalna G, et al. Epigenetic downregulation of human disabled homolog 2 switches TGF-beta from a tumor suppressor to a tumor promoter. J Clin Invest. 2010;120:2842–57.

128. Taylor MA, Lee YH, Schiemann WP. Role of TGF-beta and the tumor microenvironment during mammary tumorigenesis. Gene Expr. 2011;15:117–32.
129. Pickup M, Novitskiy S, Moses HL. The roles of TGFbeta in the tumour microenvironment. Nat Rev Cancer. 2013;13:788–99.
130. Lu S-L, Reh D, Li AG, et al. Overexpression of transforming growth factor β1 in head and neck epithelia results in inflammation, angiogenesis, and epithelial Hyperproliferation. Cancer Res. 2004;64:4405–10.
131. Rosenthal E, McCrory A, Talbert M, et al. Elevated expression of TGF-β1 in head and neck cancer–associated fibroblasts. Mol Carcinog. 2004;40:116–21.
132. Kapral M, Strzalka B, Kowalczyk M, et al. Transforming growth factor beta isoforms (TGF-beta1, TGF-beta2, TGF-beta3) messenger RNA expression in laryngeal cancer. Am J Otolaryngol. 2008;29:233–7.
133. Bae WJ, Lee SH, Rho YS, et al. Transforming growth factor beta1 enhances stemness of head and neck squamous cell carcinoma cells through activation of Wnt signaling. Oncol Lett. 2016;12:5315–20.
134. Watabe T, Miyazono K. Roles of TGF-beta family signaling in stem cell renewal and differentiation. Cell Res. 2009;19:103–15.
135. Lim YC, Oh SY, Cha YY, et al. Cancer stem cell traits in squamospheres derived from primary head and neck squamous cell carcinomas. Oral Oncol. 2011;47:83–91.
136. Takahashi K, Yamanaka S. Induction of pluripotent stem cells from mouse embryonic and adult fibroblast cultures by defined factors. Cell. 2006;126:663–76.
137. Chaffer CL, Weinberg RA. A perspective on cancer cell metastasis. Science. 2011;331:1559–64.
138. Bao S, Wu Q, McLendon RE, et al. Glioma stem cells promote radioresistance by preferential activation of the DNA damage response. Nature. 2006;444:756–60.
139. Lu SL, Herrington H, Reh D, et al. Loss of transforming growth factor-beta type II receptor promotes metastatic head-and-neck squamous cell carcinoma. Genes Dev. 2006;20:1331–42.
140. Arin MJ, Longley MA, Wang XJ, Roop DR. Focal activation of a mutant allele defines the role of stem cells in mosaic skin disorders. J Cell Biol. 2001;152:645–9.
141. Forrester E, Chytil A, Bierie B, et al. Effect of conditional knockout of the type II TGF-beta receptor gene in mammary epithelia on mammary gland development and polyomavirus middle T antigen induced tumor formation and metastasis. Cancer Res. 2005;65:2296–302.
142. Jackson EL, Willis N, Mercer K, et al. Analysis of lung tumor initiation and progression using conditional expression of oncogenic K-ras. Genes Dev. 2001;15:3243–8.
143. Bierie B, Moses HL. Transforming growth factor beta (TGF-beta) and inflammation in cancer. Cytokine Growth Factor Rev. 2010;21:49–59.
144. Yu Q, Stamenkovic I. Cell surface-localized matrix metalloproteinase-9 proteolytically activates TGF-beta and promotes tumor invasion and angiogenesis. Genes Dev. 2000;14:163–76.
145. Bian Y, Hall B, Sun ZJ, et al. Loss of TGF-beta signaling and PTEN promotes head and neck squamous cell carcinoma through cellular senescence evasion and cancer-related inflammation. Oncogene. 2012;31:3322–32.
146. Honjo Y, Bian Y, Kawakami K, et al. TGF-beta receptor I conditional knockout mice develop spontaneous squamous cell carcinoma. Cell Cycle. 2007;6:1360–6.
147. Kawakami M, Kawakami K, Kasperbauer JL, et al. Interleukin-13 receptor α2 chain in human head and neck Cancer serves as a unique diagnostic marker. Clin Cancer Res. 2003;9:6381–8.
148. Joshi BH, Kawakami K, Leland P, Puri RK. Heterogeneity in Interleukin-13 receptor expression and subunit structure in squamous cell carcinoma of head and neck: differential sensitivity to chimeric fusion proteins comprised of Interleukin-13 and a mutated form of Pseudomonas exotoxin. Clin Cancer Res. 2002;8:1948–56.
149. Osawa H, Shitara Y, Shoji H, et al. Mutation analysis of transforming growth factor beta type II receptor, Smad2, Smad3 and Smad4 in esophageal squamous cell carcinoma. Int J Oncol. 2000;17:723–8.

150. Stransky N, Egloff AM, Tward AD, et al. The mutational landscape of head and neck squamous cell carcinoma. Science. 2011;333:1157–60.
151. Hedberg ML, Goh G, Chiosea SI, et al. Genetic landscape of metastatic and recurrent head and neck squamous cell carcinoma. J Clin Invest. 2016;126:1606.
152. Morris LG, Chandramohan R, West L, et al. The molecular landscape of recurrent and metastatic head and neck cancers: insights from a precision oncology sequencing platform. JAMA Oncol. 2017;3(2):244–55.
153. Garrigue-Antar L, Munoz-Antonia T, Antonia SJ, et al. Missense mutations of the transforming growth factor beta type II receptor in human head and neck squamous carcinoma cells. Cancer Res. 1995;55:3982–7.
154. Wang D, Song H, Evans JA, et al. Mutation and downregulation of the transforming growth factor beta type II receptor gene in primary squamous cell carcinomas of the head and neck. Carcinogenesis. 1997;18:2285–90.
155. Fukai Y, Fukuchi M, Masuda N, et al. Reduced expression of transforming growth factor-beta receptors is an unfavorable prognostic factor in human esophageal squamous cell carcinoma. Int J Cancer. 2003;104:161–6.
156. Muro-Cacho CA, Anderson M, Cordero J, Munoz-Antonia T. Expression of transforming growth factor beta type II receptors in head and neck squamous cell carcinoma. Clin Cancer Res. 1999;5:1243–8.
157. Xu Y, Pasche B. TGF-beta signaling alterations and susceptibility to colorectal cancer. Hum Mol Genet. 2007;16 Spec No 1:R14–20.
158. Grady WM, Markowitz SD. Genetic and epigenetic alterations in colon cancer. Annu Rev Genomics Hum Genet. 2002;3:101–28.
159. Grady WM, Myeroff LL, Swinler SE, et al. Mutational inactivation of transforming growth factor beta receptor type II in microsatellite stable colon cancers. Cancer Res. 1999;59:320–4.
160. Chen T, Yan W, Wells RG, et al. Novel inactivating mutations of transforming growth factor-beta type I receptor gene in head-and-neck cancer metastases. Int J Cancer. 2001;93:653–61.
161. Pasche B, Kolachana P, Nafa K, et al. TβR-I(6A) is a candidate tumor susceptibility allele. Cancer Res. 1999;59:5678–82.
162. Pasche B, Knobloch TJ, Bian Y, et al. Somatic acquisition and signaling of TGFBR1*6A in cancer. JAMA. 2005;294:1634–46.
163. Pasche B, Luo Y, Rao PH, et al. Type I transforming growth factor β receptor maps to 9q22 and exhibits a polymorphism and a rare variant within a Polyalanine tract. Cancer Res. 1998;58:2727–32.
164. Chen ZG. The cancer stem cell concept in progression of head and neck cancer. J Oncol. 2009;2009:894064.
165. Chiou SH, Yu CC, Huang CY, et al. Positive correlations of Oct-4 and Nanog in oral cancer stem-like cells and high-grade oral squamous cell carcinoma. Clin Cancer Res. 2008;14:4085–95.
166. Prince ME, Sivanandan R, Kaczorowski A, et al. Identification of a subpopulation of cells with cancer stem cell properties in head and neck squamous cell carcinoma. Proc Natl Acad Sci U S A. 2007;104:973–8.
167. Schiffer M, von Gersdorff G, Bitzer M, et al. Smad proteins and transforming growth factor-beta signaling. Kidney Int Suppl. 2000;77:S45–52.
168. Hahn SA, Schutte M, Hoque ATMS, et al. DPC4, a candidate tumor suppressor gene at human chromosome 18q21.1. Science. 1996;271:350–3.
169. Eppert K, Scherer SW, Ozcelik H, et al. MADR2 maps to 18q21 and encodes a TGFbeta-regulated MAD-related protein that is functionally mutated in colorectal carcinoma. Cell. 1996;86:543–52.
170. Dimitriou R, Carr IM, West RM, et al. Genetic predisposition to fracture non-union: a case control study of a preliminary single nucleotide polymorphisms analysis of the BMP pathway. BMC Musculoskelet Disord. 2011;12

171. Snijders AM, Schmidt BL, Fridlyand J, et al. Rare amplicons implicate frequent deregulation of cell fate specification pathways in oral squamous cell carcinoma. Oncogene. 2005;24:4232–42.
172. Sparano A, Quesnelle KM, Kumar MS, et al. Genome-wide profiling of oral squamous cell carcinoma by array-based comparative genomic hybridization. Laryngoscope. 2006;116:735–41.
173. Jones JW, Raval JR, Beals TF, et al. Frequent loss of heterozygosity on chromosome arm 18q in squamous cell carcinomas. Identification of 2 regions of loss--18q11.1-q12.3 and 18q21.1-q23. Arch Otolaryngol Head Neck Surg. 1997;123:610–4.
174. Watanabe T, Wang X, Miyakawa A, et al. Mutational state of tumor suppressor genes (DCC, DPC4) and alteration on chromosome 18q21 in human oral cancer. Int J Oncol. 1997;11:1287–90.
175. Van Dyke DL, Worsham MJ, Benninger MS, et al. Recurrent cytogenetic abnormalities in squamous cell carcinomas of the head and neck region. Genes Chromosomes Cancer. 1994;9:192–206.
176. Kelker W, Van Dyke DL, Worsham MJ, et al. Loss of 18q and homozygosity for the DCC locus: possible markers for clinically aggressive squamous cell carcinoma. Anticancer Res. 1996;16:2365–72.
177. Kasamatsu A, Uzawa K, Usukura K, et al. Loss of heterozygosity in oral cancer. Oral Sci Int. 2011;8:37–43.
178. Takebayashi S, Ogawa T, Jung KY, et al. Identification of new minimally lost regions on 18q in head and neck squamous cell carcinoma. Cancer Res. 2000;60:3397–403.
179. Kim SK, Fan YH, Papadimitrakopoulou V, et al. DPC4, a candidate tumor suppressor gene, is altered infrequently in head and neck squamous cell carcinoma. Cancer Res. 1996;56:2519–21.
180. Muro-Cacho CA, Rosario-Ortiz K, Livingston S, Munoz-Antonia T. Defective transforming growth factor beta signaling pathway in head and neck squamous cell carcinoma as evidenced by the lack of expression of activated Smad2. Clin Cancer Res. 2001;7:1618–26.
181. Nagatake M, Takagi Y, Osada N, et al. Somatic in vivo alterations of the DPC4 gene at 18q21 in human lung cancers. Cancer Res. 1996;56:2718–20.
182. Bornstein S, White R, Malkoski S, et al. Smad4 loss in mice causes spontaneous head and neck cancer with increased genomic instability and inflammation. J Clin Invest. 2009;119:3408–19.
183. Malkoski SP, Wang XJ. Two sides of the story? Smad4 loss in pancreatic cancer versus head-and-neck cancer. FEBS Lett. 2012;586:1984–92.
184. Xie W, Aisner S, Baredes S, et al. Alterations of Smad expression and activation in defining 2 subtypes of human head and neck squamous cell carcinoma. Head Neck (Journal for the Sciences and Specialties of the Head and Neck). 2013;35:76–85.
185. Krisanaprakornkit S, Iamaroon A. Epithelial-mesenchymal transition in oral squamous cell carcinoma. ISRN Oncol. 2012;2012:681469.
186. Franz M, Spiegel K, Umbreit C, et al. Expression of Snail is associated with myofibroblast phenotype development in oral squamous cell carcinoma. Histochem Cell Biol. 2009;131:651–60.
187. Richter P, Umbreit C, Franz M, et al. EGF/TGFbeta1 co-stimulation of oral squamous cell carcinoma cells causes an epithelial-mesenchymal transition cell phenotype expressing laminin 332. J Oral Pathol Med. 2011;40:46–54.
188. Hwang YS, Park KK, Chung WY. Stromal transforming growth factor-beta 1 is crucial for reinforcing the invasive potential of low invasive cancer. Arch Oral Biol. 2014;59:687–94.
189. Gandalovicova A, Vomastek T, Rosel D, Brabek J. Cell polarity signaling in the plasticity of cancer cell invasiveness. Oncotarget. 2016;7:25022–49.
190. Leong HS, Robertson AE, Stoletov K, et al. Invadopodia are required for cancer cell extravasation and are a therapeutic target for metastasis. Cell Rep. 2014;8:1558–70.
191. Eckert MA, Lwin TM, Chang AT, et al. Twist1-induced invadopodia formation promotes tumor metastasis. Cancer Cell. 2011;19:372–86.

192. Sun L, Diamond ME, Ottaviano AJ, et al. Transforming growth factor-beta 1 promotes matrix metalloproteinase-9-mediated oral cancer invasion through snail expression. Mol Cancer Res. 2008;6:10–20.
193. Qiao B, Johnson NW, Gao J. Epithelial-mesenchymal transition in oral squamous cell carcinoma triggered by transforming growth factor-beta1 is Snail family-dependent and correlates with matrix metalloproteinase-2 and -9 expressions. Int J Oncol. 2010;37:663–8.
194. Vincent T, Neve EP, Johnson JR, et al. A SNAIL1-SMAD3/4 transcriptional repressor complex promotes TGF-beta mediated epithelial-mesenchymal transition. Nat Cell Biol. 2009;11:943–50.
195. Ye X, Tam WL, Shibue T, et al. Distinct EMT programs control normal mammary stem cells and tumour-initiating cells. Nature. 2015;525:256–60.
196. Cano A, Perez-Moreno MA, Rodrigo I, et al. The transcription factor snail controls epithelial-mesenchymal transitions by repressing E-cadherin expression. Nat Cell Biol. 2000;2:76–83.
197. Smith AP, Verrecchia A, Faga G, et al. A positive role for Myc in TGFbeta-induced Snail transcription and epithelial-to-mesenchymal transition. Oncogene. 2009;28:422–30.
198. Li H, Wang H, Wang F, et al. Snail involves in the transforming growth factor beta1-mediated epithelial-mesenchymal transition of retinal pigment epithelial cells. PLoS One. 2011;6:e23322.
199. Nakamura R, Kayamori K, Oue E, et al. Transforming growth factor-beta synthesized by stromal cells and cancer cells participates in bone resorption induced by oral squamous cell carcinoma. Biochem Biophys Res Commun. 2015;458:777–82.
200. Sun N, Taguchi A, Hanash S. Switching roles of TGF-beta in cancer development: implications for therapeutic target and biomarker studies. J Clin Med. 2016;5.
201. Neuzillet C, Tijeras-Raballand A, Cohen R, et al. Targeting the TGFbeta pathway for cancer therapy. Pharmacol Ther. 2015;147:22–31.
202. Akhurst RJ, Hata A. Targeting the TGFbeta signalling pathway in disease. Nat Rev Drug Discov. 2012;11:790–811.
203. Herbertz S, Sawyer JS, Stauber AJ, et al. Clinical development of galunisertib (LY2157299 monohydrate), a small molecule inhibitor of transforming growth factor-beta signaling pathway. Drug Des Devel Ther. 2015;9:4479–99.
204. Bhola NE, Balko JM, Dugger TC, et al. TGF-beta inhibition enhances chemotherapy action against triple-negative breast cancer. J Clin Invest. 2013;123:1348–58.
205. Yue L, Bartenstein M, Zhao W, et al. Preclinical efficacy of TGF-Beta receptor I kinase inhibitor, galunisertib, in myelofibrosis. Blood. 2015;126:603.
206. Kim B-G, Sergeeva O, Luo G, et al. Abstract 2647: TGF-β type I receptor inhibitor (TEW-7197) diminishes myeloma progression by multiple immunomodulatory mechanisms in combination with ixazomib. Cancer Res. 2017;77:2647.
207. Tolcher AW, Berlin JD, Cosaert J, et al. A phase 1 study of anti-TGF beta receptor type-II monoclonal antibody LY3022859 in patients with advanced solid tumors. Cancer Chemother Pharmacol. 2017;79:673–80.
208. Rice LM, Padilla CM, McLaughlin SR, et al. Fresolimumab treatment decreases biomarkers and improves clinical symptoms in systemic sclerosis patients. J Clin Invest. 2015;125:2795–807.
209. Trachtman H, Fervenza FC, Gipson DS, et al. A phase 1, single-dose study of fresolimumab, an anti-TGF-beta antibody, in treatment-resistant primary focal segmental glomerulosclerosis. Kidney Int. 2011;79:1236–43.
210. Ling H, Li X, Jha S, et al. Therapeutic role of TGF-beta-neutralizing antibody in mouse cyclosporin A nephropathy: morphologic improvement associated with functional preservation. J Am Soc Nephrol. 2003;14:377–88.
211. Morris JC, Tan AR, Olencki TE, et al. Phase I study of GC1008 (fresolimumab): a human anti-transforming growth factor-beta (TGFbeta) monoclonal antibody in patients with advanced malignant melanoma or renal cell carcinoma. PLoS One. 2014;9:e90353.
212. Boussemart L, Routier E, Mateus C, et al. Prospective study of cutaneous side-effects associated with the BRAF inhibitor vemurafenib: a study of 42 patients. Ann Oncol. 2013;24:1691–7.

213. Frouin E, Guillot B, Larrieux M, et al. Cutaneous epithelial tumors induced by vemurafenib involve the MAPK and Pi3KCA pathways but not HPV nor HPyV viral infection. PLoS One. 2014;9:e110478.
214. Bedi A, Chang X, Noonan K, et al. Inhibition of TGF-beta enhances the in vivo antitumor efficacy of EGF receptor-targeted therapy. Mol Cancer Ther. 2012;11:2429–39.
215. de Gramont A, Faivre S, Raymond E. Novel TGF-beta inhibitors ready for prime time in onco-immunology. Oncoimmunology. 2017;6:e1257453.
216. Metelli A, Wu BX, Fugle CW, et al. Surface expression of TGFbeta docking receptor GARP promotes oncogenesis and immune tolerance in breast cancer. Cancer Res. 2016;76:7106–17.
217. Li MO, Wan YY, Sanjabi S, et al. Transforming growth factor-beta regulation of immune responses. Annu Rev Immunol. 2006;24:99–146.
218. Clay TM, Hobeika AC, Mosca PJ, et al. Assays for monitoring cellular immune responses to active immunotherapy of cancer. Clin Cancer Res. 2001;7:1127–35.
219. Park BV, Freeman ZT, Ghasemzadeh A, et al. TGFbeta1-mediated SMAD3 enhances PD-1 expression on antigen-specific T cells in cancer. Cancer Discov. 2016;6:1366–81.
220. Locci M, Wu JE, Arumemi F, et al. Activin A programs the differentiation of human TFH cells. Nat Immunol. 2016;17:976–84.
221. Baas M, Besancon A, Goncalves T, et al. TGFbeta-dependent expression of PD-1 and PD-L1 controls CD8(+) T cell anergy in transplant tolerance. elife. 2016;5:e08133.
222. Hahn T, Akporiaye ET. Targeting transforming growth factor beta to enhance cancer immunotherapy. Curr Oncol. 2006;13:141–3.
223. Ghebeh H, Bakr MM, Dermime S. Cancer stem cell immunotherapy: the right bullet for the right target. Hematol Oncol Stem Cell Ther. 2008;1:1–2.
224. Hanks BA, Holtzhausen A, Evans K, et al. Combinatorial TGF-β signaling blockade and anti-CTLA-4 antibody immunotherapy in a murine BRAFV600E-PTEN−/− transgenic model of melanoma. J Clin Oncol. 2014;5s:32.
225. O'Connor-McCourt MD, Lenferink AEG, Zwaagstra J, et al. Abstract B058: development of AVID200, a novel and highly potent TGF-beta neutralizing immunotherapy. Cancer Immunol Res. 2016;4:B058.
226. Gotwals P, Cameron S, Cipolletta D, et al. Prospects for combining targeted and conventional cancer therapy with immunotherapy. Nat Rev Cancer. 2017;
227. Economopoulou P, Perisanidis C, Giotakis EI, Psyrri A. The emerging role of immunotherapy in head and neck squamous cell carcinoma (HNSCC): anti-tumor immunity and clinical applications. Ann Transl Med. 2016;4:173.
228. Ferris RL, Blumenschein G, Fayette J, et al. Nivolumab for recurrent squamous-cell carcinoma of the head and neck. N Engl J Med. 2016;375:1856–67.
229. Bauman JE, Cohen E, Ferris RL, et al. Immunotherapy of head and neck cancer: emerging clinical trials from a National Cancer Institute head and neck cancer steering committee planning meeting. Cancer. 2017;123:1259–71.
230. Bauml J, Seiwert TY, Pfister DG et al. Pembrolizumab for platinum- and cetuximab-refractory head and neck cancer: results from a single-arm, phase II study. J Clin Oncol. 2017;; JCO2016701524.
231. Chow LQM, Haddad R, Gupta S, et al. Antitumor activity of pembrolizumab in biomarker-unselected patients with recurrent and/or metastatic head and neck squamous cell carcinoma: results from the phase Ib KEYNOTE-012 expansion cohort. J Clin Oncol. 2016;34:3838–45.
232. Colak S, ten Dijke P. Targeting TGF-β signaling in cancer. Trends Cancer. 3:56–71.
233. Faivre SJ, Santoro A, Kelley RK, et al. A phase 2 study of a novel transforming growth factor-beta (TGF-beta 1) receptor I kinase inhibitor, LY2157299 monohydrate (LY), in patients with advanced hepatocellular carcinoma (HCC). J Clin Oncol 2014;32:3_suppl, LBA173–LBA173.
234. Rodon J, Carducci M, Sepulveda-Sanchez JM, et al. Pharmacokinetic, pharmacodynamic and biomarker evaluation of transforming growth factor-beta receptor I kinase inhibitor, galunisertib, in phase 1 study in patients with advanced cancer. Investig New Drugs. 2015;33:357–70.
235. Farrington DL, Yingling JM, Fill JA, et al. Development and validation of a phosphorylated SMAD ex vivo stimulation assay. Biomarkers. 2007;12:313–30.

Chapter 5
The PI3K Signaling Pathway in Head and Neck Squamous Cell Carcinoma

Alexander Y. Deneka, Jason D. Howard, and Christine H. Chung

Abstract The PI3K/PTEN/AKT/mTOR signaling axis has been intensively studied in many cancer systems. Extensive evidence suggests deregulation of this pathway plays an important role in the initiation, development, and recurrence of head and neck squamous cell carcinoma (HNSCC). A heterogeneous disease by nature, HNSCC encompasses a disparate collection of anatomical sites with complex tumor biology. Nevertheless, PI3K/PTEN/AKT/mTOR signaling has a critical role in nearly every facet of this disease. In this chapter we will provide a brief introduction to the mechanisms involved in PI3K/PTEN/AKT/mTOR signaling and how specific alterations in these signaling nodes enable HNSCC development and progression. We will also discuss differences in PI3K/PTEN/AKT/mTOR signaling with respect to human papillomavirus (HPV) status. A number of inhibitors targeting multiple nodes in this pathway have been developed, with these agents having potential application and in some cases demonstrated clinical activity in HNSCC. We will briefly review how these therapeutic agents are being evaluated and what predictive biomarkers have been established for them in HNSCC. Finally, PI3K/PTEN/AKT/mTOR signaling represents an important source of resistance to radiation and chemotherapy as well as other targeted agents. We will speculate on how PI3K/PTEN/AKT/mTOR inhibitors may increase the efficacy of these established therapies.

Keywords PI3K · PTEN · AKT · mTOR · HPV · HNSCC · Biomarkers

A. Y. Deneka (✉)
Molecular Therapeutics Program, Fox Chase Cancer Center, Philadelphia, PA, USA

Kazan Federal University, Kazan, Russia
e-mail: Alexander.Deneka@fccc.edu

J. D. Howard
Department of Oncology, Sidney Kimmel Cancer Center, Johns Hopkins Medical Institute, Baltimore, MD, USA

C. H. Chung
Department of Oncology, Sidney Kimmel Cancer Center, Johns Hopkins Medical Institute, Baltimore, MD, USA

Department of Head and Neck-Endocrine Oncology, Moffitt Cancer Center, Tampa, FL, USA

B. Burtness, E. A. Golemis (eds.), *Molecular Determinants of Head and Neck Cancer*, Current Cancer Research, https://doi.org/10.1007/978-3-319-78762-6_5

5.1 Introduction

The PI3K/PTEN/AKT/mTOR pathway is a critical signaling axis, which consolidates and regulates numerous extracellular signals required for complex, multicellular organisms. The end result of appropriate PI3K/PTEN/AKT/mTOR signaling is homeostasis: the careful balance of survival, proliferation, metabolism, growth, autophagy, cap-dependent translation, migration, apoptosis, and many other cellular requirements. Given the range of functionalities associated with this pathway, deregulation at any of its signaling nodes can have dire biological consequences. Thus, many of the proteins involved in this pathway have been established as *bona fide* oncogenes or tumor suppressors. Recent evidence suggests that at least 47% of head and neck squamous cell carcinomas (HNSCCs) have at least one molecular alteration in this pathway [1], making the PI3K pathway the most frequently mutated oncogenic pathway (30.5%) in HNSCC (Fig. 5.1) [2, 3]. In this chapter, we will summarize key features of this pathway, how these molecular alterations are associated with HNSCC development and progression, and how this differs for human papillomavirus (HPV)-related and HPV-unrelated (will be referred to as HPV-negative) head and neck cancers. In addition, we will provide some perspective regarding the translational potential of known therapeutic targets involved in this signaling network and the development of biomarkers for assessing clinical outcomes.

5.2 PI3K/PTEN/AKT/mTOR Signal Transduction

5.2.1 *Phosphoinositide 3-Kinase (PI3K)*

The intracellular transduction of extracellular stimuli often requires receptor-mediated signaling. Thus, membrane-bound receptors translate extracellular ligand binding into intracellular signaling cascades to various downstream cellular compartments (Fig. 5.2). Adaptor proteins and second messengers play an important role in correctly mediating and regulating these signals. One group of second messengers involves the class I phosphoinositide 3-kinase (PI3K) family (p110α, p100β, p110γ and p110δ). Other classes of PI3Ks include the class II and III families, comprehensively reviewed in [4]. These second messengers exert a common signaling mechanism utilized by a wide array of receptor tyrosine kinases (RTKs) and G-protein-coupled receptors (GPCRs). A functional PI3K-signaling unit contains one regulatory (typically p85-alpha or its splice variants, p55-alpha and p50-alpha; p85-beta or p85-gamma) and one catalytic (p110-alpha, beta, and delta) subunit, creating a heterodimeric kinase with enzymatic activity for lipid and protein substrates [4–6]. However, only the lipid kinase activity is required for oncogenic signaling [7]. PI3K activation is extensively regulated, particularly, by numerous RTKs

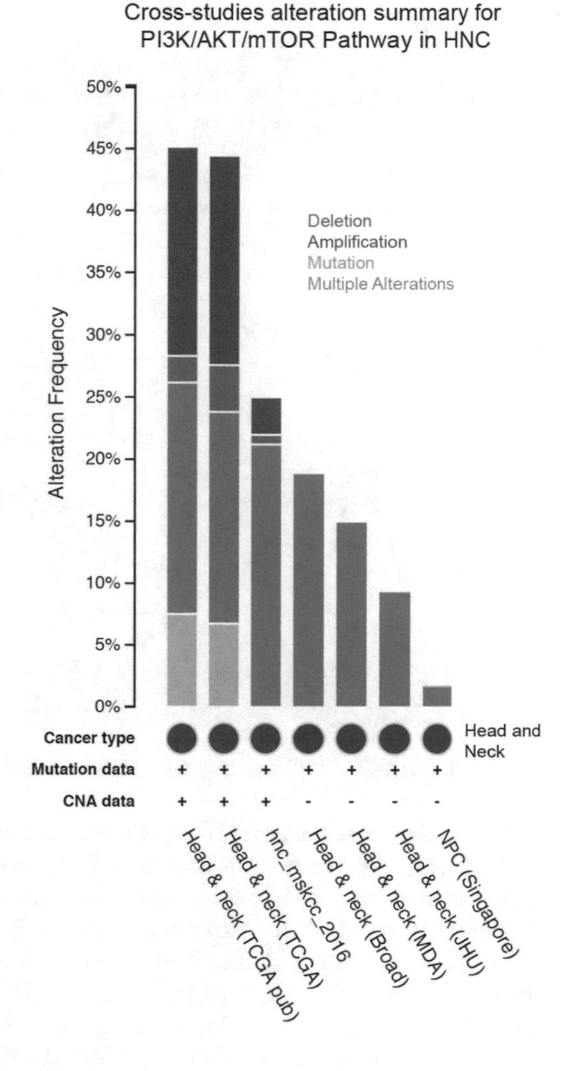

Fig. 5.1 Cross-study alteration summary for PI3K/AKT/mTOR pathway in head and neck cancer. TCGA data. Chart represents summary of deletions, amplifications, mutations, and multiple alterations in different head and neck cancer cohorts (horizontal) for *FOXO1*, *FOXO3*, *AKT1*, *AKT2*, *AKT3*, *PIK3R1*, *PIK3CA*, and *PTEN* genes

(such as the ERBB family, insulin-like, and fibroblast growth factor receptors) and G-protein-coupled receptors. When a receptor is activated, PI3K translocates to the cell membrane, where it associates with the receptor through p85 and various adaptor proteins (i.e., IRS1) [8, 9]. This binding reduces p85-dependent negative regulation of p110, initiating catalytic activity. PI3K can also be positively induced by activation of Ras, a critical member of small GTPases, which may facilitate PI3K membrane localization [10–12].

Once active, PI3K catalyzes the phosphorylation of phosphatidylinositol 4,5-bisphosphate (PI4,5P2) to phosphatidylinositol 3,4,5-trisphosphate (PIP3) [13]. The synthesis of PI3,4P2 typically follows this process, perhaps resulting from the

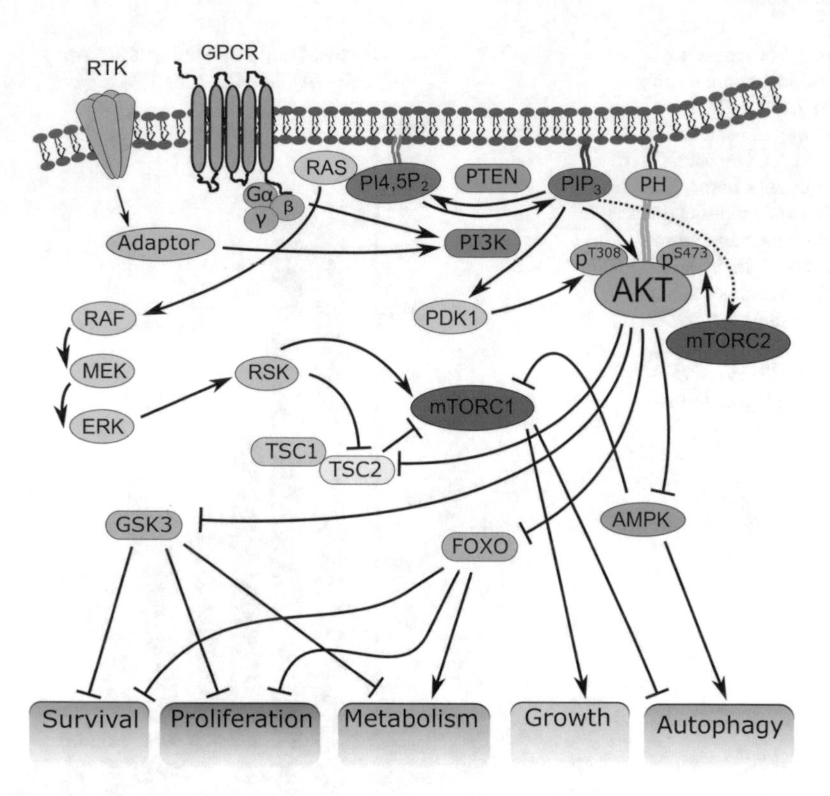

Fig. 5.2 PI3K/AKT/mTOR signaling map. See discussion in the text

action of the 5-phosphatase SH2-domain-containing inositol 5-phosphatase (SHIP) on PIP3 [14, 15]. PI3,4P2 can also be synthesized by class II PI3Ks using PI4P as substrate (reviewed in [16]). Although the specific PI3K isoform activated in a given cellular context may differ, the ultimate output is the same: relocalization of inactive AKT and its activating kinase, 3-phosphoinositide-dependent kinase 1 (PDK1), to membrane sites of PI3,4P2 or PIP3 accumulation, via engagement of the AKT pleckstrin homology (PH) domain (Fig. 5.2). PDK1-mediated AKT activation occurs via phosphorylation at threonine 308 (T308) in the activation or T-loop [17] of AKT. In parallel, AKT receives an activating phosphorylation mediated by mTORC2 at serine 473 (S473) in a C-terminal hydrophobic motif [18]. Activated AKT exerts a number of downstream effects on transcription, protein synthesis, metabolism, proliferation, autophagy, and apoptosis [19]. Although earlier studies suggested that PIP3 is exclusively localized to the plasma membrane, more recent reports have provided evidence for endomembrane pools of PIP3 and PI3,4P2 that directly contribute to AKT activation [20]. There are numerous additional cancer-relevant PI3K effectors: for example, one recent study has suggested that PI3K overexpression in an oral carcinogenesis mouse model signals through PDK1 to

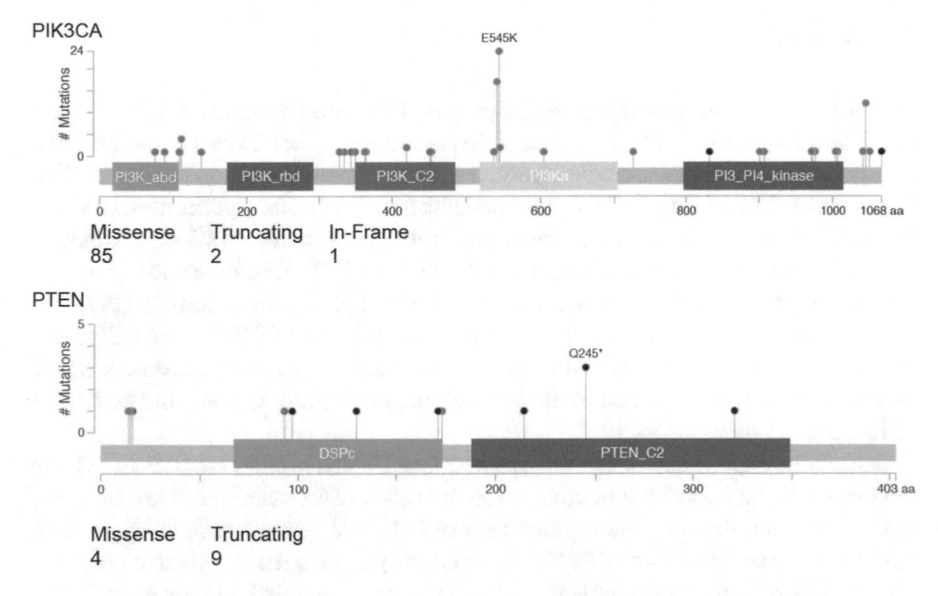

Fig. 5.3 TCGA data for point mutation rates in *PIK3CA* and *PTEN* genes. Missense mutations are indicated with green circles: truncating mutations with black. See text

enhance invasion through the activation of TGF-β signaling pathway [21]. Finally, protein phosphatases carefully balance the activity of these kinases in normal cells. Notably, the PI3K pathway is negatively regulated by phosphatase and tensin homologue (PTEN), which catalyzes the dephosphorylation of PIP3 to PIP2 [22, 23], as discussed below.

Investigations of oncogenic PI3K have been focused largely on p110α (PIK3CA). Exome sequencing projects have determined that the vast majority of cancer-related PI3K mutations are found in this isoform (Fig. 5.3) [3, 24]. It is currently estimated that p110α is mutated in 6–36% of HNSCC tumors [3, 25–30]. Unlike tumor suppressors, these activating point mutations are not found throughout the gene; rather, 80% of these cancer-associated modifications occur within three "hot spot" locations in the helical region (E542K, E545K) or the kinase (H1047R) domain. The first two mutations decouple p110 from p85, releasing the inhibitory effect of the regulatory subunit [31, 32]. The third mutation introduces a conformational change in the activation loop [31], possibly mimicking Ras-mediated activation [33, 34]. These three mutations are common in HPV-negative HNSCC tumors, whereas HPV-positive tumors present with *PIK3CA* mutations mostly in the helical domain [35]. Other cancer-specific mutations do exist within the gene; however, they have lower oncogenic activity and provide less of a selective advantage for tumorigenesis [36, 37].

5.2.2 PTEN

PTEN is a critical tumor suppressor, originally discovered because complete or partial deletion of a region of chromosome 10 encompassing *PTEN* is a common event in a number of cancers, including brain, bladder, prostate, and head and neck [38, 39]. At least 80% of Cowden's disease patients harbor heritable, germline mutations in *PTEN*, which confer this rare cancer predisposition syndrome [40–43]. Although mostly known for catalyzing the reaction of PIP3 to PIP2, this gene encodes a protein which possesses both peptide and phospholipid phosphatase activity (Fig. 5.2) [22, 23]. Loss of PTEN function causes an accumulation of PIP3 at the cell membrane. This enriched pool of PIP3 recruits AKT/mTOR pathway members (AKT isoforms, PDK1, etc.) to the membrane and inappropriately initiates the activation of this central signaling axis.

Knockout experiments have determined that PTEN, while essential for viable development, also has tumor-suppressive functions in endometrial, liver, prostate, gastrointestinal, thyroid, and thymus tissues [44, 45]. Haploinsufficiency is often sufficient to mediate a loss of PTEN function [46]. Due to the significance of these effects, PTEN function is regulated, and consequently deregulated, by a myriad of mechanisms: mutation, deletion, epigenetic silencing, transcriptional and posttranscriptional (microRNA (miRNA) regulation), posttranslational modification, and various protein–protein interactions. PTEN is the most frequently deregulated tumor suppressor associated with the PI3K/AKT/mTOR pathway and has some additional functions independent of PI3K/AKT/mTOR. For example, a loss of PTEN causes depletion of PIP2, an important membrane-associated regulator of cell polarity. This morphological modulation initiates a loss of epithelial characteristics, similar to epithelial–mesenchymal transition (EMT) [47], a hallmark of particularly aggressive cancers. Following EMT, neoplastic cells have increased cell motility and are often more resistant to standard therapy (reviewed in [48–50]). Some PTEN also localizes to the nucleus and is involved in maintaining chromosomal stability. Consequently, a loss of nuclear PTEN enhances chromosomal instability and leads to spontaneous DNA double-strand breaks [51]. As PTEN only exhibits PIP3 phosphatase activity in the cytoplasm, nuclear PTEN may affect genomic stability and cell cycle progression by lipid phosphatase-independent mechanisms [52, 53]. Due to its numerous PI3K-dependent and PI3K-independent functions, PTEN is accepted as an important tumor suppressor with loss of function often resulting in cancer.

5.2.3 AKT

AKT, also known as protein kinase B (PKB), is a critical node for mammalian signal transduction and the major effector of PI3K signaling. This vital serine/threonine protein kinase was originally discovered as the human homologue of *v-akt*, an oncogene transduced by the murine retrovirus AKT8 [54–57]. The AKT family is represented by three isoforms: AKT1, AKT2, and AKT3. AKT1 is ubiquitously expressed at high levels [55, 56, 58], while the remaining isoforms are expressed in a more tissue-specific manner. Insulin-sensitive cells, such as liver, skeletal muscle, and adipose tissue, demonstrate high levels of AKT2 expression [59, 60]. Meanwhile, AKT3 is highly expressed in the brain and testes, with lower levels of expression observed in muscle and intestinal organs [61]. While cancer-related AKT research largely focuses on AKT1, large-scale cancer sequencing projects have uncovered single nucleotide polymorphisms (SNPs) and somatic mutations associated with AKT2 and AKT3 [62, 63]. Mouse knockout models of the various AKT isoforms demonstrate specific mutant phenotypes, but all are viable [64–68]. Thus, the lack of embryonic lethality suggests that while each AKT isoform has characteristic signaling functions, they share a degree of functional compensation.

AKT kinases are comprised of an N-terminal PH domain, a flexible linker, and a C-terminal catalytic domain. While PIP3 interacts with AKT via the PH domain [14], AKT is phosphorylated by PDK1 on the C-terminal activation loop (T308) and at serine 473 (S473) by mTORC2 (Fig. 5.2; [69]) to achieve full kinase activity [70]. While these mechanisms represent canonical AKT activation, a number of PIP3-independent mechanisms also initiate AKT signaling. Activated CDC42 kinase 1 (Ack1 or TNK2) [71, 72], Src [73], protein-tyrosine kinase 6 (PTK6) [74], and serine/threonine-protein kinase 1 (TBK1) [75–77] all possess the ability to modulate AKT activity by noncanonical means. Once activated, AKT phosphorylates downstream targets, altering cell survival, growth, proliferation, metabolism, and crosstalk with other signaling pathways. The most important downstream target of AKT is mammalian target of rapamycin (mTOR), a master regulator of cell growth, metabolism, translation initiation, and ribosome biogenesis. AKT also affects cell survival by negatively regulating pro-apoptotic proteins such as FOXO and MDM2, a negative regulator of p53 [78, 79]. AKT can also enhance cell-cycle turnover by phosphorylating glycogen synthase kinase 3 (GSK-3), which stabilizes cyclin D/E, c-jun, and c-myc proteins [80–83].

Recent evidence suggests that subcellular localization is an important determinant of AKT activity and downstream signaling. In fact, two important AKT substrates (FOXO proteins and p300) are sequestered solely in the nucleus [84, 85]. Despite lacking a nuclear localization signal, AKT likely translocates to the nucleus by interacting with members of the T-cell leukemia-1 (TCL1) family of oncoproteins. These proteins are capable of complexing with AKT to serve as coactivators and nuclear shuttles [86, 87]. Increased nuclear phospho-AKT has been observed in acute myeloid leukemia [88, 89], lung [90], breast [91], thyroid [92], and prostate cancers [93]. Nuclear phospho-AKT detection has also been positively correlated with prostate cancer progression [93] and Gleason score [94]. Nuclear AKT activity may have specific oncogenic effects as promyelocytic leukemia protein (PML), which functions to dephosphorylate AKT within the nucleus and is a known tumor suppressor [95].

Due to the staggering number of pathways dependent on AKT signaling, deregulation of this enzyme by alterations in associative proteins or changes in subcellular localization can have disastrous biological consequences. For example, mosaic expression of an AKT point mutant (AKT E17K) is responsible for almost 90% of Proteus syndrome cases, the debilitating growth disorder suffered by Joseph Merrick, popularly known as "The Elephant Man" [96]. Proteus syndrome is characterized by segmental overgrowth and hyperplasia of a variety of tissues and organs, which also includes an increased risk of tumorigenesis [97, 98]. The rare nature of this crippling disease (<1 case/1 million) lies in its dependence on mosaic expression, as constitutive somatic or germline expression of this mutant would be lethal. Not surprisingly, this mutation has been detected in a variety of cancers including breast [99], urinary tract [100], and endometrial cancers [101]. Although the incidence of AKT E17K in patient tumor samples is low (1–4%), it nonetheless represents an important example of the total PI3K/PTEN/AKT/mTOR deregulation that occurs during tumorigenesis.

5.2.4 mTOR

As mentioned above, mTOR is the single most important effector of AKT signaling. Serving as the catalytic subunit of two macromolecular complexes (mTORC1 and mTORC2), mTOR is a master regulator of cell growth, via mediation of growth factor inputs and nutrient sensing. Although mTOR is shared between these two complexes, the associative proteins unique to each tune the activity of this enzyme for distinct substrates and sources of regulation [102–108]. mTORC1 consists of mTOR, Deptor, Raptor, mLST8, and PRAS40 [109, 110]. This complex is rapamycin-sensitive [111–113], and S6 K1 and 4E-BP1 are its most important downstream targets (Fig. 5.2; [114–116]). Phosphorylation of S6 K1 promotes mRNA translation by facilitating initiation and elongation complex formation at the mRNA transcript. Activation of 4E-BP1 allows eIF4E to recruit eIF4G and initiate 5′ 188 mRNA translation. Aside from protein synthesis, mTORC1 also regulates ribosome biogenesis and autophagy [117–119]. Recent studies have shown mTORC1 activation is sufficient to inhibit autophagy, which is reversible following mTORC1 inhibition [120].

mTORC2 also contains Deptor and mLST8; however, additional associative proteins include Rictor, mSIN1, and Protor [105, 121, 122]. Differential phosphorylation of AKT (T308 vs. S473) had long been understood, with PDK1 mediating T308 activation. mTORC2 is the enzyme responsible for "PDK2" activity, phosphorylating AKT at S473 and further facilitating T308 phosphorylation by PDK1. This dual phosphorylation is essential for full AKT activation [69]. Consequently, this functionality places mTORC2 in a positive feedback loop within the pathway, allowing AKT to achieve full activation. A recent study suggested a mechanism of this regulation, indicating that a PH domain within the SIN1 component of mTORC2 serves to bind to PIP3, leading to relief of autoinhibition of mTOR kinase activity within the complex [123]. PIP3 binding would therefore have the dual function of relocalizing mTORC2 to membranes where AKT is being recruited, as well as relieving conformational constraints on mTOR, allowing AKT phosphorylation; however, a separate study using intracellular compartment-specific reporters concluded that PI3K activation is dispensable for mTORC2 activity on membranes [124]. In this model, it is the relocalization of AKT to specific membranes through its PH domain that allows mTORC2 to gain access to S473.

This function was initially difficult to elucidate as mTORC2 is rapamycin-insensitive during acute treatment [111–113]. Along with AKT, mTORC2 can also activate serum- and glucocorticoid-regulated kinase (SGK) and other AGC kinases, including particularly protein kinase C (PKC) [69, 125–127] at their corresponding hydrophobic motif residues, although the corresponding motif in S6 K1 (T389) is targeted by mTORC1 [128].

Because mTOR has a central role in controlling cell growth, appropriate regulation of mTOR itself is paramount to maintaining homeostasis. Thus, it is not surprising that a number of familial cancer syndromes involve germline mutations of mTOR-negative regulators, such as PTEN in Cowden disease and TSC in tuberous sclerosis [41, 129]. Transgenic mice have also provided experimental evidence for the importance of appropriate mTOR regulation. Mice heterozygous for beclin or autophagy-related 4C (ATG4C), both critical regulators of autophagy, are prone to tumor formation due to defects in autophagosome formation [130–132]. As a negative regulator of autophagy [128], sustained mTORC1 activation has the ability to mimic these genetic modifications and enhance tumor development. Sustained mTORC2 activity is also capable of driving tumorigenesis through constitutive activation of AKT and SGK. Furthermore, the expression of Rictor is required for tumor cell line and prostate tumor growth in PTEN-deficient mice [133, 134]. Consequently, tumor-associated defects in PI3K, PTEN, or AKT all have the potential to initiate pathological mTOR signaling. However, multiple routes of deregulation may provide important biomarkers and potential targets of therapeutic intervention to alleviate the oncogenic effects of mTOR signaling in HNSCC.

5.3 PI3K/PTEN/AKT/mTOR Deregulation in HNSCC

5.3.1 Genetic Alterations of PIK3CA in HNSCC

PI3K functions are of critical importance to potentiate and regulate receptor-mediated extracellular stimuli. This vital second messenger has been intensively studied in cancer progression, including HNSCC. *PI3K*, and more specifically *PIK3CA*, is a *bona fide* oncogene in HNSCC. As mentioned above, *PIK3CA* contains activating point mutations (commonly E542, E545, or H1047) in 6–20% of HNSCC tumor samples [26, 27] (Fig. 5.3; Table 5.1). Two sequencing projects independently identified *PIK3CA* as a significantly mutated oncogene in HNSCC tumor samples [25, 28]. Stransky et al. determined 8% (6/74) of their tumor samples had *PIK3CA* activating mutations, identifying one as R115L (rare), three affecting E542–545, and two targeting H1047 [28]. Agrawal et al. reported 6% of their tumors

Table 5.1 Investigation of *PI3K/PTEN/AKT/mTOR* genetic alterations in premalignant lesions and squamous cell carcinoma of the head and neck

Gene	Samples	Type of alteration	Reference
PIK3CA (p110α)	Fresh tissue samples after surgical resection, 117 patients with HNSCC, stages I–IV 33 fresh-frozen tissue samples from oral cavity, larynx and pharynx, different stages 29 HPV(+) and 13 HPV (−) oropharyngeal carcinoma samples 115 non-lymph node metastatic HNSCC samples	Amplification	[1] [135] [136] [137]
	33 fresh-frozen tissue samples from oral cavity, larynx and pharynx, different stages 20 HPV+ and 20 HPV-laser-capture microdissected oropharyngeal carcinomas 46 brushed samples of oropharyngeal SCCC	Copy number gain	[135] [138] [139]
PI3KR1 (p85α)	92 HNSCC patients. Samples from oral cavity, oropharynx, hypopharynx, larynx, and sinonasal cavity	Missense mutation	[28]
	92 HNSCC patients. Samples from oral cavity, oropharynx, hypopharynx, larynx, and sinonasal cavity	In-frame insertion	[28]
	151 HNSCC tumors, not specified sites	Nonsense mutation	[2]
	120 matched samples, oropharynx, oral cavity, hypopharynx, larynx	Non-synonymous mutation	[140]
PTEN	151 HNSCC tumors, not specified sites DNA from 213 HNSCC patient samples. Oropharynx and oral cavity, predominantly IVA stage	Nonsense mutation	[2] [141]
	151 HNSCC tumors, not specified sites DNA from 213 HNSCC patient samples. Oropharynx and oral cavity, predominantly IVA stage 52 samples, moderately differentiated SCCC of oropharynx, hypopharynx, larynx, and lymph node mets	Missense mutation	[2] [141] [142]
	Fresh tissue samples after surgical resection, 117 patients with HNSCC, stages I–IV 19 HNSCC samples, different locations, II–IV stages	Loss of heterozygosity	[1] [143]
	151 HNSCC tumors, not specified sites	Splice-site single nucleotide polymorphism	[2]
	Fresh tissue samples after surgical resection, 117 patients with HNSCC, stages I–IV 79 oropharyngeal squamous cell carcinoma samples	Reduced expression	[1] [144]

(continued)

Table 5.1 (continued)

Gene	Samples	Type of alteration	Reference
AKT	151 HNSCC tumors, not specified sites	Missense mutation	[2]
	Fresh-frozen oral SCC samples	Copy number variations	[145]
	Fresh tissue samples after surgical resection, 117 patients with HNSCC, stages I–IV 10 different formalin-fixed, paraffin-embedded tumors of each type Formalin-fixed, paraffin-embedded samples of dysplasias; carcinomas in situ; invasive SCCs	Increased activation	[1] [146] [147]
	33 HNSCCC samples, various locations	Overexpression	[135]
mTOR	93 patients samples with laryngeal and hypopharyngeal SCC	Overexpression	[148]
	151 HNSCC tumors, not specified sites	Missense mutation	[2]

harbored *PIK3CA* mutations: all three affecting H1047 [25]. In addition to these activating point mutations, copy number gains within the *PIK3CA* locus (3q26) are extremely common [149]. Genomic studies suggest that mutations and copy variations of *PIK3CA* are detected in aggregate in 27% of HPV-negative and 37% of HPV-related HNSCC tumors [150]. Intriguingly, one recent study using reverse phase protein array (RPPA) analysis reported that although activating *PIK3CA* mutations occur more commonly in HPV-positive tumors, these tumors had lower mean levels of activated AKT and less phosphorylation of targets downstream of AKT than did HPV-negative tumors, even though the *PIK3CA*-mutated tumors had elevated activation of mTOR [136]. Relationship of specific mutation to activation requires further study.

Current evidence suggests *PIK3CA* copy number gain is an early event in HNSCC development. Oral premalignant lesions commonly possess an increase in *PIK3CA* copy number (39%) [1]. An equivalent incidence in *PIK3CA* copy number gain is reflected in a more broadly defined set of HNSCC tumors (32–37%) [1, 137]. Along with alterations in ERK/MAPK, fibroblast growth factor (FGF), and p53, deregulated PTEN and PI3K/AKT pathway members discriminate high-grade premalignant lesions from low-grade dysplasias [151]. Increased *PIK3CA* copy number is also associated with early HNSCC recurrence, but this difference is only statistically significant in patients without lymph node metastases ($p = 0.026$) [137].

5.3.2 PTEN Loss

A loss of *PI3K*-negative regulation has been observed in a number of independent HNSCC studies, as alterations in *PTEN* status are common (Fig. 5.3; Table 5.1). These include nonsense and missense mutations [2, 141, 142], loss of heterozygosity [1, 143], hemizygous deletions [152], intron [153], and splice-site single nucleotide polymorphisms [2], as well as reduced expression [1]. Given the many different genomic and proteomic alterations seen with PTEN, its dysregulation in HNSCC might be attributed to multiple molecular mechanisms.

Early efforts to catalog PTEN deregulation in HNSCC began with a screen of 19 tumors, which determined *PTEN* was mutated in three samples [143]. A loss of heterozygosity (LOH) also occurred within the *PTEN* locus (10q23) in 6 of the 15 evaluable samples. Within the mutant PTEN patients, two had stage IV disease while the third had recurrent, metastatic disease. In a larger study involving 117 HNSCCC samples, targeted analysis of PI3K/AKT/mTOR HNSCC genetic alterations detected *PTEN* LOH in 14% of the samples [1]. Three of the eight patients with *PTEN* LOH also demonstrated abnormal PTEN levels in the adjacent mucosa, suggesting both PIK3CA and PTEN deregulation are early events in HNSCC development. An additional investigation in squamous cell carcinoma of the tongue determined *PTEN* loss was evident in 29% of the tumor samples [154]. Deregulated PTEN also correlated with decreased overall survival ($p = 0.03$) and event-free survival ($p = 0.01$). While these studies were targeted in nature, *PTEN* loss was also evident in one of the two HNSCC genome sequencing projects referenced above. Stransky et al. detected *PTEN* mutations in 7% of their tumor samples [28], although *PTEN* abnormalities were not detected by Agrawal et al. [25].

PTEN downregulation may occur not only via loss of heterozygosity but also by miRNA deregulation. These short, noncoding RNAs are capable of regulating a wide variety of proteins, and thus represent oncogenes and tumor suppressors in their own right. Recent studies of HNSCC tumor samples and cell lines determined that miR-21 is overexpressed with respect to normal tissue [155, 156]. miR-21 overexpression is capable of downregulating HNSCC PTEN protein levels in vitro, causing elevation of phospho-AKT, and increasing the proliferation of immortalized keratinocytes (HaCaT) [155]. Consequently, miR-21 has been described as a proto-oncogene in HNSCC [156]. However, miR-21 is not the only miRNA capable of targeting PTEN protein expression. *miR-9* is a frequently methylated gene in HNSCC tumor samples, with miR-9 expression levels closely correlating with methylation status [157]. When miR-9 is re-expressed with the use of a demethylating agent, a significant increase in PTEN and a concomitant decrease in cell growth is observed [157]. While the connection between miR-9 and PTEN is indirect, this study does provide additional evidence for miRNA-mediated PTEN modulation in HNSCC cells.

Recent evidence has also demonstrated an association between transforming growth factor-β (TGF-β) signaling and PTEN loss in HNSCC development. The TGF-β superfamily of ligands and receptors represent a signaling pathway unified by a shared group of second messengers, the SMADs. Knockdown of a downstream effector, *SMAD4* is sufficient to promote HNSCC in mouse models [158, 159]. In contrast, conditional knockdown of TGF-β receptor 1 (*TGF-βRI*) in the oral cavity of mice only leads to early HNSCC development when combined with a topical carcinogen (DMBA) [160]. Specifically, 16 weeks after DMBA treatment, 45% of the *TGF-βRI* conditional knockdown mice develop HNSCC, but only 10% of these mice develop HNSCC within a year without DMBA treatment [161]. *TGF-βRI* knockdown tumors are characterized by an increase in AKT activity with a paradoxical upregulation of PTEN. However, when *PTEN* is conditionally knocked down in combination with *TGF-βRI* loss, the mice develop benign papillomas within 4 weeks, and after 10 weeks, 100% of the mice develop HNSCC [161]. These tumors demonstrate an overexpression of EGFR and chemokines; hyperactivation of AKT, NF-κB, and STAT3; and recruitment of tumor-promoting myeloid-derived suppressor cells (e.g., CD11b+). Treatment of these animals with rapamycin effectively prevents tumorigenesis; thus, carcinogenesis in this model is an mTOR-dependent event [162, 163].

5.4 Differences in PI3K-Dependent Signaling Based on HPV Status

5.4.1 PI3K Signaling in HPV-negative HNSCC

The most common genetic abnormality associated with HPV-negative HNSCC is a functional loss of p53 [25, 28], yet somatic ablation of this tumor suppressor in transgenic mice favors spontaneous tumor formation in the skin, rather than tumorigenesis in the oral mucosa [164]. However, when p53 loss is combined with constitutively active AKT (myrAKT, in which fusion of a myristylation sequence constitutively localizes AKT to the membrane), tumor formation within the oral cavity, palate, ventral side of the tongue, and lips is markedly increased [165]. These tumors also exhibit increased EGFR expression and potently activated NF-κB and STAT3 pathways, recapitulating the hallmarks of HPV-negative HNSCC [166]. As the PI3K/PTEN/AKT/mTOR pathway can be activated through a multitude of mechanisms, these data suggest that any manner of AKT activation, when combined with p53 loss, may synergize to initiate HNSCC development and progression. This is consistent with the observation that both PIK3CA and PTEN deregulation are early events in HNSCC, and targeting this pathway may have a role in chemoprevention for smokers.

The majority of patients with HPV-negative tumors have an extensive history of cigarette smoking, and these tumors are associated with an increased number of mutations compared to their HPV-related cohorts [25, 28]. Tobacco use is a well-defined causal link for the development of HNSCC [167], and cigarette pack-years is a predictive variable for survival even among patients with HPV-associated cancer [168]. Studies have shown that nicotine and an additional tobacco carcinogen [4-(methylnitrosamino)-1-(3-pyridyl)-1-butanone, NNK] are both capable of activating AKT by receptor-mediated signaling in normal human airway epithelial cells [169]. This mechanism has since been observed in human head and neck epithelium as well, where activated AKT is four times more likely in HNSCC-adjacent mucosa from smokers compared to nonsmokers [170]. However, these findings are somewhat difficult to interpret, since the majority of patients with tobacco-associated cancers are not active smokers at the time of treatment. Experiments with HNSCC cell lines have also determined NNK activation of AKT is PI3K-dependent [170]. Furthermore, cigarette smoke condensate (CSC) also upregulates the multidrug transporter ABCG2 in lung cancer and HNSCC cell lines [171]. After CSC treatment, these cells are more resistant to doxorubicin and have upregulated drug efflux mechanisms; the latter effect can be abrogated by PI3K or nicotinic acetylcholine receptor inhibition [171]. As tobacco use is a strong predictor of HNSCC recurrence, and HNSCC patients have a 10–14% risk of developing a second malignancy within 5 years of primary surgical treatment [168, 172], further studies of premalignant PI3K/AKT/mTOR activation may yield novel chemopreventive options to mitigate this public health challenge.

The finding of a simultaneous second primary tumor (SPT) is an independent and negative prognostic factor for patients with oral squamous cell carcinoma [173]. Early data suggested that SPT development may be prevented by 13-cis-retinoic acid (13-cRA) treatment [174, 175]. However, in a follow-up phase III clinical trial (C0590), no significant difference in SPT or recurrence could be observed between placebo and low-dose 13-cRA in early-stage HNSCC patients [176]. However, after long-term follow-up, this study demonstrated near significant survival advantage among patients who were women, or never/former smokers, and among patients whose primary tumor was surgically excised [177]. A recent retrospective study characterized 137 SNPs as predictive biomarkers for recurrence in the aforementioned placebo cohort. While 22 SNPs were significantly associated with recurrence, 15 SNPs were detected in the majority of patients who recurred [178]. Ten of these fifteen SNPs were located in *TSC1*, a negative regulator of mTOR. When these SNPs were assayed in the 13-cRA treatment group, two of the *TSC-1*-associated SNPs yielded a 43% decrease in SPT/recurrence with treatment. Variants in *PIK3CD* and *PTEN* were also associated with a decrease in SPT/recurrence risk with 13-cRA treatment [178]. Consequently, prospective analysis of early stage HNSCC for PI3K/PTEN/AKT/mTOR pathway genetic variants could increase the efficacy of 13-cRA chemoprevention.

5.4.2 PI3K Signaling in HPV-Positive HNSCC

The incidence of HPV-negative HNSCC is decreasing worldwide following successful tobacco cessation campaigns; however, the overall incidence of HNSCC remains constant due to an increase in HPV-related HNSCC [179, 180]. HPV-related HNSCC primarily arises within the oropharynx, as oral HPV infection most commonly persists in the palatine and lingual tonsils (reviewed in [181]). Mounting evidence suggests PI3K/PTEN/AKT/mTOR signaling has an important role in HPV infection and HPV-induced carcinogenesis.

Several studies have demonstrated that EGFR and PI3K signaling are required for viral entry into the cell. Pretreatment of HaCaT or cervical cancer cells (HeLa) in vitro with an EGFR inhibitor (gefitinib) is sufficient to inhibit HPV-16 endocytosis [182]. Additionally, two different PI3K inhibitors (PI-103, wortmannin) are also capable of preventing viral entry [182]. As the same mechanism was observed in cells from different anatomical sites, these data suggest a common requirement for EGFR and PI3K activity in high-risk HPV infection.

Following viral entry, the PI3K pathway continues to play an important role in HPV-related tumorigenesis. Gene expression profile analysis of HNSCC patient samples has determined that HPV-related tumors demonstrate upregulation in genes within the 3q26–29 chromosomal region [183]. This locus contains *PIK3CA*, and confirmatory analysis with RT-PCR established that *PIK3CA* is upregulated in HPV-related compared to HPV-negative tumors [184]. Further, in a study of 231 HNSCC patients, *PIK3CA* was identified as the most frequently altered oncogene (13% amplified or mutated), with highest levels in HPV-related oropharyngeal cancers [141]. Based on The Cancer Genome Atlas data, *PI3K* mutations are more common for HPV-associated than HPV-negative tumors (56%, $n = 36$) [185]. Similarly, the Stransky and Agrawal studies noted above found 6 non-synonymous mutations of *PIK3CA* in a group of 74 tumor/normal pairs [28] and in 6% of 32 tumors screened for *PIK3CA* mutations [25], respectively. The authors noted that far fewer genes were mutated per tumor in the HPV-related as compared with HPV-negative tumors (4.8 ± 3 versus 20.6 ± 16.7, $P < 0.05$). Next-generation sequencing found *PI3K* amplification or mutation in 13% of 213 HNSCC patients, with *PI3K* amplification associated with reduced PFS ($n = 109$) [141].

Immunohistochemical (IHC) analysis of HNSCC tissue samples has also documented a strong correlation between p16 upregulation (a surrogate marker for HPV infection) and eIF4E activation ($p = 0.03$), which reflects mTOR pathway activation [186, 187]. Although phospho-AKT only trended toward significance with p16 expression ($p = 0.06$), its lack of concordance could be caused by the additional signaling factors that regulate AKT activity compared to eIF4E function. This study suggested HPV infection was associated with mTOR-dependent activation of mRNA translation, including the upregulation of transformation-related and prosurvival pathway members [188, 189]. However, in a larger follow-up study, neither phospho-AKT (pAKT S473) nor phospho-S6 (mTOR target) was associated with HPV-related HNSCCs [190]. In addition, the HPV-related tumors were not associated with an activation of EGFR, as gauged by tissue microarray (TMA) analysis indicating specimens were positive by IHC for EGFR but negative for phospho-EGFR. Similar results were observed in the HPV-negative samples. Thus, the hyperactivation of selective mTOR targets noted in the prior study may arise through a different mechanism. However, detection of phosphorylated proteins such as AKT and S6 in clinical specimens is challenging due to rapid dephosphorylation and technical variations [191], which may account for conflicting results.

5.5 PI3K/AKT/mTOR Inhibition as a Therapeutic Option in HNSCC

Due to extensive preclinical evidence that PI3K/AKT/mTOR signaling represents an integral component of HNSCC signal transduction, and significantly contributes to development of acquired resistance to anticancer therapy [192], a number of clinical trials are currently underway to evaluate thc cfficacy of small molecules which inhibit key nodes of this pathway (Fig. 5.4; Table 5.2). Currently, rapamycin and its associated analogs (rapalogs) are the most investigated PI3K/AKT/mTOR-targeted agents in HNSCC clinical trials. Rapamycin is a secondary metabolite produced by *Streptomyces hygroscopicus*, isolated from a soil sample collected on Easter Island (Rapa Nui) [193]. This molecule is an allosteric inhibitor of mTOR, creating a complex with FKBP12, which binds and prevents mTOR activation via the FKBP12-rapamycin-binding (FRB) domain. As this domain is unique to mTOR, rapamycin-induced inhibition of mTORC1 is highly selective; however, as reviewed above, mTORC2 is largely uninhibited by this compound [194, 195]. To determine if the PI3K/AKT/mTOR pathway represents a viable clinical target in HPV-related HNSCC and cervical cancer, xenograft models for these cancers were treated with rapamycin and RAD001 (everolimus). Both xenograft models demonstrated a

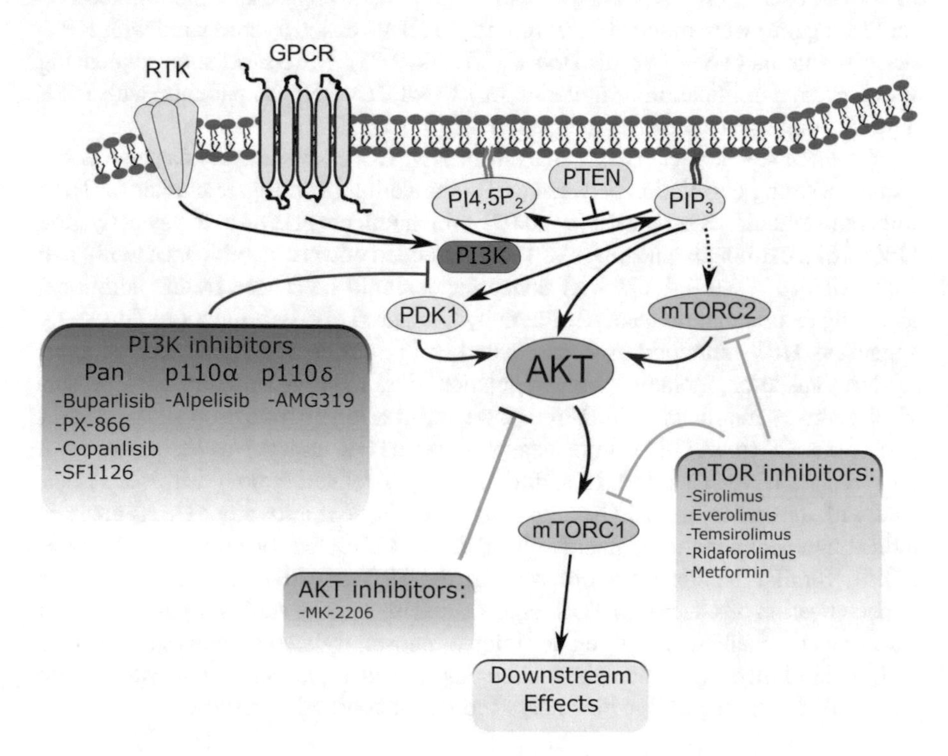

Fig. 5.4 PI3K, AKT, and mTOR signaling inhibitors currently in trials for head and neck cancers. See text for discussion

durable, cytostatic response following mTOR inhibition [190]. Similarly, Madera et al. reported that metformin treatment of HNSCC cells expressing mutated *PIK3CA* and HPV oncogenes inhibited mTOR signaling and tumor growth. This study found that metformin activity required the expression of organic cation transporter 3 (OCT3/SLC22A3), a metformin uptake transporter [196]. In summary, it is possible to conclude that mTOR inhibition represents an important therapeutic option in HNSCC patients, particularly those with HPV-positive disease.

Table 5.2 Ongoing clinical trials of agents that target PI3K/AKT/mTOR pathway in HNSCC

Target	Agent	Additional agent	Inclusion criteria	Phase	Status	Clinical trial #
PI3K	Buparlisib (BKM120)		Advanced HNSCC	2	Unknown*	NCT01527877
		Cisplatin, IMRT	High-risk LA HNSCC	1b	Active	NCT02113878
		Paclitaxel	Pt pretreated R/M HNSCC	2	Active	NCT01852292
		Cetuximab	R/M HNSCC	1/2	Active	NCT01816984
			R/M HNC	2	Active	NCT01737450
	PX-866	Docetaxel	NSCLC, HNSCC	1/2	Completed	NCT01204099
		Cetuximab	Metastatic CRC, R/M HNSCC	1/2	Completed	NCT01252628
	Copanlisib (BAY 80–6946)	Cetuximab	R/M HNSCC with PI3KCA mutation/ amplification and/or PTEN loss	1/2	Active	NCT02822482
	SF1126		R/M HNSCC with mutation in PIK3CA and/or PI3K pathway	2	Active	NCT02644122
	Alpelisib (BYL-719)	Cisplatin, IMRT	LA HNSCC	1	Active	NCT02537223
		Paclitaxel	Breast cancer and HNC	1	Active	NCT02051751
		Cetuximab, IMRT	Stage III/IVb HNSCC	1	Active	NCT02282371
			Pt therapy failed, R/M HNSCC	2	Active	NCT02145312
		Cetuximab	R/M HNSCC	1b/2	Completed	NCT01602315
	AMG319		HPV negative HNSCC	2	Active	NCT02540928
PI3K/ AKT	Perifosine (KRX-0401)		R/M HNC	2	Terminated	NCT00062387
PI3K/ PLK	Rigosertib (ON-01910)	Cisplatin, radiation	HNSCC	1	Completed	NCT02107235
AKT	MK2206		R/M HNC	2	Completed	NCT01349933

(continued)

Table 5.2 (continued)

Target	Agent	Additional agent	Inclusion criteria	Phase	Status	Clinical trial #
mTOR	Sirolimus (rapamycin)		HNSCC	1/2	Completed	NCT01195922
	Everolimus (RAD001)	Docetaxel	LA and R/M HNSCC	1/2	Terminated	NCT01313390
		Carboplatin, cetuximab	Advanced HNC	1/2	Completed	NCT01283334
		Carboplatin, paclitaxel	LA HNC not removable by surgery	1/2	Completed	NCT01333085
		Erlotinib	Recurrent HNSCC	2	Completed	NCT00942734
		Erlotinib, radiation	R/M HNSCC treated with radiation	1	Withdrawn	NCT01332279
		Cetuximab	R/M colon cancer or HNC	1	Completed	NCT01637194
		Cetuximab, cisplatin, carboplatin	R/M HNSCC	1/2	Terminated	NCT01009346
			HNSCC	2	Active	NCT01133678
			HNSCC	2	Active	NCT01051791
			HNC	2	Active	NCT01111058
		Docetaxel, cisplatin	LA HNC	1	Completed	NCT00935961
		Cisplatin, radiation	LA, inoperable HNC	1	Terminated	NCT01057277
		Cisplatin, IMRT	LA HNC	1	Terminated	NCT01058408
		Cisplatin, IMRT	LA HNC	1	Completed	NCT00858663
		Ceritinib	HNC, NSCLC	1/1b	Active	NCT02321501
			LA HNSCC	2	Active	NCT01133678
	Temsirolimus (CCI-779)		HNSCC	2	Completed	NCT01172769
		Cetuximab	R/M HNC not respond to therapy	2	Completed	NCT01256385
		Paclitaxel, carboplatin	R/M HNSCC	1/2	Active	NCT01016769
		Cisplatin, cetuximab	R/M HNSCC	1/2	Terminated	NCT01015664
			Advanced HNSCC		Completed	NCT00195299
		Erlotinib	Pt-refractory or Pt-ineligible, advanced SCC	2	Terminated	NCT01009203
		Cetuximab, cisplatin, radiation	Advanced HNC	Pilot	Withdrawn	NCT01326468
	Ridaforolimus (AP23573, MK-8669, deforolimus)		Advanced HNC, NSCLC, colon cancer	1	Terminated	NCT01212627
	Metformin (glucophage)	Paclitaxel	R/M HNSCC	2	Terminated	NCT01333852
		Cisplatin, radiation	LA HNSCC	1	Active	NCT02325401
			HNSCC	0	Terminated	NCT02402348
			HNSCC	0	Completed	NCT02083692

*Unknown: status not verified for more than 2 years

Pt = platinating agent; LA = locally advanced; R/M = recurrent or metastatic

A number of drugs have been designed to improve the pharmacokinetics of this parent compound. Temsirolimus (Torisel; Pfizer) is a water-soluble ester of rapamycin (also known as sirolimus) for oral or IV administration (reviewed in [197]). Everolimus (Afinitor; Novartis) is a hydroxyethyl ether derivative, also with increased solubility relative to the parent compound (reviewed in [198]). In a phase II study of biomarker-unselected patients with recurrent or metastatic HNSCC who failed at least one prior therapy, everolimus did not show a survival benefit as a monotherapy [199]. Temsirolimus was also assessed in combination with erlotinib in a phase II study of patients with recurrent and/or metastatic, platinum-refractory HNSCC [200]. This combination, which delivered erlotinib 150 mg daily and temsirolimus 15 mg weekly, was poorly tolerated and did not succeed in reaching its primary endpoint of improving PFS. A preclinical study utilizing in vivo xenograft models of resistant tumors assessed the effectiveness of combining the mTOR inhibitor rapamycin and the EGFR-targeting antibody cetuximab. Results of this work suggested that addition of rapamycin dramatically increased activity of cetuximab [201]: however, these findings have not been confirmed in any clinical trial to date, and excess toxicity was associated with this combination, as with the erlotinib/temsirolimus combination cited above [202]. In contrast to monotherapy or combination of mTOR inhibitors with other targeted agents, a phase II study of temsirolimus 25 mg added to a low-dose weekly carboplatin AUC 1.5 and paclitaxel 80 mg/m^2 given on days 1 and 8 of a 21-day cycle showed clinical efficacy with relatively high response rate and acceptable safety profile (NCT01016769) [203]. Among the 36 patients who were eligible for evaluation, 15 (41.7%) patients had partial response and 19 (52.3%) patients had stable disease after 2 cycles of treatment. There was no complete response, but only two patients were considered nonresponders for having adverse events during cycle 1 before the assessment of objective response after two cycles of treatment.

Ridaforolimus (AP23573, MK-8669, deforolimus) is available in oral and intravenous formulations. In combination with the Notch inhibitor MK-0752, ridaforolimus was evaluated in a phase 1 trial in patients with advanced tumors, including head and neck cancer [204]. This drug combination showed potentially promising results in the HNSCC setting, as two patients out of ten had complete or partial response; however, there were toxicity concerns. Metformin, currently used to treat type II diabetes, is also being investigated as a chemotherapeutic in HNSCC patients (NCT02325401 trial). Metformin indirectly inhibits mTORC1 by increasing intracellular AMP levels, controlling mTOR by both AMPK-dependent and AMPK-independent mechanisms [205–207]. Metformin also inhibits Ras signaling by decreasing EGFR [208], which is of interest given results with combination of mTOR and EGFR inhibition. Intriguingly, metformin has demonstrated chemopreventive activity for a number of different cancers in diabetic patients (reviewed in [209]). Thus, additional studies are warranted to determine whether this well-characterized compound will have similar chemopreventive or chemotherapeutic effects in nondiabetic patients.

Additional compounds acting upstream of mTOR are also being evaluated in HNSCC. Buparlisib is an oral PI3K inhibitor, and it inhibits the activity of all four p110 isoforms of class I PI3K, as shown by Kong et al. [210]. Analysis of tumor tissue microarrays, in association with a phase II clinical trial studying efficacy of induction chemotherapy plus cetuximab followed by cetuximab chemoradiotherapy, indicated that resistance to cetuximab may occur due to the activation of the PI3K/AKT and RAS/MAPK/ERK pathways [211]. In preclinical studies, the pan-PI3K inhibitor buparlisib in combination with cetuximab and irradiation increased tumor inhibition in comparison to cetuximab/irradiation alone in an orthotopic xenograft model of HNSCC [212]. A recent randomized phase II clinical trial (NCT01852292) has shown efficacy for buparlisib in combination with paclitaxel, compared to paclitaxel and placebo in patients with platinum-refractory recurrent or metastatic HNSCC (PFS 4.6 vs 3.5 months in buparlisib and placebo groups, respectively) [213]. However, the IC50 concentration of buparlisib is significantly higher than that of many of the PI3K inhibitors under investigation. Due to toxicity issues (elevated liver enzymes, hyperglycemia, and rash) in breast cancer trials, clinical application of PI3K inhibition will need to be studied with care, although the safety profile compares favorably to some standard agents in common use in HNSCC. Five phase 1 and/or 2 clinical trials are ongoing to evaluate the efficacy and safety of buparlisib in combination with cisplatin and IMRT, paclitaxel, or cetuximab.

Copanlisib is a highly selective and potent intravenous inhibitor of the p110α and p110δ isoforms of PI3K [214]. Copanlisib in combination with cetuximab is being evaluated in phase 1 and phase 2 trials in patients with recurrent and/or metastatic HNSCC harboring *PI3KCA* mutations/amplifications and/or PTEN loss. LY294002 is a compound that inhibits both PI3K and mTOR. It has antitumor and antiangiogenesis activity in vivo, but is not a viable drug due to poor solubility and short half-life. Therefore, SF1126 was designed as a prodrug of LY294002, by appending a small peptide tag to LY294002 to increase solubility [215]. SF1126 is currently under evaluation in a phase 2 trial in patients with recurrent or progressive HNSCC and mutations in *PIK3CA* and/or PI3K pathway genes (NCT02644122). Several isoform-specific PI3K inhibitors are currently in trials in head and neck patients. These agents are active predominantly against one of the p110 isoforms of class I PI3K and to a lesser extent against the other isoforms. Alpelisib (BYL-719, NVP-BYL719), an inhibitor of p110α isoform of PIK3CA, is being evaluated in five clinical trials (Table 5.2). Preliminary results showed encouraging antitumor activity [216], but trial results are not yet published. AMG319 specifically inhibits isoform p110δ of PI3K. This is a promising drug, because downregulation of the p110δ isoform in regulatory T cells unleashes exerts CD8+ cytotoxic T cells activity, resulting in tumor regression [217]. AMG319 is in a double-blind, placebo-controlled phase 2a trial in patients with HPV-negative HNSCC.

MK-2206 is an allosteric AKT inhibitor developed by Merck. Synergistic anticancer properties have been observed in vitro when this compound was used in combination with erlotinib (in non-small cell lung cancer (NSCLC)) or lapatinib (in breast cancer) [218]. PX-866 is a synthetic derivative of wortmannin (a potent irreversible PI3K inhibitor with antitumor activity) with antineoplastic activity and reduced liver toxicity with respect to the parent compound [219]. Aside from increased safety, PX-866 also demonstrates superior water solubility, bioavailability, and AKT inhibition. However, due to the positive feedback and compensation that can occur via mTORC2, single target inhibition with this agent led to rapid acquired resistance in preclinical and clinical trial investigations.

Additional studies have investigated dual-target inhibitors. NVP-BEZ235 is an orally available, dual PI3K/mTOR inhibitor that reversibly inhibits class I PI3K through ATP competition. This compound is unique because it simultaneously inhibits mTOR catalytic activity, as the kinase domain of mTOR is highly homologous to that of class I PI3K [220], although initial studies suggest the dual potency of NVP-BEZ235 is not equivalent for each target. In breast cancer cells, NVP-BEZ235 exerted anti-mTOR activity at lower doses (<100 nM) while dual PI3K/mTOR blockade occurred at higher concentration (>500 nM) [221].

5.6 PI3K Pathway Biomarkers

5.6.1 Activation of PI3K Pathway Prognostic Biomarkers

Ongoing studies are evaluating deregulated PI3K/AKT/mTOR pathway members as prognostic biomarkers in HNSCC. However, validated data are sparse due to small sample sizes, technical limitations, and intrinsic biological heterogeneity of HNSCC. To date, the most established prognostic biomarkers for HNSCC outcome are EGFR overexpression [222], EBV (Epstein-Barr virus) and HPV status [168, 183, 223], and the contribution of PI3K/PTEN/AKT/mTOR signaling needs to be evaluated in the context of these well-characterized biomarkers for clinical translation. Initially, investigators examined hyperphosphorylation of PI3K/AKT/mTOR pathway members to establish pathway activation. One potential prognostic biomarker is phosphorylated (phospho)-AKT [224], which was associated with poor local control in a series of 38 HNSCC patients [225]. In one study, persistent AKT upregulation associated with poor clinical outcome in oropharyngeal squamous cell carcinoma patients has been found to be independent of EGFR status, caused by PTEN loss [144], with such AKT activation forming part of a signature for cetuximab resistance [211]. eIF4E, a downstream target of mTOR, is also upregulated in many HNSCC tumor samples [226, 227]. Upregulated phosphorylated eIF4E detection in tumor-free margins is also associated with disease recurrence [228].

Phospho-protein detection in clinical samples can present a challenge as outcomes can differ depending on fixation protocol and handling time ex vivo. Thus, biomarkers utilizing total protein fractions may be preferable from a technical standpoint. One controversy in the literature involves the prognostic value of PTEN in predicting outcome following HNSCC surgery and radiotherapy. Among 140 HNSCC tissue microarray samples, PTEN-positive tumors were associated with worse locoregional control (LRC) than PTEN-negative tumors following surgery and radiation therapy (HR: 2.4) [229]. In this study, phospho-AKT was also associated with poor LRC (HR: 2.2). The authors suggested PTEN-positive tumors exhibit increased EGFR activity, and this subsequently supported PI3K/AKT/mTOR activation to provide a protective effect from ionizing radiation. However, a similar study of 147 HNSCC patients also treated with surgery and radiotherapy had very different results. In this study, the 5-year LRC-free rate for PTEN-low tumors was 52.3%, while 80.9% of PTEN-high patients were recurrence-free over the same time period ($p = 0.0007$) [230]. PTEN status did not correlate with 5-year risk of metastasis in this study ($p = 0.49$) [230]. Similar results were obtained in a study of inoperable recurrent metastatic HNSCC, where low PTEN expression was associated with worse overall survival after chemotherapy [231]. The non-concordance between these various studies highlights the difficulty in utilizing a tumor suppressor as a predictive biomarker. As PTEN haploinsufficiency can be tumorigenic [46, 232], and techniques for accurate, quantitative assessment of this target from clinical samples is lacking, further studies are required to establish this protein as a *bona fide* prognostic marker in HNSCC.

In contrast, detection of *PIK3CA* mutations is more straightforward with current technology, and it is the only oncogene in HNSCC with relatively frequent activating mutations or gene amplification (Fig. 5.3) and existing targeted therapies (Fig. 5.4). As discussed above, mutations in exon 9 and 20 are the predominant *PIK3CA* lesions associated with most cancers, and current evidence suggests these species may contribute differently to clinical outcome. Studies in breast cancer have shown that *PIK3CA* exon 9 mutations (E542/E545) are independently associated with shorter disease-free survival ($p = 0.0003$) and overall survival ($p = 0.001$) [233]. Conversely, exon 20 mutations (H1047) are associated with better overall survival [233]. Another study reported that *PIK3CA* copy number amplifications were found in 37 of 115 (32.2%) non-lymph node metastatic HNSCC samples. Theses alterations were markedly associated with cancer relapse in patients (log-rank test, $p = 0.026$) [137]. The smaller genomic datasets currently available for these mutations in HNSCC do not yet have adequate sample size to achieve the statistical power needed to delineate any prognostic significance for these *PIK3CA* alterations.

5.6.2 PI3K Pathway Members as Predictive Biomarkers for Radiation Therapy and PI3K/mTOR Inhibitor Response

The association of increased phospho-AKT and poor LRC suggests that AKT activation may be a predictive marker of radiation resistance in HNSCC. In vitro studies have indicated PI3K inhibition increases the radiosensitivity of HNSCC cell lines [225]. Future investigations of mTOR inhibition as a radiosensitizer in HNSCC treatment may establish an optimal treatment regimen and determine the maximum tolerated dose in the context of this disease.

PI3K pathway deregulation may also serve as a biomarker for response to PI3K/AKT/mTOR-targeted agents. A retrospective analysis determined the *PI3K* mutational status of solid tumors from clinical trials investigating PI3K/AKT/mTOR inhibitors [234]. Of the 1012 patients in this study, 105 were prospectively selected, and 66 of these patients harbored *PIK3CA* mutations. Although these patients had tumors arising from varying anatomical locations, those possessing an exon 20 mutation responded better to PI3K/AKT/mTOR therapies than those bearing other *PI3K* mutations (PR rate, 38% vs. 10%; $p = 0.018$). Unfortunately, an increase in progression-free survival only trended toward statistical significance (5.7 vs. 2 months, $p = 0.06$). While this study was hampered by a heterogeneous tumor population and by multiple treatment regimens, PI3K/AKT/mTOR-treatment efficacy in patients with exon 20 mutations remained an intriguing finding. Of the 66 prospectively selected patients from this study, four individuals had HNSCC, and the best responder possessed an H1047R mutation. Although these data indicate *PI3K* mutations may sensitize HNSCC tumors to PI3K/AKT/mTOR inhibitor treatment, wild-type *PI3K* status may not preclude the use of these drugs in this patient population. Subsequent studies suggested that downstream signaling in patients with HPV-positive HNSCC with mutant and wild-type *PIK3CA* differs according to activating mutations in genes encoding mTOR/S6 rather that AKT, therefore making PI3K and mTOR inhibitors more preferable in this cohort of patients [136].

5.6.3 Predictive Biomarkers for RTK Inhibitor Resistance

While the identification of predictive biomarkers for RTK inhibitor response is of paramount concern, equally important is the investigation of biomarkers for resistance. For example, additional ERBB receptors, aside from EGFR, signal through PI3K in HNSCC. A positive feedback loop has been reported between ERBB2/HER2 and ADAM12 in HNSCC cell lines [235]. This positive feedback loop is dependent on PI3K and JNK signaling. ADAM12 is a multifunctional protein possessing an intracellular domain capable of initiating second messenger signaling and an extracellular domain with protease activity cleaving extracellular matrix substrates and activating EGFR ligands [236–238]. Additionally, ADAM12 upregulation confers increased migratory and invasive phenotypes to these cells. This signaling mechanism may have clinical significance as HER2 activation and total ERBB3/HER3 expression are predictive of de novo resistance to gefitinib (an ERBB1/EGFR-targeted tyrosine kinase inhibitor (TKI)) [239]. Whether PI3K or PTEN serve as biomarkers for TKIs targeting this signaling pathway remains to be determined.

The FGFR (fibroblast growth factor receptor) family of RTKs is differentially activated by 18 FGFs. Activated FGFR RTKs signal downstream through effectors including MAPK, PI3K, p38 MAPK, JNK, STAT, and RSL2, in a context-dependent manner [240]. According to TCGA data, *FGFR1, FGFR2, FGFR3*, and *FGFR4* are altered in 10%, 2%, 2%, and 0.4% of HPV-negative HNSCC, respectively [3]. Among HPV-related HNSCC, alterations in *FGFR1* and *FGFR2* are not detected, *FGFR3* presents with 11% of mutations or fusions, and *FGFR4* have a 3% mutation rate [3]. Given this high rate of mutation in HNSCC, several studies investigated if targeting this receptor might be beneficial in the context of HNSCC. In a preclinical study, an inhibitor of FGFR1 suppressed cell growth and invasion [241], in a mechanism of action that depended on MAPK, not PI3K. Another study assessed the same inhibitor and reported no effect on proliferation rate of tumor cells but inhibition of proliferation in fibroblasts and endothelial cells in the stroma [242]. Both studies were conducted in cell lines without characterization of the mutation or copy number status of *FGFR* and *FGF* gene families, or of *PI3K/PTEN*; further studies of these potential biomarkers are required.

A strong relationship also exists between c-Met and PI3K second messenger signaling in HNSCC. C-Met activation has been reported in several subsets of HNSCC [211] and may also lead to PI3K-mediated resistance to TKIs inhibiting EGFR and other RTKs [243]. To address this concern, potent c-Met inhibitors (SU11274 and PF-2341066) have been developed. Pretreatment of HNSCC cell lines with these compounds in vitro prevented c-Met ligand-induced AKT activation [244, 245]. However, the degree of concordance between AKT inhibition and pharmacologic Met inhibition depends on which AKT phosphorylation site is studied. In 2007, AKT activation at mTORC2 phosphorylation site (S473) in HNSCC was reported to be associated with adverse patient outcomes, with 5% and 38% recurrence rates for low versus high phospho-AKT tumor levels, therefore emerging this biomarker as a predictor of response to therapy [144]. Further investigation demonstrated consistent AKT inhibition with c-Met inhibitors when utilizing the S473 phosphorylation site as an assay of AKT activity [244]. Meanwhile a similar investigation observed modest AKT inhibition across a panel of HNSCC cell lines while employing the PDK-1 phosphorylation site (T308) as a marker [245]. Due to the differential regulation of these sites, it is quite possible both observations are valid, and these parallel studies provide further insight into the PI3K/PTEN/AKT/mTOR signaling occurring downstream of c-Met. From a clinical perspective, the dual regulation of AKT may explain why combined treatments of EGFR and c-Met TKIs have potent, additive effects on HNSCC growth inhibition [245, 246]. c-Met activation following the addition of EGFR ligands has also been observed, suggesting crosstalk between these two receptors may have an important functional role [246]. As discussed above, while multiple pathways interact with c-Met, current evidence suggests PI3K/AKT/mTOR signaling is specifically capable of mediating pathologic signal transduction downstream of this receptor.

An additional pathway providing inhibitor resistance during HNSCC treatment is the TGF-β pathway. Aside from ligand-mediated signaling, non-canonical TGF-β activation can occur through the downstream pathways shared with EGFR and Met (MAPK, PI3K/AKT, and Rho GTPase) [21, 161, 247]. However, recent evidence suggests that TGF-β-induced changes in the tumor microenvironment can inhibit antibody-dependent cellular cytotoxicity (ADCC; an important source of in vivo activity of cetuximab) while simultaneously activating tumor-associated AKT signaling [248]. In this paradigm, TGF-β1 reduces the efficacy of immune-associated responses to cetuximab treatment while concurrently providing a proliferative signal to the tumor. In support of this hypothesis, HNSCC xenografts selected in vivo for cetuximab-resistance display increased TGF-β expression and TGF-β-dependent AKT activation [248]. This resistance is reversible with a TGF-β inhibitor, providing a strong preclinical rationale for this therapeutic option in HNSCCs refractory to cetuximab treatment. It is worth considering the genomic signature of PI3K/PTEN mutation as a potential biomarker for response to TGF-β inhibition, in future studies of this mechanism.

5.7 Conclusion

The PI3K/PTEN/AKT/mTOR pathway is one of the commonly deregulated signaling pathways in HNSCC. Although excellent preclinical and clinical studies have begun evaluating the therapeutic potential of this pathway in HNSCC, additional work is required to establish nodes of oncogenic dependency and addiction in this signaling network. While single-agent therapies targeting this pathway have not shown great efficacy, these compounds can enhance the efficacy of standard therapy options in use today, particularly in select patient populations. The role of PI3K/PTEN/AKT/mTOR pathway deregulation in HNSCC certainly warrants further investigation.

References

1. Pedrero JMG, Carracedo DG, Pinto CM, Zapatero AH, Rodrigo JP, Nieto CS, et al. Frequent genetic and biochemical alterations of the PI 3-K/AKT/PTEN pathway in head and neck squamous cell carcinoma. Int J Cancer. 2005;114(2):242–8.
2. Lui VW, Hedberg ML, Li H, Vangara BS, Pendleton K, Zeng Y, et al. Frequent mutation of the PI3K pathway in head and neck cancer defines predictive biomarkers. Cancer Discov. 2013;3(7):761–9.
3. Gao J, Aksoy BA, Dogrusoz U, Dresdner G, Gross B, Sumer SO, et al. Integrative analysis of complex cancer genomics and clinical profiles using the cBioPortal. Sci Signal. 2013;6(269):pl1.
4. Thorpe LM, Yuzugullu H, Zhao JJ. PI3K in cancer: divergent roles of isoforms, modes of activation and therapeutic targeting. Nat Rev Cancer. 2015;15(1):7–24.
5. Dhand R, Hiles I, Panayotou G, Roche S, Fry MJ, Gout I, et al. PI 3-kinase is a dual specificity enzyme: autoregulation by an intrinsic protein-serine kinase activity. EMBO J. 1994;13(3):522–33.
6. Foukas LC, Beeton CA, Jensen J, Phillips WA, Shepherd PR. Regulation of phosphoinositide 3-kinase by its intrinsic serine kinase activity in vivo. Mol Cell Biol. 2004;24(3):966–75.
7. Kang S, Denley A, Vanhaesebroeck B, Vogt PK. Oncogenic transformation induced by the p110beta, – , and – isoforms of class I phosphoinositide 3-kinase. Proc Natl Acad Sci. 2006;103(5):1289–94.
8. Skolnik EY, Margolis B, Mohammadi M, Lowenstein E, Fischer R, Drepps A, et al. Cloning of PI3 kinase-associated p85 utilizing a novel method for expression/cloning of target proteins for receptor tyrosine kinases. Cell. 1991;65(1):83–90.

9. Stephens L, Smrcka A, Cooke FT, Jackson TR, Sternweis PC, Hawkins PT. A novel phosphoinositide 3 kinase activity in myeloid-derived cells is activated by G protein beta gamma subunits. Cell. 1994;77(1):83–93.
10. Chan TO, Rodeck U, Chan AM, Kimmelman AC, Rittenhouse SE, Panayotou G, et al. Small GTPases and tyrosine kinases coregulate a molecular switch in the phosphoinositide 3-kinase regulatory subunit. Cancer Cell. 2002;1(2):181–91.
11. Rodriguez-Viciana P, Warne PH, Dhand R, Vanhaesebroeck B, Gout I, Fry MJ, et al. Phosphatidylinositol-3-OH kinase direct target of Ras. Nature. 1994;370(6490):527–32.
12. Rodriguez-Viciana P, Warne PH, Vanhaesebroeck B, Waterfield MD, Downward J. Activation of phosphoinositide 3-kinase by interaction with Ras and by point mutation. EMBO J. 1996;15(10):2442–51.
13. Vanhaesebroeck B, Guillermet-Guibert J, Graupera M, Bilanges B. The emerging mechanisms of isoform-specific PI3K signalling. Nat Rev Mol Cell Biol. 2010;11(5):329–41.
14. Franke TF, Kaplan DR, Cantley LC, Toker A. Direct regulation of the Akt proto-oncogene product by phosphatidylinositol-3,4-bisphosphate. Science. 1997;275(5300):665–8.
15. Guilherme A, Klarlund JK, Krystal G, Czech MP. Regulation of phosphatidylinositol 3,4,5-trisphosphate 5′-phosphatase activity by insulin. J Biol Chem. 1996;271(47):29533–6.
16. Hawkins PT, Stephens LR. Emerging evidence of signalling roles for PI(3,4)P2 in Class I and II PI3K-regulated pathways. Biochem Soc Trans. 2016;44(1):307–14.
17. Alessi DR, James SR, Downes CP, Holmes AB, Gaffney PRJ, Reese CB, et al. Characterization of a 3-phosphoinositide-dependent protein kinase which phosphorylates and activates protein kinase Bα. Curr Biol. 1997;7(4):261–9.
18. Alessi DR, Caudwell FB, Andjelkovic M, Hemmings BA, Cohen P. Molecular basis for the substrate specificity of protein kinase B; comparison with MAPKAP kinase-1 and p70 S6 kinase. FEBS Lett. 1996;399(3):333–8.
19. Manning BD, Toker A. AKT/PKB signaling: navigating the network. Cell. 2017;169(3):381–405.
20. Jethwa N, Chung GH, Lete MG, Alonso A, Byrne RD, Calleja V, et al. Endomembrane PtdIns(3,4,5)P3 activates the PI3K-Akt pathway. J Cell Sci. 2015;128(18):3456–65.
21. Du L, Chen X, Cao Y, Lu L, Zhang F, Bornstein S, et al. Overexpression of PIK3CA in murine head and neck epithelium drives tumor invasion and metastasis through PDK1 and enhanced TGFbeta signaling. Oncogene. 2016;35(35):4641–52.
22. Maehama T, Dixon JE. The tumor suppressor, PTEN/MMAC1, dephosphorylates the lipid second messenger, phosphatidylinositol 3,4,5-trisphosphate. J Biol Chem. 1998;273(22):13375–8.
23. Stambolic V, Suzuki A, de la Pompa JL, Brothers GM, Mirtsos C, Sasaki T, et al. Negative regulation of PKB/Akt-dependent cell survival by the tumor suppressor PTEN. Cell. 1998;95(1):29–39.
24. Samuels Y, Wang Z, Bardelli A, Silliman N, Ptak J, Szabo S, et al. High frequency of mutations of the PIK3CA gene in human cancers. Science. 2004;304(5670):554.
25. Agrawal N, Frederick MJ, Pickering CR, Bettegowda C, Chang K, Li RJ, et al. Exome sequencing of head and neck squamous cell carcinoma reveals inactivating mutations in NOTCH1. Science. 2011;333(6046):1154–7.
26. Murugan A, Hong N, Fukui Y, Munirajan A, Tsuchida N. Oncogenic mutations of the PIK3CA gene in head and neck squamous cell carcinomas. Int J Oncol. 2008;
27. Qiu W, Schonleben F, Li X, Ho DJ, Close LG, Manolidis S, et al. PIK3CA mutations in head and neck squamous cell carcinoma. Clin Cancer Res. 2006;12(5):1441–6.
28. Stransky N, Egloff AM, Tward AD, Kostic AD, Cibulskis K, Sivachenko A, et al. The mutational landscape of head and neck squamous cell carcinoma. Science. 2011;333(6046):1157–60.
29. Machiels J-P. Evaluation for the mutational landscape of head and neck squamous cell carcinoma. F1000 – Post-publication peer review of the biomedical literature: Faculty of 1000, Ltd.; 2014.

30. Kommineni N, Jamil K, Pingali UR, Addala L, M V, Naidu M. Association of PIK3CA gene mutations with head and neck squamous cell carcinomas. Neoplasma. 2015;62(01):72–80.
31. Huang CH, Mandelker D, Schmidt-Kittler O, Samuels Y, Velculescu VE, Kinzler KW, et al. The structure of a human p110 /p85 complex elucidates the effects of oncogenic PI3K mutations. Science. 2007;318(5857):1744–8.
32. Miled N, Yan Y, Hon WC, Perisic O, Zvelebil M, Inbar Y, et al. Mechanism of two classes of cancer mutations in the phosphoinositide 3-kinase catalytic subunit. Science. 2007;317(5835):239–42.
33. Zhao L, Vogt PK. Helical domain and kinase domain mutations in p110 of phosphatidylinositol 3-kinase induce gain of function by different mechanisms. Proc Natl Acad Sci. 2008;105(7):2652–7.
34. Sun M, Hillmann P, Hofmann BT, Hart JR, Vogt PK. Cancer-derived mutations in the regulatory subunit p85 of phosphoinositide 3-kinase function through the catalytic subunit p110. Proc Natl Acad Sci. 2010;107(35):15547–52.
35. Memorial Sloan Kettering Cancer Center. cBioPortal for Cancer Genomics. http://www.cbioportal.org/public-portal/. 2014.
36. Ikenoue T, Kanai F, Hikiba Y, Obata T, Tanaka Y, Imamura J, et al. Functional analysis of PIK3CA gene mutations in human colorectal cancer. Cancer Res. 2005;65(11):4562–7.
37. Gymnopoulos M, Elsliger MA, Vogt PK. Rare cancer-specific mutations in PIK3CA show gain of function. Proc Natl Acad Sci U S A. 2007;104(13):5569–74.
38. Bigner SH, Mark J, Mahaley MS, Bigner DD. Patterns of the early, gross chromosomal changes in malignant human gliomas. Hereditas. 2008;101(1):103–13.
39. Squarize CH, Castilho RM, Abrahao AC, Molinolo A, Lingen MW, Gutkind JS. PTEN deficiency contributes to the development and progression of head and neck cancer. Neoplasia. 2013;15(5):461–71.
40. Li J, Yen C, Liaw D, Podsypanina K, Bose S, Wang SI, et al. PTEN, a putative protein tyrosine phosphatase gene mutated in human brain, breast, and prostate cancer. Science. 1997;275(5308):1943–7.
41. Liaw D, Marsh DJ, Li J, Dahia PLM, Wang SI, Zheng Z, et al. Germline mutations of the PTEN gene in Cowden disease, an inherited breast and thyroid cancer syndrome. Nat Genet. 1997;16(1):64–7.
42. Steck PA, Pershouse MA, Jasser SA, Yung WKA, Lin H, Ligon AH, et al. Identification of a candidate tumour suppressor gene, MMAC1, at chromosome 10q23.3 that is mutated in multiple advanced cancers. Nat Genet. 1997;15(4):356–62.
43. Fackenthal JD. Male breast cancer in Cowden syndrome patients with germline PTEN mutations. J Med Genet. 2001;38(3):159–64.
44. Podsypanina K, Ellenson LH, Nemes A, Gu J, Tamura M, Yamada KM, et al. Mutation of Pten/Mmac1 in mice causes neoplasia in multiple organ systems. Proc Natl Acad Sci. 1999;96(4):1563–8.
45. Podsypanina K, Lee RT, Politis C, Hennessy I, Crane A, Puc J, et al. An inhibitor of mTOR reduces neoplasia and normalizes p70/S6 kinase activity in Pten+/– mice. Proc Natl Acad Sci. 2001;98(18):10320–5.
46. Di Cristofano A, Kotsi P, Peng YF, Cordon-Cardo C, Elkon KB, Pandolfi PP. Impaired Fas response and autoimmunity in Pten+/– mice. Science. 1999;285(5436):2122–5.
47. Martin-Belmonte F, Gassama A, Datta A, Yu W, Rescher U, Gerke V, et al. PTEN-mediated apical segregation of phosphoinositides controls epithelial morphogenesis through Cdc42. Cell. 2007;128(2):383–97.
48. Yang J, Weinberg RA. Epithelial-mesenchymal transition: at the crossroads of development and tumor metastasis. Dev Cell. 2008;14(6):818–29.
49. Ombrato L, Malanchi I. The EMT universe: space between cancer cell dissemination and metastasis initiation. Crit Rev Oncog. 2014;19(5):349–61.
50. Smith BN, Bhowmick NA. Role of EMT in metastasis and therapy resistance. J Clin Med. 2016;5(2)

51. Shen WH, Balajee AS, Wang J, Wu H, Eng C, Pandolfi PP, et al. Essential role for nuclear PTEN in maintaining chromosomal integrity. Cell. 2007;128(1):157–70.
52. Lindsay Y, McCoull D, Davidson L, Leslie NR, Fairservice A, Gray A, et al. Localization of agonist-sensitive PtdIns(3,4,5)P3 reveals a nuclear pool that is insensitive to PTEN expression. J Cell Sci. 2006;119(24):5160–8.
53. Leslie NR, Yang X, Downes CP, Weijer CJ. PtdIns(3,4,5)P3-dependent and -independent roles for PTEN in the control of cell migration. Curr Biol. 2007;17(2):115–25.
54. Bellacosa A, Testa JR, Staal S, Tsichlis P. A retroviral oncogene, akt, encoding a serine-threonine kinase containing an SH2-like region. Science. 1991;254(5029):274–7.
55. Coffer PJ, Woodgett JR. Molecular cloning and characterisation of a novel putative protein-serine kinase related to the cAMP-dependent and protein kinase C families. Eur J Biochem. 1991;201(2):475–81.
56. Jones PF, Jakubowicz T, Pitossi FJ, Maurer F, Hemmings BA. Molecular cloning and identification of a serine/threonine protein kinase of the second-messenger subfamily. Proc Natl Acad Sci. 1991;88(10):4171–5.
57. Staal SP, Hartley JW, Rowe WP. Isolation of transforming murine leukemia viruses from mice with a high incidence of spontaneous lymphoma. Proc Natl Acad Sci. 1977;74(7):3065–7.
58. Bellacosa A, Franke TF, Gonzalez-Portal ME, Datta K, Taguchi T, Gardner J, et al. Structure, expression and chromosomal mapping of c-akt: relationship to v-akt and its implications. Oncogene. 1993;8(3):745–54.
59. Jones PF, Jakubowicz T, Hemmings BA. Molecular cloning of a second form of rac protein kinase. Mol Biol Cell. 1991;2(12):1001–9.
60. Konishi H, Shinomura T, Kuroda S, Ono Y, Kikkawa U. Molecular cloning of rat RAC protein kinase α and β and their association with protein kinase Cζ. Biochem Biophys Res Commun. 1994;205(1):817–25.
61. Nakatani K, Sakaue H, Thompson DA, Weigel RJ, Roth RA. Identification of a human Akt3 (protein kinase B γ) which contains the regulatory serine phosphorylation site. Biochem Biophys Res Commun. 1999;257(3):906–10.
62. Greenman C, Stephens P, Smith R, Dalgliesh GL, Hunter C, Bignell G, et al. Patterns of somatic mutation in human cancer genomes. Nature. 2007;446(7132):153–8.
63. Sjoblom T, Jones S, Wood LD, Parsons DW, Lin J, Barber TD, et al. The consensus coding sequences of human breast and colorectal cancers. Science. 2006;314(5797):268–74.
64. Chen WS, Xu P-Z, Gottlob K, Chen M-L, Sokol K, Shiyanova T, et al. Growth retardation and increased apoptosis in mice with homozygous disruption of the akt1 gene. Genes Dev. 2001;15(17):2203–8.
65. Cho H, Mu J, Kim JK, Thorvaldsen JL, Chu Q, EBr C, et al. Insulin resistance and a diabetes mellitus-like syndrome in mice lacking the protein kinase Akt2 (PKBbeta). Science. 2001;292(5522):1728–31.
66. Cho H, Thorvaldsen JL, Chu Q, Feng F, Birnbaum MJ. Akt1/PKBα is required for normal growth but dispensable for maintenance of glucose homeostasis in mice. J Biol Chem. 2001;276(42):38349–52.
67. Easton RM, Cho H, Roovers K, Shineman DW, Mizrahi M, Forman MS, et al. Role for Akt3/protein kinase B in attainment of normal brain size. Mol Cell Biol. 2005;25(5):1869–78.
68. Tschopp O, Yang ZZ, Brodbeck D, Dummler BA, Hemmings-Mieszczak M, Watanabe T, et al. Essential role of protein kinase B gamma (PKB gamma/Akt3) in postnatal brain development but not in glucose homeostasis. Development. 2005;132(13):2943–54.
69. Sarbassov DD, Guertin DA, Siraj MA, Sabatini DM. Phosphorylation and regulation of Akt/PKB by the rictor-mTOR complex. Science. 2005;307(5712):1098–101.
70. Alessi DR, Andjelkovic M, Caudwell B, Cron P, Morrice N, Cohen P, et al. Mechanism of activation of protein kinase B by insulin and IGF-1. EMBO J. 1996;15(23):6541–51.
71. Mahajan K, Coppola D, Challa S, Fang B, Chen YA, Zhu W, et al. Ack1 mediated AKT/PKB tyrosine 176 phosphorylation regulates its activation. PLoS One. 2010;5(3):e9646.

72. Mahajan K, Mahajan NP. Shepherding AKT and androgen receptor by Ack1 tyrosine kinase. J Cell Physiol. 2010;224(2):327–33.
73. Chen R, Kim O, Yang J, Sato K, Eisenmann KM, McCarthy J, et al. Regulation of Akt/PKB activation by tyrosine phosphorylation. J Biol Chem. 2001;276(34):31858–62.
74. Zheng Y, Peng M, Wang Z, Asara JM, Tyner AL. Protein tyrosine kinase 6 directly phosphorylates AKT and promotes AKT activation in response to epidermal growth factor. Mol Cell Biol. 2010;30(17):4280–92.
75. Joung SM, Park ZY, Rani S, Takeuchi O, Akira S, Lee JY. Akt contributes to activation of the TRIF-dependent signaling pathways of TLRs by interacting with TANK-binding kinase 1. J Immunol. 2010;186(1):499–507.
76. Ou Y-H, Torres M, Ram R, Formstecher E, Roland C, Cheng T, et al. TBK1 directly engages Akt/PKB survival signaling to support oncogenic transformation. Mol Cell. 2011;41(4):458–70.
77. Xie X, Zhang D, Zhao B, Lu MK, You M, Condorelli G, et al. I B kinase and TANK-binding kinase 1 activate AKT by direct phosphorylation. Proc Natl Acad Sci. 2011;108(16):6474–9.
78. Mayo LD, Donner DB. A phosphatidylinositol 3-kinase/Akt pathway promotes translocation of Mdm2 from the cytoplasm to the nucleus. Proc Natl Acad Sci. 2001;98(20):11598–603.
79. Zhou BP, Liao Y, Xia W, Spohn B, Lee M-H, Hung M-C. Cytoplasmic localization of p21Cip1/WAF1 by Akt-induced phosphorylation in HER-2/neu-overexpressing cells. Nat Cell Biol. 2001;3(3):245–52.
80. Diehl JA, Cheng M, Roussel MF, Sherr CJ. Glycogen synthase kinase-3beta regulates cyclin D1 proteolysis and subcellular localization. Genes Dev. 1998;12(22):3499–511.
81. Wei W, Jin J, Schlisio S, Harper JW, Kaelin WG. The v-Jun point mutation allows c-Jun to escape GSK3-dependent recognition and destruction by the Fbw7 ubiquitin ligase. Cancer Cell. 2005;8(1):25–33.
82. Welcker M, Singer J, Loeb KR, Grim J, Bloecher A, Gurien-West M, et al. Multisite phosphorylation by Cdk2 and GSK3 controls cyclin E degradation. Mol Cell. 2003;12(2):381–92.
83. Yeh E, Cunningham M, Arnold H, Chasse D, Monteith T, Ivaldi G, et al. A signalling pathway controlling c-Myc degradation that impacts oncogenic transformation of human cells. Nat Cell Biol. 2004;6(4):308–18.
84. Brunet A, Bonni A, Zigmond MJ, Lin MZ, Juo P, Hu LS, et al. Akt promotes cell survival by phosphorylating and inhibiting a Forkhead transcription factor. Cell. 1999;96(6):857–68.
85. Huang WC, Chen CC. Akt phosphorylation of p300 at Ser-1834 is essential for its histone acetyltransferase and transcriptional activity. Mol Cell Biol. 2005;25(15):6592–602.
86. Laine J, Künstle G, Obata T, Sha M, Noguchi M. The protooncogene TCL1 is an Akt kinase coactivator. Mol Cell. 2000;6(2):395–407.
87. Pekarsky Y, Koval A, Hallas C, Bichi R, Tresini M, Malstrom S, et al. Tcl1 enhances Akt kinase activity and mediates its nuclear translocation. Proc Natl Acad Sci. 2000;97(7):3028–33.
88. Brandts CH, Sargin B, Rode M, Biermann C, Lindtner B, Schwäble J, et al. Constitutive activation of Akt by Flt3 internal tandem duplications is necessary for increased survival, proliferation, and myeloid transformation. Cancer Res. 2005;65(21):9643–50.
89. Cappellini A, Tabellini G, Zweyer M, Bortul R, Tazzari PL, Billi AM, et al. The phosphoinositide 3-kinase/Akt pathway regulates cell cycle progression of HL60 human leukemia cells through cytoplasmic relocalization of the cyclin-dependent kinase inhibitor p27Kip1 and control of cyclin D1 expression. Leukemia. 2003;17(11):2157–67.
90. Lee SH, Kim HS, Park WS, Kim SY, Lee KY, Kim SH, et al. Non-small cell lung cancers frequently express phosphorylated Akt; an immunohistochemical study. APMIS. 2002;110(7–8):587–92.
91. Nicholson KM, Streuli CH, Anderson NG. Autocrine signalling through erbB receptors promotes constitutive activation of protein kinase B/Akt in breast cancer cell lines. Breast Cancer Res Treat. 2003;81(2):117–28.

92. Vasko V, Saji M, Hardy E, Kruhlak M, Larin A, SAvchenko V, et al. Akt activation and localisation correlate with tumour invasion and oncogene expression in thyroid cancer. J Med Genet. 2004;41(3):161–70.
93. Van de Sande T, Roskams T, Lerut E, Joniau S, Van Poppel H, Verhoeven G, et al. High-level expression of fatty acid synthase in human prostate cancer tissues is linked to activation and nuclear localization of Akt/PKB. J Pathol. 2005;206(2):214–9.
94. Montironi R, Mazzuccheli R, Scarpelli M, Lopez-Beltran A, Fellegara G, Algaba F. Gleason grading of prostate cancer in needle biopsies or radical prostatectomy specimens: contemporary approach, current clinical significance and sources of pathology discrepancies. BJU Int. 2005;95(8):1146–52.
95. Trotman LC, Alimonti A, Scaglioni PP, Koutcher JA, Cordon-Cardo C, Pandolfi PP. Identification of a tumour suppressor network opposing nuclear Akt function. Nature. 2006;441(7092):523–7.
96. Lindhurst MJ, Sapp JC, Teer JK, Johnston JJ, Finn EM, Peters K, et al. A mosaic activating mutation inAKT1Associated with the Proteus syndrome. N Engl J Med. 2011;365(7):611–9.
97. Biesecker L. The challenges of Proteus syndrome: diagnosis and management. Eur J Hum Genet. 2006;14(11):1151–7.
98. Biesecker LG. The multifaceted challenges of Proteus syndrome. JAMA. 2001;285(17):2240.
99. Carpten JD, Faber AL, Horn C, Donoho GP, Briggs SL, Robbins CM, et al. A transforming mutation in the pleckstrin homology domain of AKT1 in cancer (AKT1-PH_E17K). Protein Data Bank, Rutgers University; 2007.
100. Askham JM, Platt F, Chambers PA, Snowden H, Taylor CF, Knowles MA. AKT1 mutations in bladder cancer: identification of a novel oncogenic mutation that can co-operate with E17K. Oncogene. 2009;29(1):150–5.
101. Cohen Y, Shalmon B, Korach J, Barshack I, Fridman E, Rechavi G. AKT1 pleckstrin homology domain E17K activating mutation in endometrial carcinoma. Gynecol Oncol. 2010;116(1):88–91.
102. Hara K, Maruki Y, Long X, Yoshino K-i, Oshiro N, Hidayat S, et al. Raptor, a binding partner of target of rapamycin (TOR), mediates TOR action. Cell. 2002;110(2):177–89.
103. Kim D-H, Sarbassov DD, Ali SM, King JE, Latek RR, Erdjument-Bromage H, et al. mTOR interacts with raptor to form a nutrient-sensitive complex that signals to the cell growth machinery. Cell. 2002;110(2):163–75.
104. Nojima H, Tokunaga C, Eguchi S, Oshiro N, Hidayat S, Yoshino K-i, et al. The mammalian target of rapamycin (mTOR) partner, raptor, binds the mTOR substrates p70 S6 kinase and 4E-BP1 through their TOR signaling (TOS) motif. J Biol Chem. 2003;278(18):15461–4.
105. Pearce Laura R, Huang X, Boudeau J, Pawłowski R, Wullschleger S, Deak M, et al. Identification of Protor as a novel Rictor-binding component of mTOR complex-2. Biochem J. 2007;405(3):513–22.
106. Sancak Y, Peterson TR, Shaul YD, Lindquist RA, Thoreen CC, Bar-Peled L, et al. The rag GTPases bind raptor and mediate amino acid signaling to mTORC1. Science. 2008;320(5882):1496–501.
107. Schalm SS, Fingar DC, Sabatini DM, Blenis J. TOS motif-mediated raptor binding regulates 4E-BP1 multisite phosphorylation and function. Curr Biol. 2003;13(10):797–806.
108. Wullschleger S, Loewith R, Oppliger W, Hall MN. Molecular organization of target of rapamycin complex 2. J Biol Chem. 2005;280(35):30697–704.
109. Sancak Y, Thoreen CC, Peterson TR, Lindquist RA, Kang SA, Spooner E, et al. PRAS40 is an insulin-regulated inhibitor of the mTORC1 protein kinase. Mol Cell. 2007;25(6):903–15.
110. Haar EV, Lee S-I, Bandhakavi S, Griffin TJ, Kim D-H. Insulin signalling to mTOR mediated by the Akt/PKB substrate PRAS40. Nat Cell Biol. 2007;9(3):316–23.
111. Dos DS, Ali SM, Kim D-H, Guertin DA, Latek RR, Erdjument-Bromage H, et al. Rictor, a novel binding partner of mTOR, defines a rapamycin-insensitive and raptor-independent pathway that regulates the cytoskeleton. Curr Biol. 2004;14(14):1296–302.

112. Jacinto E, Loewith R, Schmidt A, Lin S, Rüegg MA, Hall A, et al. Mammalian TOR complex 2 controls the actin cytoskeleton and is rapamycin insensitive. Nat Cell Biol. 2004;6(11):1122–8.
113. Loewith R, Jacinto E, Wullschleger S, Lorberg A, Crespo JL, Bonenfant D, et al. Two TOR complexes, only one of which is rapamycin sensitive, have distinct roles in cell growth control. Mol Cell. 2002;10(3):457–68.
114. Hara K, Yonezawa K, Kozlowski MT, Sugimoto T, Andrabi K, Weng Q-P, et al. Regulation of eIF-4E BP1 phosphorylation by mTOR. J Biol Chem. 1997;272(42):26457–63.
115. Holz MK, Ballif BA, Gygi SP, Blenis J. mTOR and S6K1 mediate assembly of the translation preinitiation complex through dynamic protein interchange and ordered phosphorylation events. Cell. 2005;123(4):569–80.
116. Ma XM, Yoon S-O, Richardson CJ, Jülich K, Blenis J. SKAR links pre-mRNA splicing to mTOR/S6K1-mediated enhanced translation efficiency of spliced mRNAs. Cell. 2008;133(2):303–13.
117. Mayer C, Zhao J, Yuan X, Grummt I. mTOR-dependent activation of the transcription factor TIF-IA links rRNA synthesis to nutrient availability. Genes Dev. 2004;18(4):423–34.
118. Noda T, Ohsumi Y. Tor, a phosphatidylinositol kinase homologue, controls autophagy in yeast. J Biol Chem. 1998;273(7):3963–6.
119. Thoreen CC, Kang SA, Chang JW, Liu Q, Zhang J, Gao Y, et al. An ATP-competitive mammalian target of rapamycin inhibitor reveals rapamycin-resistant functions of mTORC1. J Biol Chem. 2009;284(12):8023–32.
120. Kamada Y, Funakoshi T, Shintani T, Nagano K, Ohsumi M, Ohsumi Y. Tor-mediated induction of autophagy via an Apg1 protein kinase complex. J Cell Biol. 2000;150(6):1507–13.
121. Frias MA, Thoreen CC, Jaffe JD, Schroder W, Sculley T, Carr SA, et al. mSin1 is necessary for Akt/PKB phosphorylation, and its isoforms define three distinct mTORC2s. Curr Biol. 2006;16(18):1865–70.
122. Yang Q, Inoki K, Ikenoue T, Guan KL. Identification of Sin1 as an essential TORC2 component required for complex formation and kinase activity. Genes Dev. 2006;20(20):2820–32.
123. Liu P, Gan W, Chin YR, Ogura K, Guo J, Zhang J, et al. PtdIns(3,4,5)P3-dependent activation of the mTORC2 kinase complex. Cancer Discov. 2015;5(11):1194–209.
124. Ebner M, Sinkovics B, Szczygiel M, Ribeiro DW, Yudushkin I. Localization of mTORC2 activity inside cells. J Cell Biol. 2017;216(2):343–53.
125. Facchinetti V, Ouyang W, Wei H, Soto N, Lazorchak A, Gould C, et al. The mammalian target of rapamycin complex 2 controls folding and stability of Akt and protein kinase C. EMBO J. 2008;27(14):1932–43.
126. García-Martínez Juan M, Alessi Dario R. mTOR complex 2 (mTORC2) controls hydrophobic motif phosphorylation and activation of serum- and glucocorticoid-induced protein kinase 1 (SGK1). Biochem J. 2008;416(3):375–85.
127. Ikenoue T, Inoki K, Yang Q, Zhou X, Guan K-L. Essential function of TORC2 in PKC and Akt turn motif phosphorylation, maturation and signalling. EMBO J. 2008;27(14):1919–31.
128. Saxton RA, Sabatini DM. mTOR signaling in growth, metabolism, and disease. Cell. 2017;168(6):960–76.
129. van Slegtenhorst M, de Hoogt R, Hermans C, Nellist M, Janssen B, Verhoef S, et al. Identification of the tuberous sclerosis gene TSC1 on chromosome 9q34. Science. 1997;277(5327):805–8.
130. Mariño G, Salvador-Montoliu N, Fueyo A, Knecht E, Mizushima N, López-Otín C. Tissue-specific autophagy alterations and increased tumorigenesis in mice deficient in Atg4C/Autophagin-3. J Biol Chem. 2007;282(25):18573–83.
131. Qu X, Yu J, Bhagat G, Furuya N, Hibshoosh H, Troxel A, et al. Promotion of tumorigenesis by heterozygous disruption of the beclin 1 autophagy gene. J Clin Investig. 2003;112(12):1809–20.

132. Yue Z, Jin S, Yang C, Levine AJ, Heintz N. Beclin 1, an autophagy gene essential for early embryonic development, is a haploinsufficient tumor suppressor. Proc Natl Acad Sci. 2003;100(25):15077–82.
133. Guertin DA, Stevens DM, Saitoh M, Kinkel S, Crosby K, Sheen J-H, et al. mTOR complex 2 is required for the development of prostate cancer induced by Pten loss in mice. Cancer Cell. 2009;15(2):148–59.
134. Hietakangas V, Cohen SM. TOR complex 2 is needed for cell cycle progression and anchorage-independent growth of MCF7 and PC3 tumor cells. BMC Cancer. 2008;8(1)
135. Fenic I, Steger K, Gruber C, Arens C, Woenckhaus J. Analysis of PIK3CA and Akt/protein kinase B in head and neck squamous cell carcinoma. Oncol Rep. 2007;18(1):253–9.
136. Sewell A, Brown B, Biktasova A, Mills GB, Lu Y, Tyson DR, et al. Reverse-phase protein array profiling of oropharyngeal cancer and significance of PIK3CA mutations in HPV-associated head and neck cancer. Clin Cancer Res. 2014;20(9):2300–11.
137. Suda T, Hama T, Kondo S, Yuza Y, Yoshikawa M, Urashima M, et al. Copy number amplification of the PIK3CA gene is associated with poor prognosis in non-lymph node metastatic head and neck squamous cell carcinoma. BMC Cancer. 2012;12(1)
138. Lechner M, Frampton GM, Fenton T, Feber A, Palmer G, Jay A, et al. Targeted next-generation sequencing of head and neck squamous cell carcinoma identifies novel genetic alterations in HPV+ and HPV tumors. Genome Med. 2013;5(5):49.
139. Lin SC, Liu CJ, Ko SY, Chang HC, Liu TY, Chang KW. Copy number amplification of 3q26-27 oncogenes in microdissected oral squamous cell carcinoma and oral brushed samples from areca chewers. J Pathol. 2005;206(4):417–22.
140. Seiwert TY, Zuo Z, Keck MK, Khattri A, Pedamallu CS, Stricker T, et al. Integrative and comparative genomic analysis of HPV-positive and HPV-negative head and neck squamous cell carcinomas. Clin Cancer Res. 2015;21(3):632–41.
141. Chau NG, Li YY, Jo VY, Rabinowits G, Lorch JH, Tishler RB, et al. Incorporation of next-generation sequencing into routine clinical care to direct treatment of head and neck squamous cell carcinoma. Clin Cancer Res. 2016;22(12):2939–49.
142. Poetsch M, Lorenz G, Kleist B. Detection of new PTEN/MMAC1 mutations in head and neck squamous cell carcinomas with loss of chromosome 10. Cancer Genet Cytogenet. 2002;132(1):20–4.
143. Shao X, Tandon R, Samara G, Kanki H, Yano H, Close LG, et al. Mutational analysis of the PTEN gene in head and neck squamous cell carcinoma. Int J Cancer. 1998;77(5):684–8.
144. Yu Z, Weinberger PM, Sasaki C, Egleston BL, Speier WF 4th, Haffty B, et al. Phosphorylation of Akt (Ser473) predicts poor clinical outcome in oropharyngeal squamous cell cancer. Cancer Epidemiol Biomark Prev. 2007;16(3):553–8.
145. Pickering CR, Zhang J, Yoo SY, Bengtsson L, Moorthy S, Neskey DM, et al. Integrative genomic characterization of oral squamous cell carcinoma identifies frequent somatic drivers. Cancer Discov. 2013;3(7):770–81.
146. Segrelles C, Moral M, Lara MF, Ruiz S, Santos M, Leis H, et al. Molecular determinants of Akt-induced keratinocyte transformation. Oncogene. 2006;25(8):1174–85.
147. Amornphimoltham P, Sriuranpong V, Patel V, Benavides F, Conti CJ, Sauk J, et al. Persistent activation of the Akt pathway in head and neck squamous cell carcinoma: a potential target for UCN-01. Clin Cancer Res. 2004;10(12 Pt 1):4029–37.
148. García-Carracedo D, Villaronga MÁ, Álvarez-Teijeiro S, Hermida-Prado F, Santamaría I, Allonca E, et al. Impact of PI3K/AKT/mTOR pathway activation on the prognosis of patients with head and neck squamous cell carcinomas. Oncotarget. 2016;7(20):29780–93.
149. Woenckhaus J, Steger K, Werner E, Fenic I, Gamerdinger U, Dreyer T, et al. Genomic gain of PIK3CA and increased expression of p110alpha are associated with progression of dysplasia into invasive squamous cell carcinoma. J Pathol. 2002;198(3):335–42.
150. Chung CH, Guthrie VB, Masica DL, Tokheim C, Kang H, Richmon J, et al. Genomic alterations in head and neck squamous cell carcinoma determined by cancer gene-targeted sequencing. Ann Oncol. 2015;26(6):1216–23.

151. Tsui IFL, Poh CF, Garnis C, Rosin MP, Zhang L, Lam WL. Multiple pathways in the FGF signaling network are frequently deregulated by gene amplification in oral dysplasias. Int J Cancer. 2009;125(9):2219–28.
152. Xu B, Wang L, Borsu L, Ghossein R, Katabi N, Ganly I, et al. A proportion of primary squamous cell carcinomas of the parotid gland harbour high-risk human papillomavirus. Histopathology. 2016;69(6):921–9.
153. Hu YC, Lam KY, Tang JC, Srivastava G. Mutational analysis of the PTEN/MMAC1 gene in primary oesophageal squamous cell carcinomas. Mol Pathol. 1999;52(6):353–6.
154. Lee JI, Soria J-C, Hassan KA, El-Naggar AK, Tang X, Liu DD, et al. Loss of PTEN expression as a prognostic marker for tongue cancer. Arch Otolaryngol Head Neck Surg. 2001;127(12):1441.
155. Darido C, Georgy Smitha R, Wilanowski T, Dworkin S, Auden A, Zhao Q, et al. Targeting of the tumor suppressor GRHL3 by a miR-21-dependent proto-oncogenic network results in PTEN loss and tumorigenesis. Cancer Cell. 2011;20(5):635–48.
156. Georgy SR, Cangkrama M, Srivastava S, Partridge D, Auden A, Dworkin S, et al. Identification of a novel proto-oncogenic network in head and neck squamous cell carcinoma. JNCI (Journal of the National Cancer Institute). 2015;107(9)
157. Minor J, Wang X, Zhang F, Song J, Jimeno A, Wang X-J, et al. Methylation of microRNA-9 is a specific and sensitive biomarker for oral and oropharyngeal squamous cell carcinomas. Oral Oncol. 2012;48(1):73–8.
158. Bornstein S, White R, Malkoski S, Oka M, Han G, Cleaver T, et al. Smad4 loss in mice causes spontaneous head and neck cancer with increased genomic instability and inflammation. J Clin Investig. 2009;
159. Malkoski SP, Wang X-J. Two sides of the story? Smad4 loss in pancreatic cancer versus head-and-neck cancer. FEBS Lett. 2012;586(14):1984–92.
160. Bian Y, Terse A, Du J, Hall B, Molinolo A, Zhang P, et al. Progressive tumor formation in mice with conditional deletion of TGF- signaling in head and neck epithelia is associated with activation of the PI3K/Akt pathway. Cancer Res. 2009;69(14):5918–26.
161. Bian Y, Hall B, Sun ZJ, Molinolo A, Chen W, Gutkind JS, et al. Loss of TGF-β signaling and PTEN promotes head and neck squamous cell carcinoma through cellular senescence evasion and cancer-related inflammation. Oncogene. 2011;31(28):3322–32.
162. Sun ZJ, Zhang L, Hall B, Bian Y, Gutkind JS, Kulkarni AB. Chemopreventive and chemotherapeutic actions of mTOR inhibitor in genetically defined head and neck squamous cell carcinoma mouse model. Clin Cancer Res. 2012;18(19):5304–13.
163. Amornphimoltham P, Roth SJ, Ideker T, Silvio Gutkind J. Targeting the mTOR signaling circuitry in head and neck Cancer. Squamous cell carcinoma. Netherlands: Springer; 2017. p. 163–81.
164. Martinez-Cruz AB, Santos M, Lara MF, Segrelles C, Ruiz S, Moral M, et al. Spontaneous squamous cell carcinoma induced by the somatic inactivation of retinoblastoma and Trp53 tumor suppressors. Cancer Res. 2008;68(3):683–92.
165. Moral M, Segrelles C, Lara MF, Martinez-Cruz AB, Lorz C, Santos M, et al. Akt activation synergizes with Trp53 loss in oral epithelium to produce a novel mouse model for head and neck squamous cell carcinoma. Cancer Res. 2009;69(3):1099–108.
166. Vander Broek R, Snow GE, Chen Z, Van Waes C. Chemoprevention of head and neck squamous cell carcinoma through inhibition of NF-κB signaling. Oral Oncol. 2014;50(10):930–41.
167. Vineis P, Alavanja M, Buffler P, Fontham E, Franceschi S, Gao YT, et al. Tobacco and cancer: recent epidemiological evidence. JNCI (Journal of the National Cancer Institute). 2004;96(2):99–106.
168. Ang KK, Harris J, Wheeler R, Weber R, Rosenthal DI, Nguyen-Tân PF, et al. Human papillomavirus and survival of patients with oropharyngeal cancer. N Engl J Med. 2010;363(1):24–35.
169. West KA, Brognard J, Clark AS, Linnoila IR, Yang X, Swain SM, et al. Rapid Akt activation by nicotine and a tobacco carcinogen modulates the phenotype of normal human airway epithelial cells. J Clin Investig 2003;111(1):81–90.

170. Weber SM, Bornstein S, Li Y, Malkoski SP, Wang D, Rustgi AK, et al. Tobacco-specific carcinogen nitrosamine 4-(methylnitrosamino)-1-(3-pyridyl)-1-butanone induces AKT activation in head and neck epithelia. Int J Oncol. 2011;39(5):1193–8.
171. An Y, Kiang A, Lopez JP, Kuo SZ, Yu MA, Abhold EL, et al. Cigarette smoke promotes drug resistance and expansion of cancer stem cell-like side population. PLoS One. 2012;7(11):e47919.
172. Lin K, Patel SG, Chu PY, Matsuo JMS, Singh B, Wong RJ, et al. Second primary malignancy of the aerodigestive tract in patients treated for cancer of the oral cavity and larynx. Head Neck. 2005;27(12):1042–8.
173. Hsu S-H, Wong Y-K, Wang C-P, Wang C-C, Jiang R-S, Chen F-J, et al. Survival analysis of patients with oral squamous cell carcinoma with simultaneous second primary tumors. Head Neck. 2013;35(12):1801–7.
174. Benner SE, Pajak TF, Lippman SM, Earley C, Hong WK. Prevention of second primary tumors with isotretinoin in patients with squamous cell carcinoma of the head and neck: long-term follow-up. JNCI (Journal of the National Cancer Institute). 1994;86(2):140–1.
175. Hong WKI, Lippman SM, Itri LM, Karp DD, Lee JS, Byers RM, et al. Prevention of second primary tumors with isotretinoin in squamous-cell carcinoma of the head and neck. N Engl J Med. 1990;323(12):795–801.
176. Khuri FR, Lee JJ, Lippman SM, Kim ES, Cooper JS, Benner SE, et al. Randomized phase III trial of low-dose isotretinoin for prevention of second primary tumors in stage I and II head and neck Cancer patients. JNCI (Journal of the National Cancer Institute). 2006;98(7):441–50.
177. Bhatia AK, Lee JW, Pinto HA, Jacobs CD, Limburg PJ, Rubin P, et al. Double-blind, randomized phase 3 trial of low-dose 13-cis retinoic acid in the prevention of second primaries in head and neck cancer: long-term follow-up of a trial of the Eastern Cooperative Oncology Group-ACRIN Cancer Research Group (C0590). Cancer. 2017;123(23):4653–62.
178. Hildebrandt MAT, Lippman SM, Etzel CJ, Kim E, Lee JJ, Khuri FR, et al. Genetic variants in the PI3K/PTEN/AKT/mTOR pathway predict head and neck Cancer patient second primary tumor/recurrence risk and response to retinoid chemoprevention. Clin Cancer Res. 2012;18(13):3705–13.
179. Gillison ML, Chaturvedi AK, Anderson WF, Fakhry C. Epidemiology of human papillomavirus–positive head and neck squamous cell carcinoma. J Clin Oncol. 2015;33(29):3235–42.
180. Maritz GS, Mutemwa M. Tobacco smoking: patterns, health consequences for adults, and the long-term health of the offspring. Glob J Health Sci. 2012;4(4):62–75.
181. Howard JD, Chung CH. Biology of human papillomavirus–related oropharyngeal cancer. Semin Radiat Oncol. 2012;22(3):187–93.
182. Schelhaas M, Shah B, Holzer M, Blattmann P, Kühling L, Day PM, et al. Entry of human papillomavirus type 16 by actin-dependent, clathrin- and lipid raft-independent endocytosis. PLoS Pathog. 2012;8(4):e1002657.
183. Slebos RJC, Yi Y, Ely K, Carter J, Evjen A, Zhang X, et al. Gene expression differences associated with human papillomavirus status in head and neck squamous cell carcinoma. Clin Cancer Res. 2006;12(3):701–9.
184. Yarbrough WG, Whigham A, Brown B, Roach M, Slebos R. Phosphoinositide kinase-3 status associated with presence or absence of human papillomavirus in head and neck squamous cell carcinomas. Int J Radiat Oncol Biol Phys. 2007;69(2):S98–S101.
185. The Cancer Genome Atlas N. Comprehensive genomic characterization of head and neck squamous cell carcinomas. Nature. 2015;517(7536):576–82.
186. Fury MG, Drobnjak M, Sima CS, Asher M, Shah J, Lee N, et al. Tissue microarray evidence of association between p16 and phosphorylated eIF4E in tonsillar squamous cell carcinoma. Head Neck. 2010;33(9):1340–5.
187. Lewis JS, Chernock RD, Bishop JA. Squamous and neuroendocrine specific immunohistochemical markers in head and neck squamous cell carcinoma: a tissue microarray study. Head Neck Pathol. 2017;12(1):62–70.

188. Mamane Y, Petroulakis E, Martineau Y, Sato T-A, Larsson O, Rajasekhar VK, et al. Epigenetic activation of a subset of mRNAs by eIF4E explains its effects on cell proliferation. PLoS One. 2007;2(2):e242.
189. Rajasekhar VK, Viale A, Socci ND, Wiedmann M, Hu X, Holland EC. Oncogenic Ras and Akt signaling contribute to glioblastoma formation by differential recruitment of existing mRNAs to polysomes. Mol Cell. 2003;12(4):889–901.
190. Molinolo AA, Marsh C, El Dinali M, Gangane N, Jennison K, Hewitt S, et al. mTOR as a molecular target in HPV-associated oral and cervical squamous carcinomas. Clin Cancer Res. 2012;18(9):2558–68.
191. Yang SXND, Rubinstein L, Sherman ME, Swain SM, Tomaszewska JE, Doroshow JH. pAKT expression in paraffin-embedded xenograft tumors after fixation delays and human breast cancer by optimized immunohistochemistry. J Clin Oncol. 2012;30(Suppl):Abstr 10603.
192. Burris HA 3rd. Overcoming acquired resistance to anticancer therapy: focus on the PI3K/AKT/mTOR pathway. Cancer Chemother Pharmacol. 2013;71(4):829–42.
193. Vezina C, Kudelski A, Sehgal SN. Rapamycin (AY-22,989), a new antifungal antibiotic. I. Taxonomy of the producing streptomycete and isolation of the active principle. J Antibiot. 1975;28(10):721–6.
194. Heitman J, Movva N, Hall M. Targets for cell cycle arrest by the immunosuppressant rapamycin in yeast. Science. 1991;253(5022):905–9.
195. Zheng Y, Jiang Y. mTOR inhibitors at a glance. Mol Cell Pharmacol. 2015;7(2):15–20.
196. Madera D, Vitale-Cross L, Martin D, Schneider A, Molinolo AA, Gangane N, et al. Prevention of tumor growth driven by PIK3CA and HPV oncogenes by targeting mTOR signaling with metformin in oral squamous carcinomas expressing OCT3. Cancer Prev Res. 2015;8(3):197–207.
197. Rini BI. Temsirolimus, an inhibitor of mammalian target of rapamycin. Clin Cancer Res. 2008;14(5):1286–90.
198. Gabardi S, Baroletti SA. Everolimus: a proliferation signal inhibitor with clinical applications in organ transplantation, oncology, and cardiology. Pharmacotherapy. 2010;30(10):1044–56.
199. Geiger JL, Bauman JE, Gibson MK, Gooding WE, Varadarajan P, Kotsakis A, et al. Phase II trial of everolimus in patients with previously treated recurrent or metastatic head and neck squamous cell carcinoma. Head Neck. 2016;38(12):1759–64.
200. Bauman JE, Arias-Pulido H, Lee SJ, Fekrazad MH, Ozawa H, Fertig E, et al. A phase II study of temsirolimus and erlotinib in patients with recurrent and/or metastatic, platinum-refractory head and neck squamous cell carcinoma. Oral Oncol. 2013;49(5):461–7.
201. Wang Z, Martin D, Molinolo AA, Patel V, Iglesias-Bartolome R, Degese MS, et al. mTOR co-targeting in cetuximab resistance in head and neck cancers harboring PIK3CA and RAS mutations. J Natl Cancer Inst. 2014;106(9)
202. Burtness BMS, Marur S, Bauman JE, Golemis EA, Mehra R, Cohen SJ. Comment on "epidermal growth factor receptor is essential for Toll-Like receptor 3 signaling". Sci Signal. 2012;5(254):lc5
203. Dunn LA, Fury MG, Xiao H, Baxi SS, Sherman EJ, Korte S, et al. A phase II study of temsirolimus added to low-dose weekly carboplatin and paclitaxel for patients with recurrent and/or metastatic (R/M) head and neck squamous cell carcinoma (HNSCC). Ann Oncol. 2017;28(10):2533–8.
204. Piha-Paul SA, Munster PN, Hollebecque A, Argiles G, Dajani O, Cheng JD, et al. Results of a phase 1 trial combining ridaforolimus and MK-0752 in patients with advanced solid tumours. Eur J Cancer. 2015;51(14):1865–73. (1879–0852 (Electronic))
205. Dowling RJO, Zakikhani M, Fantus IG, Pollak M, Sonenberg N. Metformin inhibits mammalian target of rapamycin dependent translation initiation in breast cancer cells. Cancer Res. 2007;67(22):10804–12.
206. Gwinn DM, Shackelford DB, Egan DF, Mihaylova MM, Mery A, Vasquez DS, et al. AMPK phosphorylation of raptor mediates a metabolic checkpoint. Mol Cell. 2008;30(2):214–26.

207. Kalender A, Selvaraj A, Kim SY, Gulati P, Brûlé S, Viollet B, et al. Metformin, independent of AMPK, inhibits mTORC1 in a rag GTPase-dependent manner. Cell Metab. 2010;11(5):390–401.
208. Nair V, Sreevalsan S, Basha R, Abdelrahim M, Abudayyeh A, Rodrigues Hoffman A, et al. Mechanism of metformin-dependent inhibition of mammalian target of rapamycin (mTOR) and Ras activity in pancreatic cancer: role of specificity protein (Sp) transcription factors. J Biol Chem. 2014;289(40):27692–701. (1083-351X (Electronic))
209. Bo S, Benso A, Durazzo M, Ghigo E. Does use of metformin protect against cancer in Type 2 diabetes mellitus? J Endocrinol Investig. 2012;35(2):231–5.
210. Kong D, Yamori T, Yamazaki K, Dan S. In vitro multifaceted activities of a specific group of novel phosphatidylinositol 3-kinase inhibitors on hotspot mutant PIK3CA. Invest New Drugs. 2014;32(6):1134–43. (1573–0646 (Electronic))
211. Psyrri A, Lee JW, Pectasides E, Vassilakopoulou M, Kosmidis EK, Burtness BA, et al. Prognostic biomarkers in phase II trial of cetuximab-containing induction and chemoradiation in resectable HNSCC: eastern cooperative oncology group E2303. Clin Cancer Res. 2014;20(11):3023–32.
212. Bozec A, Ebran N, Radosevic-Robin N, Chamorey E, Yahia HB, Marcie S, et al. Combination of phosphotidylinositol-3-kinase targeting with cetuximab and irradiation: a preclinical study on an orthotopic xenograft model of head and neck cancer. Head Neck. 2017;39(1):151–9.
213. Soulieres D, Faivre S, Mesia R, Remenar E, Li SH, Karpenko A, et al. Buparlisib and paclitaxel in patients with platinum-pretreated recurrent or metastatic squamous cell carcinoma of the head and neck (BERIL-1): a randomised, double-blind, placebo-controlled phase 2 trial. Lancet Oncol. 2017;18(3):323–35.
214. Liu N, Rowley BR, Bull CO, Schneider C, Haegebarth A, Schatz CA, Fracasso PR, et al. BAY 80–6946 is a highly selective intravenous PI3K inhibitor with potent p110alpha and p110delta activities in tumor cell lines and xenograft models. Mol Cancer Ther. 2013;12(11):2319–30. (1538–8514 (Electronic))
215. Garlich JR, De P, Dey N, Su JD, Peng X, Miller A, Murali R, et al. A vascular targeted pan phosphoinositide 3-kinase inhibitor prodrug, SF1126, with antitumor and antiangiogenic activity. Cancer Res. 2008;68(1):206–15. (1538–7445 (Electronic))
216. Furet P, Guagnano V, Fairhurst RA, Imbach-Weese P, Bruce I, Knapp M, Fritsch C, et al. Discovery of NVP-BYL719 a potent and selective phosphatidylinositol-3 kinase alpha inhibitor selected for clinical evaluation. Bioorg Med Chem Lett. 2013;23(13):3741–8. (1464–3405 (Electronic))
217. Ali K, Soond DR, Pineiro R, Hagemann T, Pearce W, Lim EL, et al. Inactivation of PI(3) K p110delta breaks regulatory T-cell-mediated immune tolerance to cancer. Nature. 2014;510(7505):407–11. (1476–4687 (Electronic))
218. Hirai H, Sootome H, Nakatsuru Y, Miyama K, Taguchi S, Tsujioka K, et al. MK-2206, an allosteric Akt inhibitor, enhances antitumor efficacy by standard chemotherapeutic agents or molecular targeted drugs in vitro and in vivo. Mol Cancer Ther. 2010;9(7):1956–67.
219. Ihle NT, Williams R, Chow S, Chew W, Berggren MI, Paine-Murrieta G, Minion DJ, et al. Molecular pharmacology and antitumor activity of PX-866, a novel inhibitor of phosphoinositide-3-kinase signaling. Mol Cancer Ther. 2004;3(7):763–72. (1535–7163 (Print))
220. Maira SM, Stauffer F, Brueggen J, Furet P, Schnell C, Fritsch C, et al. Identification and characterization of NVP-BEZ235, a new orally available dual phosphatidylinositol 3-kinase/mammalian target of rapamycin inhibitor with potent in vivo antitumor activity. Mol Cancer Ther. 2008;7(7):1851–63.
221. Serra V, Markman B, Scaltriti M, Eichhorn PJA, Valero V, Guzman M, et al. NVP-BEZ235, a dual PI3K/mTOR inhibitor, prevents PI3K signaling and inhibits the growth of cancer cells with activating PI3K mutations. Cancer Res. 2008;68(19):8022–30.

222. Chung CH, Ely K, McGavran L, Varella-Garcia M, Parker J, Parker N, et al. Increased epidermal growth factor receptor gene copy number is associated with poor prognosis in head and neck squamous cell carcinomas. J Clin Oncol. 2006;24(25):4170–6.
223. Kang H, Kiess A, Chung CH. Emerging biomarkers in head and neck cancer in the era of genomics. Nat Rev Clin Oncol. 2015;12(1):11–26.
224. Freudlsperger C, Horn D, Weissfuss S, Weichert W, Weber KJ, Saure D, et al. Phosphorylation of AKT(Ser473) serves as an independent prognostic marker for radiosensitivity in advanced head and neck squamous cell carcinoma. Int J Cancer. 2015;136(12):2775–85.
225. Gupta AK, McKenna WG, Weber CN, Feldman MD, Goldsmith JD, Mick R, et al. Local recurrence in head and neck cancer: relationship to radiation resistance and signal transduction. Clin Cancer Res. 2002;8(3):885–92.
226. Nathan C-AO, Liu L, Li BD, Abreo FW, Nandy I, De Benedetti A. Detection of the proto-oncogene eIF4E in surgical margins may predict recurrence in head and neck cancer. Oncogene. 1997;15(5):579–84.
227. Alain T, Morita M, Fonseca BD, Yanagiya A, Siddiqui N, Bhat M, et al. eIF4E/4E-BP ratio predicts the efficacy of mTOR targeted therapies. Cancer Res. 2012;72(24):6468–76.
228. Nathan CAO, Amirghahari N, Abreo FW, Rong X, Caldito G, Jones ML, et al. Overexpressed eIF4E is functionally active in surgical margins of head and neck cancer patients via activation of the Akt/mammalian target of rapamycin pathway. Clin Cancer Res. 2004;10(17):5820–7.
229. Pattje WJ, Schuuring E, Mastik MF, Slagter-Menkema L, Schrijvers ML, Alessi S, et al. The phosphatase and tensin homologue deleted on chromosome 10 mediates radiosensitivity in head and neck cancer. Br J Cancer. 2010;102(12):1778–85.
230. Snietura M, Jaworska M, Mlynarczyk-Liszka J, Goraj-Zajac A, Piglowski W, Lange D, et al. PTEN as a prognostic and predictive marker in postoperative radiotherapy for squamous cell Cancer of the head and neck. PLoS One. 2012;7(3):e33396.
231. da Costa AA, D'Almeida Costa F, Ribeiro AR, Guimaraes AP, Chinen LT, Lopes CA, et al. Low PTEN expression is associated with worse overall survival in head and neck squamous cell carcinoma patients treated with chemotherapy and cetuximab. Int J Clin Oncol. 2015;20(2):282–9.
232. Di Cristofano A. Impaired Fas response and autoimmunity in Pten+/ mice. Science. 1999;285(5436):2122–5.
233. Barbareschi M, Buttitta F, Felicioni L, Cotrupi S, Barassi F, Del Grammastro M, et al. Different prognostic roles of mutations in the helical and kinase domains of the PIK3CA gene in breast carcinomas. Clin Cancer Res. 2007;13(20):6064–9.
234. Janku F, Wheler JJ, Naing A, Falchook GS, Hong DS, Stepanek VM, et al. PIK3CA mutation H1047R is associated with response to PI3K/AKT/mTOR signaling pathway inhibitors in early-phase clinical trials. Cancer Res. 2012;73(1):276–84.
235. Rao VH, Kandel A, Lynch D, Pena Z, Marwaha N, Deng C, et al. A positive feedback loop between HER2 and ADAM12 in human head and neck cancer cells increases migration and invasion. Oncogene. 2012;31(23):2888–98.
236. Asakura M, Kitakaze M, Takashima S, Liao Y, Ishikura F, Yoshinaka T, et al. Cardiac hypertrophy is inhibited by antagonism of ADAM12 processing of HB-EGF: metalloproteinase inhibitors as a new therapy. Nat Med. 2002;8(1):35–40.
237. Kang Q, Cao Y, Zolkiewska A. Direct interaction between the cytoplasmic tail of ADAM 12 and the Src homology 3 domain of p85α activates phosphatidylinositol 3-kinase in C2C12 cells. J Biol Chem. 2001;276(27):24466–72.
238. Roy R, Wewer UM, Zurakowski D, Pories SE, Moses MA. ADAM 12 cleaves extracellular matrix proteins and correlates with cancer status and stage. J Biol Chem. 2004;279(49):51323–30.
239. Erjala K, Sundvall M, Junttila TT, Zhang N, Savisalo M, Mali P, et al. Signaling via ErbB2 and ErbB3 associates with resistance and epidermal growth factor receptor (EGFR) amplification with sensitivity to EGFR inhibitor gefitinib in head and neck squamous cell carcinoma cells. Clin Cancer Res. 2006;12(13):4103–11.

240. Turner N, Grose R. Fibroblast growth factor signalling: from development to cancer. Nat Rev Cancer. 2010;10(2):116–29.
241. Nguyen PT, Tsunematsu T, Yanagisawa S, Kudo Y, Miyauchi M, Kamata N, et al. The FGFR1 inhibitor PD173074 induces mesenchymal-epithelial transition through the transcription factor AP-1. Br J Cancer. 2013;109(8):2248–58.
242. Sweeny L, Liu Z, Lancaster W, Hart J, Hartman YE, Rosenthal EL. Inhibition of fibroblasts reduced head and neck cancer growth by targeting fibroblast growth factor receptor. Laryngoscope. 2012;122(7):1539–44.
243. Argiris A, Ghebremichael M, Gilbert J, Lee JW, Sachidanandam K, Kolesar JM, et al. Phase III randomized, placebo-controlled trial of docetaxel with or without gefitinib in recurrent or metastatic head and neck cancer: an eastern cooperative oncology group trial. J Clin Oncol. 2013;31(11):1405–14.
244. Knowles LM, Stabile LP, Egloff AM, Rothstein ME, Thomas SM, Gubish CT, et al. HGF and c-Met participate in paracrine tumorigenic pathways in head and neck squamous cell cancer. Clin Cancer Res. 2009;15(11):3740–50.
245. Seiwert TY, Jagadeeswaran R, Faoro L, Janamanchi V, Nallasura V, El Dinali M, et al. The MET receptor tyrosine kinase is a potential novel therapeutic target for head and neck squamous cell carcinoma. Cancer Res. 2009;69(7):3021–31.
246. Xu H, Stabile LP, Gubish CT, Gooding WE, Grandis JR, Siegfried JM. Dual blockade of EGFR and c-Met abrogates redundant signaling and proliferation in head and neck carcinoma cells. Clin Cancer Res. 2011;17(13):4425–38.
247. Zhang YE. Non-Smad pathways in TGF-β signaling. Cell Res. 2009;19(1):128–39.
248. Bedi A, Chang X, Noonan K, Pham V, Bedi R, Fertig EJ, et al. Inhibition of TGF- enhances the in vivo antitumor efficacy of EGF receptor-targeted therapy. Mol Cancer Ther. 2012;11(11):2429–39.

Chapter 6
Jak/STAT Signaling in Head and Neck Cancer

Elizabeth Cedars, Daniel E. Johnson, and Jennifer R. Grandis

Abstract The Janus kinase/signal transducer and activator of transcription (Jak/STAT) pathway conveys cytokine receptor and receptor tyrosine kinase activation signals to the nucleus, leading to alteration of gene expression involved in normal cell functioning, including growth, differentiation, and cell survival. Constitutive Jak/STAT activation has been identified in many epithelial malignancies, including head and neck cancer (HNC). STAT3 activation in HNC promotes cell cycle progression and prevents apoptotic cell death, resulting in the survival and proliferation of cancer cells. There is also evidence that this pathway stimulates neovascularization and establishes a pro-inflammatory state. Inhibition of aberrant STAT3 activity has been shown to impede HNC growth in vitro and in vivo, suggesting that strategies to block STAT3 activity may be valuable therapeutic modalities. A thorough understanding of the impact of Jak/STAT signaling is necessary to develop safe and effective targeted therapies for head and neck cancer. This chapter will review the fundamentals of Jak and STAT structure and function and introduce inhibitors of the pathway that are currently in preclinical testing, under clinical investigation, or approved for use.

Keywords Head and neck cancer · Squamous cell carcinoma · Jak · STAT · Targeted therapies

6.1 Overview of the Jak/STAT Pathway

The Janus kinase/signal transducer and activator of transcription (Jak/STAT) pathway is a multi-protein mechanism that translates signals from extracellular molecules into changes in gene expression that alter cellular functions. First described in

---

E. Cedars · D. E. Johnson · J. R. Grandis (✉)
Department of Otolaryngology – Head and Neck Surgery, University of California, San Francisco, CA, USA
e-mail: Jennifer.Grandis@ucsf.edu

B. Burtness, E. A. Golemis (eds.), *Molecular Determinants of Head and Neck Cancer*, Current Cancer Research, https://doi.org/10.1007/978-3-319-78762-6_6

the context of interferon signaling and EGFR activation [22, 33, 34, 64], it is now well-established that the Jak/STAT pathway is activated by binding of ligands such as growth factors and cytokines to Jak-associated cell surface receptors, leading to nuclear translocation of STAT transcription factors that promote cellular proliferation and survival, differentiation, and development, as well as immune function and inflammation [14, 69, 90]. To date, four distinct mammalian Jak molecules (Jak1, Jak2, Jak3, Tyk2) and seven different STAT proteins (STAT1, STAT2, STAT3, STAT4, STAT5a, STAT5b, STAT6) have been identified.

6.2 Mechanism of Jak/STAT Pathway Activation

Jak/STAT complexes consist of a cell surface receptor constitutively bound to a Jak protein and an associated intracellular STAT molecule. Named for the Roman god Janus known for having two faces, Jaks are 110–140 kDa proteins containing two tyrosine kinase domains. The Jak-homology (JH) 1 domain, located at the C-terminus, encodes an active kinase; proximal to this is the JH2 domain, encoding a pseudokinase domain that lacks catalytic activity [39, 66]. N-terminal to the kinase regions, Jak proteins consist of five additional conserved regions (JH3–JH7), including a Src homology-2 (SH2) domain which spans the JH3/JH4 regions and a region with similarity to a FERM (four-point-one, ezrin, radixin, and moesin) domain, spanning JH4–JH7 [20, 39] (Fig. 6.1a). These N-terminal domains are critical for cytokine receptor binding and kinase regulation [39, 66].

Multiple STAT proteins have been evaluated by X-ray crystallography, which supports the presence of conserved regions among the several STAT subtypes. These homologous regions include an N-terminal domain, consisting of 130 amino acids forming eight alpha-helices, which mediates association with DNA in transcriptional complexes by stabilizing STAT dimer interactions; a coiled-coil domain involved in interactions with regulatory proteins; a DNA-binding domain that forms transcriptional complexes; a linker region, an SH2 domain through which inactive STAT proteins bind to receptors and active STAT monomers undergo dimerization; and a C-terminal transactivation domain that modulates transcription of target genes [8, 14, 79, 132, 155] (Fig. 6.1b). The STAT proteins may be bound and phosphorylated by receptor tyrosine kinases such as epidermal growth factor receptor (EGFR) and vascular endothelial growth factor receptor (VEGFR); or they can associate with non-receptor tyrosine kinases such as Jak [32]. G-protein-coupled receptors (GPCRs) have also been implicated in Jak-mediated STAT activation [156].

Two broad signaling mechanisms for Jak/STAT have been described. Canonical Jak/STAT pathway signaling is initiated with the binding of a ligand to a cognate cell surface receptor, resulting in receptor aggregation into homo- or heterodimers or oligomers [20, 74] (Fig. 6.2). These conformational changes bring the associated Jak C-terminal JH1 kinase domains into close proximity to each other, allowing for rapid transphosphorylation of Jak at several different sites within those domains [66]. Phosphorylated Jak proteins, now activated, phosphorylate tyrosine residues

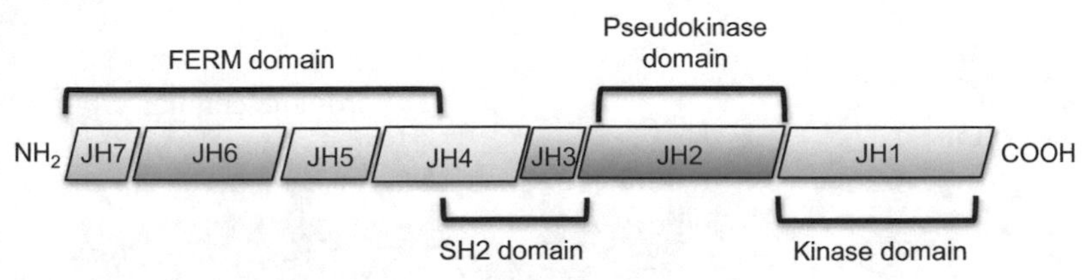

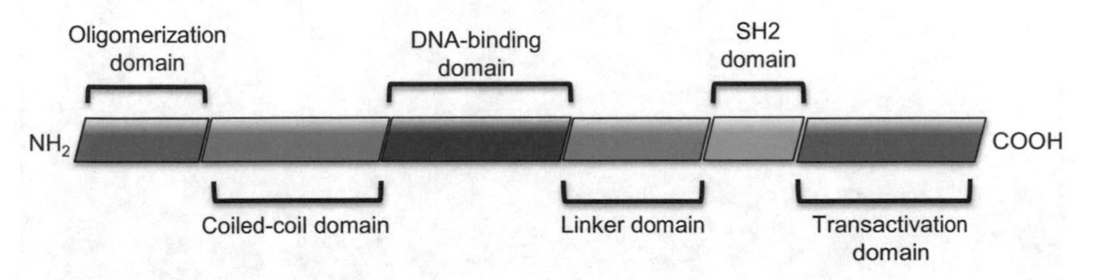

Fig. 6.1 The functional domains of the Jak and STAT proteins. (**a**) Jak consists of a kinase domain (JH1), a pseudokinase domain (JH2), as well as regions containing an SH2 domain and a FERM element (JH3–JH7). (**b**) STAT isomers all include a transactivation domain, an SH2 domain, linker domain, DNA-binding domain, coil-coil domain, and oligomerization domain

on the cytoplasmic regions of the receptor proteins. These phosphorylated tyrosine residues provide the docking site for inactive STAT proteins to bind via their SH2 domains. Once bound, the inactive STAT monomers are phosphorylated by Jak at a critical tyrosine residue adjacent to the SH2 domain. The phosphorylated STATs (pSTATs) then undergo dimerization, enabling the activated dimers to translocate to the nucleus where they induce the transcription of target genes harboring STAT-responsive elements in their promoters [22, 66, 74].

Non-canonical mechanisms of STAT signaling have also been described, initially in *Drosophila* and subsequently in mammalian cell lines. In contrast to nuclear translocation of phosphorylated/activated STAT proteins, inactive forms of STATs have also been found to be present in the nucleus [74]. Unphosphorylated STAT is associated with heterochromatin protein 1 (HP1), assisting in heterochromatin stability. Dissociation of this complex – potentially via recruitment to the cytosol through equilibration of pSTAT and unphosphorylated STAT molecules or via local activation by Jak or other kinases – leads to instability of heterochromatin and results in pSTAT-dependent transcription of regions of euchromatin [74]. Using mouse models and tumor cell lines, unphosphorylated STAT5A has been shown not only to stabilize heterochromatin but also to inhibit growth of tumors [46]. In addition to the role of unphosphorylated STAT in the nucleus, there is also evidence of cytosolic signaling by inactive STAT1 and STAT3, including binding of unphosphorylated STAT3 to NF-κB with resultant downstream activation of NF-κB target genes [150, 152].

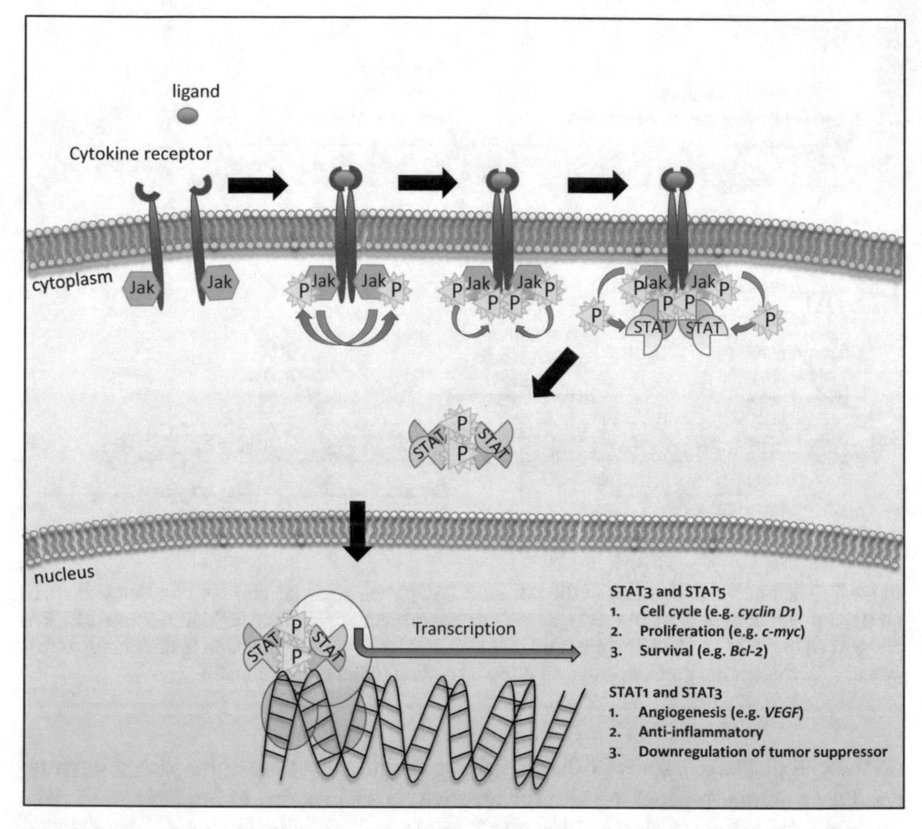

Fig. 6.2 The canonical Jak/STAT pathway. Ligand binding to a cognate cell surface receptor initiates the signaling pathway by causing receptor aggregation. This brings the constitutively bound Jak proteins into position to transphosphorylate each other via their JH1 domains, leading to activation and subsequent Jak-mediated phosphorylation of the associated receptor. These tyrosine phosphorylated receptor sites allow for docking of the SH2 domains of STAT proteins, with the docked STAT proteins subsequently undergoing phosphorylation by Jak. The phosphorylated/activated STAT proteins then dimerize in a head-to-tail fashion and translocate to the nucleus, where their DNA-binding domains allow for association with STAT response elements in the promoters of target genes, initiating transcription of the STAT-driven genes

Regulation of the Jak/STAT pathway involves not only activation by phosphorylation but also changes in STAT protein expression and modulation by additional molecules. The effects of signaling initiation, including activation-induced degradation, occur within hours [14, 74, 109]. Several STAT proteins act as inducers of their own expression, forming a positive feedback loop [74, 152]. There are also several negative regulatory mechanisms of Jak/STAT signaling, including dephosphorylation of both proteins, transportation of STAT out of the nucleus, and association with inhibitory proteins. Molecules involved in these processes include phosphatases such as SH2-containing phosphatases (SHPs) and protein tyrosine phosphatases

(PTPs), protein inhibitor of activated STAT (PIAS), and suppressor of cytokine signaling (SOCS) proteins [66, 74, 109].

6.3 Jak/STAT in Oncogenesis

The role of the Jak/STAT pathway in malignant neoplasia was elucidated initially in *Drosophila melanogaster* models of gain-of-function Jak mutations [40, 74]. In the 1980s, a mutated allele of a Jak kinase, *hop*$^{Tum\text{-}l}$, was found to lead to overactivation of the pathway and lymphoid tissue hypertrophy, leading to a leukemia-like presentation in the fruit fly. Subsequently, dysregulated Jak/STAT pathway activation has been implicated in many types of cancer [14]. Several lines of research have supported an important role for STAT signaling in tumorigenesis: tyrosine kinase pathways involved in neoplastic transformation lead to constitutive STAT signaling; dominant-negative STAT mutants prevent the downstream activation of oncogenic tyrosine kinase pathways; activated mutant STAT drives expression of genes seen in tumor transformation; increased expression of STAT is seen in cancer cell lines; and multiple STAT target genes play a role in oncogenesis by promoting cellular proliferation and survival [12]. Additional evidence supports a connection between the Jak/STAT pathway and cancer, including the important role of microRNAs in Jak-/STAT-mediated tumor progression and chemoresistance and a role for Toll-like receptor (TLR) mediation of immune suppression via STAT3 in cancers including HNC [156].

Constitutive STAT activity has been demonstrated in a wide variety of malignancies through study of human cancer cell lines and primary tumors, including hematologic malignancies such as leukemia and multiple myeloma, and solid tumors of the pancreas, breast, lung, prostate, colon, stomach, esophagus, liver, kidney, uterus, ovary, bone, brain, and head and neck [12, 14, 75, 77, 107, 108, 128, 129, 137, 138, 149, 157, 158, 160, 161]. STATs 1, 3, 5a, and 5b, in particular, have been implicated in head and neck cancer [66, 124].

Relatively few somatic, tumor-associated mutations have been found in Jak and STAT proteins. Missense mutations in Jak1 (V658F, T782M, and L783F), Jak2 (V617F), and Jak3 (V715I, A572V, V722I, and P132T) have been described primarily in myeloproliferative disorders [51, 52, 63, 73, 154]. Two additional missense mutations in Jak2 have been found in breast tumors, with one in the FERM domain and another in the JH1 kinase domain [25]. Though such activating Jak mutations have been identified in several solid tumors (breast, lung [52]), thus far – in spite of recent RNA and cDNA analyses – none have been identified in cancers of the head and neck [23]. Several activating mutations of STAT have also been identified, thus far solely in hematologic malignancies. Four mutations were identified in STAT3 in large granular lymphocytic leukemia (Y640F, D661V, D661Y, and N647I) [62], all in exon 21 coding for the SH2 domain. Additional mutations have been discovered in NKT cell lymphomas in STAT3 (S614R, G618R, and A702T) as well as STAT5B (N642H and Y665F), with these mutations also located in the SH2 region [64].

More recently, a non-SH2 mutation was identified in leukemia, H410R, located in exon 13 and corresponding to the DNA-binding domain [4]. No examples of gene duplication or deletion have so far been reported.

6.3.1 STATs in HNC

Research into the mechanisms of head and neck cancer development has shown that squamous cell carcinoma of the head and neck can arise via two distinct paths, one mediated by nonspecific mutagenic substances like tobacco and alcohol and the other by specific changes associated with human papilloma virus (HPV) infection [70]. HPV-negative and HPV-driven head and neck squamous cell carcinomas (HNSCC), though grossly similar, have different underlying etiologies and pathologic cellular mechanisms. HPV induces oncogenic changes via expression of two proteins, E6 and E7, that target tumor suppressor proteins p53 and Rb, respectively (see extended discussions in Chaps. 20 and 21). Both p53 and Rb act in part by regulating the cell cycle downstream of Jak/STAT pathway activation, though there is some evidence from the cervical cancer literature that STAT3 also may mediate E6 and E7 activity [121]. In the case of HPV-negative HNSCC, multiple STAT proteins have been implicated in alcohol- and tobacco-induced carcinogenesis [10]. As research continues, we will better understand the molecular differences underlying these distinct types of HNC and thus the necessary therapeutic differences to effectively treat both entities.

As the broader roles of STAT proteins have been elucidated, it has become possible to investigate their involvement in the origin and progression of HNSCC. In general, STAT3 and STAT5 function as potential oncogenes, regulating the expression of cell cycle genes such as *cyclin D1*, genes regulating proliferation such as *c-myc*, and genes regulating cell survival, including members of the *Bcl-2* family. STAT3 has also been demonstrated to downregulate p53, a key tumor suppressor; induce VEGF, which promotes angiogenesis; and inhibit pro-inflammatory signals mediating antitumor responses such as nitric oxide and TNF-alpha. In contrast, STAT1 acts primarily as a tumor suppressor, promoting apoptosis and countering the effects of STAT3 and STAT5 [155].

6.3.1.1 STAT1 in HNC

A pathway involving STAT1 has recently been identified though microarray analysis of the transcriptomes of oral cavity SCC as one of several mechanisms through which differentially expressed genes contribute to HNC [130]. A potential role of STAT1 in HNC was proposed even prior to this discovery, as it has been found to be present in significantly lower levels in HNSCC cells, and exogenous overexpression of STAT1 suppresses xenograft tumor growth [144]. Several distinct mechanisms have been proposed for STAT1 involvement in HNC, specifically in modulating immune pathways. IFN-γ has been shown to mediate apoptosis through STAT1 activation, triggering Fas signaling and cell cycle arrest through inducing transcription of p21 [18]. Studies of responses of HNC cells to immunotherapy have demonstrated STAT1 activation of oxidative stress-dependent pathways via accumulation of reactive oxygen species, which occurs through upregulating indoleamine 2,3-dioxygenase (IDO), suppressing heme oxygenase-1 (HO-1) [28]. Additionally, Jak2/STAT1 has been shown to mediate EGFR- and IFN-γ-induced upregulation of programmed cell death-ligand 1 (PD-L1), an inhibitor of T-cell-mediated cytotoxicity in HNSCC and other cancers. Inhibition of the Jak/STAT pathway suppressed upregulation of PD-L1 and enhanced cytotoxicity of chemotherapeutic agents [19].

The role of STAT1 in tumor suppression requires further clarification, as conflicting data exist regarding the relationship of activated STAT1 levels to cancer survival. Initial studies showed that increased levels of nuclear or phosphorylated STAT1 in HNCs correlated with improved prognosis in response to chemotherapy [67, 99]. A more recent work has shown that higher levels of S727- or Y701-phosphorylated STAT1 in HNSCC tumor specimens correlate with worse overall survival [110].

6.3.1.2 STAT3 in HNC

HNC is one of the first cancer subtypes in which STATs were shown to play a key role in human tumorigenesis. Multiple in vitro and in vivo studies have implicated STAT3 as pro-oncogenic in this disease. Increased expression and activation of STAT3 has been demonstrated in HNC cell lines and tumors. Studies of tumor cells from patients with HNC showed both constitutively active and TGF-α-induced activities of STAT3 dimers (STAT3 homodimers and STAT3/STAT1 heterodimers) [33, 34]. Moreover, increased expression and phosphorylation of STAT3 was seen not only in the tumor cells but also in nearby non-cancerous mucosal cells from HNC patients, compared to epithelium from healthy controls [35], suggesting that STAT3 activation is an early event in the development of HNSCC. Further, loss of STAT3 function impedes HNC cell proliferation. Enforced expression of dominant-negative mutant STAT3 in HNSCC cell lines has been shown to halt proliferation, trigger apoptosis, and inhibit downstream pathways of STAT activation [33–35, 57].

Several STAT3-mediated signal transduction pathways are associated with HNC tumorigenesis and growth. EGFR and its ligand TGF-α have been clearly shown to be overexpressed in HNSCC [70], and their effects are mediated by STAT3 activation, while antisense inhibition of EGFR activity in xenograft models inhibits STAT3 activation, reduces tumor cell proliferation, and triggers apoptosis [37]. Studies in HNC cell lines demonstrate that constitutively activated STAT3 leads to increased expression of cyclin D1 and Bcl-XL and results in increased proliferation and cell survival [57]. Inhibition of STAT3 in HNC and other cancers leads to reduced expression of multiple anti-apoptotic proteins and rapid induction of apoptosis [124, 155]. Relationships have also been identified between the well-known tumor suppressor p53 and STAT3 pathways, showing STAT3 inhibits p53 expression in some cancer cells [94]. A complex relationship between p53, STAT3, and NF-κB exists in HNSCC that is still being elucidated [148]. Recent evidence has also suggested that STAT3 may play a role in the production of specific microRNAs, particularly miRNA-21, a small oligonucleotide that has been shown to inhibit expression of tumor suppressor genes in many cancers, including HNSCC [11].

In addition to affecting cell cycle regulation, STAT3 has also been implicated in angiogenesis. pSTAT3 induces VEGF, a key regulator of vascular development that is integral to recruitment of vessels for tumor growth, and inhibition of STAT3 activity downregulates VEGF expression [93]. VEGF has long been known to be overexpressed in multiple cancers, with levels correlating with prognosis, and in HNSCC pSTAT3 has been identified as a direct mediator of both increased VEGF production and intratumoral microvessel density, a measure of tumor recruitment of vasculature [88].

Inflammatory pathways are also regulated, in part, by STAT3 activity, while STAT3 also is regulated by activity of inflammatory signaling cascades. For example, IL-6 is a prototypical pro-inflammatory cytokine, and IL-6/gp130 interaction leads to activation of multiple STATs including STAT3 [16, 43]. STAT3 activation in turn leads to increased IL-6 expression, and this positive feedback loop contributes to the development of an inflammatory, tumorigenic environment [16]. Importantly, microRNAs have also been found to play a significant role in tumorigenesis via the IL-6/STAT3 pathway [50]. IL-6 has been shown to induce epithelial-to-mesenchymal transition (EMT) via Jak/STAT3 in HNSCC, contributing to development of metastasis [147], and depletion of IL-6 in HNC cells leads to reduced STAT3 phosphorylation [127]. IL-6 serum levels were elevated in patients with HNSCC compared to healthy controls and correlated with advanced tumor stage and worse prognosis [27, 105]. Elevated IL-6 expression in tumor specimens is also associated with worse overall survival [31]. Thus, IL-6 autocrine/paracrine activation of Jak/STAT3 pathways may play an important role in HNC. Evidence is growing that a related cytokine, leukemia inhibitory factor (LIF), induces phosphorylation of STAT3, which may similarly play an important role in several cancers, including nasopharyngeal carcinoma (NPC) [101, 156].

Given the involvement of STAT3 activation in tumor proliferation and immune function, inhibition of mechanisms regulating STAT3 dephosphorylation may provide yet another avenue for development of HNC. STAT3 activity is negatively regulated by protein tyrosine phosphatase receptor T (PTPRT) [159]. Mutations likely to cause reduced function or loss of function of this phosphatase have been shown in analyses of whole-exome sequencing to be present in 5.6% of HNC tumors. Such mutations in the family of PTP receptors, including PTPRT, may be present in as high as 31% of HNSCC [82]. Epigenetic changes leading to reduced expression of these phosphatases, specifically hypermethylation of their promoter DNA, have also been described in many cancer types [136], including in 60% of HNSCC samples in The Cancer Genome Atlas (TCGA) [102, 103], and may also play a role in enhancing STAT3 activity in HNC. Reversal of hypermethylation in HNC cell lines was shown to reduce levels of pSTAT3 [102, 103]. Other evidence of oncogenic changes in STAT3 regulatory proteins includes the findings that *SOCS-1* is hypermethylated in as many as a third of HNSCC samples, and hypermethylation of the *SOCS-1* and *SOCS-3* genes was associated with reduced STAT3 activation; data in analysis of other cancer types has shown that SOCS-3 reduces pSTAT3 levels by occupying JAK binding sites [68, 140].

The activation of STAT3 may not only be relevant to HNC development but may provide a valuable prognostic indicator for this disease. mRNA and immunohistochemical analysis of 20 HNSCC biopsies revealed that STAT3 levels were elevated in more poorly differentiated HNSCC, while elevated STAT1 levels were found in well-differentiated tumors, suggesting that STAT3/STAT1 ratios may be indicative of tumor differentiation status [6]. Further, elevated levels of pSTAT3 correlated with nodal metastasis and higher clinical stage, as well as reduced rates of disease-specific survival in oral tongue SCC biopsies [89]. This relationship to overall survival has been replicated in additional studies of oral SCC [118].

The impact of impaired STAT3 function has been highlighted via investigation of several STAT subtypes that exist as the result of alternative splicing of mRNA or protein cleavage. In most cases, these variants result simply in a truncated protein, although the STAT3β isoform is a protein in which the C-terminal transactivation domain is replaced by a novel series of seven residues (CT7 domain) [49]. In cells transfected with STAT3β, STAT3-mediated gene expression was dramatically impaired, as measured by a luciferase reporter construct. Further, in esophageal cancer samples, high STAT3β expression correlated with longer overall survival and recurrence-free survival, supporting the hypothesis that STAT3β acts to suppress the oncogenic activity of STAT3α in spite of increased levels of pSTAT3α [157, 158, 160, 161]. Additional heterodimers composed of alternative STAT3 isoforms have shown similar inhibitory results. STAT3D, an isoform generated through experimentally induced E434A and E435A mutations, reduced pSTAT3 target gene expression and arrested cell growth, likely mediated by impaired DNA binding, while another mutant isoform, STAT3F, may work by competing for binding to kinases [89, 92]. Similarly, HNC cells transfected with a dominant-negative carboxy-truncated STAT5b (STAT5b**Δ**754) impaired cellular proliferation [72].

6.3.1.3 STAT5 in HNSCC

STAT5 activation can also contribute to the development of HNSCC, as indicated by in vitro, animal model, and clinical studies. Greater expression and phosphorylation of STAT5 was found in head and neck tumors compared to normal epithelial tissue, and tumor growth was observed in a xenograft model with constitutive STAT5b activation, suggesting a role in tumorigenesis [146]. Using cell culture and xenograft models, Koppikar et al. demonstrated a role for STAT5 in enhancing tumor growth, invasion, and EMT, as well as development of cisplatin and erlotinib resistance [61]. More recent work delineating the IL-8/miR-424-5p/STAT5 pathway has revealed a role for STAT5 activation in cell migration and invasion in oral SCC [100]. The pathway involves IL-8-induced activation of STAT5, which subsequently induces expression of SOCS-2, an inhibitor of STAT5, as well as of miR-424-5p. Increased expression of miR-424-5p was found to suppress SOCS-2 activity and led to constitutive STAT5 expression, correlating with increased tumor cell migration and invasion in oral squamous cell cancer cells.

6.4 Therapies Targeting Jak/STAT Pathways

Research into the key molecular mechanisms underlying oncogenesis allows for the development of targeted therapeutic agents that can overcome the fundamental underpinnings of human malignancies. In HNC, investigation into molecules that influence the Jak/STAT pathway is underway, with several promising leads (Fig. 6.3).

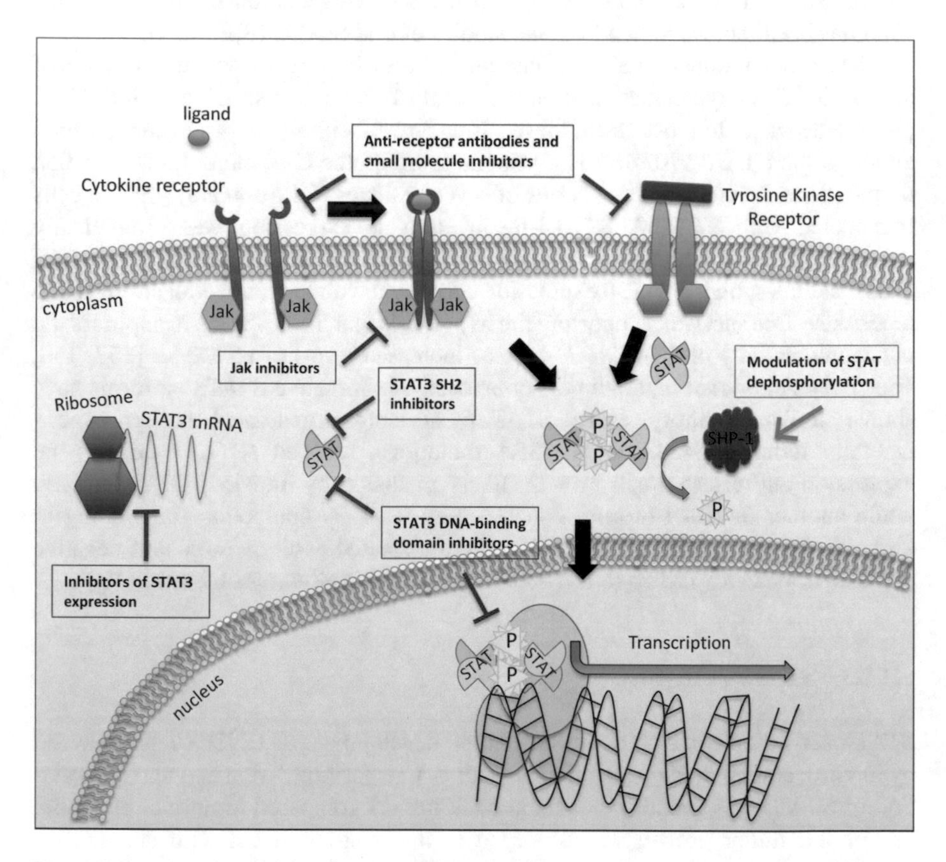

Fig. 6.3 Therapeutic mechanisms of Jak/STAT pathway disruption. Cytokine or growth factor binding to cognate receptors may be blocked by monoclonal antibodies or small molecule inhibitors. Jak activity has been targeted by small molecule inhibitors and plant derivatives. STAT3 receptor association and dimerization can be prevented with plant derivatives, small molecule inhibitors, aptamers, peptide mimics, and dominant-negative isoforms that act on the SH2 domain. DNA binding by STAT3 is prevented by platinum-based compounds, oligonucleotide decoys, and G-quartets. Destruction of STAT3 mRNA via targeting by antisense nucleotides or siRNA reduces STAT3 expression. Modulation of STAT3 phosphorylation status by induction of phosphatases is another means by which STAT3 activity is therapeutically reduced

6.4.1 *Targeting Regulators of Jak/STAT Activation*

6.4.1.1 EGFR Pathway

EGFR signaling is one of several mechanisms by which the STAT pathway is activated, and multiple studies have confirmed its early role in HNC, as described above [36, 124]. It is therefore a plausible therapeutic target. Cetuximab, an IgG1 monoclonal antibody to EGFR, was approved in 2006 by the FDA for treatment of locoregional HNC in combination with radiotherapy. It acts by binding and obstructing the ligand-binding region of EGFR with high affinity, acting to prevent ligand interaction and sterically inhibiting dimerization and autophosphorylation [76]. As a result, downstream signaling mechanisms are blocked. This and other anti-EGFR therapies are covered in detail in Chap. 2.

While anti-EGFR antibodies have been found to be effective for HNC, small molecule inhibitors have been less successful. The limited response of HNC to EGFR tyrosine kinase inhibitors (TKIs) indicates that additional, downstream pathway activation can develop, as well as upregulation of alternative receptor tyrosine kinases [13]. Constitutive STAT3 activation has been shown to contribute to therapeutic resistance to EGFR TKIs; for example, activation of the IL-6/STAT3 pathway decreased tumor response to erlotinib in non-small cell lung cancer [153]. Inhibition of STAT3 has been demonstrated to inhibit tumor growth in TKI- and cetuximab-resistant HNC xenograft models, suggesting it is a viable method to augment responses to these therapies [112]. STAT3 inhibitors are discussed in more detail below. Inhibitors of other downstream EGFR pathways, including Src kinase inhibitors or mTOR kinase inhibitors, are under investigation, as well as inhibitors of alternative receptor tyrosine kinases, including MET and HER2 [13, 164].

6.4.1.2 IL-6 Pathway

IL-6/Jak1 or Jak2 activation of STAT has been observed in multiple cancers including HNC, where IL-6 levels have been shown to be a marker of oral SCC recurrence and poor prognosis [27, 43]. Anti-IL-6 therapies have been developed, including a humanized monoclonal antibody, tocilizumab, which competes with target receptors for binding and prevents activation of IL-6-regulated pathways. In studies of renal cell carcinoma, tocilizumab increased phosphorylation of STAT1 in response to interferon and suppressed STAT3 phosphorylation in cell lines; the effect of tocilizumab with interferon in xenografts was tumor suppression [96]. Tocilizumab is FDA-approved for use in rheumatoid arthritis and juvenile idiopathic arthritis [32], but given the identification of an IL-6-mediated pathway in the development of several cancer types, clinical trials are currently focusing on evaluating efficacy in other malignancies such as ovarian and pancreatic cancer. CNTO328 (siltuximab), another monoclonal antibody to IL-6, has also been studied in several malignancies and appears promising. Early studies in lung cancer cells suggested that inhibition of Jak1 with CNTO328 reduced tyrosine phosphorylation of STAT3 and, when combined with inhibition of STAT3 serine phosphorylation with erlotinib (an EGFR inhibitor), impaired the growth of lung cancers in mice [125]. In prostate cancer, CNTO328 was found to have some biologic activity in both in vitro and in vivo studies, but little clinically relevant activity [21, 26]; it was found to stabilize disease in over 50% of patients with metastatic renal cell carcinoma [106]. Potentially beneficial results were seen in a phase 1 trial when combined with standard therapy for previously untreated multiple myeloma [117]. Further studies are needed to evaluate the potential utility of CNTO328 in HNC.

A plant steroid, guggulsterone (GS), has been demonstrated to inhibit IL-6-mediated activation of STAT3 in HNSCC and to block STAT3 activation of VEGF expression [85]. The activity of GS is due, in part, to the induction of tyrosine phosphatase SHP-1, which may act to dephosphorylate Jak2 and c-Src kinases, both activators of STAT3, in addition to direct dephosphorylation of STAT3 itself [3]. All data thus far are preclinical.

6.4.2 *Targeting Jak*

Given the handful of constitutively activating Jak mutations identified in myeloproliferative neoplasms, as well as less commonly in breast and lung tumors, Jaks represent another possible target for therapy. At this point in time, there are two FDA-approved Jak inhibitors, both used in autoimmune disorders and one also approved for use in myelofibrosis. There is active research ongoing into additional disease states that might benefit from these medications, as well as into other Jak inhibitors.

6.4.2.1 Small Molecule Jak Inhibitors

Ruxolitinib (also known as Jakafi® and INCB018424) is an orally bioavailable small molecule Jak1/Jak2 inhibitor that was approved by the FDA in 2011 for the treatment of intermediate- and high-risk myelofibrosis [41]. Though not yet in clinical use in squamous cell carcinoma, recent data suggests that SCC invasion and proliferation induced by loss of CADM1, a cell adhesion molecule that prevents STAT3 activity, can be halted by the use of ruxolitinib. This suggests that testing for loss of CADM1, and possibly STAT3 activation, may identify candidates for ruxolitinib inhibitory therapy [135].

Due to the success of ruxolitinib in myelofibrosis, other Jak inhibitors have been developed, including a Jak3 inhibitor, tofacitinib, which is currently used for inflammatory bowel disease and psoriasis [32], and a novel Jak1/Jak2 inhibitor, AZD1480. AZD1480 inhibited xenograft solid tumor growth mediated by constitutive STAT3 in an HNC model as well as in other cancers [42, 114]. In HNC cell lines, AZD1480 treatment reduced cell proliferation and pSTAT3 expression [114].

Two other Jak inhibitors, WP1066 and fedratinib, initially appeared promising in preclinical testing. However, investigation of these inhibitors has been discontinued due to concerns about lack of efficacy and neurotoxicity, respectively [32]. In contrast, clinical investigation is ongoing involving ruxolitinib for use in HNSCC (NCT03153982) in addition to multiple other tumor types; and another Jak inhibitor, pacritinib, is being trialed in several hematologic malignancies as well as in non-small cell lung cancer (NCT02342353). At an earlier stage of development, AG490 is a preclinical tool Jak2 inhibitor with evidence of benefit in multiple cancer types, including in vitro efficacy against laryngeal carcinoma cells via Jak2-mediated STAT3 phosphorylation [162] and antitumor effects in a murine ovarian cancer model [60].

6.4.2.2 Plant-Based Therapies

In addition to chemically derived therapeutic molecules, there are naturally occurring plant-based agents, including GS, that have shown evidence of inhibitory activity on the Jak/STAT pathway. Quercetin is found in many fruits, vegetables, and grains and is often used as a supplement in food products. Like other members of the flavonoid family, quercetin has been noted to have antioxidant, anticancer, and anti-inflammatory properties [32]. Further study has revealed a mechanism mediated by the Jak/STAT pathway [32]. In cervical cancer cells, administration of quercetin reduced Jak2 expression, inhibited proliferation, reduced migration, and stimulated apoptosis, likely via suppression of a Jak2/STAT5 pathway, leading to reduced xenograft tumor growth [83].

Curcumin is a plant derivative that has been used in the culinary world for centuries in the form of the spice turmeric, and research has demonstrated anti-inflammatory and antineoplastic effects [32]. Studies of microglial cells showed that administration of curcumin impaired phosphorylation of Jak1 and Jak2, as well as STAT1 and STAT3 in response to LPS and IFN-γ, likely mediated by SHP-2 phosphatase action on Jak1/Jak2 [58]. Other activities of curcumin include inhibition of NF-κB via reduction of cytokine production and release, including that of IL-6 [32]. A recent review concluded that HNSCC cell lines treated with curcumin experienced cell cycle arrest in the G2/M phase and were induced to undergo apoptosis and that some animal models experienced reduction in tumor growth, along with increased expression of pro-apoptotic proteins and reduction of cyclin D1 expression [9]. Current clinical trials of curcumin include treatment of non-small cell lung cancer (NCT02321293) and prostate cancer (NCT02064673).

Some concern has been raised that curcumin has low human bioavailability [119], so investigation of chemically derived analogues has been pursued. One analogue, FLLL12, was shown to be at least tenfold more potent than curcumin in inhibiting growth of HNC cell lines and induced apoptosis through upregulation of pro-apoptotic Bcl-2 proteins while also inhibiting the growth of xenograft tumors [5]. Both this compound and its relative FLLL11 have been shown to inhibit STAT3 phosphorylation at Tyr705 in breast and prostate cancer cells, as well as impede its DNA-binding and transcriptional activity [80]. An additional curcumin analogue, FLLL32, targets the SH2 domain of STAT3 and is described below.

6.4.3 Targeting STAT

Due to the prominent role that STATs play in HNSCC, multiple therapeutic approaches have been and continue to be pursued to target these proteins. As data suggest that among the STATs, STAT3 is the major mediator of oncogenesis, proliferation, and metastasis, this protein has been the major focus of interest thus far, though therapeutic targeting of STAT1 and STAT5 has also been investigated. The STAT protein inhibitors that have been developed to date target the STAT proteins and their activation at multiple points: (1) receptor binding and phosphorylation, (2) DNA binding and transcription, and (3) protein expression [32, 86].

6.4.3.1 Targeting the SH2 Domain

The SH2 domain in STAT proteins is the principal mediator of STAT/receptor interactions as well as STAT protein dimerization. Due to the conserved nature of the SH2 domain across multiple STAT subtypes and its fundamental role in STAT activation, targeting of the SH2 domain has attracted considerable attention as a means of inhibiting STAT activity in cancer cells.

Plant-Based Therapies

The curcumin analogue FLLL32 interacts with the SH2 domain to impede STAT3 phosphorylation at Tyr705 and resultant activation. Treatment of cisplatin-sensitive HNSCC cells with FLLL32 enhanced apoptosis in dual therapy studies and markedly sensitized cisplatin-resistant HNSCC cells to treatment with this conventional chemotherapy drug [1]. Another natural compound, cryptotanshinone, a Chinese herbal medicine, has been shown to suppress STAT3 signaling, likely through SH2 domain targeting [32]. It has been studied in multiple cancer types though not in HNC; its use in glioma cells inhibited STAT3 phosphorylation at Tyr705, reduced nuclear translocation, and downregulated expression of cyclin D1 and survivin [81]. Cryptotanshinone has not been carefully evaluated in clinical studies.

Small Molecule Inhibitors

Non-peptide small molecule inhibitors targeting the STAT3 SH2 domain have also been developed. Among these, Stattic (STAT three inhibitory compound), identified via screening of chemical libraries, was the first small molecule inhibitor [91].

Stattic has been shown to decrease tumor growth and to radiosensitize tumors in a murine xenograft model of HNC [2]. Administration of Stattic also prevented development of oral SCC in mice exposed to the chemical carcinogen 4-NQO [102, 103].

Phase 1 studies have been performed with another small molecule inhibitor, OPB-51602, a low-molecular-weight molecule that binds to the SH2 domain and inhibits phosphorylation at both Tyr705 and a second phosphorylation site Ser727 [143]. Maximum tolerated dose (MTD), pharmacokinetics, tolerability, and antitumor effects were evaluated, with 5% of patients (2/37) with non-small cell lung cancer showing a partial response [143], though with poor tolerance for the continuous dosing administered. Additional phase 1 investigation for hematologic malignancies found similar MTD and toxicities, though with no clinical benefit for the population studied [95]. Further studies evaluating different dosing strategies may be useful. A similar compound, OPB-31121, was investigated in patients with advanced solid tumors and yielded partial response (tumor shrinkage) in 2 of 25 patients, with stable disease in 8 [97], though results in hepatocellular carcinoma and hematologic malignancies were less beneficial [32, 98]. Continued investigations into its optimal dosing and into cancer types yielding the best response are needed.

Other SH2-interacting, small molecule inhibitors have shown promise in some cancer cell line models. Database screening utilizing computational docking and estimation of binding affinity helped to identify a novel small molecule inhibitor of STAT3, STA-21, which in breast carcinoma cells inhibited dimerization and DNA binding and inhibited the survival of cells with constitutive STAT3 activity [126]. S3I-201 is another chemical compound, identified by computational modeling as binding to the SH2 domain of STAT3. In human breast carcinoma cells and transformed mouse fibroblasts, S3I-201 inhibited cell growth and induced apoptosis [123]. Further study in prostate cancer cell lines showed that S3I-201, similar to the Jak2 inhibitor AG490, inhibited IL-6-/Jak2-mediated STAT3 activation and decreased expression of STAT3 as well as upregulated expression of pro-apoptotic proteins [38].

Repurposing of small molecules currently used as medications for other purposes may provide another additional STAT inhibitor for evaluation in HNC. In models of polycystic kidney disease, a condition also found to entail elevated STAT3 activation [142], the anti-helminthic medication pyrimethamine was found to inhibit STAT3 phosphorylation [131]. High-throughput screening of compounds identified pyrimethamine as a potentially therapeutic medication for acute myeloid leukemia, showing selective and potent action in inducing apoptosis and differentiation of both murine and human AML cells [120]. A phase 1 trial of pyrimethamine in patients with chronic lymphocytic leukemia/small lymphocytic lymphoma is currently recruiting (NCT01066663).

Mimics as Inhibitors

Peptidomimetic inhibitors are stabilized derivatives of peptide inhibitory molecules and are commonly based on peptide sequences present in the target protein or an interacting partner of the target protein. ISS-610 and S31-M2001 were developed from short amino acid sequences near the SH2 domain of STAT3 and inhibit SH2-mediated protein interactions [32]. S31-M2001 has shown evidence of tumor growth inhibition and reduction of STAT3-mediated cell survival and migration [122]. Another compound, STX-0119, similarly derived from virtual screens of a target peptide sequence in the SH2 domain, downregulated multiple STAT3 target genes and suppressed tumor growth in mouse models of glioma [7]. Prodrugs of phosphopeptide mimetics have also been investigated, with design of a modified p-Tyr mimic showing increased binding affinity to the SH2 domain, as well as inhibition of STAT3 phosphorylation and pSTAT3 nuclear translocation in several cancer cell lines [87]. Further data suggests that such drugs may also lead to reduced angiogenesis [32]. While multiple peptidomimetic compounds have been identified and there is some evidence of efficacy in tumor cells, thus far, none have moved beyond preclinical testing [91].

6.4.3.2 Targeting the DNA-Binding Domains

Platinum Compounds

Platinum-based compounds, of which cisplatin is the classic example, act both by DNA alkylation and by direct effects on proteins. Several novel platinum-containing compounds preferentially bind the N-terminal domain of STAT proteins [133]. CPA-7, one of the most potent of these, inhibited STAT3 DNA binding in vitro and caused tumor regression in a colon cancer mouse model [133]. Studies in prostate cancer cell lines and xenografts showed not only reduction of activated STAT3 but also increased T-cell responses with administration of CPA-7 [78], suggesting enhanced immune-mediated anticancer activity. Another platinum compound, IS3 295, has been evaluated in cell lines, and administration led to induction of apoptosis and concomitant downregulation of the STAT3 target genes *cyclin D1* and $Bcl\text{-}X_L$ in transformed mouse fibroblasts and in human breast cancer cells [134].

Designer Molecules

Inhibitors directed against the DNA-binding domain of STAT3 have been rationally designed to suppress STAT3-mediated gene transcription. A synthetic analogue of the STAT3 helix 2 region present in the N-terminal domain, STAT3-Hel2A-2, was derived and found to bind to STAT3 but not STAT1 in HEK293 cells as detected by FRET analysis and induced apoptosis in breast cancer cells [132]. Another small molecule, inS3-54, was identified via virtual screening as targeting the STAT3 DNA-binding region and similarly appears to impair proliferation in breast and lung cancer cell lines but not fibroblast or epithelial cells [47].

Oligonucleotide Inhibitors

Short strands of nucleotides have been utilized to target STAT cellular activity. Leong and colleagues generated a 15-base pair double-stranded oligonucleotide mimicking the STAT3 response element found in the promoter of STAT3 target genes [71]. This STAT3 decoy oligonucleotide competitively inhibited the binding of pSTAT to promoter elements, suppressing the expression of STAT3 target genes. Other oligonucleotides have been developed into complex structures that physically impede STAT3 binding [53].

Early studies of the STAT3 decoy oligonucleotide showed selectivity for STAT3 in vitro and inhibited the survival and proliferation of HNC cells [71] and led to regression of tumors in HNC murine xenograft models [145]. As toxicity was shown to be minimal in primates [116], the decoy was next studied in a phase 0 clinical trial in HNC patients. Immediately prior to surgical resection, patients received either a single intratumoral injection of the decoy or saline control vehicle. Injection of the STAT3 decoy was found to significantly reduce the expression of the STAT3 target genes cyclin D1 and Bcl-X_L relative to saline control, and no adverse effects were seen [115]. Systemic delivery of a modified, cyclic form of the STAT3 decoy inhibited HNC murine xenograft tumor growth and STAT3 target gene expression without effecting STAT3 phosphorylation or total levels [115]. Dosing and safety of this modified decoy were then investigated following systemic administration in a mouse model, and no toxicities were noted after intravenous administration at two different doses, despite a significant reduction in tumor volume with either dose [113]. A modification of this approach, wherein decoy was coupled to the ligand for TLR9, induced cancer regression in mouse models of acute myeloid leukemia [157, 158, 160, 161].

In an additional approach, oligonucleotides have been used to form tertiary structures that impair the DNA binding of STAT3. These guanine-rich RNA or DNA molecules, or "G-quartets," consist of a 4-nucleotide cyclical formation with each nucleotide forming two hydrogen bonds, one to each neighbor. The G-quartets can undergo further higher-order folding to form complex structures that occupy the DNA-binding site of proteins such as the HIV-1 integrase enzyme. T40214 is a G-quartet oligonucleotide that was found to bind primarily to the STAT3 SH2 domain. In cell lines, T40214 inhibits the transcription of the STAT3-regulated genes *Bcl-X*$_L$ and *Mcl-1* [53]. Further, HNSCC mouse xenograft models showed reduced tumor growth with the use of T40214 and enhanced reduction in tumor size when treated with T40214 + paclitaxel [54]. These results were supported by similar inhibition of T40214 on STAT3-regulated gene expression. Similar results were seen in NSCLC xenograft mouse models, where T40214 reduced STAT3 target gene expression, cell survival, angiogenesis, and proliferation [141], as well as in a T-cell leukemia model where delivery of T40214 to transgenic mice impaired STAT3-DNA complex formation and reduced tumor burden [44]. Current data support the use of G-quartet oligonucleotides in further studies focusing on optimization of selectivity and bioavailability.

6.4.3.3 Inhibition of STAT Expression

While blocking STAT phosphorylation or DNA binding can be an effective therapeutic strategy, reducing production of the STAT protein via silencing represents an alternative approach. Antisense oligonucleotides and small interfering RNA (siRNA) strategies aimed at achieving highly specific targeting and destruction of STAT mRNA transcripts have been described in preclinical studies in cell lines and xenografts of multiple cancer types [17, 29, 48, 104, 157, 158, 60, 161]. Antisense RNA and siRNA methods differ both in delivery to the target cell and by their mechanism of inhibition.

Antisense RNA

Antisense RNA is introduced in the form of a stable single strand of RNA, generally a sequence of 18–21 nucleotides, which then binds complementary mRNA sequences either in the cytoplasm or nucleus. Silencing ensues through one of several mechanisms, including steric inhibition of translation via binding to the 5′ untranslated region of the mRNA, targeting of the double-stranded region for degradation by endogenous nucleases such as RNase H, or induction of alternative splicing pathways that lead to desired isoforms. All of these mechanisms result in decreased expression of the aberrant protein [24, 139]. Some of the limitations to clinical application of antisense RNA have been substantially overcome through the introduction of chemical modifications that improve the pharmacologic properties of the antisense molecule. For example, the use of phosphorothioate – a replacement of a non-bridging oxygen in a phosphate group with a sulfur atom – was determined to significantly improve stability of antisense oligonucleotides [139], and virtually all oligonucleotides in clinical use now incorporate phosphorothioate residues [15]. Recently, a phosphorothioate STAT3 oligonucleotide was introduced into human endothelial cell models and found to inhibit cell proliferation while inducing apoptosis and impeding angiogenesis [59]. Although STAT3 antisense therapies are still in the early phases of clinical trials [45], multiple antisense therapeutics directed toward other gene targets involved in cancer have undergone phase 1–3 trials [15, 24, 139].

siRNA

Small interfering RNAs (siRNAs) are delivered in double-stranded form and utilize existing nuclear proteins to bind to their target sequences of mRNA. siRNAs consist of 19–22 base pair sequences, which associate with the RNA-induced silencing complex (RISC), releasing the "passenger" RNA strand and incorporating the functional "guide" strand into the complex, bound to the complementary target site [139]. The RISC mechanism functions with endogenous small RNAs as well. Bound mRNA is either enzymatically cleaved by the associated RISC ribonuclease, or expression is simply repressed through its incorporation with the complex [24, 139].

The use of siRNAs has been shown to be a formidable method for reducing STAT3 expression in murine models of several malignancies including laryngeal, breast, pancreatic, and colorectal cancers [30, 48, 65, 104]. The approach has also been effective in targeting STAT5: the use of siRNA in esophageal cancer cell lines reduced the expression of Bcl-2, cyclin D1, and STAT5 and impaired proliferation [151]. In a hepatocellular carcinoma xenograft model, treatment with siRNA to STAT5 led to growth inhibition and apoptosis [163].

Toxicities known to be associated with oligonucleotide treatments include coagulopathy, hepatic injury, splenomegaly, and lymphoid hyperplasia, mediated by either direct action of the oligonucleotide binding (less common) or by off-target, non-antisense effects triggering inflammatory responses [24, 111]. However, studies have shown good tolerability of oligonucleotide therapies in multiple animal models including primates [116] and have been developed into promising therapies. Still, although the stability and efficacy of these modalities have been improved significantly, limitations in drug delivery have been a barrier to clinical application of these approaches to reducing STAT expression [55]. Several barriers have been addressed by means such as chemical modifications and antibody targeting to cell-specific receptors [84], as well as via nanoparticle delivery [56], but some still remain, including the inefficiency of endocytosis, permeability of tumor vasculature, and binding to extracellular matrix proteins [15, 139].

6.5 Conclusions

Evidence from cell lines and from human tumors demonstrates the relevance of the Jak/STAT pathway, and particularly of STAT3, in the pathogenesis of HNC. As a result, STAT3 and other members of this signaling pathway have been identified as plausible therapeutic targets in the treatment of HNC. Targets include cytokine and growth factor receptors upstream of the Jak/STAT pathway, modulators of Jak activation and direct Jak inhibitors, STAT3 inhibitors targeting multiple domains, and molecules that prevent STAT expression. While many compounds have been identified with inhibitory effects on STAT3, few have advanced to clinical testing, and even fewer are approved for use.

Several upstream inhibitors of STAT activity are promising therapeutic agents, including an FDA-approved anti-EGFR therapy for HNC, cetuximab. Additional anti-EGFR antibodies and small molecule inhibitors are under investigation, as well as anti-IL-6 antibodies. Stimulating modulators of phosphorylation such as phosphatases provide another potential approach to regulating STAT3 activation. Multiple small molecule inhibitors of Jaks are being investigated in clinical trials, with two currently FDA-approved for autoimmune disease, though none as yet has a role in the treatment of HNC. Several plant-derived Jak inhibitors are also being evaluated as potential therapies. Multiple agents directed at the STAT3 SH2 domain have been developed to interfere with receptor association and dimerization, including plant-based molecules, small molecule inhibitors, peptide mimics, and dominant-negative STAT isoforms. Platinum-based compounds, designer molecules, oligonucleotide decoys, and G-quartets have been developed to target STAT3 DNA-binding activity. Lastly, antisense RNA and siRNA have both been investigated to decrease STAT3 expression.

Of the current investigatory molecules, thus far EGFR small molecule inhibitors, anti-IL-6 antibodies, curcumin inhibition of Jak, small molecule inhibitors of the STAT3 SH2 domain, and STAT3 antisense therapies have progressed to clinical trials, though not all in HNC patients. However, these trials provide the groundwork for potential investigation for HNC treatment, as well. Limitations to clinical application of these therapies include lack of target selectivity and efficacy. Additionally, as tumor heterogeneity is common in HNC, more than one oncogenic pathway may be implicated in each patient. As personalized medicine advances, improved patient selection may assist with identifying potential roles for each of these therapeutics, alone or in combination with conventional chemoradiation. Overall, clinical application of Jak/STAT pathway inhibitors will benefit from continued detection and elucidation of Jak/STAT pathway dysregulation in HNC tumors, leading to more effective targeted approaches.

References

1. Abuzeid WM, Davis S, Tang AL, Saunders L, Brenner JC, Lin J, Fuchs JR, Light E, Bradford CR, Prince ME, Carey TE. Sensitization of head and neck cancer to cisplatin through the use of a novel curcumin analog. Arch Otolaryngol Head Neck Surg. 2011;137(5):499–507.
2. Adachi M, Cui C, Dodge CT, Bhayani MK, Lai SY. Targeting STAT3 inhibits growth and enhances radiosensitivity in head and neck squamous cell carcinoma. Oral Oncol. 2012;48(12):1220–6.
3. Ahn KS, Sethi G, Sung B, Goel A, Ralhan R, Aggarwal BB. Guggulsterone, a farnesoid X receptor antagonist, inhibits constitutive and inducible STAT3 activation through induction of a protein tyrosine phosphatase SHP-1. Cancer Res. 2008;68(11):4406–15.
4. Andersson E, Kuusanmäki H, Bortoluzzi S, Lagström S, Parsons A, Rajala H, van Adrichem A, Eldfors S, Olson T, Clemente MJ, Laasonen A, Ellonen P, Heckman C, Loughran TP, Maciejewski JP, Mustjoki S. Activating somatic mutations outside the SH2-domain of STAT3 in LGL leukemia. Leukemia. 2016;30(5):1204–8.
5. Anisuzzaman AS, Haque A, Rahman MA, Wang D, Fuchs JR, Hurwitz S, Liu Y, Sica G, Khuri FR, Chen ZG, Shin DM, Amin AR. Preclinical in vitro, in vivo, and pharmacokinetic evaluations of FLLL12 for the prevention and treatment of head and neck cancers. Cancer Prev Res (Phila). 2016;9(1):63–73.
6. Arany I, Chen SH, Megyesi JK, Adler-Storthz K, Chen Z, Rajaraman S, Ember IA, Tyring SK, Brysk MM. Differentiation-dependent expression of signal transducers and activators of transcription (STATs) might modify responses to growth factors in the cancers of the head and neck. Cancer Lett. 2003;199(1):83–9.
7. Ashizawa T, Akiyama Y, Miyata H, Iizuka A, Komiyama M, Kume A, Omiya M, Sugino T, Asai A, Hayashi N, Mitsuya K, Nakasu Y, Yamaguchi K. Effect of the STAT3 inhibitor STX-0119 on the proliferation of a temozolomide-resistant glioblastoma cell line. Int J Oncol. 2014;45(1):411–8.

8. Becker S, Groner B, Muller CW. Three-dimensional structure of the Stat3beta homodimer bound to DNA. Nature. 1998;394(6689):145–51.
9. Borges GÁ, Rêgo DF, Assad DX, Coletta RD, De Luca Canto G, Guerra EN. In vivo and in vitro effects of curcumin on head and neck carcinoma: a systematic review. J Oral Pathol Med. 2016. https://doi.org/10.1111/jop.12455.
10. Bose P, Brockton NT, Dort JC. Head and neck cancer: from anatomy to biology. Int J Cancer. 2013;133(9):2013–23.
11. Bourguignon LY, Earle C, Wong G, Spevak CC, Krueger K. Stem cell marker (Nanog) and Stat-3 signaling promote MicroRNA-21 expression and chemoresistance in hyaluronan/CD44-activated head and neck squamous cell carcinoma cells. Oncogene. 2012;31(2):149–60.
12. Bowman T, Garcia R, Turkson J, Jove R. STATs in oncogenesis. Oncogene. 2000;19(21):2474–88. Review
13. Burtness B, Bauman JE, Galloway T. Novel targets in HPV-negative head and neck cancer: overcoming resistance to EGFR inhibition. Lancet Oncol. 2013;14(8):e302–9.
14. Calo V, Migliavacca M, Bazan V, Macaluso M, Buscemi M, Gebbia N, Russo A. STAT proteins: from normal control of cellular events to tumorigenesis. J Cell Physiol. 2003;197(2):157–68.
15. Castanotto D, Stein CA. Antisense oligonucleotides in cancer. Curr Opin Oncol. 2014;26(6):584–9.
16. Chang Q, Daly L, Bromberg J. The IL-6 feed-forward loop: a driver of tumorigenesis. Semin Immunol. 2014;26(1):48–53.
17. Chang L, Gong F, Cai H, Li Z, Cui Y. Combined RNAi targeting human Stat3 and ADAM9 as gene therapy for non-small cell lung cancer. Oncol Lett. 2016;11(2):1242–50.
18. Chin YE, Kitagawa M, Su WC, You ZH, Iwamoto Y, Fu XY. Cell growth arrest and induction of cyclin-dependent kinase inhibitor p21 WAF1/CIP1 mediated by STAT1. Science. 1996;272(5262):719–22.
19. Concha-Benavente F, Srivastava RM, Trivedi S, Lei Y, Chandran U, Seethala RR, Freeman GJ, Ferris RL. Identification of the cell-intrinsic and -extrinsic pathways downstream of EGFR and IFNγ that induce PD-L1 expression in head and neck cancer. Cancer Res. 2016;76(5):1031–43.
20. Constantinescu SN, Girardot M, Pecquet C. Mining for JAK-STAT mutations in cancer. Trends Biochem Sci. 2008;33(3):122–31.
21. Culig Z, Puhr M. Interleukin-6 and prostate cancer: current developments and unsolved questions. Mol Cell Endocrinol. 2018;462(Pt A):25–30.
22. Darnell JE Jr, Kerr IM, Stark GR. Jak-STAT pathways and transcriptional activation in response to IFNs and other extracellular signaling proteins. Science. 1994;264(5164):1415–21.
23. De Carvalho TG, De Carvalho AC, Maia DC, Ogawa JK, Carvalho AL, Vettore AL. Search for mutations in signaling pathways in head and neck squamous cell carcinoma. Oncol Rep. 2013;30(1):334–40.
24. Dean NM, Bennett CF. Antisense oligonucleotide-based therapeutics for cancer. Oncogene. 2003;22(56):9087–96.
25. Ding L, Ellis MJ, Li S, Larson DE, Chen K, et al. Genome remodelling in a basal-like breast cancer metastasis and xenograft. Nature. 2010;464(7291):999–1005.
26. Dorff TB, Goldman B, Pinski JK, Mack PC, Lara PN Jr, Van Veldhuizen PJ Jr, Quinn DI, Vogelzang NJ, Thompson IM Jr, Hussain MH. Clinical and correlative results of SWOG S0354: a phase II trial of CNTO328 (siltuximab), a monoclonal antibody against interleukin-6, in chemotherapy-pretreated patients with castration-resistant prostate cancer. Clin Cancer Res. 2010;16(11):3028–34.
27. Duffy SA, Taylor JM, Terrell JE, Islam M, Li Y, Fowler KE, Wolf GT, Teknos TN. Interleukin-6 predicts recurrence and survival among head and neck cancer patients. Cancer. 2008;113(4):750–7.

28. El Jamal SM, Taylor EB, Abd Elmageed ZY, Alamodi AA, Selimovic D, Alkhateeb A, Hannig M, Hassan SY, Santourlidis S, Friedlander PL, Haikel Y, Vijaykumar S, Kandil E, Hassan M. Interferon gamma-induced apoptosis of head and neck squamous cell carcinoma is connected to indoleamine-2,3-dioxygenase via mitochondrial and ER stress-associated pathways. Cell Div. 2016;11:11.
29. Gao X, Sun J, Huang C, Hu X, Jiang N, Lu C. RNAi-mediated silencing of NOX4 inhibited the invasion of gastric cancer cells through JAK2/STAT3 signaling. Am J Transl Res. 2017;9(10):4440–9.
30. Gao LF, Wen LJ, Yu H, Zhang L, Meng Y, Shao YT, Xu DQ, Zhao XJ. Knockdown of Stat3 expression using RNAi inhibits growth of laryngeal tumors in vivo. Acta Pharmacol Sin. 2006;27(3):347–52.
31. Gao J, Zhao S, Halstensen TS. Increased interleukin-6 expression is associated with poor prognosis and acquired cisplatin resistance in head and neck squamous cell carcinoma. Oncol Rep. 2016;35(6):3265–74.
32. Geiger JL, Grandis JR, Bauman JE. The STAT3 pathway as a therapeutic target in head and neck cancer: barriers and innovations. Oral Oncol. 2016;56:84–92.
33. Grandis JR, Chakraborty A, Zeng Q, Melhem MF, Tweardy DJ. Downmodulation of TGF-alpha protein expression with antisense oligonucleotides inhibits proliferation of head and neck squamous carcinoma but not normal mucosal epithelial cells. J Cell Biochem. 1998a;69(1):55–62.
34. Grandis JR, Drenning SD, Chakraborty A, Zhou MY, Zeng Q, Pitt AS, Tweardy DJ. Requirement of Stat3 but not Stat1 activation for epidermal growth factor receptor- mediated cell growth in vitro. J Clin Invest. 1998b;102(7):1385–92.
35. Grandis JR, Drenning SD, Zeng Q, Watkins SC, Melhem MF, Endo S, Johnson DE, Huang L, He Y, Kim JD. Constitutive activation of Stat3 signaling abrogates apoptosis in squamous cell carcinogenesis in vivo. Proc Natl Acad Sci U S A. 2000a;97(8):4227–32.
36. Grandis JR, Tweardy DJ. Elevated levels of transforming growth factor alpha and epidermal growth factor receptor messenger RNA are early markers of carcinogenesis in head and neck cancer. Cancer Res. 1993;53(15):3579–84.
37. Grandis JR, Zeng Q, Drenning SD. Epidermal growth factor receptor-mediated stat3 signaling blocks apoptosis in head and neck cancer. Laryngoscope. 2000b;110(5 Pt 1):868–74.
38. Gurbuz V, Konac E, Varol N, Yilmaz A, Gurocak S, Menevse S, Sozen S. Effects of AG490 and S3I-201 on regulation of the JAK/STAT3 signaling pathway in relation to angiogenesis in TRAIL-resistant prostate cancer cells in vitro. Oncol Lett. 2014;7(3):755–63.
39. Haan C, Kreis S, Margue C, Behrmann I. Jaks and cytokine receptors – an intimate relationship. Biochem Pharmacol. 2006;72:1538–46.
40. Harrison DA, Binari R, Nahreini TS, Gilman M, Perrimon N. Activation of a Drosophila Janus kinase (JAK) causes hematopoietic neoplasia and developmental defects. EMBO J. 1995;14(12):2857–65.
41. Harrison C, Kiladjian JJ, Al-Ali HK, Gisslinger H, Waltzman R, Stalbovskaya V, McQuitty M, Hunter DS, Levy R, Knoops L, Cervantes F, Vannucchi AM, Barbui T, Barosi G. JAK Inhibition with Ruxolitinib versus Best Available Therapy for Myelofibrosis. New England Journal of Medicine. 2012;366(9):787–98.
42. Hedvat M, Huszar D, Herrmann A, Gozgit JM, Schroeder A, Sheehy A, Buettner R, Proia D, Kowolik CM, Xin H, Armstrong B, Bebernitz G, Weng S, Wang L, Ye M, McEachern K, Chen H, Morosini D, Bell K, Alimzhanov M, Ioannidis S, McCoon P, Cao ZA, Yu H, Jove R, Zinda M. The JAK2 inhibitor AZD1480 potently blocks Stat3 signaling and oncogenesis in solid tumors. Cancer Cell. 2009;16(6):487–97.
43. Heinrich PC, Behrmann I, Haan S, Hermanns HM, Müller-Newen G, Schaper F. Principles of interleukin (IL)-6-type cytokine signalling and its regulation. Biochem J. 2003;374(Pt 1):1–20.

44. Hillion J, Belton AM, Shah SN, Turkson J, Jing N, Tweardy DJ, di Cello F, Huso DL, Resar LM. Nanoparticle delivery of inhibitory signal transducer and activator of transcription 3 G-quartet oligonucleotides blocks tumor growth in HMGA1 transgenic model of T-cell leukemia. Leuk Lymphoma. 2014;55(5):1194–7.
45. Hong D, Kurzrock R, Kim Y, Woessner R, Younes A, Nemunaitis J, Fowler N, Zhou T, Schmidt J, Jo M, Lee SJ, Yamashita M, Hughes SG, Fayad L, Piha-Paul S, Nadella MV, Mohseni M, Lawson D, Reimer C, Blakey DC, Xiao X, Hsu J, Revenko A, Monia BP, MacLeod AR. AZD9150, a next-generation antisense oligonucleotide inhibitor of STAT3 with early evidence of clinical activity in lymphoma and lung cancer. Sci Transl Med. 2015;7(314):314ra185.
46. Hu X, Dutta P, Tsurumi A, Li J, Wang J, Land H, Li WX. Unphosphorylated STAT5A stabilizes heterochromatin and suppresses tumor growth. Proc Natl Acad Sci U S A. 2013;110(25):10213–8.
47. Huang W, Dong Z, Wang F, Peng H, Liu JY, Zhang JT. A small molecule compound targeting STAT3 DNA-binding domain inhibits cancer cell proliferation, migration, and invasion. ACS Chem Biol. 2014;9(5):1188–96.
48. Huang C, Jiang T, Zhu L, Liu J, Cao J, Huang KJ, Qiu ZJ. STAT3-targeting RNA interference inhibits pancreatic cancer angiogenesis in vitro and in vivo. Int J Oncol. 2011;38(6):1637–44.
49. Huang Y, Qiu J, Dong S, Redell MS, Poli V, Mancini MA, Tweardy DJ. Stat3 isoforms, alpha and beta, demonstrate distinct intracellular dynamics with prolonged nuclear retention of Stat3beta mapping to its unique C-terminal end. J Biol Chem. 2007;282(48):34958–67.
50. Iliopoulos D, Hirsch HA, Struhl K. An epigenetic switch involving NF-kappaB, Lin28, Let-7 MicroRNA, and IL6 links inflammation to cell transformation. Cell. 2009;139(4):693–706.
51. James C, Ugo V, Le Couedic JP, Staerk J, Delhommeau F, Lacout C, Garcon L, Raslova H, Berger R, Bennaceur-Griscelli A, Villeval JL, Constantinescu SN, Casadevall N, Vainchenker W. A unique clonal JAK2 mutation leading to constitutive signalling causes polycythaemia vera. Nature. 2005;434(7037):1144–8.
55. Jeong EG, Kim MS, Nam HK, Min CK, Lee S, Chung YJ, Yoo NJ, Lee SH. Somatic mutations of JAK1 and JAK3 in acute leukemias and solid cancers. Clin Cancer Res. 2008;14(12):3716–21.
53 Jing N, Li Y, Xu X, Sha W, Li P, Feng L, Tweardy DJ. Targeting Stat3 with G-quartet oligodeoxynucleotides in human cancer cells. DNA Cell Biol. 2003;22(11):685–96.
54. Jing N, Zhu Q, Yuan P, Li Y, Mao L, Tweardy DJ. Targeting signal transducer and activator of transcription 3 with G-quartet oligonucleotides: a potential novel therapy for head and neck cancer. Mol Cancer Ther. 2006;5(2):279–86.
55. Juliano R, Bauman J, Kang H, Ming X. Biological barriers to therapy with antisense and siRNA oligonucleotides. Mol Pharm. 2009;6(3):686–95.
56. Kamrani Moghaddam L, Ramezani Paschepari S, Zaimy MA, Abdalaian A, Jebali A. The inhibition of epidermal growth factor receptor signaling by hexagonal selenium nanoparticles modified by SiRNA. Cancer Gene Ther. 2016;23(9):321–5.
57. Kijima T, Niwa H, Steinman RA, Drenning SD, Gooding WE, Wentzel AL, Xi S, Grandis JR. STAT3 activation abrogates growth factor dependence and contributes to head and neck squamous cell carcinoma tumor growth in vivo. Cell Growth Differ. 2002;13(8):355–62.
58. Kim HY, Park EJ, Joe EH, Jou I. Curcumin suppresses Janus kinase-STAT inflammatory signaling through activation of Src homology 2 domain-containing tyrosine phosphatase 2 in brain microglia. J Immunol. 2003;171(11):6072–9.
59. Klein JD, Sano D, Sen M, Myers JN, Grandis JR, Kim S. STAT3 oligonucleotide inhibits tumor angiogenesis in preclinical models of squamous cell carcinoma. PLoS One. 2014;9(1):e81819.
60. Kobayashi A, Tanizaki Y, Kimura A, Ishida Y, Nosaka M, Toujima S, Kuninaka Y, Minami S, Ino K, Kondo T. AG490, a Jak2 inhibitor, suppressed the progression of murine ovarian cancer. European Journal of Pharmacology. 2015;766:63–75.

61. Koppikar P, Lui VW, Man D, Xi S, Chai RL, Nelson E, Tobey AB, Grandis JR. Constitutive activation of signal transducer and activator of transcription 5 contributes to tumor growth, epithelial-mesenchymal transition, and resistance to epidermal growth factor receptor targeting. Clin Cancer Res. 2008;14(23):7682–90.
62. Koskela HL, Eldfors S, Ellonen P, van Adrichem AJ, Kuusanmäki H, Andersson EI, Lagström S, Clemente MJ, Olson T, Jalkanen SE, Majumder MM, Almusa H, Edgren H, Lepistö M, Mattila P, Guinta K, Koistinen P, Kuittinen T, Penttinen K, Parsons A, Knowles J, Saarela J, Wennerberg K, Kallioniemi O, Porkka K, Loughran TP Jr, Heckman CA, Maciejewski JP, Mustjoki S. Somatic STAT3 mutations in large granular lymphocytic leukemia. N Engl J Med. 2012;366(20):1905–13.
63. Kralovics R, Passamonti F, Buser AS, Teo SS, Tiedt R, Passweg JR, Tichelli A, Cazzola M, Skoda RC. A gain-of-function mutation of JAK2 in myeloproliferative disorders. N Engl J Med. 2005;352(17):1779–90.
64. Küçük C, Jiang B, Hu X, Zhang W, Chan JK, Xiao W, Lack N, Alkan C, Williams JC, Avery KN, Kavak P, Scuto A, Sen E, Gaulard P, Staudt L, Iqbal J, Zhang W, Cornish A, Gong Q, Yang Q, Sun H, d'Amore F, Leppä S, Liu W, Fu K, de Leval L, McKeithan T, Chan WC. Activating mutations of STAT5B and STAT3 in lymphomas derived from γδ-T or NK cells. Nat Commun. 2015;6:6025.
65. Kunigal S, Lakka SS, Sodadasu PK, Estes N, Rao JS. Stat3-siRNA induces Fas-mediated apoptosis in vitro and in vivo in breast cancer. Int J Oncol. 2009;34(5):1209–20.
66. Lai SY, Johnson FM. Defining the role of the JAK-STAT pathway in head and neck and thoracic malignancies: implications for future therapeutic approaches. Drug Resist Updat. 2010;13(3):67–78.
67. Laimer K, Spizzo G, Obrist P, Gastl G, Brunhuber T, Schäfer G, Norer B, Rasse M, Haffner MC, Doppler W. STAT1 activation in squamous cell cancer of the oral cavity: a potential predictive marker of response to adjuvant chemotherapy. Cancer. 2007;110(2):326–33.
68. Lee TL, Yeh J, Van Waes C, Chen Z. Epigenetic modification of SOCS-1 differentially regulates STAT3 activation in response to interleukin-6 receptor and epidermal growth factor receptor signaling through JAK and/or MEK in head and neck squamous cell carcinomas. Mol Cancer Ther. 2006;5(1):8–19.
69. Leeman RJ, Lui VW, Grandis JR. STAT3 as a therapeutic target in head and neck cancer. Expert Opin Biol Ther. 2006;6(3):231–41.
70. Leemans CR, Braakhuis BJ, Brakenhoff RH. The molecular biology of head and neck cancer. Nat Rev Cancer. 2011;11(1):9–22.
71. Leong PL, Andrews GA, Johnson DE, Dyer KF, Xi S, Mai JC, Robbins PD, Gadiparthi S, Burke NA, Watkins SF, Grandis JR. Targeted inhibition of Stat3 with a decoy oligonucleotide abrogates head and neck cancer cell growth. Proc Natl Acad Sci U S A. 2003;100(7):4138–43.
72. Leong PL, Xi S, Drenning SD, Dyer KF, Wentzel AL, Lerner EC, Smithgall TE, Grandis JR. Differential function of STAT5 isoforms in head and neck cancer growth control. Oncogene. 2002;21(18):2846–53.
73. Levine RL, Wadleigh M, Cools J, Ebert BL, Wernig G, Huntly BJ, Boggon TJ, Wlodarska I, Clark JJ, Moore S, Adelsperger J, Koo S, Lee JC, Gabriel S, Mercher T, D'Andrea A, Frohling S, Dohner K, Marynen P, Vandenberghe P, Mesa RA, Tefferi A, Griffin JD, Eck MJ, Sellers WR, Meyerson M, Golub TR, Lee SJ, Gilliland DG. Activating mutation in the tyrosine kinase JAK2 in polycythemia vera, essential thrombocythemia, and myeloid metaplasia with myelofibrosis. Cancer Cell. 2005;7(4):387–97.
74. Li WX. Canonical and non-canonical JAK-STAT signaling. Trends Cell Biol. 2008;18(11):545–51.
75. Li S, Priceman SJ, Xin H, Zhang W, Deng J, Liu Y, Huang J, Zhu W, Chen M, Hu W, Deng X, Zhang J, Yu H, He G. Icaritin inhibits JAK/STAT3 signaling and growth of renal cell carcinoma. PLoS One. 2013;8(12):e81657.
76. Li S, Schmitz KR, Jeffrey PD, Wiltzius JJ, Kussie P, Ferguson KM. Structural basis for inhibition of the epidermal growth factor receptor by cetuximab. Cancer Cell. 2005;7(4):301–11.

77. Liang Z, Gao LH, Cao LJ, Feng DY, Cao Y, Luo QZ, Yu P, Li M. Detection of STAT2 in early stage of cervical premalignancy and in cervical cancer. Asian Pac J Trop Med. 2012;5(9):738–42.
78. Liang M, Zhan F, Zhao J, Li Q, Wuyang J, Mu G, Li D, Zhang Y, Huang X. CPA-7 influences immune profile and elicits anti-prostate cancer effects by inhibiting activated STAT3. BMC Cancer. 2016;16:504.
79. Lim CP, Cao X. Structure, function, and regulation of STAT proteins. Mol BioSyst. 2006;2(11):536–50.
80. Lin L, Hutzen B, Ball S, Foust E, Sobo M, Deangelis S, Pandit B, Friedman L, Li C, Li PK, Fuchs J, Lin J. New curcumin analogues exhibit enhanced growth-suppressive activity and inhibit AKT and signal transducer and activator of transcription 3 phosphorylation in breast and prostate cancer cells. Cancer Sci. 2009;100(9):1719–27.
81. Lu L, Li C, Li D, Wang Y, Zhou C, Shao W, Peng J, You Y, Zhang X, Shen X. Cryptotanshinone inhibits human glioma cell proliferation by suppressing STAT3 signaling. Mol Cell Biochem. 2013;381(1–2):273–82.
82. Lui VW, Peyser ND, Ng PK, Hritz J, Zeng Y, Lu Y, Li H, Wang L, Gilbert BR, General IJ, Bahar I, Ju Z, Wang Z, Pendleton KP, Xiao X, Du Y, Vries JK, Hammerman PS, Garraway LA, Mills GB, Johnson DE, Grandis JR. Frequent mutation of receptor protein tyrosine phosphatases provides a mechanism for STAT3 hyperactivation in head and neck cancer. Proc Natl Acad Sci U S A. 2014;111(3):1114–9.
83. Luo CL, Liu YQ, Wang P, Song CH, Wang KJ, Dai LP, Zhang JY, Ye H. The effect of quercetin nanoparticle on cervical cancer progression by inducing apoptosis, autophagy and anti-proliferation via JAK2 suppression. Biomed Pharmacother. 2016;82:595–605.
84. Ma Y, Kowolik CM, Swiderski PM, Kortylewski M, Yu H, Horne DA, Jove R, Caballero OL, Simpson AJ, Lee FT, Pillay V, Scott AM. Humanized Lewis-Y specific antibody based delivery of STAT3 siRNA. ACS Chem Biol. 2011;6(9):962–70.
85. Macha MA, Matta A, Chauhan SS, Siu KW, Ralhan R. Guggulsterone (GS) inhibits smokeless tobacco and nicotine-induced NF-κB and STAT3 pathways in head and neck cancer cells. Carcinogenesis. 2011;32(3):368–80.
86. Mali SB. Review of STAT3 (Signal Transducers and Activators of Transcription) in head and neck cancer. Oral Oncol. 2015;51(6):565–9.
87. Mandal PK, Gao F, Lu Z, Ren Z, Ramesh R, Birtwistle JS, Kaluarachchi KK, Chen X, Bast RC Jr, Liao WS, McMurray JS. Potent and selective phosphopeptide mimetic prodrugs targeted to the Src homology 2 (SH2) domain of signal transducer and activator of transcription 3. J Med Chem. 2011;54(10):3549–63.
88. Masuda M, Ruan HY, Ito A, Nakashima T, Toh S, Wakasaki T, Yasumatsu R, Kutratomi Y, Komune S, Weinstein IB. Signal transducers and activators of transcription 3 up-regulates vascular endothelial growth factor production and tumor angiogenesis in head and neck squamous cell carcinoma. Oral Oncol. 2007;43(8):785–90.
89. Masuda M, Suzui M, Yasumatu R, Nakashima T, Kuratomi Y, Azuma K, Tomita K, Komiyama S, Weinstein IB. Constitutive activation of signal transducers and activators of transcription 3 correlates with cyclin D1 overexpression and may provide a novel prognostic marker in head and neck squamous cell carcinoma. Cancer Res. 2002;62(12):3351–5.
90. Mertens C, Darnell JE Jr. SnapShot: JAK-STAT signaling. Cell. 2007;131(3):612.
91. Munoz J, Dhillon N, Janku F, Watowich SS, Hong DS. STAT3 inhibitors: finding a home in lymphoma and leukemia. Oncologist. 2014;19(5):536–44. Published online 2014 Apr 4
92. Nakajima K, Yamanaka Y, Nakae K, Kojima H, Ichiba M, Kiuchi N, Kitaoka T, Fukada T, Hibi M, Hirano T. A central role for Stat3 in IL-6-induced regulation of growth and differentiation in M1 leukemia cells. EMBO J. 1996;15(14):3651–8.
93. Niu G, Wright KL, Huang M, Song L, Haura E, Turkson J, Zhang S, Wang T, Sinibaldi D, Coppola D, Heller R, Ellis LM, Karras J, Bromberg J, Pardoll D, Jove R, Yu H. Constitutive Stat3 activity up-regulates VEGF expression and tumor angiogenesis. Oncogene. 2002;21:2000–8.

94. Niu G, Wright KL, Ma Y, Wright GM, Huang M, Irby R, Briggs J, Karras J, Cress WD, Pardoll D, Jove R, Chen J, Yu H. Role of Stat3 in regulating p53 expression and function. Mol Cell Biol. 2005;25(17):7432–40.
95. Ogura M, Uchida T, Terui Y, Hayakawa F, Kobayashi Y, Taniwaki M, Takamatsu Y, Naoe T, Tobinai K, Munakata W, Yamauchi T, Kageyama A, Yuasa M, Motoyama M, Tsunoda T, Hatake K. Phase I study of OPB-51602, an oral inhibitor of signal transducer and activator of transcription 3, in patients with relapsed/refractory hematological malignancies. Cancer Sci. 2015;106(7):896–901.
96. Oguro T, Ishibashi K, Sugino T, Hashimoto K, Tomita S, Takahashi N, Yanagida T, Haga N, Aikawa K, Suzutani T, Yamaguchi O, Kojima Y. Humanised antihuman IL-6R antibody with interferon inhibits renal cell carcinoma cell growth in vitro and in vivo through suppressed SOCS3 expression. Eur J Cancer. 2013;49(7):1715–24.
97. Oh DY, Lee SH, Han SW, Kim MJ, Kim TM, Kim TY, Heo DS, Yuasa M, Yanagihara Y, Bang YJ. Phase I study of OPB-31121, an oral STAT3 inhibitor, in patients with advanced solid tumors. Cancer Res Treat. 2015;47(4):607–15.
98. Okusaka T, Ueno H, Ikeda M, Mitsunaga S, Ozaka M, Ishii H, Yokosuka O, Ooka Y, Yoshimoto R, Yanagihara Y, Okita K. Phase 1 and pharmacological trial of OPB-31121, a signal transducer and activator of transcription-3 inhibitor, in patients with advanced hepatocellular carcinoma. Hepatol Res. 2015;45(13):1283–91.
99. Pappa E, Nikitakis N, Vlachodimitropoulos D, Avgoustidis D, Oktseloglou V, Papadogeorgakis N. Phosphorylated signal transducer and activator of transcription-1 immunohistochemical expression is associated with improved survival in patients with oral squamous cell carcinoma. J Oral Maxillofac Surg. 2014;72(1):211–21.
100. Peng HY, Jiang SS, Hsiao JR, Hsiao M, Hsu YM, Wu GH, Chang WM, Chang JY, Jin SL, Shiah SG. IL-8 induces miR-424-5p expression and modulates SOCS2/STAT5 signaling pathway in oral squamous cell carcinoma. Mol Oncol. 2016;10(6):895–909.
101. Peñuelas S, Anido J, Prieto-Sánchez RM, Folch G, Barba I, Cuartas I, García-Dorado D, Poca MA, Sahuquillo J, Baselga J, Seoane J. TGF-beta increases glioma-initiating cell self-renewal through the induction of LIF in human glioblastoma. Cancer Cell. 2009;15(4):315–27. https://doi.org/10.1016/j.ccr.2009.02.011.
102. Peyser ND, Freilino M, Wang L, Zeng Y, Li H, Johnson DE, Grandis JR. Frequent promoter hypermethylation of PTPRT increases STAT3 activation and sensitivity to STAT3 inhibition in head and neck cancer. Oncogene. 2016a;35(9):1163–9.
103. Peyser ND, Wang L, Zeng Y, Acquafondata M, Freilino M, Li H, Sen M, Gooding WE, Satake M, Wang Z, Johnson DE, Grandis JR. STAT3 as a chemoprevention target in carcinogen-induced head and neck squamous cell carcinoma. Cancer Prev Res (Phila). 2016b;9(8):657–63.
104. Qian WF, Guan WX, Gao Y, Tan JF, Qiao ZM, Huang H, Xia CL. Inhibition of STAT3 by RNA interference suppresses angiogenesis in colorectal carcinoma. Braz J Med Biol Res. 2011;44(12):1222–30.
105. Riedel F, Zaiss I, Herzog D, Gotte K, Naim R, Hormann K. Serum levels of interleukin-6 in patients with primary head and neck squamous cell carcinoma. Anticancer Res. 2005;25(4):2761–5.
106. Rossi JF, Négrier S, James ND, Kocak I, Hawkins R, Davis H, Prabhakar U, Qin X, Mulders P, Berns B. A phase I/II study of siltuximab (CNTO 328), an anti-interleukin-6 monoclonal antibody, in metastatic renal cell cancer. Br J Cancer. 2010;103(8):1154–62.
107. Sahu RP, Srivastava SK. The role of STAT-3 in the induction of apoptosis in pancreatic cancer cells by benzyl isothiocyanate. J Natl Cancer Inst. 2009;101(3):176–93.
108. Schaefer LK, Ren Z, Fuller GN, Schaefer TS. Constitutive activation of Stat3alpha in brain tumors: localization to tumor endothelial cells and activation by the endothelial tyrosine kinase receptor (VEGFR-2). Oncogene. 2002;21(13):2058–65.
109. Schindler C, Levy DE, Decker T. JAK-STAT signaling: from interferons to cytokines. J Biol Chem. 2007;282(28):20059–63.

110. Schmitt NC, Trivedi S, Ferris RL. STAT1 activation is enhanced by cisplatin and variably affected by EGFR inhibition in HNSCC cells. Mol Cancer Ther. 2015;14(9):2103–11.
111. Sen M, Grandis JR. Nucleic acid-based approaches to STAT inhibition. JAKSTAT. 2012;1(4):285–91.
112. Sen M, Joyce S, Panahandeh M, Li C, Thomas SM, Maxwell J, Wang L, Gooding WE, Johnson DE, Grandis JR. Targeting Stat3 abrogates EGFR inhibitor resistance in cancer. Clin Cancer Res. 2012b;18(18):4986–96.
113. Sen M, Paul K, Freilino ML, Li H, Li C, Johnson DE, Wang L, Eiseman J, Grandis JR. Systemic administration of a cyclic signal transducer and activator of transcription 3 (STAT3) decoy oligonucleotide inhibits tumor growth without inducing toxicological effects. Mol Med. 2014;20:46–56.
114. Sen M, Pollock NI, Black J, DeGrave KA, Wheeler S, Freilino ML, Joyce S, Lui VW, Zeng Y, Chiosea SI, Grandis JR. JAK kinase inhibition abrogates STAT3 activation and head and neck squamous cell carcinoma tumor growth. Neoplasia. 2015;17(3):256–64.
115. Sen M, Thomas SM, Kim S, Yeh JI, Ferris RL, Johnson JT, Duvvuri U, Lee J, Sahu N, Joyce S, Freilino ML, Shi H, Li C, Ly D, Rapireddy S, Etter JP, Li PK, Wang L, Chiosea S, Seethala RR, Gooding WE, Chen X, Kaminski N, Pandit K, Johnson DE, Grandis JR. First-in-human trial of a STAT3 decoy oligonucleotide in head and neck tumors: implications for cancer therapy. Cancer Discov. 2012a;2(8):694–705.
116. Sen M, Tosca PJ, Zwayer C, Ryan MJ, Johnson JD, Knostman KA, Giclas PC, Peggins JO, Tomaszewski JE, McMurray TP, Li C, Leibowitz MS, Ferris RL, Gooding WE, Thomas SM, Johnson DE, Grandis JR. Lack of toxicity of a STAT3 decoy oligonucleotide. Cancer Chemother Pharmacol. 2009;63(6):983–95. https://doi.org/10.1007/s00280-008-0823-6.
117. Shah JJ, Feng L, Thomas SK, Berkova Z, Weber DM, Wang M, Qazilbash MH, Champlin RE, Mendoza TR, Cleeland C, Orlowski RZ. Siltuximab (CNTO 328) with lenalidomide, bortezomib and dexamethasone in newly-diagnosed, previously untreated multiple myeloma: an open-label phase I trial. Blood Cancer J. 2016;6:e396.
118. Shah NG, Trivedi TI, Tankshali RA, Goswami JV, Jetly DH, Shukla SN, Shah PM, Verma RJ. Prognostic significance of molecular markers in oral squamous cell carcinoma: a multivariate analysis. Head Neck. 2009;31(12):1544–56.
119. Sharma RA, Euden SA, Platton SL, Cooke DN, Shafayat A, Hewitt HR, Marczylo TH, Morgan B, Hemingway D, Plummer SM, Pirmohamed M, Gescher AJ, Steward WP. Phase I clinical trial of oral curcumin: biomarkers of systemic activity and compliance. Clin Cancer Res. 2004;10(20):6847–54.
120. Sharma A, Jyotsana N, Lai CK, Chaturvedi A, Gabdoulline R, Görlich K, Murphy C, Blanchard JE, Ganser A, Brown E, Hassell JA, Humphries RK, Morgan M, Heuser M. Pyrimethamine as a potent and selective inhibitor of acute myeloid leukemia identified by high-throughput drug screening. Curr Cancer Drug Targets. 2016;16(9):818–28.
121. Shukla S, Mahata S, Shishodia G, Pandey A, Tyagi A, Vishnoi K, Basir SF, Das BC, Bharti AC. Functional regulatory role of STAT3 in HPV16-mediated cervical carcinogenesis. PLoS One. 2013;8(7):e67849.
122. Siddiquee KA, Gunning PT, Glenn M, Katt WP, Zhang S, Schrock C, Sebti SM, Jove R, Hamilton AD, Turkson J. An oxazole-based small-molecule Stat3 inhibitor modulates Stat3 stability and processing and induces antitumor cell effects. ACS Chem Biol. 2007a;2(12):787–98.
123. Siddiquee K, Zhang S, Guida WC, Blaskovich MA, Greedy B, Lawrence HR, Yip ML, Jove R, McLaughlin MM, Lawrence NJ, Sebti SM, Turkson J. Selective chemical probe inhibitor of Stat3, identified through structure-based virtual screening, induces antitumor activity. Proc Natl Acad Sci U S A. 2007b;104(18):7391–6.
124. Song JI, Grandis JR. STAT signaling in head and neck cancer. Oncogene. 2000;19(21):2489–95. Review
125. Song L, Rawal B, Nemeth JA, Haura EB. JAK1 activates STAT3 activity in non-small-cell lung cancer cells and IL-6 neutralizing antibodies can suppress JAK1-STAT3 signaling. Mol Cancer Ther. 2011;10(3):481–94.

126. Song H, Wang R, Wang S, Lin J. A low-molecular-weight compound discovered through virtual database screening inhibits Stat3 function in breast cancer cells. Proc Natl Acad Sci U S A. 2005;102(13):4700–5.
127. Sriuranpong V, Park JI, Amornphimoltham P, Patel V, Nelkin BD, Gutkind JS. Epidermal growth factor receptor-independent constitutive activation of STAT3 in head and neck squamous cell carcinoma is mediated by the autocrine/paracrine stimulation of the interleukin 6/gp130 cytokine system. Cancer Res. 2003;63(11):2948–56.
128. Subramaniam KS, Omar IS, Kwong SC, Mohamed Z, Woo YL, Mat Adenan NA, Chung I. Cancer-associated fibroblasts promote endometrial cancer growth via activation of interleukin-6/STAT-3/c-Myc pathway. Am J Cancer Res. 2016;6(2):200–13.
129. Suh YA, Jo SY, Lee HY, Lee C. Inhibition of IL-6/STAT3 axis and targeting Axl and Tyro3 receptor tyrosine kinases by apigenin circumvent taxol resistance in ovarian cancer cells. Int J Oncol. 2015;46(3):1405–11.
130. Sun Y, Sang Z, Jiang Q, Ding X, Yu Y. Transcriptomic characterization of differential gene expression in oral squamous cell carcinoma: a meta-analysis of publicly available microarray data sets. Tumour Biol. 2016. https://doi.org/10.1007/s13277-016-5439-6.
131. Takakura A, Nelson EA, Haque N, Humphreys BD, Zandi-Nejad K, Frank DA, Zhou J. Pyrimethamine inhibits adult polycystic kidney disease by modulating STAT signaling pathways. Hum Mol Genet. 2011;20(21):4143–54.
132. Timofeeva OA, Gaponenko V, Lockett SJ, Tarasov SG, Jiang S, Michejda CJ, Perantoni AO, Tarasova NI. Rationally designed inhibitors identify STAT3 N-domain as a promising anticancer drug target. ACS Chem Biol. 2007;2(12):799–809.
133. Turkson J, Kim JS, Zhang S, Yuan J, Huang M, Glenn M, Haura E, Sebti S, Hamilton AD, Jove R. Novel peptidomimetic inhibitors of signal transducer and activator of transcription 3 dimerization and biological activity. Mol Cancer Ther. 2004b;3(3):261–9.
134. Turkson J, Zhang S, Mora LB, Burns A, Sebti S, Jove R. A novel platinum compound inhibits constitutive Stat3 signaling and induces cell cycle arrest and apoptosis of malignant cells. J Biol Chem. 2005;280(38):32979–88.
135. Vallath S, Sage EK, Kolluri KK, Lourenco SN, Teixeira VS, Chimalapati S, George PJ, Janes SM, Giangreco A. CADM1 inhibits squamous cell carcinoma progression by reducing STAT3 activity. Sci Rep. 2016;6:24006. https://doi.org/10.1038/srep24006.
136. Veeriah S, Brennan C, Meng S, Singh B, Fagin JA, Solit DB, Paty PB, Rohle D, Vivanco I, Chmielecki J, Pao W, Ladanyi M, Gerald WL, Liau L, Cloughesy TC, Mischel PS, Sander C, Taylor B, Schultz N, Major J, Heguy A, Fang F, Mellinghoff IK, Chan TA. The tyrosine phosphatase PTPRD is a tumor suppressor that is frequently inactivated and mutated in glioblastoma and other human cancers. Proc Natl Acad Sci U S A. 2009;106(23):9435–40.
137. Wang Y, Qu A, Wang H. Signal transducer and activator of transcription 4 in liver diseases. Int J Biol Sci. 2015;11(4):448–55.
138. Wang Z, Si X, Xu A, Meng X, Gao S, Qi Y, Zhu L, Li T, Li W, Dong L. Activation of STAT3 in human gastric cancer cells via interleukin (IL)-6-type cytokine signaling correlates with clinical implications. PLoS One. 2013;8(10):e75788.
139. Watts JK, Corey DR. Silencing disease genes in the laboratory and the clinic. J Pathol. 2012;226(2):365–79.
140. Weber A, Hengge UR, Bardenheuer W, Tischoff I, Sommerer F, Markwarth A, Dietz A, Wittekind C, Tannapfel A. SOCS-3 is frequently methylated in head and neck squamous cell carcinoma and its precursor lesions and causes growth inhibition. Oncogene. 2005;24(44):6699–708.
141. Weerasinghe P, Garcia GE, Zhu Q, Yuan P, Feng L, Mao L, Jing N. Inhibition of Stat3 activation and tumor growth suppression of non-small cell lung cancer by G-quartet oligonucleotides. Int J Oncol. 2007;31(1):129–36.
142. Weimbs T, Olsan EE, Talbot JJ. Regulation of STATs by polycystin-1 and their role in polycystic kidney disease. JAKSTAT. 2013;2(2):e23650.

143. Wong AL, Soo RA, Tan DS, Lee SC, Lim JS, Marban PC, Kong LR, Lee YJ, Wang LZ, Thuya WL, Soong R, Yee MQ, Chin TM, Cordero MT, Asuncion BR, Pang B, Pervaiz S, Hirpara JL, Sinha A, Xu WW, Yuasa M, Tsunoda T, Motoyama M, Yamauchi T, Goh BC. Phase I and biomarker study of OPB-51602, a novel signal transducer and activator of transcription (STAT) 3 inhibitor, in patients with refractory solid malignancies. Ann Oncol. 2015;26(5):998–1005.
144. Xi S, Dyer KF, Kimak M, Zhang Q, Gooding WE, Chaillet JR, Chai RL, Ferrell RE, Zamboni B, Hunt J, Grandis JR. Decreased STAT1 expression by promoter methylation in squamous cell carcinogenesis. J Natl Cancer Inst. 2006;98(3):181–9.
145. Xi S, Gooding WE, Grandis JR. In vivo antitumor efficacy of STAT3 blockade using a transcription factor decoy approach: implications for cancer therapy. Oncogene. 2005;24(6):970–9.
146. Xi S, Zhang Q, Gooding WE, Smithgall TE, Grandis JR. Constitutive activation of Stat5b contributes to carcinogenesis in vivo. Cancer Res. 2003;63(20):6763–71.
147. Yadav A, Kumar B, Datta J, Teknos TN, Kumar P. IL-6 promotes head and neck tumor metastasis by inducing epithelial-mesenchymal transition via the JAK-STAT3-SNAIL signaling pathway. Mol Cancer Res. 2011;9(12):1658–67.
148. Yan B, Li H, Yang X, Shao J, Jang M, Guan D, Zou S, Van Waes C, Chen Z, Zhan M. Unraveling regulatory programs for NF-kappaB, p53 and microRNAs in head and neck squamous cell carcinoma. PLoS One. 2013;8(9):e73656.
149. Yan S, Zhou C, Zhang W, Zhang G, Zhao X, Yang S, Wang Y, Lu N, Zhu H, Xu N. Beta-catenin/TCF pathway upregulates STAT3 expression in human esophageal squamous cell carcinoma. Cancer Lett. 2008;271(1):85–97.
150. Yang J, Chatterjee-Kishore M, Staugaitis SM, Nguyen H, Schlessinger K, Levy DE, Stark GR. Novel roles of unphosphorylated STAT3 in oncogenesis and transcriptional regulation. Cancer Res. 2005;65(3):939–47.
151. Yang Q, Li M, Wang T, Xu H, Zang W, Zhao G. Effect of STAT5 silenced by siRNA on proliferation apoptosis and invasion of esophageal carcinoma cell line Eca-109. Diagn Pathol. 2013;8:132.
152. Yang J, Stark GR. Roles of unphosphorylated STATs in signaling. Cell Res. 2008;18(4):443–51. https://doi.org/10.1038/cr.2008.41.
153. Yao Z, Fenoglio S, Gao DC, Camiolo M, Stiles B, Lindsted T, Schlederere M, Johns C, Altoriki N, Mittal V, Kenner L, Sordella R. TGF-beta IL-6 axis mediates selective and adaptive mechanisms of resistance to molecular targeted therapy in lung cancer. Proc Natl Acad Sci U S A. 2010;107(35):15535–40.
154. Yin C, Sandoval C, Baeg GH. Identification of mutant alleles of JAK3 in pediatric patients with acute lymphoblastic leukemia. Leuk Lymphoma. 2015;56(5):1502–6.
155. Yu H, Jove R. The STATs of cancer--new molecular targets come of age. Nat Rev Cancer. 2004;4(2):97–105.
156. Yu H, Lee H, Herrmann A, Buettner R, Jove R. Revisiting STAT3 signalling in cancer: new and unexpected biological functions. Nat Rev Cancer. 2014;14(11):736–46.
157. Zhang HF, Chen Y, Wu C, Wu ZY, Tweardy DJ, Alshareef A, Liao LD, Xue YJ, Wu JY, Chen B, Xu XE, Gopal K, Gupta N, Li EM, Xu LY, Lai R. The opposing function of STAT3 as an oncoprotein and tumor suppressor is dictated by the expression status of STAT3β in esophageal squamous cell carcinoma. Clin Cancer Res. 2016b;22(3):691–703.
158. Zhang J, Du J, Liu Q, Zhang Y. Down-regulation of STAT3 expression using vector-based RNA interference promotes apoptosis in Hepatocarcinoma cells. Artif Cells Nanomed Biotechnol. 2016c;44(5):1201–5.
159. Zhang X, Guo A, Yu J, Possemato A, Chen Y, Zheng W, Polakiewicz RD, Kinzler KW, Vogelstein B, Velculescu VE, Wang ZJ. Identification of STAT3 as a substrate of receptor protein tyrosine phosphatase T. Proc Natl Acad Sci U S A. 2007;104(10):4060–4. Epub 2007 Feb 21

160. Zhang Q, Hossain DM, Duttagupta P, Moreira D, Zhao X, Won H, Buettner R, Nechaev S, Majka M, Zhang B, Cai Q, Swiderski P, Kuo YH, Forman S, Marcucci G, Kortylewski M. Serum-resistant CpG-STAT3 decoy for targeting survival and immune checkpoint signaling in acute myeloid leukemia. Blood. 2016a;127(13):1687–700.
161. Zhang Y, Ma Q, Liu T, Guan G, Zhang K, Chen J, Jia N, Yan S, Chen G, Liu S, Jiang K, Lu Y, Wen Y, Zhao H, Zhou Y, Fan Q, Qiu X. Interleukin-6 suppression reduces tumour self-seeding by circulating tumour cells in a human osteosarcoma nude mouse model. Oncotarget. 2016d;7(1):446–58.
162. Zhang H, Zhang D, Luan X, Xie G, Pan X. Inhibition of the signal transducers and activators of transcription (STAT) 3 signalling pathway by AG490 in laryngeal carcinoma cells. J Int Med Res. 2010;38(5):1673–81.
163. Zhang L, Zhao Z, Feng Z, Yin N, Lu G, Shan B. RNA interference-mediated silencing of Stat5 induces apoptosis and growth suppression of hepatocellular carcinoma cells. Neoplasma. 2012;59(3):302–9. https://doi.org/10.4149/neo_2012_039.
164. Zibelman M, Mehra R. Overview of current treatment options and investigational targeted therapies for locally advanced squamous cell carcinoma of the head and neck. Am J Clin Oncol. 2016;39(4):396–406.

Chapter 7
Molecular Regulation of Cell Cycle and Cell Cycle-Targeted Therapies in Head and Neck Squamous Cell Carcinoma (HNSCC)

Elena V. Demidova, Waleed Iqbal, and Sanjeevani Arora

Abstract Head and neck squamous cell carcinomas (HNSCCs) are clinically challenging. The molecular mechanisms and genetic changes that drive HNSCCs have been studied with the aim of developing better therapeutic strategies involving novel molecular targets. Genomic studies have identified mutations in genes that mediate cell cycle, and key differences in cell cycle regulation differentiate both human papillomavirus (HPV)-associated and HPV-negative HNSCC cases from normal tissue. Some of these differences may nominate specific therapeutic targets and impact treatment response in HNSCC. For example, one of the most frequent cell cycle alterations in HPV (−) HNSCC is the disruption of the p53 (*TP53*) pathway (over 80% of tumors, based on data in TCGA and other studies), which is involved in cell cycle control, DNA damage signaling, and overall maintenance of genome stability. Other frequent alterations disrupt the cell cycle regulator *CDKN2A* (28% alteration frequency), which encodes p16, an inhibitor of cell cycle kinases CDK4 and CDK6, and alters expression of *CCND1*, resulting in inactivation of the tumor suppressor Rb. Other mutations found less commonly in patients target elements of the cell cycle checkpoint and DNA damage response machinery. Such

E. V. Demidova
Programs in Cancer Prevention and Control, Fox Chase Cancer Center, Philadelphia, PA, USA

Kazan Federal University, Kazan, Russia

W. Iqbal
Programs in Molecular Therapeutics, Fox Chase Cancer Center, Philadelphia, PA, USA

Cancer Biology Program, Drexel Univeristy College Of Medicine, Philadelphia, PA, USA

S. Arora (✉)
Programs in Cancer Prevention and Control, Fox Chase Cancer Center, Philadelphia, PA, USA
e-mail: sanjeevani.arora@fccc.edu

B. Burtness, E. A. Golemis (eds.), *Molecular Determinants of Head and Neck Cancer*, Current Cancer Research, https://doi.org/10.1007/978-3-319-78762-6_7

observations and a growing recognition of the importance of cell cycle regulatory defects in HNSCC response to typically DNA-damaging chemotherapies and radiation therapy have rationalized the development of novel cell cycle-targeted therapies for HNSCCs. We here provide a general overview of the process of cell cycle control, cell cycle checkpoints, and how these are dysregulated in HNSCC and other cancers and discuss current cell cycle-targeted therapies in development and in clinical trials for HNSCC. The ultimate goal of these efforts is to develop new, potent therapeutic agents and to identify patient subpopulations that will be more responsive to cell cycle-targeted therapies.

Keywords Cell cycle inhibitors · Cancer treatment · Head and neck squamous cell carcinomas · Cyclin-dependent kinases · Cyclins · WEE1 · CHK1 · ATR · p16 · p53 · Aurora kinase · Plk1 · Human papillomavirus (HPV)-associated · HPV-negative (−)

7.1 Introduction

Head and neck squamous cell carcinoma (HNSCC) treatment options have long employed classical chemotherapies, including DNA-damaging agents such as platinum compounds, that are associated with high toxicity and incomplete efficacy in advanced tumors. These limitations have motivated studies of the genetic and protein expression profiles of HNSCCs to help identify novel targets and design treatment strategies for better prognosis and overall survival [1, 2]. Genetic analyses have identified prominent distinctions between subclasses of HNSCC tumors. The most prominent subclasses are human papillomavirus (HPV)-positive or HPV-associated tumors (caused by infection with high-risk HPV strains) and HPV-negative (HPV(−)) tumors [3, 4]. HPV encodes two oncoproteins, E6 and E7, that inactivate cell cycle regulators, p53 and retinoblastoma protein (RB), respectively. This causes deregulation of the cell cycle and the onset of HPV-mediated tumorigenesis [1, 2, 5–7]. In contrast, molecular analysis of HNSCCs has shown that cell cycle regulators, such as *CDKN2A*, *TP53*, *CCND1*, and *CDK2AP1*, are frequently mutated and can be early events in head and neck carcinogenesis, particularly in HPV(−) tumors [8–10].

The normal cell cycle is divided into active phases of DNA synthesis (S phase) and cell partitioning in mitosis (M phase), separated by gap periods G1 and G2, during which the cells prepare for the successful completion of S and M phase, respectively. Progression through the cell cycle is coordinated by regulatory cyclins expressed in distinct cell cycle phases: specific cyclins, complexed with cyclin-dependent kinases (CDKs), phosphorylate a large number of substrates on Ser/Thr residues to control expression and activity of proteins required for each phase [11]. Cell cycle checkpoints guarantee faithful completion of this progression [12]. Of particular relevance to cancer therapy, cell cycle proteins maintain critical connections with pathways that assist with DNA repair, DNA damage response (DDR), and maintenance of genomic stability (discussed at length in Sect. 7.2) [13].

Deregulation of the cell cycle and its components is a hallmark of tumor formation and progression. Thus, the cell cycle is a rational target for actively dividing cancer cells. Standard chemotherapies for HNSCCs include DNA-damaging agents such as the platinum compounds (cisplatin and carboplatin) [14–16] and mitotic inhibitors such as taxanes (paclitaxel and docetaxel) [17–20]. In addition, the use of radiation therapy typically causes DNA damage and triggers cell cycle checkpoints [21, 22]. Drugs that target key molecular components of the cell cycle may offer increased tumor specificity (reviewed in [23]). Several agents targeting the cell cycle have recently completed clinical trials in advanced and/or metastatic breast cancer. The leading dual CDK4/6 inhibitors in clinical use or development are palbociclib (PD-0332991, Pfizer) [24, 25] and ribociclib (LEE011; Novartis) [26, 27], both FDA-approved for estrogen-receptor positive, human epidermal growth factor receptor 2-negative (ER+/HER2−) metastatic breast cancer; and abemaciclib (LY2835219, Lilly) [28, 29]. The results of clinical trials in advanced breast cancer (ER+/HER-) with cell cycle inhibitors have been practice-changing: the combination of palbociclib with letrozole led to an average median progression-free survival (PFS) of 22 months (NCT00721409, NCT01740427) [30–32] and with fulvestrant of 9.2 months (NCT01942135) [33–36]. For ribociclib in combination with letrozole, the median duration of follow-up was 15.3 months. After 18 months, the PFS rate was 63.0% in patients with measurable disease at baseline versus 42.2% in the fulvestrant plus placebo group, and the overall response rates were 52.7% and 37.1%, respectively (NCT01958021) [37]. Based on these exciting results, clinical trials in other cancer types such as HNSCC have been initiated. These are discussed further in Sect. 7.4 of this chapter.

In addition to CDK inhibitors such as palbociclib [24, 25], therapeutic inhibitors of other kinases are also in clinical development for targets which regulate mitotic entry and progression. These targets include polo-like kinase-1 (PLK1) and Aurora kinases [38–45]; negative cell cycle regulators, such as the dual-specificity kinases WEE1 and MYT1 [46–52]; and proteins associated with DNA damage checkpoint pathways, ataxia telangiectasia-mutated (ATM) and Rad3-related (ATR)/checkpoint Kinase 1 (CHK1) (ATR/CHK1) [53–59]. In Sect. 7.2, we discuss the basic biology of the key cell cycle and related proteins. The biology of these proteins rationalizes the use of agents that target cell cycle-related proteins in HNSCC therapy. In Sect. 7.3, we discuss frequent alterations in cell cycle and related genes in HNSCCs. This section also discusses the different mechanisms of cell cycle deregulation that are observed in HPV-associated and HPV (−) HNSCCs and the impact of cell cycle alterations (with a focus on *TP53*) in response to conventional therapy (platinum and radiation). Finally, in Sect. 7.4, we discuss the development and evaluation of cell cycle inhibitors that are promising for the treatment of HNSCC. Table 7.1 is a comprehensive list of current clinical trials with various cell cycle-targeted agents for HNSCC.

Table 7.1 List of current clinical trials for cell cycle targets in HNSCC

ID number	Trial name	Drugs tested	Main target	Notes
NCT02526316	Pilot study of concurrent CDDP-based chemotherapy combined with vaccination therapy with the P16_37-63 peptide in patients with HPV– and p16INK4a-positive cancer	P16_37-63, Montanide	$p16^{INK4A}$ (recovering suppressor function)	HPV-induced cancers (including HPV+ HNSCC), phase I, ongoing
NCT02187783	Modular phase II study to link targeted therapy to patients with pathway-activated tumors: module 8 – LEE011 for patients with CDK4/6 pathway-activated tumors	Ribociclib (LEE011)	CDK4/6 pathway (inhibition)	Patients with solid tumors or hematological malignancy that have been pre-identified as having relevant CDK4/6, cyclin D1/3, or p16 aberrations. Patients must have received at least one prior and have no remaining standard therapy options anticipated to result in a durable response, phase II, ongoing
NCT02686008	A pilot pharmacodynamic study to assess the antiproliferative activity of the PARP inhibitor olaparib in patients with HPV+ and HPV– HNSCC	Olaparib	PARP (inhibition)	Phase I, not open for recruiting participants
NCT01695122	Efficacy evaluation of the combination of valproic acid and standard platinum-based chemoradiation in patients with locally advanced HNSCC	Valproic acid, standard platinum-based chemoradiation	HDAC (inhibition)	Valproic acid activates apoptosis pathways, cell differentiation, downregulating expression of growth factors and promoting radiosensitization. Head and neck cancer, oral cavity cancer, oropharyngeal cancer, phase II, completed, no results posted
NCT00387127	A randomized, double-blind, placebo-controlled, multicenter, phase II study of oral lapatinib in combination with concurrent radiotherapy and CDDP versus radiotherapy and CDDP alone, in subjects with stage III, IVA, B HNSCC	Lapatinib, concurrent radiotherapy, CDDP vs. radiotherapy, CDDP	Tyrosine kinase (inhibition in HER2/neu and EGFR pathways)	Lapatinib is a dual tyrosine kinase inhibitor, which interrupts the HER2/neu and EGFR pathways. Advanced HNSCCs, phase II, completed, overall survival increased

NCT02499120	A randomized, multicenter, double-blind phase II study of palbociclib plus cetuximab versus cetuximab for the treatment of HPV−, cetuximab-naïve patients with recurrent/metastatic HNSCC after failure of one prior platinum-containing chemotherapy regimen	Palbociclib, Cetuximab vs. Cetuximab only	CDK4 and 6 (inhibition), RB (phosphorylation blocking), EGFR (MAb, inhibition)	Patients with recurrent/metastatic HPV− HNSCC after failure of chemotherapy with platinum agents, phase II, recruiting participants
NCT03065062	Phase I study of the CDK4/6 inhibitor palbociclib (PD-0332991) in combination with the PI3K/mTOR inhibitor gedatolisib (PF-05212384) for patients with advanced squamous cell lung, pancreatic, head and neck, and other solid tumors	Palbociclib, Gedatolisib	CDK4 and 6 (inhibition), PI3K/mTOR (inhibition)	Lung cancer squamous cell, solid tumors, head and neck cancer, pancreatic cancer. Study of combination of drugs in tumors that might have a specific change in the phosphatidylinositol-3 phosphate pathway. Phase I, recruiting participants
NCT03024489	A phase I/II dose-escalation study of the CDK4/6 inhibitor palbociclib in combination with cetuximab and IMRT for locally advanced unresectable HNSCC	Palbociclib, Cetuximab, IMRT	CDK4 and 6 (inhibition), RB (phosphorylation blocking), EGFR (inhibition)	Patients with locally advanced, unresectable HNSCC. Phase I and II, not open for recruiting participants
NCT00899054	An open label, multicenter phase I/II study of selective CDKs inhibitor P276-00 in combination with radiation in subjects with recurrent and/or locally advanced HNSCC	Riviciclib (P267-00), radiation	CDK1, Cdk4/cyclin D1, Cdk1/cyclin B, Cdk9/cyclin T1, serine/threonine kinases (inhibition)	Patients with advanced HNSCC, phase I and II, completed, no results posted
NCT00824343	An open label, multicenter phase II study to evaluate efficacy and safety of P276-00 in Indian subjects with recurrent, metastatic, or unresectable locally advanced HNSCC	Riviciclib (P267-00)	CDK1, Cdk4/cyclin D1, Cdk1/cyclin B, Cdk9/cyclin T1, serine/threonine kinases (inhibition)	Recurrent, metastatic or unresectable locally advanced HNSCC. Phase II, completed, no results posted
NCT01051791	Phase II study of RAD001 for treatment of refractory, recurrent, locally advanced HNSCC	Everolimus (RAD001)	mTOR pathway (inhibition)	HNSCC, phase II, ongoing

(continued)

Table 7.1 (continued)

ID number	Trial name	Drugs tested	Main target	Notes
NCT00176241	A phase II trial of postoperative radiation, CDDP, and panitumumab in locally advanced head and neck cancer	Panitumumab, CDDP, postoperative radiation	EGFR (MAb, inhibition), apoptosis induction	Patients with pathologic stage III or IVA squamous cell carcinoma of the oral cavity, larynx, hypopharynx, or HPV− oropharynx, without gross residual tumor, featuring high-risk factors. Probability of PFS at 2 year – 70% (average), up to 90 months for cohort; individual patients, up to 24 months. Phase II, ongoing
NCT00281866	Genotypic-based pharmacodynamic evaluation of erlotinib (Tarceva™, OSI Pharmaceuticals, Uniondale, NY) in patients with HNSCC	Erlotinib hydrochloride	Tyrosine kinase (EGFR inhibition)	Patients with HNSCC, phase II, completed, no results posted
NCT00099021	A phase IIa cancer prevention trial of the PPARγ agonist pioglitazone in oral leukoplakia	Pioglitazone hydrochloride (Actos)	PPARγ (stimulation)	Prevention study, HNSCC patients with oral leukoplakia. Phase II, completed, partial overall response, 15/21 patients; clinical response, 17/21
NCT02429037	Recombinant adenoviral human p53 gene combined with radio- and chemotherapy in treatment of unresectable, locally advanced head and neck cancer – an open-labeled randomized phase II study	rAd-p53, radiation, CDDP	P53 (recovering suppressor function)	Gene therapy drug, advanced HNSCC, phase II, not yet recruiting participants
NCT02178072	A pilot study to assess the activity of demethylation therapy in patients with HPV+ compared with HPV− HNSCC	5-azacitadine (Vidaza)	DNA – methyltransferase (inhibition), cytotoxicity	Nucleotide analogue, patients with HPV+ and HPV− HNSCC, phase II, recruiting participants
NCT01115790	A phase I study of LY2606368 in patients with advanced cancer	Prexasertib	CHK1 (inhibition)	Several types of squamous cell carcinomas, including HNSCC, phase I, completed, no results posted
NCT02508246	A phase I clinical trial of AZD1775 in combination with neoadjuvant weekly docetaxel and CDDP prior to surgery in HNSCC	AZD1775, CDDP, docetaxel	WEE1 (inhibition)	Patients with HNSCC prior to surgery, phase I, recruiting participants

NCT03028766	A phase I trial of WEE1 inhibition with chemotherapy and radiotherapy as adjuvant treatment and a window of opportunity trial with CDDP in patients with head and neck cancer	AZD1775, CDDP, radiation	WEE1 (inhibition)	Patients with hypopharynx squamous cell carcinoma, oral cavity squamous cell carcinoma, larynx cancer, phase I, not yet recruiting participants
NCT02585973	Phase Ib trial of dose-escalating AZD1775 in combination with concurrent radiation and CDDP for intermediate and high-risk HNSCC	AZD1775, CDDP, radiation	WEE1 (inhibition)	Patients with HNSCC, phase I, recruiting participants
NCT00997906	A randomized phase II/III study of concurrent CDDP-radiotherapy with or without induction chemotherapy using gemcitabine, carboplatin, and paclitaxel in locally advanced nasopharyngeal cancer	CDDP, radiotherapy with/without carboplatin, gemcitabine, hydrochloride, paclitaxel	Inducing apoptosis, microtubule stimulating, suppressing their organization (division suppression)	HNSCC, phase II and III, ongoing
NCT00588770	A phase III randomized trial of chemotherapy with or without bevacizumab in patients with recurrent or metastatic HNSCC	Bevacizumab with/without carboplatin, CDDP, docetaxel, 5-fluorouracil	VEGF (MAb, inhibition)	Patients with HNSCC, phase III, ongoing
NCT02734537	Phase II randomized trial of adjuvant radiotherapy with or without CDDP for p53 mutated, surgically resected HNSCC	CDDP with/without IMRT	Inducing apoptosis	Patients with surgically resected HNSCC with p53 mutations, phase II, recruiting participants
NCT01898494	Phase II randomized trial of transoral surgical resection followed by low-dose or standard-dose IMRT in resectable p16+ locally advanced oropharynx cancer	Surgery, IMRT, CDDP, carboplatin	Inducing apoptosis	Patients with HPV+ advanced squamous cell carcinomas of the oropharynx, phase II, recruiting participants

CDDP, cisplatin; *EGFR*, epidermal growth factor receptor; *HDAC*, histone deacetylase; *HER2/neu*, receptor tyrosine-protein kinase erbB-2; *HNSCC*, head and neck squamous cell carcinoma; *HPV*, human papilloma virus; *HPV+*, human papilloma virus positive; *HPV−*, human papilloma virus negative; *IMRT*, intensity-modulated radiotherapy; *MAb*, monoclonal antibody; *P16_37-63*, peptide p16 vaccine; *PARP*, poly ADP-ribose polymerase; *PFS*, progression-free survival; *PPARɤ*, peroxisome proliferator-activated receptor gamma; *rAd-p53*, adenoviral vector with inserted p53; *VEGF*, vascular endothelial growth factor

7.2 Cell Cycle Controls: Key Regulators and Checkpoints

Movement through the cell cycle is strongly regulated by specialized groups of proteins, with cyclins and CDKs as the main drivers. In normal cell cycle progression, CDKs are allosterically activated after binding to cyclins, which form active heterodimers. The cyclins are positive regulators of cell cycle progression, and their temporal expression, localization, and degradation are tightly regulated ([60] and reviewed in [61]). There are more than 15 families of cyclins in humans (including A, B, C, D, E, F, H, K, L, T, and Y), with functions ranging from cell cycle regulation to control of DNA synthesis and repair, transcription, proteolysis, RNA splicing, and neuronal differentiation. Progression through each cell cycle phase is dominated by specific cyclins, such as cyclin D in G1 phase, cyclins E and A in S phase, and cyclins A and B in G2/M phase [62, 63].

Of particular importance to HNSCC, elimination of cell cycle arrest mediated by the tumor suppressors p53 and Rb is commonly associated with the disease [8–10]. The genomic profile of HNSCC is discussed in detail in Chap. 11 of this book. A number of studies have observed genetic alterations directly targeting *TP53* (~46–73%) and RB1 (~5%), as well as changes targeting genes encoding proteins that interact with RB1 or p53 (e.g., *CCND1, CDKN2A*), in ways that are predicted to reduce or eliminate tumor suppressor function [8–10, 64–70]. In HPV-associated HNSCCs, HPV oncoprotein-induced degradation of the p53 and Rb proteins (discussed in Sect. 7.3.1 of this chapter) performs a similar disruptive role [8–10]. Cdc25, a protein phosphatase that dephosphorylates within the activation loop of the kinase domain, helps activate the CDK2/Cyclin E (during G1 and S) and CDK1/Cyclin B (during G2 and M) complexes ([71, 72] and reviewed in [73]). Negative cell cycle regulators include Wee1 and Myt1; these employ inhibitory phosphorylation to restrict the activity of M-phase cyclin-CDK complexes [46–52] and CDK inhibitors (CKIs) such as the INK4 proteins, which include p16, that restrict the activity of CDK4/6 [74, 75]. Figure 7.1 and the following discussion provide a brief overview of key factors regulating progression and arrest, specifically relevant to HNSCC biology and therapies; for extended discussion of cell cycle, see reviews [2, 21, 22, 46, 61, 76–83].

7.2.1 G1 Phase

The early events in the G1 phase of the cell cycle are controlled by interphase CDKs (CDK4 and 6) that assemble with mitogen-regulated D-type cyclins (*CCND1, CCND2, or CCND3*), forming a CDK4/6-cyclin D complex. D-type cyclin production is stimulated by mitogenic signals, and the withdrawal of these signals can arrest cells in G1/G0 phase. The synthesis, assembly, stability, and nuclear export of D-type cyclins are tightly regulated by signals from multiple integral signal transduction pathways such as the RAS/RAF/ERK and phosphoinositide 3-kinase (PI3K)/AKT pathways [60, 74] (reviewed in [81–83]).

CDK4 and CDK6 and their interacting D-class cyclins (particularly *CCND1*) have emerged as key proteins that integrate the main signaling events related to cell cycle progression and are particularly relevant to HNSCC. In G1, active CDK4/6-cyclin D complexes phosphorylate and partially inactivate members of the retinoblastoma (Rb) family, including Rb, RbL1 (p107), and RbL2 (p130) ([84, 85] and reviewed in [86–88]). Active Rb family members directly bind transcription activation domains of the E2F family of proteins (Rb-E2F complex) and restrict cell cycle progression by preventing the transcription of cell cycle genes that contain E2F sites [89, 90]. Rb-deficient cells do not require cyclin D-dependent kinase activity during the cell cycle [91], a point critical for understanding the key events in both HPV(−) and HPV-associated HNSCC.

Later in G1, CDK4/6-cyclin D complexes drive the hyperphosphorylation of RB, releasing Rb from the Rb–E2F complex and promoting transcription of S-phase-promoting genes such as the E-type cyclins (*CCNE1* and *CCNE2)* [89–91]. Binding of E2F to the Rb pocket and C-terminal domains is regulated by multiple phosphorylation sites [92]. Hyperphosphorylation of Rb leads to conformational changes that inhibit binding with E2F domains. For G1- to S-phase progression, both complexes, CDK2/cyclin E and CDK4/cyclin D, are required for completely hyperphosphorylating Rb. CDK4/cyclin D acts earlier than CDK2/cyclin E, and partial phosphorylation of Rb by the CDK4 complex is essential for CDK2/cyclin E phosphorylation. The CDK4 complex preferentially phosphorylates Rb at sites S807/S811, which facilitates further phosphorylation (priming phosphorylation) by the CDK2 complex. The CDK2/Cyclin E complex phosphorylates sites S608/S612 in Rb, which facilitates the recruitment of the prolyl-isomerase, Pin1. Pin1 then mediates the phosphorylation of Ser/Thr-Pro motifs in Rb, which enhances hyperphosphorylation of Rb. Additionally, the phosphorylation of other sites (such as T373, S788/S795, and T821/T826) by CDKs 4 and 2 also contributes to dissociation of Rb from E2F [92]. Shortly before the G1/S transition, the CDK2-cyclin E complex also phosphorylates and inactivates E3 ubiquitin ligase anaphase-promoting complex/cyclosome-Cdh1 (APC^{CDH1}) [93], completely committing cells to S-phase entry and the initiation of DNA synthesis [94, 95].

7.2.2 S Phase

Once cells have entered S phase, the CDK2-cyclin E complex is inactivated by the degradation of cyclin E by the Skp, cullin, F-box-containing (SCF) ubiquitinating complex [96]. Continuing Rb inactivation permits the transcription of A-type (CCNA1, CCNA2) and subsequently B-type (CCNB1, CCNB2, and CCNB3) cyclins, with an expression of these cyclins continuing to rise in late S phase and in early G2 phase. Cellular function during S phase is dominated by DNA replication. At the end of S phase, the CDK2-cyclin A complex phosphorylates and inactivates E2F, causing a transition to the G2 phase [97, 98].

7.2.3 G2 and M Phase

During the S/G2 transition, A-type cyclins are degraded, while B-type cyclin levels continue to increase. Activation of the CDK1-cyclin B complex largely governs the entry into mitosis. Active CDK1 is restricted in G2 by inhibitory phosphorylation mediated by the Wee1 and Myt1 kinases, which are reversed by the Cdc25 phosphatases [99]. Cdc25 dual-specificity phosphatases activate CDK complexes [71, 72] (reviewed in [100]). There are three known mammalian isoforms of cdc25: Cdc25 A, B, and C. These Cdc25 phosphatases dephosphorylate conserved Thr14 and Tyr15 inhibitory sites in the ATP-binding loop of Cdk1 [48, 100–102]. Further, Cdk-activating kinase (CAK) phosphorylates the cyclin-Cdk complex at Thr161. The CAK complex is comprised of Cdk7, cyclin B, and MAT1 [103–105]. Binding of CKIs (such as p16, p27, or p21/WAF1) regulates CAKs indirectly, by binding CDKs and causing conformation changes that prevent their interaction with CAKs [106].

Active Cdk1-cyclin B complexes drive phosphorylation events that inactivate Wee1 and Myt1, in a feed-forward loop that also involves the mitotic kinases polo-like kinase 1 (PLK1) and Aurora A (AURKA). Aurora A (AURKA) expression peaks in the late G2 phase of the cell cycle [107], and in this phase it is primarily involved in centrosome maturation and physically associates with PLK1. PLK1 is exclusively present in centrosomes and kinetochores in the G2 phase, where it regulates centrosome maturation, bipolar spindle formation, chromosome arm resolution, chromosome alignment and segregation, and cytokinesis [108]. AURKA, PLK1, Cdc25B, and Cdc25C are heavily recruited to centrosomes in late G2, leading to centrosome maturation and recruitment of the Cdk1-cyclin B complex. Following its activation at the G2/M boundary, AURKA activates PLK1 by phosphorylation of a highly conserved residue (threonine 210) in the T-loop of the PLK1 kinase domain [41]. In parallel, PLK1 and AURKA also promote phosphorylation of Cdc25 and B-type cyclin, thereby further promoting the formation of B-type cyclin-CDK1 complexes [109]. Cdk1-cyclin B- and PLK1-dependent phosphorylation of Wee1 [110] promotes its degradation and of Myt1 [111] decreases its kinase activity; once Wee1 and Myt1 activities are lost, the cell enters M phase.

The activity of AURKA and PLK1 at the centrosome and radiating microtubule asters promotes formation and function of the bipolar spindle; another Aurora kinase, AURKB, is active at the centromere/kinetochore and supports appropriate DNA alignment in metaphase. After alignment of chromosomes has been achieved,

Fig. 7.1 (continued) PLK1 and Aurora kinases. Aurora A activates PLK1 and they together play an important role in checkpoint reversal and mitotic entry. Also, spindle assembly checkpoint in M phase with multiple components is not shown. *The bottom part of each cell cycle phase* lists traditional chemotherapy agents and current agents in clinical trials targeting specific proteins in the corresponding cell cycle phase for treating HNSCCs. *In the figure*, *solid arrows* show activation; *dotted arrows*, downstream impact of events such as replicative senescence, mitogenic signals, transcription of S-phase proteins, DNA damage, and no repair in mitosis; *arrows with blunt ends*, inhibition. P, phosphorylation; PP, hyper phosphorylation; pRB, phosphorylated form of Retinoblastoma protein

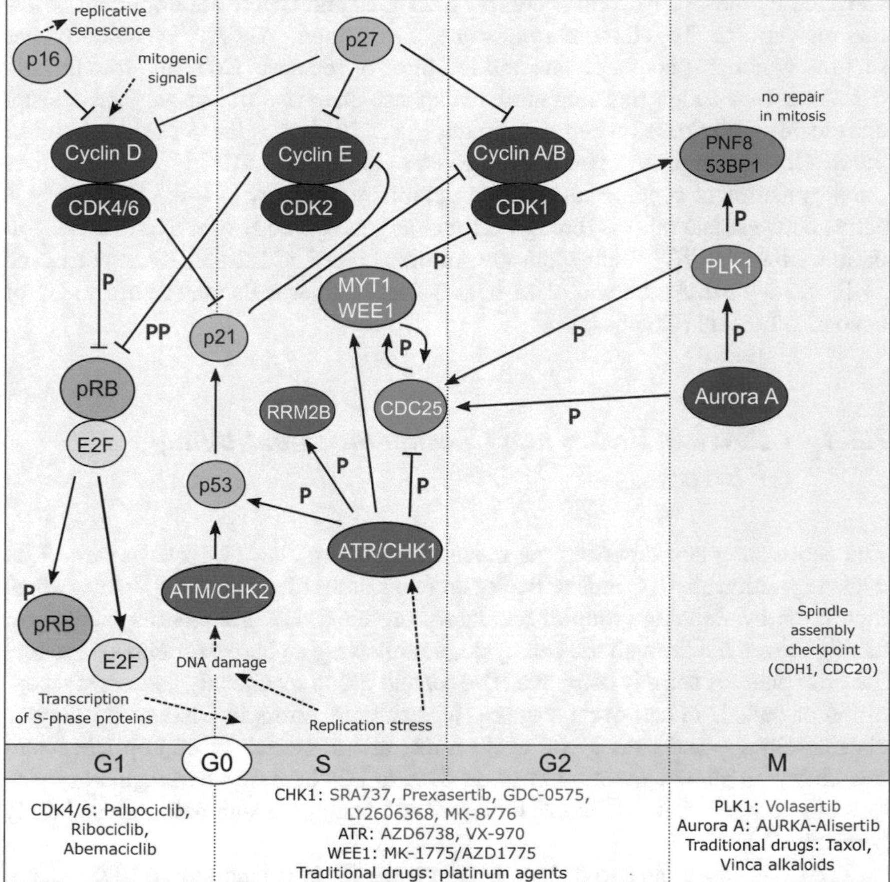

Fig. 7.1 Cell cycle regulatory pathways and key therapeutic targets. The figure presents the main cell cycle regulatory pathways and the current targeted therapy drugs for some of the main cell cycle players. The scheme is divided into four parts according to four cell cycle phases – G1, S, G2, and M. The main regulatory proteins in each phase are cyclins (pink) and cyclin-dependent kinases (CDKs, purple) complexes. *G1 phase*: mitogenic signals stimulate cyclin D/CDK4/6 complex for S-phase progression that is achieved by phosphorylation of RB family of proteins. Such phosphorylation prevents pRB growth inhibitory properties and activates the E2F family of transcription factors to transcribe S phase-relevant genes. DNA damage signals in the G1 phase majorly activate the ATM/CHK2/P53 pathway, and if repaired, the cell can transition to the S phase. *S phase*: replication stress in the S phase activates checkpoint kinases of the ATR/CHK1 complex leading to Wee1/Myt1 complex activation – which prevents replication origin firing (CDK1 activity inhibition), inhibitory phosphorylation of CDC25 (which helps to activate CDK complexes) or CDK activator kinases (not shown), and other factors such as RRM2B (to increase deoxynucleotide pools). *G2/M phases*: The main complex in this phase is cyclin 1 with CDK A and/or B. CDK1 levels are regulated in G2 by inhibitory phosphorylation of Wee1 and Myt1 kinases. Also, the above cyclin-CDK complex is driven by regulatory mitotic kinases such as

an E3 ubiquitin ligase – the anaphase-promoting complex/cyclosome (APC/C) – is activated by the Cdk1-cyclin B complex. CDC20 and CDH1 are important cofactors of APC/C. To allow chromosome segregation, APC/C^{Cdc20} ubiquitylates securin, which is a co-chaperone and inhibitor of separase. Cohesin proteins hold the sister chromatids together until metaphase-anaphase transition. These sister chromatids are released when the separase protein cleaves the cohesin subunit of the SCC1 cohesin protein (further reviewed in [112, 113]). APC/C^{Cdc20} also causes the degradation of cyclin B and thereby inhibits the activity of Cdk1. APC/C^{Cdh1} is active between late mitosis through G1 and targets A- and B-type cyclin for degradation. Other APC/C^{Cdh1} substrates are Aurora A and B, Cdc20, PLK1, securin, and SKP2 [114–116]. At the end of M phase, the daughter cells may return to G1 or become quiescent (G0 phase).

7.2.4 Cell Cycle Brakes and Checkpoints: Controlling for Errors

The above summary describes the machinery driving the cell cycle forward. Cell cycle regulation also depends critically on two classes of negative regulators, which involve an overlapping group of regulatory proteins. One group is responsible for pacing movement through the cell cycle, not allowing a phase to commence before the prior phase is entirely complete. The second group specifically addresses recognition of defects in cell cycle stages – for example, errors in DNA or misaligned chromosomes – and causes cell cycle arrest and activation of appropriate repair machinery, to allow correction of these defects. As these negative regulatory proteins also are relevant to HNSCC biology and therapy, we summarize these briefly here (Fig. 7.1).

The CKIs constitute two distinct families, the INK4 (INhibitors of CDK4) family, with four members (p16^{INK4a}, p15^{INK4b}, p18^{INK4c}, p19^{INK4d}) that exclusively and directly bind to and inhibit D-type cyclin-dependent kinases (CDK4 and CDK6), and the CIP/KIP family, with three members (p21$^{CIP1/WAF1}$, p27^{KIP1}, p57^{KIP2}) that inhibit the activity of all CDKs [75]. The p16 INK4a/ARF locus has been implicated as an important target in cancers based on genetic alterations in human cancers and murine models [117–119]. However, the other INK4 or CIP/ KIP loci do not appear to be involved in carcinogenesis [117, 120, 121]. p21$^{CIP1/WAF1}$ (encoded by the p53 responsive gene, *CDKN1A*) is a universal cell cycle inhibitor that is directly controlled by p53 [122–124].

In G1 phase, DNA damage activates ataxia telangiectasia-mutated (ATM)/checkpoint kinase 2 (CHK2) and ATM- and Rad3-related (ATR)/checkpoint kinase 1 (CHK1) to a lesser degree. Activation of this signaling cascade leads to p53-p21-mediated G1/S-checkpoint signaling and results in G1-arrest [22, 125]. During G1 phase, DNA damage such as DNA double-strand breaks (DSBs, the most lethal form of DNA damage) is primarily repaired by the proteins involved in nonhomologous end-joining (NHEJ) pathways [54]. Not surprisingly, the Rb proteins, p16^{INK4a}

and p53, are potent tumor suppressors that are frequently altered in cancers; mutation or viral-protein-mediated degradation with loss of RB and p53 is the most common molecular alterations in HPV(−) and HPV-associated HNSCCs, respectively.

In S phase, DNA is proofread and DNA damage is repaired/resolved by proteins involved in intra-S-phase checkpoint signaling. The S-phase checkpoint regulates progression through S phase and is mediated by the activity of sensory kinases (ATM/ATR) and checkpoint-signaling kinases (CHK2/CHK1) [54]. These kinases engage effector proteins from various DNA repair pathways to repair DNA damage and also activate CKIs such as p21 and/or inhibit the phosphatase cdc25. Inhibition of cdc25 prevents the activation of B-type cyclin-Cdk1, thus preventing S/G2 transition [81, 99, 102]. Exclusively in S phase, homologous recombination (HR) pathways can mediate error-free repair utilizing the sister chromatid as a template for repair. Interestingly, recent evidence also suggests that CDK activity plays specific roles in HR repair; Cdk1 and 2 can phosphorylate BRCA2 thereby modulating its interaction with the HR protein, RAD51 [126, 127], promoting end-resection and stimulating HR. *BRCA*-mutated tumors are highly sensitive to poly (ADP-ribose) polymerase inhibitors (PARPi) and Cdk1 inhibition may compromise HR. The combination of PARPi and Cdk1 inhibitors sensitized BRCA-proficient human tumor xenografts in a mouse model of lung adenocarcinoma, and this combination has been proposed as a therapeutic strategy [77, 128].

Genomic instability is a hallmark of cancer, and, not surprisingly, cancer cells often have dysregulation of the cell cycle as well as dysregulation of DNA damage response (DDR) signaling, increased replication stress, and loss of certain DDR pathways; these abnormalities all lead to increased levels of endogenous DNA damage. Unrepaired DNA damage can trigger checkpoint activity (p53, ATM/ATR, CHK1/2, gamma (γ)-H2AX), Cdk2 activity, and replication stress [54, 57, 76, 111, 129–131]. Replication stress can result in fork slowing and fork stalling, with extended regions of RPA (replication protein A)-coated single-stranded DNA. Such single-stranded DNA leads to the activation of signaling cascades mainly involving the ATR kinase, ATRIP (ATR-Interacting protein), CHK1, and DNA-PK signaling [132]. ATR- and CHK1-kinase led signaling events prevent replication-fork collapse and new-origin firing. Of note, PARP binding to CHK1 at stalled replication forks is needed for checkpoint activation [133]. ATR signaling also increases levels of ribonucleotide reductase M2 (RRM2, which generates deoxyribonucleotides from ribonucleotides) and the activity of fork-remodeling enzymes such as SMARCAL1 (to prevent generation of DSBs) [83]. Thus, dysregulation or inhibition of the ATR-CHK1 pathway leads to increased replication stress and DSBs. Increased replication stress can also select for the inactivation of certain DDR pathways, mainly loss of p53 and mutations in the DDR kinases. Replication stress may also enhance intra-tumor heterogeneity with the emergence of DDR-defective clones within the tumor [76, 111]. Thus, targeting ongoing vulnerabilities in the S phase of the cell cycle has been a long-standing basis of tumor therapy. Understanding the intimate relationship between cell cycle regulators and DDRs has identified new therapeutic targets for HNSCC (Fig. 7.1).

The DNA damage checkpoint in G2 is largely mediated by the ATR/CHK1 signaling cascade. Activation of this pathway can drive inactivation of Cdc25, Cdk1 inhibition (via inhibitory phosphorylation mediated by Wee1 and Myt1), and G2 arrest, preventing entry into mitosis. Activation of the G2/M checkpoint prevents forced entry of damaged cells into mitosis, avoiding senescence or mitotic catastrophe. Mitotic catastrophe results in cell death either in M phase or in the subsequent G1 phase of the cell cycle (reviewed in [23, 134]). The hallmarks of mitotic catastrophe include large nonviable cells with several micronuclei (nuclear envelopes around individual clusters of unsegregated chromosomes), regions of uncondensed chromatin, abnormal duplication of centrosomes, multipolar mitoses, abnormal segregation of chromosomes, depletion of centrosomal proteins that perturbs microtubule organization, and inhibition of some proteins that contribute to G2 checkpoints (including ATM, ATR, CHK1 and 2, PLK1, PIN1, MLH1, p53, and p53-inducible checkpoint regulators) [135–137]. After repair, PLK1 and Aurora A can mediate reversal of checkpoint arrest and drive mitotic entry [138–142].

In mitosis, several factors prevent DNA repair to avoid promotion of genomic instability [143]. DNA breaks during mitosis are marked, but not repaired. During mitosis, double-stranded DNA ends are recognized by the Mre11/Rad50/Nbs1 (MRN) complex, which is recruited to sites of DNA damage. ATM kinase is activated, which leads to the phosphorylation of H2AX to γH2AX, and recruitment of MDC1 (mediator of DNA damage checkpoint) to the damaged sites, which in turn recruits DNA repair proteins. However, there is no activation of downstream DDR signals such as recruitment of 53BP1 (TP53-binding protein), ubiquitination of histones, or recruitment of DDR components that require ubiquitinated histones as a recruitment signal (BRCA1, RAP80, and others) [143]. Inhibitory Cdk1/cyclin B- and PLK1-dependent phosphorylation of the E3 ubiquitin ligase protein RNF8 (at T198) and 53BP1 (at sites T1609 and S1618) prevents the recruitment of these DSB repair proteins during mitosis [144]. Further, Plk1- and Cdk1-dependent phosphorylation of the DNA repair protein, XRCC4 (X-ray repair cross complementing 4 protein, at serine 326 in the C-terminus), prevents NHEJ repair during mitosis. As XRCC4 also stimulates the activity of other repair proteins such as DNA ligase IV and polynucleotide kinase/phosphatase (PNKP) for correct ligation of DNA breaks, their action is also inhibited during mitosis [145]. This tight control of mitotic DSB repair prevents the production of dicentric chromosomes and aneuploidy [146, 147].

Another important checkpoint ensures the precise segregation of the chromosomes during mitosis, regulated by the spindle assembly checkpoint (SAC)[23]. The SAC delays the metaphase-anaphase transition until all chromosomes are properly aligned on the mitotic spindle. However, if the SAC fails or is inactivated [148], cells still continue with segregation of chromatids and mitotic exit, termed as adaptation or mitotic slippage. This can lead to aneuploidy and genetic instability. The segregation of chromosomes is delayed by proteins in the mitotic checkpoint complex (MCC), which include MPS1, BUB1, BUBR1, and others until all chromosomes are aligned [40, 149]. Specifically, Aurora family kinases play an important role in the regulation of the SAC. Aurora B is an important and active participant of the catalytic subunit of the chromosome passenger complex (CPC) in SAC, as it phosphorylates unattached kinetochores and ATM (phosphorylation site – S1403) and recruits SAC kinase MPS1. ATM phosphorylation by Aurora B in response to DNA damage also leads to the phosphorylation of other SAC components, such as BUB1 (S314) and MAD1 (S214) and activation of mitotic checkpoint arrest. Also, CHK1 assists Aurora B phosphorylation activity in kinetochores [150, 151].

7.3 Changes in Cell Cycle Regulators Associated with HNSCC Pathogenesis

Changes affecting regulators of cell cycle progression observed in HNSCC include overexpression of cyclins, loss of tumor suppressor genes, and mutations in checkpoint proteins. The disease biology has been studied in relation to HPV status and tobacco and/or alcohol exposure. HPV-associated HNSCCs are more treatment responsive than HPV(−) HNSCCs [4, 152, 153]. We here discuss the biology, molecular, and genetic changes associated with deregulation of the cell cycle found in HPV-associated versus HPV(−) HNSCC (Fig. 7.2).

7.3.1 Biology, Chromosomal, and Genomic Changes

Somatic mutations, chromosomal arrangements, and structural alterations in cell cycle regulators are commonly detected in HNSCCs (Fig. 7.2, Tables 7.2 (mutations) and 7.3 (structural alterations or copy number alterations or CNA)) [65, 68, 70, 154–159]. Structural perturbations are observed regardless of HNSCC HPV status. Several differences in these alterations are in cell cycle regulatory genes between HPV (−) and HPV-associated tumors (Fig. 7.2). It is also interesting to note that the cell cycle genes that are frequently altered in HNSCC connect functionally (Fig. 7.3). As early as 2002, molecular differences in HPV-associated and HPV(−) HNSCCs were appreciated [160]. Below we review the HPV-related specific alterations related to the two genetic subclasses of HNSCC.

7.3.1.1 *Cell Cycle Alterations in HPV-Associated HNSCC* (Fig. 7.2a)

As discussed above, prior studies have identified alterations in several cell cycle genes, which were confirmed by The Cancer Genome Atlas (TCGA) study and work from other groups [4, 64, 65, 153, 161–163]. Further studies with additional HPV-associated HNSCC tumors are needed to extend our understanding of the

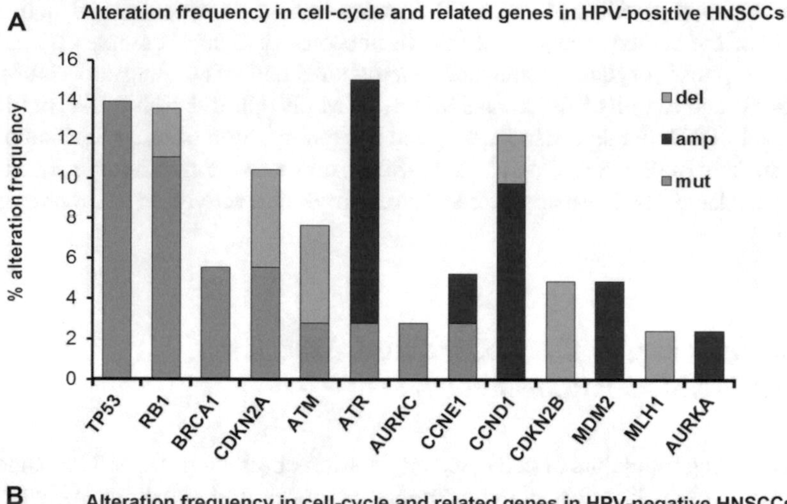

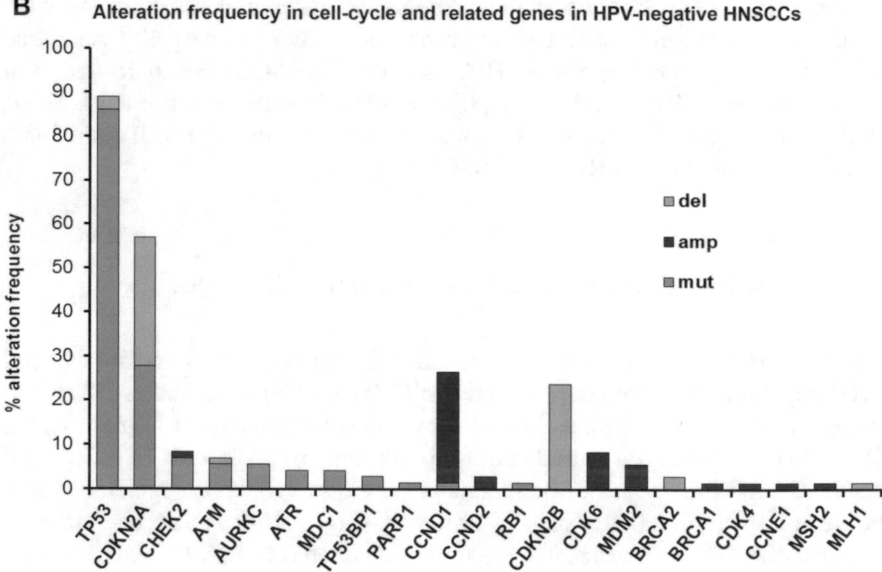

Fig. 7.2 Alteration frequency of cell cycle and related genes in HPV-associated (**a**) and HPV-negative (**b**) HNSCCs. The data was downloaded from The Cancer Genome Atlas (TCGA) database (06/16/2017). The studies determined HPV− status by p16 testing. Mut, mutation; del, deletion; amp, amplification; amp*, amplification of the same gene but in another cytoband

biology of this disease. The TCGA database uses a classification of HNSCC based on expression profiles, as proposed by Chung et al., which are the [1] classical, [2] basal, [3] mesenchymal, and [4] atypical groups, with HPV-associated HNSCC included in the atypical group and characterized by upregulation of the cycle and DNA replication genes [64]. Leemans et al. have proposed a classification of tumors as [1] HPV-associated tumors, [2] HPV(−) tumors with high chromosome instability (high CIN), or [3] HPV(−) tumors with low CIN [164].

Table 7.2 Mining of TCGA database for mutations in HPV-positive and HPV-negative HNSCCs

HPV-positive			
Gene	# Mut	#	% Freq
TP53	5	5	13.89
RB1	4	4	11.11
BRCA1	3	2	5.56
CDKN2A	2	2	5.56
ATM	1	1	2.78
ATR	1	1	2.78
AURKC	1	1	2.78
CCNE1	1	1	2.78
HPV-negative			
Gene	# Mut	#	% Freq
TP53	76	62	86.11
CDKN2A	20	20	27.78
CHEK2	5	5	6.94
ATM	5	4	5.56
AURKC	4	4	5.56
ATR	3	3	4.17
MDC1	3	3	4.17
CDKN	2	2	2.78
PARP1	1	1	1.39
CCND1	1	1	1.39
PRB1	1	1	1.39

HPV status was determined by p16 testing. Total number of patients in the TCGA provisional cohort, 528 (512 samples are profiled); HPV-positive, 41 (36 samples are profiled); HPV-negative, 73 (72 samples are profiled); N/A, 414. Data was downloaded on 06/16/2017. The table shows frequently mutated cell cycle genes in HNSCC. # Mut, number of mutations in the tested gene; #, number of samples carrying a mutation in the tested gene; Freq, frequency of alteration

HPV-associated tumors usually have wild-type *TP53* and *CDKN2A* genes. The cancer-promoting action of HPV can be attributed to the HPV viral proteins, E6 and E7, which act synergistically to induce carcinogenesis by targeting and degrading the cell cycle regulators p53 and Rb, respectively. E6 and E7 are transcribed together as part of a viral polycistronic E6/E7 full-length mRNA. Two alternatively spliced transcripts, E6*I and E6*II, can also be produced and have been found in high-risk strains (high-risk and low-risk HPV strains have been extensively reviewed in Chap. 20 of this book). The E6 protein interacts with E3 ubiquitin-ligase E6-associated protein (E6AP), forming a complex that polyubiquitinates p53 and causes its subsequent proteasomal degradation. Only the E6 protein from high-risk HPV strains has been found to degrade p53. Although the mechanism is not understood, it has been speculated that it might be due to differences in E6 localization or expression levels, among high- vs. low-risk strains [165]. The HPV E7 protein can associate with phospho Rb (pRb) and related pocket proteins p107 and p130, and through another interaction with cullin 2 ubiquitin ligase complex, the protein causes proteasomal

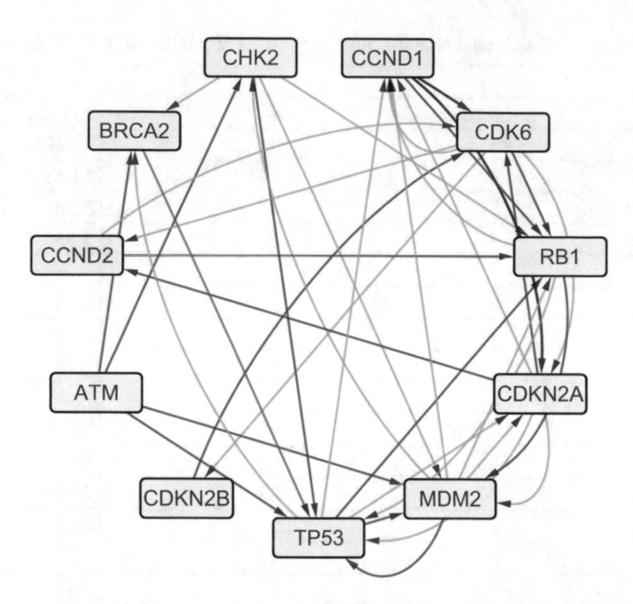

Fig. 7.3 An interactive network of cell cycle-related genes that are frequently mutated in HNSCCs. The network represents interactions within a group of related cell cycle genes (yellow boxes), which are frequently altered in HNSCCs (Fig. 7.2). Purple lines represent direct interaction; gray lines, physical association; blue lines, indirect association; and black lines, colocalization. Data was downloaded from the BioGRID database (06/18/2017) and analyzed by Cytoscape v. 3.5.1.

degradation of pRb. The E7 protein from high-risk types has a higher affinity for pRb (compared to E6), making it better able to degrade pRb. E7 promotes E2F transcription by binding directly to E2F1 and enhancing its transcriptional activity, and additionally, E7 can bind and inhibit E2F6, a negative feedback repressor of E2F1 activation [166]. The E6 and E7 viral proteins have also been shown to interact with histone-modifying enzymes such as histone acetylases and deacetylases and modulate their transcriptional coactivator activity. For example, E6 protein inhibits the histone acetyltransferase p300/CBP-mediated acetylation of p53 and thus inhibits p53-dependent gene activation [167]. E6 and E7 from high-risk types cooperate to uncouple centrosome duplication from the cell cycle, inducing genetic instability. As a result, cell division carries the risk of chromosomal missegregation, and increases in copy number alterations can be observed [168]. In additional activities, E7 also impacts cell division through the inhibition of p21 and p27, which leads to the activation of Cdk2 [169]. E6 also activates telomerase reverse transcriptase (TERT) through its interactions with SP1, MYC, nuclear transcription factor, X box-binding protein-123, and UBE3A (ubiquitin-protein ligase 3A), thereby promoting immortalization of the HPV-infected cells. Another HPV protein, E5, exhibits weak oncogenic properties as an independent gene, but it enhances the interactions of E6 and E7 and their downregulation of p21 and p27. Emerging evidence suggests that the E5, E6, and E7 proteins also help HPV-infected cells escape immune cells, based on mechanisms reviewed in [169].

HPV-DNA may be integrated into the DNA of the infected cells, may be present in the non-integrated form, or both. HPV integration sites are distributed throughout the genome, apparently at random. HPV integration usually occurs in the presence of unrepaired double-stranded DNA breaks (thought to be the main factor of integration in cervical carcinomas) [170]. HPV integration also leads to high and stable expression of E6 and E7 and deregulates cell cycle and DDR genes, such as *RAD51B* and *TP63* [170].

Overall, there are relatively few other alterations in cell cycle genes in HPV-associated HNSCCs. The TCGA has mined data from 528 HNSCC cases, out of which 41 cases are HPV-associated (confirmed by p16 testing, data accessed on 06/16/2017). A previously published study with the initial TCGA HNSCC cohort (279 cases) showed *TP53* mutations were found in 84% of 243 HPV (−) patients, while only 3% of HPV-associated ($n = 36$) had the mutation [69]. Similarly, although *CCND1* was amplified in 31% of 243 HPV (−) patients, only 3% of HPV-associated cases had the mutation.

HNSCCs overall (without the HPV classification) had a 72% rate of *TP53* mutations in TCGA (and 41% in COSMIC data) [64, 171, 172]. Analysis of the 41 HPV-associated cases in the current (referenced on 06/16/2017) provisional TCGA data set (of 528 cases overall) suggests that *RB1* mutations are present in ~10–11% of the HPV-associated cases. By contrast, another study which focused on genomic differences between HPV-associated and HPV(−) HNSCC cases, in the 52 (out of 120 total) HPV-associated cases, the *RB1* mutation frequency was 6% (similar to the published TCGA data at ~5%) [64, 65]. *CDKN2A* can be inactivated due to either inactivating mutations or hypermethylation [64, 156]. TCGA studies have also predominantly identified focal amplifications in E2F1 (~20% cases) [173]. Of relevance to cell cycle control, amplifications in *ATR* (~10–12%), mutations or deletions in *ATM* (~8%), and amplifications in *MDM2* (~5%) and *AURKA* (<5%) are also observed in these tumors (see Fig. 7.2a). Mutations have also been reported in *SLX4* (8.33%), *FANCD2* (5.56%), *BRCA1* (5.56%), *FANCA* (2.78%), *FANCM* (2.78%), and *FANCI* (2.78%) and also amplifications in *FANCG* (2.44%) [64]. These genetic alterations in cell cycle and DDR genes may have therapeutic implications for targeted agents, which are currently in development and/or in clinical trials (Table 7.1, also see Sect. 7.4).

7.3.1.2 *Biology and Cell Cycle Alterations in HPV(−) HNSCC* (Fig. 7.2b)

Previous studies have identified genetic alterations in tumor suppressors, *CDKN2A* [174], and *TP53* [175], as key events in the pathogenesis of HPV(−) HNSCCs. The TCGA analysis confirmed these findings; *TP53* (80–87%) and *CDKN2A* (50–60%) are very commonly mutated in HPV(−) tumors [64]. Analysis of TCGA studies for HNSCC tumors with available data for copy number variation and mutational profile ($n = 73$, with 72 confirmed HPV (−)) also confirmed genetic changes in other cell cycle regulatory genes including *CHEK2* (7%), *ATM* (5.5%), *AURKC* (5.5%), and *ATR* (4%), and CNAs in *CDKN2A* (29.17%, deletion), *CCND1* (25%, amplification), *CDKN2B* (23.61%, deletion), and *ATR* (19.44%, amplification) are very commonly mutated in HPV(−) tumors [141] (see Table 7.2 and Fig. 7.2b). Also, changes in other cell cycle and DDR-related genes such as the Fanconi anemia genes in HPV(−) tumors have been detected. The frequency of mutations observed is *FANCG* at 2.78%, while *BRIP1, ERCC4, SLX4, AURKA, RAD51C*, and *PALB2* are all altered at the frequency of 1.39% each. Amplification frequencies are *FANCG*, 6.94%, and *FANCC*, 2.78%, while *BRCA1, FANCM, FANCL, FANCI, FANCA, FANCF*, and *ERCC4* are all altered at the frequency of less than 2% each. Finally, deletions have been observed in *BRCA2* (2.78%) and *FANCD2* (1.39%) [64]. Also, promoter hypermethylation of *CDKN2A* is commonly observed with loss of gene expression in most HPV(−) HNSCCs. The *TP53* mutations related to exogenous damage from alcohol and cigarette smoke are usually frameshift mutations [159].

Due to the high frequency of *TP53* mutations, several studies have suggested improved survival for patients with WT *TP53* compared to mutated *TP53*. WT *TP53* status may be a universal marker for a better outcome for all HNSCCs [176]. The relationship of *TP53* and prognosis is reviewed extensively in Chap. 9 of this book; we here focus on the impact on cell cycle and discuss survival and response to therapy (Sect. 7.3.2).

Evaluation of *CDKN2A* status has been discordant among different groups, with TCGA reporting 50–60% alterations, while other studies range genomic alterations from 4% to 74% [65, 67–70]. Studies of *CDKN2A* are complicated by difficulties in sequencing this gene, due to its high GC content [177]. Beck et al. reported that patients with low mRNA expression of *CDKN2A* (RNA-Seq studies) had reduced survival ($p = 0.037$) [178]. This is consistent with previous work indicating improved survival for p16-positive non-oropharyngeal squamous cell carcinoma patients [162]. Interestingly, TCGA studies have also observed increased amplifications (~30%) in *CCND1* (confirming earlier results) [64, 65, 69, 178]. Beck et al. in their studies from the TCGA data set also reported that cases with high mRNA expression of *CCND1* correlated with reduced survival and co-occurred frequently with *CDKN2A* deletions [178].

A recent study of the TCGA HNSCC cohort has identified that microRNA let-7c, which is a cell cycle regulator, is frequently inactivated in both HPV-associated and HPV(−) HNSCC. Reduced expression of this microRNA is associated with increased expression of important cell cycle genes such as *CDK4*, *CDK6*, *E2F1*, and *PLK1* [64, 179]. Heritable forms of HNSCC have been associated with mutations in Lynch syndrome and Fanconi anemia genes [180]. Another study also concluded that BRCA1 and MDM2 (E3 ubiquitin ligase, p53 inhibitor) could be potential biomarkers for HNSCC [181]. The TCGA data suggests that compared to the 27 anatomical tumor sites studied, the mutation load in HNSCCs is highly significant (placed 9 out of 27 sites). To summarize (both HPV-associated and HPV(−) HNSCCs), loss of heterozygosity in chromosomal regions that contain cell cycle genes such as 9p21 (*CDKN2A*) [174] and 17p13 (*TP53*) [175] and amplification of 11q13 (*CCND1*) [182] may drive HNSCC tumorigenesis. *TP53* mutations are considered the earliest and most frequent alteration in HNSCC, with mutations found in 50–80% of the cases [64, 175]. pRb is targeted early in HNSCC carcinogenesis through the inactivation of the *CDKN2A* gene, with mutations and copy number changes. *CCND1* (encodes for cyclin D1) is amplified or overexpressed in HNSCCs. In addition, *TP53* mutations, p16 loss, and overexpression of cyclin D1 are associated with reduced survival in HNSCCs. The increased rate of *TP53* mutation has been associated with tobacco and alcohol use in HNSCC cases. With the advent of next-generation sequencing, large-scale HNSCC studies have been performed (such as the TCGA) which have supported and confirmed the findings from the previous studies [64].

7.3.2 Cell Cycle Changes Impact Survival and Response to Standard HNSCC Therapies

In HNSCC, *TP53* mutations, as well as loss of p53 functionality, can affect treatment response and prognosis in patients. A 2007 study by Poeta et al. analyzed 420 patient samples (not differentiated by HPV status but likely predominantly HPV-unrelated) from the Eastern Cooperative Oncology Group (E4393) study to identify *TP53* mutations and their impact on survival and prognosis in HNSCC [176]. Of the 420 tumor samples, 43% were from the oral cavity, 21% from the larynx, 22% from the oropharynx, and 7.6% from the hypopharynx; 5.2% were unknown or belonged to other regions, and 0.7% had multiple tumor sites. According to this study, patients with any *TP53* mutation had a worse median overall survival (3.2 years) than median overall survival for those with WT *TP53* (5.4 years). Patients with disruptive *TP53* mutations had a lower median overall survival at 2 years rather than 3.9 years for nondisruptive *TP53* mutations. In HPV-associated HNSCC tumors, loss of p53 activity results due to the degradation of protein rather than a mutation in *TP53*; it is speculated that the better treatment response and prognosis (independent of treatment modality) are linked to residual wild-type activity [1]. It is also important to note that certain HNSCCs such as nasopharyngeal carcinomas (NPC) have lower rates of *TP53* mutations (as in the case of HPV-associated HNSCCs) and, in general, up to 95% of these overexpress WT p53: these cancers differ from other HNSCCs in incidence rates, epidemiology, and treatment [183, 184]. A study analyzing NPC tumors and matched normal tissue showed that loss of p53 function is a result of ΔNp63 activity (inhibitor of p53) and silencing of p14 (which normally inhibits the p53 inhibitor, MDM2) through hypermethylation [185, 186].

TP53 mutations (Fig. 7.2 and Tables 7.2 and 7.3) are often associated with resistance to DNA-damaging therapy (radiation and platinum chemotherapy). Mechanistically, in response to DNA damage, p53 is stabilized by phosphorylation at S15 (conferred by the kinases ATM and ATR) and at S37 (by ATR) and promotes cell cycle arrest by increasing expression of the CDK inhibitory proteins p21 and p27 and other actions. Mutations causing inactivation or loss of p53 lead to loss of checkpoint integrity, which results in genomic instability, as cells can undergo rep-

Table 7.3 Mining of TCGA data for copy number alterations in HPV-negative and HPV-positive HNSCCs

HPV-positive				
Gene	Cytoband	NA	#	% Freq
ATR	3q23	MP	5	12.2
CCND1	11q13	MP	4	9.76
CDKN2A	9p21	EL	2	4.88
CDKN2B	9p21	EL	2	4.88
ATM	11q22-q23	EL	2	4.88
MDM2	12q14.3-q15	MP	2	4.88
CCND1	11q13	EL	1	2.44
CCNE1	19q12	MP	1	2.44
RB1	13q14.2	EL	1	2.44
MLH1	3p21.3	EL	1	2.44
AURKA	20q13	MP	1	2.44
HPV-negative				
Gene	Cytoband	NA	#	% Freq
CDKN2A	9p21	EL	21	29.17
CCND1	11q13	MP	18	25
CDKN2B	9p21	EL	17	23.61
ATR	3q23	MP	14	19.44
CDK6	7q21-q22	MP	6	8.33
MDM2	12q14.3-q15	MP	4	5.56
TP53	17p13.1	EL	2	2.78
CCND2	12p13	MP	2	2.78
PARP1	1q41-q42	MP	2	2.78
BRCA2	13q12.3	EL	2	2.78
ATM	11q22-q23	EL	1	1.39
CHEK2	22q12.1	MP	1	1.39
BRCA1	17q21	MP	1	1.39
CDK4	12q14	MP	1	1.39
CCNE1	19q12	MP	1	1.39
MSH2	2p21	MP	1	1.39
MLH1	3p21.3	EL	1	1.39

HPV status determined by p16 testing was analyzed. Total number of patients in the TCGA provisional cohort, 528 (522 samples are profiled); HPV-positive, 41 (41 samples are profiled); HPV-negative, 73 (72 samples are profiled); N/A, 414. Data was downloaded on 06/16/2017. The table shows frequently mutated cell cycle genes in HNSCC. CNA, copy number alteration; AMP, amplification; DEL, deletion; #, number of samples carrying a mutation in the tested gene; Freq, frequency of alteration

lication with damaged DNA, resulting in further gene mutations and amplifications [187, 188]. P53 can also activate the apoptosis pathway if DNA damage cannot be repaired. In this case, p53 transcriptionally activates pro-apoptotic proteins of the Bcl-2 family, which include Bax, as well as BH3-only protein members (Bid, Bad, Puma, and Noxa) [189]. This leads to the permeabilization of the mitochondrial membrane and the downstream activation of the apoptotic machinery [190]. Hence, defects in *TP53* activity result in failure of cells to activate apoptotic pathways and are associated with resistance to therapy with DNA-damaging treatments such as radiation and platinum agents [9, 191].

The p53 protein has multiple domains with defined functions, such as the N-terminal transactivation (TA) domain (residues 1–73), a proline-rich region [63–97], the central DNA-binding core domain (residues 94–312), the C-terminus oligomerization domain (residues 324–355), and an unstructured basic domain (residues 360–393) [192]. Of the total 528 HNSCC cases in the TCGA, sequence data are available for 504: among these, 243 cases have missense *TP53* mutations that are found primarily (239/243) in the central DNA-binding domain; 173 are truncating mutations that are spread throughout the p53 structure, and 10 are in-frame deletions that are also localized to the DNA-binding domain. Among cases with defined HPV status, one HPV-associated case had a truncating mutation (DNA-binding domain) and four HPV-associated cases had missense mutations (three in DNA-binding domain and one in the C-terminal) [64]. Mutations in the oligomerization domain can prevent tetramerization of the p53 tetramer, which is essential for its in vivo activity. The majority of *TP53* mutations in the DNA-binding domain prevent p53 binding to its target DNA sequences [193]. Importantly, because p53 is a tetramer, a damaging point mutation affecting DNA binding in one of the alleles can act in a dominant negative manner to inhibit the transcriptional activity of p53 [154].

TP53 mutations can be classified as loss of function (LOF) or gain of function (GOF) mutations, with both classes altering p53 function in a pathogenic manner. Deletions or LOF mutations in *TP53* lead to an inability to arrest at G1, causing affected cells to rely on the G2 checkpoint arrest to maintain genomic stability. However, such cells eventually lose the ability to arrest in G2 and are forced to enter mitosis with unrepaired DNA damage, leading to mitotic catastrophe [193]. *TP53* GOF mutations alter the DNA-binding and protein interaction abilities in a manner which promotes oncogenesis (as reviewed in [194]). *TP53* with GOF mutations such as R282W and R723H can regulate p63 and p67 and promote perinuclear localization of these proteins, suppressing their activity. Similarly, p53 with GOF mutations R175H and R273H can interact and inhibit the tumor suppressor, Kruppel-like-factor 17 (KLF17). These mutations along with R248W can also transactivate oncogenes such as CYP3A4, promoting drug resistance. Mutations such as R282W and R248Q have been shown to cause an increase in mir-155 levels, which is important for posttranscriptional control of gene expression that promotes cellular transformation and invasion [194]. H1299 (p53-null human non-small-cell lung carcinoma) cells were used to show that p53 mutants (R175H, conformational mutant and R248W and R273H, DNA-binding site mutants) have a dominant negative effect on WT p53 when co-expressed. These mutants inhibited the ability of p53 to transactivate target genes and negated WT p53's ability to suppress proliferation and induce cell cycle arrest. Additionally, these mutants did not increase the expression of the inhibitor MDM2, nor were they degraded by MDM2 [195]. Such mutations have been identified in HNSCCs, according to the TCGA data [64].

There are differences in treatment outcomes with DNA-damaging agents in tumors with wild-type p53 protein that can be attributed to mutations targeting proteins that directly or indirectly impact the function of p53. For example, mutations in downstream p53 apoptotic effectors such as the apoptotic proteins, Apaf1 and Casp9, result in increased cancer cell proliferation and reduced sensitivity to DNA damage and thereby increased resistance to therapy [196]. The effect of *TP53* mutations on cisplatin resistance has been extensively studied in cell lines. Cell lines with R72 polymorphism were more resistant due to mutations affecting nuclear localization signal of p53 [197]. p53 mutations such as F134C, A161S, Y236C, R175H, C176F, and C238F were shown to cause cisplatin resistance in vitro [198]. Interestingly, cisplatin resistance due to *TP53* mutation or loss of function can be overcome by CHK1 or CHK2 inhibition, which may suggest some therapeutic combinations [199]. However, some studies have produced conflicting data, where cisplatin sensitivity was observed with *TP53* mutations. In a 2003 study [200], 23 HNSCC cell lines were evaluated, where 13 had *TP53* mutations, and these correlated with cisplatin sensitivity (in vitro IC50 studies). Other studies by Hoffman et al. [201] and Andrews et al. [202] also found that cell lines with p53 mutations had increased susceptibility to cisplatin-induced apoptosis. Despite conflicting data, as discussed above, both preclinical and clinical studies support the use of cisplatin as a radiosensitizer in *TP53*-mutated HNSCCs [203–207].

CDKN2A is another cell cycle protein that might play an important role in HNSCC treatment response. In HPV(−) HNSCCs, *CDK2NA* deletions are common, while in HPV-associated cancers, CDKN2A function is generally intact [156]. CDKN2NA/B inactivation has been observed in leukemia patients (with greater inactivation observed in patients who relapse after therapy) [208] and was also confirmed in mouse models of lymphoblastic leukemia [209]. Cyclin D1 expression can also vary among HNSCC tumors; in a study that analyzed 48 HNSCC tumors, 21 (43.75%) had weak, 18 (37.5%) had intermediate, and 9 (18.8%) had strong expression, and there was a significant correlation between tumor stage but not tumor grade and cyclin D1 expression [210]. For HNSCCs that have an intermediate or strong expression of *CCND1*, resistance to receptor tyrosine kinase inhibitors (such as gefitinib) has been studied in HNSCC cell lines [211]. A 2015 study analyzed 94 HPV(−)HNSCC tumor samples from the oral cavity (44%), tongue (22%), and the glottis (17%). The authors reported that inhibitory CDK4/6 phosphorylation of Rb1 on threonine 356 was associated with worse prognosis and reduced overall survival (OS). The authors speculated that if high phosopho-Rb1 at threonine 356 predicts poor outcome, then increased CDK4/6 activity should also predict survival and may be used as a detectable biomarker. CDK4/6 activity can be influenced by loss of function mutations in *CDKN2A* (p16) or mRNA overexpression and/or gain of function mutations of *CDK4, CDK6,* and *CCND1* (cyclin D1) [178]. Additionally, elevated expression of DNA repair proteins in tumors can also lead to therapeutic resistance to radiation therapy and platinum-based chemotherapy [212]. Overexpression of the DNA repair proteins XPF and ECCR1 has been observed in HNSCC tumors and has been reported to cause resistance to cytotoxic drugs such as cisplatin [213].

7.4 Key Cell Cycle Targets and Current Clinical Trials

A number of cell cycle regulatory proteins have been taken forward for preclinical and clinical evaluation in HNSCCs.

7.4.1 *Conventional Chemotherapy and Limitations*

Chemotherapy has been the most widely used form of treatment for metastatic or recurrent HNSCC and also improves outcomes in combination with radiation for locally advanced HNSCC. Conventional chemotherapy agents for treating HNSCC have been platinum agents (such as cisplatin and carboplatin) and taxanes (such as docetaxel and paclitaxel) [214]. Other conventional agents are methotrexate and 5-fluorouracil (5-FU). However, nonspecific activity, treatment resistance, and toxicities associated with conventional chemotherapy agents have fueled the development of novel cell cycle-targeted therapies for HNSCC [14, 15, 17, 83, 215].

7.4.2 *Cyclin-Dependent Kinase (CDK) Inhibitors*

The first CDK inhibitors were relatively nonspecific and were termed "pan-CDK" inhibitors (reviewed in [216]). The pan-CDK inhibitor alvocidib (flavopiridol, Tolero Pharmaceuticals) was a first-generation CDK inhibitor (selective CDK9 inhibitor, with some activity against CDK4/6) that demonstrated the initial in vivo antitumor activity against multiple human tumor xenografts of prostate carcinoma, head and neck cancer, non-Hodgkin's lymphoma, and leukemia [217]. However, flavopiridol did not exhibit any durable responses in phase II studies with multiple solid tumor types (reviewed extensively in [216, 218]). In parallel, seliciclib (roscovitine, Cyclacel) was the first purine-based CDK inhibitor and was among the first clinically evaluated in a phase I trial. This was a relatively more selective inhibitor of CDK4 and CDK6 with relatively low affinity [218–220]. Subsequent studies suggest that it also inhibits CDK7 and CDK9 [82, 216, 221]. However, these early pan-CDK inhibitors failed due to various reasons including lack of clear understanding of mechanistic action of the inhibitor (such as which CDK was being inhibited in vivo and how, which confounded the ability to develop these agents for targeted therapies and for combination with other agents); low potency and selectivity; inadequate or inappropriate patient selection; inhibition of CDKs critical for normal cell proliferation and survival; and, above all, lack of therapeutic window, leading to toxicities such as diarrhea, myelosuppression, anemia, and nausea [218].

As discussed in the introduction to this chapter, potent and selective CDK4/CDK6 inhibitors have been developed, such as palbociclib (PD-0332991, Pfizer), ribociclib (LEE011, Novartis), and abemaciclib (LY2835219, Eli Lilly) [30–37]. The rationale for these agents arises from preclinical studies that showed that inhibition of CDK4 or CDK6 is not essential for viability: knockout mice are viable, and severe phenotypes are not observed because of functional compensation between CDK4 and CDK6 [222, 223]. However, CDK4/6 double-knockout mice were anemic in the later stages of embryonic development, with only some cells that were still able to proliferate normally. Further, embryonic fibroblasts lacking CDK4 and CDK6 can only enter S phase with reduced efficacy, which was proposed to be due to CDK2 interaction with cyclin D [222]. These results suggested inhibition of CDK4/6 would be potent in rapidly growing tumors, particularly as tumors usually bypass cell cycle regulation and G1-/S-phase arrest by overexpression of cyclin D and CDK4/6 complexes.

The development of selective CDK4/6 inhibitors has now made it possible to selectively prevent the G1-to-S-phase transition [223]. These inhibitors also have reduced toxicities when compared to traditional chemotherapies [223]. The CDK4/6 inhibitor palbociclib inhibits the phosphorylation of Rb protein, which depends on CDK4/cyclin D activity. Mitogenic signaling leads to enhanced cyclin D expression (transcriptional and posttranscriptional) through the Ras/RAF/MAPK pathway or the PI3K/AKT pathway. Since the D-type cyclins respond to mitogenic signals, combination strategies of CDK4/6 inhibitors with inhibitors of mitogenic signaling pathways such as MEK/ERK or PI3K/AKT are also currently being preclinically and clinically assessed.

The PI3K-dependent activation of AKT causes inhibition of glycogen synthase kinase3-beta (GSK3-β) which, when active, phosphorylates cyclin D, leading to its nuclear export as well as degradation. Hormone and growth factor signaling can also lead to a transcriptional and translational increase of cyclin D. In ER+ breast cancer cells, overexpression of cyclin D/CDKs is observed along with other E2F-dependent genes important for cell proliferation and division [224]. Recently, CDK4/6 inhibitors have been shown to be more effective when used in combination with ER-signaling inhibitors (letrozole, an aromatase inhibitor) [225, 226], or with the PI3K inhibitor, BYL-719, in ER+/HER2- breast cancer [227].

CDK4/6 inhibitors have much potential in cancer types such as HNSCC, in which the integrity of G1/S checkpoints is commonly perturbed [7, 82, 216]. Recently, CDK4/6 inhibition with palbociclib has been tested as a therapeutic strategy in HNSCC [7, 82, 216]. A randomized, phase II study of palbociclib and cetuximab (Erbitux, Eli Lilly, EGFR inhibitor) versus cetuximab is currently being conducted in HPV(−), cetuximab-naïve patients with recurrent or metastatic platinum-refractory HNSCC (NCT02499120). Palbociclib (125 mg daily) is also being evaluated in combination with carboplatin for recurrent or metastatic HNSCC (NCT03194373). A phase I/II dose-escalation study of palbociclib in combination with cetuximab and intensity-modulated radiation therapy (IMRT) for locally advanced unresectable HNSCCs (NCT0302448) is also currently ongoing.

A phase I trial is currently recruiting patients with advanced solid tumors including HNSCC to test the combination of CDK4/6 inhibitor (palbociclib) in combination with the dual PI3K/mTOR inhibitor, gedatolisib (PF-05212384, Pfizer

Inc.) (NCT03065062). A phase I and II trial (NCT0089905) to determine the maximum tolerated dose (MTD) and dose-limiting toxicity of CDK1, 4, and 9 inhibitor, riviciclib (P276-00, Piramal Life Sciences), in combination with radiation therapy was recently completed and is awaiting results. Similarly, another phase II trial with riviciclib was recently completed in Indian subjects with recurrent, metastatic, or unresectable locally advanced HNSCC. All patients were given 144 mg/m^2/day of the inhibitor (NCT00824343). Palbociclib is also being tested in the setting of incurable HNSCC in a phase II trial in comparison to the standard of care. This is a biomarker-enrichment trial referred to as UPSTREAM (NCT03088059); here, patients with recurrent or metastatic platinum-refractory HNSCCs will be allocated in biomarker-defined cohorts. Patients who are p16 negative, with *CCND1* amplification, will be randomized between palbociclib or the current standard of care for HNSCC.

7.4.3 Mitotic Inhibitors: Cytotoxic Chemotherapy (Microtubule Inhibitors) Versus Targeted Kinase Inhibitors

Microtubule inhibitors (MI), such as taxanes (paclitaxel, docetaxel) and vinca alkaloids (vinblastine, vincristine), impact the SAC, induce mitotic arrest, and are frequently employed in HNSCC [17, 83]. The benefit from the addition of docetaxel to a platinum and 5-FU regimen has been reported from multiple trials in HNSCC [214, 228].

The taxanes stabilize pre-existing microtubules, while the vinca alkaloids prevent microtubule polymerization. However, as mentioned in Sect. 7.2.4 of this chapter, cancer cells can often undergo adaptations such as mitotic slippage, which causes the escape of both mitotic arrest and apoptosis. Cells that escape and enter the subsequent G1 phase can become tetraploid or senescent or undergo apoptosis depending on the p53 status of the cells. The cells that continue to divide will lead to increased genomic instability and resistance to MIs. Another limitation of MI is the impact on normal cells, particularly neurons [23]. Side effects include neutropenia, toxicity of hematopoietic cells and myelosuppression, and peripheral neuropathy. The other limitations include overexpression of drug efflux pumps that prevent drug accumulation inside tumor cells, lack of activity on non-mitotic or slowly growing tumor cells, as well as delay in cell death by influencing the levels and activity of apoptotic regulators by antimitotic drugs (regulators such as Mcl-1, Bcl-xL, and Bcl-2) [229]. Thus, the limitations such as increased toxicities and side effects associated with traditional MI agents have fueled the development of novel cell cycle-targeted therapies for HNSCC.

Currently, clinical trials are ongoing for therapies that target mitotic kinases, Aurora and PLK1. Multiple specific (AURKA, AURKB, and AURKC) and pan-Aurora kinase inhibitors have been developed. Alisertib (MLN8237, Takeda Pharmaceuticals) is a highly specific, second-generation AURKA inhibitor [230]. A phase II study with alisertib in patients with platinum-resistant or refractory ovarian cancer has shown modest single-agent antitumor activity [231, 232]. The clinical data for Aurora kinase inhibitors in HNSCC are limited. A phase II study of alisertib was conducted with 249 patients with advanced solid tumors. There were 55 HNSCCs cases assessed for safety and 45 for drug response. For patients with HNSCC, median PFS was 2.7 months. An objective response was observed in 4/47 or 9% of the patients; 3 were HPV-negative, and 1 had unknown HPV status. Four HNSCC patients discontinued treatment because of adverse effects such as fatigue and febrile neutropenia; additionally, two deaths were observed within 30 days of the last dose of alisertib in the HNSCC group. This trial concluded that alisertib as an oral single agent showed positive antitumor activity and tolerable safety profile, at least in patients with advanced breast cancer and small-cell lung cancer [233].

A currently ongoing phase I study (NCT01540682) is testing the combination of an AURKA inhibitor (alisertib) with the EGFR inhibitor cetuximab and definitive radiation. The study is evaluating correlative biomarkers from AURKA and EGFR signaling on pre- and post-therapy biopsy specimens to establish proof of mechanism for this novel combination. Another phase I trial using the EffTox design (a Bayesian adaptive dose-finding trial design that jointly scrutinizes binary efficacy and toxicity outcomes) [234] will evaluate activity and safety of alisertib, in combination with the selective VEGFR inhibitor, pazopanib, in patients with advanced, previously treated non-hematologic solid tumors (NCT01639911). Another combination strategy being explored preclinically in colorectal cancer models combines AURKA kinase inhibitors with MEK inhibitors and may also be useful in HNSCC [235, 236].

The AURKB inhibitor barasertib (AZD1152, AstraZeneca) has shown some promise in acute myeloid leukemia (AML) [237]. Barasertib was used in a phase I dose-escalation study in acute myeloid leukemia (n = 22) patients, in combination with a low-dose cytosine arabinose (LDAC), with about 45% overall response rate [238]. In the phase II trial of this study, barasertib as a single agent was compared with LDAC as a single agent. From the 74 patients in this study, median OS with AZD1152 was 8.2 months and 4.5 months with LDAC [239]. Another AURKB inhibitor, AT9283 (Astex Therapeutics), was used in a phase I study in patients with relapsed/refractory leukemia or myelofibrosis (n = 48). However, these patients showed no objective response, despite evidence of AURKB inhibition [240]. In a phase II study of a pan-Aurora inhibitor, danusertib (PHA-739358, Nerviano Medical Sciences), comprising of 223 patients (advanced or metastatic breast, ovarian, colorectal, pancreatic, small-cell, and non-small-cell lung cancers), it was estimated that the PFS at 4 months was 18.4% in breast cancer, 12.1% in ovarian cancer, 10.0% in pancreatic cancer, 10.4% in non-small-cell lung carcinoma (all histotypes), 16.1% in squamous non-small-cell lung carcinoma, and 0% in small-cell lung carcinoma [241]. In another phase II study with danusertib, 88 patients with metastatic prostate cancer were divided into two groups and treated with different doses (330 and 500 mg/m^2, respectively). The results showed SD in 8 (for 330 mg/m^2) and 13 (500 mg/m^2) patients, respectively. Additionally, overall 13.6% treated patients had SD for ≥6 months. Danusertib was estimated as generally well-tolerated [242]. These agents have not yet been assessed in HNSCC but may be useful in this setting.

As discussed in Sect. 7.2 of this chapter, the PLK1 mitotic kinase is activated by AURKA, which phosphorylates PLK1 at T210. Additionally, a PLK1-T210D phospho-mimetic mutant was shown to be able to overcome AURKA-dependent checkpoint recovery [41]. Therefore, targeting PLK1 instead of or addition to AURKA can be promising in HNSCCs that have PLK1 overexpression or aberrant activation. In a preclinical study in NPC cells (which overexpress PLK1), it was shown that combining AURKA inhibitor (MK-5108) or AURKB inhibitor (barasertib) with a PLK1 inhibitor (BI 2536) increased mitotic defects such as mitotic slippage and induced metaphase arrest [65].

Volasertib (Boehringer Ingelheim) is a potent and selective PLK1 inhibitor [243]. A phase I trial treated 65 patients at doses of 12–450 mg. Reversible hematological toxicity was the main adverse effect. Three patients achieved partial response (PR), while SD was reported in 40% patients. Two patients remained progression-free for over 1 year [244, 245]. There are no current volasertib HNSCC trials. The next generation of inhibitors targeting the mitotic kinases may be more effective if they are combined to target cancer-specific vulnerabilities such as pathway defects, genomic instability, cell cycle checkpoints, etc. Such strategies may be synthetic lethal and may produce the desired therapeutic effect. Another strategy under clinical testing (NCT01954316) is against PLK4 (CFI-400945) [246], a key regulator of centriole duplication. This is a phase I trial in advanced cancers to evaluate the highest dose level that does not lead to unacceptable toxicity and measure the pharmacokinetic profile. PLK4 is also dysregulated in cancer cells and promotes genetic instability. CFI-400945 inhibits PLK4 and other kinases such as AURKB. Other mitotic kinases governing centrosome dynamics and mitotic spindle function are also potential targets such as MASTL, Haspin, and Nek [139].

7.4.4 Targeting ATR/CHK1/WEE1 Signaling

The ablation of G2-DNA damage checkpoint by drugs targeting the ATR/CHK1/WEE1 pathway is being actively tested in preclinical and clinical models for multiple cancers including HNSCC (see Table 7.1). These agents alone or in combination with other DNA-damaging therapies are being tested with promising results in p53-mutant tumors [247–249]. However, a study recently also showed that WEE1 inhibitors could sensitize independent of p53 status. Drugs with these G2-DNA damage checkpoint activities include ATR (AZD6738 (AstraZeneca) and VX-970 (Vertex)), CHK1 (GDC-0575 (Genentech), prexasertib (LY2606368, Eli Lilly), and MK-8776 (Merck)) and WEE1 inhibitors (AZD1775, AstraZeneca) (reviewed in [76]).

In a phase I dose-escalation trial, the side effects and optimal dose of the ATR kinase inhibitor, VX-970, in combination with cisplatin and radiation therapy are being assessed in patients with locally advanced HPV(−) HNSCC (NCT02567422). The primary outcome measures are frequency and grade of toxic response, including the establishment of dose-limiting toxicities, and assessment of the highest doses of cisplatin and VX-790 that can be safely combined with radiation therapy. Secondary outcome measures are metabolic and objective response rates. Pharmacokinetic properties of VX-790 will also be assessed. Another currently recruiting modular phase I study (NCT02264678) is to assess the safety, tolerability, pharmacokinetics, and preliminary antitumor activity of the ATR inhibitor AZD6738 in combination with either cytotoxic chemotherapy (carboplatin) and/or the DNA damage repair inhibitor (PARP inhibitor, olaparib (AstraZeneca)) or durvalumab (MED14736, AstraZeneca, anti PD-L1) in patients with advanced solid malignancies including HNSCCs.

A recent study showed that a combination of a CHK1/2 inhibitor (prexasertib, Eli Lilly) with cetuximab and radiation therapy enhanced cytotoxicity in both HPV-associated and HPV(−) HNSCC preclinical models [250]. A phase I trial of the CHK1 inhibitor SRA737 for patients with advanced cancer, including HNSCC, is currently recruiting participants (NCT02797964). A phase I trial with prexasertib for advanced cancers (including HNSCC) was recently completed, but results have not been posted (NCT01115790).

Overexpression of WEE1 has been reported in several cancers such as breast, ovarian, colorectal, gastric and malignant melanoma [77]. Initial preclinical studies showed that WEE1 inhibitors in cancer cells with mutations in the *CDKN2A* locus that lead to abrogation of the G1 to M checkpoints could cause synthetic lethality [251]. In a recent study with HNSCC p53 mutant cells, cisplatin treatment led to a G2-arrest: a combination of cisplatin with WEE1 inhibition using MK-1775 (now renamed AZD1775) abrogated the G2 checkpoint and increased the sensitivity of these HNSCC cells to cisplatin [252].

A phase I study (NCT01748825) was performed to determine the MTD of AZD1775 in patients with solid refractory tumors, with 1 of 25 patients having HNSCC. Patients treated with AZD1775 had an increase in unrepaired DNA damage reflected by elevated γH2AX levels and confirmed an on-target activity, i.e., reduction in pY15-Cdk levels [253]. An interesting outcome of this study was the observation of AZD1775 single-agent activity in the two patients (one HNSCC and one ovarian cancer patient) carrying *BRCA* mutations. A partial response was observed in the patient with HNSCC carrying the *BRCA* mutation.

Another phase I trial of AZD1775 in locally advanced HNSCC is a single-arm dose-escalation trial to determine the MTD in combination with the standard of care chemotherapy cisplatin and radiation (NCT02585973). A secondary outcome measure of this trial is to study the toxicity profile and objective response rate. Another phase I trial is currently recruiting patients to assess the side effects and optimal dose of AZD1775. This is in a presurgery setting with AZD1775, cisplatin, and docetaxel for patients with borderline resectable stage III–IVB HNSCC (NCT02508246). A phase II trial is currently testing AZD1775 alone or in combination with cisplatin for recurrent or metastatic HNSCC (NCT02196168). This study has recently completed and is awaiting results. This study evaluated the WEE1 inhibitor in combination with cisplatin and compared it to a placebo and cisplatin arm. The primary objectives were to assess the overall response rate and efficacy of AZD1775. The efficacy was measured by protein status of p53. The study also explored predictive and pharmacodynamic biomarkers.

7.5 Conclusions and Future Directions

New classes of cell cycle inhibitors are showing clinical promise as monotherapy and in combination with traditional chemotherapy and/or radiation and with other targeted agents. The results from ongoing preclinical and clinical studies will further highlight the factors that govern their efficacy in HNSCC and potential therapeutic biomarkers that could be further developed. Development of novel cell cycle targets would be further facilitated by a better definition of tumor-specific requirements for growth and survival. Identification of these tumor-specific susceptibilities can lead to the development of synthetic lethal combinations to enhance tumor cell killing. In the setting of HNSCC, the most clinically relevant distinction is the HPV status. The similarities and differences between HPV-positive and HPV-negative subtypes are an ongoing area of study in the effort to develop personalized therapies for HNSCC patients. Further studies are needed to evaluate synthetic lethal interactions between genetic changes (e.g., drugs combinations for ATR/CHK1/WEE1 inhibition) in HNSCC and other cancers. Therapeutic strategies involving correction of cell cycle-related genomic alterations such as *TP53* mutations in HNSCC via CRISPR-CAS9 genome editing methods may also be possible in the future [254]. The first CRISPR genome editing human trial commenced in 2016, delivering gene-edited PD-1 knockout engineered T-cells into a patient with aggressive lung cancer [255]. The results from this first trial are being evaluated and more trials will be starting this year. The development of new technologies combined with an advanced understanding of disease biology and genetics should help optimize the use of cell cycle-targeting agents.

References

1. Dok R, Nuyts S. HPV positive head and neck cancers: molecular pathogenesis and evolving treatment strategies. Cancers (Basel). 2016;8:pii: E41. https://doi.org/10.3390/cancers8040041.
2. Jenkins G, O'Byrne KJ, Panizza B, Richard DJ. Genome stability pathways in head and neck cancers. Int J Genomics. 2013;2013:464720. https://doi.org/10.1155/2013/464720.
3. Gillison ML, Lowy DR. A causal role for human papillomavirus in head and neck cancer. Lancet. 2004;363:1488–9. https://doi.org/10.1016/S0140-6736(04)16194-1.
4. Ang KK, et al. Human papillomavirus and survival of patients with oropharyngeal cancer. N Engl J Med. 2010;363:24–35. https://doi.org/10.1056/NEJMoa0912217.
5. Feldman R, et al. Molecular profiling of head and neck squamous cell carcinoma. Head Neck. 2016;38(Suppl 1):E1625–38. https://doi.org/10.1002/hed.24290.

6. Bingham HG, Copeland EM, Hackett R, Caffee HH. Breast cancer in a patient with silicone breast implants after 13 years. Ann Plast Surg. 1988;20:236–7.
7. Riaz N, Morris LG, Lee W, Chan TA. Unraveling the molecular genetics of head and neck cancer through genome-wide approaches. Genes Dis. 2014;1:75–86. https://doi.org/10.1016/j.gendis.2014.07.002.
8. Deshpande AM, Wong DT. Molecular mechanisms of head and neck cancer. Expert Rev Anticancer Ther. 2008;8:799–809. https://doi.org/10.1586/14737140.8.5.799.
9. Rothenberg SM, Ellisen LW. The molecular pathogenesis of head and neck squamous cell carcinoma. J Clin Invest. 2012;122:1951–7.
10. Worsham MJ, Ali H, Dragovic J, Schweitzer VP. Molecular characterization of head and neck cancer: how close to personalized targeted therapy? Mol Diagn Ther. 2012;16:209–22. https://doi.org/10.2165/11635330-000000000-00000.
11. Morgan DO. Cyclin-dependent kinases: engines, clocks, and microprocessors. Annu Rev Cell Dev Biol. 1997;13:261–91. https://doi.org/10.1146/annurev.cellbio.13.1.261.
12. Hartwell LH, Weinert TA. Checkpoints: controls that ensure the order of cell cycle events. Science. 1989;246:629–34.
13. Shapiro GI, Harper JW. Anticancer drug targets: cell cycle and checkpoint control. J Clin Invest. 1999;104:1645–53. https://doi.org/10.1172/JCI9054.
14. Dasari S, Tchounwou PB. Cisplatin in cancer therapy: molecular mechanisms of action. Eur J Pharmacol. 2014;740:364–78. https://doi.org/10.1016/j.ejphar.2014.07.025.
15. Florea AM, Busselberg D. Cisplatin as an anti-tumor drug: cellular mechanisms of activity, drug resistance and induced side effects. Cancers (Basel). 2011;3:1351–71. https://doi.org/10.3390/cancers3011351.
16. Rozencweig M, von Hoff DD, Slavik M, Muggia FM. Cis-diamminedichloroplatinum (II). A new anticancer drug. Ann Intern Med. 1977;86:803–12.
17. Weaver B, How A. Taxol/paclitaxel kills cancer cells. Mol Biol Cell. 2014;25:2677–81. https://doi.org/10.1091/mbc.E14-04-0916.
18. Herbst RS, Khuri FR. Mode of action of docetaxel – a basis for combination with novel anticancer agents. Cancer Treat Rev. 2003;29:407–15.
19. Bissery MC, Nohynek G, Sanderink GJ, Lavelle F. Docetaxel (Taxotere): a review of preclinical and clinical experience. Part I: preclinical experience. Anticancer Drugs. 1995;6:339–55., 363–338.
20. Altmann KH. Microtubule-stabilizing agents: a growing class of important anticancer drugs. Curr Opin Chem Biol. 2001;5:424–31.
21. Gabrielli B, Brooks K, Pavey S. Defective cell cycle checkpoints as targets for anti-cancer therapies. Front Pharmacol. 2012;3:9. https://doi.org/10.3389/fphar.2012.00009.
22. Kastan MB, Bartek J. Cell-cycle checkpoints and cancer. Nature. 2004;432:316–23. https://doi.org/10.1038/nature03097.
23. Visconti R, Della Monica R, Grieco D. Cell cycle checkpoint in cancer: a therapeutically targetable double-edged sword. J Exp Clin Cancer Res. 2016;35:153. https://doi.org/10.1186/s13046-016-0433-9.
24. Loibl S, Turner NC, Ro J, Cristofanilli M, Iwata H, Im SA, Masuda N, Loi S, André F, Harbeck N, Verma S, Folkerd E, Puyana Theall K, Hoffman J, Zhang K, Bartlett CH, Dowsett M. Palbociclib Combined with Fulvestrant in Premenopausal Women with Advanced Breast Cancer and Prior Progression on Endocrine Therapy: PALOMA-3 Results. Oncologist. 2017 Sep;22(9):1028-1038. doi: 10.1634/theoncologist.2017-0072. Epub 2017 Jun 26.
25. Verma S, Bartlett CH, Schnell P, DeMichele AM, Loi S, Ro J, Colleoni M, Iwata H, Harbeck N, Cristofanilli M, Zhang K, Thiele A, Turner NC, Rugo HS. Palbociclib in Combination With Fulvestrant in Women With Hormone Receptor-Positive/HER2-Negative Advanced Metastatic Breast Cancer: Detailed Safety Analysis From a Multicenter, Randomized, Placebo-Controlled, Phase III Study (PALOMA-3). Oncologist. 2016 Oct;21(10):1165-1175. Epub 2016 Jul 1.
26. Hortobagyi G, Stemmer S, Burris H, et al. First-line ribociclib plus letrozole for postmenopausal women with HR+, HER2-, advanced breast cancer: first results from the phase III

MONALEESA-2 study. Presented at the European Society for Medical Oncology (ESMO) Congress, Copenhagen, Denmark (October 8, 2016).
27. Shah A, Bloomquist E, Tang S, Fu W, Bi Y, Liu Q, Yu J, Zhao P, Palmby TR, Goldberg KB, CJG C, Patel P, Alebachew E, Tilley A, Pierce WF, Ibrahim A, Blumenthal GM, Sridhara R, Beaver JA, Pazdur R. FDA Approval: ribociclib for the treatment of postmenopausal women with hormone receptor-positive, HER2-Negative Advanced or Metastatic Breast Cancer. Clin Cancer Res. 2018. https://doi.org/10.1158/1078-0432.CCR-17-2369. clincanres.2369.2017. [Epub ahead of print]. PMID:29437768.
28. Dempsey JA, et al. AACR Annual Meeting; April 6–10, 2013; Washington, DC. Abstract LB122.
29. Gelbert LM, et al. AACR-NCI-EORTC International Conference: Molecular Targets and Cancer Therapeutics; November 12–16, 2011; San Francisco, CA. Abstract B233.
30. Bell T, et al. Impact of palbociclib plus letrozole on pain severity and pain interference with daily activities in patients with estrogen receptor-positive/human epidermal growth factor receptor 2-negative advanced breast cancer as first-line treatment. Curr Med Res Opin. 2016;32:959–65. https://doi.org/10.1185/03007995.2016.1157060.
31. Finn RS, et al. The cyclin-dependent kinase 4/6 inhibitor palbociclib in combination with letrozole versus letrozole alone as first-line treatment of oestrogen receptor-positive, HER2-negative, advanced breast cancer (PALOMA-1/TRIO-18): a randomised phase 2 study. Lancet Oncol. 2015;16:25–35. https://doi.org/10.1016/S1470-2045(14)71159-3.
32. Finn RS, et al. Palbociclib and letrozole in advanced breast cancer. N Engl J Med. 2016;375:1925–36. https://doi.org/10.1056/NEJMoa1607303.
33. Cristofanilli M, et al. Fulvestrant plus palbociclib versus fulvestrant plus placebo for treatment of hormone-receptor-positive, HER2-negative metastatic breast cancer that progressed on previous endocrine therapy (PALOMA-3): final analysis of the multicentre, double-blind, phase 3 randomised controlled trial. Lancet Oncol. 2016;17:425–39. https://doi.org/10.1016/S1470-2045(15)00613-0.
34. Harbeck N, Iyer S, Turner N, Cristofanilli M, Ro J, André F, Loi S, Verma S, Iwata H, Bhattacharyya H, Puyana Theall K, Bartlett CH, Loibl S. Quality of life with palbociclib plus fulvestrant in previously treated hormone receptor-positive, HER2-negative metastatic breast cancer: patient-reported outcomes from the PALOMA-3 trial. Ann Oncol. 2016;27(6):1047–54. https://doi.org/10.1093/annonc/mdw139. Epub 2016 Mar 30.
35. Turner NC, et al. Palbociclib in hormone-receptor-positive advanced breast cancer. N Engl J Med. 2015;373:209–19. https://doi.org/10.1056/NEJMoa1505270.
36. Verma S, Bartlett CH, Schnell P, DeMichele AM, Loi S, Ro J, Colleoni M, Iwata H, Harbeck N, Cristofanilli M, Zhang K, Thiele A, Turner NC, Rugo HS. Palbociclib in combination with fulvestrant in women with hormone receptor-positive/HER2-Negative Advanced Metastatic Breast Cancer: detailed safety analysis from a multicenter, randomized, placebo-controlled, phase III study (PALOMA-3). Oncologist. 2016;21(10):1165–75. Epub 2016 Jul 1
37. Hortobagyi GN, et al. Ribociclib as first-line therapy for HR-positive, advanced breast cancer. N Engl J Med. 2016;375:1738–48. https://doi.org/10.1056/NEJMoa1609709.
38. de Carcer G, Manning G, Malumbres M. From Plk1 to Plk5: functional evolution of polo-like kinases. Cell Cycle. 2011;10:2255–62. https://doi.org/10.4161/cc.10.14.16494.
39. Glover DM, Hagan IM, Tavares AA. Polo-like kinases: a team that plays throughout mitosis. Genes Dev. 1998;12:3777–87.
40. Kops GJ, Weaver BA, Cleveland DW. On the road to cancer: aneuploidy and the mitotic checkpoint. Nat Rev Cancer. 2005;5:773–85. https://doi.org/10.1038/nrc1714.
41. Macurek L, et al. Polo-like kinase-1 is activated by aurora A to promote checkpoint recovery. Nature. 2008;455:119–23. https://doi.org/10.1038/nature07185.
42. Asteriti IA, De Mattia F, Guarguaglini G. Cross-talk between AURKA and Plk1 in mitotic entry and spindle assembly. Front Oncol. 2015;5:283. https://doi.org/10.3389/fonc.2015.00283.
43. Nikonova AS, Astsaturov I, Serebriiskii IG, Dunbrack RL Jr, Golemis EA. Aurora A kinase (AURKA) in normal and pathological cell division. Cell Mol Life Sci. 2013;70:661–87. https://doi.org/10.1007/s00018-012-1073-7.

44. Wirtz-Peitz F, Nishimura T, Knoblich JA. Linking cell cycle to asymmetric division: aurora-A phosphorylates the Par complex to regulate Numb localization. Cell. 2008;135:161–73. https://doi.org/10.1016/j.cell.2008.07.049.
45. Chan CS, Botstein D. Isolation and characterization of chromosome-gain and increase-in-ploidy mutants in yeast. Genetics. 1993;135:677–91.
46. Rhind N, Russell P. Signaling pathways that regulate cell division. Cold Spring Harb Perspect Biol. 2012;4:a005942. https://doi.org/10.1101/cshperspect.a005942.
47. Wells NJ, et al. The C-terminal domain of the Cdc2 inhibitory kinase Myt1 interacts with Cdc2 complexes and is required for inhibition of G(2)/M progression. J Cell Sci. 1999;112(Pt 19):3361–71.
48. Booher RN, Holman PS, Fattaey A. Human Myt1 is a cell cycle-regulated kinase that inhibits Cdc2 but not Cdk2 activity. J Biol Chem. 1997;272:22300–6.
49. Liu F, Rothblum-Oviatt C, Ryan CE, Piwnica-Worms H. Overproduction of human Myt1 kinase induces a G2 cell cycle delay by interfering with the intracellular trafficking of Cdc2-cyclin B1 complexes. Mol Cell Biol. 1999;19:5113–23.
50. Den Haese GJ, Walworth N, Carr AM, Gould KL. The Wee1 protein kinase regulates T14 phosphorylation of fission yeast Cdc2. Mol Biol Cell. 1995;6:371–85.
51. Coleman TR, Dunphy WG. Cdc2 regulatory factors. Curr Opin Cell Biol. 1994;6:877–82.
52. Rowley R, Hudson J, Young PG. The wee1 protein kinase is required for radiation-induced mitotic delay. Nature. 1992;356:353–5. https://doi.org/10.1038/356353a0.
53. Manic G, Obrist F, Sistigu A, Vitale I. Trial watch: targeting ATM-CHK2 and ATR-CHK1 pathways for anticancer therapy. Mol Cell Oncol. 2015;2:e1012976. https://doi.org/10.1080/23723556.2015.1012976.
54. Smith J, Tho LM, Xu N, Gillespie DA. The ATM-Chk2 and ATR-Chk1 pathways in DNA damage signaling and cancer. Adv Cancer Res. 2010;108:73–112. https://doi.org/10.1016/B978-0-12-380888-2.00003-0.
55. Keith CT, Schreiber SL. PIK-related kinases: DNA repair, recombination, and cell cycle checkpoints. Science. 1995;270:50–1.
56. Lecona E, Fernandez-Capetillo O. Replication stress and cancer: it takes two to tango. Exp Cell Res. 2014;329:26–34. https://doi.org/10.1016/j.yexcr.2014.09.019.
57. Mazouzi A, Velimezi G, Loizou JI. DNA replication stress: causes, resolution and disease. Exp Cell Res. 2014;329:85–93. https://doi.org/10.1016/j.yexcr.2014.09.030.
58. Wallace MD, Southard TL, Schimenti KJ, Schimenti JC. Role of DNA damage response pathways in preventing carcinogenesis caused by intrinsic replication stress. Oncogene. 2014;33:3688–95. https://doi.org/10.1038/onc.2013.339.
59. Reinhardt HC, Yaffe MB. Kinases that control the cell cycle in response to DNA damage: Chk1, Chk2, and MK2. Curr Opin Cell Biol. 2009;21:245–55. https://doi.org/10.1016/j.ceb.2009.01.018.
60. van den Heuvel S, Harlow E. Distinct roles for cyclin-dependent kinases in cell cycle control. Science. 1993;262:2050–4.
61. Malumbres M, Barbacid M. Mammalian cyclin-dependent kinases. Trends Biochem Sci. 2005;30:630–41. https://doi.org/10.1016/j.tibs.2005.09.005.
62. Gopinathan L, Ratnacaram CK, Kaldis P. Established and novel Cdk/cyclin complexes regulating the cell cycle and development. Results Probl Cell Differ. 2011;53:365–89. https://doi.org/10.1007/978-3-642-19065-0_16.
63. Lim S, Kaldis P. Cdks, cyclins and CKIs: roles beyond cell cycle regulation. Development. 2013;140:3079–93. https://doi.org/10.1242/dev.091744.
64. Cancer Genome Atlas Network. Comprehensive genomic characterization of head and neck squamous cell carcinomas. Nature. 2015;517:576–82. https://doi.org/10.1038/nature14129.
65. Seiwert TY, et al. Integrative and comparative genomic analysis of HPV-positive and HPV-negative head and neck squamous cell carcinomas. Clin Cancer Res. 2015;21:632–41. https://doi.org/10.1158/1078-0432.CCR-13-3310.
66. Lin DC, et al. The genomic landscape of nasopharyngeal carcinoma. Nat Genet. 2014;46:866–71. https://doi.org/10.1038/ng.3006.

67. Pickering CR, et al. Squamous cell carcinoma of the oral tongue in young non-smokers is genomically similar to tumors in older smokers. Clin Cancer Res. 2014;20:3842–8. https://doi.org/10.1158/1078-0432.CCR-14-0565.
68. Pickering CR, et al. Integrative genomic characterization of oral squamous cell carcinoma identifies frequent somatic drivers. Cancer Discov. 2013;3:770–81. https://doi.org/10.1158/2159-8290.CD-12-0537.
69. Stransky N, et al. The mutational landscape of head and neck squamous cell carcinoma. Science. 2011;333:1157–60. https://doi.org/10.1126/science.1208130.
70. Agrawal N, et al. Exome sequencing of head and neck squamous cell carcinoma reveals inactivating mutations in NOTCH1. Science. 2011;333:1154–7. https://doi.org/10.1126/science.1206923.
71. Russell P, Nurse P. cdc25+ functions as an inducer in the mitotic control of fission yeast. Cell. 1986;45:145–53.
72. Pines J. Four-dimensional control of the cell cycle. Nat Cell Biol. 1999;1:E73–9. https://doi.org/10.1038/11041.
73. Donzelli M, Draetta GF. Regulating mammalian checkpoints through Cdc25 inactivation. EMBO Rep. 2003;4:671–7. https://doi.org/10.1038/sj.embor.embor887.
74. Serrano M, Hannon GJ, Beach D. A new regulatory motif in cell-cycle control causing specific inhibition of cyclin D/CDK4. Nature. 1993;366:704–7. https://doi.org/10.1038/366704a0.
75. Sherr CJ, Roberts JM. Inhibitors of mammalian G1 cyclin-dependent kinases. Genes Dev. 1995;9:1149–63.
76. O'Connor MJ. Targeting the DNA damage response in cancer. Mol Cell. 2015;60:547–60. https://doi.org/10.1016/j.molcel.2015.10.040.
77. Geenen JJJ, Schellens JHM. Molecular pathways: targeting the protein kinase Wee1 in cancer. Clin Cancer Res. 2017;23:4540–4. https://doi.org/10.1158/1078-0432.CCR-17-0520.
78. Matheson CJ, Backos DS, Reigan P. Targeting WEE1 kinase in cancer. Trends Pharmacol Sci. 2016;37:872–81. https://doi.org/10.1016/j.tips.2016.06.006.
79. Otto T, Sicinski P. Cell cycle proteins as promising targets in cancer therapy. Nat Rev Cancer. 2017;17:93–115. https://doi.org/10.1038/nrc.2016.138.
80. Besson A, Dowdy SF, Roberts JM. CDK inhibitors: cell cycle regulators and beyond. Dev Cell. 2008;14:159–69. https://doi.org/10.1016/j.devcel.2008.01.013.
81. Malumbres M, Barbacid M. Cell cycle, CDKs and cancer: a changing paradigm. Nat Rev Cancer. 2009;9:153–66. https://doi.org/10.1038/nrc2602.
82. Sherr CJ, Beach D, Shapiro GI. Targeting CDK4 and CDK6: from discovery to therapy. Cancer Discov. 2016;6:353–67. https://doi.org/10.1158/2159-8290.CD-15-0894.
83. Sherr CJ, Bartek AJ. Cell cycle–targeted cancer therapies. Annu Rev Cancer Biol. 2017;1:41–57. https://doi.org/10.1146/annurev-cancerbio-040716-075628.
84. Ewen ME, et al. Functional interactions of the retinoblastoma protein with mammalian D-type cyclins. Cell. 1993;73:487–97.
85. DeCaprio JA, et al. The product of the retinoblastoma susceptibility gene has properties of a cell cycle regulatory element. Cell. 1989;58:1085–95.
86. Weinberg RA. The retinoblastoma protein and cell cycle control. Cell. 1995;81:323–30.
87. Sherr CJ. The ins and outs of RB: coupling gene expression to the cell cycle clock. Trends Cell Biol. 1994;4:15–8.
88. Cobrinik D, Dowdy SF, Hinds PW, Mittnacht S, Weinberg RA. The retinoblastoma protein and the regulation of cell cycling. Trends Biochem Sci. 1992;17:312–5.
89. Chellappan SP, Hiebert S, Mudryj M, Horowitz JM, Nevins JR. The E2F transcription factor is a cellular target for the RB protein. Cell. 1991;65:1053–61.
90. Dyson N. The regulation of E2F by pRB-family proteins. Genes Dev. 1998;12:2245–62.
91. Lukas J, Bartkova J, Rohde M, Strauss M, Bartek J. Cyclin D1 is dispensable for G1 control in retinoblastoma gene-deficient cells independently of cdk4 activity. Mol Cell Biol. 1995;15:2600–11.
92. Rubin SM. Deciphering the retinoblastoma protein phosphorylation code. Trends Biochem Sci. 2013;38:12–9. https://doi.org/10.1016/j.tibs.2012.10.007.

93. Cappell SD, Chung M, Jaimovich A, Spencer SL, Meyer T. Irreversible APC(Cdh1) inactivation underlies the point of no return for cell-cycle entry. Cell. 2016;166:167–80. https://doi.org/10.1016/j.cell.2016.05.077.
94. Eguren M, Manchado E, Malumbres M. Non-mitotic functions of the anaphase-promoting complex. Semin Cell Dev Biol. 2011;22:572–8. https://doi.org/10.1016/j.semcdb.2011.03.010.
95. Rape M, Kirschner MW. Autonomous regulation of the anaphase-promoting complex couples mitosis to S-phase entry. Nature. 2004;432:588–95. https://doi.org/10.1038/nature03023.
96. Vodermaier HC. APC/C and SCF: controlling each other and the cell cycle. Curr Biol. 2004;14:R787–96. https://doi.org/10.1016/j.cub.2004.09.020.
97. Sherr CJ, Roberts JM. CDK inhibitors: positive and negative regulators of G1-phase progression. Genes Dev. 1999;13:1501–12.
98. Kitagawa M, et al. Phosphorylation of E2F-1 by cyclin A-cdk2. Oncogene. 1995;10:229–36.
99. Perry JA, Kornbluth S. Cdc25 and Wee1: analogous opposites? Cell Div. 2007;2:12. https://doi.org/10.1186/1747-1028-2-12.
100. Boutros R, Lobjois V, Ducommun B. CDC25 phosphatases in cancer cells: key players? Good targets? Nat Rev Cancer. 2007;7:495–507. https://doi.org/10.1038/nrc2169.
101. Heald R, McLoughlin M, McKeon F. Human wee1 maintains mitotic timing by protecting the nucleus from cytoplasmically activated Cdc2 kinase. Cell. 1993;74:463–74.
102. Kristjansdottir K, Rudolph J. Cdc25 phosphatases and cancer. Chem Biol. 2004;11:1043–51. https://doi.org/10.1016/j.chembiol.2004.07.007.
103. Lolli G, Johnson LN. CAK-Cyclin-dependent Activating Kinase: a key kinase in cell cycle control and a target for drugs? Cell Cycle. 2005;4:572–7.
104. Draetta GF. Cell cycle: will the real Cdk-activating kinase please stand up. Curr Biol. 1997;7:R50–2.
105. Shiekhattar R, et al. Cdk-activating kinase complex is a component of human transcription factor TFIIH. Nature. 1995;374:283–7. https://doi.org/10.1038/374283a0.
106. Kaldis P. The cdk-activating kinase (CAK): from yeast to mammals. Cell Mol Life Sci. 1999;55:284–96. https://doi.org/10.1007/s000180050290.
107. Palazzo RE, Vogel JM, Schnackenberg BJ, Hull DR, Wu X. Centrosome maturation. Curr Top Dev Biol. 2000;49:449–70.
108. Lera RF, et al. Decoding polo-like kinase 1 signaling along the kinetochore-centromere axis. Nat Chem Biol. 2016;12:411–8. https://doi.org/10.1038/nchembio.2060.
109. Dutertre S, et al. Phosphorylation of CDC25B by aurora-A at the centrosome contributes to the G2-M transition. J Cell Sci. 2004;117:2523–31. https://doi.org/10.1242/jcs.01108.
110. Lindqvist A, Rodriguez-Bravo V, Medema RH. The decision to enter mitosis: feedback and redundancy in the mitotic entry network. J Cell Biol. 2009;185:193–202. https://doi.org/10.1083/jcb.200812045.
111. Shaltiel IA, Krenning L, Bruinsma W, Medema RH. The same, only different - DNA damage checkpoints and their reversal throughout the cell cycle. J Cell Sci. 2015;128:607–20. https://doi.org/10.1242/jcs.163766.
112. Nasmyth K. Disseminating the genome: joining, resolving, and separating sister chromatids during mitosis and meiosis. Annu Rev Genet. 2001;35:673–745. https://doi.org/10.1146/annurev.genet.35.102401.091334.
113. Peters JM. The anaphase promoting complex/cyclosome: a machine designed to destroy. Nat Rev Mol Cell Biol. 2006;7:644–56. https://doi.org/10.1038/nrm1988.
114. Engelbert D, Schnerch D, Baumgarten A, Wasch R. The ubiquitin ligase APC(Cdh1) is required to maintain genome integrity in primary human cells. Oncogene. 2008;27:907–17. https://doi.org/10.1038/sj.onc.1210703.
115. Garcia-Higuera I, et al. Genomic stability and tumour suppression by the APC/C cofactor Cdh1. Nat Cell Biol. 2008;10:802–11. https://doi.org/10.1038/ncb1742.
116. Li M, et al. The adaptor protein of the anaphase promoting complex Cdh1 is essential in maintaining replicative lifespan and in learning and memory. Nat Cell Biol. 2008;10:1083–9. https://doi.org/10.1038/ncb1768.

117. Ruas M, Peters G. The p16INK4a/CDKN2A tumor suppressor and its relatives. Biochim Biophys Acta. 1998;1378:F115–77.
118. Sherr CJ. Tumor surveillance via the ARF-p53 pathway. Genes Dev. 1998;12:2984–91.
119. Sharpless NE, DePinho RA. The INK4A/ARF locus and its two gene products. Curr Opin Genet Dev. 1999;9:22–30.
120. Hirama T, Koeffler HP. Role of the cyclin-dependent kinase inhibitors in the development of cancer. Blood. 1995;86:841–54.
121. Drexler HG. Review of alterations of the cyclin-dependent kinase inhibitor INK4 family genes p15, p16, p18 and p19 in human leukemia-lymphoma cells. Leukemia. 1998;12:845–59.
122. Xiong Y, et al. p21 is a universal inhibitor of cyclin kinases. Nature. 1993;366:701–4. https://doi.org/10.1038/366701a0.
123. el-Deiry WS, et al. WAF1, a potential mediator of p53 tumor suppression. Cell. 1993;75:817–25.
124. El-Deiry WS. p21(WAF1) mediates cell-cycle inhibition, relevant to cancer suppression and therapy. Cancer Res. 2016;76:5189–91. https://doi.org/10.1158/0008-5472.CAN-16-2055.
125. Warfel NA, El-Deiry WS. p21WAF1 and tumourigenesis: 20 years after. Curr Opin Oncol. 2013;25:52–8. https://doi.org/10.1097/CCO.0b013e32835b639e.
126. Esashi F, et al. CDK-dependent phosphorylation of BRCA2 as a regulatory mechanism for recombinational repair. Nature. 2005;434:598–604. https://doi.org/10.1038/nature03404.
127. Wohlbold L, Fisher RP. Behind the wheel and under the hood: functions of cyclin-dependent kinases in response to DNA damage. DNA Repair (Amst). 2009;8:1018–24. https://doi.org/10.1016/j.dnarep.2009.04.009.
128. Johnson N, et al. Compromised CDK1 activity sensitizes BRCA-proficient cancers to PARP inhibition. Nat Med. 2011;17:875–82. https://doi.org/10.1038/nm.2377.
129. Foster SS, De S, Johnson LK, Petrini JH, Stracker TH. Cell cycle- and DNA repair pathway-specific effects of apoptosis on tumor suppression. Proc Natl Acad Sci U S A. 2012;109:9953–8. https://doi.org/10.1073/pnas.1120476109.
130. Kaufmann WK, Paules RS. DNA damage and cell cycle checkpoints. FASEB J. 1996;10:238–47.
131. Langerak P, Russell P. Regulatory networks integrating cell cycle control with DNA damage checkpoints and double-strand break repair. Philos Trans R Soc Lond B Biol Sci. 2011;366:3562–71. https://doi.org/10.1098/rstb.2011.0070.
132. Buisson R, Boisvert JL, Benes CH, Zou L. Distinct but concerted roles of ATR, DNA-PK, and Chk1 in countering replication stress during S phase. Mol Cell. 2015;59:1011–24. https://doi.org/10.1016/j.molcel.2015.07.029.
133. Min W, et al. Poly(ADP-ribose) binding to Chk1 at stalled replication forks is required for S-phase checkpoint activation. Nat Commun. 2013;4:2993. https://doi.org/10.1038/ncomms3993.
134. Vitale I, Galluzzi L, Castedo M, Kroemer G. Mitotic catastrophe: a mechanism for avoiding genomic instability. Nat Rev Mol Cell Biol. 2011;12:385–92. https://doi.org/10.1038/nrm3115.
135. Castedo M, et al. Cell death by mitotic catastrophe: a molecular definition. Oncogene. 2004;23:2825–37. https://doi.org/10.1038/sj.onc.1207528.
136. Roninson IB, Broude EV, Chang BD. If not apoptosis, then what? Treatment-induced senescence and mitotic catastrophe in tumor cells. Drug Resist Updat. 2001;4:303–13. https://doi.org/10.1054/drup.2001.0213.
137. Vakifahmetoglu H, Olsson M, Zhivotovsky B. Death through a tragedy: mitotic catastrophe. Cell Death Differ. 2008;15:1153–62. https://doi.org/10.1038/cdd.2008.47.
138. Aarts M, Linardopoulos S, Turner NC. Tumour selective targeting of cell cycle kinases for cancer treatment. Curr Opin Pharmacol. 2013;13:529–35. https://doi.org/10.1016/j.coph.2013.03.012.
139. Dominguez-Brauer C, et al. Targeting mitosis in cancer: emerging strategies. Mol Cell. 2015;60:524–36. https://doi.org/10.1016/j.molcel.2015.11.006.

140. Keen N, Taylor S. Aurora-kinase inhibitors as anticancer agents. Nat Rev Cancer. 2004;4:927–36. https://doi.org/10.1038/nrc1502.
141. Strebhardt K, Ullrich A. Targeting polo-like kinase 1 for cancer therapy. Nat Rev Cancer. 2006;6:321–30. https://doi.org/10.1038/nrc1841.
142. Nigg EA. Mitotic kinases as regulators of cell division and its checkpoints. Nat Rev Mol Cell Biol. 2001;2:21–32. https://doi.org/10.1038/35048096.
143. Heijink AM, Krajewska M, van Vugt MA. The DNA damage response during mitosis. Mutat Res. 2013;750:45–55. https://doi.org/10.1016/j.mrfmmm.2013.07.003.
144. Orthwein A, et al. Mitosis inhibits DNA double-strand break repair to guard against telomere fusions. Science. 2014;344:189–93. https://doi.org/10.1126/science.1248024.
145. Mani RS, et al. Dual modes of interaction between XRCC4 and polynucleotide kinase/phosphatase: implications for nonhomologous end joining. J Biol Chem. 2010;285:37619–29. https://doi.org/10.1074/jbc.M109.058719.
146. Lees-Miller SP. DNA double strand break repair in mitosis is suppressed by phosphorylation of XRCC4. PLoS Genet. 2014;10:e1004598. https://doi.org/10.1371/journal.pgen.1004598.
147. Hustedt N, Durocher D. The control of DNA repair by the cell cycle. Nat Cell Biol. 2016;19:1–9. https://doi.org/10.1038/ncb3452.
148. Rossio V, Galati E, Piatti S. Adapt or die: how eukaryotic cells respond to prolonged activation of the spindle assembly checkpoint. Biochem Soc Trans. 2010;38:1645–9. https://doi.org/10.1042/BST0381645.
149. Thompson RC, Dripps DJ, Eisenberg SP. Interleukin-1 receptor antagonist (IL-1ra) as a probe and as a treatment for IL-1 mediated disease. Int J Immunopharmacol. 1992;14:475–80.
150. Lawrence KS, Engebrecht J. The spindle assembly checkpoint: more than just keeping track of the spindle. Trends Cell Mol Biol. 2015;10:141–50.
151. Musacchio A. The molecular biology of spindle assembly checkpoint signaling dynamics. Curr Biol. 2015;25:R1002–18. https://doi.org/10.1016/j.cub.2015.08.051.
152. O'Sullivan B, et al. Outcomes of HPV-related oropharyngeal cancer patients treated by radiotherapy alone using altered fractionation. Radiother Oncol. 2012;103:49–56. https://doi.org/10.1016/j.radonc.2012.02.009.
153. Richards L. Human papillomavirus-a powerful predictor of survival in patients with oropharyngeal cancer. Nat Rev Clin Oncol. 2010;7:481. https://doi.org/10.1038/nrclinonc.2010.123.
154. Zhou G, Liu Z, Myers JN. TP53 mutations in head and neck squamous cell carcinoma and their impact on disease progression and treatment response. J Cell Biochem. 2016;117:2682–92. https://doi.org/10.1002/jcb.25592.
155. Suh Y, Amelio I, Guerrero Urbano T, Tavassoli M. Clinical update on cancer: molecular oncology of head and neck cancer. Cell Death Dis. 2014;5:e1018. https://doi.org/10.1038/cddis.2013.548.
156. Aung KL, Siu LL. Genomically personalized therapy in head and neck cancer. Cancers Head Neck. 2016;1:2. https://doi.org/10.1186/s41199-016-0004-y.
157. Chung CH, et al. Genomic alterations in head and neck squamous cell carcinoma determined by cancer gene-targeted sequencing. Ann Oncol. 2015;26:1216–23. https://doi.org/10.1093/annonc/mdv109.
158. Guerrero-Preston R, et al. Key tumor suppressor genes inactivated by "greater promoter" methylation and somatic mutations in head and neck cancer. Epigenetics. 2014;9:1031–46. https://doi.org/10.4161/epi.29025.
159. Lim AM, et al. Differential mechanisms of CDKN2A (p16) alteration in oral tongue squamous cell carcinomas and correlation with patient outcome. Int J Cancer. 2014;135:887–95. https://doi.org/10.1002/ijc.28727.
160. Wiest T, Schwarz E, Enders C, Flechtenmacher C, Bosch FX. Involvement of intact HPV16 E6/E7 gene expression in head and neck cancers with unaltered p53 status and perturbed pRb cell cycle control. Oncogene. 2002;21:1510–7. https://doi.org/10.1038/sj.onc.1205214.
161. Bhatia A, Burtness B. Human papillomavirus-associated oropharyngeal cancer: defining risk groups and clinical trials. J Clin Oncol. 2015;33:3243–50. https://doi.org/10.1200/JCO.2015.61.2358.

162. Chung CH, et al. p16 protein expression and human papillomavirus status as prognostic biomarkers of nonoropharyngeal head and neck squamous cell carcinoma. J Clin Oncol. 2014;32:3930–8. https://doi.org/10.1200/JCO.2013.54.5228.
163. Lechner M, et al. Targeted next-generation sequencing of head and neck squamous cell carcinoma identifies novel genetic alterations in HPV+ and HPV- tumors. Genome Med. 2013;5:49. https://doi.org/10.1186/gm453.
164. Leemans CR, Braakhuis BJ, Brakenhoff RH. The molecular biology of head and neck cancer. Nat Rev Cancer. 2011;11:9–22. https://doi.org/10.1038/nrc2982.
165. Mesplede T, et al. p53 degradation activity, expression, and subcellular localization of E6 proteins from 29 human papillomavirus genotypes. J Virol. 2012;86:94–107. https://doi.org/10.1128/JVI.00751-11.
166. McLaughlin-Drubin ME, Munger K. The human papillomavirus E7 oncoprotein. Virology. 2009;384:335–44. https://doi.org/10.1016/j.virol.2008.10.006.
167. Durzynska J, Lesniewicz K, Poreba E. Human papillomaviruses in epigenetic regulations. Mutat Res. 2017;772:36–50. https://doi.org/10.1016/j.mrrev.2016.09.006.
168. Duensing S, et al. The human papillomavirus type 16 E6 and E7 oncoproteins cooperate to induce mitotic defects and genomic instability by uncoupling centrosome duplication from the cell division cycle. Proc Natl Acad Sci U S A. 2000;97:10002–7. https://doi.org/10.1073/pnas.170093297.
169. Sano D, Oridate N. The molecular mechanism of human papillomavirus-induced carcinogenesis in head and neck squamous cell carcinoma. Int J Clin Oncol. 2016;21:819–26. https://doi.org/10.1007/s10147-016-1005-x.
170. Romanczuk H, Howley PM. Disruption of either the E1 or the E2 regulatory gene of human papillomavirus type 16 increases viral immortalization capacity. Proc Natl Acad Sci U S A. 1992;89:3159–63.
171. Chow YP, et al. Exome sequencing identifies potentially druggable mutations in nasopharyngeal carcinoma. Sci Rep. 2017;7:42980. https://doi.org/10.1038/srep42980.
172. Forbes SA, et al. COSMIC: somatic cancer genetics at high-resolution. Nucleic Acids Res. 2017;45:D777–83. https://doi.org/10.1093/nar/gkw1121.
173. Beck TN, Golemis EA. Genomic insights into head and neck cancer. Cancers Head Neck. 2016;1:1. https://doi.org/10.1038/nature14129.
174. Riese U, et al. Tumor suppressor gene p16 (CDKN2A) mutation status and promoter inactivation in head and neck cancer. Int J Mol Med. 1999;4:61–5.
175. Greenblatt MS, Bennett WP, Hollstein M, Harris CC. Mutations in the p53 tumor suppressor gene: clues to cancer etiology and molecular pathogenesis. Cancer Res. 1994;54:4855–78.
176. Poeta ML, et al. TP53 mutations and survival in squamous-cell carcinoma of the head and neck. N Engl J Med. 2007;357:2552–61. https://doi.org/10.1056/NEJMoa073770.
177. Wong SQ, et al. Targeted-capture massively-parallel sequencing enables robust detection of clinically informative mutations from formalin-fixed tumours. Sci Rep. 2013;3:3494. https://doi.org/10.1038/srep03494.
178. Beck TN, et al. Phospho-T356RB1 predicts survival in HPV-negative squamous cell carcinoma of the head and neck. Oncotarget. 2015;6:18863–74. https://doi.org/10.18632/oncotarget.4321.
179. Liu Z, et al. Knocking down CDK4 mediates the elevation of let-7c suppressing cell growth in nasopharyngeal carcinoma. BMC Cancer. 2014;14:274. https://doi.org/10.1186/1471-2407-14-274.
180. Birkeland AC, Ludwig ML, Spector ME, Brenner JC. The potential for tumor suppressor gene therapy in head and neck cancer. Discov Med. 2016;21:41–7.
181. Bhowmik A, et al. BRCA1 and MDM2 as independent blood-based biomarkers of head and neck cancer. Tumour Biol. 2016. https://doi.org/10.1007/s13277-016-5359-5.
182. Wang X, et al. Amplification and overexpression of the cyclin D1 gene in head and neck squamous cell carcinoma. Clin Mol Pathol. 1995;48:M256–9.

183. Sheu LF, Chen A, Lee HS, Hsu HY, Yu DS. Cooperative interactions among p53, bcl-2 and Epstein-Barr virus latent membrane protein 1 in nasopharyngeal carcinoma cells. Pathol Int. 2004;54:475–85. https://doi.org/10.1111/j.1440-1827.2004.01654.x.
184. Niedobitek G, et al. P53 overexpression and Epstein-Barr virus infection in undifferentiated and squamous cell nasopharyngeal carcinomas. J Pathol. 1993;170:457–61. https://doi.org/10.1002/path.1711700409.
185. Crook T, Nicholls JM, Brooks L, O'Nions J, Allday MJ. High level expression of deltaN-p63: a mechanism for the inactivation of p53 in undifferentiated nasopharyngeal carcinoma (NPC)? Oncogene. 2000;19:3439–44. https://doi.org/10.1038/sj.onc.1203656.
186. Kwong J, et al. Promoter hypermethylation of multiple genes in nasopharyngeal carcinoma. Clin Cancer Res. 2002;8:131–7.
187. Aas T, et al. Specific P53 mutations are associated with de novo resistance to doxorubicin in breast cancer patients. Nat Med. 1996;2:811–4.
188. Pellegata NS, Antoniono RJ, Redpath JL, Stanbridge EJ. DNA damage and p53-mediated cell cycle arrest: a reevaluation. Proc Natl Acad Sci U S A. 1996;93:15209–14.
189. Fridman JS, Lowe SW. Control of apoptosis by p53. Oncogene. 2003;22:9030–40. https://doi.org/10.1038/sj.onc.1207116.
190. Chipuk JE, et al. Direct activation of Bax by p53 mediates mitochondrial membrane permeabilization and apoptosis. Science. 2004;303:1010–4. https://doi.org/10.1126/science.1092734.
191. Harbour JW, Dean DC. Rb function in cell-cycle regulation and apoptosis. Nat Cell Biol. 2000;2:E65–7. https://doi.org/10.1038/35008695.
192. Laptenko O, Prives C. Transcriptional regulation by p53: one protein, many possibilities. Cell Death Differ. 2006;13:951–61. https://doi.org/10.1038/sj.cdd.4401916.
193. Varley J, Germline M. TP53 mutations and Li-Fraumeni syndrome. Hum Mutat. 2003;21:313–20. https://doi.org/10.1002/humu.10185.
194. Zhang Y, Coillie SV, Fang JY, Xu J. Gain of function of mutant p53: R282W on the peak? Oncogene. 2016;5:e196. https://doi.org/10.1038/oncsis.2016.8.
195. Willis A, Jung EJ, Wakefield T, Chen X. Mutant p53 exerts a dominant negative effect by preventing wild-type p53 from binding to the promoter of its target genes. Oncogene. 2004;23:2330–8. https://doi.org/10.1038/sj.onc.1207396.
196. Soengas MS, et al. Apaf-1 and caspase-9 in p53-dependent apoptosis and tumor inhibition. Science. 1999;284:156–9.
197. Mandic R, et al. Reduced cisplatin sensitivity of head and neck squamous cell carcinoma cell lines correlates with mutations affecting the COOH-terminal nuclear localization signal of p53. Clin Cancer Res. 2005;11:6845–52. https://doi.org/10.1158/1078-0432.CCR-05-0378.
198. Osman AA, et al. Evolutionary action score of TP53 coding variants is predictive of platinum response in head and neck cancer patients. Cancer Res. 2015;75:1205–15. https://doi.org/10.1158/0008-5472.CAN-14-2729.
199. Gadhikar MA, et al. Chk1/2 inhibition overcomes the cisplatin resistance of head and neck cancer cells secondary to the loss of functional p53. Mol Cancer Ther. 2013;12:1860–73. https://doi.org/10.1158/1535-7163.MCT-13-0157.
200. Bradford CR, et al. P53 mutation correlates with cisplatin sensitivity in head and neck squamous cell carcinoma lines. Head Neck. 2003;25:654–61. https://doi.org/10.1002/hed.10274.
201. Hoffmann TK, et al. Alterations in the p53 pathway and their association with radio- and chemosensitivity in head and neck squamous cell carcinoma. Oral Oncol. 2008;44:1100–9. https://doi.org/10.1016/j.oraloncology.2008.02.006.
202. Andrews GA, et al. Mutation of p53 in head and neck squamous cell carcinoma correlates with Bcl-2 expression and increased susceptibility to cisplatin-induced apoptosis. Head Neck. 2004;26:870–7. https://doi.org/10.1002/hed.20029.
203. Ekshyyan O, et al. Comparison of radiosensitizing effects of the mammalian target of rapamycin inhibitor CCI-779 to cisplatin in experimental models of head and neck squamous cell carcinoma. Mol Cancer Ther. 2009;8:2255–65. https://doi.org/10.1158/1535-7163.MCT-08-1184.

204. Tokalov SV, Abolmaali N. Radiosensitization of p53-deficient lung cancer cells by pre-treatment with cytostatic compounds. Anticancer Res. 2012;32:1239–43.
205. Adelstein DJ, et al. An intergroup phase III comparison of standard radiation therapy and two schedules of concurrent chemoradiotherapy in patients with unresectable squamous cell head and neck cancer. J Clin Oncol. 2003;21:92–8. https://doi.org/10.1200/JCO.2003.01.008.
206. Cooper JS, et al. Postoperative concurrent radiotherapy and chemotherapy for high-risk squamous-cell carcinoma of the head and neck. N Engl J Med. 2004;350:1937–44. https://doi.org/10.1056/NEJMoa032646.
207. Forastiere AA, et al. Concurrent chemotherapy and radiotherapy for organ preservation in advanced laryngeal cancer. N Engl J Med. 2003;349:2091–8. https://doi.org/10.1056/NEJMoa031317.
208. Mullighan CG, Williams RT, Downing JR, Sherr CJ. Failure of CDKN2A/B (INK4A/B--ARF)-mediated tumor suppression and resistance to targeted therapy in acute lymphoblastic leukemia induced by BCR-ABL. Genes Dev. 2008;22:1411–5. https://doi.org/10.1101/gad.1673908.
209. Williams RT, den Besten W, Sherr CJ. Cytokine-dependent imatinib resistance in mouse BCR-ABL+, Arf-null lymphoblastic leukemia. Genes Dev. 2007;21:2283–7. https://doi.org/10.1101/gad.1588607.
210. Dhingra V, Verma J, Misra V, Srivastav S, Hasan F. Evaluation of cyclin D1 expression in head and neck squamous cell carcinoma. J Clin Diagn Res. 2017;11:EC01–4. https://doi.org/10.7860/JCDR/2017/21760.9329.
211. Kalish LH, et al. Deregulated cyclin D1 expression is associated with decreased efficacy of the selective epidermal growth factor receptor tyrosine kinase inhibitor gefitinib in head and neck squamous cell carcinoma cell lines. Clin Cancer Res. 2004;10:7764–74. https://doi.org/10.1158/1078-0432.CCR-04-0012.
212. Brockstein BE, Vokes EE, Yoo DS, Posner MR, Brizel DM, Ross ME. Methods to overcome radiation resistance in head and neck cancer. https://www.uptodate.com/contents/methods-to-overcome-radiation-resistance-in-head-and-neck-cancer. UpToDate, Wolters Kluwer Health – 2012-04-20.
213. Pendleton KP, Grandis JR. Cisplatin-based chemotherapy options for recurrent and/or metastatic squamous cell cancer of the head and neck. Clin Med Insights Ther. 2013. https://doi.org/10.4137/CMT.S10409.
214. Posner MR. Paradigm shift in the treatment of head and neck cancer: the role of neoadjuvant chemotherapy. Oncologist. 2005;10(Suppl 3):11–9. https://doi.org/10.1634/theoncologist.10-90003-11.
215. Zhu H, et al. Molecular mechanisms of cisplatin resistance in cervical cancer. Drug Des Devel Ther. 2016;10:1885–95. https://doi.org/10.2147/DDDT.S106412.
216. Shapiro GI. Cyclin-dependent kinase pathways as targets for cancer treatment. J Clin Oncol. 2006;24:1770–83. https://doi.org/10.1200/JCO.2005.03.7689.
217. Sedlacek H, et al. Flavopiridol (L86 8275; NSC 649890), a new kinase inhibitor for tumor therapy. Int J Oncol. 1996;9:1143–68.
218. Asghar U, Witkiewicz AK, Turner NC, Knudsen ES. The history and future of targeting cyclin-dependent kinases in cancer therapy. Nat Rev Drug Discov. 2015;14:130–46. https://doi.org/10.1038/nrd4504.
219. Le Tourneau C, et al. Phase I evaluation of seliciclib (R-roscovitine), a novel oral cyclin-dependent kinase inhibitor, in patients with advanced malignancies. Eur J Cancer. 2010;46:3243–50. https://doi.org/10.1016/j.ejca.2010.08.001.
220. Meijer L, Raymond E. Roscovitine and other purines as kinase inhibitors. From starfish oocytes to clinical trials. Acc Chem Res. 2003;36:417–25. https://doi.org/10.1021/ar0201198.
221. Whittaker SR, Walton MI, Garrett MD, Workman P. The Cyclin-dependent kinase inhibitor CYC202 (R-roscovitine) inhibits retinoblastoma protein phosphorylation, causes loss of Cyclin D1, and activates the mitogen-activated protein kinase pathway. Cancer Res. 2004;64:262–72.

222. Malumbres M, et al. Mammalian cells cycle without the D-type cyclin-dependent kinases Cdk4 and Cdk6. Cell. 2004;118:493–504. https://doi.org/10.1016/j.cell.2004.08.002.
223. O'Leary B, Finn RS, Turner NC. Treating cancer with selective CDK4/6 inhibitors. Nat Rev Clin Oncol. 2016;13:417–30. https://doi.org/10.1038/nrclinonc.2016.26.
224. VanArsdale T, Boshoff C, Arndt KT, Abraham RT. Molecular pathways: targeting the cyclin D-CDK4/6 axis for cancer treatment. Clin Cancer Res. 2015;21:2905–10. https://doi.org/10.1158/1078-0432.CCR-14-0816.
225. Finn RS, et al. PD 0332991, a selective cyclin D kinase 4/6 inhibitor, preferentially inhibits proliferation of luminal estrogen receptor-positive human breast cancer cell lines in vitro. Breast Cancer Res. 2009;11:R77. https://doi.org/10.1186/bcr2419.
226. Lukas J, Bartkova J, Bartek J. Convergence of mitogenic signalling cascades from diverse classes of receptors at the cyclin D-cyclin-dependent kinase-pRb-controlled G1 checkpoint. Mol Cell Biol. 1996;16:6917–25.
227. Kim S, et al. Abstract PR02: LEE011: an orally bioavailable, selective small molecule inhibitor of CDK4/6– reactivating Rb in cancer. Mol Cancer Ther. 2013;12:PR02. https://doi.org/10.1158/1535-7163.targ-13-pr02.
228. Schrijvers D, Vermorken JB. Role of taxoids in head and neck cancer. Oncologist. 2000;5:199–208.
229. Marzo I, Naval J. Antimitotic drugs in cancer chemotherapy: promises and pitfalls. Biochem Pharmacol. 2013;86:703–10. https://doi.org/10.1016/j.bcp.2013.07.010.
230. Fathi AT, et al. Phase I study of the aurora A kinase inhibitor alisertib with induction chemotherapy in patients with acute myeloid leukemia. Haematologica. 2017;102:719–27. https://doi.org/10.3324/haematol.2016.158394.
231. Ding YH, et al. Alisertib, an Aurora kinase A inhibitor, induces apoptosis and autophagy but inhibits epithelial to mesenchymal transition in human epithelial ovarian cancer cells. Drug Des Devel Ther. 2015;9:425–64. https://doi.org/10.2147/DDDT.S74062.
232. Matulonis UA, et al. Phase II study of MLN8237 (alisertib), an investigational Aurora A kinase inhibitor, in patients with platinum-resistant or -refractory epithelial ovarian, fallopian tube, or primary peritoneal carcinoma. Gynecol Oncol. 2012;127:63–9. https://doi.org/10.1016/j.ygyno.2012.06.040.
233. Melichar B, et al. Safety and activity of alisertib, an investigational aurora kinase A inhibitor, in patients with breast cancer, small-cell lung cancer, non-small-cell lung cancer, head and neck squamous-cell carcinoma, and gastro-oesophageal adenocarcinoma: a five-arm phase 2 study. Lancet Oncol. 2015;16:395–405. https://doi.org/10.1016/S1470-2045(15)70051-3.
234. Thall PF, Cook JD. Dose-finding based on efficacy-toxicity trade-offs. Biometrics. 2004;60:684–93. https://doi.org/10.1111/j.0006-341X.2004.00218.x.
235. Hoellein A, et al. Aurora kinase inhibition overcomes cetuximab resistance in squamous cell cancer of the head and neck. Oncotarget. 2011;2:599–609. https://doi.org/10.18632/oncotarget.311.
236. Collins S, Blair D, Zarycki J, Szynal C, Gangolli E, Vincent P, Chakravarty A, Ecsedy J. Abstract 3738: a rationale for combining the targeted investigational agents TAK-733, a MEK1/2 inhibitor, with alisertib (MLN8237), an Aurora A kinase inhibitor, for cancer therapy. Cancer Res. 2012;72(8 Suppl):3738. https://doi.org/10.1158/1538-7445.AM2012-3738.
237. Yu MG, Zheng HY. Acute myeloid leukemia: advancements in diagnosis and treatment. Chin Med J (Engl). 2017;130:211–8. https://doi.org/10.4103/0366-6999.198004.
238. Kantarjian HM, et al. Phase I study assessing the safety and tolerability of barasertib (AZD1152) with low-dose cytosine arabinoside in elderly patients with AML. Clin Lymphoma Myeloma Leuk. 2013;13:559–67. https://doi.org/10.1016/j.clml.2013.03.019.
239. Kantarjian HM, et al. Stage I of a phase 2 study assessing the efficacy, safety, and tolerability of barasertib (AZD1152) versus low-dose cytosine arabinoside in elderly patients with acute myeloid leukemia. Cancer. 2013;119:2611–9. https://doi.org/10.1002/cncr.28113.

240. Foran J, et al. A phase I and pharmacodynamic study of AT9283, a small-molecule inhibitor of aurora kinases in patients with relapsed/refractory leukemia or myelofibrosis. Clin Lymphoma Myeloma Leuk. 2014;14:223–30. https://doi.org/10.1016/j.clml.2013.11.001.
241. Schoffski P, et al. Efficacy and safety of biweekly i.v. administrations of the Aurora kinase inhibitor danusertib hydrochloride in independent cohorts of patients with advanced or metastatic breast, ovarian, colorectal, pancreatic, small-cell and non-small-cell lung cancer: a multi-tumour, multi-institutional phase II study. Ann Oncol. 2015;26:598–607. https://doi.org/10.1093/annonc/mdu566.
242. Meulenbeld HJ, et al. Randomized phase II study of danusertib in patients with metastatic castration-resistant prostate cancer after docetaxel failure. BJU Int. 2013;111:44–52. https://doi.org/10.1111/j.1464-410X.2012.11404.x.
243. Liu Z, Sun Q, Wang X. PLK1, a potential target for cancer therapy. Transl Oncol. 2017;10:22–32. https://doi.org/10.1016/j.tranon.2016.10.003.
244. Schoffski P, et al. A phase I, dose-escalation study of the novel Polo-like kinase inhibitor volasertib (BI 6727) in patients with advanced solid tumours. Eur J Cancer. 2012;48:179–86. https://doi.org/10.1016/j.ejca.2011.11.001.
245. Gutteridge RE, Ndiaye MA, Liu X, Ahmad N. Plk1 inhibitors in cancer therapy: from laboratory to clinics. Mol Cancer Ther. 2016;15:1427–35. https://doi.org/10.1158/1535-7163.MCT-15-0897.
246. Yu B, Yu Z, Qi PP, Yu DQ, Liu HM. Discovery of orally active anticancer candidate CFI-400945 derived from biologically promising spirooxindoles: success and challenges. Eur J Med Chem. 2015;95:35–40. https://doi.org/10.1016/j.ejmech.2015.03.020.
247. Hirai H, et al. Small-molecule inhibition of Wee1 kinase by MK-1775 selectively sensitizes p53-deficient tumor cells to DNA-damaging agents. Mol Cancer Ther. 2009;8:2992–3000. https://doi.org/10.1158/1535-7163.MCT-09-0463.
248. Sorensen CS, Syljuasen RG. Safeguarding genome integrity: the checkpoint kinases ATR, CHK1 and WEE1 restrain CDK activity during normal DNA replication. Nucleic Acids Res. 2012;40:477–86. https://doi.org/10.1093/nar/gkr697.
249. Van Linden AA, et al. Inhibition of Wee1 sensitizes cancer cells to antimetabolite chemotherapeutics in vitro and in vivo, independent of p53 functionality. Mol Cancer Ther. 2013;12:2675–84. https://doi.org/10.1158/1535-7163.MCT-13-0424.
250. Zeng L, Beggs RR, Cooper TS, Weaver AN, Yang ES. Combining Chk1/2 inhibition with cetuximab and radiation enhances in vitro and in vivo cytotoxicity in head and neck squamous cell carcinoma. Mol Cancer Ther. 2017;16:591–600. https://doi.org/10.1158/1535-7163.MCT-16-0352.
251. Kato S, et al. Cyclin-dependent kinase pathway aberrations in diverse malignancies: clinical and molecular characteristics. Cell Cycle. 2015;14:1252–9. https://doi.org/10.1080/15384101.2015.1014149.
252. Osman AA, et al. Wee-1 kinase inhibition overcomes cisplatin resistance associated with high-risk TP53 mutations in head and neck cancer through mitotic arrest followed by senescence. Mol Cancer Ther. 2015;14:608–19. https://doi.org/10.1158/1535-7163.MCT-14-0735-T.
253. Do K, et al. Phase I study of single-agent AZD1775 (MK-1775), a Wee1 kinase inhibitor, in patients with refractory solid tumors. J Clin Oncol. 2015;33:3409–15. https://doi.org/10.1200/JCO.2014.60.4009.
254. Ledford H. CRISPR, the disruptor. Nature. 2015;522:20–4. https://doi.org/10.1038/522020a.
255. Cyranoski D. CRISPR gene-editing tested in a person for the first time. Nature. 2016;539:479. https://doi.org/10.1038/nature.2016.20988.

Chapter 8
Role of the NOTCH Signaling Pathway in Head and Neck Cancer

Adrian D. Schubert, Fernando T. Zamuner, Nyall R. London Jr, Alex Zhavoronkov, Ranee Mehra, Mohammad O. Hoque, Atul Bedi, Rajani Ravi, Elana J. Fertig, David Sidransky, Daria A. Gaykalova, and Evgeny Izumchenko

Abstract The NOTCH signaling cascade has been implicated in multiple cellular functions, such as cell proliferation, differentiation, and survival. Dysregulation of the NOTCH pathway is associated with the progression of several types of malignant tumors, including head and neck squamous cell carcinoma (HNSCC). Accumulating data suggest that NOTCH is one of the most frequently altered pathways in HNSCC. Given the importance of NOTCH signaling in regulating tumor cell behavior, several NOTCH-targeted strategies are currently being developed and tested in preclinical and clinical settings. However, the precise role of the NOTCH pathway in head and neck malignancies remains incompletely defined and controversial. In most tumor types, *NOTCH1* has been reported as an oncogene. However, early characterization of the genomic landscape found that inactivating mutations of *NOTCH1* frequently occur in HNSCC, suggesting that *NOTCH1* is a tumor suppressor. More recent evidence indicates that NOTCH signaling may be activated in a

---

Adrian D. Schubert and Fernando T. Zamuner are contributed equally to this work.
Daria A. Gaykalova and Evgeny Izumchenko are Co-senior authors.

A. D. Schubert · F. T. Zamuner · N. R. London Jr · M. O. Hoque · A. Bedi · R. Ravi
D. Sidransky · D. A. Gaykalova · E. Izumchenko (✉)
Johns Hopkins University, School of Medicine, Department of Otolaryngology-Head & Neck Cancer Research, Baltimore, MD, USA
e-mail: eizumch1@jhmi.edu

A. Zhavoronkov
Insilico Medicine, Inc., Emerging Technology Centers, Johns Hopkins University at Eastern, Baltimore, MD, USA

R. Mehra
Department of Oncology, Johns Hopkins University School of Medicine, Baltimore, MD, USA

E. J. Fertig
Johns Hopkins University, School of Medicine, Department of Oncology-Division of Biostatistics and Bioinformatics, Baltimore, MD, USA

B. Burtness, E. A. Golemis (eds.), *Molecular Determinants of Head and Neck Cancer*, Current Cancer Research, https://doi.org/10.1007/978-3-319-78762-6_8

subset of HNSCC tumors, similar to other tumor types, indicating a more complex function in HNSCC. This overview will summarize the evidence for oncogenic and tumor suppressor roles of the NOTCH signaling pathway in HNSCC, discuss recent studies that aid in interpretation of these contradictory findings, and describe potential therapeutic opportunities and future directions.

Keywords Head and neck cancer · Notch receptors · Notch signaling pathway · γ-secretase inhibitor · Mutation · Oncogene · Tumor suppressor

8.1 Introduction

Head and neck squamous cell carcinoma (HNSCC) is the sixth most common cancer in the world, resulting in over 300,000 deaths worldwide and with an annual incidence of over 50,000 cases in the United States [1, 2]. HNSCC is a heterogeneous disease at both the clinical and molecular levels; this term includes squamous cell carcinomas (SCCs) of the oral cavity, nasal cavity and sinuses, pharynx, and larynx. Although major HNSCC-associated risk factors are predominantly related to tobacco and alcohol consumption, a subset of cancers of the oropharynx results from infection with the human papillomavirus (HPV), as discussed in other chapters of this work (Chaps. 20 and 21). Due to the asymptomatic course of early-stage disease and the absence of the routine screening techniques, HNSCC is often present as clinically advanced disease upon diagnosis. Despite the improved understanding of disease pathogenesis and advances in molecular diagnosis, the 5-year overall survival for locally advanced disease remains poor, at approximately 50% [3–6].

Like other solid malignancies, HNSCC is thought to progress through a series of genetic alterations. The landscape of genetic variants in HNSCC is highly heterogeneous and differs for HPV-unrelated and HPV-related tumors.

Several signaling proteins, including the tumor suppressors TP53, FAT1, and CDKN2A and the growth-promoting CCND1, HRAS, and PIK3CA [7–9], were shown to be commonly dysregulated in HNSCC through genetic and epigenetic alterations, supporting their critical role in HNSCC tumorigenesis. NOTCH1, also identified as a target of frequent mutation in these studies [7–9], is particularly noteworthy. *NOTCH1* is one of the four paralogous *NOTCH* genes in humans, which encode four heterodimeric transmembrane receptor signaling proteins (NOTCH1-4). Large-scale sequencing studies have indicated that potentially inactivating mutations in *NOTCH1* occurred in 15–19% of US HNSCC patients [7–10], making *NOTCH1* the second most frequently mutated gene in HNSCC after *TP53* [7–9].

The NOTCH signaling pathway has been linked to multiple cellular functions associated with cancer development, including regulation of self-renewal capacity, proliferation, survival, cell fate determination, and apoptosis [11, 12]. NOTCH1 chromosomal translocation and overexpression of its constitutively active forms were initially identified in the development of certain types of leukemia [13]. Abnormally elevated expression and activating mutations in NOTCH receptors (predominantly NOTCH1) and their ligands (delta-like protein (DLL) 1, DLL3, DLL4, JAGGED1, and JAGGED2) were later reported in a number of solid human malignancies, including HNSCC [7, 9, 14–16]. Due to its multifaceted effects in promoting angiogenesis, epithelial-mesenchymal transition (EMT), and chemoresistance [17–20] in many tumor types, the NOTCH signaling pathway is considered as a potential therapeutic target for cancer treatment.

Although NOTCH1 has a well-established role as an oncogene in T-cell leukemia [21], its role in HNSCC is more complex. The location of the missense mutations initially identified in ligand-binding domains and frequent truncating mutations has led to the suggestion that NOTCH1 acts as a tumor suppressor in HNSCC. Therefore, NOTCH1 was considered an unlikely therapeutic target for patients with head and neck cancers due to generalized difficulty in restoring the activity of proteins with loss-of-function mutations. However, subsequent studies of HNSCC in Chinese patients showed mutation rates of *NOTCH1* about 50% with a considerable portion of potentially activating mutations [22, 23]. While these observations underscore the critical role of NOTCH1 pathway activity in the development of head and neck neoplasms, they add another twist to the already complicated picture of NOTCH signaling in head and neck cancers. A growing body of work indicates that downstream effects of NOTCH1 activation in malignant initiation and progression are highly context-dependent and vary with cell lineage, pathology, and stage of differentiation [24, 25]. Recent work has shown that NOTCH1 pathway activation in subsets of HNSCCs is associated with a poor prognosis [15, 19, 22, 26–28]. The nature of the switch that determines whether NOTCH1 acts as a tumor promoter or as a tumor suppressor has been the subject of intense and even controversial research.

Given the importance of NOTCH signaling in regulating tumor cell behavior, clinical trials of NOTCH pathway inhibitors in patients with solid tumors have been undertaken (such as phase 1/2 studies of gamma secretase inhibitors [LY3039478, RO4929097 and PF-03084014] and anti-DLL4 or anti-NOTCH antibodies [enoticumab, demcizumab, tarextumab]) [29–36], and several approaches are under preclinical evaluation [37]. However, the overall response rates to NOTCH-targeted therapy from clinical trials in patients with melanoma, sarcoma, lung, breast, pancreatic and prostate cancer remain suboptimal. Factors contributing to limited success with these drugs include the off-target effects of some classes of NOTCH inhibitor and inability to identify those patients who will most likely respond to anti-NOTCH therapy [37–40]. NOTCH pathway inhibition underlies a potential therapeutic strategy in HNSCC, but further biological insight will be necessary before NOTCH pathway inhibition can be optimally exploited for effective treatment of patients with head and neck cancer.

This article will summarize the current understanding of the role of the NOTCH pathway in HNSCC, discuss the controversy regarding its biological function, and review signaling aspects that will impact strategies for treating HNSCC patients.

8.2 NOTCH1 Domain Architecture

The NOTCH1 protein is one of the four NOTCH transmembrane receptor paralogs, which display both overlapping and unique functions [41]. Mature NOTCH receptors consist of an extracellular ligand-binding domain and an intracellular domain, which subsequently translocate to the nucleus and mediate target gene transcription upon ligand-dependent activation. The extracellular domain of NOTCH1 contains 36 epidermal growth factor (EGF)-like repeats. Repeats 11–13 constrain the "ligand-binding" domain, which is essential for direct interactions between NOTCH1 and its ligands [42, 43] (Fig. 8.1). Since NOTCH1 ligand-mediated receptor activation relies on these interactions, mutations in this region may inhibit ligand-receptor interactions and subsequently signal transduction. Although little is known about the contribution of EGF repeats to NOTCH1 function, it was shown that the Abruptex region (EGF repeats 24–29) may regulate the ligand-binding activity through interactions with the ligand-binding domain [44]. It was reported that the integrity of the Abruptex region is required for suppression of NOTCH1 activity and that mutations within this region enhance NOTCH1 signaling [44, 45]. LIN12-NOTCH repeats (LNR) and N- and C-terminal heterodimerization domains (HDDs) together form the negative regulatory region (NRR) (Fig. 8.1). The membrane-proximal NRR acts as a receptor activation switch and prevents ligand-independent activity: mutations in this area can lead to NOTCH1 ligand-independent signaling [46]. The NOTCH1 intracellular domain (NICD) contains structural regions essential for interaction with downstream effector proteins and consists of a *r*ecombination signal-binding protein – Jkappa-*a*ssociated *m*odule (RAM), an *ank*yrin repeat domain (ANK), a *t*rans*a*ctivation *d*omain (TAD), and a region rich in *p*roline, *g*lutamine, *s*erine, and *t*hreonine residues (PEST) (Fig. 8.1). NICD, post-translationally cleaved from full-length NOTCH1, is a core component of NOTCH1 nuclear complexes that also include the DNA-binding factor CSL (an acronym for CBF1, Suppressor of Hairless, Lag-1; also known as CBF1/RBPJ) and coactivator proteins of the MAML family [47]. The RAM region constitutes a high-affinity binding module for CSL, and the ANK domain together with the CSL creates a composite-binding site for MAML recruitment into these complexes [47, 48]. Therefore, mutations in the RAM/ANK module may disrupt proper nuclear

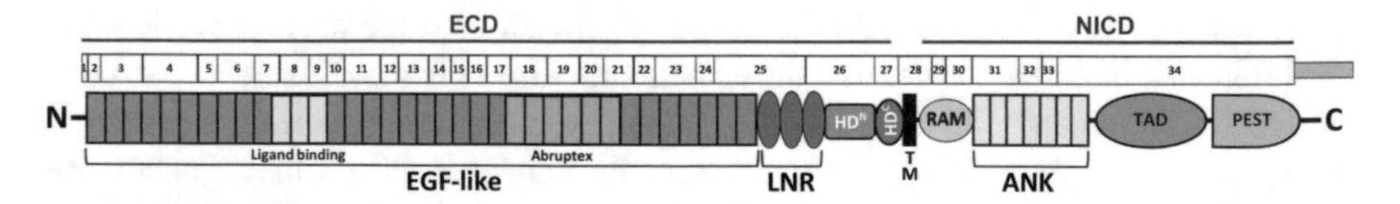

Fig. 8.1 *Schematic representation of the NOTCH1 structure*. Thirty-four exons are shown in the top, and the protein domains are displayed at the bottom. LNR LIN12-NOTCH repeats, HD heterodimerization domain, TM transmembrane domain, ANK ankyrin repeats, TAD transactivation domain, PEST proline (P), glutamine (E), serine (S), and threonine (T) residues domain. ECD extracellular domain, NICD NOTCH intracellular domain

complex assembly and subsequently prevent transcription of NOTCH1-dependent genes [23]. The TAD/PEST region is essential for degradation of NOTCH1 and its cleaved derivative by proteolysis [49]. Mutations in this region are usually associated with gain of function [50, 51], emphasizing the important functional role of regulated NICD degradation. Although the overall spectrum of *NOTCH1* mutations in HNSCC varies in patients of different ethnicities, the majority of genetic changes cluster around the extracellular "ligand-binding" domain, suggesting that disruption of the architecture of this region plays a prominent role NOTCH1-associated tumorigenesis [22, 23].

8.3 Canonical Ligand-Dependent NOTCH1 Signaling

The understanding of the complex NOTCH signaling network is rapidly evolving. Although the number of NOTCH1-auxiliary proteins and genes known to be directly and indirectly regulated by NOTCH1 is quickly expanding due to extensive ongoing research into the finer mechanistic detail of NOTCH signaling, the core NOTCH pathway is now understood in reasonable detail. Canonical NOTCH pathway activation involves a proteolysis-mediated release of the NICD, its translocation to the nucleus, and association with a DNA-bound protein complex (Fig. 8.2). The NOTCH receptor is activated by juxtacrine interaction with the members of Delta/Serrate/Lag-2 (DSL) family: JAG1, JAG2, DLL1, DLL3, and DLL4 [52, 53]. This receptor-ligand interaction elicits changes in the configuration of the NOTCH extracellular domain that expose cleavage site (S2), located within the NRR just before the transmembrane domain, to metalloprotease TACE (also known as ADAM17). Cleavage by TACE removes the extracellular domain from the outer portion of the membrane (at site S2; Fig. 8.2). Removal of the ectodomain results in the formation of a membrane-tethered intermediate form of the NICD with a transmembrane moiety. Subsequently, γ-secretase complex cleaves the NICD of the receptor at the transmembrane domain (at site S3; Fig. 8.2), releasing a transcriptionally active NICD, which is now free to translocate to the nucleus. In the nucleus, NICD binds to and activates the DNA-binding protein CSL and recruits other coactivators such as MAML and the MED8 mediator complex, leading to expression of downstream target genes (such as *HES* and *HEY* family members, *NFκB*, *MYC*, *CCND1*, *BCL2*, and *CCR7*). Which genes are expressed varies in a cell type-specific manner [47, 48, 54]. As NOTCH signaling modulates a variety of fundamental cellular processes, it is a topic of very active research. The network of proteins that have recognized ability to modify the output from NOTCH receptors by regulating ligand-mediated activation, receptor proteolysis, and target selection is rapidly growing, and the molecular details of the current understanding of the complex NOTCH signaling (including the noncanonical role for NOTCH pathway in oncogenesis) have been well summarized in recent reviews [43, 55–57].

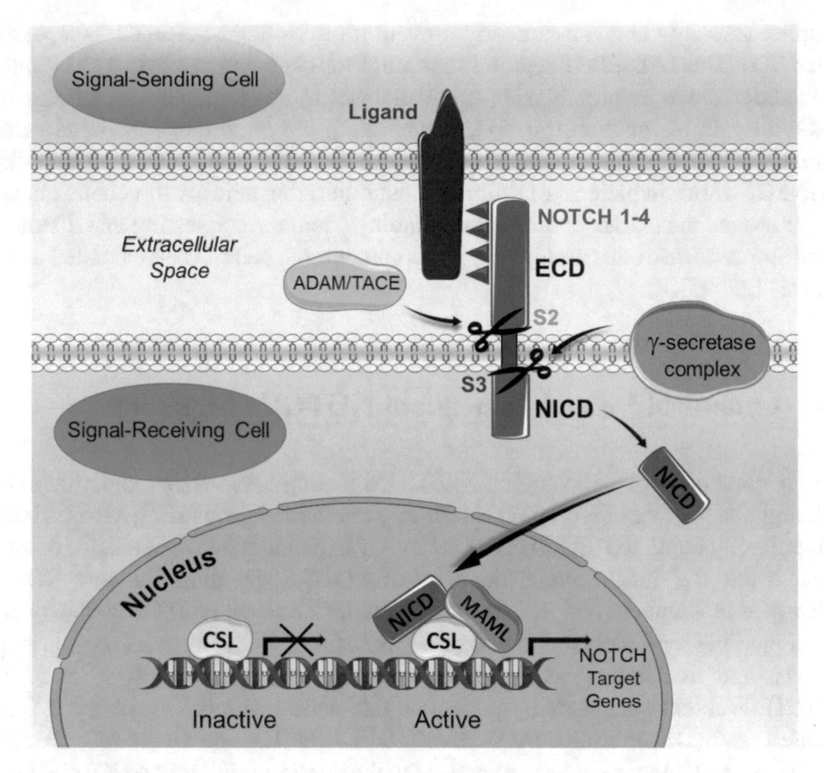

Fig. 8.2 *Canonical NOTCH1 signaling activation*. Ligand binding initiates two successive proteolytic cleavages (S2 and S3). ADAM/TACE proteinase mediates the first cleavage that occurs at the S2 site of the extracellular domain. The S2 cleavage allows access to the γ-secretase complex, which is responsible for the second proteolytic cleavage at the S3 site within the transmembrane domain. Released NICD domain translocates to the nucleus, where it interacts with the CSL transcription factor. Docking of the ankyrin repeat domain of NOTCH to the CSL protein generates a composite-binding surface for the recruitment of MAML. These interactions convert CSL from a transcriptional repressor to an activator by displacing corepressors and histone deacetylases and recruiting histone acetyltransferases. MAML, in turn, recruits additional coactivators, which leads to expression of the NOTCH target genes

8.4 The Role of NOTCH Signaling in HNSCC: An Oncogene or Tumor Suppressor?

Although the link between dysregulation of NOTCH1 signaling and human cancer is strong and compelling [58], only a few studies have directly examined the function of NOTCH1 activation in the context of HNSCC. The biological effects of NOTCH signaling in this disease are still not fully understood and whether it acts as a tumor promoter or as a tumor suppressor remains controversial. While the spectrum of predominantly inactivating mutations in NOTCH1 suggests it is a tumor suppressor [7, 9, 10], other studies report that activation of NOTCH can enhance proliferation, inhibit apoptosis, and promote angiogenesis, suggesting that in some cases, NOTCH1 can behave as an oncogene. While the precise mechanisms that explain this paradox are not yet elucidated, the dual roles of aberrant NOTCH signaling indicate its complex biological function in the development and progression of HNSCC.

8.5 The NOTCH Pathway as a Tumor Promoter in HNSCC and Evaluation of NOTCH-Inhibiting Therapies

Abnormal expression of NOTCH pathway components and the potential involvement of NOTCH signaling activation in HNSCC tumorigenesis have been reported by multiple studies over the last decade. For example, JAG1-triggered NOTCH1 activation in the HNSCC SCC9 cell line resulted in endothelial sprouting of co-cultured endothelial cells in vitro, significantly enhanced neovascularization and tumor growth in vivo, and also correlated with blood vessel density in primary human HNSCC tissues, supporting the role of NOTCH signaling in activating the angiogenic process [19]. Other independent studies also indicated that NOTCH1 protein expression correlated with increased microvessel density [59] and that downregulation of NOTCH1 coincided with cetuximab-induced inhibition of angiogenesis in vitro and in vivo [60], further supporting the putative role of NOTCH signaling in mediating tumor vascularization [61]. NOTCH receptors, their ligands, and the downstream signaling components were first found to be overexpressed in HNSCC by gene profiling of four HNSCC cases versus their matching normal tissues in 2000 [62]. A series of latter studies based on the gene expression array, RT-PCR analysis, and immunohistochemistry of the central NOTCH pathway genes supported a role for upregulation of NOTCH1 signaling in the progression, lymph node metastasis, and unfavorable prognosis of patients with HNSCC neoplasms from multiple anatomical sites [26, 27, 59, 63–73]. Moreover, it was shown that NOTCH1 activation in several HNSCC-derived cell lines in vitro induced a marked increase in cell growth, migration, and invasion [69, 71], whereas NOTCH1 knockdown caused an anti-tumorigenic effect [26, 67, 69, 70].

More recently, comprehensive analyses of immunohistochemistry, NOTCH mutations, copy number variation, and gene expression of the NOTCH signaling pathway genes in primary HNSCCs revealed that nearly one-third of tumors (14/44) exhibited activated NOTCH signaling through overexpression of NOTCH ligands or receptors [15]. Consistent with this finding, the NOTCH1 pathway was found to be activated in a subset of HNSCC patients with distinctive peripheral (only at the outermost layer of tumor mass) or non-peripheral staining patterns of cleaved NICD [28]. Interestingly, another study also showed that NOTCH1 protein localizes to the invasive front of oral squamous cell carcinoma (OSCC) specimens [65]. Such localization may suggest that NOTCH1 is involved in the invasive ability of OSCC [27, 59], potentially through inactivation of DTX1, a positive regulator of the NOTCH signaling pathway [73].

Notably, high levels of NOTCH1 expression in a subset of HNSCC tumor specimens were closely associated with clinical resistance to cisplatin therapy [20, 64], consistent with observations in other solid tumors [70, 74, 75]. Given the limited curative success of cisplatin-based chemotherapy against locally advanced or metastatic HNSCC, these studies position NOTCH expression as a potential molecular biomarker for response to cytotoxic chemotherapy. Although mechanisms underlying this correlation in HNSCC remain to be further investigated, it is believed that the failure of cisplatin-based regimens and subsequent tumor relapse can be partially attributed to the presence of highly tumorigenic, chemoresistant cancer stem cells (CSCs), which possess unique features of self-renewal, unlimited proliferation and asymmetric cell division [76]. Emerging evidence, both in vitro and in vivo, has shown that activation of NOTCH signaling is involved in the control of CSC self-renewal in HNSCC and that knockdown or blockade of NOTCH1 in HNSCC cell lines augments the chemosensitizing effects of cisplatin [72, 77–79]. Taken together with data from studies of human embryonic stem cells, these findings suggest that NOTCH1 may act as a key regulator of CSC-associated chemoresistance and tumorigenicity by altering the NANOG and OCT4 transcription network [80, 81].

Nearly 20% of patients with OSCC harbor premalignant lesions showing signs of dysplasia, often visually identified as leukoplakia or erythroplakia [82]. As some of these lesions evolve to malignant neoplasms [82, 83], they represent intermediate steps in OSCC progression [84, 85]. This multistep process from normal epithelium to early premalignant change and fully invasive squamous cell carcinoma provides a rational framework for studying the timing of molecular alterations underlying OSCC tumorigenesis. Although relatively few molecular markers associated with the progression of oral dysplasia to OSCC are currently recognized, upregulated expression of NOTCH1 receptors and their ligands was seen in these precancerous dysplastic lesions [19, 65, 86, 87]. Moreover, expression of NOTCH1, NICD, JAG1, and HES1 was gradually increased from normal to dysplastic to malignant tissue [19, 65, 87], suggesting that activation of NOTCH signaling is an early event in OSCC tumor progression. Recently, a high rate of potentially activating NOTCH1 mutations was reported in oral dysplastic lesions and invasive OSCCs resected from Chinese patients [22, 23], further supporting an important role of NOTCH1 in the progression of early neoplasms.

Of note, *NOTCH1* is the most frequently altered gene in adenoid cystic carcinoma (ACC), a relatively rare tumor of the head and neck, with four mutations identified in 3 (12%) of 25 tumors tested (including one nonsense and one nonsynonymous substitution, as well as two indel alterations) [88]. In ACC, *NOTCH1* mutations were proposed to be associated with higher NICD expression, worse prognosis, and likely better response to brontictuzumab (a monoclonal antibody that binds to NOTCH1 on the cell surface, thereby inhibiting NOTCH1-mediated signaling), as demonstrated by an index patient with *NOTCH1*-mutant tumor and xenograft model in vivo [89].

Several studies have focused on the potential utility of downregulating NOTCH signaling as an anticancer therapy for HNSCC. A number of studies reported that γ-secretase inhibition results in downregulation of the NICD downstream targets and reduces proliferation, migration, EMT-dependent invasiveness, and chemoresistance in HNSCC cell lines in vitro and in vivo [26, 65, 69, 72, 86, 90]. Currently, several γ-secretase inhibitors (GSIs) have been developed (including PF-03084014, GSI-1, GSI-IX (DAPT), MK-0752, BMS-906024, LY3039478,

LY450139, LY900009, MRK-003, and RO4929097, among others [91–96]) and tested in preclinical studies and phase I/II trials in patients with advanced solid tumors, either as single agents or in combination with targeted therapeutics or chemotherapy [40]. These studies have suffered from the use of drugs with poor pharmacokinetics (PK) for on-target effects and the lack of patient selection for NOTCH activation. For instance, a recent phase I study reported that only 2 out of 15 HNSCC patients (13%) treated with the combination of GSI MK-0752, a γ-secretase inhibitor, and the mTOR inhibitor ridaforolimus achieved partial response [39, 97], which might have been due to the effect induced by the ridaforolimus. Taken together, the overall clinical experience indicates that although some patients with advanced disease experience cancer remissions, most patients do not respond to γ-secretase inhibitor therapy.

Monoclonal antibodies against NOTCH1 receptors or ligands (such as enoticumab, demcizumab, tarextumab, and brontictuzumab) [29–31] and other molecules with non-GSI modes of action that block NOTCH signaling (such as the γ-secretase modulators E2012 and JLK6, nonsteroidal anti-inflammatory drugs (NSAID) flurbiprofen/flurizan, the endoplasmic reticulum-exporting inhibitor molecule FLI-06, or TACE inhibitors INCB7839/XL784) are currently being developed and evaluated preclinically or in early-stage clinical trials [31, 37, 95, 98–100]. However, further improvement in clinical success is contingent upon identification of molecular biomarkers that will allow preselecting patients more likely to benefit from NOTCH1 inhibitors. These patients will likely be those in which NOTCH1 is altered and oncogenic, and therefore further studies are required to determine the molecular features that define this function.

8.6 NOTCH as a Tumor Suppressor

Although investigation of the role of NOTCH signaling during HNSCC tumorigenesis has focused mainly on its role as an oncogene, a tumor suppressor function for NOTCH1 in squamous epithelial tumors is well supported by multiple in vivo models of various types of cancer and in vitro studies with normal keratinocytes from the skin or esophagus [101–105]. Studies indicating that NOTCH exerts tumor suppressor function and can inhibit carcinogenesis in certain tissues have been recently thoroughly reviewed [57].

Prior to the recent next-generation sequencing analyses of the HNSCC mutational landscape, only a few studies pointed to an association between loss of NOTCH activity and HNSCC carcinogenesis. In one study, overexpression of NICD in a tongue cancer cell line resulted in growth suppression in vitro and in vivo accompanied by G0/G1 cell cycle arrest and apoptosis [106]. Similar results were reported in a laryngeal cancer cell line, where overexpression of NOTCH1 suppressed proliferation, causing cell cycle arrest in the G0/G1 phase and inducing apoptosis [107]. In a later study, immunohistochemical analysis of NOTCH1 in 56 cases of oral epithelial specimens, including OSCC and oral intraepithelial neoplasm (OIN), showed reduced NOTCH1 expression in cancer compared to normal basal cells within the same specimen [108]. The downregulation of NOTCH1 in OSCC was also confirmed by microarray and Western blot analyses. Although the mechanisms underlying redundant and non-overlapping roles of NOTCH proteins in tumorigenesis of solid malignancies are complex and not yet well understood, in this study, neither NOTCH2 nor NOTCH3 was upregulated in OSCC, suggesting that quantitative compensation of NOTCH1 function in OSCC is unlikely [108]. Furthermore, the same study reported that knockdown of *NOTCH1* in normal human fibroblast cells (HFS) resulted in dysplastic stratified epithelium formation in vitro [108].

With the advancement of next-generation sequencing technologies, several independent groups performed whole-exome sequencing of HNSCC tumors, predominantly from Caucasian patients, reporting that up to 15% of HNSCCs carry *NOTCH1* mutations [7–9, 15, 109]. This observation was later confirmed by The Cancer Genome Atlas (TCGA) study, which found *NOTCH1* mutations with a prevalence of approximately 19% in the 279 HNSCC tumors analyzed, confirming that *NOTCH1* is one of the most frequently mutated genes discovered in HNSCC [10]. In these large-scale studies, the location of missense mutations in the ligand-binding domain as well as frequent nonsense and frameshift aberrations [7] led to the hypothesis that *NOTCH1* is likely a tumor suppressor. Subsequently, it was demonstrated that expression of cleaved NICD or full-length wild-type *NOTCH1* in mutant HNSCC cell lines inhibited in vitro proliferation and growth in mice [8], functionally supporting the hypothesis that the NOTCH pathway has a tumor suppressor effect in HNSCC. Of note, mutations in FBXW7 (a component of the SCF ubiquitin ligase complex and a negative regulator of NOTCH1) within a hotspot known to block NOTCH1 degradation may result in sustained NOTCH signaling, adding another level of complexity to the functional status of NOTCH1 in these tumors [7].

Interestingly, in a clinical trial of GSI for the treatment of Alzheimer disease, an increased risk of nonmelanoma skin cancers was reported in a subset of patient who received higher doses of the drug [110, 111], consistent with previous observations that NOTCH1 acts as a tumor suppressor in the epidermis [112]. The emergence of skin cancers with GSI administration further supports evidence that GSIs are more effective against tumors with NOTCH1-activating mutations or malignancies where NOTCH1 has lost its tumor-suppressive function (via inactivating mutations or other molecular mechanisms) [113]. Since ablation of NOTCH1 signaling in a case where it plays a tumor-suppressive role may induce tumorigenic effects [112], until the molecular signature for better selection of patients likely to respond to GSIs is available, close attention must be paid.

While findings that loss-of-function mutations are recurrent in clinical samples suggests that NOTCH1 functions as a tumor suppressor in a subset of HNSCC tumors, continued delineation of the mechanisms underlying NOTCH-mediated tumor suppression will be necessary for developing new therapeutic strategies that enable the tumor function of NOTCH to be effectively harnessed to improve patient benefits. For example, development of NOTCH-activating antibodies that facilitate TACE-mediated S2 cleavage [114] or targeting NOTCH's negative regulators such as DTX1 or FBXW7 could be a potential therapeutic strategy to restore NOTCH pathway activity [73].

8.7 NOTCH Functional Heterogeneity: A Potential Key to Conflicting Results

Although the role of NOTCH signaling in HNSCC remains controversial, several recent studies provide some explanations for the contradictory findings and shed new light on the complex molecular alterations of the NOTCH signaling pathway in HNSCC.

The vast majority of mutations identified in high-throughput sequencing studies are located around the ligand-binding domain, indicating that disrupting the NOTCH1-ligand interaction may be the most prevalent cause for NOTCH1-related tumorigenesis in the Western populations [23]. Overall, most genetic alterations identified in Caucasians have been located in areas associated with the loss-of-function phenotype, a mutational spectrum fundamentally different from that observed in Chinese patients, where significant proportion of alterations are located in domains likely to result in activation of NOTCH signaling. Interestingly, the NOTCH1 mutation frequency was considerably higher (over 50%) in Chinese HNSCC patients and strongly associated with poor prognosis and shorter survival [16, 22, 23]. On the other hand, the rate of *NOTCH1* mutations in Japanese OSCC patients was on par with that seen in Caucasians, with most of the NOTCH1 alterations predicted to be inactivating [115]. Although further studies with much larger independent cohorts from similar ethnic backgrounds will be needed to validate these observations, these studies emphasize that NOTCH1 may employ different tumorigenic mechanisms in patients from distinct ethnic and geographic areas.

A recent comprehensive genetic, epigenetic, and transcriptomic analysis of the NOTCH signaling pathway in 44 predominantly Caucasian patients with HNSCC revealed that *NOTCH1*-mutant tumors display lack of NOTCH pathway activation due to the loss-of-function mutations, whereas a large subset of tumors with wild-type *NOTCH1* exhibit increased expression or gene copy number of either the receptor or ligands, as well as downstream pathway activation [15]. This bimodal pattern of NOTCH pathway alterations suggests that NOTCH signaling may be more common in HNSCC than previously thought but through different mechanisms dependent on inherited genetic background, environmental factors, and etiology.

More recently, immunohistochemical analysis of transcriptionally active NICD in 79 tumors from Caucasian patients with known *NOTCH1* mutation status revealed that NICD is expressed in three distinct patterns. This study suggests that NOTCH dysregulation may differ from tumor to tumor, with NOTCH1 functioning as a tumor suppressor in a subset of tumors with negative staining, and as an oncogene in another subgroup (tumors with positive nonperipheral expression), while retaining normal function in the remainder (tumors with positive peripheral staining pattern) [28]. Interestingly, about half of the *NOTCH1* mutants expressed cleaved NOTCH1 despite having missense mutations in the EGF-like ligand-binding domain. While it is possible that these patients were heterozygous for *NOTCH1* mutations, further studies are necessary to clarify this observation.

Furthermore, a few recent studies have indicated a possible effect of HPV infection on NOTCH1 signaling patterns in HNSCC. It was reported that *NOTCH1*-mutant tumors [28, 116] and tumors that lacked downstream NICD staining are more likely to be HPV-negative [28]. Another study revealed that while NOTCH1 drives angiogenesis in HPV-negative head and neck cancers, angiogenesis in HPV-positive tumors is most likely NOTCH1 independent [61]. Although the significance of these findings is not yet clear [117], these preliminary observations suggest that the oncogenic or tumor-suppressive role of NOTCH1 is also dependent on the tumor HPV genotype. In studies in another HPV-induced tumor type, using cervical cancer cell lines, NOTCH1 was shown to be able to interact with the HPV-encoded oncoproteins E6 and E7 to either induce transformation [118] or to protect against HPV-induced oncogenesis [119]. Therefore, it is possible that the NOTCH1-E6/E7 interaction also mediates NOTCH1 signaling in HPV-positive HNSCCs. However, this regulatory axis remains a subject of future investigation.

Taken together, these recent studies aid in interpreting the apparent inconsistencies of existing literature and further recognize that the biological role of NOTCH could be contextual. HNSCC is considered to be a vastly heterogeneous disease, and the functional role of the NOTCH pathway is likely dependent on the genetic background of the individual tumor. While NOTCH undoubtedly plays a complex biological role in HNSCC, further elucidation of the mechanisms by which NOTCH signaling mediates its tumor-suppressive or tumor-promoting activities will better focus therapeutic interventions.

8.8 Current Challenges and Future Directions

Historically, most of the functional studies of NOTCH1 activation in the context of HNSCC were performed in vitro or in cell line-derived xenograft models. Mouse xenografts of human tumor cell lines diverge substantially from the actual tumors from which they were derived [120], do not represent the heterogeneity of human malignancy that occurs among individuals on a population basis, and are unlikely to accurately recapitulate the processes that occur during disease development. Hence, there is a need for more relevant experimental systems that better reflect the features of individual tumors to functionally test the role of NOTCH signaling and its inhibition in a preclinical setting. Patient-derived xenograft (PDX) models, established from tumor tissue samples directly implanted into immunodeficient mice, faithfully maintain mutational status, gene expression patterns, DNA copy number alterations, and other important biological features of human malignancies [121]. Continuous generation and analysis of in vivo models that truly recapitulate HNSCC may allow characterization of the molecular features that define the contextual functional role of NOTCH signaling. Furthermore, these in vivo systems may provide a powerful modality for preclinical evaluation of therapeutics that perturb the NOTCH pathway in HNSCC and other solid malignancies.

Even morphologically similar HNSCC cancers are extremely heterogeneous at the genetic and epigenetic level, arise in different histological sites, and respond to treatments with very diverse pathogenic mechanisms. To date, the functionality of NOTCH signaling has been analyzed in HNSCCs that represent a clinically diverse set of cancers from different anatomical sites that may or may not contain oncogenic HPV, and from patients that have had varied levels of tobacco smoke exposure, impacting mutational burden. Therefore, the population-based interpatient genomic heterogeneity may underlie, in part, the contradictory evidence of NOTCH signaling in cancer. Further studies using more homogeneous cohorts of well-annotated tumors may aid in a fuller understanding of the function of NOTCH in HNSCC.

Remarkable advances in high-throughput genomic technologies and decreasing sequencing cost have allowed rapid generation of high-quality genetic, transcriptomic, and epigenetic data from hundreds of HNSCC tumors. Profiling this data with constantly emerging new integrative analytical approaches [122, 123] may further enhance identification of potentially targetable components of NOTCH signaling pathway and clinically relevant predictors of response to current and novel NOTCH-targeted therapies. Development of novel bioinformatic methods to account for inter-tumor heterogeneity [124] and pathway dysregulation [123, 125, 126] is critical to evaluating the complex role of NOTCH in these samples.

It has been proposed that in some cases, *NOTCH1* mutation might underlie progression of oral premalignant lesions into a primary invasive carcinoma [23]. However, the risk of invasive transformation associated with *NOTCH1* mutations or NOTCH pathway dysregulation is yet to be determined. Similar to *TP53*, *NOTCH1* mutations may exist in benign oral lesions for many years without progression to malignancy [127], and it is difficult to obtain longitudinal collection of oral dysplasia and subsequent OSCCs that developed in the same patients. Nevertheless, continued investigation of the NOTCH pathway in samples collected across the entire developmental path (from oral dysplasia to carcinoma in situ (CIS) and subsequently to invasive carcinoma) is needed to further validate the role of NOTCH signaling in OSCC evolution. Such studies may have considerable implications for risk assessment and treatment of patients with OSCC and other cancers of the head and neck arising from different anatomical sites.

8.9 Conclusion

The majority of patients with advanced HNSCC become resistant to repeated lines of traditional chemotherapy, and addition of cetuximab (a monoclonal antibody targeting epidermal growth factor receptor and the only FDA-approved targeted therapy available for this disease) offers on average only a modest survival benefit. Therefore, there is an urgent need for new therapeutic strategies for patients with HNSCC. It is apparent that dysregulation of the NOTCH signaling cascade plays an important role in proliferation, invasiveness, angiogenesis, stem cell maintenance, and chemoresistance of head and neck tumors. However, while dysregulation of NOTCH pathway activity promotes cancer growth in some circumstances, it mediates cell death and tumor suppression in others. It has often been reported in past studies of many different tissue types that the ability of the NOTCH pathway to carry out different biologic outputs in cancers is profoundly dependent on the cellular context [128]. It seems that HNSCC is no exception.

Initially viewed as a simple linear sequence of events, NOTCH signaling is now appreciated to involve a highly interconnected and complex network of protein modifiers that can affect its activity (Fig. 8.3). Moreover, crosstalk with components of other signaling pathways, such as receptor tyrosine kinases (i.e., VEGFR2) [129, 130], Wnt [131, 132], and TGF-β [133], may also augment, inhibit, or modulate

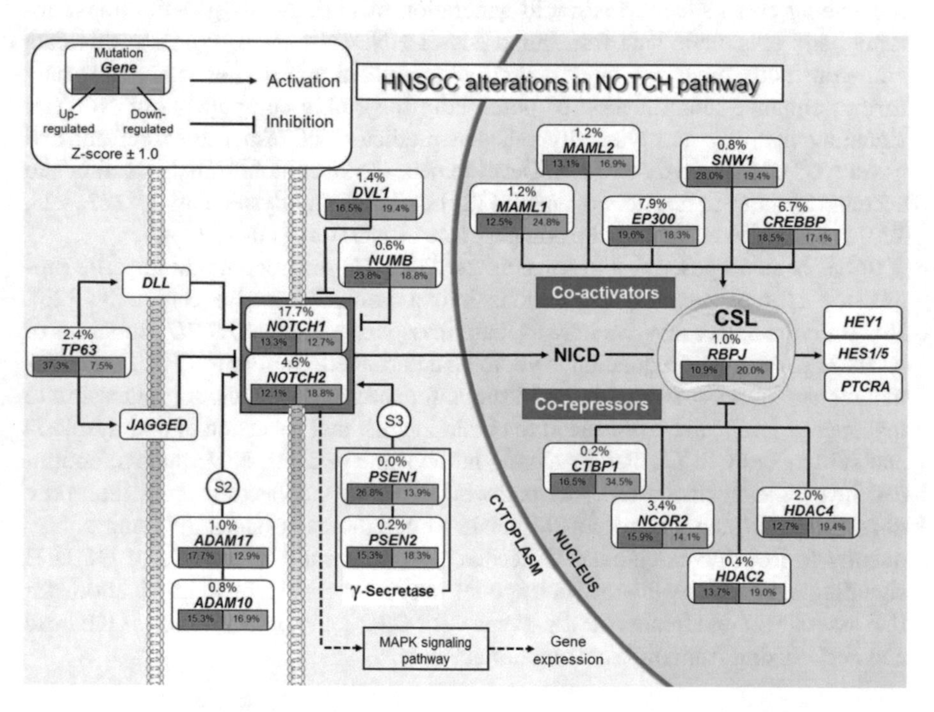

Fig. 8.3 *Dysregulation of the key NOTCH1 pathway genes in HNSCC*. The canonical NOTCH1 signaling pathway consisting of 44 key proteins annotated by KEGG [134] pathway collection was constructed based on a provisional TCGA HNSCC cohort (n = 528) using the CBioPortal database. Activating or inhibitory symbols are based on the predicted effects of genomic alterations in 44 genes analyzed. The frequency of mutations (top) and the gene expression (bottom) are summarized for each gene in the pathway. Gene expression alterations (Z-score ± 1) are shown separately for upregulation (red) and downregulation (blue). For simplicity, only genes that are dysregulated on the gene expression level in more than 10% of the TCGA HNSCC patient cohort are shown

NOTCH pathway activity and subsequently influence the outcome of this fundamental signaling axis. Recent advancements in high-throughput technologies and integrative analytical tools have brought a better understanding of NOTCH1 signaling in the context of head and neck cancers. As of 2017, cBioPortal analysis (Fig. 8.3) of the provisional TCGA HNSCC cohort shows that mutations in the key components of the NOTCH1 pathway occur in 46% of the patients, whereas gene expression changes are seen in virtually 100% of tumors, highlighting the importance of NOTCH pathway dysregulation in pathogenesis of HNSCC. Given the role of NOTCH signaling in regulating behavior of tumor cells, it is no surprise that several NOTCH-targeted treatment paradigms are currently being developed and tested in preclinical and clinical settings. Finally, developing drugs with better PK and pharmacodynamic (PD) profiles and better patient selection is needed before the NOTCH pathway can be optimally exploited for the treatment of patients with HNSCC and other cancers driven by NOTCH pathway alterations.

Conflicts of Interest The authors declare no conflict of interest.

References

1. Torre LA, et al. Global cancer statistics, 2012. CA Cancer J Clin. 2015;65(2):87–108.
2. Siegel RL, Miller KD, Jemal A. Cancer statistics, 2015. CA Cancer J Clin. 2015;65(1):5–29.
3. Califano J, et al. Unknown primary head and neck squamous cell carcinoma: molecular identification of the site of origin. J Natl Cancer Inst. 1999;91(7):599–604.
4. Cianchetti M, et al. Diagnostic evaluation of squamous cell carcinoma metastatic to cervical lymph nodes from an unknown head and neck primary site. Laryngoscope. 2009;119(12):2348–54.
5. Nagao T, et al. Oral cancer screening as an integral part of general health screening in Tokoname City, Japan. J Med Screen. 2000;7(4):203–8.
6. Subramanian S, et al. Cost-effectiveness of oral cancer screening: results from a cluster randomized controlled trial in India. Bull World Health Organ. 2009;87(3):200–6.
7. Agrawal N, et al. Exome sequencing of head and neck squamous cell carcinoma reveals inactivating mutations in NOTCH1. Science. 2011;333(6046):1154–7.
8. Pickering CR, et al. Integrative genomic characterization of oral squamous cell carcinoma identifies frequent somatic drivers. Cancer Discov. 2013;3(7):770–81.
9. Stransky N, et al. The mutational landscape of head and neck squamous cell carcinoma. Science. 2011;333(6046):1157–60.
10. Cancer Genome Atlas N. Comprehensive genomic characterization of head and neck squamous cell carcinomas. Nature. 2015;517(7536):576–82.
11. Artavanis-Tsakonas S, Rand MD, Lake RJ. Notch signaling: cell fate control and signal integration in development. Science. 1999;284(5415):770–6.
12. Hori K, Sen A, Artavanis-Tsakonas S. Notch signaling at a glance. J Cell Sci. 2013;126(Pt 10):2135–40.
13. Ellisen LW, et al. TAN-1, the human homolog of the Drosophila notch gene, is broken by chromosomal translocations in T lymphoblastic neoplasms. Cell. 1991;66(4):649–61.
14. Ntziachristos P, et al. From fly wings to targeted cancer therapies: a centennial for notch signaling. Cancer Cell. 2014;25(3):318–34.

15. Sun W, et al. Activation of the NOTCH pathway in head and neck cancer. Cancer Res. 2014;74(4):1091–104.
16. Liu YF, et al. Somatic mutations and genetic variants of NOTCH1 in head and neck squamous cell carcinoma occurrence and development. Sci Rep. 2016;6:24014.
17. Capaccione KM, Pine SR. The Notch signaling pathway as a mediator of tumor survival. Carcinogenesis. 2013;34(7):1420–30.
18. Wang Z, et al. Emerging role of Notch in stem cells and cancer. Cancer Lett. 2009;279(1):8–12.
19. Zeng Q, et al. Crosstalk between tumor and endothelial cells promotes tumor angiogenesis by MAPK activation of Notch signaling. Cancer Cell. 2005;8(1):13–23.
20. Gu F, et al. Expression of Stat3 and Notch1 is associated with cisplatin resistance in head and neck squamous cell carcinoma. Oncol Rep. 2010;23(3):671–6.
21. Ferrando AA. The role of NOTCH1 signaling in T-ALL. Hematology Am Soc Hematol Educ Program. 2009;2009:353–61.
22. Song X, et al. Common and complex Notch1 mutations in Chinese oral squamous cell carcinoma. Clin Cancer Res. 2014;20:701–10.
23. Izumchenko E, et al. Notch1 mutations are drivers of oral tumorigenesis. Cancer Prev Res (Phila). 2015;8(4):277–86.
24. Egloff AM, Grandis JR. Molecular pathways: context-dependent approaches to Notch targeting as cancer therapy. Clin Cancer Res. 2012;18(19):5188–95.
25. Louvi A, Artavanis-Tsakonas S. Notch and disease: a growing field. Semin Cell Dev Biol. 2012;23(4):473–80.
26. Hijioka H, et al. Upregulation of Notch pathway molecules in oral squamous cell carcinoma. Int J Oncol. 2010;36(4):817–22.
27. Zhang TH, et al. Activation of Notch signaling in human tongue carcinoma. J Oral Pathol Med. 2011;40(1):37–45.
28. Rettig EM, et al. Cleaved NOTCH1 expression pattern in head and neck squamous cell carcinoma is associated with NOTCH1 mutation, HPV status, and high-risk features. Cancer Prev Res (Phila). 2015;8(4):287–95.
29. Chiorean EG, et al. A phase I first-in-human study of Enoticumab (REGN421), a fully Human Delta-like ligand 4 (Dll4) monoclonal antibody in patients with advanced solid tumors. Clin Cancer Res. 2015;21(12):2695–703.
30. Smith DC, et al. A phase I dose escalation and expansion study of the anticancer stem cell agent demcizumab (anti-DLL4) in patients with previously treated solid tumors. Clin Cancer Res. 2014;20(24):6295–303.
31. Takebe N, Nguyen D, Yang SX. Targeting notch signaling pathway in cancer: clinical development advances and challenges. Pharmacol Ther. 2014;141(2):140–9.
32. Lee SM, et al. Phase 2 study of RO4929097, a gamma-secretase inhibitor, in metastatic melanoma: SWOG 0933. Cancer. 2015;121(3):432–40.
33. Diaz-Padilla I, et al. A phase Ib combination study of RO4929097, a gamma-secretase inhibitor, and temsirolimus in patients with advanced solid tumors. Investig New Drugs. 2013;31(5):1182–91.
34. Locatelli MA, et al. Phase I study of the gamma secretase inhibitor PF-03084014 in combination with docetaxel in patients with advanced triple-negative breast cancer. Oncotarget. 2017;8(2):2320–8.
35. Papayannidis C, et al. A phase 1 study of the novel gamma-secretase inhibitor PF-03084014 in patients with T-cell acute lymphoblastic leukemia and T-cell lymphoblastic lymphoma. Blood Cancer J. 2015;5:e350.
36. Messersmith WA, et al. A phase I, dose-finding study in patients with advanced solid malignancies of the oral gamma-secretase inhibitor PF-03084014. Clin Cancer Res. 2015;21(1):60–7.
37. Andersson ER, Lendahl U. Therapeutic modulation of Notch signalling – are we there yet? Nat Rev Drug Discov. 2014;13(5):357–78.
38. Diaz-Padilla I, et al. A phase II study of single-agent RO4929097, a gamma-secretase inhibitor of Notch signaling, in patients with recurrent platinum-resistant epithelial ovarian cancer:

a study of the Princess Margaret, Chicago and California phase II consortia. Gynecol Oncol. 2015;137(2):216–22.
39. Piha-Paul SA, et al. Results of a phase 1 trial combining ridaforolimus and MK-0752 in patients with advanced solid tumours. Eur J Cancer. 2015;51(14):1865–73.
40. Yuan X, et al. Notch signaling: an emerging therapeutic target for cancer treatment. Cancer Lett. 2015;369(1):20–7.
41. Shimizu K, et al. Functional diversity among Notch1, Notch2, and Notch3 receptors. Biochem Biophys Res Commun. 2002;291(4):775–9.
42. Kopan R, Ilagan MX. The canonical Notch signaling pathway: unfolding the activation mechanism. Cell. 2009;137(2):216–33.
43. Chillakuri CR, et al. Notch receptor-ligand binding and activation: insights from molecular studies. Semin Cell Dev Biol. 2012;23(4):421–8.
44. Pei Z, Baker NE. Competition between Delta and the Abruptex domain of Notch. BMC Dev Biol. 2008;8:4.
45. de Celis JF, Bray SJ. The Abruptex domain of Notch regulates negative interactions between Notch, its ligands and fringe. Development. 2000;127(6):1291–302.
46. Malecki MJ, et al. Leukemia-associated mutations within the NOTCH1 heterodimerization domain fall into at least two distinct mechanistic classes. Mol Cell Biol. 2006;26(12):4642–51.
47. Choi SH, et al. Conformational locking upon cooperative assembly of notch transcription complexes. Structure. 2012;20(2):340–9.
48. Arnett KL, et al. Structural and mechanistic insights into cooperative assembly of dimeric Notch transcription complexes. Nat Struct Mol Biol. 2010;17(11):1312–7.
49. Sulis ML, et al. NOTCH1 extracellular juxtamembrane expansion mutations in T-ALL. Blood. 2008;112(3):733–40.
50. Weng AP, et al. Activating mutations of NOTCH1 in human T cell acute lymphoblastic leukemia. Science. 2004;306(5694):269–71.
51. Baldus CD, et al. Prognostic implications of NOTCH1 and FBXW7 mutations in adult acute T-lymphoblastic leukemia. Haematologica. 2009;94(10):1383–90.
52. Fiuza UM, Arias AM. Cell and molecular biology of Notch. J Endocrinol. 2007;194(3):459–74.
53. Bray SJ. Notch signalling: a simple pathway becomes complex. Nat Rev Mol Cell Biol. 2006;7(9):678–89.
54. Chesworth BM, et al. Reliability and validity of two versions of the upper extremity functional index. Physiother Can. 2014;66(3):243–53.
55. Zhang M, et al. Does Notch play a tumor suppressor role across diverse squamous cell carcinomas? Cancer Med. 2016;5(8):2048–60.
56. Yap LF, et al. The opposing roles of NOTCH signalling in head and neck cancer: a mini review. Oral Dis. 2015;21(7):850–7.
57. Nowell CS, Radtke F. Notch as a tumour suppressor. Nat Rev Cancer. 2017;17(3):145–59.
58. Ranganathan P, Weaver KL, Capobianco AJ. Notch signalling in solid tumours: a little bit of everything but not all the time. Nat Rev Cancer. 2011;11(5):338–51.
59. Joo YH, et al. Relationship between vascular endothelial growth factor and Notch1 expression and lymphatic metastasis in tongue cancer. Otolaryngol Head Neck Surg. 2009;140(4):512–8.
60. Wang W-M, et al. Epidermal growth factor receptor inhibition reduces angiogenesis via hypoxia-inducible factor-1α and Notch1 in head neck squamous cell carcinoma. PLoS One. 2015;10(2):e0119723.
61. Troy JD, et al. Expression of EGFR, VEGF, and NOTCH1 suggest differences in tumor angiogenesis in HPV-positive and HPV-negative head and neck squamous cell carcinoma. Head Neck Pathol. 2013;7(4):344–55.
62. Leethanakul C, et al. Distinct pattern of expression of differentiation and growth-related genes in squamous cell carcinomas of the head and neck revealed by the use of laser capture microdissection and cDNA arrays. Oncogene. 2000;19(March):3220–4.

63. Ha PK, et al. A transcriptional progression model for head and neck cancer. Clin Cancer Res. 2003;9(8):3058–64.
64. Zhang ZP, et al. Correlation of Notch1 expression and activation to cisplatin-sensitivity of head and neck squamous cell carcinoma. Ai Zheng. 2009;28(2):100–3.
65. Yoshida R, et al. The pathological significance of Notch1 in oral squamous cell carcinoma. Lab Investig. 2013;93(10):1068–81.
66. Snijders AM, et al. Rare amplicons implicate frequent deregulation of cell fate specification pathways in oral squamous cell carcinoma. Oncogene. 2005;24(26):4232–42.
67. Li D, et al. Notch1 overexpression associates with poor prognosis in human laryngeal squamous cell carcinoma. Ann Otol Rhinol Laryngol. 2014;123(10):705–10.
68. Lin JT, et al. Association of high levels of Jagged-1 and Notch-1 expression with poor prognosis in head and neck cancer. Ann Surg Oncol. 2010;17(11):2976–83.
69. Inamura N, et al. Notch1 regulates invasion and metastasis of head and neck squamous cell carcinoma by inducing EMT through c-Myc. Auris Nasus Larynx. 2016;44(4):447–57.
70. Dai MY, et al. Downregulation of Notch1 induces apoptosis and inhibits cell proliferation and metastasis in laryngeal squamous cell carcinoma. Oncol Rep. 2015;34(6):3111–9.
71. Weaver AN, et al. Notch Signaling activation is associated with patient mortality and increased FGF1-mediated invasion in squamous cell carcinoma of the oral cavity. Mol Cancer Res. 2016;14(9):883–91.
72. Zhao ZL, et al. NOTCH1 inhibition enhances the efficacy of conventional chemotherapeutic agents by targeting head neck cancer stem cell. Sci Rep. 2016;6:24704.
73. Gaykalova DA, et al. Integrative computational analysis of transcriptional and epigenetic alterations implicates DTX1 as a putative tumor suppressor gene in HNSCC. Oncotarget. 2017;8(9):15349–63.
74. Stylianou S, Clarke RB, Brennan K. Aberrant activation of notch signaling in human breast cancer. Cancer Res. 2006;66(3):1517–25.
75. Xie XQ, et al. Dysregulation of mRNA profile in cisplatin-resistant gastric cancer cell line SGC7901. World J Gastroenterol. 2017;23(7):1189–202.
76. Zhang Z, Filho MS, Nor JE. The biology of head and neck cancer stem cells. Oral Oncol. 2012;48(1):1–9.
77. Lee SH, et al. Notch1 signaling contributes to stemness in head and neck squamous cell carcinoma. Lab Invest. 2016;96(5):508–16.
78. Lee SH, et al. Epigallocatechin-3-gallate attenuates head and neck cancer stem cell traits through suppression of Notch pathway. Eur J Cancer. 2013;49(15):3210–8.
79. Upadhyay P, et al. Notch pathway activation is essential for maintenance of stem-like cells in early tongue cancer. Oncotarget. 2016;7(31):50437–49.
80. Yu X, et al. Notch signaling activation in human embryonic stem cells is required for embryonic, but not trophoblastic, lineage commitment. Cell Stem Cell. 2008;2(5):461–71.
81. Shi G, Jin Y. Role of Oct4 in maintaining and regaining stem cell pluripotency. Stem Cell Res Ther. 2010;1(5):39.
82. Neville BW, Day TA. Oral cancer and precancerous lesions. CA Cancer J Clin. 2002;52(4):195–215.
83. Haya-Fernandez MC, et al. The prevalence of oral leukoplakia in 138 patients with oral squamous cell carcinoma. Oral Dis. 2004;10(6):346–8.
84. Mehanna HM, et al. Treatment and follow-up of oral dysplasia – a systematic review and meta-analysis. Head Neck. 2009;31(12):1600–9.
85. Silverman S Jr, Gorsky M, Lozada F. Oral leukoplakia and malignant transformation. A follow-up study of 257 patients. Cancer. 1984;53(3):563–8.
86. Lee SH, et al. TNFalpha enhances cancer stem cell-like phenotype via Notch-Hes1 activation in oral squamous cell carcinoma cells. Biochem Biophys Res Commun. 2012;424(1):58–64.
87. Gokulan R, Halagowder D. Expression pattern of Notch intracellular domain (NICD) and Hes-1 in preneoplastic and neoplastic human oral squamous epithelium: their correlation with c-Myc, clinicopathological factors and prognosis in oral cancer. Med Oncol. 2014;31(8):126.

88. Rettig EM, et al. Whole-genome sequencing of salivary gland adenoid cystic carcinoma. Cancer Prev Res (Phila). 2016;9(4):265–74.
89. Ferrarotto R, et al. Activating NOTCH1 mutations define a distinct subgroup of patients with adenoid cystic carcinoma who have poor prognosis, propensity to bone and liver metastasis, and potential responsiveness to Notch1 inhibitors. J Clin Oncol. 2017;35(3):352–60.
90. Yao J, et al. Gamma-secretase inhibitors exerts antitumor activity via down-regulation of Notch and nuclear factor kappa B in human tongue carcinoma cells. Oral Dis. 2007;13(6):555–63.
91. Wu CX, et al. Notch inhibitor PF-03084014 inhibits hepatocellular carcinoma growth and metastasis via suppression of cancer stemness due to reduced activation of Notch1-Stat3. Mol Cancer Ther. 2017;16(8):1531–43.
92. Gavai AV, et al. Discovery of clinical candidate BMS-906024: a potent pan-notch inhibitor for the treatment of leukemia and solid tumors. ACS Med Chem Lett. 2015;6(5):523–7.
93. Mohamed AA, et al. Synergistic activity with NOTCH inhibition and androgen ablation in ERG-positive prostate cancer cells. Mol Cancer Res. 2017;15(10):1308–17.
94. Barat S, et al. Gamma-Secretase inhibitor IX (GSI) impairs concomitant activation of Notch and wnt-beta-catenin pathways in CD44+ gastric Cancer stem cells. Stem Cells Transl Med. 2017;6(3):819–29.
95. De Kloe GE, De Strooper B. Small molecules that inhibit Notch signaling. Methods Mol Biol. 2014;1187:311–22.
96. Pant S, et al. A first-in-human phase I study of the oral Notch inhibitor, LY900009, in patients with advanced cancer. Eur J Cancer. 2016;56:1–9.
97. Bossi P, Alfieri S. Investigational drugs for head and neck cancer. Expert Opin Investig Drugs. 2016;25(7):797–810.
98. Kramer A, et al. Small molecules intercept Notch signaling and the early secretory pathway. Nat Chem Biol. 2013;9(11):731–8.
99. Borgegard T, et al. First and second generation gamma-secretase modulators (GSMs) modulate amyloid-beta (Abeta) peptide production through different mechanisms. J Biol Chem. 2012;287(15):11810–9.
100. Kumar R, Juillerat-Jeanneret L, Golshayan D. Notch antagonists: potential modulators of Cancer and inflammatory diseases. J Med Chem. 2016;59(17):7719–37.
101. Nicolas M, et al. Notch1 functions as a tumor suppressor in mouse skin. Nat Genet. 2003;33(3):416–21.
102. Naganuma S, et al. Notch receptor inhibition reveals the importance of cyclin D1 and Wnt signaling in invasive esophageal squamous cell carcinoma. Am J Cancer Res. 2012;2(4):459–75.
103. Nguyen BC, et al. Cross-regulation between Notch and p63 in keratinocyte commitment to differentiation. Genes Dev. 2006;20(8):1028–42.
104. Rangarajan A, et al. Notch signaling is a direct determinant of keratinocyte growth arrest and entry into differentiation. EMBO J. 2001;20(13):3427–36.
105. Ohashi S, et al. NOTCH1 and NOTCH3 coordinate esophageal squamous differentiation through a CSL-dependent transcriptional network. Gastroenterology. 2010;139(6):2113–23.
106. Duan L, et al. Growth suppression induced by Notch1 activation involves Wnt-beta-catenin down-regulation in human tongue carcinoma cells. Biol Cell. 2006;98(8):479–90.
107. Jiao J, et al. Potential role of Notch1 signaling pathway in laryngeal squamous cell carcinoma cell line Hep-2 involving proliferation inhibition, cell cycle arrest, cell apoptosis, and cell migration. Oncol Rep. 2009;22(4):815–23.
108. Sakamoto K, et al. Reduction of NOTCH1 expression pertains to maturation abnormalities of keratinocytes in squamous neoplasms. Lab Investig. 2012;92(5):688–702.
109. Gaykalova DA, et al. Novel insight into mutational landscape of head and neck squamous cell carcinoma. PLoS One. 2014;9(3):e93102.
110. Coric V, et al. Safety and tolerability of the gamma-secretase inhibitor avagacestat in a phase 2 study of mild to moderate Alzheimer disease. Arch Neurol. 2012;69(11):1430–40.
111. Doody RS, et al. Peripheral and central effects of gamma-secretase inhibition by semagacestat in Alzheimer's disease. Alzheimers Res Ther. 2015;7(1):36.

112. Radtke F, Raj K. The role of Notch in tumorigenesis: oncogene or tumour suppressor? Nat Rev Cancer. 2003;3(10):756–67.
113. Olsauskas-Kuprys R, Zlobin A, Osipo C. Gamma secretase inhibitors of Notch signaling. Onco Targets Ther. 2013;6:943–55.
114. Li K, et al. Modulation of Notch signaling by antibodies specific for the extracellular negative regulatory region of NOTCH3. J Biol Chem. 2008;283(12):8046–54.
115. Aoyama K, et al. Frequent mutations in NOTCH1 ligand-binding regions in Japanese oral squamous cell carcinoma. Biochem Biophys Res Commun. 2014;452(4):980–5.
116. Seiwert TY, et al. Integrative and comparative genomic analysis of HPV-positive and HPV-negative head and neck squamous cell carcinomas. Clin Cancer Res. 2015;21(3):632–41.
117. Zhong R, et al. Notch1 activation or loss promotes HPV-induced oral tumorigenesis. Cancer Res. 2015;75(18):3958–69.
118. Weijzen S, et al. HPV16 E6 and E7 oncoproteins regulate Notch-1 expression and cooperate to induce transformation. J Cell Physiol. 2003;194(3):356–62.
119. Talora C, et al. Specific down-modulation of Notch1 signaling in cervical cancer cells is required for sustained HPV-E6/E7 expression and late steps of malignant transformation. Genes Dev. 2002;16(17):2252–63.
120. Sausville EA, Burger AM. Contributions of human tumor xenografts to anticancer drug development. Cancer Res. 2006;66(7):3351–4, discussion 3354.
121. Izumchenko E, et al. Patient-derived xenografts as tools in pharmaceutical development. Clin Pharmacol Ther. 2016;99(6):612–21.
122. Kagohara L, et al. Epigenetic regulation of gene expression in cancer: techniques, resources, and analysis. Brief Funct Genomics. 2018;17(1):49–63.
123. Ozerov IV, et al. In silico Pathway Activation Network Decomposition Analysis (iPANDA) as a method for biomarker development. Nat Commun. 2016;7:13427.
124. Mroz EA, Rocco JW. MATH, a novel measure of intratumor genetic heterogeneity, is high in poor-outcome classes of head and neck squamous cell carcinoma. Oral Oncol. 2013;49(3):211–5.
125. Afsari B, Geman D, Fertig EJ. Learning dysregulated pathways in cancers from differential variability analysis. Cancer Inform. 2014;13(Suppl 5):61–7.
126. Makarev E, et al. In silico analysis of pathways activation landscape in oral squamous cell carcinoma and oral leukoplakia. Cell Death Discov. 2017;3:17022.
127. Ogmundsdottir HM, Bjornsson J, Holbrook WP. Role of TP53 in the progression of pre-malignant and malignant oral mucosal lesions. A follow-up study of 144 patients. J Oral Pathol Med. 2009;38(7):565–71.
128. Hori K, et al. Synergy between the ESCRT-III complex and Deltex defines a ligand-independent Notch signal. J Cell Biol. 2011;195(6):1005–15.
129. Li JL, et al. DLL4-Notch signaling mediates tumor resistance to anti-VEGF therapy in vivo. Cancer Res. 2011;71(18):6073–83.
130. Doroquez DB, Rebay I. Signal integration during development: mechanisms of EGFR and Notch pathway function and cross-talk. Crit Rev Biochem Mol Biol. 2006;41(6):339–85.
131. Hayward P, et al. Notch modulates Wnt signalling by associating with Armadillo/beta-catenin and regulating its transcriptional activity. Development. 2005;132(8):1819–30.
132. Shin M, Nagai H, Sheng G. Notch mediates Wnt and BMP signals in the early separation of smooth muscle progenitors and blood/endothelial common progenitors. Development. 2009;136(4):595–603.
133. Blokzijl A, et al. Cross-talk between the Notch and TGF-beta signaling pathways mediated by interaction of the Notch intracellular domain with Smad3. J Cell Biol. 2003;163(4):723–8.
134. Kanehisa M, Goto S. KEGG: Kyoto encyclopedia of genes and genomes. Nucleic Acids Res. 2000;28(1):27–30.

Chapter 9
P53 in Head and Neck Squamous Cell Carcinoma

Janaki Parameswaran and Barbara Burtness

Abstract *TP53* is the most commonly mutated gene in head and neck cancer. Mutations in *TP53* are associated with poor prognosis; approximately 50% of patients with locally advanced disease and nearly all patients with metastatic disease succumb to their illness. Novel and more effective treatment strategies are needed for these patients. However, due to the numerous intracellular roles of p53, and to the presence of both gain-of-function and loss-of-function mutations, targeting p53 has been challenging. Here, we review the p53 pathway and its role in the pathogenesis, prognosis, and treatment of head and neck squamous cell carcinoma.

Keywords P53 · Head and neck squamous cell carcinoma · Tumor suppressor

9.1 Introduction

Since its discovery in 1979, *TP53* (encoding the p53 tumor suppressor protein) has emerged as one of the most important and commonly mutated genes in human cancers. P53 plays a crucial role in cellular response to stress, determining cell fate in the settings of nutrient deprivation, hypoxia, metabolic disturbance, hyperproliferative signals, and DNA damage [1]. Under both normal physiologic and pathological conditions, p53 activation can lead to cell cycle arrest, apoptosis, and senescence. It is, therefore, considered a central tumor suppressor and "guardian of the genome" [2]. Altered p53 function contributes to tumor development and progression by influencing the cell cycle [3], angiogenesis [4–6], tumor-microenvironment interaction

J. Parameswaran
H. Lee Moffitt Cancer Center and Research Institute, Department of Head and Neck-Endocrine Oncology, Tampa, FL, USA

B. Burtness (✉)
Department of Internal Medicine, Yale University School of Medicine, New Haven, CT, USA
e-mail: barbara.burtness@yale.edu

B. Burtness, E. A. Golemis (eds.), *Molecular Determinants of Head and Neck Cancer*, Current Cancer Research, https://doi.org/10.1007/978-3-319-78762-6_9

[7, 8], and tumor invasion [9, 10]. Mutations in *TP53* occur in over 50% of malignancies [11]. Both loss-of-function (LOF) and gain-of-function (GOF) mutations have been described and result in various degrees of altered wild-type (WT) function [12, 13]. In head and neck squamous cell carcinoma (HNSCC), *TP53* mutation is the most common genetic alteration, occurring in 50–70% of cases [14, 15].

HNSCC can be broadly categorized by anatomic origin. The term "HNSCC" will refer to squamous cancers originating in the oral cavity, oropharynx, hypopharynx, and larynx, unless otherwise specified. HNSCC can also be classified by human papillomavirus (HPV) status, with an increasing proportion of cancers that are now associated with chronic HPV infection (see Chaps. 20 and 21). *TP53* mutations are found predominantly in HPV-negative disease, including in premalignant lesions [16, 17], suggesting a role in pathogenesis. In established tumors, *TP53* mutations are an independent risk factor for poor prognosis [18] and occur in approximately 70–80% of HPV-negative and less than 10% of HPV-related cases [19–21]. In HPV-related disease, the E6 viral protein targets p53 for proteasomal degradation; this likely has similar effects to an inactivating LOF mutation in *TP53*, with the exception that in response to DNA damage, levels of WT p53 can rise in HPV-associated disease, while WT p53 cannot be synthesized in the context of biallelic mutation [22].

Despite the frequency of *TP53* mutation, mutant p53 remains undruggable and creates a barrier to improved outcome. New treatment strategies that either target abnormal p53 or take advantage of vulnerabilities associated with *TP53* mutation are needed.

9.2 p53 Structure

TP53 is located on chromosome 17p13 and is approximately 22,000 bp in length. It contains 2 promoters (P1 and P2) and 11 exons. The gene encodes multiple isoforms that result from varying use of promoters, alternative splicing, and/or altering the site of mRNA initiation. To date, 9 mRNAs that encode 12 different isoforms have been defined [23, 24]. The most abundant isoform in humans is p53α, typically referred to as p53. Although p53 isoforms can be differentially expressed in malignancy, in HNSCC, there are no reports of significant difference in isoform expression between normal and cancerous tissue. Full-length p53 is 393 amino acids long, consisting of an N-terminal transactivation domain (TAD), a proline-rich region (PRR), a large DNA-binding domain (DBD), a tetramerization domain (TD), and a C-terminal regulatory domain (CTD) (Fig. 9.1) [25].

The TAD interacts with transcriptional coactivators and corepressors, and its activity is regulated by phosphorylation via a variety of kinases. The TAD can be divided into two domains, AD1 (amino acids 20–40) and AD2 (amino acids 40–61). AD1 plays an important role in response to DNA damage [26], and AD2 is involved in apoptosis [27]. Both TAD1 and TAD2 undergo conformational change upon

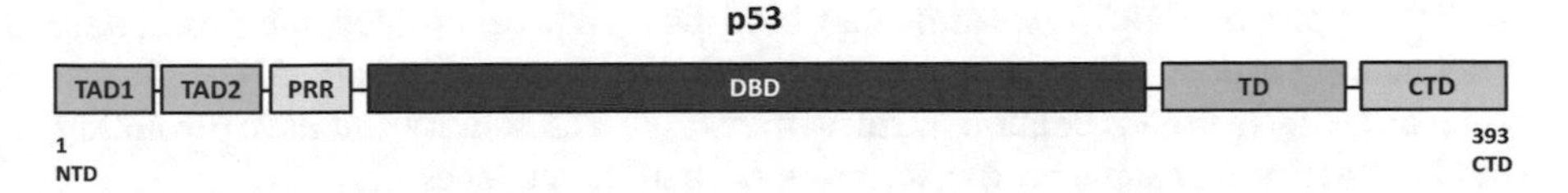

Fig. 9.1 From N-terminus to C-terminus p53 is made up of five main domains: the transactivation domain consisting of activation domain 1 (AD1) and activation domain 2 (AD2), proline-rich region (PRR), DNA-binding domain (DBD), tetramerization domain (TD), and the C-terminal domain (CTD)

binding to their target proteins, which include coactivators P300 and CBP and key inhibitors MDM2 and MDM4 [28–30].

The DNA-binding domain (amino acids 96–293), encoded by exons 5–8, is the largest domain of p53. Made up of two β-turn loops, L2 and L3, and a loop-sheet-helix motif [31], it is a highly conserved "core" region of the protein [32] that allows p53 to bind promoters of a variety of genes and regulate their transcription. Recent analysis of the International Agency for Research on Cancer (IARC) database and of the COSMIC database suggests that 80–90% of mutations associated with malignancy occur in the DBD [33].

In its main role as a transcription factor, p53 functions as a homotetramer [34]. The TD (amino acids 324–356) enables the formation of p53 homotetramers, as well as heterotetramers with p53 isoforms or mutant p53 [35]. Tetramers can exist in inactive (less DNA binding) and active (more DNA binding) forms, and the TD enables this shift in conformation by affecting the p53 tertiary and quaternary structure [36]. The TD also plays a role in the degradation of p53, as ubiquitination targets the oligomeric form [37, 38]. The TD also impacts p53 intracellular localization, as it contains a nuclear export signal which becomes hidden when p53 tetramerizes, allowing p53 to remain in the nucleus to regulate transcription [39].

Lastly, the CTD (amino acids 364–393) enables p53 DNA binding in a sequence-dependent manner [40] and helps stabilize DNA/p53 binding [41]. Within the CTD are two highly conserved motifs at residues 381-–383 and 391–393 [42]. Lysines 381 and 382 are either acetylated or ubiquitinated by p300/CBP or MDM2, respectively, and play a major role in determining p53 transcriptional capacity [43–47].

9.3 P63 and P73

The p53 family also includes two additional proteins, p63 and p73, which were identified in 1997 [48, 49]. These proteins are evolutionarily more ancient paralogues of p53 and resemble p53 in structure [50]. They have some overlapping functions with p53 (cell cycle regulation, apoptosis) but also play an important role in squamous cell differentiation. P63 is located on chromosome 3q28, and p73 is on chromosome 1p36. Like p53, p63 and p73 contain TADs, DNA-binding domains, and tetramerization domains and have alternative promoters and splicing sites which render multiple different protein isoforms. Most notably, they can encode protein variants that are full length (TAp63 or TAp73) or that lack TADs (ΔNp63 or ΔNp73).

Many ΔNp63 or ΔNp73 isoforms can bind DNA but cannot activate promoters, engendering a dominant negative-like phenotype. However, certain ΔN isoforms, such as DeltaNp73beta, contain intrinsic transcriptional activity and have the ability to either activate or suppress transcription of their target genes [51, 52].

P63 has 12 known isoforms that result from alternate promoter usage and alternate splicing [53]. It mainly exists in an inactive homodimerized state and is rarely mutated in human cancers. Interestingly, isoform ΔNp63α, an important regulator of keratinocyte differentiation [54], can be overexpressed in squamous cell carcinomas (SCC), including those originating in the head and neck, due to chromosomal amplification [55]. ΔNp63α may contribute to oncogenesis via stimulatory and inhibitory transcriptional regulation. Depending on the context, it can have dominant negative effects by blocking p53 response elements (e.g., influencing keratinocyte differentiation [54]) or activating transcription (e.g., of β-catenin-dependent genes [56] and VEGF [57]).

P73 mainly exists as an active homotetramer, and the TP73 gene is also rarely mutated in cancer. It has 14 known isoforms and an additional 15 theoretical isoforms. Its predominant isoform in cancer can vary, and its role in tumor development can differ depending on whether the isoform is pro- or anti-apoptotic. TAp73 isoforms bind to DNA via p53 response elements (p53RE), activating p53 target genes that induced cell cycle arrest or apoptosis. ΔNp73 isoforms can behave in a dominant negative manner (as mentioned above) by competing for p53 DNA-binding sites or protein interactions, leading to an anti-apoptotic effect [58]. In general, squamous cell carcinomas are associated with TAp73 isoform overexpression [55]. Additionally, patients with the TP73 G4C14-to-A4T14 (GC/AT) polymorphism may have a slightly increased risk for developing HPV-related oropharyngeal squamous cell carcinoma [59]. This polymorphism results in a dinucleotide substitution, which alters amino acid 4 from G to A and amino acid 14 from C to T. Due to monoallelic expression of p73, either the GC or AT allele is expressed. The polymorphism occurs in exon 2, upstream of p73's start codon, and can theoretically alter p73 gene expression, but mechanism of carcinogenesis is unknown [49].

9.4 p53 Regulation

Transcriptional and posttranslational modifications affect p53's activity and cellular localization. Over 30 proteins bind to p53 [60]. The E3 ubiquitin ligases MDM2 and MDM4 are two of those key proteins that regulate abundance of p53 by promoting its rate of ubiquitin-targeted degradation and limiting its transcriptional activity [3, 61–63]. MDM2 and MDM4 are structurally paralogous and contain four conserved regions: an N-terminal p53 binding domain which interacts with p53, a zinc finger domain, an acidic domain, and a C-terminal ring structure [64].

MDM4 heterodimerizes with and stabilizes MDM2 [65]. Based on differential expression of MDM4 and MDM2 in mouse models, it has been proposed that the MDM2 homodimer functions in slowly proliferating and terminally differentiated

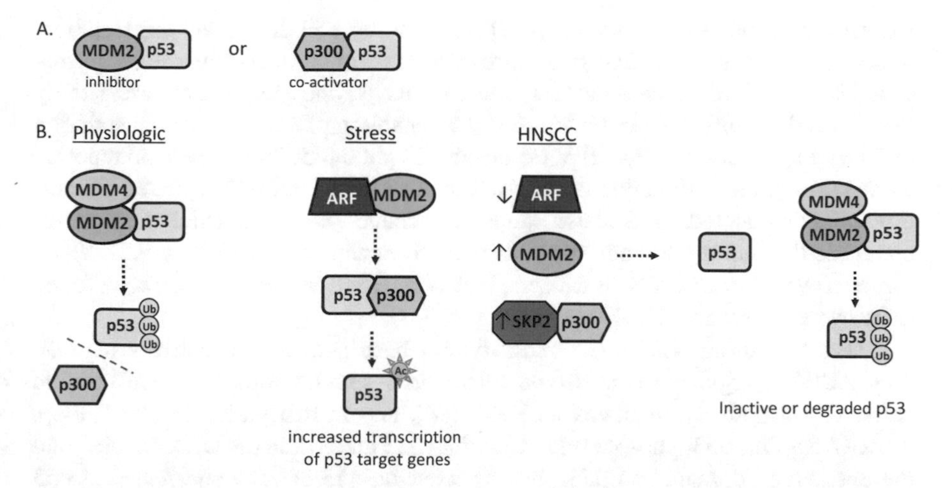

Fig. 9.2 Regulation of p53 in HNSCC. (**a**) p53 can either bind its inhibitor MDM2 or its key transcriptional co-activator p300. (**b**) Under normal physiologic conditions, MDM2 inhibits p53 function by targeting it to proteasomal degradation and preventing binding of p300. MDM4 stabilizes MDM2. Under cellular stress (e.g., DNA damage), numerous interactions contribute to p53 activation including ARF inhibition of MDM2. When ARF binds MDM2, p53 is free to bind p300. Dysregulation of the p53 pathway can contribute to oncogenesis. In HNSCC, studies have found decreased ARF, increased MDM2, and increased SKP2 (binds p300 and prevents p300-p53 interaction) which can lead to inappropriate inhibition of p53

cells, whereas MDM4 enhances MDM2 activity in highly proliferating cells [66, 67]. Interestingly, a growing body of evidence suggests MDM4 overexpression correlates with poor prognosis in a variety of cancers [68–71]. In HNSCC, MDM4 was found to be overexpressed in 50% of cases, based on immunohistochemistry (IHC) [72].

In response to stress, there are a variety of mechanisms that disrupt the MDM2-p53 interaction, including ADP-ribosylation factor (ARF) inactivation of MDM2. ARF is one of the two tumor suppressors transcribed from the p16INK4A-ARF locus (CDKN2A) [73]. ARF binds and sequesters MDM2, thereby increasing p53 levels and activity (Fig. 9.2) [74, 75]. Under pathological conditions, reduced ARF expression [76] or increased MDM2 expression leads to inhibition of p53, which can contribute to oncogenesis via loss of cell cycle checkpoints. HNSCC has been associated with alterations in both MDM2 and ARF expression. MDM2 is overexpressed in up to 46% of cases [77]. ARF is underexpressed (via deletion or methylation) in 30% of oral SCC cases, and ARF underexpression correlates with increased recurrence in retrospective analysis [78]. Moreover, LOH of 9p21 (site of ARF) was found to occur early during HNSCC development, suggesting a role in pathogenesis [17]. Interestingly, as opposed to oral cancers which are typically HPV-negative and more likely to have *TP53* mutation, HPV-positive oropharyngeal cancers have been shown to be associated with *CDKN2A* hypermethylation and increased ARF expression [79].

Another mechanism to disrupt the MDM2-p53 interaction is phosphorylation of Thr18 on p53, which prevents binding of MDM2 [80] and increases binding affinity for the transcriptional coactivator p300 [81]. During p53-dependent transcription, p300 interaction with the TAD of tetrameric p53 stabilizes the tetrameric conformation. P300 also loosens chromatin structure via its histone acetyltransferase activity [82, 83] and recruits RNA polymerase II [84], enabling p53 to regulate transcription of its target genes. Although HPV E6 inactivates p300 [85, 86], there is no reported evidence of p300 dysfunction in HPV-negative HNSCC. P300 is regulated by multiple proteins including S-phase kinase-associated protein 2 (SKP2), a proto-oncogene that binds p300 and prevents p300 from activating p53 (Fig. 9.2). SKP2 can be elevated in a variety of cancers, including HNSCC, where it has been found to be overexpressed in 24/47 cases [87].

Recently, Aurora Kinase A (AURKA) has been shown to negatively regulate p53. AURKA regulates centrosome maturation, spindle formation, and mitotic entry. P53 negatively regulates AURKA [88], and in a negative feedback loop, AURKA inhibits p53, phosphorylating serines 215 and 315 in the DNA-binding and tetramerization domains on p53, thereby reducing p53 activity and fostering p53 degradation [89, 90]. Further support of the inhibitory effect of AURKA on p53 comes from in vitro data showing AURKA inhibition stabilizes p53's downstream target, CDKN1A/p21 [91]. Pathologic upregulation of AURKA occurs in a variety of tumors. In HNSCC, high AURKA levels correlate with poor prognosis and cisplatin resistance [92–94]. Alisertib, a small molecule that is primarily an AURKA inhibitor, has demonstrated a 9% monotherapy response rate in previously treated HNSCC [95]. Clinical response rates to AURKA inhibition in p53 mutant vs. WT tumors have not been established in HNSCC.

9.5 P53 Function

Posttranslational modification (including phosphorylation, acetylation, and methylation) by stress-induced kinases allows p53 to bind DNA at p53 response elements and induce the transcription of genes affecting apoptosis, cell cycle progression, and DNA repair. Following DNA damage, p53 triggers the G1/S checkpoint by inducing expression of Waf/CIP1, which leads to inhibition of CDK2 and G1 arrest. P53 also has established roles in nucleotide excision, base excision, and double-stranded and mismatch repair. In this capacity, it affects the accessibility of chromatin to repair enzymes by modulating the helicase activity of TFIIH [96, 97], and it increases transcription of p48 and XPC, which increase cellular ability to locate and target DNA damage for nucleotide excision repair. P53 also regulates transcription of several other DNA repair and damage response genes including *RAD51, MSH2, OGG1, MUTYH*, and *APE1* [98–103]. Cells that have mutant *TP53* have impaired control over the G1 checkpoint and DNA repair, contributing to oncogenesis.

P53 also interacts directly with several proteins that influence cell survival including Bcl-2 and Bcl-2-associated protein X (Bax), structural paralogues in the BCL-2 family. Bcl-2 promotes cell survival, whereas Bax induces apoptosis (Fig. 9.3). When apoptosis is triggered, Bax induces mitochondrial membrane permeability

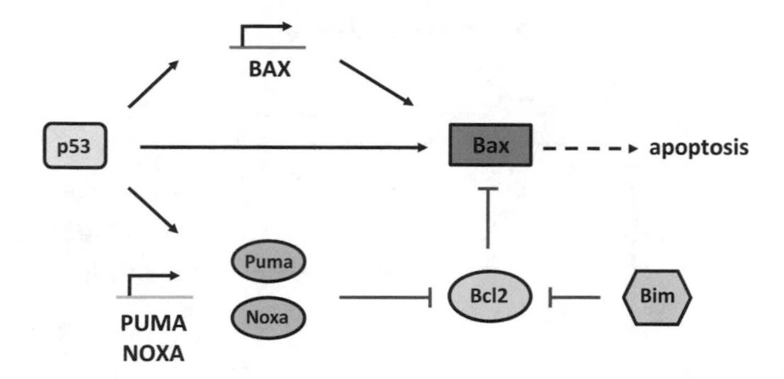

Fig. 9.3 P53 activation of Bax. One of the pathways by which p53 leads to apoptosis is via activation of Bax. P53 can increase transcription of *BAX*, directly activate Bax, and increase transcription of *PUMA* and *NOXA*. Puma and Noxa inhibit Bcl2, a negative regulator of Bax

and ultimately caspase activation. Bax is inhibited by Bcl-2, which is negatively regulated by Puma, Noxa, and Bim. P53 can lead to apoptosis by increasing transcription of *BAX*, directly activating Bax, or increasing transcription of *PUMA* and *NOXA* [104, 105].

Bcl2 is detected via IHC in about 13–21% of HNSCC [106, 107]. Correlation of Bcl-2 expression and prognosis in HNSCC varies with stage and treatment modality. In a study analyzing BCL-2 expression using IHC in 400 samples, Bcl-2 overexpression correlated with poorly differentiated, advanced disease but improved locoregional tumor control (relative risk 0.51, 95% CI 0.31–0.85, $p = 0.009$) and overall survival (relative risk 0.37, 95% CI 0.22–0.62, $p = 0.0002$) in multivariate analysis [106]. Pena et al. [108] also found that increased Bcl-2 expression detected by IHC was associated with improved prognosis in patients with stage IV locally advanced disease. Interestingly, in patients with early-stage HNSCC treated with radiation therapy (RT), Bcl-2 expression correlated with worse disease-free and 5-year overall survival [107], with a potential mechanism being impaired apoptosis during RT. None of the above studies correlated bcl-2 and p53 status.

9.6 Single Nucleotide Polymorphisms (SNP) of the p53 Pathway

A common SNP in *TP53* occurs in codon 72, located in exon 4 in a proline-rich region important for p53 regulation of apoptosis [109]. In 1988, it was discovered that amino acid 72, which typically codes for arginine (R72), can also code for proline (P72) [110]. In HNSCC, there is no clear correlation between P72 SNP and risk of cancer development. However, patients with oral and laryngeal cancer with P72/P72 may have younger age of tumor onset [111]. Additionally, based on a small sample of patients, those with P72/P72 appear to have significantly worse

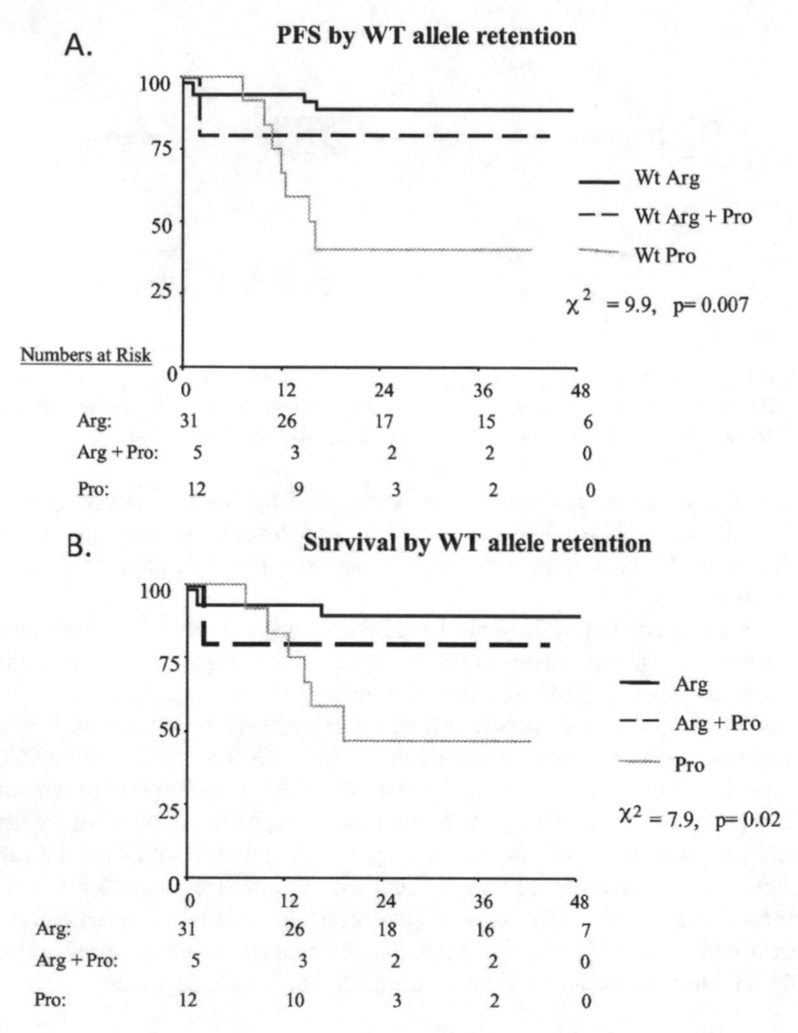

Fig. 9.4 Polymorphism in WT TP53 codon 72 influences the clinical outcome of combined modality treatment in HNSCC. Estimates of progression-free survival (PFS) and overall survival in HNSCC based on findings from Sullivan et al. [112] WT p53 R72/R72 = Wt Arg, WT p53 R72/P72 = Wt Arg + Pro, WT P72/P72 = Wt Pro. (**a**) Presence of at least one R72 allele is associated with improved progression-free survival. (**b**) Presence of at least one R72 allele is associated with improved overall survival

progression-free survival and overall survival after concurrent chemoradiation with cisplatin compared to those with R72/R72 or R72/P72 (Fig. 9.4) [112].

Another SNP in *TP53* occurs at amino acid 47, which can code for proline (predominant form) or serine. In response to UV radiation, p38 MAP kinase phosphorylates serine residues on p53, thereby stabilizing p53 and leading to apoptosis [113]. When S47 is present, p38 may have reduced ability to bind the p53 proline-rich region [114]. In vitro data suggest that the S47 SNP confers resistance to treatment with cisplatin, the main chemotherapy used in HNSCC treatment [115], but there are no data on this SNP in HNSCC. S47 is found in less than 5% of African-Americans and confers a five-fold reduced ability to induce apoptosis, versus 0% of Americans of European descent, which may contribute to racial disparities in response to treatment [116].

In addition to SNPs in p53, the T309G SNP in *MDM2* can impact p53 functionality. *MDM2* SNP 309GG stabilizes *MDM2* mRNA and consequently increases MDM2 protein levels and degradation of *TP53*. Patients with SNP 309GG have a 9-year earlier cancer onset in both hereditary and sporadic tumors [117]. This polymorphism may also may confer risk for nasopharyngeal cancer, but not risk for other head and neck cancer types [118]. In patients treated with platinum-based therapy, those with SNP 309GG or TG had worse survival than TT patients [119].

9.7 *TP53* Mutations in HNSCC

Mutations affecting TP53 activity that are related to malignancy can be categorized as germ line or somatic. Germline mutations in *TP53* are uncommon driver mutations for HNSCC and, when they occur, are associated with Li-Fraumeni, Bloom, or Werner syndromes. Li-Fraumeni is due predominantly to *TP53* germline mutations in the DNA-binding domain; these are often missense mutations that lead to defective protein production [120]. Li-Fraumeni patients are therefore predisposed to malignancy at a young age (mean age of onset 25) and often have multiple malignancies in their lifetime [121]. Based on a case series, HNSCC accounts for about 3% of the cancers found in these patients [122].

Bloom and Werner syndromes indirectly affect p53 function and are due to mutations that affect helicases. The Bloom syndrome protein (BLM) and the Werner syndrome protein (WRN) are two of five RecQ helicases that belong to the SF2 helicase superfamily. Mutations in both genes significantly predispose to cancer. BLM has been shown to interact with p53 in yeast two-hybrid systems and likely causes p53 transactivation [123, 124]. Mutations in *BLM* lead to severe growth retardation, UV light sensitivity, and genomic instability (chromosome breakage and short telomeres) [125]. In a series of 100 individual cancer specimens in patients with Bloom syndrome, 10 were in the head and neck (excluding the skin): 2 auditory canal, 4 tongue, 1 tonsil, and 3 larynx/epiglottis [126]. WRN binds to the CTD of p53 and stimulates p53-dependent apoptosis [127]; mutant WRN can lead to reduced apoptotic function of p53 [128]. Mutations in *WRN* lead to osteoporosis, atherosclerosis, type 2 diabetes mellitus, and cataracts. There are reports of Werner syndrome patients with head and neck cancer, and in a primarily Japanese population, about 3% of head and neck cancers were associated with *WRN* mutation [129].

Much more common are somatic *TP53* mutations that occur early in tumor growth. Somatic *TP53* mutations fall into several distinct functional categories: while most lead to loss-of-function (LOF), or have dominant-negative effects, some provide new oncogenic properties to the protein, in a gain-of-function (GOF). In

1995, Brennan et al. [130] sequenced *TP53* in 129 tumors from patients with HNSCC treated at Johns Hopkins. Most patients were smokers or had at least a 20 pack-year smoking history. *TP53* mutations were found in 54 of the 129 patients. Most of the observed mutations were substitutions, and 7% were frameshift (either insertion or deletion).

Several other studies have shown that function-disabling *TP53* mutations occur predominantly in HPV-negative disease, including that of Gillison et al. [131] Their study compared rates of *TP53* mutation in HPV-associated and HPV-unassociated head and neck tumors. They evaluated 253 samples, 200 from the primary tumor at time of diagnosis and 53 from local biopsies at time of recurrence. In the oropharyngeal subset, *TP53* mutation was present in 10% of HPV-associated disease and 67% of HPV-unassociated disease.

In 2015, The Cancer Genome Atlas Group reported results of a multi-platform genomic characterization of 279 HNSCCs (62% oral, 26% laryngeal, and 12% oropharyngeal) [15]. Tumors were classified as HPV-positive or HPV-negative using RNA sequencing of viral genes E6 and E7. Among their findings were differences in rates of *TP53* mutation between HPV-related and negative tumors: 86% of HPV-negative tumors vs. 1/36 (2.7%) of HPV-positive tumors had *TP53* mutations. Tinhofer et al. [132] similarly reported *TP53* mutation rates of 4% and 67% in HPV-positive and HPV-negative HNSCC, respectively, based on a sample of 179 specimens. More recently, in 2016, Morris et al. published data on the molecular landscape of recurrent and metastatic head and neck cancers treated at Memorial Sloan Kettering Cancer Center [133]. Next-generation sequencing on 53 locally advanced and metastatic treatment-resistant HNSCCs (21 HPV-positive, 30 HPV-negative) again confirmed that *TP53* mutation is much more common in HPV-negative than HPV-positive disease (72% vs. 15%); however, the *TP53* mutation rate in HPV-positive cancers was higher than in the larger studies reported above. Moreover, HPV-associated recurrent/metastatic tumors had higher rates of *TP53* mutation (15% = 3/20), compared to primary HPV-related tumors (3% = 1/36). Based on correlation with clinical data, these authors suggest that *TP53* mutation in a subset of HPV-positive tumors is associated with smoking, occurs later in tumor evolution, and reflects poorer prognosis. Further studies addressing prognosis of *TP53* mutations are described in the following section.

9.8 *TP53* Mutations and Prognosis in HNSCC

It has been well-established in HNSCC that, in contrast to *TP53* genotype, p53 expression detected by IHC does not correlate with prognosis. This discrepancy between the prognostic import of genotype and protein expression arises because a subset of *TP53* mutations affect p53 function but not expression, or lead to prolonged p53 half-life, including GOF and dominant-negative mutations. In 1997, Bradford et al. [134] analyzed *TP53* mutation versus p53 expression in tumors from the VA Laryngeal Cancer Cooperative Study [135]. In the 44 laryngeal cancers analyzed, there was poor correlation between p53 expression and *TP53* exon 5–8 mutational status. Sixteen patients had overexpression of p53 without detectable mutation on sequencing. Conversely, seven patients with normal p53 expression on IHC had detectable mutations on sequencing. Additionally, there was a trend toward decreased survival in patients with *TP53* mutation (HR 1.83, 95% CI 0.9–3.72, $p = 0.09$), suggesting that p53 status was an independent poor prognostic factor in HNSCC.

To prognosticate the effects of *TP53* mutation in HNSCC, many classification systems have been developed. In 2007, Poeta et al. [18] associated *TP53* mutation with poor survival in patients with resected HNSCC of all stages (Fig. 9.5), enrolled in the Eastern Cooperative Oncology Group E4393 study. This study had two objectives: [1] to determine the value of molecular detection of cancer cells at tumor margins and [2] to determine the incidence of *TP53* mutation and its impact on survival in HNSCC. All patients had newly diagnosed or recurrent HNSCC, underwent surgical resection with negative margins, and received risk-based adjuvant therapy. Of the 420 patient samples analyzed, 43% were from the oral cavity, 21% from the larynx, 22% from the oropharynx, 7.6% from the hypopharynx, and 5.2% others or unknown.

Tumors were analyzed by GeneChip p53 assay, which detects mutations in exons 2–11 and is able to detect some, but not all, frameshift mutations. The investigators classified mutations as disruptive or non-disruptive [32]. Mutations were classified as disruptive if they occurred in the L2–L3 region of the DBD or produced stop codons in any region. Non-disruptive mutations occurred outside the L2–L3 region and were not truncation mutations. Patients who had *TP53* mutations had a median overall survival of 3.2 years, and those with WT *TP53* had a median overall survival of 5.4 years. Furthermore, patients with disruptive *TP53* mutations had worse prognosis compared to patients with non-disruptive mutations, with median overall survival of 2 years vs. 3.9 years. The hazard ratio for death was 1.32 (95% CI 1.01–1.73, $p = 0.04$) with any *TP53* mutation and 1.69 (95% CI 1.20–2.36, $p = 0.003$) with a disruptive *TP53* mutation. The adverse effect of disruptive *TP53* mutations was observed in both locally advanced and early-stage disease.

Since missense mutations lead to dominant-negative or GOF effects, using sequence location alone to predict the effect of *TP53* mutations can be difficult. To better classify p53 mutations and determine which mutations adversely affect prognosis in HNSCC, Neskey et al. [136] used an evolutionary trace (ET) approach. Every nucleotide was assigned a level of functional sensitivity defined by the degree of phylogenetic divergence resulting from substitutions in that region. The ET functional sensitivity in addition to conservativeness of the amino acid change was used to calculate an evolutionary action (EA) score. TCGA data were used to classify mutations into high risk and low risk. High-risk mutations (including R175H, C238F, G245D) had increased likelihood for lung metastasis and worse overall survival. The method was validated using an internal data set from M.D. Anderson, and results were confirmed using mouse models [136].

Using data from the patients enrolled in E4393, Masica et al. [137] compared 15 different methods of predicting the effect of *TP53* mutations on survival. For the

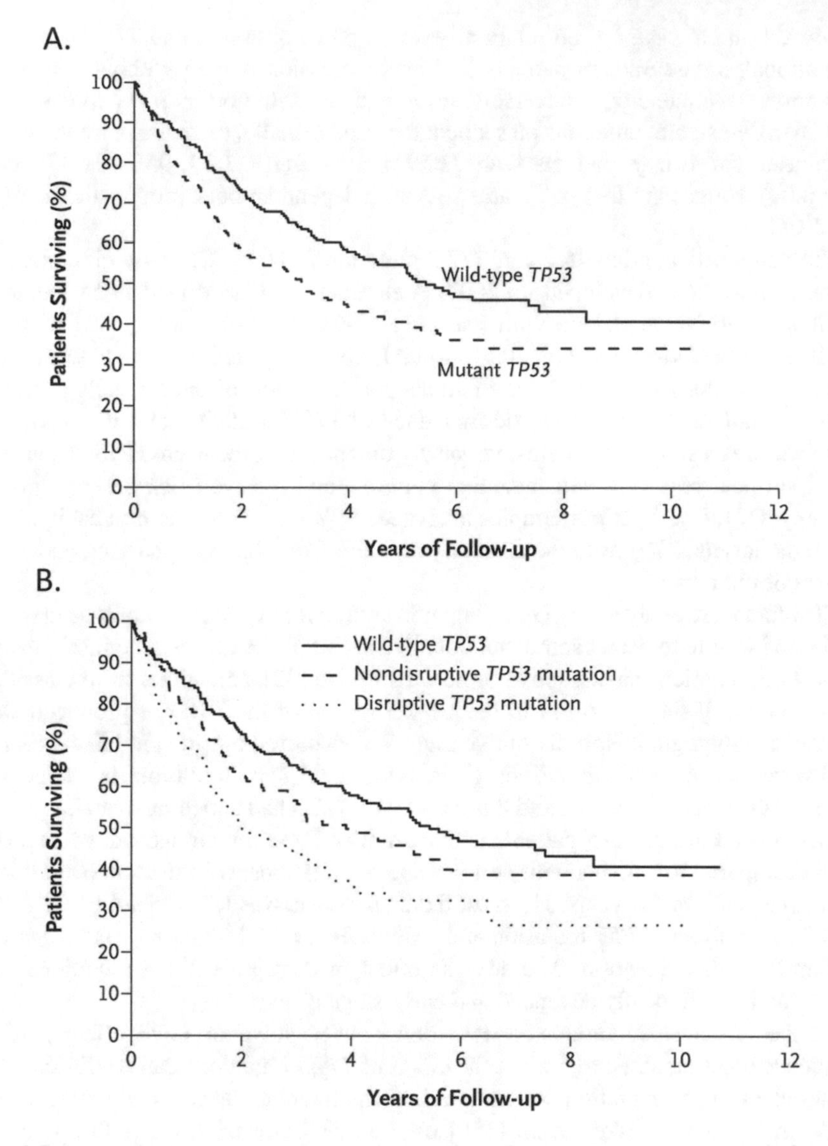

Fig. 9.5 Kaplan-Meier estimates of overall survival in HNSCC from Poeta et al. [18] Patients included in the study underwent surgical resection with negative margins, followed by risk-based adjuvant therapy. (**a**) Among the 196 patients with wild-type *TP53* (of whom 99 died) and the 224 patients with mutant *TP53* (of whom 133 died), the median survival among patients with mutant *TP53* was 3.2 years, as compared with 5.4 years among patients with wild-type *TP53*. (**b**) Among the 139 patients with non-disruptive *TP53* mutations (of whom 76 died) and the 85 patients with disruptive *TP53* mutations (of whom 57 died), median survival among patients with disruptive mutations was 2.0 years, whereas that among patients with non-disruptive mutations was 3.9 years. Disruptive mutations were defined as non-conservative mutations located inside the key DNA-binding domain (the L2–L3 region) or stop codons in any region; non-disruptive mutations were defined as conservative or non-conservative mutations (excluding stop codons) outside the L2–L3 region

analysis, the Poeta algorithm was included, but the Neskey algorithm was not. A total of 420 patient samples were analyzed, of which 224 were *TP53* mutated. The classification system used by Poeta was superior to the remaining 14 algorithms and was the only method to correlate statistically with survival.

Lastly, in patients with p53 mutant disease, there have been attempts to correlate minimal residual disease (MRD) at surgical margin sites with prognosis. No studies have shown statistical significance. This may be due to high rate of false-positive *TP53* mutation detection or presence of *TP53* mutations in secondary early precursor lesions [138]. In 2009, data were published using the LigAmp *TP53* assay to assess MRD at surgical margin sites from patients enrolled in E4393 and others in RTOG 9614. Ninety-five samples contained a missense *TP53* mutation found in at least 1 other sample: 85 had negative margins and 10 had positive margins (defined by severe dysplasia or invasive carcinoma). There was no statistical correlation between *TP53* positive margin status and either cancer-specific or overall survival. A high false-positive rate with the LigAmp *TP53* assay (32.5% of patients without local recurrence were found to have positive margins) may have contributed to the failure to identify a correlation.

9.9 P53 in Nasopharyngeal Carcinomas

Nasopharyngeal carcinoma (NPC) is an uncommon type of squamous cell carcinoma that differs from other HPV-negative HNSCCs in terms of incidence, epidemiology, and treatment. NPC has an incidence of about 80,000 people worldwide per year [139], with the highest prevalence in Southeast Asia. It is classified by WHO into the following types: keratinizing SCC and nonkeratinizing well-differentiated, nonkeratinizing poorly differentiated, and basaloid SCC [140]. Nearly all nonkeratinizing NPCs are associated with EBV [141]. More recently, NPC has also been found to be associated with HPV, especially in the non-endemic, keratinizing type [142, 143].

Unlike other HPV-negative squamous cell carcinomas of the head and neck, NPC has a low rate of *TP53* mutation. In fact, up to 95% of NPCs overexpress WT p53 [144–146]. One study examined 38 tumors for mutations in *TP53* exons 4–8 and found no mutations that led to amino acid change [147]. When 12 NPC tumors from Hunan province of China were sequenced, only 1 had a *TP53* mutation [148]. A third study sequenced *TP53* exons 5–8 in 28 tumors derived from non-endemic cases (mostly Caucasian and African-American) and found only 3 patients with non-silent C→T substitutions [149].

9.10 P53 and Treatment Response

TP53 mutation correlates with worse outcome in HNSCC. This may in part be a consequence of varied response to standard treatment, which relies on surgery and the DNA-damaging modalities of radiation therapy (RT) and cisplatin-based chemotherapy.

The effect of *TP53* mutation on radiosensitivity can vary with cancer cell type [150]. In HNSCC, *TP53* mutations are typically associated with radioresistance based on preclinical and clinical data. In 1996, Koch et al. sequenced 110 HNSCCs from patients treated with radiation only or surgery followed by adjuvant radiation [151]. *TP53* status was determined by sequencing exons 5–9. Multivariate analysis showed that *TP53* mutation correlated with worse locoregional control. In 2001, Alsner et al. [152] evaluated samples from 90 patients with HNSCC, most of whom had locally advanced disease and received radiation as their only treatment. Forty of ninety patients had *TP53* mutations (mostly missense, as determined by exon 5–9 sequencing), which were associated with worse locoregional control rates and reduced DFS and overall survival.

In 2012, Skinner et al. [153] sequenced *TP53* exons 2–12 in tumor samples from 74 patients who were treated with surgery and adjuvant radiation. Mutations were classified as disruptive or non-disruptive based on the Poeta criteria. Patients with disruptive *TP53* mutations had increased risk of locoregional recurrence and worse overall survival. At 5 years, freedom from locoregional recurrence was 41% with disruptive *TP53* mutation, 64% with non-disruptive *TP53* mutation, and 76% with WT *TP53*. The 5-year overall survival rates for patients with disruptive mutations, non-disruptive mutations, and WT *TP53* were 19%, 41%, and 52%, respectively. These authors went on to evaluate 38 HNSCC cell lines and demonstrated that cells with disruptive *TP53* mutations were radioresistant and implicated a mechanism of impaired senescence.

Although it is fairly clear that *TP53* mutations typically confer radioresistance, their effect on cisplatin sensitivity can vary. In 2003, Bradford et al. [154] evaluated 23 HNSCC cell lines, of which 13 were *TP53* mutant as determined by SSCP and p53 GeneChip analysis. *TP53* mutation correlated with cisplatin sensitivity in vitro based on IC_{50} values. Similarly, data from Hoffmann et al. [155] and Andrews et al. [156] on HNSCC cell lines suggested cisplatin sensitivity in *TP53* mutant cell lines. Hoffman's team compared 26 p53 mutant cell lines to 3 p53 WT cell lines. There was a trend toward increased cisplatin sensitivity in p53 mutant cell lines, but this was not statistically significant ($p = 0.13$), which may have been due to the presence of an additional *TP53* WT allele in 11/26 cell lines. Of note, the one cell line that was HPV-positive was highly resistant to cisplatin. Lastly, Andrews et al. [156] found that induction of mutant p53 in the USMSCC74B cell lines via transfection of the temperature-sensitive p53 construct LTRp53cG (Val 135) resulted in increased susceptibility to cisplatin-induced apoptosis.

However, other studies [157–159] correlate a subset of *TP53* variants or mutations with cisplatin resistance. Mandic et al. [157] demonstrated that cell lines containing the *TP53* polymorphism R72/R72 were resistant to cisplatin if they had mutations affecting the p53 nuclear localization signal (corresponding to amino acids 305–322). Osman et al. [158] used three cell lines with lack of p53 expression and transfected them with seven different *TP53* mutant constructs that were either high (R175H, C176F, C238F, G245D) or low risk (F134C, A161S, Y236C) for poor outcome based on the evolutionary action score. They found that cells transfected with high-risk mutant constructs had reduced cisplatin sensitivity compared to low-risk constructs and the parental cell line. Of note, two of the mutations were previously included in studies that concluded *TP53* mutations correlate with cisplatin sensitivity: the C176F mutation was included in Bradford's analysis, and the C238F mutation was included in Hoffman's analysis. Conflicting data on the effect of these two mutations on cisplatin sensitivity may be related to differences in cisplatin doses, varied presence of R72 SNP in *TP53* which can confer cisplatin resistance [160], or presence of background mutations in a multitude of other genes affecting the cell cycle or cell survival.

Despite varying data on the effect of *TP53* mutation on cisplatin sensitivity, both in vitro [161, 162] and clinical data [163–165] support the use of cisplatin as a radiosensitizer in *TP53*-mutated HNSCC. In patients with HNSCC associated with disruptive *TP53* mutation, the addition of cisplatin to radiation appears to improve survival compared to radiation alone. A small study conducted by Fallai et al. [166] analyzed tumors from 78 patients with oropharyngeal cancer who participated in the ORO93-01 and MEDUSA trials. The ORO 93-01 trial was a phase III trial randomizing patients with locally advanced HNSCC to definitive RT, accelerated split course RT, or standard RT with carboplatin and 5FU. The MEDUSA trial randomized patients with locally advanced HNSCC to receive radiation with concomitant boost with either low-dose daily cisplatin or three cycles of cisplatin/5FU every 3 weeks. Twenty-two samples had disruptive *TP53* mutations, another twenty-two samples had non-disruptive *TP53* mutations, and thirty-five samples were *TP53* WT. Although sample sizes were small, they found that when a platinum agent was added to radiation, 5-year OS was nearly doubled in both the *TP53* mutant and WT groups (42% vs. 22% if mutant and 30% vs. 16% if WT).

9.11 New Treatments

Treatment of patients with *TP53* mutant tumors has been challenging across a variety of cancers including HNSCC, as clinically translatable methods of restoring normal p53 function in the face of disruptive mutations have not been developed. Since patients with *TP53* mutations have worse outcome and are at higher risk of relapse following definitive surgery, the EA3132 [NCT02734537] trial was designed to determine whether *TP53* mutant disease should be treated more aggressively. For patients with locally advanced disease (stage III or IV), standard of care typically involves adjuvant radiation for patients who undergo primary resection of their tumor. Those with positive surgical margins or extracapsular spread of their disease from lymph nodes benefit from addition of chemotherapy to radiation [167, 168]. The EA3132 trial will assess *TP53* status in patients with locally advanced disease who had surgical resection without the recognized indications for adjuvant chemotherapy, stratify on the basis of disruptive *TP53* mutation, and randomize patients to either postoperative radiation or concurrent chemoradiation with cisplatin.

Other recent approaches in p53 mutant disease involve the use of cell cycle checkpoint kinase inhibitors (discussed in more detail in Chap. 7). Since cells with p53 mutations lose their ability to control the G1/S checkpoint, they rely more heavily on the G2/M checkpoint [169]. The G2/M checkpoint is mainly regulated by the CDK1/cyclin B1 complex: when the complex is activated, the cell enters mitosis, and when it is inactivated, the cell arrests in G2. Cdc25 and WEE1 regulate activation and inactivation of CDK1, respectively. If WEE1 is inhibited, cells are predisposed to inappropriate mitotic entry, especially those with an already defective G1 checkpoint (i.e., those with *TP53* mutation). Similarly, WEE1 is also involved in duration of S-phase via interactions with CDK2 and may have a role in DNA repair. Therefore, inhibition of WEE1 may lead to mitotic catastrophe if cells that are unable to repair DNA damage enter mitosis [170, 171].

A phase I clinical trial combining AZD-1775 with either cisplatin, carboplatin, or gemcitabine in advanced solid tumors showed that overall, 53% of patients had stable disease and 10% had partial responses [172]. Tumor samples were available in 52 patients: 19 with *TP53*-mutations and 33 with WT *TP53*. Patients with *TP53* mutations had numerically increased partial response rates compared those with wild-type *TP53* (21% vs. 12%), but larger sample size is required to determine statistical significance. The same group also conducted a phase II trial of AZD-1775 plus carboplatin in *TP53*-mutant ovarian cancer patients that relapsed or were refractory to first-line platin-based chemotherapy [173]. The overall response rate was 43%, and one patient had a prolonged partial response, suggesting that WEE1 inhibition can enhance the efficacy of platinum agents, at least in ovarian cancer.

There are several small molecule-based therapies that target either WT or mutant p53 proteins. COTI-2 is an orally available third-generation thiosemicarbazone developed by Critical Outcome Technologies Inc. Initial preclinical data showed that COTI-2 leads to apoptosis in a variety of *TP53* mutant cancer cell lines and reduces tumor growth in xenograft mouse models [174]. Although the structure remains proprietary, COTI-2 is believed to restore p53 activity and negatively regulate the PI3K/AKT/mTOR pathway, which can be overexpressed in malignancy and promotes cell survival. In HNSCC, preliminary in vitro data from M.D. Anderson show that COTI-2 has activity in both p53 WT and mutant cell lines leading to increased p21 expression and also inhibits tumor growth in mouse orthotopic models of oral cancer [175]. COTI-2 is currently being evaluated in a phase I trial in gynecological malignancies, with planned phase II expansion that includes HPV-negative recurrent/metastatic head and neck cancer patients [NCT02433626].

Other small molecules in various stages of testing/development for restoration of p53 function include PRIMA-1^{MET} [176], CP31398 [177, 178], MIRA-1 [179], RETRA [180], nutlins, and RITA (reactivation of p53 and induction of tumor cell apoptosis). Nutlins and RITA prevent p53 degradation, and in vitro data show effectiveness in reducing the growth of *TP53* WT HNSCC cell lines when used either alone or in combination with cisplatin [181, 182]. RITA binds p53 and induces a conformational change that prevents MDM2 binding [183], whereas nutlins bind MDM2 and disrupt the MDM2-p53 interaction [184]. Recently Nutlin-3 has been shown to sensitize p53 WT esophageal squamous cancer cells (ECA-109) to irradiation in vitro and in xenograft mouse models created with ECA-109 cells [185]. Resistance mechanisms to both nutlins and RITA have been described. Although nutlins do not bind p53, selection for nutlin resistance can lead to acquisition of mutations in p53 that alter its transcriptional activity [186] and may contribute to nutlin resistance. Resistance to RITA may be partially attributed to NF-κB RelA/p65 phosphorylation status: reduced phosphorylation of RelA/p65 at S536 and increased phosphorylation at S276 and S468 can reduce p65 activity and potential for cell death. Resistance may also be attributable to increased export of RITA via upregulation of ABC transporter ABCC6 expression [187].

Lastly, gene therapy using adenoviral vectors can be used in both p53 WT and mutant HNSCC. Vectors can restore or boost p53 function when they contain WT *TP53* (e.g., Ad-p53). They can also be designed to induce cell lysis in p53-deficient cells (e.g., Onyx-015, ColoAd1). These agents have been undergoing evaluation for several years [188]. In a phase III trial using Ad-p53 in refractory HNSCC, patients received intratumoral injections with Ad-p53 or methotrexate. Patients with WT p53 or low expression of mutant p53 had similar outcomes with Ad-p53 and methotrexate, with a trend to improved survival in the Ad-p53 group [189]. In *TP53*-mutant disease, the *TP53* antisense agent ONYX-15 underwent preclinical and clinical testing [190]. However, due to a requirement for BID injections and lack of impact on median overall survival in the initial study, there are no new trials with ONYX-15.

Although not directly targeting p53 mutant cells, over the past few years, there have been exciting advances in the use of immune checkpoint inhibitors in malignancy regardless of *TP53* mutational state. In 2016, two agents received FDA approval to treat recurrent or refractory HNSCC that has progressed despite platin-based therapy—the humanized anti-PD-1 antibody pembrolizumab and the human IgG4 anti-PD-1 antibody nivolumab. Data supporting their use in HNSCC are presented in Chap. 14. The appeal of these agents includes the potential for long-term survival and a different toxicity profile compared to chemotherapy, predominantly autoimmune in nature. Higher PD-L1 expression corresponds to improved response rates to PD-1 inhibition in several cancers including HNSCC [191, 192]. P53 has been shown to activate miR-34a, which binds PD-L1 mRNA and prevents PD-L1 expression on the cell membrane [193, 194]. Furthermore, analysis tumors from the TCGA showed that *TP53* mutations were associated with lower levels of mir-34a and higher PDL-1 expression in non-small cell lung cancer [194]. Further investigation into correlation of *TP53* mutational status and response to PD-1 or PD-L1 inhibition and patient selection and resistance mechanisms are needed in HNSCC.

Despite current treatment options of immune checkpoint inhibition, chemotherapy, radiation, and surgery, treating p53 mutant HNSCC remains challenging. We are unable to effectively target mutant p53 HNSCC, and many questions regarding treatment remain. Does locally advanced p53 mutant disease benefit from treatment intensification or modification? Can immune checkpoint inhibition in the adjuvant setting overcome the poor prognosis conferred by the mutation? Should we continue our attempts to target p53 or develop treatments that will be effective regardless of p53 status? How can analysis of synthetic lethal and pathway target approaches exploit the abnormalities in cell cycle control in p53 mutant cancer? We eagerly await results of ongoing clinical and laboratory research, which may provide the answers.

References

1. Bieging KT, Mello SS, Attardi LD. Unravelling mechanisms of p53-mediated tumour suppression. Nat Rev Cancer. 2014;14:359–70.
2. Lane DP. Cancer. p53, guardian of the genome. Nature. 1992;358:15–6.
3. Brown CJ, Lain S, Verma CS, et al. Awakening guardian angels: drugging the p53 pathway. Nat Rev Cancer. 2009;9:862–73.
4. Ravi R, Mookerjee B, Bhujwalla ZM, et al. Regulation of tumor angiogenesis by p53-induced degradation of hypoxia-inducible factor 1α. Genes Dev. 2000;14:34–44.
5. Farhang Ghahremani M, Goossens S, Nittner D, et al. p53 promotes VEGF expression and angiogenesis in the absence of an intact p21-Rb pathway. Cell Death Differ. 2013;20:888–97.
6. Teodoro JG, Evans SK, Green MR. Inhibition of tumor angiogenesis by p53: a new role for the guardian of the genome. J Mol Med (Berl). 2007;85:1175–86.
7. Huang Y, Yu P, Li W, et al. p53 regulates mesenchymal stem cell-mediated tumor suppression in a tumor microenvironment through immune modulation. Oncogene. 2014;33:3830–8.
8. Palumbo A Jr, Da Costa Nde O, Bonamino MH, et al. Genetic instability in the tumor microenvironment: a new look at an old neighbor. Mol Cancer. 2015;14:145.
9. Muller PAJ, Vousden KH, Norman JC. p53 and its mutants in tumor cell migration and invasion. J Cell Biol. 2011;192:209.
10. Powell E, Piwnica-Worms D, Piwnica-Worms H. Contribution of p53 to metastasis. Cancer Discov. 2014;4:405.
11. Soussi T. p53 alterations in human cancer: more questions than answers. Oncogene. 2007;26:2145–56.
12. Muller PAJ, Vousden KH. p53 mutations in cancer. Nat Cell Biol. 2013;15:2–8.
13. Peltonen JK, Helppi HM, Pääkkö P, et al. p53 in head and neck cancer: functional consequences and environmental implications of TP53 mutations. Head Neck Oncol. 2010;2:36.
14. Stransky N, Egloff AM, Tward AD, et al. The mutational landscape of head and neck squamous cell carcinoma. Science. 2011;333:1157–60.
15. The Cancer Genome Atlas N. Comprehensive genomic characterization of head and neck squamous cell carcinomas. Nature. 2015;517:576–82.
16. Shin DM, Kim J, Ro JY, et al. Activation of p53 gene expression in premalignant lesions during head and neck tumorigenesis. Cancer Res. 1994;54:321–6.
17. Califano J, van der Riet P, Westra W, et al. Genetic progression model for head and neck cancer: implications for field cancerization. Cancer Res. 1996;56:2488–92.
18. Poeta ML, Manola J, Goldwasser MA, et al. TP53 mutations and survival in squamous-cell carcinoma of the head and neck. N Engl J Med. 2007;357:2552–61.
19. Hayes DN, Waes C, Seiwert TY. Genetic landscape of human papillomavirus-associated head and neck cancer and comparison to tobacco-related tumors. J Clin Oncol. 2015;33:3227.
20. Beck TN, Golemis EA. Genomic insights into head and neck cancer. Cancers Head Neck. 2016;1:1.

21. Maruyama H, Yasui T, Ishikawa-Fujiwara T, et al. Human papillomavirus and p53 mutations in head and neck squamous cell carcinoma among Japanese population. Cancer Sci. 2014;105:409–17.
22. Hafkamp HC, Speel EJM, Haesevoets A, et al. A subset of head and neck squamous cell carcinomas exhibits integration of HPV 16/18 DNA and overexpression of p16INK4A and p53 in the absence of mutations in p53 exons 5–8. Int J Cancer. 2003;107:394–400.
23. Khoury MP, Bourdon JC. The isoforms of the p53 protein. Cold Spring Harb Perspect Biol. 2010;2:a000927.
24. Khoury MP, Bourdon J-C. p53 isoforms: an intracellular microprocessor? Genes Cancer. 2011;2:453–65.
25. Joerger AC, Fersht AR. The tumor suppressor p53: from structures to drug discovery. Cold Spring Harb Perspect Biol. 2010;2:a000919.
26. Raj N, Attardi LD. The transactivation domains of the p53 protein. Cold Spring Harb Perspect Med. 2016. https://doi.org/10.1101/cshperspect.a026047.
27. Puca R, Nardinocchi L, Givol D, et al. Regulation of p53 activity by HIPK2: molecular mechanisms and therapeutical implications in human cancer cells. Oncogene. 2010;29:4378–87.
28. Miller Jenkins LM, Yamaguchi H, Hayashi R, et al. Two distinct motifs within the p53 transactivation domain bind to the Taz2 domain of p300 and are differentially affected by phosphorylation. Biochemistry. 2009;48:1244–55.
29. Shan B, Li DW, Bruschweiler-Li L, et al. Competitive binding between dynamic p53 transactivation subdomains to human MDM2 protein: implications for regulating the p53.MDM2/MDMX interaction. J Biol Chem. 2012;287:30376–84.
30. Lee CW, Martinez-Yamout MA, Dyson HJ, et al. Structure of the p53 transactivation domain in complex with the nuclear receptor coactivator binding domain of CREB binding protein. Biochemistry. 2010;49:9964–71.
31. Cho Y, Gorina S, Jeffrey PD, et al. Crystal structure of a p53 tumor suppressor-DNA complex: understanding tumorigenic mutations. Science. 1994;265:346–55.
32. Friend S. p53: a glimpse at the puppet behind the shadow play. Science. 1994;265:334–5.
33. Hainaut P, Pfeifer GP. Somatic TP53 mutations in the era of genome sequencing. Cold Spring Harb Perspect Med. 2016;6:a026179.
34. McLure KG, Lee PW. How p53 binds DNA as a tetramer. EMBO J. 1998;17:3342–50.
35. Chene P. The role of tetramerization in p53 function. Oncogene. 2001;20:2611–7.
36. Halazonetis TD, Kandil AN. Conformational shifts propagate from the oligomerization domain of p53 to its tetrameric DNA binding domain and restore DNA binding to select p53 mutants. EMBO J. 1993;12:5057–64.
37. Kubbutat MH, Ludwig RL, Ashcroft M, et al. Regulation of Mdm2-directed degradation by the C terminus of p53. Mol Cell Biol. 1998;18:5690–8.
38. Maki CG. Oligomerization is required for p53 to be efficiently ubiquitinated by MDM2. J Biol Chem. 1999;274:16531–5.
39. Stommel JM, Marchenko ND, Jimenez GS, et al. A leucine-rich nuclear export signal in the p53 tetramerization domain: regulation of subcellular localization and p53 activity by NES masking. EMBO J. 1999;18:1660–72.
40. Hupp TR, Meek DW, Midgley CA, et al. Regulation of the specific DNA binding function of p53. Cell. 1992;71:875–86.
41. Laptenko O, Shiff I, Freed-Pastor W, et al. The p53 C terminus controls site-specific DNA binding and promotes structural changes within the central DNA binding domain. Mol Cell. 2015;57:1034–46.
42. Laptenko O, Tong DR, Manfredi J, et al. The tail that wags the dog: how the disordered C-terminal domain controls the transcriptional activities of the p53 tumor-suppressor protein. Trends Biochem Sci. 2016;41:1022–34.
43. Gu W, Roeder RG. Activation of p53 sequence-specific DNA binding by acetylation of the p53 C-terminal domain. Cell. 90:595–606.

44. Sakaguchi K, Herrera JE, Saito S, et al. DNA damage activates p53 through a phosphorylation-acetylation cascade. Genes Dev. 1998;12:2831–41.
45. Liu L, Scolnick DM, Trievel RC, et al. p53 sites acetylated in vitro by PCAF and p300 are acetylated in vivo in response to DNA damage. Mol Cell Biol. 1999;19:1202–9.
46. Rodriguez MS, Desterro JMP, Lain S, et al. Multiple C-terminal lysine residues target p53 for ubiquitin-proteasome-mediated degradation. Mol Cell Biol. 2000;20:8458–67.
47. Nakamura S, Roth JA, Mukhopadhyay T. Multiple lysine mutations in the C-terminal domain of p53 interfere with MDM2-dependent protein degradation and ubiquitination. Mol Cell Biol. 2000;20:9391–8.
48. Yang A, Kaghad M, Wang Y, et al. p63, a p53 homolog at 3q27-29, encodes multiple products with transactivating, death-inducing, and dominant-negative activities. Mol Cell. 1998;2:305–16.
49. Kaghad M, Bonnet H, Yang A, et al. Monoallelically expressed gene related to p53 at 1p36, a region frequently deleted in neuroblastoma and other human cancers. Cell. 90:809–19.
50. Dotsch V, Bernassola F, Coutandin D, et al. p63 and p73, the ancestors of p53. Cold Spring Harb Perspect Biol. 2010;2:a004887.
51. Liu G, Nozell S, Xiao H, et al. DeltaNp73beta is active in transactivation and growth suppression. Mol Cell Biol. 2004;24:487–501.
52. Helton ES, Zhu J, Chen X. The unique NH2-terminally deleted (DeltaN) residues, the PXXP motif, and the PPXY motif are required for the transcriptional activity of the DeltaN variant of p63. J Biol Chem. 2006;281:2533–42.
53. Du J, Romano RA, Si H, et al. Epidermal overexpression of transgenic ΔNp63 promotes type 2 immune and myeloid inflammatory responses and hyperplasia via NF-κB activation. J Pathol. 2014;232:356–68. https://doi.org/10.1002/path.4302.
54. King KE, Ponnamperuma RM, Yamashita T, et al. deltaNp63alpha functions as both a positive and a negative transcriptional regulator and blocks in vitro differentiation of murine keratinocytes. Oncogene. 2003;22:3635–44.
55. Deyoung MP, Ellisen LW. p63 and p73 in human cancer: defining the network. Oncogene. 2007;26:5169–83.
56. Patturajan M, Nomoto S, Sommer M, et al. ΔNp63 induces β-catenin nuclear accumulation and signaling. Cancer Cell. 1:369–79.
57. Senoo M, Matsumura Y, Habu S. TAp63gamma (p51A) and dNp63alpha (p73L), two major isoforms of the p63 gene, exert opposite effects on the vascular endothelial growth factor (VEGF) gene expression. Oncogene. 2002;21:2455–65.
58. Murray-Zmijewski F, Lane DP, Bourdon JC. p53//p63//p73 isoforms: an orchestra of isoforms to harmonise cell differentiation and response to stress. Cell Death Differ. 2006;13:962–72.
59. Chen X, Sturgis EM, Etzel CJ, et al. p73 G4C14-to-A4T14 polymorphism and risk of human papillomavirus associated squamous cell carcinoma of the oropharynx in never smokers and never drinkers. Cancer. 2008;113:3307–14.
60. Inoue K, Fry EA, Frazier DP. Transcription factors that interact with p53 and Mdm2. Int J Cancer. 2016;138:1577–85.
61. Momand J, Zambetti GP, Olson DC, et al. The mdm-2 oncogene product forms a complex with the p53 protein and inhibits p53-mediated transactivation. Cell. 1992;69:1237–45.
62. Oliner JD, Pietenpol JA, Thiagalingam S, et al. Oncoprotein MDM2 conceals the activation domain of tumour suppressor p53. Nature. 1993;362:857–60.
63. Haupt Y, Maya R, Kazaz A, et al. Mdm2 promotes the rapid degradation of p53. Nature. 1997;387:296–9.
64. Cheok CF, Verma CS, Baselga J, et al. Translating p53 into the clinic. Nat Rev Clin Oncol. 2011;8:25–37.
65. Sharp DA, Kratowicz SA, Sank MJ, et al. Stabilization of the MDM2 oncoprotein by interaction with the structurally related MDMX protein. J Biol Chem. 1999;274:38189–96.
66. Valentin-Vega YA, Box N, Terzian T, et al. Mdm4 loss in the intestinal epithelium leads to compartmentalized cell death but no tissue abnormalities. Differentiation. 2009;77:442–9.

67. Marine JC, Jochemsen AG. MDMX (MDM4), a promising target for p53 reactivation therapy and beyond. Cold Spring Harb Perspect Med. 2016;6:a026237.
68. Bartel F, Schulz J, Bohnke A, et al. Significance of HDMX-S (or MDM4) mRNA splice variant overexpression and HDMX gene amplification on primary soft tissue sarcoma prognosis. Int J Cancer. 2005;117:469–75.
69. Danovi D, Meulmeester E, Pasini D, et al. Amplification of Mdmx (or Mdm4) directly contributes to tumor formation by inhibiting p53 tumor suppressor activity. Mol Cell Biol. 2004;24:5835–43.
70. Ramos YF, Stad R, Attema J, et al. Aberrant expression of HDMX proteins in tumor cells correlates with wild-type p53. Cancer Res. 2001;61:1839–42.
71. Riemenschneider MJ, Knobbe CB, Reifenberger G. Refined mapping of 1q32 amplicons in malignant gliomas confirms MDM4 as the main amplification target. Int J Cancer. 2003;104:752–7.
72. Valentin-Vega YA, Barboza JA, Chau GP, et al. Overexpression of the p53 inhibitor MDM4 in head and neck squamous carcinomas. Hum Pathol. 2007;38:1553–62.
73. Hipfner DR, Cohen SM. Connecting proliferation and apoptosis in development and disease. Nat Rev Mol Cell Biol. 2004;5:805–15.
74. Zhang Y, Xiong Y, Yarbrough WG. ARF promotes MDM2 degradation and stabilizes p53: ARF-INK4a locus deletion impairs both the Rb and p53 tumor suppression pathways. Cell. 1998;92:725–34.
75. Weber JD, Taylor LJ, Roussel MF, et al. Nucleolar Arf sequesters Mdm2 and activates p53. Nat Cell Biol. 1999;1:20–6.
76. Ozenne P, Eymin B, Brambilla E, et al. The ARF tumor suppressor: structure, functions and status in cancer. Int J Cancer. 2010;127:2239–47.
77. Millon R, Muller D, Schultz I, et al. Loss of MDM2 expression in human head and neck squamous cell carcinomas and clinical significance. Oral Oncol. 2001;37:620–31.
78. Sailasree R, Abhilash A, Sathyan KM, et al. Differential roles of p16INK4A and p14ARF genes in prognosis of oral carcinoma. Cancer Epidemiol Biomark Prev. 2008;17:414–20.
79. Schlecht NF, Ben-Dayan M, Anayannis N, et al. Epigenetic changes in the CDKN2A locus are associated with differential expression of P16INK4A and P14ARF in HPV-positive oropharyngeal squamous cell carcinoma. Cancer Med. 2015;4:342–53.
80. Sakaguchi K, Saito S, Higashimoto Y, et al. Damage-mediated phosphorylation of human p53 threonine 18 through a cascade mediated by a casein 1-like kinase. Effect on Mdm2 binding. J Biol Chem. 2000;275:9278–83.
81. Dornan D, Hupp TR. Inhibition of p53-dependent transcription by BOX-I phospho-peptide mimetics that bind to p300. EMBO Rep. 2001;2:139–44.
82. Bannister AJ, Kouzarides T. The CBP co-activator is a histone acetyltransferase. Nature. 1996;384:641–3.
83. Ogryzko VV, Schiltz RL, Russanova V, et al. The transcriptional coactivators p300 and CBP are histone acetyltransferases. Cell. 1996;87:953–9.
84. Kee BL, Arias J, Montminy MR. Adaptor-mediated recruitment of RNA polymerase II to a signal-dependent activator. J Biol Chem. 1996;271(5):2373.
85. Patel D, Huang SM, Baglia LA, et al. The E6 protein of human papillomavirus type 16 binds to and inhibits co-activation by CBP and p300. EMBO J. 1999;18:5061–72.
86. Zimmermann H, Degenkolbe R, Bernard HU, et al. The human papillomavirus type 16 E6 oncoprotein can down-regulate p53 activity by targeting the transcriptional coactivator CBP/p300. J Virol. 1999;73:6209–19.
87. Carracedo DG, Astudillo A, Rodrigo JP, et al. Skp2, p27kip1 and EGFR assessment in head and neck squamous cell carcinoma: prognostic implications. Oncol Rep. 2008;20:589–95.
88. Wu CC, Yang TY, Yu CT, et al. p53 negatively regulates Aurora A via both transcriptional and posttranslational regulation. Cell Cycle. 2012;11:3433–42.
89. Katayama H, Sasai K, Kawai H, et al. Phosphorylation by aurora kinase A induces Mdm2-mediated destabilization and inhibition of p53. Nat Genet. 2004;36:55–62.

90. Liu Q, Kaneko S, Yang L, et al. Aurora-A abrogation of p53 DNA binding and transactivation activity by phosphorylation of serine 215. J Biol Chem. 2004;279:52175–82.
91. Marxer M, Ma HT, Man WY, et al. p53 deficiency enhances mitotic arrest and slippage induced by pharmacological inhibition of Aurora kinases. Oncogene. 2014;33:3550–60.
92. Mehra R, Serebriiskii IG, Burtness B, et al. Aurora kinases in head and neck cancer. Lancet Oncol. 2013;14:e425–35.
93. Reiter R, Gais P, Jütting U, et al. Aurora kinase a messenger RNA overexpression is correlated with tumor progression and shortened survival in head and neck squamous cell carcinoma. Clin Cancer Res. 2006;12:5136.
94. Li Y, Zhang J. AURKA is a predictor of chemotherapy response and prognosis for patients with advanced oral squamous cell carcinoma. Tumour Biol. 2015;36:3557–64.
95. Melichar B, Adenis A, Lockhart AC, et al. Safety and activity of alisertib, an investigational aurora kinase A inhibitor, in patients with breast cancer, small-cell lung cancer, non-small-cell lung cancer, head and neck squamous-cell carcinoma, and gastro-oesophageal adenocarcinoma: a five-arm phase 2 study. Lancet Oncol. 2015;16:395–405.
96. Wang XW, Yeh H, Schaeffer L, et al. p53 modulation of TFIIH-associated nucleotide excision repair activity. Nat Genet. 1995;10:188–95.
97. Leveillard T, Andera L, Bissonnette N, et al. Functional interactions between p53 and the TFIIH complex are affected by tumour-associated mutations. EMBO J. 1996;15:1615–24.
98. Arias-Lopez C, Lazaro-Trueba I, Kerr P, et al. p53 modulates homologous recombination by transcriptional regulation of the RAD51 gene. EMBO Rep. 2006;7:219–24.
99. Gatz SA, Wiesmuller L. p53 in recombination and repair. Cell Death Differ. 2006;13:1003–16.
100. Zink D, Mayr C, Janz C, et al. Association of p53 and MSH2 with recombinative repair complexes during S phase. Oncogene. 2002;21:4788–800.
101. Achanta G, Huang P. Role of p53 in sensing oxidative DNA damage in response to reactive oxygen species-generating agents. Cancer Res. 2004;64:6233–9.
102. Oka S, Leon J, Tsuchimoto D, et al. MUTYH, an adenine DNA glycosylase, mediates p53 tumor suppression via PARP-dependent cell death. Oncogene. 2014;3:e121.
103. Zhou J, Ahn J, Wilson SH, et al. A role for p53 in base excision repair. EMBO J. 2001;20:914–23.
104. Hemann MT, Lowe SW. The p53-Bcl-2 connection. Cell Death Differ. 2006;13:1256–9.
105. Nakano K, Vousden KH. PUMA, a novel proapoptotic gene, is induced by p53. Mol Cell. 2001;7:683–94.
106. Wilson GD, Saunders MI, Dische S, et al. bcl-2 expression in head and neck cancer: an enigmatic prognostic marker. Int J Radiat Oncol Biol Phys. 2001;49:435–41.
107. Gallo O, Boddi V, Calzolari A, et al. bcl-2 protein expression correlates with recurrence and survival in early stage head and neck cancer treated by radiotherapy. Clin Cancer Res. 1996;2:261–7.
108. Pena JC, Thompson CB, Recant W, et al. Bcl-xL and Bcl-2 expression in squamous cell carcinoma of the head and neck. Cancer. 1999;85:164–70.
109. Grochola LF, Zeron-Medina J, Mériaux S, et al. Single-nucleotide polymorphisms in the p53 signaling pathway. Cold Spring Harb Perspect Biol. 2010;2:a001032.
110. Buchman VL, Chumakov PM, Ninkina NN, et al. A variation in the structure of the protein-coding region of the human p53 gene. Gene. 1988;70:245–52.
111. Shen H, Zheng Y, Sturgis EM, et al. p53 codon 72 polymorphism and risk of squamous cell carcinoma of the head and neck: a case-control study. Cancer Lett. 2002;183:123–30.
112. Sullivan A, Syed N, Gasco M, et al. Polymorphism in wild-type p53 modulates response to chemotherapy in vitro and in vivo. Oncogene. 2004;23:3328–37.
113. Bulavin DV, Saito S, Hollander MC, et al. Phosphorylation of human p53 by p38 kinase coordinates N-terminal phosphorylation and apoptosis in response to UV radiation. EMBO J. 1999;18:6845–54.
114. Li X, Dumont P, Della Pietra A, et al. The codon 47 polymorphism in p53 is functionally significant. J Biol Chem. 2005;280:24245–51.

115. Basu S, Barnoud T, Kung CP, et al. The African-specific S47 polymorphism of p53 alters chemosensitivity. Cell Cycle. 2016;15:2557–60.
116. Felley-Bosco E, Weston A, Cawley HM, et al. Functional studies of a germ-line polymorphism at codon 47 within the p53 gene. Am J Hum Genet. 1993;53:752–9.
117. Yu H, Huang YJ, Liu Z, et al. Effects of MDM2 promoter polymorphisms and p53 codon 72 polymorphism on risk and age at onset of squamous cell carcinoma of the head and neck. Mol Carcinog. 2011;50:697–706.
118. Zhou J, Yang Y, Zhang D, et al. Association of the recurrence of vocal leukoplakia with MDM2-309 variants over a 2-year period: a prospective study. Acta Otolaryngol. 2016;136:95–9.
119. Vivenza D, Gasco M, Monteverde M, et al. MDM2 309 polymorphism predicts outcome in platinum-treated locally advanced head and neck cancer. Oral Oncol. 2012;48(7):602.
120. Malkin D. Li-fraumeni syndrome. Genes Cancer. 2011;2:475–84.
121. Schneider KZK, Nichols KE, et al. Li-Fraumeni syndrome. In: Pagon RA, Adam MP, Ardinger HH, et al. GeneReviews® [Internet], 1999 Jan 19 [Updated 2013 Apr 11].
122. Bougeard G, Renaux-Petel M, Flaman JM, et al. Revisiting Li-Fraumeni syndrome from TP53 mutation carriers. J Clin Oncol. 2015;33:2345–52.
123. Sengupta S, Linke SP, Pedeux R, et al. BLM helicase-dependent transport of p53 to sites of stalled DNA replication forks modulates homologous recombination. EMBO J. 2003;22:1210–22.
124. Wang XW, Tseng A, Ellis NA, et al. Functional interaction of p53 and BLM DNA helicase in apoptosis. J Biol Chem. 2001;276:32948–55.
125. Arora H, Chacon AH, Choudhary S, et al. Bloom syndrome. Int J Dermatol. 2014;53:798–802.
126. Gorlin RJ, et al. Syndromes of the head and neck: Oxford University Press, New York, NY; 2001.
127. Blander G, Kipnis J, Leal JF, et al. Physical and functional interaction between p53 and the Werner's syndrome protein. J Biol Chem. 1999;274:29463–9.
128. Spillare EA, Robles AI, Wang XW, et al. p53-mediated apoptosis is attenuated in Werner syndrome cells. Genes Dev. 1999;13:1355–60.
129. Lauper JM, Krause A, Vaughan TL, et al. Spectrum and risk of neoplasia in Werner syndrome: a systematic review. PLoS One. 2013;8:e59709.
130. Brennan JA, Mao L, Hruban RH, et al. Molecular assessment of histopathological staging in squamous-cell carcinoma of the head and neck. N Engl J Med. 1995;332:429–35.
131. Gillison ML, Koch WM, Capone RB, et al. Evidence for a causal association between human papillomavirus and a subset of head and neck cancers. J Natl Cancer Inst. 2000;92:709–20.
132. Tinhofer I, Budach V, Saki M, et al. Targeted next-generation sequencing of locally advanced squamous cell carcinomas of the head and neck reveals druggable targets for improving adjuvant chemoradiation. Eur J Cancer. 2016;57:78–86.
133. Morris LT, Chandramohan R, West L, et al. The molecular landscape of recurrent and metastatic head and neck cancers: insights from a precision oncology sequencing platform. JAMA Oncol. 2017;3(2):244-255
134. Bradford CR, Zhu S, Poore J, et al. p53 mutation as a prognostic marker in advanced laryngeal carcinoma. Department of Veterans Affairs Laryngeal Cancer Cooperative Study Group. Arch Otolaryngol Head Neck Surg. 1997;123:605–9.
135. Group* TDoVALCS. Induction chemotherapy plus radiation compared with surgery plus radiation in patients with advanced laryngeal cancer. N Engl J Med. 1991;324:1685–90.
136. Neskey DM, Osman AA, Ow TJ, et al. Evolutionary action score of TP53 identifies high-risk mutations associated with decreased survival and increased distant metastases in head and neck cancer. Cancer Res. 2015;75:1527–36.
137. Masica DL, Li S, Douville C, et al. Predicting survival in head and neck squamous cell carcinoma from TP53 mutation. Hum Genet. 2015;134:497–507.
138. van Houten VM, Tabor MP, van den Brekel MW, et al. Mutated p53 as a molecular marker for the diagnosis of head and neck cancer. J Pathol. 2002;198:476–86.

139. Ferlay J, Soerjomataram I, Dikshit R, et al. Cancer incidence and mortality worldwide: sources, methods and major patterns in GLOBOCAN 2012. Int J Cancer. 2015;136:E359–86.
140. Shanmugaratnam K, Sobin LH. The World Health Organization histological classification of tumours of the upper respiratory tract and ear. A commentary on the second edition. Cancer. 1993;71:2689–97.
141. Vasef MA, Ferlito A, Weiss LM. Nasopharyngeal carcinoma, with emphasis on its relationship to Epstein-Barr virus. Ann Otol Rhinol Laryngol. 1997;106:348–56.
142. Singhi AD, Califano J, Westra WH. High-risk human papillomavirus in nasopharyngeal carcinoma. Head Neck. 2012;34:213–8.
143. Maxwell JH, Kumar B, Feng FY, et al. HPV-positive/p16-positive/EBV-negative nasopharyngeal carcinoma in white North Americans. Head Neck. 2010;32:562–7.
144. Sheu LF, Chen A, Lee HS, et al. Cooperative interactions among p53, bcl-2 and Epstein-Barr virus latent membrane protein 1 in nasopharyngeal carcinoma cells. Pathol Int. 2004;54:475–85.
145. Niedobitek G, Agathanggelou A, Barber P, et al. P53 overexpression and Epstein-Barr virus infection in undifferentiated and squamous cell nasopharyngeal carcinomas. J Pathol. 1993;170:457–61.
146. Murono S, Yoshizaki T, Park CS, et al. Association of Epstein-Barr virus infection with p53 protein accumulation but not bcl-2 protein in nasopharyngeal carcinoma. Histopathology. 1999;34:432–8.
147. Lo KW, Mok CH, Huang DP, et al. p53 mutation in human nasopharyngeal carcinomas. Anticancer Res. 1992;12:1957–63.
148. Sun Y, Hegamyer G, Cheng YJ, et al. An infrequent point mutation of the p53 gene in human nasopharyngeal carcinoma. Proc Natl Acad Sci U S A. 1992;89:6516–20.
149. Van Tornout JM, Spruck CH 3rd, Shibata A, et al. Presence of p53 mutations in primary nasopharyngeal carcinoma (NPC) in non-Asians of Los Angeles, California, a low-risk population for NPC. Cancer Epidemiol Biomark Prev. 1997;6:493–7.
150. Gudkov AV, Komarova EA. The role of p53 in determining sensitivity to radiotherapy. Nat Rev Cancer. 2003;3:117–29.
151. Koch WM, Brennan JA, Zahurak M, et al. p53 mutation and locoregional treatment failure in head and neck squamous cell carcinoma. J Natl Cancer Inst. 1996;88:1580–6.
152. Alsner J, Sørensen SB, Overgaard J. TP53 mutation is related to poor prognosis after radiotherapy, but not surgery, in squamous cell carcinoma of the head and neck. Radiother Oncol. 2001;59:179–85.
153. Skinner HD, Sandulache VC, Ow TJ, et al. TP53 disruptive mutations lead to head and neck cancer treatment failure through inhibition of radiation-induced senescence. Clin Cancer Res. 2012;18:290–300.
154. Bradford CR, Zhu S, Ogawa H, et al. P53 mutation correlates with cisplatin sensitivity in head and neck squamous cell carcinoma lines. Head Neck. 2003;25:654–61.
155. Hoffmann TK, Sonkoly E, Hauser U, et al. Alterations in the p53 pathway and their association with radio- and chemosensitivity in head and neck squamous cell carcinoma. Oral Oncol. 2008;44:1100–9.
156. Andrews GA, Xi S, Pomerantz RG, et al. Mutation of p53 in head and neck squamous cell carcinoma correlates with Bcl-2 expression and increased susceptibility to cisplatin-induced apoptosis. Head Neck. 2004;26:870–7.
157. Mandic R, Schamberger CJ, Muller JF, et al. Reduced cisplatin sensitivity of head and neck squamous cell carcinoma cell lines correlates with mutations affecting the COOH-terminal nuclear localization signal of p53. Clin Cancer Res. 2005;11:6845–52.
158. Osman AA, Neskey DM, Katsonis P, et al. Evolutionary action score of TP53 coding variants is predictive of platinum response in head and neck cancer patients. Cancer Res. 2015;75:1205–15.
159. Gadhikar MA, Sciuto MR, Alves MVO, et al. Chk1/2 inhibition overcomes the cisplatin resistance of head and neck cancer cells secondary to the loss of functional p53. Mol Cancer Ther. 2013;12:1860.

160. Bergamaschi D, Gasco M, Hiller L, et al. p53 polymorphism influences response in cancer chemotherapy via modulation of p73-dependent apoptosis. Cancer Cell. 2003;3:387–402.
161. Ekshyyan O, Rong Y, Rong X, et al. Comparison of radiosensitizing effects of the mammalian target of rapamycin inhibitor CCI-779 to cisplatin in experimental models of head and neck squamous cell carcinoma. Mol Cancer Ther. 2009;8:2255–65.
162. Tokalov SV, Abolmaali N. Radiosensitization of p53-deficient lung cancer cells by pretreatment with cytostatic compounds. Anticancer Res. 2012;32:1239–43.
163. Adelstein DJ, Li Y, Adams GL, et al. An intergroup phase III comparison of standard radiation therapy and two schedules of concurrent chemoradiotherapy in patients with unresectable squamous cell head and neck cancer. J Clin Oncol. 2003;21:92–8.
164. Cooper JS, Pajak TF, Forastiere AA, et al. Postoperative concurrent radiotherapy and chemotherapy for high-risk squamous-cell carcinoma of the head and neck. N Engl J Med. 2004;350:1937–44.
165. Forastiere AA, Goepfert H, Maor M, et al. Concurrent chemotherapy and radiotherapy for organ preservation in advanced laryngeal cancer. N Engl J Med. 2003;349:2091–8.
166. Fallai C, Perrone F, Licitra L, et al. Oropharyngeal squamous cell carcinoma treated with radiotherapy or radiochemotherapy: prognostic role of TP53 and HPV status. Int J Radiat Oncol Biol Phys. 2009;75:1053–9.
167. Cooper JS, Zhang Q, Pajak TF, et al. Long-term follow-up of the RTOG 9501/intergroup phase III trial: postoperative concurrent radiation therapy and chemotherapy in high-risk squamous cell carcinoma of the head & neck. Int J Radiat Oncol Biol Phys. 2012;84:1198–205.
168. Bernier J, Cooper JS, Pajak TF, et al. Defining risk levels in locally advanced head and neck cancers: a comparative analysis of concurrent postoperative radiation plus chemotherapy trials of the EORTC (#22931) and RTOG (# 9501). Head Neck. 2005;27:843–50.
169. Wang Y, Li J, Booher RN, et al. Radiosensitization of p53 mutant cells by PD0166285, a novel G(2) checkpoint abrogator. Cancer Res. 2001;61:8211–7.
170. Do K, Wilsker D, Ji J, et al. Phase I study of single-agent AZD1775 (MK-1775), a Wee1 kinase inhibitor, in patients with refractory solid tumors. J Clin Oncol. 2015;33:3409–15.
171. Suzanne L, Jos HB, Jan HMS. Abrogation of the G2 checkpoint by inhibition of Wee-1 kinase results in sensitization of p53-deficient tumor cells to DNA-damaging agents. Curr Clin Pharmacol. 2010;5:186–91.
172. Leijen S, van Geel RM, Pavlick AC, et al. Phase I study evaluating WEE1 inhibitor AZD1775 as monotherapy and in combination with gemcitabine, cisplatin, or carboplatin in patients with advanced solid tumors. J Clin Oncol. 2016;34:4371–80.
173. Leijen S, van Geel RM, Sonke GS, et al. Phase II study of WEE1 inhibitor AZD1775 plus carboplatin in patients with TP53-mutated ovarian cancer refractory or resistant to first-line therapy within 3 months. J Clin Oncol. 2016;34:4354–61.
174. Salim KY, Maleki Vareki S, Danter WR, et al. COTI-2, a novel small molecule that is active against multiple human cancer cell lines in vitro and in vivo. Oncotarget. 2016;7:41363–79.
175. Silver NL, Osman AA, Patel AA, et al. A novel third generation thiosemicarbazone, COTI-2, is highly effective in killing head and neck squamous cell carcinomas (HNSCC) bearing a variety of TP53 mutations. Int J Radiat Oncol Biol Phys. 94:942.
176. Lambert JMR, Gorzov P, Veprintsev DB, et al. PRIMA-1 reactivates mutant p53 by covalent binding to the core domain. Cancer Cell. 2009;15:376–88.
177. Wang W, Takimoto R, Rastinejad F, et al. Stabilization of p53 by CP-31398 inhibits ubiquitination without altering phosphorylation at serine 15 or 20 or MDM2 binding. Mol Cell Biol. 2003;23:2171–81.
178. Wischhusen J, Naumann U, Ohgaki H, et al. CP-31398, a novel p53-stabilizing agent, induces p53-dependent and p53-independent glioma cell death. Oncogene. 22:8233–8245, 0000.
179. Bykov VJ, Issaeva N, Zache N, et al. Reactivation of mutant p53 and induction of apoptosis in human tumor cells by maleimide analogs. J Biol Chem. 2005;280:30384–91.

180. Kravchenko JE, Ilyinskaya GV, Komarov PG, et al. Small-molecule RETRA suppresses mutant p53-bearing cancer cells through a p73-dependent salvage pathway. Proc Natl Acad Sci U S A. 2008;105:6302–7.
181. Roh JL, Kang SK, Minn I, et al. p53-reactivating small molecules induce apoptosis and enhance chemotherapeutic cytotoxicity in head and neck squamous cell carcinoma. Oral Oncol. 2011;47:8–15.
182. Roh JL, Ko JH, Moon SJ, et al. The p53-reactivating small-molecule RITA enhances cisplatin-induced cytotoxicity and apoptosis in head and neck cancer. Cancer Lett. 2012;325:35–41.
183. Issaeva N, Bozko P, Enge M, et al. Small molecule RITA binds to p53, blocks p53-HDM-2 interaction and activates p53 function in tumors. Nat Med. 2004;10:1321–8.
184. Vassilev LT, Vu BT, Graves B, et al. In vivo activation of the p53 pathway by small-molecule antagonists of MDM2. Science. 2004;303:844–8.
185. He T, Guo J, Song H, et al. Nutlin-3, an antagonist of MDM2, enhances the radiosensitivity of esophageal squamous cancer with wild-type p53. Pathol Oncol Res. 2018;24:75–81.
186. Aziz MH, Shen H, Maki CG. Acquisition of p53 mutations in response to the non-genotoxic p53 activator Nutlin-3. Oncogene. 2011;30:4678–86.
187. Bu Y, Cai G, Shen Y, et al. Targeting NF-kappaB RelA/p65 phosphorylation overcomes RITA resistance. Cancer Lett. 2016;383:261–71.
188. Garber K. China approves world's first oncolytic virus therapy for cancer treatment. J Natl Cancer Inst. 2006;98:298–300.
189. Nemunaitis J, Clayman G, Agarwala SS, et al. Biomarkers predict p53 gene therapy efficacy in recurrent squamous cell carcinoma of the head and neck. Clin Cancer Res. 2009;15:7719–25.
190. Senior K. ONYX-015 phase II clinical trial results. Lancet Oncol. 2001;2:3.
191. Ferris RL, Blumenschein G Jr, Fayette J, et al. Nivolumab for recurrent squamous-cell carcinoma of the head and neck. N Engl J Med. 2016;375:1856.
192. Chow LQ, Haddad R, Gupta S, et al. Antitumor activity of pembrolizumab in biomarker-unselected patients with recurrent and/or metastatic head and neck squamous cell carcinoma: results from the phase Ib KEYNOTE-012 expansion cohort. J Clin Oncol. 2016;34:3838.
193. Okada N, Lin CP, Ribeiro MC, et al. A positive feedback between p53 and miR-34 miRNAs mediates tumor suppression. Genes Dev. 2014;28:438–50.
194. Cortez MA, Ivan C, Valdecanas D, et al. PDL1 regulation by p53 via miR-34. J Natl Cancer Inst. 2016;108:1–9.

Chapter 10
APOBEC as an Endogenous Mutagen in Cancers of the Head and Neck

Tomoaki Sasaki, Natalia Issaeva, Wendell G. Yarbrough, and Karen S. Anderson

Abstract Apolipoprotein B mRNA-editing enzyme catalytic polypeptide-like 3 (APOBEC3) proteins are a family of cytidine deaminases that play important roles in diverse physiological processes in humans. APOBEC3-driven cytidine deamination results in a C-to-U conversion in target nucleotides, classifying this biological activity as a DNA-/RNA-editing mechanism. In recent years, biochemical, cellular, and bioinformatics studies have supported a role for APOBEC3 proteins in the etiology of human cancers, based on their ability to mutate genomic DNA. In this chapter, we provide a thorough review of recent studies that have implicated APOBEC3 in a number of diverse cancers, including head and neck cancers. These studies suggest that APOBEC3-dependent mutations are most associated with squamous cell carcinomas of the head and neck and may be linked to human papillomavirus (HPV) expression and production of neoantigens. Cell-based and structural evidence corroborates the initial bioinformatics data on APOBEC3 function. In conclusion, we raise several prospects for targeted therapeutic avenues including potential immunotherapy options that can leverage neoantigens generated from APOBEC activity.

Keywords APOBEC3 · Cytidine deaminases · Mutations · Mutational burden · SCCHN · Head and neck squamous cell carcinoma · Neoantigens · DNA editing · Human papillomavirus · Targeted therapy · Immunotherapy

10.1 Introduction: Enzymatic Function of APOBEC Family Proteins

Apolipoprotein B mRNA-editing enzyme catalytic polypeptide-like (APOBEC) proteins are a group of DNA and RNA mutators that play important physiological roles in vivo. In humans, APOBEC proteins consist of 11 gene products including the founding family member, APOBEC1, the activation-induced deaminase (AID),

---

T. Sasaki · N. Issaeva · W. G. Yarbrough · K. S. Anderson (✉)
Yale University School of Medicine, New Haven, CT, USA
e-mail: karen.anderson@yale.edu

B. Burtness, E. A. Golemis (eds.), *Molecular Determinants of Head and Neck Cancer*, Current Cancer Research, https://doi.org/10.1007/978-3-319-78762-6_10

and the recently discovered APOBEC3A-3H (APOBEC3) enzymes. APOBEC1 functions as a mRNA-editing enzyme, which acts on apolipoprotein B (*apoB*) mRNA to introduce a stop codon in some transcripts. This allows one gene to form two distinct *apoB* gene products (apoB48 and apoB100), both of which play an indispensable role in lipid metabolism [1–3]. This was the first known example of RNA editing in vivo. In subsequent studies, AID was found to contribute to antibody diversification through mutation of the variable regions of antibody genes, serving as a crucial component of class switch recombination and somatic hypermutation [4–6]. The recently discovered APOBEC3 enzymes are closely conserved paralogs, which cluster in tandem on chromosome 22 [7]. Collectively, they function in cellular defense against both extracellular viruses and intragenic infectious elements (reviewed in Chiu and Greene) [8].

All APOBEC enzymes catalyze a single chemical reaction in vivo: the cytidine deamination of single-stranded DNA (ssDNA) or RNA. This results in the replacement of the 4′-amino moiety on cytosine with a carbonyl group leading to a base modification (Fig. 10.1). This change in sequence from a cytidine to a uridine is not encoded in the genome, classifying this activity as a DNA-editing mechanism. Target substrates of APOBEC enzymes are diverse. As noted above, APOBEC1 targets ApoB mRNA (and also DNA) in hepatocytes [3, 9], while AID functions on immunoglobulin genes in germinal centers of B cells [5]. Some of the APOBEC3 enzymes have been defined as hypermutating ssDNA of several extracellular retroviruses and intragenic elements [8]. Key questions in the field remain how APOBEC proteins properly locate their substrates among the myriad of nucleic acids existing in the cell and how this biological activity is regulated to finely control this process.

APOBEC enzymes are composed of either single or tandemly arranged zinc fingers known as zinc-dependent deaminase (ZDD) domains, which are characterized by the primary amino acid sequence H-X-$\underline{\mathbf{E}}$-$X_{23–28}$-P-C-$X_{2–4}$-C (Fig. 10.2) [10]. The catalytic residue in this highly conserved domain is the glutamate, which is important in driving the proton shuttling mechanism that is ultimately responsible for the replacement of the 4′-amino group with a carbonyl moiety (Fig. 10.1b). In addition, the histidine and cysteine residues together coordinate the zinc metal, which properly positions a water molecule that is central for electron movement during catalysis (Fig. 10.1b). Single-amino-acid mutagenesis resulting in removal of any of these residues resulted in the complete loss of enzymatic activity, in several family members [3, 11, 12]. For those APOBECs containing two zinc finger domains, the C-terminal ZDD is primarily responsible for chemical catalysis [11–14]. The N-terminal domain is instead thought to play an accessory role in RNA binding, encapsulation into virions, and oligomerization, though the role of this particular domain is still in debate, due to the limited availability of structural information (discussed below) [15–18].

Interestingly, the spectrum of nucleic acids in which these base modifications occur varies with each APOBEC family member. For instance, APOBEC1 drives C-to-U conversions in messenger RNA as well as ssDNA [9], whereas APOBEC3 enzymes are exclusively ssDNA mutators [8]. Even among APOBEC3s, the pri-

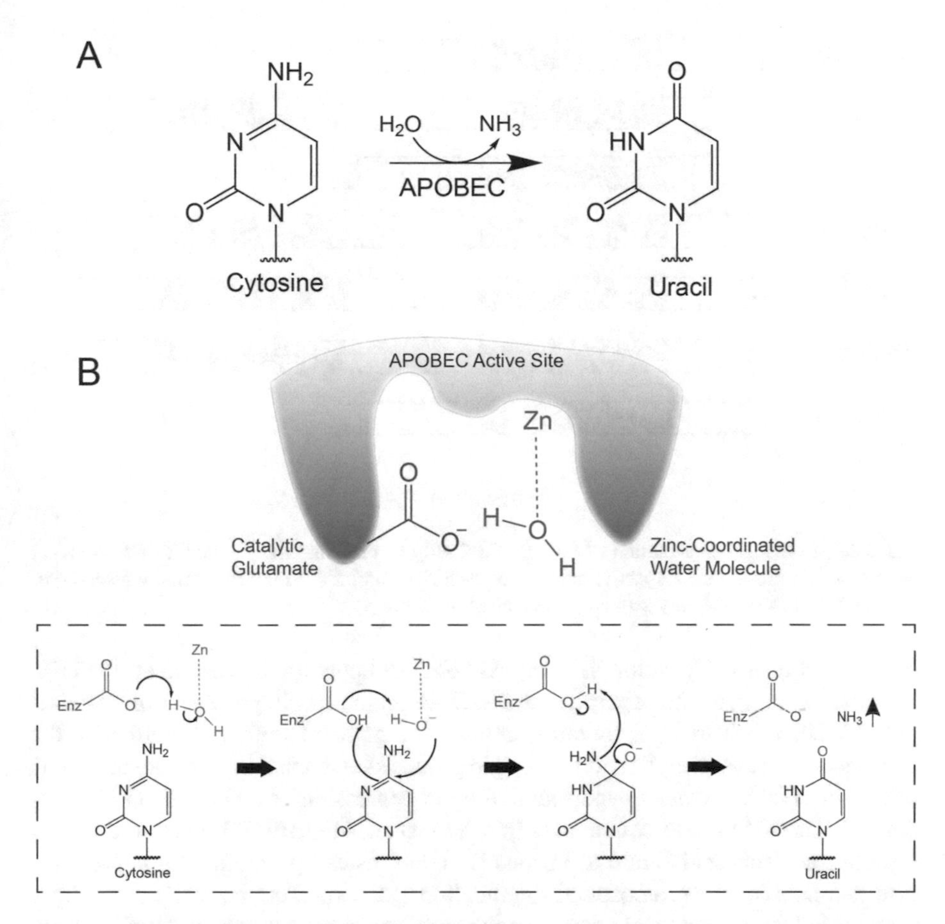

Fig. 10.1 Overall schematic of (**a**) an APOBEC enzyme-catalyzed deamination reaction and (**b**) the catalytic mechanism at the active site. The arrows indicated in the mechanism represent electron pushing, and the dotted line represents zinc coordination of the water molecule at the active site. Note the presence of a tetrahedral intermediate involved in the proton shuttling electron-pushing mechanism

mary sequence context of the deaminated cytosine is unique to each APOBEC3 enzyme. APOBEC3G catalyzes cytidine deamination within a 5'–CC<u>C</u>A–3′ motif (the deaminated cytosine underlined) [19], while the recently characterized APOBEC3A and APOBEC3B enzymes require a 5'–T<u>C</u>–3′ dinucleotide motif [20]. As such, each independent APOBEC enzyme has the capability of recognizing different substrates in vivo, and this may explain their divergent roles in physiological and pathological processes in humans.

Among APOBEC3 proteins, APOBEC3G has been well characterized in the context of human immunodeficiency virus (HIV) infection. APOBEC3G (also known as CEM15) was initially discovered in a subtractive hybridization screen, as an antiviral enzyme for HIV infection, and its activity was shown to be counteracted

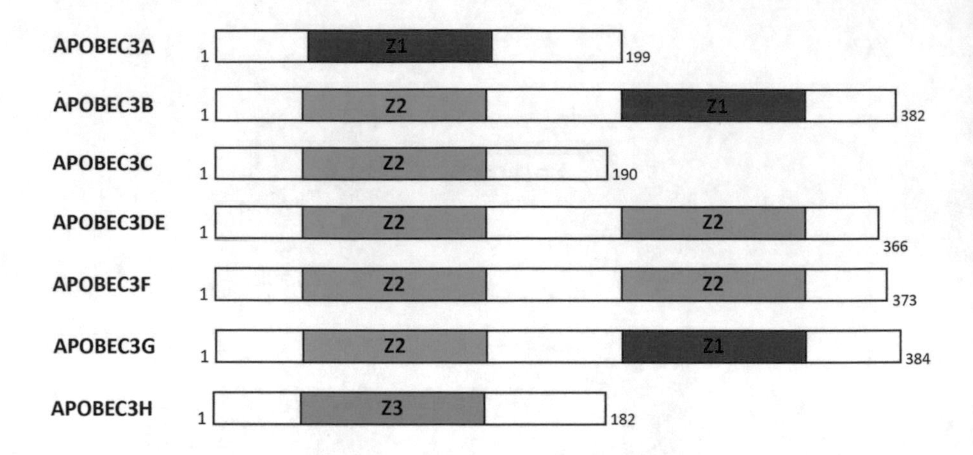

Fig. 10.2 Domain representation of APOBEC3 family of proteins. Z1, Z2, and Z3 notations on each protein represent the respective subtype of each ZDD motif. All ZDD motifs are conserved in that they contain the primary sequence denoted at the bottom

by the viral infectivity factor, Vif [21]. APOBEC3G potently hypermutates the HIV genome by acting on the exposed strand ssDNA regions during reverse transcription [11, 12, 22, 23]. This hypermutation of the viral genome renders it ineffective for subsequent rounds of replication. Multiple groups showed that in the absence of Vif, retroviral cDNA becomes hypermutated upon expression of APOBEC3G [11, 12, 22, 23]. The Vif protein counteracts this activity of APOBEC3G by excluding the protein from encapsulation into virions [24] and instead promoting its ubiquitination and subsequent proteosomal degradation [25, 26]. This coevolution of host restriction factors and viral accessory proteins has been dubbed an "evolutionary arms race."

Since the discovery that APOBEC3G hypermutates retroviral cDNA, other APOBEC proteins have been recognized as antiviral agents. For instance, the closely related APOBEC3F is also an antiretroviral factor inhibiting HIV-1 replication [27]. APOBEC3DE exhibits antiretroviral activity against HIV-1 and the related simian immunodeficiency virus (SIV) [28]. The more distantly related APOBEC3B has also been identified as a potent antiviral agent against HIV [29], SIV [30], murine leukemia virus [31], and intragenic infectious elements such as retrotransposons [32–35]. Thus, diverse APOBEC3 enzymes are thought to collectively function through potent hypermutation of many viruses and harmful intragenic elements.

10.2 Molecular Features of APOBEC3 Cytidine Deaminases

The first structural studies on APOBEC3s were conducted in 2008, with the C-terminal ZDD of APOBEC3G structure solved at a high resolution using solution nuclear magnetic resonance (NMR) spectroscopy (Fig. 10.3) [36]. These studies revealed that the catalytic domain resembled that of known cytidine deaminases, consisting of a beta-stranded core surrounded by several alpha helices and loop regions. The active site of the enzyme is located on one face of the beta-sheet core, between the α2 and α3 helices. Coordinates for the zinc metal were assigned to a region spanning the α2-β3-α3 motif, indicating that this represents the active site where the catalytic activity takes place. The integration of surface charge analyses, NMR titration studies and computational modeling suggested a binding groove for ssDNA. Subsequent crystal structures of the APOBEC3G catalytic domain have confirmed this overall architecture of the catalytic domain [37–39].

The first crystal structure for the catalytic domain of APOBEC3B, the leading candidate for APOBEC3 involvement in human cancers, was solved in 2010 by Shi et al. [14] The crystallized structure showed similarities to previously solved APOBEC3G catalytic domain structures, including features unique to these structures. However, this protein crystallized in a closed conformation in contrast to other APOBEC3s, suggesting that conformational changes may occur during catalysis. Single-stranded DNA binding potentially drives this conformational change to an open, active conformation. Importantly, the structure also reveals that the bound zinc molecule mediates important interactions between zinc-coordinating amino

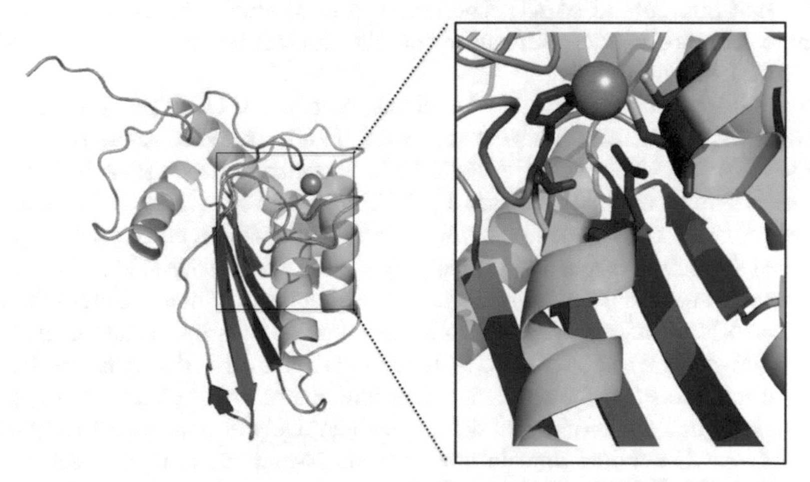

Fig. 10.3 Solution nuclear magnetic resonance (NMR) structure of the catalytic domain of APOBEC3G (PDB ID: 2JYW). The left shows an overall structure of the domain consisting of its active site characterized by an antiparallel beta-sheet core surrounded by alpha helices and loop regions. On the right is the zoomed version of the structure focusing on the active site where red-highlighted residues are cysteines, histidine, and glutamate important for the catalytic mechanism

acid side chains and a water molecule/cryoprotectant (depending on crystallization conditions), providing insight into the mechanistic requirements of efficient APOBEC catalysis. Similar molecular features have also been observed for APOBEC3A [40], but with the distinguishing factor that this single-domain protein crystallized in a dimeric form in which the individual active sites were connected by a single groove through the dimer interface.

Until recently, structures with bound substrate remained elusive, due in part to the difficulty in obtaining high-resolution crystal structures of APOBEC3s. However, in 2017, two studies reported DNA-bound crystal structures of APOBEC3A [41, 42]. Both studies revealed extensive contacts of APOBEC active site residues to the deaminated cytosine and the base directly 5′ to the cytosine. The structural data from Shi and colleagues pointed toward a U-shaped DNA-binding conformation where these two bases were flipped out, making extensive hydrogen bonding interactions in a groove containing the active site, while the succeeding bases on the 3′ end stacked onto each other distal from this site [41]. These data suggest that the base at the −1 position (preceding the deaminated base) is an important residue in determining selectivity for DNA substrates. Comparison of the DNA-bound APOBEC3A structure with the *apo* structure solved by a different group shows a slight change in conformation, in which DNA binding leads to a more open conformation [42]. While the positions of the catalytic residues remain unchanged, the surrounding loop regions known to govern DNA binding and substrate selectivity shift to accommodate for the binding of single-stranded DNA. Shi and colleagues additionally solved the DNA-bound structure of APOBEC3B, though the construct had to be modified to optimize solubility [41]. These first-in-class studies provide the basis for a more comprehensive evaluation of the molecular mechanisms driving APOBEC3 function.

No full-length structures of the double-domain APOBEC3s (APOBEC3B, APOBEC3F, APOBEC3G) have been solved. However, the structure of the noncatalytic N-terminal domain of APOBEC3G, important in DNA/RNA binding and oligomerization, has been determined [43]. This crystal structure provides compelling evidence for the unique role of the N-terminal domain in nucleic acid binding, mediated through an extensive positively charged surface on one side of the crystallized protein and through oligomerization. The dimerization interface is required for DNA and RNA binding. While these studies provide a partial picture of how the N-terminal domains of APOBEC3s function in vivo, a full-length structure is required to understand the respective contributions of both domains. Open questions include how the N-terminal domain cooperatively or noncooperatively interacts with the C-terminal domain and how the N-terminal domain regulates the catalytic activity driven by its C-terminal counterpart.

10.3 APOBEC3 Proteins and Cancer

Until recently, studies on APOBECs have been confined to characterization of their activity on viruses and intragenic elements. However, with the increasingly common use of next-generation sequencing technology, researchers have been able to comprehensively study cancer genomes at the molecular level. Publicly shared sequence databases such as The Cancer Genome Atlas (TCGA) project facilitate this effort by allowing researchers to easily access and analyze DNA sequence information of annotated clinical samples. These analyses can be used in an attempt to understand the etiologies of diverse human cancers, particularly on the respective contributions of endogenous and exogenous sources of mutation in shaping cancer genomes [44, 45]. Using this unprecedented capability to closely examine cancer genomes, researchers have begun to test the hypothesis that APOBEC proteins may in some cases deaminate the host chromosomal DNA, leading to mutagenesis and cancer progression.

Evidence uncovered from the comprehensive sequencing and initial analyses of the mutational signatures in 21 breast cancer genomes led to the discovery of unique mutational patterns marked by local hypermutation in a strand-coordinated fashion [44]. The authors termed the phenomenon in these regions "kataegis" ("rain showers," in Greek) to describe the concentrated and localized fashion in which these mutations occur [44]. Suggestively, these regions were characterized by base substitutions of primarily cytosines. Given the activity of APOBECs as cytidine deaminases and that APOBEC-induced mutations occur in this particular sequence context in both a sporadic and clustered fashion, the authors proposed that APOBEC family members were the likely cause of kataegis [44]. This study was the first comprehensive analysis suggesting that APOBECs may play a role in mutagenesis in a cancer.

Expanding on this study [44], the authors created a comprehensive collection of mutational signatures compiled from ~5 million mutations in over 7000 human cancer genome sequences [45]. Examination of the trinucleotide motif containing the mutated base in the central position defined 21 distinct mutational signatures arising from various exogenous and endogenous sources. Excitingly, this study revealed two signatures, Signature 2 and Signature 13, containing an enrichment of C-to-T and C-to-G mutations in the TC dinucleotide context, suggesting that the source of these mutation types could be APOBEC3 enzymes.

In a parallel effort, Burns and colleagues combined TCGA data with evidence from cellular and biochemical experiments to suggest that APOBEC3B is the enzymatic source of mutations in breast cancer [20]. The authors found evidence of *APOBEC3* upregulation at the mRNA level in RNA-Seq data from the TCGA, which they confirmed by qRT-PCR of primary breast tumors and breast cancer cell lines. They directly demonstrated that APOBEC3B upregulation increased mutational load and triggered cell cycle aberrations, cell death, DNA fragmentation, and DNA double-stranded breaks. Knockdown of *APOBEC3B* expression using small hairpin RNAs (shRNAs) reversed these phenotypes. These experiments convincingly showed that *APOBEC3B* overexpression is important in driving hypermutation and downstream cellular consequences in breast cancer.

APOBEC3-dependent mutagenesis has now been implicated in multiple cancer types, including head and neck cancer, lung cancer, breast cancer, and serous ovarian carcinoma [20, 46–49]. For instance, Roberts et al. identified APOBEC-induced clustered mutations from analyses of approximately 40 whole genome sequencing datasets and 2680 whole exome datasets representative of multiple cancers using various sequence repositories, including TCGA [49]. Interestingly, clustered mutations preferentially occurred at chromosomal rearrangement breakpoints, suggesting that exposed ssDNA during rearrangement is a substrate requirement for APOBEC activity [49, 50]. In addition, the study found that mRNA levels of *APOBEC3A* and *APOBEC3B* correlated with the frequency of *APOBEC* mutations identified in the exome sequence for individual tumors. The study conclusively identified *APOBEC* mutations in cervical, bladder, lung, breast, and head and neck cancers as accounting for more than half of the total mutational load in some clinical samples [49]. In a parallel effort, Burns et al. mined bioinformatics data of over 4800 exome sequences from TCGA and identified a positive correlation between APOBEC3B expression and the presence of APOBEC signature mutations [47]. Similar to the study conducted by Roberts et al., the authors identified a high enrichment of *APOBEC* mutations in six different cancers including lung adenocarcinoma, lung squamous cell carcinoma, and cervical, bladder, breast, and head and neck cancers [47]. These and other studies suggested that APOBEC proteins may be ubiquitous drivers of cancers of diverse origins and may shed light on the molecular mechanisms underlying processes such as tumor subclonal evolution and mutational accumulation in vivo.

A highly debated topic in the field is the respective contribution of each APOBEC enzyme in driving cancer induction and progression. Initial data suggested that APOBEC3B is the primary driver, as APOBEC3B overexpression, both at the cellular and organismal level, correlated with a higher incidence of total APOBEC mutations and resulted in a number of downstream consequences indicative of cancer hallmarks such as the induction of double-stranded DNA breaks [20, 47]. Some findings suggest that more than one APOBEC enzyme may contribute to mutations in cancer. For example, a common *APOBEC3A-3B* deletion polymorphism [51] that fuses the *APOBEC3A* and *3B* genes in a manner that removes the coding region of *APOBEC3B* has been linked to increased breast cancer risk in patients. [52, 53] This polymorphism has also been found to result in higher APOBEC-driven mutations and subsequent mutational burden [54] and associated with both risk and enrichment of APOBEC mutations in breast cancer [55]. Furthermore, additional evidence from yeast model studies have found an abundance of APOBEC3A mutational signatures in the background of APOBEC3B-induced mutations, suggesting that APOBEC3A also plays an important role in shaping the mutational landscape [46]. Taken together, these studies highlight the role of multiple APOBEC3 enzymes, including APOBEC3B, in driving the characteristic APOBEC mutational signatures and *kataegis*.

10.4 Evidence and Consequences of APOBEC-Induced Driver Mutations

APOBEC-induced somatic mutations manifest as a complex combination of T-to-C transitions and T-to-G transversions in cancer genomes, accumulated over time [45, 56]. Typically, APOBEC mutations occur in a specific TCW trinucleotide sequence context (the deaminated cytosine is underlined; W represents either A or T) [49]. Moreover, APOBEC mutations can occur in both a sporadic or clustered (kataegis) fashion, further distinguishing them as unique in a cancer genome. Together, the type and sequence context of these mutations allow bioinformatic pipelines to efficiently identify APOBEC activity in vivo.

Given that APOBEC proteins catalyze a C-to-U conversion in ssDNA, single-strand intermediates from diverse biological processes including DNA repair and replication can potentially serve as substrates for APOBEC activity [50, 57–60]. As APOBEC mutation clusters occur most frequently at rearrangement breakpoints, multiple studies have examined whether APOBEC activity at these breakpoints results in the mutational shower phenotype [50, 57]. Studies in yeast have suggested that intermediates from lagging strand synthesis in DNA replication serve as endogenous substrates of APOBEC mutation [58]. This hypothesis was further strengthened by the observation that genetic perturbation of replication fork stabilizing proteins and chemical induction of replication stress both resulted in an increase in APOBEC-induced mutations, likely due to more exposed ssDNA that can serve as APOBEC substrates [58]. Interestingly, APOBEC enzymatic activity seems to have a spatiotemporal dependency during genome replication, such that early replicating, gene-dense regions are uniquely and preferentially deaminated by APOBEC [61]. Since sites of high APOBEC activity are characterized by high DNA fragility and increased chromosome breakage, exposed ssDNA in the genome is a likely source of the most physiological substrates for APOBEC activity.

Once APOBEC catalysis produces a deoxyuridine moiety in ssDNA, various components of cellular DNA repair machinery, including uracil DNA glycosylase (UDG), detect these lesions. The resulting C-to-U conversion can be directly replicated, resulting in the incorporation of a thymidine (a C-to-T transition). However, if acted on by UDG, which generates an abasic site [62], DNA repair mediated by the base excision repair (BER) or mismatch repair (MMR) machinery can correct these lesions, or the REV1 translesion polymerase can change the base to a guanosine after replication over the abasic site (a C-to-G transversion) [63, 64]. The frequency of C-to-T transition and C-to-G transversions is heavily dependent on the activities of DNA repair enzymes and polymerases [60]. Thus, the coordination of APOBEC enzymes and downstream repair mechanisms result in the mutational spectra observed in vivo.

10.5 Aberrant APOBEC Expression and Evidence of APOBEC Activity in Clinical Samples of Head and Neck Cancer

From the original studies by Burns et al. [20] and Roberts et al. [49] which identified head and neck cancer as one of the cancers showing an elevated level of APOBEC signatures, several groups have explored the role of APOBEC-induced mutagenesis in this cancer type. Mounting clinical evidence suggests that there is a strong association between APOBEC overexpression and induction of head and neck cancers. This is consistent with the notable observation that APOBEC3B has been most strongly associated with a number of squamous cell carcinomas originating in diverse organs, specifically through evidence uncovered by the TCGA database elucidating a bias toward C-to-T transition mutations in these clinical samples [65].

Single mutation analysis from 510 head and neck squamous cell carcinoma (HNSCC) sequences showed that the APOBEC-driven C-to-T mutation is the most common mutation occurring in the trinucleotide TCW motif [65]. *APOBEC3B* mRNA is overexpressed in oral squamous cell carcinomas (OSCCs) compared to normal oral squamous cell samples [65]. In immunohistochemical analysis, OSCC tissues stained intensely for APOBEC3B protein expression compared to controls [65]. Interestingly, not all clinical samples of OSCC had high APOBEC3B nuclear staining [65], in contrast to previous data suggesting predominantly nuclear localization of APOBEC3B [20]; whether this has regulatory significance is not known.

These findings have further been extended to other aerodigestive squamous cell carcinomas. Whole genome and whole exome sequencing studies indicated that approximately half of 192 clinical esophageal squamous cell carcinomas contained APOBEC mutational signatures [66]. An independent sequencing study using clinical samples from a distinct sub-Saharan African population also identified an enrichment of APOBEC signatures, albeit in the presence of other mutational signatures such as those caused by smoking [67]. These studies suggest that APOBEC3 enzymes may play an important role in the tumorigenesis and subclonal evolution of squamous cell carcinomas from various origins. While studies examining APOBEC involvement are currently restricted to these two types, additional sequencing efforts and studies with clinical samples and cell lines may reveal APOBEC signatures in other squamous cell carcinoma types.

10.6 The Role of Human Papillomavirus and Overexpression of APOBEC in Patients and in Experimental Cell Lines

Virally-driven HNSCCs have begun to be appreciated as distinct cancers driven by separate mechanisms as compared to non-viral HNSCCs. For instance, in the case of HIV-infected patients, the underlying biology is vastly different from that of HIV-null HNSCCs, particularly through a distinct pattern of *TP53* mutations in these cancers [68, 69]. The same observation of differences in tumor biology applies to human papillomavirus (HPV) in which a systematic review of 60 studies reveals that an estimated 25% of head and neck squamous cell carcinomas (HNSCCs) are linked to HPV infection (discussed in Chaps. 20 and 21), with the frequency increasing in recent years, and confirmed as a driver mechanism [70]. The underlying pathology, clinical presentation, and therapeutic approaches are distinct when considering virally-induced HNSCC in comparison to HPV(-) cases. Patients with HPV-related HNSCC often are diagnosed at a younger age and are less likely to have a history of tobacco exposure and known risk factors that include sexual transmission, whereas HPV(-) HNSCC is typically correlated with a history of smoking and often consumption of alcohol [71].

Interestingly, recent studies suggest a link between APOBEC-mediated mutagenesis and HPV-driven cancers. In an early study, well before the first studies linking APOBEC enzymes to cancer, Vartanian and colleagues identified APOBEC-driven mutations in the genomes of two subtypes of HPV virus – HPV1a and HPV16 – which were present in clinical samples [72]. This observation was independently supported through transient transfection experiments where *APOBEC3A*, *APOBEC3B*, and *APOBEC3H* were ectopically expressed in the presence of HPV DNA which resulted in an increase in APOBEC-induced 5’-TC-3′ mutations [72]. Several APOBEC3 family members were capable of hypermutating the viral E2 gene of the HPV16 genome [73]. A follow-up study using a HPV16 pseudovirion production system found that APOBEC3A and APOBEC3C together contribute to a significant reduction in infectivity, implying functional importance of this hypermutation phenotype [74]. These observations suggest APOBEC3s act as restriction factors in reducing HPV infectivity [74, 75], which is not surprising given the role of APOBEC3 enzymes in the context of antiviral defense.

There also is evidence that the host DNA can be preferentially deaminated in HPV-related HNSCCs, in comparison to HPV-unrelated HNSCCs, based on the enrichment of APOBEC-induced mutagenesis in HPV-related head and neck cancers in cancer genome databases [76–78]. In a pivotal study, Henderson et al. analyzed TCGA data for 299 HNSCC specimens and found APOBEC signatures to be disproportionately enriched in HPV+ HNSCC genomes [76]. Importantly, targeted sequencing of the *PIK3CA* gene – highly mutated in HNSCC (see Chap. 5) – showed an enrichment of mutations characteristic of APOBEC activity in the sequences encoding helical domain of the PIK3CA protein, resulting in appearance of E542K and E545K mutational variants [76]. These mutations are a result of C-to-T transition mutations in the complementary strand, which is likely driven by APOBEC activity [76]. These helical domain mutations are gain-of-function mutations [79, 80]. Mechanistically, these mutations require interactions between GTP-Ras and the PI3K protein that promote downstream activation [80]. The helical domain mutation E545K results in increased phosphorylation of Akt, downstream of PI3K activation, and is linked to poor prognosis in patients [79]. These observations serve as a link between APOBEC-induced mutations and downstream consequences of those mutations, such as proto-oncogene activation.

To probe more deeply into selectively elevated APOBEC activity in HPV+ clinical samples, several studies have investigated the relationship between HPV infection and subsequent changes to APOBEC expression levels. HPV infection upregulated *APOBEC3B* mRNA expression and expression of the APOBEC enzymes through the activity of the oncogenic HPV E6 protein, which is known to bind and stimulate the degradation of p53 [78]. Furthermore, transduction and knockdown experiments of the oncogenic E6 protein in HPV-positive cell lines together supported the hypothesis that this protein is a crucial and required factor in the upregulation of APOBEC3s [78]. These studies support a model in which viral infection serves as a trigger that drives higher APOBEC expression and thus APOBEC signatures in vivo.

10.7 5-Azacytidine Treatment in HPV Positive HNSCC Cell Lines and the Role of APOBEC3B

5-Azacytidine (5-aza) and the analog 5-aza-2′-deoxycytidine (decitabine) are synthetic cytidine analogs that cause DNA demethylation and are used in the clinic to treat myelodysplastic syndromes and acute myeloid leukemia (AML) [81, 82]. HPV-positive HNSCC cell lines and primary cells are more sensitive than HPV-negative cells to 5-aza [83]. These studies uncovered several mechanisms of 5-aza-induced toxicity in HPV-associated HNSCC [83], including the induction of transcription-dependent DNA double-strand breaks (DSBs) that were reliant on high levels of APOBEC3B expression. Importantly, *APOBEC3B* knockdown not only inhibited formation of DNA DSBs after 5-aza treatment but also significantly increased resistance of HPV-positive cells to 5-aza treatment (manuscript under preparation). Remarkably, despite protecting from 5-aza-induced DSBs and cellular toxicity, depletion of APOBEC3B in untreated HPV-positive head and neck cancer cells inhibited clonogenic survival. Together, these data suggest that HPV-positive cells depend on APOBEC3B activity for clonogenic survival but that this dependence can be used as an Achilles heel in a synthetic lethal-like strategy, where demethylation leads to DSB mediated by APOBEC3B.

10.8 Conclusions and Future Prospects in Targeted Therapeutics

Based on evidence from numerous recent studies, APOBEC3 proteins may serve as an important endogenous source of de novo mutations. Considering that mutations accumulate over many years, we hypothesize that APOBEC3 proteins may contribute to a relatively large portion of the total mutational load present in a given tumor. As studies reveal more about the downstream consequences of APOBEC3-induced mutations in key proto-oncogenes and tumor suppressor genes, it may be beneficial to develop small molecule inhibitors of APOBEC3 proteins as therapeutics in various cancers, including HNSCC. Further structural and biochemical work will provide an efficacious framework to develop selective and potent inhibitors of APOBEC3s to prevent off-target effects. An interesting question moving forward would be the compensatory effect of other APOBEC3 proteins. For instance, if specific inhibition of APOBEC3B is achieved, can the other APOBEC3s compensate for the loss of APOBEC3B activity? Would there be an advantage of developing pan-APOBEC inhibitors, or might these disrupt in vivo family roles in antiviral defense and adaptive immunity?

Given that APOBEC mutagenesis has been shown to induce replication stress through the induction of abasic sites [84], this could potentially represent a unique case of synthetic lethality in which APOBEC-driven cancers are sensitized to DNA damage response inhibitors. Indeed, cell lines expressing a high level of APOBEC3A and APOBEC3B were sensitive and susceptible to ataxia-telangiectasia and Rad3-related protein (ATR) inhibition [84]. Further experiments have shown that ATR inhibition resulted in increased abasic sites at replication sites, suggesting that these cells are dependent on ATR-mediated DNA damage response pathways to keep APOBEC-mediated mutagenesis under control. This study therefore suggests that those cancers driven by APOBEC activity could be selectively targeted using unique therapeutic interventions.

Recent studies have found that higher PD-L1 expression is associated with higher incidence of APOBEC-induced kataegis [85] and elevated *APOBEC* mRNA expression levels [86]. As high PD-L1 expression is often correlated with higher success rates of immune checkpoint inhibitor treatment regimens [87], this raises the question of whether tumors containing a high prevalence of APOBEC mutations can be predictive biomarkers for immunotherapy. In addition, the presence of neoantigens has been proposed to be one of the central mechanisms by which the immune system distinguishes cancerous from noncancerous cells [88]. APOBEC3 proteins, as potent DNA mutators contributing to significant mutational loads in cancer genomes, could therefore be thought of as potential sources of neoantigens. As response to anti-PD-L1, treatment has been well correlated with the presence of specific neoantigens [89]; perhaps tumors with high levels of APOBEC signatures are sound candidates for immunotherapy.

In closing, we have proposed a potential role for APOBEC proteins in the context of human cancers, particularly through potent deamination of genomic DNA. Mounting evidence suggests that HNSCC seems to be one cancer in which APOBEC3 proteins contribute to a significant proportion of mutational load, and thus an understanding of this mutagenic process would be of utmost importance. The knowledge accrued from biochemical, cellular, and bioinformatics studies are crucial in identifying important mechanistic details that drive the identification of potential therapeutic avenues involving APOBEC3 proteins.

References

1. Teng B, Burant CF, Davidson NO. Molecular cloning of an apolipoprotein B messenger RNA editing protein. Science (New York, NY). 1993;260:1816–9.
2. Navaratnam N, et al. The p27 catalytic subunit of the apolipoprotein B mRNA editing enzyme is a cytidine deaminase. J Biol Chem. 1993;268:20709–12.
3. Driscoll DM, Zhang Q. Expression and characterization of p27, the catalytic subunit of the apolipoprotein B mRNA editing enzyme. J Biol Chem. 1994;269:19843–7.
4. Petersen-Mahrt SK, Harris RS, Neuberger MS. AID mutates E. coli suggesting a DNA deamination mechanism for antibody diversification. Nature. 2002;418:99–103. https://doi.org/10.1038/nature00862.
5. Muramatsu M, et al. Specific expression of activation-induced cytidine deaminase (AID), a novel member of the RNA-editing deaminase family in germinal center B cells. J Biol Chem. 1999;274:18470–6.
6. Muramatsu M, et al. Class switch recombination and hypermutation require activation-induced cytidine deaminase (AID), a potential RNA editing enzyme. Cell. 2000;102:553–63.
7. Jarmuz A, et al. An anthropoid-specific locus of orphan C to U RNA-editing enzymes on chromosome 22. Genomics. 2002;79:285–96. https://doi.org/10.1006/geno.2002.6718.
8. Chiu YL, Greene WC. Multifaceted antiviral actions of APOBEC3 cytidine deaminases. Trends Immunol. 2006;27:291–7. https://doi.org/10.1016/j.it.2006.04.003.
9. Harris RS, Petersen-Mahrt SK, Neuberger MS. RNA editing enzyme APOBEC1 and some of its homologs can act as DNA mutators. Mol Cell. 2002;10:1247–53.
10. Salter JD, Bennett RP, Smith HC. The APOBEC protein family: united by structure, divergent in function. Trends Biochem Sci. 2016;41:578–94. https://doi.org/10.1016/j.tibs.2016.05.001.
11. Zhang H, et al. The cytidine deaminase CEM15 induces hypermutation in newly synthesized HIV-1 DNA. Nature. 2003;424:94–8. https://doi.org/10.1038/nature01707.
12. Mangeat B, et al. Broad antiretroviral defence by human APOBEC3G through lethal editing of nascent reverse transcripts. Nature. 2003;424:99–103. https://doi.org/10.1038/nature01709.
13. Hakata Y, Landau NR. Reversed functional organization of mouse and human APOBEC3 cytidine deaminase domains. J Biol Chem. 2006;281:36624–31. https://doi.org/10.1074/jbc.M604980200.
14. Shi K, Carpenter MA, Kurahashi K, Harris RS, Aihara H. Crystal structure of the DNA deaminase APOBEC3B catalytic domain. J Biol Chem. 2015;290:28120–30. https://doi.org/10.1074/jbc.M115.679951.
15. Belanger K, Savoie M, Rosales Gerpe MC, Couture JF, Langlois MA. Binding of RNA by APOBEC3G controls deamination-independent restriction of retroviruses. Nucleic Acids Res. 2013;41:7438–52. https://doi.org/10.1093/nar/gkt527.
16. Navarro F, et al. Complementary function of the two catalytic domains of APOBEC3G. Virology. 2005;333:374–86. https://doi.org/10.1016/j.virol.2005.01.011.
17. Huthoff H, Autore F, Gallois-Montbrun S, Fraternali F, Malim MH. RNA-dependent oligomerization of APOBEC3G is required for restriction of HIV-1. PLoS Pathog. 2009;5:e1000330. https://doi.org/10.1371/journal.ppat.1000330.
18. Opi S, et al. Monomeric APOBEC3G is catalytically active and has antiviral activity. J Virol. 2006;80:4673–82. https://doi.org/10.1128/jvi.80.10.4673-4682.2006.
19. Yu Q, et al. Single-strand specificity of APOBEC3G accounts for minus-strand deamination of the HIV genome. Nat Struct Mol Biol. 2004;11:435–42. https://doi.org/10.1038/nsmb758.
20. Burns MB, et al. APOBEC3B is an enzymatic source of mutation in breast cancer. Nature. 2013;494:366–70. https://doi.org/10.1038/nature11881.
21. Sheehy AM, Gaddis NC, Choi JD, Malim MH. Isolation of a human gene that inhibits HIV-1 infection and is suppressed by the viral Vif protein. Nature. 2002;418:646–50. https://doi.org/10.1038/nature00939.

22. Lecossier D, Bouchonnet F, Clavel F, Hance AJ. Hypermutation of HIV-1 DNA in the absence of the Vif protein. Science (New York, NY). 2003;300:1112. https://doi.org/10.1126/science.1083338.
23. Harris RS, et al. DNA deamination mediates innate immunity to retroviral infection. Cell. 2003;113:803–9.
24. Mariani R, et al. Species-specific exclusion of APOBEC3G from HIV-1 virions by Vif. Cell. 2003;114:21–31.
25. Mehle A, et al. Vif overcomes the innate antiviral activity of APOBEC3G by promoting its degradation in the ubiquitin-proteasome pathway. J Biol Chem. 2004;279:7792–8. https://doi.org/10.1074/jbc.M313093200.
26. Shirakawa K, et al. Ubiquitination of APOBEC3 proteins by the Vif-Cullin5-ElonginB-ElonginC complex. Virology. 2006;344:263–6. https://doi.org/10.1016/j.virol.2005.10.028.
27. Zheng YH, et al. Human APOBEC3F is another host factor that blocks human immunodeficiency virus type 1 replication. J Virol. 2004;78:6073–6. https://doi.org/10.1128/jvi.78.11.6073-6076.2004.
28. Dang Y, Wang X, Esselman WJ, Zheng YH. Identification of APOBEC3DE as another antiretroviral factor from the human APOBEC family. J Virol. 2006;80:10522–33. https://doi.org/10.1128/jvi.01123-06.
29. Doehle BP, Schafer A, Wiegand HL, Bogerd HP, Cullen BR. Differential sensitivity of murine leukemia virus to APOBEC3-mediated inhibition is governed by virion exclusion. J Virol. 2005;79:8201–7. https://doi.org/10.1128/jvi.79.13.8201-8207.2005.
30. Yu Q, et al. APOBEC3B and APOBEC3C are potent inhibitors of simian immunodeficiency virus replication. J Biol Chem. 2004;279:53379–86. https://doi.org/10.1074/jbc.M408802200.
31. Doehle BP, Schafer A, Cullen BR. Human APOBEC3B is a potent inhibitor of HIV-1 infectivity and is resistant to HIV-1 Vif. Virology. 2005;339:281–8. https://doi.org/10.1016/j.virol.2005.06.005.
32. Bogerd HP, et al. Cellular inhibitors of long interspersed element 1 and Alu retrotransposition. Proc Natl Acad Sci U S A. 2006;103:8780–5. https://doi.org/10.1073/pnas.0603313103.
33. Bogerd HP, Wiegand HL, Doehle BP, Lueders KK, Cullen BR. APOBEC3A and APOBEC3B are potent inhibitors of LTR-retrotransposon function in human cells. Nucleic Acids Res. 2006;34:89–95. https://doi.org/10.1093/nar/gkj416.
34. Muckenfuss H, et al. APOBEC3 proteins inhibit human LINE-1 retrotransposition. J Biol Chem. 2006;281:22161–72. https://doi.org/10.1074/jbc.M601716200.
35. Stenglein MD, Harris RS. APOBEC3B and APOBEC3F inhibit L1 retrotransposition by a DNA deamination-independent mechanism. J Biol Chem. 2006;281:16837–41. https://doi.org/10.1074/jbc.M602367200.
36. Chen KM, et al. Structure of the DNA deaminase domain of the HIV-1 restriction factor APOBEC3G. Nature. 2008;452:116–9. https://doi.org/10.1038/nature06638.
37. Holden LG, et al. Crystal structure of the anti-viral APOBEC3G catalytic domain and functional implications. Nature. 2008;456:121–4. https://doi.org/10.1038/nature07357.
38. Harjes E, et al. An extended structure of the APOBEC3G catalytic domain suggests a unique holoenzyme model. J Mol Biol. 2009;389:819–32. https://doi.org/10.1016/j.jmb.2009.04.031.
39. Shandilya SM, et al. Crystal structure of the APOBEC3G catalytic domain reveals potential oligomerization interfaces. Structure (London, England : 1993). 2010;18:28–38. https://doi.org/10.1016/j.str.2009.10.016.
40. Bohn MF, et al. The ssDNA mutator APOBEC3A is regulated by cooperative dimerization. Structure (London, England: 1993). 2015;23:903–11. https://doi.org/10.1016/j.str.2015.03.016.
41. Shi K, et al. Structural basis for targeted DNA cytosine deamination and mutagenesis by APOBEC3A and APOBEC3B. Nat Struct Mol Biol. 2017;24:131–9. https://doi.org/10.1038/nsmb.3344.

42. Kouno T, et al. Crystal structure of APOBEC3A bound to single-stranded DNA reveals structural basis for cytidine deamination and specificity. Nat Commun. 2017;8:15024. https://doi.org/10.1038/ncomms15024.
43. Xiao X, Li SX, Yang H, Chen XS. Crystal structures of APOBEC3G N-domain alone and its complex with DNA. Nat Commun. 2016;7:12193. https://doi.org/10.1038/ncomms12193.
44. Nik-Zainal S, et al. Mutational processes molding the genomes of 21 breast cancers. Cell. 2012;149:979–93. https://doi.org/10.1016/j.cell.2012.04.024.
45. Alexandrov LB, et al. Signatures of mutational processes in human cancer. Nature. 2013;500:415–21. https://doi.org/10.1038/nature12477.
46. Chan K, et al. An APOBEC3A hypermutation signature is distinguishable from the signature of background mutagenesis by APOBEC3B in human cancers. Nat Genet. 2015;47:1067. https://doi.org/10.1038/ng.3378.
47. Burns MB, Temiz NA, Harris RS. Evidence for APOBEC3B mutagenesis in multiple human cancers. Nat Genet. 2013;45:977–83. https://doi.org/10.1038/ng.2701.
48. Leonard B, et al. APOBEC3B upregulation and genomic mutation patterns in serous ovarian carcinoma. Cancer Res. 2013;73:7222–31. https://doi.org/10.1158/0008-5472.can-13-1753.
49. Roberts SA, et al. An APOBEC cytidine deaminase mutagenesis pattern is widespread in human cancers. Nat Genet. 2013;45:970–6. https://doi.org/10.1038/ng.2702.
50. Taylor BJ, et al. DNA deaminases induce break-associated mutation showers with implication of APOBEC3B and 3A in breast cancer kataegis. eLife. 2013;2:e00534. https://doi.org/10.7554/eLife.00534.
51. Kidd JM, Newman TL, Tuzun E, Kaul R, Eichler EE. Population stratification of a common APOBEC gene deletion polymorphism. PLoS Genet. 2007;3:e63. https://doi.org/10.1371/journal.pgen.0030063.
52. Long J, et al. A common deletion in the APOBEC3 genes and breast cancer risk. J Natl Cancer Inst. 2013;105:573–9. https://doi.org/10.1093/jnci/djt018.
53. Xuan D, et al. APOBEC3 deletion polymorphism is associated with breast cancer risk among women of European ancestry. Carcinogenesis. 2013;34:2240–3. https://doi.org/10.1093/carcin/bgt185.
54. Nik-Zainal S, et al. Association of a germline copy number polymorphism of APOBEC3A and APOBEC3B with burden of putative APOBEC-dependent mutations in breast cancer. Nat Genet. 2014;46:487–91. https://doi.org/10.1038/ng.2955.
55. Middlebrooks CD, et al. Association of germline variants in the APOBEC3 region with cancer risk and enrichment with APOBEC-signature mutations in tumors. Nat Genet. 2016;48:1330–8. https://doi.org/10.1038/ng.3670.
56. Harris RS. Cancer mutation signatures, DNA damage mechanisms, and potential clinical implications. Genome Med. 2013;5:87. https://doi.org/10.1186/gm490.
57. Roberts SA, et al. Clustered mutations in yeast and in human cancers can arise from damaged long single-strand DNA regions. Mol Cell. 2012;46:424–35. https://doi.org/10.1016/j.molcel.2012.03.030.
58. Hoopes JI, et al. APOBEC3A and APOBEC3B preferentially deaminate the lagging strand template during DNA replication. Cell Rep. 2016;14:1273–82. https://doi.org/10.1016/j.celrep.2016.01.021.
59. Sakofsky CJ, et al. Break-induced replication is a source of mutation clusters underlying kataegis. Cell Rep. 2014;7:1640–8. https://doi.org/10.1016/j.celrep.2014.04.053.
60. Chen J, Miller BF, Furano AV. Repair of naturally occurring mismatches can induce mutations in flanking DNA. eLife. 2014;3:e02001. https://doi.org/10.7554/eLife.02001.
61. Kazanov MD, et al. APOBEC-induced cancer mutations are uniquely enriched in early-replicating, gene-dense, and active chromatin regions. Cell Rep. 2015;13:1103–9. https://doi.org/10.1016/j.celrep.2015.09.077.
62. Wilson DM 3rd, Bohr VA. The mechanics of base excision repair, and its relationship to aging and disease. DNA Repair. 2007;6:544–59. https://doi.org/10.1016/j.dnarep.2006.10.017.

63. Ross AL, Sale JE. The catalytic activity of REV1 is employed during immunoglobulin gene diversification in DT40. Mol Immunol. 2006;43:1587–94. https://doi.org/10.1016/j.molimm.2005.09.017.
64. Jansen JG, et al. Strand-biased defect in C/G transversions in hypermutating immunoglobulin genes in Rev1-deficient mice. J Exp Med. 2006;203:319–23. https://doi.org/10.1084/jem.20052227.
65. Fanourakis G, et al. Evidence for APOBEC3B mRNA and protein expression in oral squamous cell carcinomas. Exp Mol Pathol. 2016;101:314–9. https://doi.org/10.1016/j.yexmp.2016.11.001.
66. Zhang L, et al. Genomic analyses reveal mutational signatures and frequently altered genes in esophageal squamous cell carcinoma. Am J Hum Genet. 2015;96:597–611. https://doi.org/10.1016/j.ajhg.2015.02.017.
67. Liu W, et al. Subtyping sub-Saharan esophageal squamous cell carcinoma by comprehensive molecular analysis. JCI Insight. 2016;1:e88755. https://doi.org/10.1172/jci.insight.88755.
68. Gleber-Netto FO, et al. Distinct pattern of TP53 mutations in human immunodeficiency virus-related head and neck squamous cell carcinoma. Cancer. 2017;124:84. https://doi.org/10.1002/cncr.31063.
69. Burtness B. The tumor genome in human immunodeficiency virus-related head and neck cancer: exploitable targets? Cancer. 2017. https://doi.org/10.1002/cncr.31059.
70. Kreimer AR, Clifford GM, Boyle P, Franceschi S. Human papillomavirus types in head and neck squamous cell carcinomas worldwide: a systematic review. Cancer Epidemiol Biomark Prev. 2005;14:467–75. https://doi.org/10.1158/1055-9965.epi-04-0551.
71. Gillison ML, et al. Distinct risk factor profiles for human papillomavirus type 16-positive and human papillomavirus type 16-negative head and neck cancers. J Natl Cancer Inst. 2008;100:407–20. https://doi.org/10.1093/jnci/djn025.
72. Vartanian JP, Guetard D, Henry M, Wain-Hobson S. Evidence for editing of human papillomavirus DNA by APOBEC3 in benign and precancerous lesions. Science (New York, NY). 2008;320:230–3. https://doi.org/10.1126/science.1153201.
73. Wang Z, et al. APOBEC3 deaminases induce hypermutation in human papillomavirus 16 DNA upon beta interferon stimulation. J Virol. 2014;88:1308–17. https://doi.org/10.1128/jvi.03091-13.
74. Ahasan MM, et al. APOBEC3A and 3C decrease human papillomavirus 16 pseudovirion infectivity. Biochem Biophys Res Commun. 2015;457:295–9. https://doi.org/10.1016/j.bbrc.2014.12.103.
75. Warren CJ, et al. APOBEC3A functions as a restriction factor of human papillomavirus. J Virol. 2015;89:688–702. https://doi.org/10.1128/jvi.02383-14.
76. Henderson S, Chakravarthy A, Su X, Boshoff C, Fenton TR. APOBEC-mediated cytosine deamination links PIK3CA helical domain mutations to human papillomavirus-driven tumor development. Cell Rep. 2014;7:1833–41. https://doi.org/10.1016/j.celrep.2014.05.012.
77. Henderson S, Chakravarthy A, Fenton T. When defense turns into attack: Antiviral cytidine deaminases linked to somatic mutagenesis in HPV-associated cancer. Mol Cell Oncol. 2014;1:e29914. https://doi.org/10.4161/mco.29914.
78. Vieira VC, et al. Human papillomavirus E6 triggers upregulation of the antiviral and cancer genomic DNA deaminase APOBEC3B. mBio. 2014;5. https://doi.org/10.1128/mBio.02234-14.
79. Barretina J, et al. Subtype-specific genomic alterations define new targets for soft-tissue sarcoma therapy. Nat Genet. 2010;42:715–21. https://doi.org/10.1038/ng.619.
80. Zhao L, Vogt PK. Helical domain and kinase domain mutations in p110alpha of phosphatidylinositol 3-kinase induce gain of function by different mechanisms. Proc Natl Acad Sci U S A. 2008;105:2652–7. https://doi.org/10.1073/pnas.0712169105.
81. Adelstein D, et al. NCCN guidelines insights: head and neck cancers, version 2.2017. J Natl Compr Cancer Netw: JNCCN. 2017;15:761–70. https://doi.org/10.6004/jnccn.2017.0101.

82. Scott L, Azacitidine J. A review in myelodysplastic syndromes and acute myeloid leukaemia. Drugs. 2016;76:889–900. https://doi.org/10.1007/s40265-016-0585-0.
83. Biktasova A, et al. Demethylation therapy as a targeted treatment for human papillomavirus-associated head and neck cancer. Clin Cancer Res. 2017;23:7276. https://doi.org/10.1158/1078-0432.ccr-17-1438.
84. Buisson R, Lawrence MS, Benes CH, Zou L. APOBEC3A and APOBEC3B activities render cancer cells susceptible to ATR inhibition. Cancer Res. 2017;77:4567–78. https://doi.org/10.1158/0008-5472.can-16-3389.
85. Boichard A, Tsigelny IF, Kurzrock R. High expression of PD-1 ligands is associated with kataegis mutational signature and APOBEC3 alterations. Oncoimmunology. 2017;6:e1284719. https://doi.org/10.1080/2162402x.2017.1284719.
86. Mullane SA, et al. Correlation of apobec mrna expression with overall survival and pd-l1 expression in urothelial carcinoma. Sci Rep. 2016;6:27702. https://doi.org/10.1038/srep27702.
87. Patel SP, Kurzrock R. PD-L1 expression as a predictive biomarker in cancer immunotherapy. Mol Cancer Ther. 2015;14:847–56. https://doi.org/10.1158/1535-7163.mct-14-0983.
88. Schumacher TN, Schreiber RD. Neoantigens in cancer immunotherapy. Science (New York, NY). 2015;348:69–74. https://doi.org/10.1126/science.aaa4971.
89. Rizvi NA, et al. Cancer immunology. Mutational landscape determines sensitivity to PD-1 blockade in non-small cell lung cancer. Science (New York, NY). 2015;348:124–8. https://doi.org/10.1126/science.aaa1348.

Chapter 11
The Genome-Wide Molecular Landscape of HPV-Driven and HPV-Negative Head and Neck Squamous Cell Carcinoma

Farhoud Faraji, Adrian D. Schubert, Luciane T. Kagohara, Marietta Tan, Yanxun Xu, Munfarid Zaidi, Jean-Philippe Fortin, Carole Fakhry, Evgeny Izumchenko, Daria A. Gaykalova, and Elana J. Fertig

Abstract Recent advances in sequencing technology have enabled unprecedented genome-wide characterization of head and neck squamous cell carcinoma (HNSCC). Integrated analyses of publicly available multiplatform high-throughput data have uncovered the vast genomic, epigenetic, and transcriptional diversity of HNSCC. Recognition of human papillomavirus (HPV) involvement in HNSCC carcinogenesis has resulted in the categorization of two HNSCC subtypes (HPV-driven and HPV-negative) with distinct etiologies, molecular properties, clinical features, and prognostic outcomes. Differences in the molecular landscapes of HPV-driven and HPV-negative HNSCC occur genome-wide and encompass changes in genomic, epigenetic, and transcriptional landscapes. Even within each subtype, HNSCC tumors have substantial inter-tumor and intra-tumor molecular heterogeneity. Improving

F. Faraji · A. D. Schubert · M. Tan · M. Zaidi · C. Fakhry
E. Izumchenko · D. A. Gaykalova (✉)
Department of Otolaryngology-Head and Neck Surgery, Johns Hopkins School of Medicine, Baltimore, MD, USA
e-mail: dgaykal1@jhmi.edu

L. T. Kagohara · E. J. Fertig (✉)
Department of Oncology and Division of Biostatistics and Bioinformatics, Sidney Kimmel Comprehensive Cancer Center, Johns Hopkins School of Medicine, Baltimore, MD, USA
e-mail: ejfertig@jhmi.edu

Y. Xu
Department of Applied Mathematics and Statistics, Whiting School of Engineering, Johns Hopkins University, Baltimore, MD, USA

J.-P. Fortin
Department of Biostatistics, Epidemiology, and Informatics, Perelman School of Medicine, University of Pennsylvania, Philadelphia, PA, USA

B. Burtness, E. A. Golemis (eds.), *Molecular Determinants of Head and Neck Cancer*, Current Cancer Research, https://doi.org/10.1007/978-3-319-78762-6_11

the understanding of the underlying biological function of these complex molecular landscapes through emerging cross-platform genomic analyses is essential to developing more effective diagnostic and therapeutic strategies for HNSCC.

Keywords Head and neck squamous cell carcinoma · Human papillomavirus · Genomics · Mutation · Epigenetics · Transcriptomics · Heterogeneity · Alternative splicing · Gene fusion · Immunotherapy

11.1 Introduction

Head and neck squamous cell carcinoma (HNSCC) comprises a heterogeneous group of tumors arising from the stratified squamous epithelium of the upper aerodigestive tract. HNSCC occurs in a variety of anatomic sites, including the oral cavity, oropharynx, hypopharynx, and larynx (Fig. 11.1). It is the fifth most common cancer worldwide, with 600,000 new cases and 300,000 deaths from this disease occurring annually [1, 2].

HNSCC is divided into two main subtypes based on its etiology: human papillomavirus-related (HPV-positive) and human papillomavirus-unrelated (HPV-negative). HPV-positive disease is driven by human papillomavirus and occurs predominantly in the oropharynx [3]. In contrast, HPV-negative disease is driven by chemical mutagenesis associated with tobacco and alcohol use [4], and tumors occur in numerous anatomic sites in the head and neck region. In this case, intensive exposure to the tobacco and alcohol-related mutagens results in mutagenesis of broad areas of the upper aerodigestive tract epithelium as a precursor to tumorigenesis (Fig. 11.1a). This process is known as "field cancerization" [5–9] and primes all anatomic sites to genomic alterations in HPV-negative HNSCC. In contrast, HPV infection alters cells in tonsillar crypts [10] promoting HNSCC carcinogenesis in the oropharynx (Fig. 11.1b).

HPV-positive HNSCC tends to arise in a younger patient population, is more sensitive to treatment, and exhibits significantly more favorable survival [11]. Overall survival rates at 3 years are estimated at 82% in locally advanced HPV-positive HNSCC compared to only 57% for locally advanced HPV-negative HNSCC [12]. The differences in etiology and outcomes have resulted in the classification of HPV-positive and HPV-negative HNSCC as two clinically distinct diseases [3, 13]. Clinical testing using immunohistochemical staining of p16 and PCR-based detection of HPV nucleotide sequences can both distinguish HPV-positive and HPV-negative disease [12, 14]. Further characterizing the molecular differences in these cancers from high-throughput genomic data can implicate molecular drivers by HPV status.

Modern high-throughput microarray and sequencing technologies have enabled unprecedented characterization of genomic, epigenetic, and transcriptional landscapes. Application of these technologies to HNSCC has revealed numerous disease-specific molecular alterations that are associated with disease initiation and

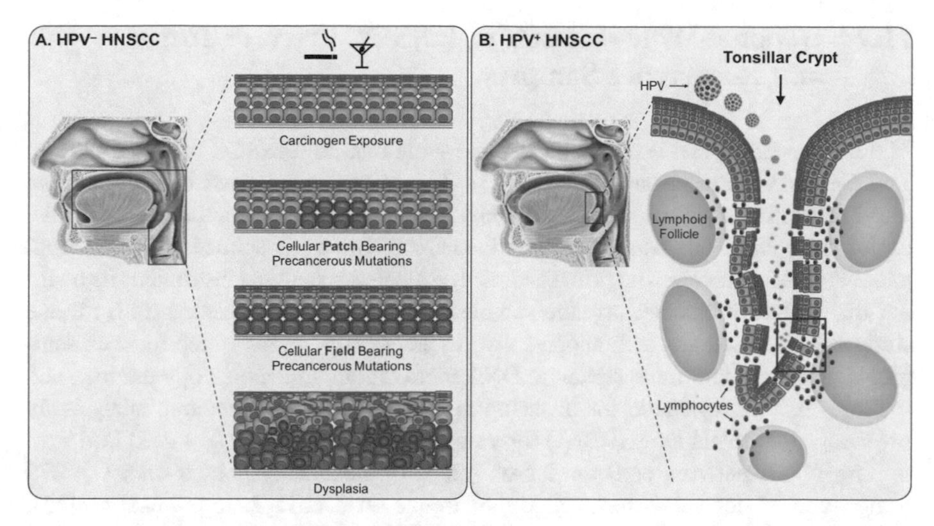

Fig. 11.1 Mode of risk factor exposure and predisposition to genomic alterations in anatomic sites that develop HNSCC. (**a**) HPV-negative HNSCC can arise at any region in the aerodigestive tract including the oral cavity, oropharynx, hypopharynx, and larynx (region in red box). Tobacco and alcohol consumption results in the exposure of mutagens to broad areas of the aerodigestive epithelia and primes this region for pervasive, random genomic alterations in a process called "field cancerization." (**b**) HPV preferentially infects the palatine tonsils, lingual tonsils, and base of the tongue (small blue box). HPV gains access to basal keratinocyte progenitors through natural interruptions in the lymphoid-associated tonsillar crypt epithelium (bold blue box) and initiates a well-established progression of molecular alterations to promote carcinogenesis

progression. HPV-positive and HPV-negative HNSCC have pervasive molecular differences detectable in the genome, epigenome, transcriptome, and proteome. However, even within each of these cancers, the molecular alterations do not appear to coalesce into common patterns, manifested by the significant inter- and intra-tumor heterogeneity. A refined biological understanding of the molecular profiles of HNSCC is thus essential to develop disease-specific treatment modalities to improve outcomes for individual patients in each subtype.

In this chapter, we review findings from genome-wide and molecular studies profiling HNSCC. We describe the somatic differences in chromatin, DNA methylation, and transcriptional profile that distinguish HPV-positive and HPV-negative HNSCC. We also describe the use of genomic methods to quantify substantial inter-tumor and intra-tumor heterogeneity. These findings have greatly enhanced the understanding of the molecular progression of carcinogenesis. Still, further multiplatform integration of high-throughput data and profiling tumors with emerging high-throughput technologies are essential to the identification of biomarkers and therapeutically actionable targets for precision therapy of all forms of HNSCC.

11.2 Genome-Wide Datasets of HNSCC Primary Tumor and Recurrence Samples

The largest centralized resource of genome-wide data for HNSCC is available from The Cancer Genome Atlas (TCGA) [15]. This project performed high-throughput multiplatform analysis of 530 primary HNSCC tumors and 44 normal samples. All tumor samples in this cohort were derived from surgical specimens resected from patients who had received no prior treatment, collected at multiple institutions throughout the United States. Because the samples for TCGA were collected from diverse study sites, p16 staining technologies used to assess HPV status varied for each sample. High-throughput transcriptional, DNA methylation, mutation, copy number, and miRNA data are available for all samples and array-based proteomic analysis by reverse-phase protein array (RPPA) for a smaller subset of tumors ($n = 212$) [15].

The TCGA network performed cross-platform data analysis on a subset of 279 tumors and 37 normal tissues, called the "freeze set," for its 2015 publication [15]. In this analysis, 36 (13%) of the samples were HPV-positive, determined from consensus calls that combine immunohistochemical p16 staining data with the presence of HPV sequence in the high-throughput sequencing data. This integration is the only standardized call of HPV status across diverse institutions and is available only for samples in the freeze set.

In addition to genomic data, TCGA contains clinical annotation for all samples, including smoking status and overall survival. The samples in TCGA are from a wide range of surgical cohorts and thus lack standardized treatment regimens or matched patient characteristics. Moreover, 70% of HPV-negative and 89% of HPV-positive HNSCC samples in TCGA are derived from male patients. While similar to gender distributions of HNSCC in the general population [16], these biases diminish the assessment of the molecular landscape of women with HNSCC. Therefore, biomarker studies developed from TCGA require substantial additional validation on independent cohorts to be sufficiently powered for clinical translation [15, 17].

Although TCGA is an unprecedented resource for genomic analyses of HNSCC, this database has two primary weaknesses: its use of adjacent normal samples as reference controls and retrospective study design. All normal samples in TCGA are from matched non-cancer tissues adjacent to tumor. While the genomic profiles of the normal tissues are homogenous and distinct from that tumor of tumors, matched normal samples may pose challenges for genomic analyses. Histopathologically normal-appearing adjacent tissue also has genetic changes relative to tissues that are unaffected by cancer because of field cancerization [18] (Fig. 11.1a). Moreover, HNSCC arises in several distinct anatomic sites of the aerodigestive tract (including the oral cavity, oropharynx, hypopharynx, and larynx). Each of these sites may have distinct molecular landscapes. Therefore, the normal cohort of samples in TCGA, predominantly collected for oral cavity samples, may be insufficient in power to discriminate subsite-specific molecular signatures from those that occur during carcinogenesis.

Beyond TCGA, numerous other sources of high-throughput genomic data from HNSCC tumors are available in the public domain (reviewed in [19, 20]). Prior to TCGA, Califano et al. [21–24] used microarray technology for cross-platform analyses of HNSCC. Landmark studies by Stransky et al. [25] and Agarwal et al. [26] also profiled the mutational landscape of HNSCC with next-generation sequencing technologies. Some of the tumors from these earlier studies are now also included in TCGA. All of these cohorts contain larger groups of HPV-negative than HPV-positive tumors. Keck et al. [27] and Guo et al. [28] contributed high-throughput transcriptional data from larger HPV-positive HNSCC cohorts. Similarly, Seiwert et al. [29] performed targeted sequencing on a panel of 617 cancer-associated genes in 120 matched tumor/normal samples, 51 of which were HPV-positive. To address the limitations of adjacent normal samples in TCGA, some of the above cohorts used samples from uvulopalatoplasty as non-cancer affected normal samples [21, 28, 30]. The data from all of these genomic studies are available in the public domain. This availability has enabled other studies to perform robust cross-study validation of genomic biomarkers [31, 32].

Similar to TCGA, the vast majority of public domain genomic datasets are from pretreated tumor or biopsy samples. Recently, Schmitz et al. [33] performed transcriptional profiling of a cohort of 12 patients before and after patients received cetuximab treatment, and Bossi et al. [34] performed transcriptional profiling of a cohort of 40 posttreatment recurrent or metastatic HNSCC samples prior to complete cetuximab response. Morris et al. [35] performed copy number analysis and targeted sequencing of recurrent and/or metastatic HNSCC samples. Hedberg et al. [36] extended analysis of HNSCC metastasis with whole exome sequencing of 13 HNSCC patients and synchronous lymph node metastases and 10 HNSCC patients with recurrent tumors. Expression profiling has also been performed on primary and metastatic samples with microarrays [37, 38] and more recently with single-cell RNA-sequencing [39]. These studies enable unprecedented characterization of the molecular mechanisms underlying treatment response and poor clinical outcomes. Recurrent and metastatic samples of HPV-negative HNSCC are most readily available for profiling, given the higher likelihood of recurrence in HPV-negative HNSCC. Nonetheless, the unique study design of Morris et al. [35] enabled profiling of HPV-positive recurrence and metastasis.

Studies assaying cell lines for genomic alterations associated with therapeutic sensitivity can serve as useful complements to the genome-wide analysis of data derived from primary tumors. Chung et al. [40] contributed the largest gene expression dataset with measurements of cetuximab sensitivity in both HPV-positive and HPV-negative HNSCC cell lines, with a subset of these samples containing paired pre- and posttreatment data [41]. The Cancer Cell Line Encyclopedia [42] and the Sanger Cell Lines Project [43] contain cross-platform genomic data with sensitivity to a wider range of therapeutic agents for HPV-negative HNSCC cell lines. To our knowledge, there are currently no similar public domain data resources for mouse models or patient-derived xenograft (PDX) models of HNSCC.

11.3 Somatic Mutational Landscape of HPV-Positive and HPV-Negative HNSCC

The mutational landscape of HNSCC has been extensively characterized, with analyses of high-throughput DNA data performed over the last decade. As a result, the critical genetic alterations that are responsible for the progression of both HPV-negative and HPV-positive carcinogenesis are well defined (Fig. 11.2). Molecular analyses demonstrated that tissues spanning the spectrum from benign hyperplasia to dysplasia and to carcinoma in situ exhibited progressively greater mutational burdens [18].

Early next-generation sequencing (NGS) studies using exome sequencing identified a two- to four-fold lower mutation rate in HPV-positive than in HPV-negative HNSCC [25, 26, 44]. However, the differences in mutational rate were more modest in the analyses of mutations performed with whole exome sequencing in TCGA [15]. Among HPV-negative tumors, higher mutation rates are observed in primary tumors from patients diagnosed with nodal metastasis relative to patients who later develop recurrence [36]. Field cancerization during early carcinogenesis (Fig. 11.1a) explains the relatively high rate of second primary tumors and their high mutation rate in HPV-negative HNSCC, a phenomenon not observed in non-smokers with HPV-positive HNSCC [45, 46].

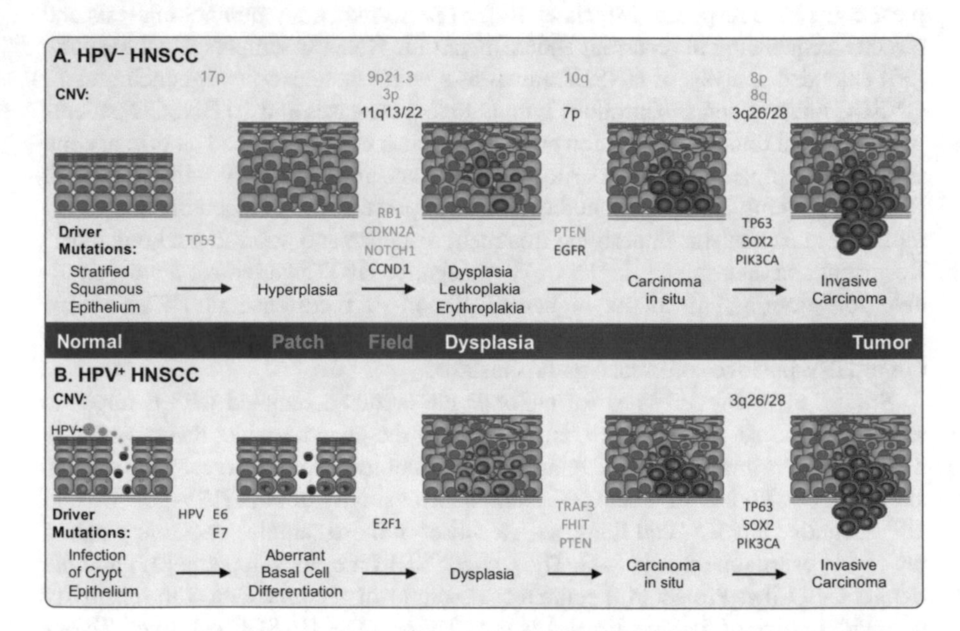

Fig. 11.2 Genetic alterations in multistage tumor progression of HNSCC. Summary of molecular and genetic alterations occurring throughout stages of (**a**) HPV-negative and (**b**) HPV-positive HNSCC carcinogenesis. Genes and loci in red are upregulated, activated, or amplified. Genes in green undergo loss of function mutation or deletion. Patch and field refer to precancerous mutation events that occur exclusively in HPV-negative HNSCC. CNV refers to copy number variation

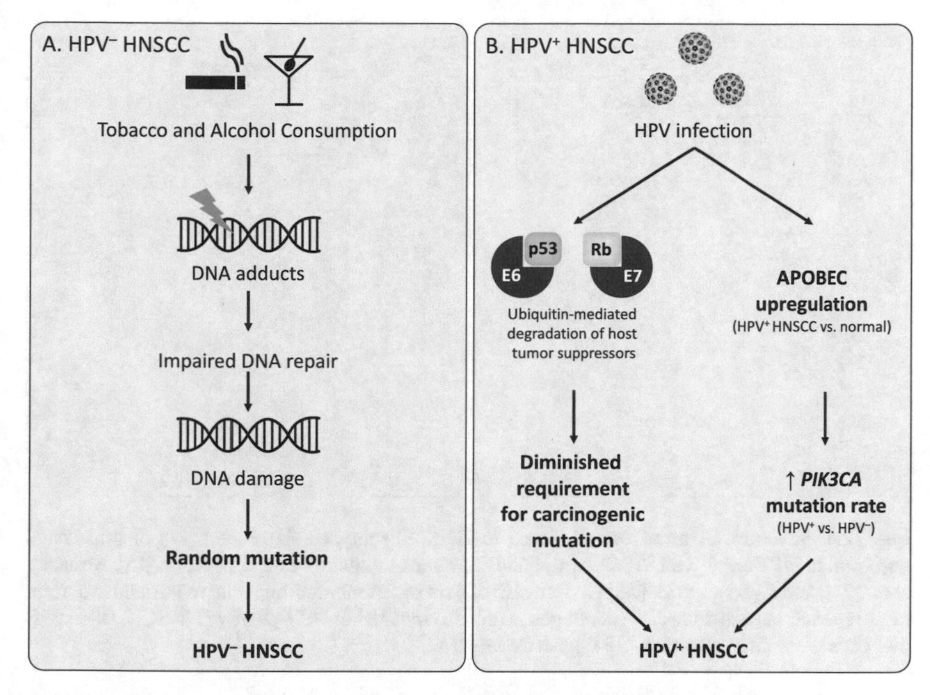

Fig. 11.3 Tobacco and alcohol consumption results in pervasive mutations in HPV-negative HNSCC and HPV viral infection mutations in HPV-positive HNSCC. (**a**) Tobacco and alcohol consumption exposes aerodigestive epithelia to mutagens that lead to the addition of DNA adducts which impede DNA repair and ultimately result in somatic mutation. (**b**) Infection of aerodigestive epithelia with human papillomavirus results in the expression of E6 and E7 oncoproteins, which induce the ubiquitin-mediated degradation of host tumor suppressor proteins p53 and Rb. In addition, host cellular response to HPV results in dysregulations of APOBEC family proteins, which result in somatic mutagenesis

The high mutational rate in HPV-negative HNSCC is readily attributable to its etiology (Fig. 11.3a). Smoking causes pervasive DNA damage and mutations, which subsequently leads to carcinogenesis. On the other hand, infection of high-risk HPV genotypes results in the expression of E6 and E7, viral oncogenes that inactivate key tumor suppressive molecular pathways (Fig. 11.3b). While the activity of viral oncoproteins may explain the lower somatic mutational burden observed in HPV-positive HNSCC tumors, HPV infection also results in dysregulation of APOBEC family cytosine deaminases, key factors in viral immunity that mutate viral DNA and restrict viral replication [47]. Hyperactive APOBEC can also cause mutation of the host genome, as seen in several virally-induced cancers [47]. Moreover, APOBEC mutation signatures are also observed in other tumors, including HPV-negative HNSCC [48].

Genomic studies have consistently concluded that distinct sets of genes are mutated in HPV-positive and HPV-negative HNSCC [15, 25, 26, 44], with the most frequently mutated genes summarized in Fig. 11.4a and b, respectively. Comparisons

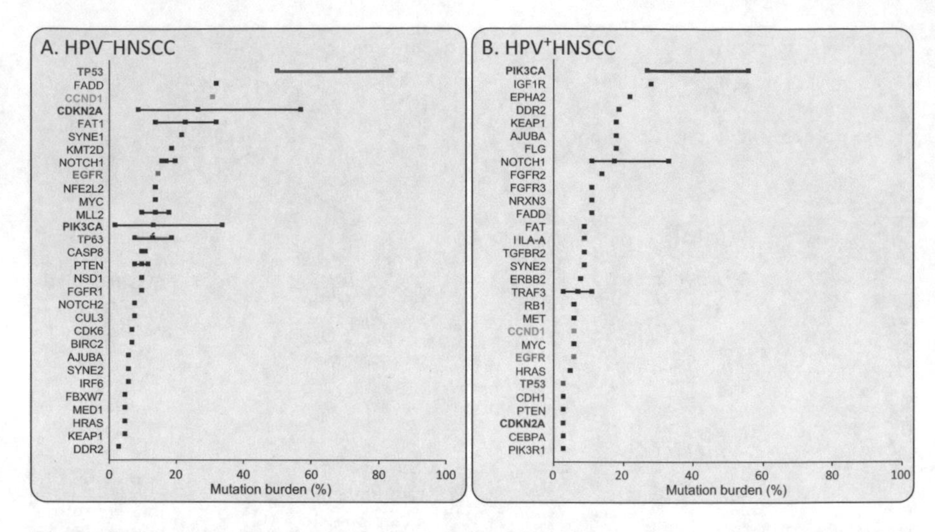

Fig. 11.4 Spectrum of mutations observed in HNSCC subtypes. (**a**) Forest plot of mutational spectrum in HPV-negative HNSCC estimated from mean mutation rates in TCGA [15], Stransky et al. [25], and Agarwal et al. [26] datasets. Error bars represent minimum and maximum mutation rates reported in each dataset. Colored genes are also mutated in HPV-positive HNSCC. (**b**) Forest plot of mutational spectrum in HPV-positive HNSCC

of these mutational landscapes to those of squamous cell carcinoma of the uterine cervix, the prototypical HPV-driven tumor, and squamous cell carcinoma of the lung, another tobacco-related tumor, demonstrate that changes in the mutational patterns between HPV-negative and HPV-positive tumors are associated with differences in their etiology [49, 50].

Genes commonly mutated in HPV-negative HNSCC include *TP53*, *CDKN2A*, *MLL2*, *CUL3*, *NSD1*, *PIK3CA*, *FAT1*, *AJUBA*, and *NOTCH* [15, 25, 26] (Fig. 11.4a). This list of genes has significant overlap with commonly mutated genes in other tobacco-induced tumors [51, 52]. Recent analysis of over 5000 sequenced tumors identified tobacco-related mutational signatures characterized primarily by nucleotide transitions, in which a purine is substituted by another purine (A↔G) or a pyrimidine by another pyrimidine (C↔T) [41, 44]. These tobacco-related mutational signatures are enriched in HPV-negative HNSCC [51].

HPV-positive HNSCC tumors most commonly have mutations in *PIK3CA*, *DDX3X*, *FGFR2*, *FGFR3*, *KRAS*, *MLL2/3*, and *NOTCH1* [15, 25, 26] (Fig. 11.4b). Notably, such a mutational pattern is also seen in HPV-driven cervical tumors [53]. In contrast to HPV-negative HNSCC, the mutational pattern of HPV-positive tumors is characterized by cytosine-to-thymidine (C > T) mutations at TpC sites [54, 55]. The enrichment of this mutational pattern in HPV-positive HNSCC results from dysregulation of APOBEC family proteins (Fig. 11.3b). Whereas smoking-related signatures are unique to HPV-negative HNSCC, APOBEC mutation signatures can also be observed in HPV-negative HNSCC at approximately similar weights [48].

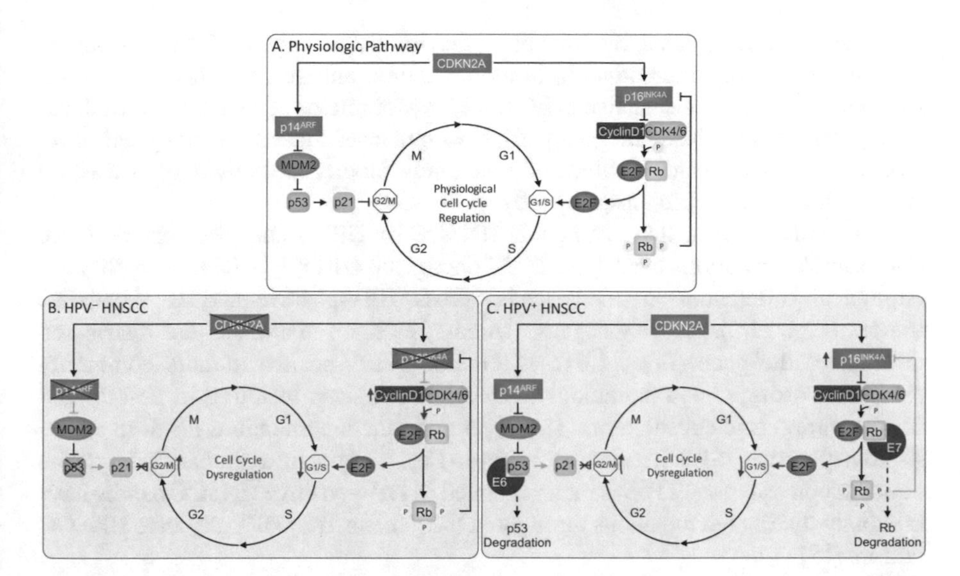

Fig. 11.5 Distinct alterations in both HPV-negative and HPV-positive HNSCC perturb the cell cycle to enable carcinogenesis. (**a**) Model depicting gene regulation of the cell cycle. Blue text represents tumor suppressor proteins, and red text represents proto-oncoproteins. *CDKN2A* encodes p14^{ARF} and p16^{INK4A} proteins. p14^{ARF} inhibits MDM2, which disinhibits p53 to activate p21 and prevent progression through the G2/M checkpoint into mitosis. p16^{INK4A} inhibits the CyclinD1/CDK4 and CyclinD1/CDK6 complexes. These complexes catalyze phosphorylation of the retinoblastoma protein (Rb), inducing Rb to release E2F family transcription factors to enter the nucleus and activate transcription of S-phase-promoting genes. Phosphorylation of Rb also results in feedback inhibition of p16^{INK4A} expression. (**b**) Model depicting mechanism by which common mutations in HPV-negative HNSCC result in cell cycle dysregulation. Frequent mutations in HPV-negative HNSCC include inactivating mutations of *CDKN2A* and *TP53* (encodes p53) and amplification of *CCND1* (encodes for Cyclin D1). *CDKN2A* mutation results in absent or nonfunctional p14^{ARF} and p16^{INK4A} protein expression, disinhibiting MDM2 and the CyclinD1/CDK4 and CyclinD1/CDK6 complexes, respectively. Disinhibition of MDM2 results in unregulated inhibition of p53 and thus diminished G2/M checkpoint blockade. Disinhibition of CyclinD1/CDK4 and CyclinD1/CDK6 results in overactive promotion of G1/S progression. In addition, mutation of *TP53* prevents signaling to p21 to prevent G2/M checkpoint blockade. (**c**) Model depicting mechanism HPV oncoprotein-mediated cell cycle dysregulation in HPV-positive HNSCC. HPV E6 oncoprotein binds and targets p53 for degradation, resulting in loss of G2/M checkpoint regulation. HPV E7 oncoprotein binds and targets Rb for degradation, resulting in the nuclear translocation of E2F and promotion of S-phase transition. In addition, downregulation of Rb results in loss of feedback inhibition and overexpression of p16^{INK4A}

Although distinct, mechanisms by which the sets of mutated genes in both HPV-negative and HPV-positive HNSCC contribute to carcinogenesis are well characterized (Fig. 11.5). Frequently mutated genes in HPV-negative tumors (including *TP53*, *CDKN2A*, and *CCND1*) play roles in cell cycle dynamics, resulting in aberrant cellular growth that predisposes to carcinogenesis. On the other hand, the HPV viral proteins themselves play critical roles in cell cycle dynamics and promote HPV-positive HNSCC carcinogenesis. The classic mechanism of action for viral E6 and E7 proteins is the ubiquitin-mediated proteolysis of p53 [56] and pRB [57] tumor suppressor proteins, respectively, inducing dysregulated cell cycle progression, disruption of differentiation, immortalization, and genomic instability [58]. Beyond these classic mechanisms, E6 and E7 oncoproteins perturb diverse molecular and cellular circuits to ultimately drive oncogenesis. Indeed, recent system-level data suggest that E6 and E7 bind and potentially modify the activity of at least 83 and 254 host cellular proteins, respectively [28, 59].

Mutational patterns that distinguish HNSCC by HPV status also vary by gene. For example, mutations in a key HNSCC oncogenic driver, *PIK3CA*, exemplify the etiological differences in oncogenesis across HPV-positive and HPV-negative tumors (Fig. 11.6). In HPV-negative tumors, *PIK3CA* mutations are distributed throughout the gene (Fig. 11.6a). In contrast, HPV-positive tumors commonly exhibit recurring C > T mutations at two hotspot sites, resulting in E542K and E545K amino acid substitutions. These gain-of-function mutations result in a constitutively active PIK3CA kinase [54, 60] (Fig. 11.6b). In a similar fashion, the tumor suppressor gene *TP53* is rarely mutated in HPV-positive HNSCC but displays randomly distributed mutations throughout the gene in most HPV-negative HNSCC tumors [15].

The mutational landscape has also been characterized in HNSCC recurrence and metastasis. As described above, Hedberg et al. [36] performed whole exome sequencing on synchronous nodal metastasis and samples from metachronous recurrences. Only one of the profiled samples was from HPV-positive HNSCC, consistent with its lower rate of recurrence relative to HPV-negative HNSCC. This study observed new mutations in *C17orf104* and *ITPR3* in synchronous nodal metastasis samples and new *DDR2* mutations in metachronous recurrent samples [36]. Morris et al. [35] found a similar mutational frequency and copy number-altered fraction of the genome in recurrent HPV-positive and HPV-negative HNSCC. Recurrent and metastatic HPV-positive HNSCC also had higher rates of *TP53* and lower rates of *PIK3CA* mutations than primary HPV-positive tumors. A recent study with single-cell RNA-sequencing found transcriptional profiles of malignant cells in lymph node metastases matched the transcriptional profiles of malignant cells in the primary oral cavity tumor for the same patient but varied from the cell types in their respective microenvironments [39].

Despite the advances in knowledge resulting from genome-wide analyses, both disease biomarkers and actionable target genes for clinical management remain elusive in HNSCC. One hurdle to clinically actionable advances is that most tumors have mutations in tumor suppressors, instead of oncogenes. In Agarwal et al. [26], 89% of samples have alterations reported in tumor suppressors, while only 18% of samples have alterations reported in oncogenes. This tendency toward tumor suppressors is observed in both HPV-positive and HPV-negative HNSCC [15]. Because tumor suppressor genes (TSG) are typically downregulated or deleted in cancer, they are not readily detectable in lower-throughput biomarker tests. Therefore, tumor suppressors have limited utility as diagnostic or prognostic markers. Unlike oncogenes characterized by hotspot mutations, TSGs are known for a wide spectrum of mutations throughout the exons, which are hard to detect using standard PCR or array technologies [61].

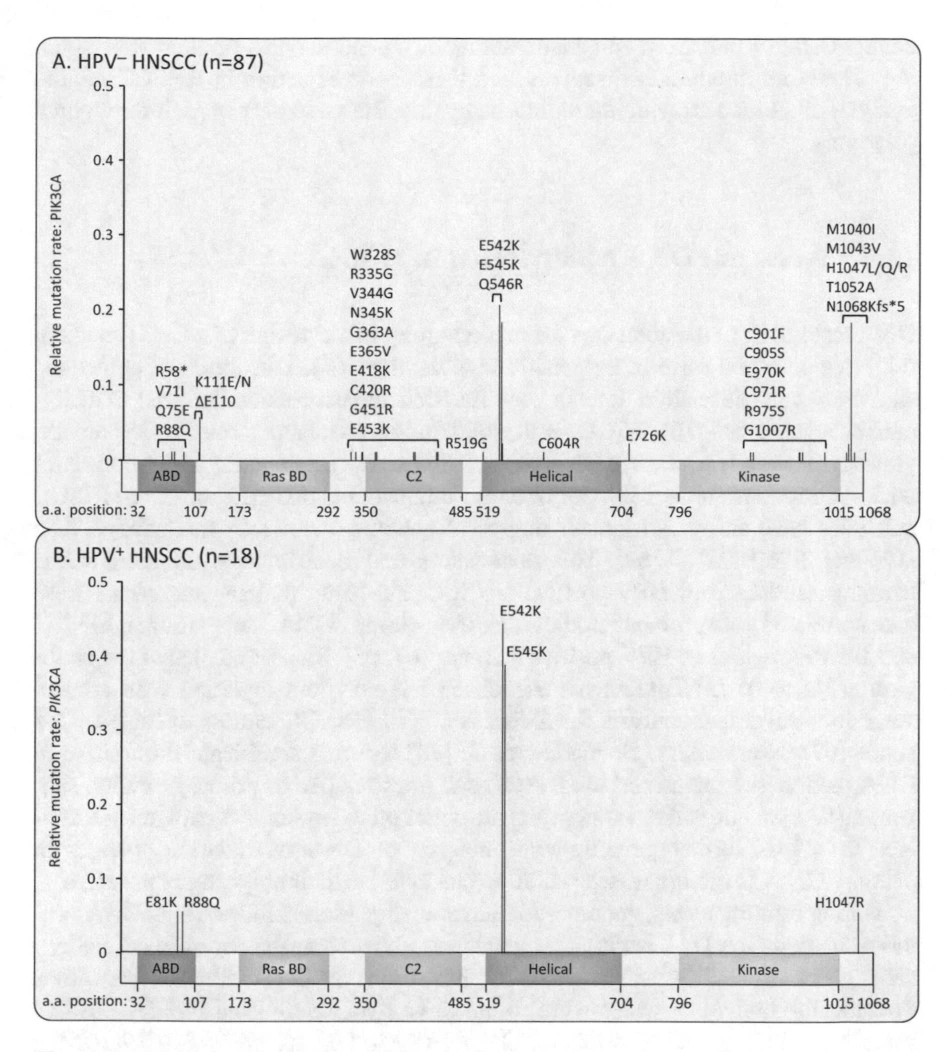

Fig. 11.6 Distribution of somatic *PIK3CA* mutations in HPV-negative and HPV-positive HNSCC tumors in TCGA. (**a**) Mutations are distributed throughout *PIK3CA* in HPV-negative HNSCC. (**b**) Greater than 80% of *PIK3CA* mutations occur at two mutational hotspots in the helical domain (E542 and E545) in HPV-positive HNSCC. ABD, adaptor-binding domain; Ras BD, Ras-binding domain; a.a., amino acid

Tumor suppressors are also not typically actionable therapeutic targets. Most targeted therapeutics block molecular function and therefore are insufficient to counteract the effect of loss-of-function mutations in tumor suppressors. Direct substitution or repair of gene function with drugs remains elusive. Current research efforts are aimed at evaluating the feasibility of targeting downstream tumor suppressor effectors that are upregulated upon tumor suppressor inactivation [62]. Determining these functional tumor suppressor effectors in different HNSCC subtypes requires further system-level analyses. Besides mutagenesis, functional changes often result from epigenetic alterations, gene fusions, or alternative splicing. Therefore, biomarker discovery and therapeutic selection in HNSCC require analysis of a broad array of alterations using data from a variety of high-throughput platforms.

11.4 Aberrant DNA Methylation in HNSCC

DNA methylation (the addition of a methyl group to a cytosine of a CpG dinucleotide) is a common form of epigenetic modification [63]. Disruption of epigenetic signatures can cause alterations in gene function and expression that lead to malignant changes in cells [64, 65]. Overall, solid tumors are characterized by hypomethylation, an overall decrease in DNA methylation across the genome, which correlates with genomic instability [66]. Sartor et al. demonstrated that HPV-positive HNSCC cell lines have an overall greater degree of genome-wide DNA methylation than HPV-negative HNSCC [67]. This observation may contribute to the more stable genomic landscape of HPV-positive HNSCC and better patient outcomes. These data correlated with previous studies identifying higher DNA methylation in LINE-1 and LUMA regions in HPV-positive samples [67, 68]. Similar studies of the methylation status of LINE elements associated DNA hypomethylation with clinical stage in a univariate analysis for a cohort of 119 HNSCC tumors of mixed HPV status [69]. Nonetheless, Hennessey et al. [70] reported significant differences in DNA methylation measured in HNSCC cell lines relative to primary tumors, suggesting that patient-derived xenografts are better model systems for epigenetic studies. Therefore, further genome-wide analysis of DNA methylation changes in primary HNSCC tumors was essential to characterize their epigenetic landscape.

During tumorigenesis, genome-wide hypomethylation is followed by DNA promoter methylation [71]. Multiple studies have shown that the list of hypermethylated genes in HPV-positive HNSCC has little overlap with HPV-negative. Specifically, individual genes were found to be hypermethylated in HPV-positive samples, including *GATA4*, *GRIA4*, *IRX4*, *ALX4*, *CUTL2*, *HOXA7*, *FBX039*, *IGSF4*, *PLOD2*, *SLITRK3*, *GALR1*, *PDX1*, *RUNX2*, *IRS1*, *and CCNA1* [15, 67, 72]. *ALDH1A2*, *OSR2*, *CHFR*, *ENPP5*, *FUZ*, *HPDL*, *HHEX*, *ICA1*, *IDUA*, *ITPKB*, *PIP5K1B*, *RBP5*, *VILL*, *ZNF141*, *ZNF420*, *KRT8*, and *CDKN2A* exhibited increased methylation in HPV-negative cells. Notably, *CDKN2A* codes for tumor suppressor proteins, $p14^{ARF}$ and $p16^{INK4A}$, highly expressed hallmarks of HPV-positive HNSCC. Promoter methylation often results in transcriptional silencing of tumor suppressor genes in HNSCC [73]. Therefore, *CDKN2A* methylation in HPV-negative HNSCC is consistent with observations of downregulated expression of $p14^{ARF}$ and $p16^{INK4A}$ in this subtype [67, 72, 74], in direct contrast to its overexpression in HPV-positive HNSCC.

The landscape of DNA methylation, as measured with microarrays, varies significantly in both HPV-negative and HPV-positive HNSCC tumors relative to nor-

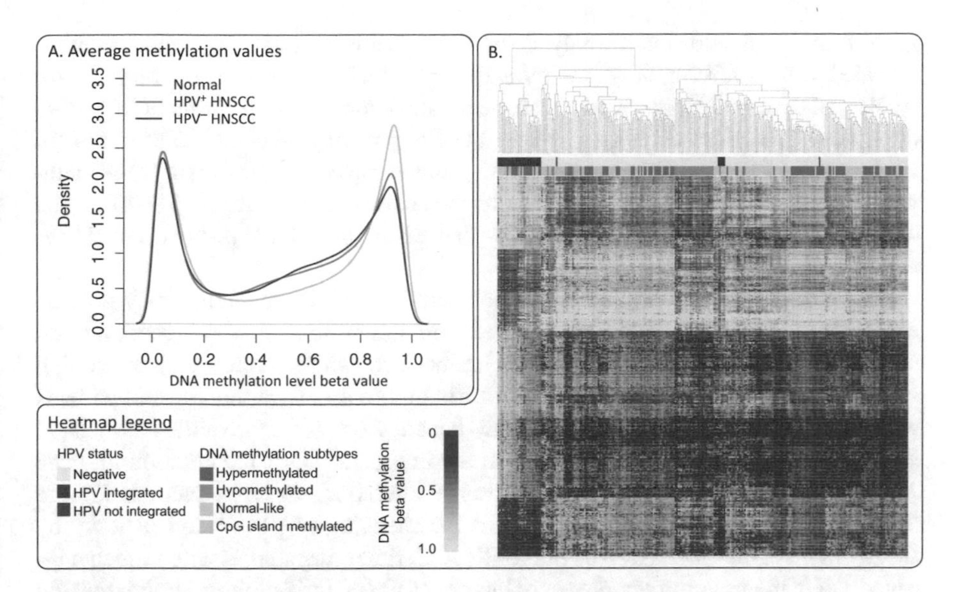

Fig. 11.7 DNA Methylation landscape in normal tissue, HPV-negative, and HPV-positive HNSCC. (**a**) Distribution of DNA methylation beta values for all probes on the DNA methylation microarrays. Beta values near zero represent absence of methylation and near one represent methylation. Distributions are averaged for all TCGA normal (gray), HPV-positive (blue), and HPV-negative (red) samples. (**b**) Heatmap of differentially methylated probes (FDR below 0.05 and the difference of mean beta values more than 0.2) in Illumina 450 K DNA methylation data for TCGA HNSCC tumor samples. Horizontal axis represents samples; vertical access represents DNA methylation array probes. Indicator bars for HPV status and DNA methylation subtypes demonstrate independence of HPV-positive and HPV-negative disease with genome-wide methylation clusters in TCGA

mal samples (Fig. 11.7). Genome-wide comparisons of $n = 4$ HPV-positive and $n = 4$ HPV-negative tumors in one study revealed 1168 differentially methylated genes – 686 hypermethylated and 467 hypomethylated – in HPV-positive HNSCC samples relative to HPV-negative HNSCC [75]. A separate study identified 360 differentially methylated CpG loci in $n = 31$ HPV-positive compared to $n = 65$ HPV-negative oropharyngeal squamous cell carcinoma (SCC) [76]. Overall, recent studies report a greater level of differential DNA methylation in HPV-positive compared to HPV-negative HNSCC [67, 76, 77].

TCGA identified four distinct DNA methylation clusters that were highly correlated with mRNA subtypes, smoking status, and tumor site [15]. These clusters were named based on the relative DNA methylation values compared to normal samples: hypermethylated, hypomethylated, CpG island methylated, and normal-like. Hypermethylated and CpG island methylated clusters were associated with smokers and oral cavity tumors, while hypomethylated and normal-like clusters were enriched in samples from nonsmokers and from the larynx as anatomic site. The hypomethylated cluster was significantly enriched for *NSD1*-altered and *NOTCH1* wild-type samples. The unsupervised clusters were not associated with

HPV status, but additional analysis of HPV-positive vs HPV-negative samples revealed 3014 differentially methylated CpG loci (1129 genes) (Fig. 11.7b) (FDR < 0.05 and difference of mean beta values more than 0.2), including 1898 hypermethylated loci (635 genes) and 1116 hypomethylated loci (520 genes) in HPV-positive samples. Other unsupervised studies have found different DNA methylation subtypes in HNSCC depending on the analysis technique [21, 77–79], some of which include methylation subtypes that distinguish HPV-positive and HPV-negative disease.

The DNA methylation gene signatures derived from these studies may provide more robust biomarkers than data derived from mutation analyses. Indeed, a DNA methylation-based prognostic signature has been shown to identify patients at high risk of treatment failure who may benefit from more intensive therapy and patients with more favorable prognosis who may benefit from lower toxicities associated with treatment de-escalation [72]. Given the impact of DNA methylation on gene expression [64, 65], many of these pervasive alterations to DNA methylation are likely to be functionally relevant, such as the silencing of $p14^{ARF}$ and $p16^{INK4A}$ by *CDKN2A* methylation in HPV-negative HNSCC. Emerging studies with sequencing-based DNA methylation analyses enable better characterization of strand-specific methylation and locus-specific methylation, not available to previous array-based studies. Integrated analyses with DNA alterations and transcriptional data [15, 21, 24, 79–81] will help better elucidate the function of these epigenetic alterations in HNSCC.

11.5 Gene Fusions Suggest Precision Medicine Strategies that Differ for HPV-Negative and HPV-Positive HNSCC

Gene structure can be altered in the absence of point mutations or epigenetic modification of DNA. With the advent of next-generation sequencing, gene fusions and alternative splicing events have been increasingly recognized as altering gene function in solid tumors [15, 82]. Given that such events are unique to tumors, they also have the potential to be utilized as biomarkers for noninvasive diagnosis, surveillance, or prognosis [83–85].

To date, two independent HNSCC cohorts have been analyzed for genome-wide fusion and alternative splicing events [15, 28]. Analysis of non-contiguous transcriptomic segments from RNA-sequencing data in TCGA with the MapSplice alignment algorithm [86] revealed 13,759 predicted fusion events in $n = 279$ HNSCC samples ($n = 45$ fusions/sample) [15]. Although 7119 events fused two genes, the general level of gene expression and rate of event recurrence between multiple tumor samples were both low [15]. Therefore, only a small subset of these fusion events were hypothesized to be functional drivers in HNSCC.

Gene fusion analysis has identified potentially actionable targets in HNSCC. Secondary analysis of pan-cancer TCGA RNA-sequencing data revealed multiple novel and recurrent fusions of the protein kinases *FGFR3*, *NTRK2*, and *NTRK3* [87]. *FGFR3* was the second most commonly fused gene after *RET* [87, 88]. Of 20 tumor types analyzed by TCGA, *FGFR3* fusion was found in eight different tumor types, including HNSCC, and was predominantly fused to *TACC3* [83, 87, 89, 90].

FGFR3-TACC3 fusion results in a constitutively activated tyrosine kinase that activates the mitogen-activated protein kinase pathway, suggestive of oncogenic activity [85, 91]. *FGFR3-TACC3* fusion also impacts activity of the p53 pathway, the AKT/mTOR/PTEN pathway, and cell cycle control (*CNNE1*, *CDK2–4*, etc.) and may potentially represent a novel mechanism of acquired resistance in EGFR-dependent cancers [89, 92, 93]. This putative oncogenic activity may potentially be pharmacologically inhibited with small molecule FGFR3 inhibitors [91, 94, 95]. Interestingly, the *FGFR3-TACC3* fusion may be specific to HPV-positive HNSCC, as it was detected in 2 of 36 (5.6%) HPV-positive samples in the TCGA dataset, but none of the 243 HPV-negative samples [15]. In further support of this specificity, *FGFR3-TACC3* fusion was also identified in RNA-sequencing data from an independent cohort of $n = 47$ HPV-positive HNSCC tumors (2%) [28]. Other gene-gene fusions identified in HPV-positive tumors from both this cohort and TCGA include *KRT14-KRT16*, *TFG-GPR128*, *ZNF750-TBCD*, and *CASZ1-CTNNBIP1* [15, 28].

Conversely, the *NTRK2-PAN3* and *NTRK3-LYN* fusions were detected in 2 of 243 (0.8%) HPV-negative HNSCC TCGA tumors and thus appeared to be specific to HPV-negative HNSCC [15]. The *JMJD7-PLA2G4B* gene fusion is another potential factor in HPV-negative HNSCC tumor progression [96]. These and other high interest recurrent gene fusions mechanistically linked to tumorigenesis may be useful in the development of cancer-specific biomarkers and the development of targeted therapy against the fused proteins.

11.6 Transcriptional Profiling Enables Characterization of Splice Variation in HNSCC

Aberrations in alternative splicing comprise another mechanism that can result in the production of non-canonical protein products and promote oncogenic function in cancer. Dysregulation of alternative splicing has been described for diverse tumor types, including HNSCC [97–99]. Similar to the detection of fused gene products, the development of RNA-sequencing and whole-transcriptome analysis methods have facilitated the elucidation of tumor-specific alternative splicing [100–103].

Alternative splicing in tumor suppressors genes *CDKN2A* (coding the $p16^{INK4A}$ protein), *TP53*, *TP63*, *TP73*, *FHIT*, *CSMD1*, *CD44*, and others were the first events discovered in HNSCC [98, 104–109]. Early microarray studies enabled genome-wide characterization of splice variants [99] but were limited to characterizing events in gene loci represented on the microarray. Analysis of RNA-Seq data enables more unbiased splice variant discovery, because short reads may span distinct gene isoforms [15, 28, 110]. Such genome-wide studies confirmed the prevalent alternative splicing of *TP63* in HNSCC [15, 99, 100], as well as the presence of alternatively spliced *AKT3*, *LAMA3*, *DST*, and *KLK12* genes [15, 99, 110, 111]. HPV-positive HNSCC samples demonstrated the prevalence of alternative splicing in *FKBP6*, *MEI1*, *SYCP2*, and *STAG2* genes [110]. Pan-cancer analyses of RNA-Seq data from TCGA revealed the diversity of disease-specific splicing events in *CAB39L*, *GSN*, *TNC*, *MYBBP1*, *KIFC3*, *CEP164*, and *ACPP*, which were detected in HPV-negative HNSCC [112].

Aberrant splicing in HNSCC can be related to aberrant phosphorylation of *SRPK2* (a splicing kinase), as well as differential expression of *SRSF3* (a proto-oncogenic splicing factor), *ESRP1/2* (the splicing regulatory proteins), and a diverse set of other RNA binding proteins, such as *CELF1* [113–116]. Analyses of mutation data from TCGA data demonstrated that genes encoding components of the splice machinery are more commonly altered in HPV-negative (18%) HNSCC samples than HPV-positive (8%) HNSCC samples [100]. This study found significant differences in alterative splicing events per gene, but not the total number of genes with alternative splicing events, between HPV-positive and HPV-negative subtypes [100].

Assembly of complete gene isoforms from short reads remains an active research area in bioinformatics. In some cases, short reads may not align to a unique isoform, which can limit the accuracy of characterization of splice variants in analyses of RNA-Seq data. Therefore, emerging long-read sequencing technologies will be essential to more accurately characterize gene isoform usage in HNSCC tumors.

11.7 Noncoding RNA

Noncoding RNAs (ncRNAs) are a class of protein noncoding transcripts with broad regulatory roles, including regulation of gene expression. ncRNAs include long noncoding RNAs (lncRNAs), long intergenic noncoding RNAs (lincRNAs), microRNAs (miRNAs), enhancer-like RNAs (eRNAs), and Piwi-RNAs (piRNAs) [117]. Initially thought to have no relevant functionality, these noncoding transcripts are currently known to control gene expression by inhibiting transcription of their targets via induction of mRNA degradation, DNA promoter competition, mRNA elongation inhibition, attraction of chromatin modifier to specific gene loci, and interference with the splicing process [118]. miRNAs represent the most studied group of ncRNAs for HNSCC and other cancers. lncRNAs are increasingly studied, but few key insights have been unveiled in the context of HNSCC carcinogenesis and prognosis [119, 120]. Only one study aimed to determine the expression profile of piRNAs in HNSCC [121].

MicroRNAs are important regulators of gene expression. These short polynucleotides (~22 base pairs in length) bind to mRNA targets with imperfect sequence complementarity and block transcription or induce transcript degradation [122].

One single miRNA can target and regulate hundreds of transcripts, resulting in each miRNA possessing pleiotropic effects [122]. Moreover, regulatory functions of miRNAs have been found to be tissue and disease specific, though the mechanisms underlying these context specificities remain to be clarified [123].

Chang et al. [124] identified a panel of miRNA as differentially expressed between HNSCC tumor and normal samples. This panel included one miRNA downregulated in tumor relative to normal (miR-494), suggestive of a potentially tumor suppressive function in cancer. In addition, several miRNAs were found to be upregulated, including miR-21, miR-7, miR-18, miR-29c, miR-142-3p, miR-155, and miR-146b.

Changes in miRNA expression have also been correlated with tumor progression and prognosis. Perhaps the best studied miRNA, miR-21, is upregulated in HNSCC and other tumor types (esophageal, prostate, breast), and high expression of miR-21 is correlated with poor survival [125]. In addition to miR-21, dysregulation of other miRNAs has been identified to predict prognosis in HNSCC. Loss of miR-375 expression correlates with higher risk of death, suggesting that miR-375 may be a prognostic marker [126]. Aberrant expression of a wide variety of miRNAs is also associated with clinical prognostic markers such as locoregional recurrence (mir-205 and let-7d) [127], disease aggressiveness and metastasis (mir-200 family) [128], and cell migration and invasion (mir-26a and mir-7) [129, 130]. Other miRNAs have no association with HNSCC clinical and pathological features but are differentially expressed at specific anatomic sites. For example, a panel of eight miRNAs are specifically dysregulated in laryngeal HPV-negative SCC: miR-21-3p and miR-106b-3p (upregulated) and let-7f-5p, miR-10a-5p, miR-125a-5p, miR-144-3p, miR-195-5p, and miR-203 (downregulated) [131].

The miRNA expression disparities between HPV-positive and HPV-negative HNSCC are not well characterized. Using a quantitative PCR-based miRNA array, Howard et al. [132] found that miR-449a is overexpressed in HPV-positive HNSCC relative to HPV-negative and confirmed the same expression pattern for miR-499 in miRNA-Seq expression data from TCGA. There is a clear difference between the repertoires of miRNAs expressed in these two HNSCC types, according to findings from Miller et al. [133] and Lajer et al. [134]. Both groups analyzed the miRNA expression profile in oropharyngeal SCC samples, identifying miR-199a-3p/miR-199b-3p, miR-143, and miR-145 to be commonly downregulated in HPV-positive HNSCC and other miRNAs showing decreased expression in at least one of the studies. The number of upregulated miRNAs was limited in both studies, and miR-320a, miR-222-3p, and miR-93-5p [133] and miR-195 and miR-363 [134] showed increased levels in the presence of viral infection. All the studies are helping to characterize the differences between HPV-positive and HPV-negative HNSCC, although all lack further investigation on the mRNA targets regulated by the aberrantly expressed miRNAs.

lncRNAs encompass transcripts of at least 200 nucleotides in length. In contrast to miRNAs that directly bind to their targets, lncRNAs regulate gene expression through a distinct variety of mechanisms. Among their many functions, lncRNAs silence genes by binding to the DNA promoter or enhancer regions and recruiting chromatin modifier or transcriptional regulatory complexes which bind directly to DNA binding sites to promote or block transcription factor (TF) activity [135]. The role of lncRNAs in HNSCC is poorly understood, with just a few differentially expressed lncRNAs associated with this tumor type: *GAS5*, *MEG3*, *H19*, *UCA1*, and *PCAT-1*. Aberrant expression of *MEG3* [136] and *UCA1* [137] has been observed in tongue SCC, with elevated *UCA1* also associated with the presence of lymph node metastasis. Using the data from TCGA, Zou et al. found 9681 lncRNAs, 232 miRNAs, and 61 piRNAs differentially expressed between HNSCC and the paired normal tissue [121]. This ncRNA landscape was associated with mutations in some relevant genes for HNSCC tumorigenesis (*TP53*, *CASP8*, *CDKN2A*, *PRDM9*, *Cyclin E*, *NOTCH*, and *FBXW7*) and copy number variations (3p, 5p, 7p, and 18q deletion and 3q and 7q amplification). This study was also the first one to detect aberrant expression of piRNAs in HNSCC. piR-34,736 and piR-36,318 have been previously described to be deregulated in breast cancers and were found by Zou et al. in HNSCC as well [121].

Despite these early insights, the precise role of ncRNAs in regulating HNSCC progression remains to be uncovered. Furthermore, little is known about the differential role of ncRNA in HNSCC with respect to HPV status. A more detailed understanding of the mechanism by which ncRNAs impact tumorigenesis and progression in both HPV-positive and HPV-negative HNSCC may unveil more reliable prognostic biomarkers [120] and novel therapeutic targets [138].

11.8 Transcriptional HNSCC Subtypes

Early gene expression profiling of HNSCC tumors found four dominant subtypes associated with distinct survival characteristics [139]. However, at the time of discovery, the drivers responsible for the discrepancies between these subtypes were unknown. Later, when HPV was established as a driver of HNSCC carcinogenesis, the gene expression profiles from HPV-positive HNSCC samples were determined to fall into one of the four dominant gene expression subtypes. Walter et al. [140] also confirmed that these subtypes are robust in several independent cohorts of samples. Upon hierarchical clustering of HNSCC tumors, the HPV-positive grouping of samples was labeled as "atypical," relative to the gene expression profiles of "classical" HPV-negative HNSCC.

Further annotation of the genes defining these groups has further determined additional subtypes: one displaying features that resemble basal tumors (the "basal" subtype) and another subtype with greater epithelial-mesenchymal transition (EMT) activity ("mesenchymal"). Single-cell RNA-sequencing data attributes differences between basal and mesenchymal subtypes in oral cavity tumors to differences in the composition of cell types within these tumors, and does not infer significant expression differences between the malignant cells within basal and mesenchymal tumors [39]. Transcriptional analysis of larger cohorts of HPV-positive HNSCC in Keck et al. [27] found further subclustering among this relatively homogeneous cohort of samples (Fig. 11.8). One of the two HPV-positive clusters shares similar gene

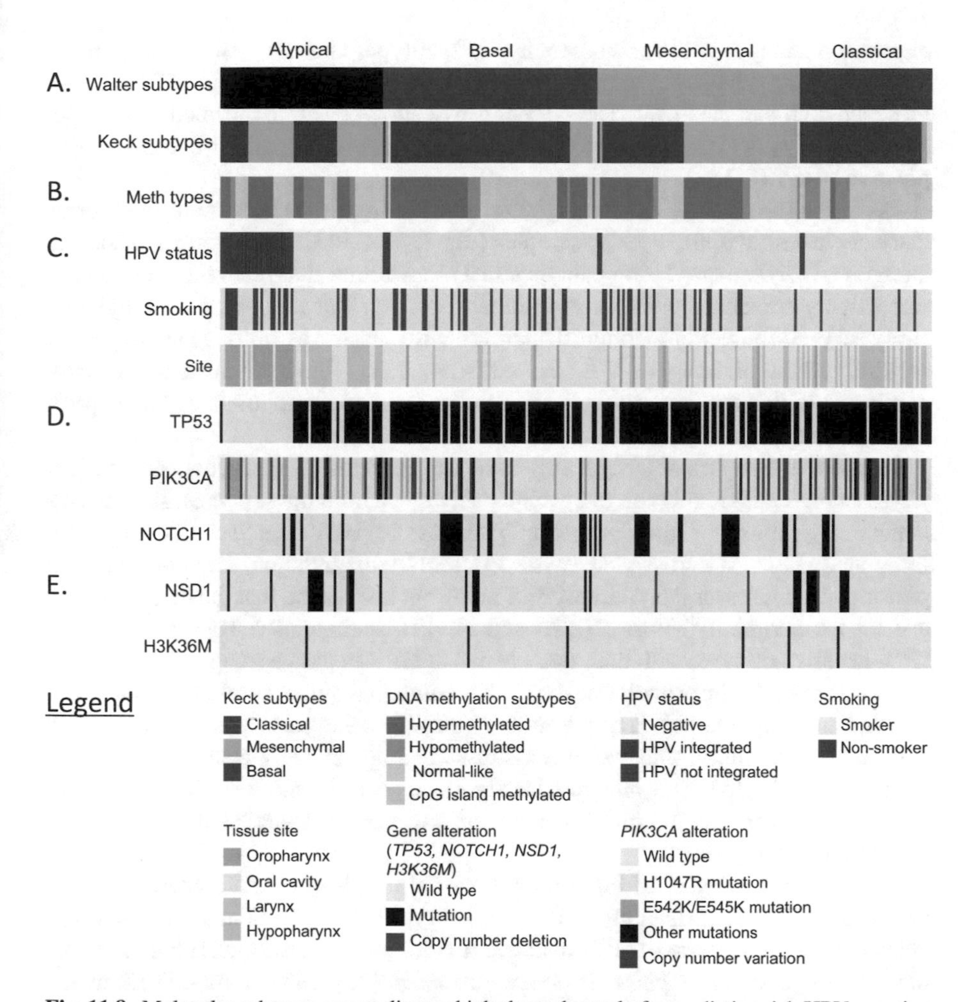

Fig. 11.8 Molecular subtypes across diverse high-throughput platforms distinguish HPV-negative and HPV-positive HNSCC. (**a**) Gene expression subtypes with Walter et al. [140] and Keck et al. [27] subtypes for TCGA samples. Atypical samples from Walter et al. are split between classical and mesenchymal subtypes in Keck et al., while the remaining basal, classical, and mesenchymal subtypes from Walter et al. are consistent with subtype calls in Keck et al. (**b**) DNA methylation subtypes reflect pervasive hypermethylation, hypomethylation, normal-like methylation, and CpG island methylation in HNSCC tumors. The hypermethylation subtype is enriched in the mesenchymal gene expression subtype and CpG island methylation subtype in the basal gene expression subtype. (**c**) Clinical attributes of TCGA samples. HPV status and oropharynx site are both enriched in the Walter et al. atypical subtype. HPV integration is associated with the HPV-positive mesenchymal Keck et al. gene expression subtype. Tumors in the larynx are split between atypical and basal subtypes, while oral cavity tumors occur predominately in basal and classical subtypes. HPV-positive tumors are enriched in nonsmoking individuals. (**d**) Commonly mutated genes distinguish HPV-negative and HNSCC-positive tumors. (**e**) Mutations in *NSD1* and *H3K36M* genes define a new epigenetic subtype in HPV-negative tumors, uniformly distributed among expression and DNA methylation subtypes

expression changes with the "mesenchymal" subtype. These mesenchymal samples are associated with greater immune activity in both HPV-positive and HPV-negative mesenchymal samples [27]. The HPV-positive subtypes are confirmed in an independent study [141] and associated with tumor-specific chromatin structure in HPV-positive HNSCC [142].

As discussed above, DNA methylation also distinguishes HPV status and forms distinct clusters within HNSCC samples [21, 77, 78, 143]. Hierarchical clustering analysis of DNA methylation subtypes in TCGA found that methylation subtypes were highly correlated with transcriptional subtypes. Nonetheless, the methylation subtypes predominantly distinguish samples with respect to smoking status, tumor subsite, and the presence of *NOTCH1* mutations [15]. These subclusters were later confirmed with a new algorithm for integrated DNA methylation and gene expression analysis [79, 143].

Integrative analyses of DNA methylation and gene expression subclusters further suggest epigenetic regulation of specific pathways within the subtypes, such as epigenetic regulation of the Hedgehog pathway in HPV-negative HNSCC [21]. Integrated statistics across gene expression, DNA methylation, and copy number performed on gene targets of transcription factors also found that the activity of the critical transcription factors STAT3 and NF-κB distinguishes HPV-positive and HPV-negative disease, with both more active in HPV-negative samples [22]. TCGA also performed bioinformatics analyses that combined gene expression, copy number, and pathway interaction data found that cell signaling pathways associated with cell death, immunity, and oxidative stress distinguish HPV-positive and HPV-negative tumors [15]. New integrated analyses that span all these data modalities are essential to determine the molecular drivers and potential therapeutic targets specific to these subtypes.

More recent analysis of high-throughput data with novel computational tools revealed a still greater diversity of HNSCC subtypes. Analysis of mutation data suggests that a new subtype of HPV-negative tumors may result from mutations to the histone H3 gene that impairs histone modification [78]. Histone modifications are associated with differences in chromatin structure. These changes alter genes that are accessible for expression and therefore may play a critical functional role in HNSCC carcinogenesis. Further collection of datasets resulting from applying rapidly emerging technologies such as ChIP-Seq or ATAC-Seq to HNSCC tumors is essential to better establish the functional role of chromatin in various disease subtypes. Figure 11.8 summarizes subtypes across all data modalities in HNSCC TCGA data.

11.9 Genomic Data Distinguishes HPV-Positive HNSCC Tumors with and without Viral Integration in the Human Genome

Data from next-generation sequencing technologies can reveal differences in the molecular landscape of the HPV virus itself in HPV-positive HNSCC tumor samples, informing the pathological action of the virus. Because sequencing techniques are agnostic to the genome, whole genome sequencing and RNA-sequencing of HPV-positive HNSCC can simultaneously measure alterations in human and HPV genomes. Sequencing reads that span both genomes indicate sites where the HPV virus integrates into the human genome, and samples that contain these reads are considered to have undergone HPV viral integration. A recent analysis of viral integration sites in both whole genome sequencing and RNA-sequencing data determined that a subset of HPV-positive HNSCC in TCGA did not have HPV integration in the genome [144]. Prior to this study, HPV was believed to be integrated in the genome in all HPV-positive HNSCC tumors. However, only roughly half of HPV-positive HNSCC tumors (39–71%) have virus integrated into the host genome [142, 147–148]. The sequencing-based studies of HPV viral integration are subject to false negatives if regions of the genome spanning these integration sites are not covered. Nonetheless, there is a wide read coverage of the HPV genome from the high-throughput assays in TCGA tumors, and similar phenomena are observed in cervical cancer [15, 49]. Using the available TCGA data, independent investigators [147, 148] found the presence of HPV-positive samples, where both episomal and integrated HPV are most likely present in the same cells, suggesting the presence of the chimeric episomes formed upon the secondary excision of the HPV from the human genome [148]. Such episomes can carry multiple copies of HPV genome as well as pieces of human genome [144, 148] and are a topic of active investigation.

HPV-positive tumors without genomically integrated virus are associated with the HPV-positive mesenchymal subtype from Keck et al. [27] (Fig. 11.8). This observation is consistent with identification of gene expression and copy number subtypes that distinguish HPV integration and viral E2/E4/E5 expression [141]. A study of mutations with HPV-positive HNSCC by Hajek et al. [149] found that alterations in *TRAF3* and *CYLD* are associated with lack of HPV integration. Moreover, recent data suggest that samples with episomal and integrated HPV genome differ on the chromatin level [142], where genes with tumor-specific H3K27ac-associated enrichment are significantly associated with HPV-KRT [141] and with the classical subtypes of HPV-positive HNSCC [27] as well as with sites of HPV integration [142]. The clinical relevance of HPV-positive HNSCC tumors that lack viral integration and their molecular landscapes is a current area of active investigation [150].

11.10 Inter- and Intra-tumor Heterogeneity Are Pervasive in HNSCC Subtypes

Carcinogen-induced somatic mutations are randomly distributed throughout the genome and across numerous adjacent tumor cells, causing a high degree of genomic heterogeneity in HPV-negative HNSCC tumors. Dysregulation of APOBEC in HPV-positive HNSCC is another source of mutation, which may also give rise to genomic heterogeneity in that disease [55, 60]. Tumor heterogeneity refers to transformed clonal cells that, with tumor progression, have developed distinct novel mutations resulting in the rise of multiple clones within a single tumor [151, 152]. The diversification of clonal populations in the tumor has been linked to worse prognosis [153, 154]. Moreover, specific therapies may eliminate certain clonal populations and select for the outgrowth of resistant clones [155]. This mechanism of therapeutic resistance occurs particularly commonly for targeted therapies, such as cetuximab [156].

Tumor heterogeneity has become an active area of research in cancer overall and HNSCC in particular [154, 157]. Tumor heterogeneity can complicate precise diagnosis with potential sampling error and may contribute to therapy failure and recurrence. New pathway dysregulation algorithms are in development to characterize functional implications of inter-tumor heterogeneity [158, 159]. These algorithms quantify the dysregulation in a single pathway comparing the variation of gene expression in all of the genes associated with that pathway for samples from one phenotype relative to the variation of gene expression in the same pathway genes for samples from another phenotype. For example, applying expression variation analysis (EVA) to gene expression data assessed using microarrays confirmed that pathways have higher variation in gene expression and are therefore more significantly dysregulated in HNSCC tumors relative to normal samples [158]. This observation was further confirmed in RNA-sequencing data from TCGA in Fig. 11.9a. These algorithms revealed that heterogeneity of pathway dysregulation increases in tumor populations with worse prognosis [159]. These methods have not yet been applied to compare the relative heterogeneity of pathway dysregulation between HPV-negative and HPV-positive tumors.

Mutant allele tumor heterogeneity (MATH) is a scoring system established in HNSCC for the measurement of tumor heterogeneity based on sequencing data [161]. MATH estimates the mutant-allele fraction (MAF) as the percentage of mutated reads versus total reads in next-generation sequencing data. This method assumes that wider distributions of MAF values across genes reflect greater genomic heterogeneity. Application of the MATH algorithm to sequencing data in TCGA identified greater intra-tumor heterogeneity in HPV-negative relative to HPV-positive tumors (Fig. 11.9b). Higher MATH scores were also associated with increased mortality independent of other clinicopathological measures, including HPV status, TNM class, or tumor site. Therefore, estimates of heterogeneity with algorithms such as MATH may serve as prognostic factors in HNSCC [156, 159].

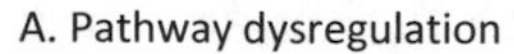

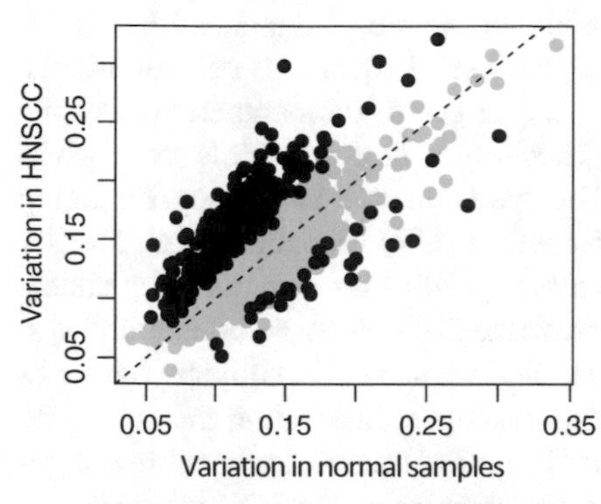

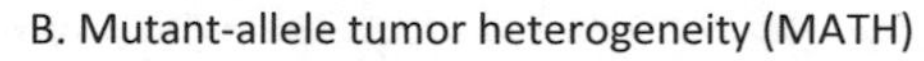

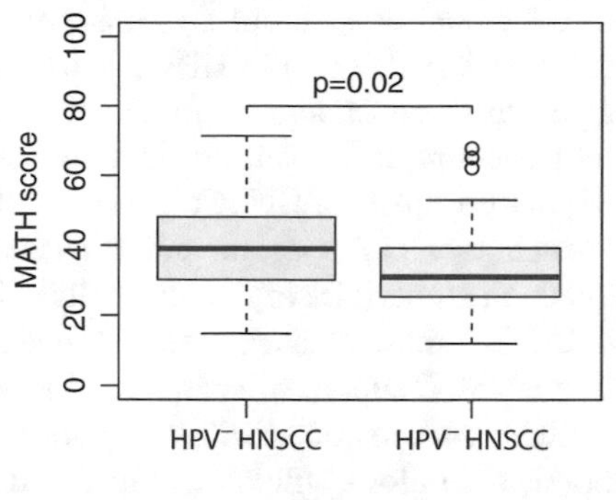

Fig. 11.9 Measures of tumor heterogeneity in HPV-negative and HPV-positive HNSCC. (**a**) Variation of gene expression profiles for genes in canonical pathways from MSigDB [160] for normal samples (x-axis) and HNSCC tumor samples (y-axis) from TCGA. Significantly dysregulated pathways in HNSCC samples relative to normals from expression variation analysis [158] are indicated in black, with all remaining pathways colored in gray. (**b**) Distribution of MATH [154] scores of intra-tumor heterogeneity of mutational profiles in HPV-negative and HPV-positive HNSCC tumors. P-value is reported as a t-test of MATH scores between 36 HPV-positive and 243 HPV-negative samples in the TCGA freeze set

Genetic tumor heterogeneity arises during tumor formation as sequences of mutations occur in subpopulations of tumor cells, resulting in heterogeneity within the same tumor and across different patients' tumors. As a result, tumor cell populations are composed of different subpopulations of cells, with each of the subpopulations – or "subclones" – defined by a unique genome. While MATH estimates the degree of heterogeneity in a tumor sample, it does not estimate the composition of the subpopulations from which the tumor arose. Single-cell RNA-sequencing is emerging as a promising new technology to measure intra-tumor heterogeneity and subclones directly and has recently been applied to study both inter- and intra-tumor heterogeneities in HNSCC primary tumors [39]. However, technical challenges in performing single-cell analysis on archival samples currently limit single-cell sequencing of large cohorts of HNSCC tumors. Therefore, new computational algorithms are emerging to estimate tumor subpopulations based on the data generated by next-generation sequencing [162, 163]. Analysis of the samples in Hedberg et al. [36] with the PyClone algorithm [163] has found that genetic alterations in metastatic samples present at the time of diagnosis were largely transmitted from primary tumors, while recurrences were more frequently derived from new subclones.

The large number of distinct cell types that comprise a tumor also contributes to intra-tumor heterogeneity. Recent advances in computational microdissection techniques enable estimation of cellular composition based on gene expression or DNA methylation data [164]. Mandal et al. [165] characterized the immune landscape of HNSCC using transcriptomic data from TCGA. They found that the degree of immune infiltration in HPV-negative HNSCC is lower than that of HPV-positive malignancies and is associated with poorer clinical outcomes. In a low-throughput study, Badoual et al. [166] also observed a higher degree of tumor infiltration by regulatory T cells (T_{reg}) in HPV-positive oropharyngeal squamous cell carcinoma, suggesting these tumors could benefit from T_{reg}-targeted immunotherapy. Despite recent encouraging results from immune checkpoint inhibitor trials in HNSCC [167], only a subset of HNSCC patients will benefit from immunotherapy. Emerging sequencing technologies, including T cell receptor (TCR)-sequencing, can directly measure the repertoire of immune cells in a tumor [168]. Although not yet available for HNSCC, these data are essential to characterize differences in the immune system in HNSCC subtypes. Therefore, future studies are needed to compare estimates of cellular compositions with immunotherapy response. Integrating this data with genomic alterations will lay a critical foundation for subsequent clinical investigation and biomarker discovery of immunotherapy-responsive HNSCC subtypes.

11.11 Discussion

Advances in genome-wide interrogation coupled with large-scale consortium efforts have begun to clarify the molecular landscapes of HPV-positive and HPV-negative head and neck squamous cell carcinoma. Observations of the clinical behavior and epidemiological features have long hinted at the possibility of two distinct HNSCC entities based on HPV status [3, 11, 16]. Comprehensive molecular profiling of large, multi-institutional datasets firmly establishes that HPV-positive and HPV-negative HNSCC are etiologically distinct neoplastic entities with divergent mutational, transcriptional, and epigenomic landscapes [15, 27, 29].

Fundamental differences based on HPV status shed light into the clinical behavior of HNSCC. While *TP53* mutations were once thought to be a hallmark of all HNSCC, it is now apparent that *TP53* mutation occurs primarily in HPV-negative tumors and rarely in HPV-positive disease. Further, mechanisms of HPV-driven carcinogenesis suggest a plausible explanation for the lack of requirement of *TP53* mutation in HPV-positive HNSCC: the viral oncoprotein E6 induces degradation of p53. Such findings provide insight into readily observed differences in the clinical behavior of HNSCC with respect to HPV status. For example, HPV-positive HNSCC may be more responsive to cytotoxic chemoradiotherapy because these tumors remain *TP53* wild type.

The recognition of the clinical impact of HPV status has initiated fundamental changes in diagnostic and therapeutic paradigms for HNSCC [13, 169]. For example, since the advent of platinum-based chemoradiotherapy, cetuximab has been identified as a targeted therapeutic [170, 171]. Genome-wide single-nucleotide resolution views of HNSCC offer perhaps the most promise toward better understanding the molecular pathophysiology of this disease. These insights hold the potential toward more accurate methods to assess prognosis and the development targeted treatments. However, the predominance of HNSCC tumors driven by silencing of tumor suppressor genes has limited further application of targeted therapies.

In addition to targeted therapeutics, genomic data have critical implications for emerging immunotherapies for HNSCC. Initial clinical trials demonstrated that 18% of HNSCC tumors are responsive to PD-1 inhibitors, independent of the HPV subtype [172]. Long-term response and optimal immunotherapy strategies in HNSCC subtypes are currently an active area of clinical research [173], reviewed extensively by Ferris [174] (and see Chap. 14). Genomic data have great potential to provide biomarkers for immunotherapy treatment selection. In colorectal cancers, total mutational burden is a predictive biomarker of immunotherapy response [175]. However, in HNSCC tumors, immunotherapy response is similar among HPV-positive and HPV-negative tumors suggesting that the total mutational burden will be a less potent biomarker. Rather, increased tumor heterogeneity is associated with a more diverse repertoire of immune cells, which is also observed in response to immunotherapy in non-small cell lung cancer [176]. Similarly, the mesenchymal subtype is also enriched for immune pathways in both HPV-positive and HPV-negative HNSCC [27]. Further genomic studies are critical for developing accurate biomarkers to enable patient selection for immunotherapy and combination therapeutic strategies for nonresponsive HNSCC tumors.

Large public domain datasets of whole-genome microarray and next-generation sequencing technologies have enabled unprecedented characterization of genomic landscape in HPV-positive and HPV-negative HNSCC. Advances to these technologies are rapidly evolving and will continue to improve the identification of biomarkers and therapeutic targets specific to individual tumors in all disease subtypes. Emerging technologies to query the chromatin structure and function, including ChIP-seq, ATAC-seq, and Hi-C, are essential to characterize the role that DNA structure plays in inducing the molecular landscape of HNSCC. These data will both refine the clinical attributes of new epigenetic subtypes of HNSCC and be critical to selecting tumors for emerging epigenetic therapies. These epigenetic landscapes will be enhanced through sequencing-based profiling of DNA methylation, which has higher resolution for genomic local and strand-specific methylation than microarrays used in TCGA.

Similarly, sequencing technologies with long reads will span entire genes, enabling far more accurate measurements of gene isoform usage specific to HNSCC tumors than bioinformatics analyses of short-read RNA-sequencing data. These gene isoforms may provide promising new therapeutic targets to currently untargetable tumors. Emerging single-cell DNA and RNA technologies will be essential to quantifying intra-tumor heterogeneity and subclones to better improve therapeutic response and predict mechanisms of acquired therapeutic resistance for combination therapeutic strategies at the time of diagnosis. Finally, emerging TCR-sequencing technologies will characterize the immune repertoire within tumors to serve as personalized biomarkers for immunotherapy. Integrating current high-throughput genomic data with data from these emerging sequencing-based technologies will provide unprecedented molecular characterization of cells that comprise individual tumors, ideally improving accuracy of biomarker development and precision medicine for future HNSCC patients.

Conflicts of Interest The authors have no relevant conflicts of interest to declare.

References

1. Siegel RL, Miller KD, Jemal A. Cancer statistics, 2016. CA Cancer J Clin. 2016;66(1):7–30.
2. Torre LA, et al. Global cancer statistics, 2012. CA Cancer J Clin. 2015;65(2):87–108.
3. Gillison ML, et al. Distinct risk factor profiles for human papillomavirus type 16-positive and human papillomavirus type 16-negative head and neck cancers. J Natl Cancer Inst. 2008;100(6):407–20.
4. Hennessey PT, Westra WH, Califano JA. Human papillomavirus and head and neck squamous cell carcinoma: recent evidence and clinical implications. J Dent Res. 2009;88(4):300–6.
5. Hayashi M, et al. Paired box 5 methylation detection by droplet digital PCR for ultra-sensitive deep surgical margins analysis of head and neck squamous cell carcinoma. Cancer Prev Res (Phila). 2015;8(11):1017–26.
6. Roh JL, et al. Tissue imprint for molecular mapping of deep surgical margins in patients with head and neck squamous cell carcinoma. Head Neck. 2012;34(11):1529–36.
7. Santhi WS, et al. Oncogenic microRNAs as biomarkers of oral tumorigenesis and minimal residual disease. Oral Oncol. 2013;49(6):567–75.
8. Pena Murillo C, et al. The utility of molecular diagnostics to predict recurrence of head and neck carcinoma. Br J Cancer. 2012;107(7):1138–43.
9. Yi HJ, et al. The role of molecular margins as prognostic factors in laryngeal carcinoma in Chinese patients. Acta Otolaryngol. 2012;132(8):874–8.
10. Faraji F, et al. Molecular mechanisms of human papillomavirus-related carcinogenesis in head and neck cancer. Microbes infect. 2017;19:464.
11. Fakhry C, et al. Improved survival of patients with human papillomavirus-positive head and neck squamous cell carcinoma in a prospective clinical trial. J Natl Cancer Inst. 2008;100(4):261–9.
12. Ang KK, et al. Human papillomavirus and survival of patients with oropharyngeal cancer. N Engl J Med. 2010;363(1):24–35.
13. Lydiatt WM, et al. Head and Neck cancers-major changes in the American Joint Committee on cancer eighth edition cancer staging manual. CA Cancer J Clin. 2017;67(2):122–37.
14. Lingen MW, et al. Low etiologic fraction for high-risk human papillomavirus in oral cavity squamous cell carcinomas. Oral Oncol. 2013;49(1):1–8.
15. Cancer Genome Atlas Network. Comprehensive genomic characterization of head and neck squamous cell carcinomas. Nature. 2015;517(7536):576–82.
16. Chaturvedi AK, et al. Incidence trends for human papillomavirus-related and -unrelated oral squamous cell carcinomas in the United States. J Clin Oncol. 2008;26(4):612–9.
17. Kang H, Kiess A, Chung CH. Emerging biomarkers in head and neck cancer in the era of genomics. Nat Rev Clin Oncol. 2015;12(1):11–26.
18. Califano J, et al. Genetic progression model for head and neck cancer: implications for field cancerization. Cancer Res. 1996;56(11):2488–92.
19. Beck TN, Golemis EA. Genomic insights into head and neck cancer. Cancer Head Neck. 2016;1(1):1.
20. Li H, et al. Genomic analysis of head and neck squamous cell carcinoma cell lines and human tumors: a rational approach to preclinical model selection. Mol Cancer Res. 2014;12(4):571–82.
21. Fertig EJ, et al. Preferential activation of the hedgehog pathway by epigenetic modulations in HPV negative HNSCC identified with meta-pathway analysis. PLoS One. 2013;8(11):e78127.
22. Gaykalova DA, et al. NF-kappaB and stat3 transcription factor signatures differentiate HPV-positive and HPV-negative head and neck squamous cell carcinoma. Int J Cancer. 2015;137(8):1879–89.
23. Rathi KS, et al. Correcting transcription factor gene sets for copy number and promoter methylation variations. Drug Dev Res. 2014;75(6):343–7.
24. Sun W, et al. Activation of the NOTCH pathway in head and neck cancer. Cancer Res. 2014;74(4):1091–104.

25. Stransky N, et al. The mutational landscape of head and neck squamous cell carcinoma. Science. 2011;333(6046):1157–60.
26. Agrawal N, et al. Exome sequencing of head and neck squamous cell carcinoma reveals inactivating mutations in NOTCH1. Science. 2011;333(6046):1154–7.
27. Keck MK, et al. Integrative analysis of head and neck cancer identifies two biologically distinct HPV and three non-HPV subtypes. Clin Cancer Res. 2015;21(4):870–81.
28. Guo T, et al. Characterization of functionally active gene fusions in human papillomavirus related oropharyngeal squamous cell carcinoma. Int J Cancer. 2016;139(2):373–82.
29. Seiwert TY, et al. Integrative and comparative genomic analysis of HPV-positive and HPV-negative head and neck squamous cell carcinomas. Clin Cancer Res. 2015;21(3):632–41.
30. Han J, et al. Identification of potential therapeutic targets in human head & neck squamous cell carcinoma. Head Neck Oncol. 2009;1:27.
31. Reddy RB, et al. Meta-analyses of microarray datasets identifies ANO1 and FADD as prognostic markers of head and neck cancer. PLoS One. 2016;11(1):e0147409.
32. De Cecco L, et al. Comprehensive gene expression meta-analysis of head and neck squamous cell carcinoma microarray data defines a robust survival predictor. Ann Oncol. 2014;25(8):1628–35.
33. Schmitz S, et al. Cetuximab promotes epithelial to mesenchymal transition and cancer associated fibroblasts in patients with head and neck cancer. Oncotarget. 2015;6(33):34288–99.
34. Bossi P, et al. Functional genomics uncover the biology behind the responsiveness of head and neck squamous cell cancer patients to cetuximab. Clin Cancer Res. 2016;22(15):3961–70.
35. Morris LG, et al. The molecular landscape of recurrent and metastatic head and neck cancers: insights from a precision oncology sequencing platform. JAMA Oncol. 2016;3:244–255.
36. Hedberg ML, et al. Genetic landscape of metastatic and recurrent head and neck squamous cell carcinoma. J Clin Invest. 2016;126(1):169–80.
37. Roepman P, et al. Maintenance of head and neck tumor gene expression profiles upon lymph node metastasis. Cancer Res. 2006;66(23):11110–4.
38. Colella S, et al. Molecular signatures of metastasis in head and neck cancer. Head Neck. 2008;30(10):1273–83.
39. Puram SV, et al. Single-cell transcriptomic analysis of primary and metastatic tumor ecosystems in head and neck cancer. In: Cell, vol. 171; 2017. p. 1611.
40. Cheng H, et al. Decreased SMAD4 expression is associated with induction of epithelial-to-mesenchymal transition and cetuximab resistance in head and neck squamous cell carcinoma. Cancer Biol Ther. 2015;16(8):1252–8.
41. Fertig EJ, et al. CoGAPS matrix factorization algorithm identifies transcriptional changes in AP-2alpha target genes in feedback from therapeutic inhibition of the EGFR network. Oncotarget. 2016;7(45):73845–64.
42. Barretina J, et al. The Cancer Cell Line Encyclopedia enables predictive modelling of anticancer drug sensitivity. Nature. 2012;483(7391):603–307.
43. Iorio F, et al. A landscape of pharmacogenomic interactions in Cancer. Cell. 2016;166(3):740–54.
44. Gaykalova DA, et al. Novel insight into mutational landscape of head and neck squamous cell carcinoma. PLoS One. 2014;9(3):e93102.
45. Xu CC, et al. HPV status and second primary tumours in oropharyngeal squamous cell carcinoma. J Otolaryngol Head Neck Surg. 2013;42:36.
46. Diaz DA, et al. Head and neck second primary cancer rates in the human papillomavirus era: a population-based analysis. Head Neck. 2016;38 Suppl 1:E873–83.
47. Henderson S, Fenton T. APOBEC3 genes: retroviral restriction factors to cancer drivers. Trends Mol Med. 2015;21(5):274–84.
48. Supek F, Lehner B. Clustered mutation signatures reveal that error-prone DNA repair targets mutations to active genes. Cell. 2017;170(3):534–547 e23.
49. Cancer Genome Atlas Research Network, et al. Integrated genomic and molecular characterization of cervical cancer. Nature. 2017;543(7645):378–84.

50. Hoadley KA, et al. Multiplatform analysis of 12 cancer types reveals molecular classification within and across tissues of origin. Cell. 2014;158(4):929–44.
51. Alexandrov LB, et al. Mutational signatures associated with tobacco smoking in human cancer. Science. 2016;354(6312):618–22.
52. Pleasance ED, et al. A small-cell lung cancer genome with complex signatures of tobacco exposure. Nature. 2010;463(7278):184–90.
53. The Cancer Genome Atlas Research Network. Integrated genomic and molecular characterization of cervical cancer. Nature. 2017;543(7645):378–84.
54. Hayes DN, Van Waes C, Seiwert TY. Genetic landscape of human papillomavirus-associated head and neck cancer and comparison to tobacco-related tumors. J Clin Oncol. 2015;33(29):3227–34.
55. Henderson S, et al. APOBEC-mediated cytosine deamination links PIK3CA helical domain mutations to human papillomavirus-driven tumor development. Cell Rep. 2014;7(6):1833–41.
56. Scheffner M, et al. The E6 oncoprotein encoded by human papillomavirus types 16 and 18 promotes the degradation of p53. Cell. 1990;63(6):1129–36.
57. Boyer SN, Wazer DE, Band V. E7 protein of human papilloma virus-16 induces degradation of retinoblastoma protein through the ubiquitin-proteasome pathway. Cancer Res. 1996;56(20):4620–4.
58. Sherman L, et al. Inhibition of serum- and calcium-induced differentiation of human keratinocytes by HPV16 E6 oncoprotein: role of p53 inactivation. Virology. 1997;237(2):296–306.
59. Guirimand T, Delmotte S, Navratil V. VirHostNet 2.0: surfing on the web of virus/host molecular interactions data. Nucleic Acids Res. 2015;43(Database issue):D583–7.
60. Roberts SA, et al. An APOBEC cytidine deaminase mutagenesis pattern is widespread in human cancers. Nat Genet. 2013;45(9):970–6.
61. Springer S, et al. A combination of molecular markers and clinical features improve the classification of pancreatic cysts. Gastroenterology. 2015;149(6):1501–10.
62. Vogelstein B, et al. Cancer genome landscapes. Science. 2013;339(6127):1546–58.
63. Razin A, Riggs AD. DNA methylation and gene function. Science. 1980;210(4470):604–10.
64. Sharma S, Kelly TK, Jones PA. Epigenetics in cancer. Carcinogenesis. 2010;31(1):27–36.
65. Esteller M. Cancer epigenomics: DNA methylomes and histone-modification maps. Nat Rev Genet. 2007;8(4):286–98.
66. Pikor L, et al. The detection and implication of genome instability in cancer. Cancer Metastasis Rev. 2013;32(3–4):341–52.
67. Sartor MA, et al. Genome-wide methylation and expression differences in HPV(+) and HPV(−) squamous cell carcinoma cell lines are consistent with divergent mechanisms of carcinogenesis. Epigenetics. 2011;6(6):777–87.
68. Poage GM, et al. Global hypomethylation identifies Loci targeted for hypermethylation in head and neck cancer. Clin Cancer Res. 2011;17(11):3579–89.
69. Smith IM, et al. DNA global hypomethylation in squamous cell head and neck cancer associated with smoking, alcohol consumption and stage. Int J Cancer. 2007;121(8):1724–8.
70. Hennessey PT, et al. Promoter methylation in head and neck squamous cell carcinoma cell lines is significantly different than methylation in primary tumors and xenografts. PLoS One. 2011;6(5):e20584.
71. Baylin SB, Jones PA. Epigenetic determinants of cancer. Cold Spring Harb Perspect Biol. 2016;8(9)
72. Kostareli E, et al. HPV-related methylation signature predicts survival in oropharyngeal squamous cell carcinomas. J Clin Invest. 2013;123(6):2488–501.
73. Ha PK, Califano JA. Promoter methylation and inactivation of tumour-suppressor genes in oral squamous-cell carcinoma. Lancet Oncol. 2006;7(1):77–82.
74. Gaykalova DA, et al. Outlier analysis defines zinc finger gene family DNA methylation in tumors and saliva of head and neck cancer patients. PLoS One. 2015;10(11):e0142148.
75. Worsham MJ, et al. Epigenetic modulation of signal transduction pathways in HPV-associated HNSCC. Otolaryngol Head Neck Surg. 2013;149(3):409–16.

76. Lleras RA, et al. Unique DNA methylation loci distinguish anatomic site and HPV status in head and neck squamous cell carcinoma. Clin Cancer Res. 2013;19(19):5444–55.
77. Colacino JA, et al. Comprehensive analysis of DNA methylation in head and neck squamous cell carcinoma indicates differences by survival and clinicopathologic characteristics. PLoS One. 2013;8(1):e54742.
78. Papillon-Cavanagh S, et al. Impaired H3K36 methylation defines a subset of head and neck squamous cell carcinomas. Nat Genet. 2017;49(2):180–5.
79. Brennan K, et al. Identification of an atypical etiological head and neck squamous carcinoma subtype featuring the CpG island methylator phenotype. EBioMedicine. 2017;17:223–36.
80. Stansfield JC, et al. Toward signaling-driven biomarkers immune to normal tissue contamination. Cancer Inform. 2016;15:15–21.
81. Gaykalova DA, et al. Integrative computational analysis of transcriptional and epigenetic alterations implicates DTX1 as a putative tumor suppressor gene in HNSCC. Oncotarget. 2017;8:15349.
82. Maher CA, et al. Transcriptome sequencing to detect gene fusions in cancer. Nature. 2009;458(7234):97–101.
83. Singh D, et al. Transforming fusions of FGFR and TACC genes in human glioblastoma. Science. 2012;337(6099):1231–5.
84. Wang R, et al. FGFR1/3 tyrosine kinase fusions define a unique molecular subtype of non-small cell lung cancer. Clin Cancer Res. 2014;20(15):4107–14.
85. Acquaviva J, et al. FGFR3 translocations in bladder cancer: differential sensitivity to HSP90 inhibition based on drug metabolism. Mol Cancer Res. 2014;12(7):1042–54.
86. Wang K, et al. MapSplice: accurate mapping of RNA-seq reads for splice junction discovery. Nucleic Acids Res. 2010;38(18):e178.
87. Stransky N, et al. The landscape of kinase fusions in cancer. Nat Commun. 2014;5:4846.
88. Wu YM, et al. Identification of targetable FGFR gene fusions in diverse cancers. Cancer Discov. 2013;3(6):636–47.
89. Daly C, et al. FGFR3-TACC3 fusion proteins act as naturally occurring drivers of tumor resistance by functionally substituting for EGFR/ERK signaling. Oncogene. 2017;36(4):471–81.
90. Majewski IJ, et al. Identification of recurrent FGFR3 fusion genes in lung cancer through kinome-centred RNA sequencing. J Pathol. 2013;230(3):270–6.
91. Di Stefano AL, et al. Detection, characterization, and inhibition of FGFR-TACC fusions in IDH wild-type glioma. Clin Cancer Res. 2015;21(14):3307–17.
92. Helsten T, et al. The FGFR landscape in cancer: analysis of 4,853 tumors by next-generation sequencing. Clin Cancer Res. 2016;22(1):259–67.
93. Costa R, et al. FGFR3-TACC3 fusion in solid tumors: mini review. Oncotarget. 2016;7(34):55924–38.
94. Yuan L, et al. Recurrent FGFR3-TACC3 fusion gene in nasopharyngeal carcinoma. Cancer Biol Ther. 2014;15(12):1613–21.
95. Nelson KN, et al. Oncogenic gene fusion FGFR3-TACC3 is regulated by tyrosine phosphorylation. Mol Cancer Res. 2016;14(5):458–69.
96. Cheng Y, et al. A novel read-through transcript JMJD7-PLA2G4B regulates head and neck squamous cell carcinoma cell proliferation and survival. Oncotarget. 2017;8(2):1972–82.
97. Rocco JW, et al. p63 mediates survival in squamous cell carcinoma by suppression of p73-dependent apoptosis. Cancer Cell. 2006;9(1):45–56.
98. Liggett WH Jr, et al. p16 and p16 beta are potent growth suppressors of head and neck squamous carcinoma cells in vitro. Cancer Res. 1996;56(18):4119–23.
99. Li R, et al. Expression microarray analysis reveals alternative splicing of LAMA3 and DST genes in head and neck squamous cell carcinoma. PLoS One. 2014;9(3):e91263.
100. Afsari B, et al. Splice Expression Variation Analysis (SEVA) for differential gene isoform usage in cancer. bioRxiv. 2016. https://doi.org/10.1101/091637.
101. Song L, Sabunciyan S, Florea L. CLASS2: accurate and efficient splice variant annotation from RNA-seq reads. Nucleic Acids Res. 2016;44(10):e98.

102. Reis EM, et al. Large-scale transcriptome analyses reveal new genetic marker candidates of head, neck, and thyroid cancer. Cancer Res. 2005;65(5):1693–9.
103. Chen P, et al. Comprehensive exon array data processing method for quantitative analysis of alternative spliced variants. Nucleic Acids Res. 2011;39(18):e123.
104. Muller M, et al. One, two, three – p53, p63, p73 and chemosensitivity. Drug Resist Updat. 2006;9(6):288–306.
105. Mao L, et al. Frequent abnormalities of FHIT, a candidate tumor suppressor gene, in head and neck cancer cell lines. Cancer Res. 1996;56(22):5128–31.
106. Frost GI, et al. HYAL1LUCA-1, a candidate tumor suppressor gene on chromosome 3p21.3, is inactivated in head and neck squamous cell carcinomas by aberrant splicing of pre-mRNA. Oncogene. 2000;19(7):870–7.
107. Cengiz B, et al. Tumor-specific mutation and downregulation of ING5 detected in oral squamous cell carcinoma. Int J Cancer. 2010;127(9):2088–94.
108. Richter TM, Tong BD, Scholnick SB. Epigenetic inactivation and aberrant transcription of CSMD1 in squamous cell carcinoma cell lines. Cancer Cell Int. 2005;5:29.
109. Assimakopoulos D, et al. The role of CD44 in the development and prognosis of head and neck squamous cell carcinomas. Histol Histopathol. 2002;17(4):1269–81.
110. Guo T, et al. A novel functional splice variant of AKT3 defined by analysis of alternative splice expression in HPV-positive oropharyngeal cancers. Cancer Res. 2017;77(19):5248–58.
111. Moller-Levet CS, et al. Exon array analysis of head and neck cancers identifies a hypoxia related splice variant of LAMA3 associated with a poor prognosis. PLoS Comput Biol. 2009;5(11):e1000571.
112. Sebestyen E, Zawisza M, Eyras E. Detection of recurrent alternative splicing switches in tumor samples reveals novel signatures of cancer. Nucleic Acids Res. 2015;43(3):1345–56.
113. Radhakrishnan A, et al. Dysregulation of splicing proteins in head and neck squamous cell carcinoma. Cancer Biol Ther. 2016;17(2):219–29.
114. Ishii H, et al. Epithelial splicing regulatory proteins 1 (ESRP1) and 2 (ESRP2) suppress cancer cell motility via different mechanisms. J Biol Chem. 2014;289(40):27386–99.
115. Peiqi L, et al. Expression of SRSF3 is correlated with carcinogenesis and progression of oral squamous cell carcinoma. Int J Med Sci. 2016;13(7):533–9.
116. House RP, et al. RNA-binding protein CELF1 promotes tumor growth and alters gene expression in oral squamous cell carcinoma. Oncotarget. 2015;6(41):43620–34.
117. Veneziano D, et al. Noncoding RNA: current deep sequencing data analysis approaches and challenges. Hum Mutat. 2016;37(12):1283–98.
118. Guil S, Esteller M. RNA-RNA interactions in gene regulation: the coding and noncoding players. Trends Biochem Sci. 2015;40(5):248–56.
119. Ma X, et al. LncRNAs as an intermediate in HPV16 promoting myeloid-derived suppressor cell recruitment of head and neck squamous cell carcinoma. Oncotarget. 2017;8(26):42061–75.
120. Cao W, et al. A three-lncRNA signature derived from the Atlas of ncRNA in cancer (TANRIC) database predicts the survival of patients with head and neck squamous cell carcinoma. Oral Oncol. 2017;65:94–101.
121. Zou AE, et al. The non-coding landscape of head and neck squamous cell carcinoma. Oncotarget. 2016;7(32):51211–22.
122. Garzon R, et al. MicroRNA expression and function in cancer. Trends Mol Med. 2006;12(12):580–7.
123. Guo Z, et al. Genome-wide survey of tissue-specific microRNA and transcription factor regulatory networks in 12 tissues. Sci Rep. 2014;4:5150.
124. Chang SS, et al. MicroRNA alterations in head and neck squamous cell carcinoma. Int J Cancer. 2008;123(12):2791–7.
125. Hedback N, et al. MiR-21 expression in the tumor stroma of oral squamous cell carcinoma: an independent biomarker of disease free survival. PLoS One. 2014;9(4):e95193.

126. Harris T, et al. Low-level expression of miR-375 correlates with poor outcome and metastasis while altering the invasive properties of head and neck squamous cell carcinomas. Am J Pathol. 2012;180(3):917–28.
127. Avissar M, et al. MicroRNA expression ratio is predictive of head and neck squamous cell carcinoma. Clin Cancer Res. 2009;15(8):2850–5.
128. Kita Y, et al. Epigenetically regulated microRNAs and their prospect in cancer diagnosis. Expert Rev Mol Diagn. 2014;14(6):673–83.
129. Yu L, et al. miR-26a inhibits invasion and metastasis of nasopharyngeal cancer by targeting EZH2. Oncol Lett. 2013;5(4):1223–8.
130. Kalinowski FC, et al. Regulation of epidermal growth factor receptor signaling and erlotinib sensitivity in head and neck cancer cells by miR-7. PLoS One. 2012;7(10):e47067.
131. Lu ZM, et al. Micro-ribonucleic acid expression profiling and bioinformatic target gene analyses in laryngeal carcinoma. Onco Targets Ther. 2014;7:525–33.
132. Howard J, et al. miRNA array analysis determines miR-205 is overexpressed in head and neck squamous cell carcinoma and enhances cellular proliferation. J Cancer Res Ther. 2013;1:153–62.
133. Miller DL, et al. Identification of a human papillomavirus-associated oncogenic miRNA panel in human oropharyngeal squamous cell carcinoma validated by bioinformatics analysis of the Cancer Genome Atlas. Am J Pathol. 2015;185(3):679–92.
134. Lajer CB, et al. The role of miRNAs in human papilloma virus (HPV)-associated cancers: bridging between HPV-related head and neck cancer and cervical cancer. Br J Cancer. 2012;106(9):1526–34.
135. Rinn JL, Chang HY. Genome regulation by long noncoding RNAs. Annu Rev Biochem. 2012;81:145–66.
136. Jia LF, et al. Expression, regulation and roles of miR-26a and MEG3 in tongue squamous cell carcinoma. Int J Cancer. 2014;135(10):2282–93.
137. Fang Z, et al. Increased expression of the long non-coding RNA UCA1 in tongue squamous cell carcinomas: a possible correlation with cancer metastasis. Oral Surg Oral Med Oral Pathol Oral Radiol. 2014;117(1):89–95.
138. Matsui M, Corey DR. Non-coding RNAs as drug targets. Nat Rev Drug Discov. 2017;16(3):167–79.
139. Chung CH, et al. Molecular classification of head and neck squamous cell carcinomas using patterns of gene expression. Cancer Cell. 2004;5(5):489–500.
140. Walter V, et al. Molecular subtypes in head and neck cancer exhibit distinct patterns of chromosomal gain and loss of canonical cancer genes. PLoS One. 2013;8(2):e56823.
141. Zhang Y, et al. Subtypes of HPV-positive head and neck cancers are associated with HPV characteristics, copy number alterations, PIK3CA mutation, and pathway signatures. Clin Cancer Res. 2016;22(18):4735–45.
142. Kelley DZ, et al. Integrated analysis of whole-genome ChIP-Seq and RNA-Seq data of primary head and neck tumor samples associates HPV integration sites with open chromatin marks. Cancer Res. 2017;77:6538.
143. Gevaert O, Tibshirani R, Plevritis SK. Pancancer analysis of DNA methylation-driven genes using MethylMix. Genome Biol. 2015;16(1):17.
144. Parfenov M, et al. Characterization of HPV and host genome interactions in primary head and neck cancers. Proc Natl Acad Sci U S A. 2014;111(43):15544–9.
145. Guo T, et al. A novel functional splice variant of AKT3 defined by analysis of alternative splice expression in HPV-positive oropharyngeal cancers. Cancer Res. 2017;77(19):5248–58.
146. Olthof NC, et al. Comprehensive analysis of HPV16 integration in OSCC reveals no significant impact of physical status on viral oncogene and virally disrupted human gene expression. PLoS One. 2014;9(2):e88718.
147. Hajek M, et al. TRAF3/CYLD mutations identify a distinct subset of human papillomavirus-associated head and neck squamous cell carcinoma. Cancer. 2017;123(10):1778–90.

148. Nulton TJ, et al. Analysis of The Cancer Genome Atlas sequencing data reveals novel properties of the human papillomavirus 16 genome in head and neck squamous cell carcinoma. Oncotarget. 2017;8(11):17684–99.
149. Hajek M, et al. TRAF3/CYLD mutations identify a distinct subset of human papilloma virus-associated head and neck squamous cell carcinoma. Cancer. 2017;123:1778.
150. Koneva LA, et al. HPV integration in HNSCC correlates with survival outcomes, immune response signatures, and candidate drivers. Mol Cancer Res. 2017;
151. Nowell PC. The clonal evolution of tumor cell populations. Science. 1976;194(4260):23–8.
152. Marusyk A, Polyak K. Tumor heterogeneity: causes and consequences. Biochim Biophys Acta. 2010;1805(1):105–17.
153. Marusyk A, Almendro V, Polyak K. Intra-tumour heterogeneity: a looking glass for cancer? Nat Rev Cancer. 2012;12(5):323–34.
154. Mroz EA, et al. Intra-tumor genetic heterogeneity and mortality in head and neck cancer: analysis of data from the Cancer Genome Atlas. PLoS Med. 2015;12(2):e1001786.
155. Greaves M, Maley CC. Clonal evolution in cancer. Nature. 2012;481(7381):306–13.
156. Sok JC, et al. Mutant epidermal growth factor receptor (EGFRvIII) contributes to head and neck cancer growth and resistance to EGFR targeting. Clin Cancer Res. 2006;12(17):5064–73.
157. Rocco JW. Mutant Allele Tumor Heterogeneity (MATH) and head and neck squamous cell carcinoma. Head Neck Pathol. 2015;9(1):1–5.
158. Afsari B, Geman D, Fertig EJ. Learning dysregulated pathways in cancers from differential variability analysis. Cancer Inform. 2014;13(Suppl 5):61–7.
159. Eddy JA, et al. Identifying tightly regulated and variably expressed networks by Differential Rank Conservation (DIRAC). PLoS Comput Biol. 2010;6(5):e1000792.
160. Subramanian A, et al. Gene set enrichment analysis: a knowledge-based approach for interpreting genome-wide expression profiles. Proc Natl Acad Sci U S A. 2005;102(43):15545–50.
161. Mroz EA, Rocco JW. MATH, a novel measure of intratumor genetic heterogeneity, is high in poor-outcome classes of head and neck squamous cell carcinoma. Oral Oncol. 2013;49(3):211–5.
162. Xu Y, et al. MAD bayes for tumor heterogeneity – feature allocation with exponential family sampling. J Am Stat Assoc. 2015;110(510):503–14.
163. Roth A, et al. PyClone: statistical inference of clonal population structure in cancer. Nat Methods. 2014;11(4):396–8.
164. Hackl H, et al. Computational genomics tools for dissecting tumour-immune cell interactions. Nat Rev Genet. 2016;17(8):441–58.
165. Mandal R, et al. The head and neck cancer immune landscape and its immunotherapeutic implications. JCI Insight. 2016;1(17):e89829.
166. Badoual C, et al. PD-1–expressing tumor-infiltrating T cells are a favorable prognostic biomarker in HPV-associated head and neck Cancer. Cancer Res. 2013;73(1):128.
167. Ferris RL, et al. Nivolumab for recurrent squamous-cell carcinoma of the head and neck. N Engl J Med. 2016;375(19):1856–67.
168. Calis JJ, Rosenberg BR. Characterizing immune repertoires by high throughput sequencing: strategies and applications. Trends Immunol. 2014;35(12):581–90.
169. Kelly JR, Husain ZA, Burtness B. Treatment de-intensification strategies for head and neck cancer. Eur J Cancer. 2016;68:125–33.
170. Adelstein DJ, et al. An intergroup phase III comparison of standard radiation therapy and two schedules of concurrent chemoradiotherapy in patients with unresectable squamous cell head and neck cancer. J Clin Oncol. 2003;21(1):92–8.
171. Bonner JA, et al. Radiotherapy plus cetuximab for squamous-cell carcinoma of the head and neck. N Engl J Med. 2006;354(6):567–78.
172. Seiwert TY, et al. Safety and clinical activity of pembrolizumab for treatment of recurrent or metastatic squamous cell carcinoma of the head and neck (KEYNOTE-012): an open-label, multicentre, phase 1b trial. Lancet Oncol. 2016;17(7):956–65.

173. Mehra R, et al. Efficacy and safety of pembrolizumab in recurrent/metastatic head and neck squamous cell carcinoma (R/M HNSCC): pooled analyses after long-term follow-up in KEYNOTE-012. In ASCO Annual Meeting. J Clin Oncol. 2016;34
174. Ferris RL. Immunology and immunotherapy of head and neck cancer. J Clin Oncol. 2015;33(29):3293–304.
175. Le DT, et al. PD-1 blockade in tumors with mismatch-repair deficiency. N Engl J Med. 2015;372(26):2509–20.
176. McGranahan N, et al. Clonal neoantigens elicit T cell immunoreactivity and sensitivity to immune checkpoint blockade. Science. 2016;351(6280):1463–9.

Chapter 12
Epigenetic Changes and Epigenetic Targets in Head and Neck Cancer

Suraj Peri, Andrew J. Andrews, Aarti Bhatia, and Ranee Mehra

Abstract Epigenetic changes are both inheritable and reversible, affecting the spatial conformation of DNA and its transcriptional activity. The most common classes of epigenetic regulation include modification of DNA (typically by methylation), or modification of the histones that form nucleosomes (typically by methylation, acetylation, or phosphorylation). Epigenetic changes can influence gene expression patterns without making permanent changes in DNA. In this article, we discuss characteristic changes in the epigenetic modification of tumor DNA that occurs in squamous cell carcinomas of the head and neck (SCCHN), which controls the selective induction and repression of genes relevant to the disease pathology. We also describe key proteins that mediate epigenetic control of gene expression, and emerging therapeutic approaches to target epigenetic control systems.

Keywords Epigenetics · DNA methylation · Histone methylation · Histone acetylation · TCGA · NSD1 · NSD2 · H3K36 · H3K27 · Tobacco · Smoking

S. Peri (✉)
Biostatistics and Bioinformatics Division, Fox Chase Cancer Center, Philadelphia, PA, USA
e-mail: Suraj.Peri@fccc.edu

A. J. Andrews
Program in Cancer Biology, Fox Chase Cancer Center, Philadelphia, PA, USA

A. Bhatia
Department of Internal Medicine, Yale School of Medicine and Yale Cancer Center, New Haven, CT, USA

R. Mehra
Department of Oncology, Johns Hopkins University School of Medicine, Baltimore, MD, USA

B. Burtness, E. A. Golemis (eds.), *Molecular Determinants of Head and Neck Cancer*, Current Cancer Research, https://doi.org/10.1007/978-3-319-78762-6_12

12.1 Introduction

There are two main subclasses of squamous cell carcinomas of the head and neck (SCCHN), based on association with human papillomavirus (HPV-positive), versus tumors arising in the absence of human papillomavirus (HPV-negative). HPV-positive SCCHN commonly arises in the oropharynx, while HPV-negative SCCHN arises in the oral cavity, oropharynx, hypopharynx, and larynx. In the past several years, genomic analyses have illustrated the genetic heterogeneity of squamous cell head and neck cancers (SCCHNs) [1–3]. There is a higher burden of mutations in HPV-negative SCCHN, because these tumors typically arise in the context of heavy use of the mutagens tobacco and alcohol. Genomic analyses have shown a diverse set of mutations, particularly in HPV-negative tumors, with the greatest commonality the mutation of *TP53* [4–6]. The overall diversity of mutations and paucity of driver mutations makes it often difficult to identify a therapeutically targetable driver mutation. Other chapters in this volume discuss some candidate driver mutations, such as in PI3K. In contrast, analyses of transcriptional profiles of SCCHN tumors indicate a greater degree of commonality among tumors, identifying a small number of subclasses. These transcriptional patterns reflect characteristic changes in the epigenetic modification of tumor DNA that occur in SCCHN and control the selective induction and repression of genes relevant to the disease pathology [7–9]. Some of the modifications, and the enzymes responsible for their introduction, represent promising candidates for therapeutic development. In this chapter, we will review potential epigenetic modifications of interest and their relevance in SCCHN. We will focus on two classes of modification that influence both transcription and response to DNA-damaging therapies: modifications targeting DNA and modifications targeting histones.

12.2 DNA Methylation Controls

Epigenetic changes are inheritable and reversible, affecting the spatial conformation of DNA and its transcriptional activity, thus ensuring the maintenance of stability and integrity of the DNA. This process leads to changes in the phenotype without changing the sequence of DNA bases, allowing the generation of diverse forms of gene expression and resulting in different phenotypes. Epigenetic changes occur during development of the organism and are reproduced during DNA replication. Epigenetic deregulation is a hallmark of several diseases including cancer, where epigenetic silencing of tumor suppressors through hypermethylation and hypomethylation of proto-oncogenes is a contributing factor in tumorigenesis[10–13]. Furthermore, the posttranslational modification of histones often accompanies DNA methylation changes, leading to changes in chromatin state that help achieve and sustain exact regulation of gene expression [14].

Earlier studies addressing methylation of small numbers of genes have been replaced by more global studies, which have relied on the development of high-throughput methods to profile DNA methylation. These technologies, including high-density microarrays and sequencing technologies, experimental approaches based on enzyme digestion, affinity enrichment, and bisulfite methods that allow comprehensive mapping of methylation profiles with single-base resolution and the establishment of general "rules" for methylation control of gene expression [15, 16]. We note that the applicability of these methods was enhanced by the simultaneous development of statistical methods for normalization, correction of batch effects, and identification of differentially methylated regions (DMR), which allow users to analyze and interpret both microarray and sequencing data [17]. Similarly, software tools that allow visual inspection and interpretation of methylation data have contributed toward comprehensive understanding of methylation profiles in the genome [18–21].

DNA methylation is the most widely studied epigenetic mechanism in mammals. The process of methylation involves covalent transfer of a methyl moiety from S-adenosylmethionine to the 5th carbon position of cytosine (5-mC) within the CG dinucleotides that are commonly found dispersed along the genome, commonly in arrays called CpG islands (CGI) (Fig. 12.1a) [22]. These short interspersed groups of nucleotides are enriched in gene-rich regions and play a role in the regulation of gene transcription. A family of enzymes known as DNA methyltransferases (DNMTs) catalyze the methylation of DNA. Among these, DNMT1 recognizes hemimethylated sites and thus predominantly maintains and ensures transmission of epigenetic marks during DNA replication. DNMT3a and DNMT3b are de novo methyltransferases responsible for establishing DNA methylation patterns during early development [23, 24] (Fig. 12.1b). DNMT3L is a catalytically inactive methyltransferase lacking methyltransferase motifs; however, it is required for activity of DNMT3a and 3b in various contexts [25, 26].

An unperturbed epigenetic maintenance and regulation of gene expression requires systematic balance between methylation and demethylation of DNA during various stages of development. Demethylation can happen passively, for example, from failure of maintenance by DNMT1 after DNA replication, or actively, through replication-independent enzymatic processes. Active DNA demethylation is a multistep process where enzymatic conversion of 5-mC to 5-hydroxymethylcytosine (5-hmC) is catalyzed by the ten-eleven translocation (TET) dioxygenases. Subsequently, 5-hmC is converted to 5-formylcytosine (5-fC) and 5-carboxylcytosine (5-caC). These are efficiently removed by the thymine DNA glycosylase (TDG) leaving an abasic site that is subsequently repaired through the base excision repair (BER) pathway [27, 28] (Fig. 12.1a).

Methylation at CpG islands (CGIs) is often associated with inhibition of gene expression. Several distinct mechanisms contribute to this inhibition. Gene expression can be modulated through an indirect mechanism, discussed further below, based on chromatin modification. However, direct inhibition of transcription can be achieved through recruitment of proteins that bind methylated DNA (such as methyl-CpG-binding protein 2, MeCP2) to gene promoters, which then compete for binding with transcription factors [29]. Methylation of the insulin 2 (*INS2)* promoter leads to the binding of MeCP2 and possibly other methylated DNA binding proteins, resulting in decreased binding of the transcriptional activators ATF2 and CREB, which further suggests a mechanism for silencing of the insulin gene [29]. Methylated DNA binding proteins such as MeCP2 contain transcriptional repression domains (TRDs) that recruit histone deacetylase complexes, which help reinforce a suppressed chromatin configuration at the transcription initiation sites, ensuring gene quiescence [30] (Fig. 12.1c).

The relationship between DNA methylation in CGIs and target gene expression is complex. Genome-wide analyses have also shown extensive methylation in both promoter and non-promoter regions, including in gene body and intergenic regions [31, 32]. While the promoter CGI methylation typically is associated with repressed

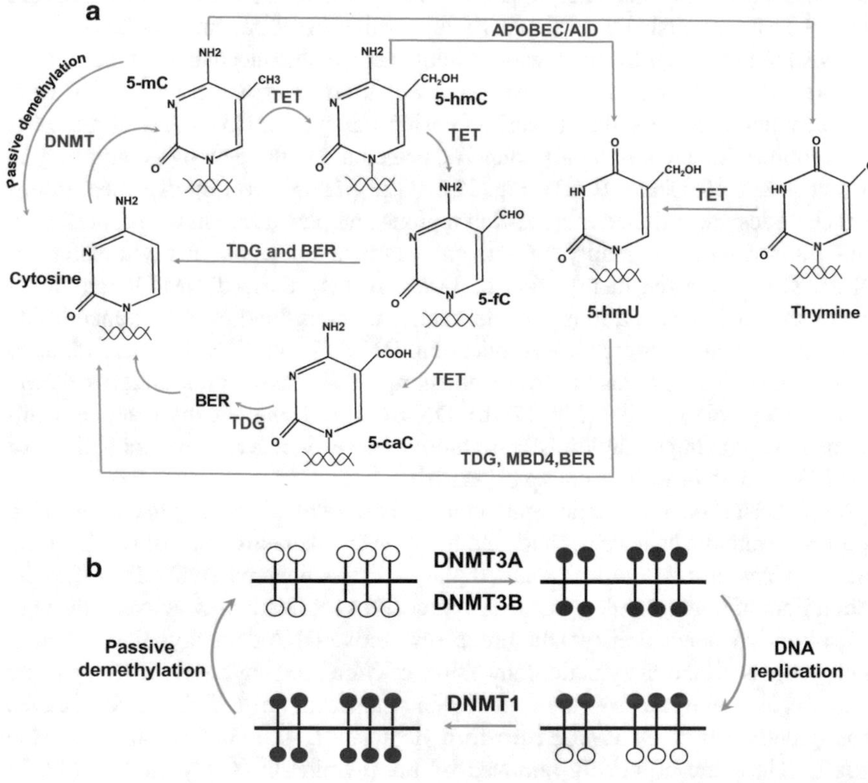

Fig. 12.1 DNA methylation mechanisms. (**a**) Various states of cytosine modifications resulting from methylation by DNMT and demethylation by TET and TDG demethylases. Activation-induced deaminase (AID) and AID-Apolipoprotein B RNA-editing catalytic component (APOBEC) family of enzymes modify 5-mC and 5-hmC to thymine and 5-hmU respectively. 5-hmU can be modified to cytosine mediated by TDG, MBD4 and base excision repair (BER) pathway (**b**) Models showing DNA methylation catalyzed by de novo methyltransferases DNMT3A/DNMT3B and maintenance methyltransferase DNMT1. De novo methyltransferases methylate CpG dinucleotides; in contrast, the maintenance methyltransferase DNMT1 methylates hemimethylated CpG sites on newly synthesized DNA after replication. The absence of appropriate methylation maintenance can lead to passive demethylation associated with loss of 5-mC over multiple replication cycles. Red and white lollipops indicate methylated and unmethylated CpGs, respectively (**c**) Model showing promoter methylation resulting in restrictive transcription and expression of the gene. Methyl-CpG-binding domain (MBD) proteins compete with transcription factors (TF) and RNA polymerase II (RNA pol II) that are necessary for active transcription. E1–3 indicate exons (**d**) Model showing gene body methylation. Methylation of CpGs in the gene body controls spurious transcription initiation by restricting binding by TFs and RNA pol II

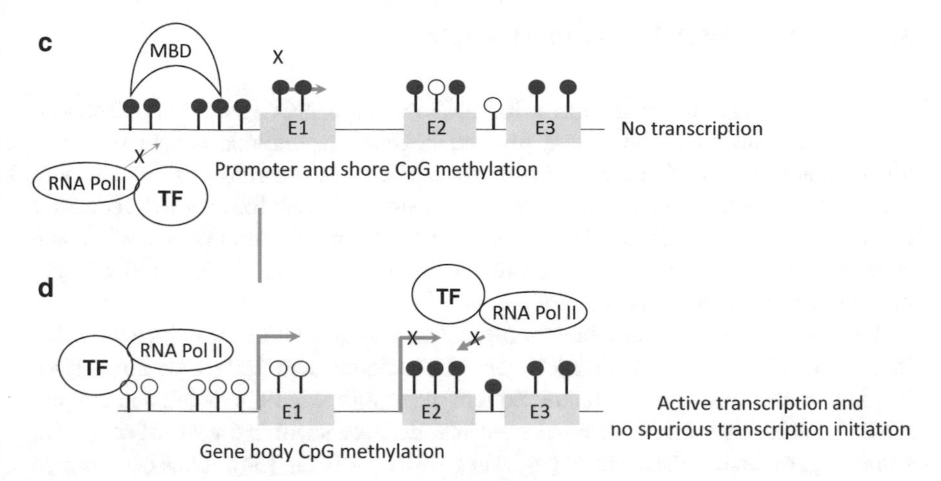

Fig. 12.1 (continued)

target gene expression, methylation of CGIs in the gene body is positively correlated with gene expression [33, 34]. The relationship between gene body methylation and gene expression is poorly understood. Several lines of evidence also show the role of gene body methylation in regulating transcription from alternative promoters [32]. It has also been shown to affect alternative splicing; for example, DNA methylation can prevent binding of the DNA binding protein CTCF to exon 5 of CD45, regulating exon inclusion [35–38].

The contrasting effect of methylation in promoter CGI versus in the gene body on gene expression has been described as a DNA methylation paradox [36, 39]. Some recent evidence suggests that intragenic DNA methylation provides protection from spurious transcription initiation through unauthentic RNA polymerase activity and cryptic intragenic transcription initiation [40]. In contrast to gene repression by promoter DNA methylation, gene body methylation correlates with transcriptional strength, providing an indirect indication that it is associated with transcriptional elongation. Hypermethylation of the gene body is mediated by DNMT3b, which recognizes H3K36me3, a histone mark for RNA pol II transcriptional elongation [41], catalyzed by the SETD2 histone methyltransferase [40] (Fig. 12.1c, d).

In addition to promoter CGIs and gene body regions, DNA methylation can also occur in low-density CGI “shores” – that is, sequences in flanking regions of CGIs (~2 kb). The methylation patterns in these regions are highly conserved and primarily reflect tissue specific modifications, with majority (70%) of these are associated with reprogramming in various progenitor cells [42–44]. In normal, untransformed cells, nearly 60% of the gene promoters associated with CpG islands exist in an unmethylated state: of these, nearly 6% of these are methylated in a tissue-specific manner during early development or tissue differentiation [37, 45]. Besides gene silencing on a localized level, DNA methylation exerts its function in development through genomic imprinting of larger regions of chromosomes. For example, as part of the dosage control of the XX chromosomes in females, hypermethylation of one X chromosome leads to monoallelic expression [46].

12.3 DNA Methylation in SCCHN

Tobacco usage is a major risk factor for development of SCCHN [47–49]. Tobacco products contain carcinogens and pro-carcinogens that include benzo(A)pyrene, nitrosamines, and arylamines. In tumorigenesis, these carcinogens induce both genetic and epigenetic changes and induce considerable risk for over a decade after cessation of smoking. These changes precede the formation of tumors, are in some cases associated with chronic obstructive pulmonary disease (COPD), [50–53] and may serve to promote tumorigenesis.

For example, some studies have shown tobacco smoke alters the epigenetic modification of CpG sites of over 1400 genes (associated with 2623 CpG sites) [54]. While risk of many diseases related to tobacco smoke can be reduced and comparable to risk levels for these diseases seen in nonsmokers within 5 years of cessation, some epigenetic modifications at CpG sites are long-lasting [54]. It has been shown that CpG sites at island shores, gene bodies, DNAse1 hypersensitive sites, and enhancers are more susceptible to methylation changes upon environmental exposures to tobacco and other agents than are CpG islands at promoters, which are more stable in the context of environmental perturbations [54–56]. A study that evaluated the differential methylation status of CpG sites among current and former smokers using buccal mucosa samples indicated significant differences in CpG sites of genes CYP1B1, involved in the pathway of detoxification for xenobiotic agents, and PARVA, an adhesion regulator [57]. In addition to the tobacco usage and alcohol consumption that are the most common risk factor for SCCHN, other factors such as chewing betel quid (prevalent in Southeast Asian countries) [58, 59] and EBV infection, known to induce epigenetic modifications with oncogenic consequences [60], are associated with nasopharyngeal cancers and can impact methylation [61].

In tumorigenesis, the cancer epigenome is characterized by increased DNA methylation at promoter and regulatory sites [37, 62], while decreased global methylation is a hallmark event. Increased promoter methylation leads to repressive chromatin state and impaired expression of tumor suppressor genes involved in cell cycle regulation, DNA repair, apoptosis, differentiation, drug resistance, angiogenesis, and metastasis. Promoter methylation is prevalent in many cancers and can serve as a robust biomarker for early detection, particularly aberrant methylation of $p16^{INK4a}$ in early lung neoplastic lesions [63]. Methylation changes have been identified in the cancer-associated genes *HOXA9*, *EDNRB*, and *DCC* from the saliva of individuals from high-risk populations, suggesting methylation changes accompany early stage of tumorigenesis of SCCHN [64].

Epigenetic alterations influence tumorigenesis through both direct and indirect interactions with other genetic factors, including mutations and copy number changes. Loss of heterozygosity, supporting Knudson's two-hit theory for cancer initiation, is an example of a direct interaction. For example, one of the two alleles in *CDKN2A*/p16 (in colon cancer) and *BRCA1* (in breast cancer) is inactivated by promoter hypermethylation, accompanied by loss of function of the protein encoded by the second allele through somatic or inherited mutation; this is a significant contributing factor to tumorigenesis [65–69]. As an example of indirect interaction, the epigenetic silencing of the DNA repair proteins *MLH1* and *MGMT* results in an increased transitional mutation rate in *KRAS* oncogene and the *TP53* tumor suppressor gene [65, 70–72]. In SCCHN, promoter methylation in the tumor suppressor genes *CDKN2A/p16* and *p14*^{ARF} is frequent [73, 74] and associated with loss of expression of these proteins and the aggressive nature of the disease [73]. The *CDKN2A* encoded p16, a key tumor suppressor protein, critical for HPV-negative SCCHN, binds and inhibits CDK4/CDK6 kinases. This inhibitory interaction blocks cell cycle progression from G1 to S phase by abrogating CDK-mediated phosphorylation of RB protein, blocking oncogenic transcription by the E2F transcription factor. Subsequent studies have found methylation changes in many genes involved in cell cycle regulation (*CDKN2A/p16ink4a, p15ink4b, p14ARF*), DNA repair (*MLH1, MGMT*), cell adhesion (*E-cadherin, H-cadherin, APC*), apoptosis (*DAPK, TMS1,* and *CASP8*), and angiogenesis (*THBS-1* and *p73*) and *RARB2* [75]. Promoter methylation of CHFR, encoding a mitotic checkpoint regulator, is associated with late stage disease in SCCHN [64, 75, 76].

Several studies of SCCHN using high-density arrays have shown extensive hypomethylation of DNA and hypermethylation at the CpG sites at the regulatory regions of genes that may downregulate the transcription of affected genes associated with risk factors including tobacco usage and alcohol consumption [8, 39]. Interestingly, the frequencies of promoter methylation among these genes have been reported to differ significantly among SCCHN patients in different regions of the world [38, 39]. These differences in part may reflect varying risk factors across different countries, although differences in experimental methods and sample size used in the experimental evaluation may be contributing factors to observed differences. However, studies from multiple series indicate E-cadherin (*CDH1*), *CDKN2A*/p16, and MGMT are most frequently found to be deregulated epigenetically (51%, 30%, and 30%, respectively) [75, 77].

Apart from epigenetic changes observed in tumor tissues, the prognostic potential of epigenetic phenotype has been reported. Studies on saliva samples from SCCHN patients versus from healthy nonsmoking individuals show that CpG methylation in promoters of *RASSF1A, DAPK1*, and *p16* genes can be identified with very high accuracy in the SCCHN patients, suggesting diagnostic potential [78]. Similarly, promoter methyation in genes such as the metalloprotease *TIMP3* and cyclin *CCNA1* serve as prediction markers associated with development of recurrence and secondary primary tumors, a common feature of SCCHN [79, 80]. Such markers may assist in developing patient-specific treatment regimens.

12.4 DNA Methylation Differences in HPV-Positive Versus HPV-Negative Disease

Methylation patterns in SCCHN are associated with HPV status [81–87]. Several studies have shown differences in methylation signatures between HPV-positive and HPV-negative tumors in SCCHN. Unsupervised clustering analyses revealed differences in CpG methylation patterns between HPV-positive and HPV-negative tumors, while Gene Set Enrichment Analysis (GSEA) found that HPV-negative tumors show enrichment of hypermethylation in CpG sites of genes *RASSF1A, CHFR, RUNX3*, and *APC* and those involved in cell cycle (*p16, CDK10*) and the JAK-STAT signaling pathway (*STAT5A, JAK3, OSM, MPL*, and *EPO*), with these CpG sites hypomethylated in HPV-positive tumors. In addition, using high-density arrays including the Illumina HumanMethylation 450 bead array platform, Esposti et al. profiled 364 SCCHN cases to identify differentially methylated regions (DMRs) between HPV-negative and HPV-positive tumors and found that at least 60% of the DMRs that are hypermethylated in HPV-negative cases are hypomethylated in HPV-positive tumors [88].

The molecular etiologic differences in SCCHN are also evident in epigenetic profiles [83]. High-throughput profiling arrays have made it possible to identify widespread CpG island promoter methylation. This "CpG island methylator phenotype" (CIMP), first identified in colon cancer [89], has now been found in many other solid tumors and in leukemia [90]. Several groups have shown the importance of CIMP as a prognosticator indicative of improved clinical outcomes [91–93]. Although the mechanistic underpinnings responsible for CIMP are not fully understood, significance of its association with *BRAF*^{V600E} mutation [94], loss of isocitrate dehydrogenase 1 and 2 (*IDH1* and *IDH2*) [92], and mutations in *TET2* have been identified [93, 94]. CIMP was identified as a relevant phenomenon in a study of HPV-negative oral cancers, with a CIMP-high phenotype associated with worse prognosis [97].

One hypothesis for the formation of mammalian cytosine methylation machinery is that it evolved as a defense mechanism against endoparasitic sequences such as transposon and transposable elements: the idea being that the cell would better be able to restrict reactivation of endoparasitic sequences through hypermethylation, thus preventing translocations, chromosomal instability, and gene disruption [37, 98]. This idea is compatible with the observed common hypermethylation of long and short interspersed nuclear elements (LINE and SINE), and Alu repeat regions in normal cell, which are transposable elements reversed transcribed into genomes [98–100]. In addition to gene-specific CpG islands, LINE and Alu repeat hypomethylation is observed in HPV-negative tumors. This is associated with mobilization of their transposition due to increased expression of L1 reverse transcriptase and leads to genomic instability measured by loss of heterozygosity [86, 100].

Among the largest studies of SCCHN methylation profiling, The Cancer Genome Atlas (TCGA) project profiled over 279 cases, including 36 HPV-positive cases, and found an epigenetic signature specific to HPV-driven cases. Analysis of this dataset has helped to identify potential future targets of interest, both as biomarkers of prognosis and of therapeutic modulation. Other studies also have shown the hypermethylation of clinically relevant polycomb repressive complex 2 (PRC2) target genes implicated in tumor progression and metastasis [87, 102]. Studies of other HPV-associated cancers, such as cervical cancers, have similarly detected differences in methylation of CGIs in these virally associated malignancies [103].

12.5 Chromatin Modifications

Histones play a critical role in maintaining the epigenetic state of the cell by regulating access to DNA. This regulation is in part accomplished by the posttranslational modification (PTM) of histones. PTMs can alter the accessibility of DNA by altering the interaction between histone and DNA and by recruiting proteins that bind specific patterns of PTMs. The three most common PTMs are methylation, acetylation, and phosphorylation. As more than 120 amino acids can be modified on a single nucleosome, this provides a tremendous amount of information content. Monitoring these modifications is typically done with antibodies, but this approach may be responsible for some of the discrepancy found in the literature. There is significant cross-reactivity between antibodies and multiple PTMs, which complicates many of these studies [104].

Trimethylation of several critical residues within histone H3, such as H3K4, H3K9, H3K27, H3K36, and H3K79 (H3K4me3, H3K9me3, H3K36me3, and H3K79me3), has been observed in carcinogenesis. H3K4me3 and H3K36me3 are typically found on active genes near the promoter and placed there by the Set1 family methyltransferases, while H3K9me3 is catalyzed by the Suv39 family of methyltransferases and is active at satellite repeats and telomeres [105–107]. Other methylation sites such as H3K79 methylated by Dot1L are found at both active and silent promoters and may play a role in heterochromatin [108, 109]. These sites of methylation are critical for regulation of gene expression. Increases in H3K4me3 in oral cancer have also been observed, but their functional role is less well understood [110]. The methylation site with the best understood impact in oral cancer is H3K27me3, which regulates homeobox genes in tongue squamous cell carcinoma (SCC) cells. Proteomic studies on oral cancer have been limited, but suggest that H4K79me2 may correlate with poor survival [111].

Indirect modulation of gene expression by methylation of CpG islands is mediated by recognition and binding of methyl-CpG-binding proteins including MeCP1, MeCP2, MBDs, and Kaiso [112, 113]. These methyl-binding proteins, as part of larger repressor complexes (NuRD, NoRC, mSin3a, and SWI-SNF), recruit histone deacetylases (HDACs) and histone methyltransferases (HMTase). These enzymes modify specific lysine residues of histones H3 and H4, resulting in a compact chromatin state that reduces access of transcription factors. In contrast, unmethylated CpG islands are in an open state, enriched for chromatin with trimethylation of histone H3K4 by Cfp1, a component of SETD1 regulatory complex [114].

Histone acetylation is the most common and well-studied modification in cancer. Many of the residues on histones that can be acetylated can also be methylated. Generally, acetylation is thought to be an activation mark and methylation a repressive mark, but this is not always the case. At any given promoter, residues such as H3K9 and H3K27 can be either acetylated or methylated but not both. In esophageal squamous cell carcinoma (ESCC), acetylation is typically decreased and methylation levels are higher [110]. However, H3K27 acetylation has been reported to increase the expression of colon cancer associated transcript 1 (*CCAT1*), which can function as a binding scaffold for polycomb repressive complex 2 (PRC2) [115].

Histone acetylation is accomplished by enzymes referred to as lysine acetyltransferases or KATs, and their removal is accomplished by histone (or pan) deacetylases (HDACs). The majority of these enzymes can act on multiple residues, which has complicated their understanding. Given that the majority of cancers have aberrant levels of acetylation, HDAC inhibitors have been extensively assessed in many cancers, including oral cancers. It has been shown that the transcription of HDAC6 (a class IIb HDAC) is increased oral squamous cell carcinoma, compared to normal tissue of the same type [116, 117]. This is unusual, as HDAC6 expression typically correlates with better survival [117].

Other modifications such as histone phosphorylation may also play an important role in gene regulation, but their role in SCCHN is less well understood. While the likelihood that these modifications are altered in the disease state is high, the function of these sites is less well understood. At its core, cancer is a multidimensional problem that will require a holistic understanding of changes to both protein and DNA. As technology and computational approaches improve, so will our ability to describe the complex disease that is cancer.

12.6 Genetic Changes Affecting Expression and Function of Epigenetic Regulators in SCCHN

Importantly, comprehensive genome and exome sequencing studies in cancers have shown that in addition to several driver genes, mutations in "epigenetic modifiers" contribute to tumorigenesis [118]. Comprehensive mapping mutation profiles in SCCHN identified genomic alterations and mutations in several DNA and chromatin modifier genes including *DNMTs, MLL2*, and *NSD* gene family members. Mutation profiling in 512 SCCHN tumors by TCGA identifies several chromatin-modifying classes of proteins including histone methyltransferases, demethyltransferases, and acetyltransferases. Among these, MLL2 (KMT2D) and MLL3 (KMT2C) are paralogous histone methyltransferases that specifically methylate lysine residue 4 located on the tail of histone H3 (H3K4). High-frequency (16%) loss of function mutations including protein-altering nonsense, splice site, insertion, and deletion variants were identified in the TCGA head and neck cancer sequencing project. Similarly, MLL3 was found to be mutated at higher frequency (9%) with similar protein function-altering mutations suggesting a tumor-suppressive function for both genes. While the loss of MLL2 in human cells leads to genomic instability and transcriptional stress, such loss in murine models is associated with impaired B-cell differentiation, class switching, and B-cell malignancies [119, 120]. In the case of MLL1-rearranged leukemias, MLL2 is required for survival and proliferation [121].

In SCCHN, among the histone methyltransferases, *NSD1, MLL2*, and *MLL3* are mutated at high frequencies (>12, 16, and 9%, respectively). Of these, the *NSD1* and *NSD2* (MMSET) genes, which belong to family of nuclear receptor binding SET domain proteins, are mutated with high frequency (> 12%) in SCCHN cases [1–3]. NSD1 and NSD2 are histone methyltransferases that catalyze transfer of di- or trimethyl groups, depending on context, to lysine at position 36 on histone H3 (H3K36). Similar to the subset of renal cell carcinoma tumors with mutations in the histone methyltransferase *SETD2*, associated with hypomethylation [5], *NSD1* and *NSD2* gene mutations are associated with a genome-wide hypomethylation phenotype in SCCHN tumors [2]. It has been shown that *NSD1* is required for embryonic development and loss is embryonic lethal [122]. Inherited germline mutations in *NSD1* cause Sotos syndrome and in the case of NSD2 cause Wolf-Hirschhorn syndrome [123]. As methyltransferases, NSD1 and NSD2 catalyze methylation of histones (H3K36, H4K20) and non-histone targets such as the inflammation and survival-regulating factor NFkB [124–126]; altered methylation of such additional non-histone targets may contribute to some of their activity in SCCHN.

Among histone demethylases, *JMJD1C, KDM6A*, and *KDM5B* are mutated at lower frequencies (< 5%) and have distinct site specificities for demethylating lysine residues on histone tails (H3K9, H3K27, and H3K4, respectively). Similarly, among histone acetyltransferases, EP300 and its paralog CBP that are transcriptional co-activators are more frequently mutated in SCCHN (8% and 7%, respectively). Acetylation by p300 results in relaxed chromatin structure and is responsible for activation of genes in the vicinity. Depending on the cellular context, p300 can act as a transcriptional cofactor for a variety of nuclear proteins with various oncogenic and tumor suppressor roles, including MYC, BRCA1, SMAD, and p53, and thus regulate cellular pathways in a context and cell-type-dependent manner [127].

12.7 Therapeutics Targeting Epigenetic Controls

Targeting epigenetic modifications for developing novel therapeutics offers significant advantage. While recovering the function of genes lost due to mutations or copy loss is a difficult therapeutic approach, epigenetic modifications can be reversible through small molecule inhibitors. Therapeutic targeting of methylation effectively enables reactivation of tumor suppressor genes that have been silenced due to promoter hypermethylation. Earlier studies have shown that complete loss of transcription upon *CDKN2A/p16* promoter methylation can be reversed with the DNMT inhibitor 5-deoxy-azacytidine [128–130]. These therapeutic inhibitors are not toxic to normal cells, and therefore development of inhibitors targeting DNA methylation is an attractive therapeutic approach.

Currently two different classes of epigenetic drugs are used in the clinic. These include inhibitors of DNMT and of HDAC, both of which modulate transcriptional silencing. Among the DNMT inhibitors are cytidine analogues including 5-azacytidine and 5-aza-2′-deoxycytidine (decitabine) [131–133]. These are currently therapeutic options for treating myelodysplastic syndromes (MDS) [134, 135]. Several HDAC inhibitors are approved for the treatment of various hematologic malignancies, with these drugs including belinostat, panobinostat, vorinostat, valproic acid, romidepsin, phenylbutyric acid, and others [136]. The US Food and Drug Administration (FDA) approved the HDAC inhibitors suberoylanilide hydroxamic acid (vorinostat) and romidepsin for patients with progressive, persistent, or recurrent T-cell lymphomas [137]. Another FDA-approved drug, panobinostat, is used in combination with bortezomib and dexamethasone for the treatment of patients with multiple myeloma [138] and belinostat for the treatment of patients with peripheral T-cell lymphoma [139].

The success of these drugs in hematologic malignancies encouraged investigators and clinicians to study them for the treatment of solid tumors. In vitro studies have shown that decitabine rescues cisplatin resistance in SCCHN [140] and combination therapies using decitabine and carboplatin have been tested in trials [141]. In addition, an antitumor effect of 5-azacytidine has been noted to a greater degree in HPV-positive cell lines and xenografts versus the HPV-negative counterparts. This is associated with a decreased expression of HPV-related genes and increased stabilization of p53 [142]. This has provided the rationale for a window of opportunity study with 5-azacytidine, which is currently ongoing (NCT02178072).

In addition, several HDAC inhibitors which target class I and class II HDACs are under study for therapeutic management of SCCHN. Mammalian HDACs can be classified primarily into four classes that vary in structure, function, expression, and subcellular localization. Class I HDACs are ubiquitously expressed nuclear localized proteins that are homologous to yeast Rpd3. Class II HDACs (which can be subdivided into classes IIa and IIb) are selectively expressed, localize to both nucleus and cytoplasm, and can also deacetylate non-histone proteins [143, 144]. Via increased acetylation of histones, these agents are thought to regulate transcription of key genes related to tumorigenesis and induce cell cycle arrest [145]. However, the exact mechanisms of activity are still under investigation. Among these HDAC-targeting drugs, resminostat, romidepsin, and SAHA are under investigation in phase I and II trials [146–148]. A phase I study of the dual EGFR and HDAC inhibitor CUDC-101, in conjunction with bolus dosing of cisplatin and radiation, was conducted among patients with HPV-negative, locally advanced SCCHN. While pharmacodynamic markers indicated that HDAC inhibition was achieved, toxicity resulting in patient discontinuation of the study agent suggested that alternative treatment schedules needed to be explored [149].

There is a growing body of literature regarding the potential interactions between epigenetic targeting agents and immune checkpoint inhibitors, such as PD-1 inhibitors [150, 151]. Several hypotheses to justify combination therapies include the following: (1) upregulation of tumor-related antigens, (2) induction of the interferon

Table 12.1 Current clinical trials of epigenetic treatments in SCCHN

Trial	n
A Phase IB/II Study with Azacitidine, Durvalumab, and Tremelimumab in Recurrent and/or Metastatic Head and Neck Cancer Patients (NCT03019003)	59
Window Trial 5-aza in SCCHN (NCT02178072)	20
Pre-operative Mocetinostat (MGCD0103) and Durvalumab (MEDI4736) (PRIMED) for Squamous Cell Carcinoma of the Oral Cavity (PRIMED-001)	12

pathway, and (3) upregulation of PD-L1 [151]. Currently, combinations of epigenetic and immune modulating agents are under study in several solid tumors, including head and neck cancers (Table 12.1).

In addition to DNMT and HDAC inhibitors, a novel class of inhibitors that target histone methyltransferases (HMTase) are showing promise in the clinic. The chromatin modifier enhancer of zeste homolog 2 (EZH2) is an HMTase which is the enzymatic subunit of polycomb repressor complex 2 (PRC2) that methylates lysine 27 of histone H3 (H3K27) to promote transcriptional silencing [125, 153] and is associated with different cancers [154]. Mutations in the PRC2 complex or SWI-SNF chromatin remodeling complex that antagonizes polycomb function or mutations in H3K27 sites cause loss of H3K27 epigenetic marks. Currently, the EZH2 inhibitor tazemetostat is in trials for the treatment of hematologic malignancies and select solid tumors including mesothelioma [154, 155]. In SCCHN, the frequent mutation of NSD1 and NSD2, and the fact that these mutations are associated with favorable prognosis in laryngeal cancer [156], is relevant given their role in the H3K36 methylation that regulates gene transcription and DNA repair mechanisms. Although currently there are no FDA-approved inhibitors for this class of genes, efforts are underway to develop novel compounds directed to these targets [157].

While epigenetic silencing is typically pro-oncogenic, some studies have identified situations in which specific promoter hypermethylation is beneficial in therapeutic management of cancer. For instance, promoter hypermethylation of MGMT, a DNA repair gene that inhibits killing of tumor cells by alkylating agents, predicts increased disease-free and overall survival after therapy with alkylating agents [158]. Hence, therapeutic strategies that rely on broadly modifying the activity of methylating or acetylating proteins are likely to induce at least some negative consequences.

12.8 Molecular Subtypes Related to Chromatin Modifications, and Integrative Analyses in SCCHN

Cancer is a multistep process that proceeds through accumulation of genomic alterations and subsequent evolutionary selection of phenotypic changes [159]. Genomic mutations, copy number changes, and epigenetic changes affecting DNA and chromatin impact expression of genes and thereby alter many molecular functions and biochemical processes. Although these genomic profiles differ significantly among patients, the functional impact of cumulative changes is more similar in ultimate impact on the tumor transcriptome, suggesting that understanding how epigenetic reprogramming leads to deregulation of transcriptome will provide critical insights into essential cancer biology.

Early studies primarily used gene expression data to characterize SCCHN. For example, Chung et al. profiled 60 SCCHN tumor samples using cDNA microarrays and defined 4 distinct subtypes [160]. Unsupervised hierarchical clustering of the expression data led to identification of four subtypes including basal, mesenchymal, classical, and atypical that are characterized by significant differences in recurrence-free survival, which have been verified by subsequent studies [2, 161] and further extended [162]. Multiple investigators have developed expression-based signatures to predict specific clinical behaviors of SCCHN, such as lymph node metastasis, level of hypoxia, or radiosensitivity [163–167]. Roepman et al. developed a 102-gene signature that is predictive of the propensity for lymph node metastasis, which theoretically could be used to guide decisions regarding lymph node dissections [165]. This predictor, however, was only able to correctly determine lymph node status in 61 of 82 patients, and the authors noted that this was of only incremental improvement to clinical decision-making. Onken et al. developed an expression signature for nodal metastasis from a mouse model of oral cavity cancer [168]. Interestingly, this signature could predict nodal metastasis development for human oral cavity cancers in a training set and a small validation set.

Efforts to study the prognostic value of expression signatures in selecting patients for strategies to overcome resistance to treatment such as radiotherapy resulted in some success [163, 166, 167]. For example, a hypoxic signature was used to retrospectively analyze data from the DAHANCA 5 trial, a randomized trial of radiotherapy ± nimorazole (a hypoxic radiosensitizer) in SCCHN [169]. This group found that the benefit of nimorazole on local control and DFS was limited to patients whose tumors were deemed to be more hypoxic by their expression signature [170]. This suggests that their signature may be predictive and could help select patients for future clinical trials using hypoxic sensitizers.

The initial TCGA SCCHN study profiled 279 untreated SCCHN patients including 36 HPV-related and 243 HPV-negative cases. The large sample size provides strong basis for molecular classification. Analyses using gene expression profiles from this dataset confirmed previously reported gene expression subtypes (atypical, mesenchymal, basal, and classical) [160, 161]. These analyses assist in the identification of subtypes and associate specific molecular portraits to the subtypes and help in designing novel therapies for patients.

Subsequently, availability of comprehensive molecular data has allowed researchers to combine multiple datasets. Integrated analysis of these multiple datasets involves classifying the molecular data from different platforms into subtypes with prognostic importance and clinical applicability. The complex genomic data from cancer genome sequencing efforts motivated the development of novel statistical methods for integrative genomic data analyses. Using these methods, data from multiple platforms such as sequencing, copy number, gene expression, and DNA methylation can be combined simultaneously for clustering to identify novel subtypes [171–175]. Among these, the iCluster and iClusterPlus algorithms use data from different genomic platforms and define a joint latent variable model to identify sets of molecular subclasses. Application of these methods to the initial TCGA dataset of 279 SCCHN cases resulted in identification of several molecular subtypes with prognostic and functional significance. For example, using TCGA data and integrative methods, Peri et al. identified 5 novel subtypes that classified 256 SCCHN tumors by disease site and HPV status. These subtypes are associated with differences in overall survival [156]. Among the larger group of SCCHN tumors, this work classified the subset of laryngeal cancers into two subtypes associated with differences in overall survival. These two novel laryngeal subtypes also differ in the methylation profiles, where the better survival group shows a marked genome-wide hypomethylation phenotype associated with *NSD1* mutations. The association between *NSD1* mutations and global DNA backbone methylation has been shown using germline DNA from Sotos syndrome patients [176]. As noted earlier, the NSD family of proteins are HMTases that mediate methylation of H3K36, which has an established relationship to activity of DNA methyltransferases [14, 177, 178]. Another study showed that decreased dimethylation of lysine in H3 histones (H3K36me2) in SCCHN is associated with damaging NSD1 mutations [179]. The ability to risk stratify patients based on this signature could be a potential mechanism to identify patients at high risk of recurrence who may benefit from novel investigational adjuvant therapies.

12.9 Future Directions

The currently approved DNMT inhibitors are limited in their clinical utility in solid tumors due to poor bioavailability and non-specific epigenetic modulation of both non-cancerous and malignant cells, resulting in significant toxicities. Other novel or second-generation molecules have been developed to overcome some of these issues. MG98 is an antisense oligonucleotide to DNMT1 that does not affect DNMT3 [180]. In bladder and colon cancer cell lines, it was able to induce p16 reexpression, and in lung and colon xenograft models in nude mice, it showed tumor inhibition. In a phase I study in 33 patients with advanced solid tumors, the drug had clinical activity and a tolerable toxicity profile and suppressed DNMT1 expression in the majority of patients [181]. RG108 is another small molecule DNMT inhibitor which has been studied as a conjugate to procainamide [182]. Procainamide has an affinity for the CpG-enriched regions of DNA and was used to concentrate RG108 in these regions. In leukemia cell lines, RG108 reactivated the tumor suppressor genes p16 and TIMP3 [183]. In comparison with other DNMT inhibitors, RG108 was the only one with no genotoxic or cytotoxic effects, lending it a potentially unique appeal for further clinical development.

Another area of active research involves microRNAs (miRNAs), which are a class of noncoding RNA molecules that have either tumor suppressor or proto-oncogenic effects [184]. These are susceptible to dysfunction which can contribute to tumorigenesis via genetic and epigenetic dysregulation [185]. Kozaki et al. studied 18 oral cancer cell lines and found that in most cases, 4 of the tumor suppressor miRNAs located around CpG islands (miR-34b, miR-137, miR-193a, and miR-203) were silenced via hypermethylation [186]. In contrast, miR-21 is overexpressed in most tumor types and is known to enhance nodal metastasis and tumor invasiveness in oral cancers [187], functioning as an oncogene. Several of the identified miRNAs confer a cisplatin-resistant profile HNSCC cell line [188–194]. Medina and colleagues generated mice that overexpress miR-21 and found that it led them to develop a malignant lymphoid-like phenotype [195]. Interestingly, the tumors completely resolved with pharmacologic intervention. This could have important therapeutic implications to enhance chemosensitivity in squamous cell head and neck cancers, where cisplatin is the mainstay of treatment.

12.10 Conclusions

Epigenetic regulation clearly has a role in the carcinogenesis of head and neck cancers. Future work will need to address factors related to recurrence and treatment resistance. In addition, with the explosion of immunotherapeutics, there is a role for further study of DNA methylation inhibitors in conjunction with immune checkpoint inhibition therapies. Due to lack of high-quality predictive biomarkers for early detection of SCCHN, epigenetic markers can be used as early diagnostic markers as they can be detected with high specificity through cost-effective experimental methods with high sensitivity. In the case of SCCHN screening, DNA obtained from saliva or buccal samples for analysis of zinc finger family methylation has provided proof for detecting carcinoma [196]. The prognostic role of the NSD proteins is intriguing and provides a rationale to study tailored treatment approaches for laryngeal cancer based on the NSD1/NSD2 mutation profile. It also suggests that there is a fundamental difference between laryngeal cancers and other types of SCCHN (for instance, affecting the oral cavity or the oropharynx), as although NSD mutations occur in these tumor subsites, they are not prognostic. This has broad implications for better parsing of the relationship of epigenetic modifications for different subsites: at present, such distinctions are not typically made, but as more tumor samples become available and are genomically profiled, this should become possible. The emerging role of epigenetic studies in SCCHN provides a tremendous opportunity for basic and translational research toward improving cancer burden among head and neck cancer patients.

References

1. Agrawal N, et al. Exome sequencing of head and neck squamous cell carcinoma reveals inactivating mutations in NOTCH1. Science. 2011;333:1154–7. https://doi.org/10.1126/science.1206923.
2. Cancer Genome Atlas, N. Comprehensive genomic characterization of head and neck squamous cell carcinomas. Nature. 2015;517:576–82. https://doi.org/10.1038/nature14129.
3. Stransky N, et al. The mutational landscape of head and neck squamous cell carcinoma. Science. 2011;333:1157–60. https://doi.org/10.1126/science.1208130.
4. Beck TN, Golemis EA. Genomic insights into head and neck cancer. Cancers Head Neck. 2016;1. https://doi.org/10.1186/s41199-016-0003-z.
5. Cancer Genome Atlas Research, N. Comprehensive molecular characterization of clear cell renal cell carcinoma. Nature. 2013;499:43–9. https://doi.org/10.1038/nature12222.
6. Seiwert TY, et al. Integrative and comparative genomic analysis of HPV-positive and HPV-negative head and neck squamous cell carcinomas. Clin Cancer Res. 2015;21:632–41. https://doi.org/10.1158/1078-0432.CCR-13-3310.
7. Ha PK, Califano JA. Promoter methylation and inactivation of tumour-suppressor genes in oral squamous-cell carcinoma. Lancet Oncol. 2006;7:77–82. https://doi.org/10.1016/S1470-2045(05)70540-4.
8. Smith IM, Mydlarz WK, Mithani SK, Califano JA. DNA global hypomethylation in squamous cell head and neck cancer associated with smoking, alcohol consumption and stage. Int J Cancer. 2007;121:1724–8. https://doi.org/10.1002/ijc.22889.
9. Tokumaru Y, et al. Inverse correlation between cyclin A1 hypermethylation and p53 mutation in head and neck cancer identified by reversal of epigenetic silencing. Cancer Res. 2004;64:5982–7. https://doi.org/10.1158/0008-5472.CAN-04-0993.
10. Baylin SB, Jones PA. A decade of exploring the cancer epigenome – biological and translational implications. Nat Rev Cancer. 2011;11:726–34. https://doi.org/10.1038/nrc3130.
11. Dawson MA, Kouzarides T. Cancer epigenetics: from mechanism to therapy. Cell. 2012;150:12–27. https://doi.org/10.1016/j.cell.2012.06.013.
12. Feinberg AP, Tycko B. The history of cancer epigenetics. Nat Rev Cancer. 2004;4:143–53. https://doi.org/10.1038/nrc1279.
13. Robertson KD. DNA methylation and human disease. Nat Rev Genet. 2005;6:597–610. https://doi.org/10.1038/nrg1655.
14. Cedar H, Bergman Y. Linking DNA methylation and histone modification: patterns and paradigms. Nat Rev Genet. 2009;10:295–304. https://doi.org/10.1038/nrg2540.
15. Schumacher A, et al. Microarray-based DNA methylation profiling: technology and applications. Nucleic Acids Res. 2006;34:528–42. https://doi.org/10.1093/nar/gkj461.
16. Yong WS, Hsu FM, Chen PY. Profiling genome-wide DNA methylation. Epigenetics Chromatin. 2016;9:26. https://doi.org/10.1186/s13072-016-0075-3.
17. Bock C. Analysing and interpreting DNA methylation data. Nat Rev Genet. 2012;13:705–19. https://doi.org/10.1038/nrg3273.
18. Chelaru F, Smith L, Goldstein N, Bravo HC. Epiviz: interactive visual analytics for functional genomics data. Nat Methods. 2014;11:938–40. https://doi.org/10.1038/nmeth.3038.
19. Kent WJ, Zweig AS, Barber G, Hinrichs AS, Karolchik D. BigWig and BigBed: enabling browsing of large distributed datasets. Bioinformatics. 2010;26:2204–7. https://doi.org/10.1093/bioinformatics/btq351.
20. Raney BJ, et al. Track data hubs enable visualization of user-defined genome-wide annotations on the UCSC genome browser. Bioinformatics. 2014;30:1003–5. https://doi.org/10.1093/bioinformatics/btt637.
21. Thorvaldsdottir H, Robinson JT, Mesirov JP. Integrative Genomics Viewer (IGV): high-performance genomics data visualization and exploration. Brief Bioinform. 2013;14:178–92. https://doi.org/10.1093/bib/bbs017.

22. Bellacosa A, Drohat AC. Role of base excision repair in maintaining the genetic and epigenetic integrity of CpG sites. DNA Repair. 2015;32:33–42. https://doi.org/10.1016/j.dnarep.2015.04.011.
23. Chen T, Li E. Structure and function of eukaryotic DNA methyltransferases. Curr Top Dev Biol. 2004;60:55–89. https://doi.org/10.1016/S0070-2153(04)60003-2.
24. Okano M, Bell DW, Haber DA, Li E. DNA methyltransferases Dnmt3a and Dnmt3b are essential for de novo methylation and mammalian development. Cell. 1999;99:247–57.
25. Chen ZX, Mann JR, Hsieh CL, Riggs AD, Chedin F. Physical and functional interactions between the human DNMT3L protein and members of the de novo methyltransferase family. J Cell Biochem. 2005;95:902–17. https://doi.org/10.1002/jcb.20447.
26. Suetake I, Shinozaki F, Miyagawa J, Takeshima H, Tajima S. DNMT3L stimulates the DNA methylation activity of Dnmt3a and Dnmt3b through a direct interaction. J Biol Chem. 2004;279:27816–23. https://doi.org/10.1074/jbc.M400181200.
27. Kohli RM, Zhang Y. TET enzymes, TDG and the dynamics of DNA demethylation. Nature. 2013;502:472–9. https://doi.org/10.1038/nature12750.
28. Pastor WA, Aravind L, Rao A. TETonic shift: biological roles of TET proteins in DNA demethylation and transcription. Nat Rev Mol Cell Biol. 2013;14:341–56. https://doi.org/10.1038/nrm3589.
29. Kuroda A, et al. Insulin gene expression is regulated by DNA methylation. PLoS One. 2009;4:e6953. https://doi.org/10.1371/journal.pone.0006953.
30. Nan X, et al. Transcriptional repression by the methyl-CpG-binding protein MeCP2 involves a histone deacetylase complex. Nature. 1998;393:386–9. https://doi.org/10.1038/30764.
31. Lister R, et al. Human DNA methylomes at base resolution show widespread epigenomic differences. Nature. 2009;462:315–22. https://doi.org/10.1038/nature08514.
32. Maunakea AK, et al. Conserved role of intragenic DNA methylation in regulating alternative promoters. Nature. 2010;466:253–7. https://doi.org/10.1038/nature09165.
33. Yang X, et al. Gene body methylation can alter gene expression and is a therapeutic target in cancer. Cancer Cell. 2014;26:577–90. https://doi.org/10.1016/j.ccr.2014.07.028.
34. Hellman A, Chess A. Gene body-specific methylation on the active X chromosome. Science. 2007;315:1141–3. https://doi.org/10.1126/science.1136352.
35. Shukla S, et al. CTCF-promoted RNA polymerase II pausing links DNA methylation to splicing. Nature. 2011;479:74–9. https://doi.org/10.1038/nature10442.
36. Jones P, The A. DNA methylation paradox. Trends Genet. 1999;15:34–7.
37. Portela A, Esteller M. Epigenetic modifications and human disease. Nat Biotechnol. 2010;28:1057–68. https://doi.org/10.1038/nbt.1685.
38. Zilberman D, Gehring M, Tran RK, Ballinger T, Henikoff S. Genome-wide analysis of Arabidopsis thaliana DNA methylation uncovers an interdependence between methylation and transcription. Nat Genet. 2007;39:61–9. https://doi.org/10.1038/ng1929.
39. Jones PA. Functions of DNA methylation: islands, start sites, gene bodies and beyond. Nat Rev Genet. 2012;13:484–92. https://doi.org/10.1038/nrg3230.
40. Neri F, et al. Intragenic DNA methylation prevents spurious transcription initiation. Nature. 2017;543:72–7. https://doi.org/10.1038/nature21373.
41. Wagner EJ, Carpenter PB. Understanding the language of Lys36 methylation at histone H3. Nat Rev Mol Cell Biol. 2012;13:115–26. https://doi.org/10.1038/nrm3274.
42. Doi A, et al. Differential methylation of tissue- and cancer-specific CpG island shores distinguishes human induced pluripotent stem cells, embryonic stem cells and fibroblasts. Nat Genet. 2009;41:1350–3. https://doi.org/10.1038/ng.471.
43. Irizarry RA, et al. The human colon cancer methylome shows similar hypo- and hypermethylation at conserved tissue-specific CpG island shores. Nat Genet. 2009;41:178–86. https://doi.org/10.1038/ng.298.
44. Ji H, et al. Comprehensive methylome map of lineage commitment from haematopoietic progenitors. Nature. 2010;467:338–42. https://doi.org/10.1038/nature09367.
45. Straussman R, et al. Developmental programming of CpG island methylation profiles in the human genome. Nat Struct Mol Biol. 2009;16:564–71. https://doi.org/10.1038/nsmb.1594.

46. Kacem S, Feil R. Chromatin mechanisms in genomic imprinting. Mamm Genome. 2009;20:544–56. https://doi.org/10.1007/s00335-009-9223-4.
47. Hashibe M, et al. Alcohol drinking in never users of tobacco, cigarette smoking in never drinkers, and the risk of head and neck cancer: pooled analysis in the international head and neck Cancer epidemiology consortium. J Natl Cancer Inst. 2007;99:777–89. https://doi.org/10.1093/jnci/djk179.
48. Maasland DH, van den Brandt PA, Kremer B, Goldbohm RA, Schouten LJ. Alcohol consumption, cigarette smoking and the risk of subtypes of head-neck cancer: results from the Netherlands Cohort Study. BMC Cancer. 2014;14:187. https://doi.org/10.1186/1471-2407-14-187.
49. Shaw R, Beasley N. Aetiology and risk factors for head and neck cancer: United Kingdom National Multidisciplinary Guidelines. J Laryngol Otol. 2016;130:S9–S12. https://doi.org/10.1017/S0022215116000360.
50. Busch R, et al. Differential DNA methylation marks and gene comethylation of COPD in African-Americans with COPD exacerbations. Respir Res. 2016;17:143. https://doi.org/10.1186/s12931-016-0459-8.
51. Hecht SS. Tobacco smoke carcinogens and lung cancer. J Natl Cancer Inst. 1999;91:1194–210.
52. Ligthart S, et al. DNA methylation signatures of chronic low-grade inflammation are associated with complex diseases. Genome Biol. 2016;17:255. https://doi.org/10.1186/s13059-016-1119-5.
53. Sundar IK, et al. DNA methylation profiling in peripheral lung tissues of smokers and patients with COPD. Clin Epigenetics. 2017;9:38. https://doi.org/10.1186/s13148-017-0335-5.
54. Joehanes R, et al. Epigenetic signatures of cigarette smoking. Circ Cardiovasc Genet. 2016;9:436–47. https://doi.org/10.1161/CIRCGENETICS.116.001506.
55. Feil R, Fraga MF. Epigenetics and the environment: emerging patterns and implications. Nat Rev Genet. 2012;13:97–109. https://doi.org/10.1038/nrg3142.
56. Ziller MJ, et al. Charting a dynamic DNA methylation landscape of the human genome. Nature. 2013;500:477–81. https://doi.org/10.1038/nature12433.
57. Wan ES, et al. Smoking-associated site-specific differential methylation in buccal mucosa in the COPDGene study. Am J Respir Cell Mol Biol. 2015;53:246–54. https://doi.org/10.1165/rcmb.2014-0103OC.
58. Sankaranarayanan R, Masuyer E, Swaminathan R, Ferlay J, Whelan S. Head and neck cancer: a global perspective on epidemiology and prognosis. Anticancer Res. 1998;18:4779–86.
59. Wang TH, Hsia SM, Shih YH, Shieh TM. Association of smoking, alcohol use, and betel quid chewing with epigenetic aberrations in cancers. Int J Mol Sci. 2017;18. https://doi.org/10.3390/ijms18061210.
60. Scott RS. Epstein-Barr virus: a master epigenetic manipulator. Curr Opin Virol. 2017;26:74–80. https://doi.org/10.1016/j.coviro.2017.07.017.
61. Chen CJ, et al. Multiple risk factors of nasopharyngeal carcinoma: Epstein-Barr virus, malarial infection, cigarette smoking and familial tendency. Anticancer Res. 1990;10:547–53.
62. Goelz SE, Vogelstein B, Hamilton SR, Feinberg AP. Hypomethylation of DNA from benign and malignant human colon neoplasms. Science. 1985;228:187–90.
63. Belinsky SA, et al. Aberrant methylation of p16(INK4a) is an early event in lung cancer and a potential biomarker for early diagnosis. Proc Natl Acad Sci USA. 1998;95:11891–6.
64. Arantes LM, de Carvalho AC, Melendez ME, Carvalho AL, Goloni-Bertollo EM. Methylation as a biomarker for head and neck cancer. Oral Oncol. 2014;50:587–92. https://doi.org/10.1016/j.oraloncology.2014.02.015.
65. Jones PA, Laird PW. Cancer epigenetics comes of age. Nat Genet. 1999;21:163–7. https://doi.org/10.1038/5947.
66. Sadikovic B, Al-Romaih K, Squire JA, Zielenska M. Cause and consequences of genetic and epigenetic alterations in human cancer. Curr Genomics. 2008;9:394–408. https://doi.org/10.2174/138920208785699580.
67. Burri N, et al. Methylation silencing and mutations of the p14ARF and p16INK4a genes in colon cancer. Lab Invest. 2001;81:217–29.

68. Esteller M, et al. Promoter hypermethylation and BRCA1 inactivation in sporadic breast and ovarian tumors. J Natl Cancer Inst. 2000;92:564–9.
69. Myohanen SK, Baylin SB, Herman JG. Hypermethylation can selectively silence individual p16ink4A alleles in neoplasia. Cancer Res. 1998;58:591–3.
70. Esteller M, et al. Promoter hypermethylation of the DNA repair gene O(6)-methylguanine-DNA methyltransferase is associated with the presence of G:C to A:T transition mutations in p53 in human colorectal tumorigenesis. Cancer Res. 2001;61:4689–92.
71. Esteller M, et al. Inactivation of the DNA repair gene O6-methylguanine-DNA methyltransferase by promoter hypermethylation is associated with G to A mutations in K-ras in colorectal tumorigenesis. Cancer Res. 2000;60:2368–71.
72. Shen H, Laird PW. Interplay between the cancer genome and epigenome. Cell. 2013;153:38–55. https://doi.org/10.1016/j.cell.2013.03.008.
73. Ishida E, et al. Promotor hypermethylation of p14ARF is a key alteration for progression of oral squamous cell carcinoma. Oral Oncol. 2005;41:614–22. https://doi.org/10.1016/j.oraloncology.2005.02.003.
74. Pierini S, et al. Promoter hypermethylation of CDKN2A, MGMT, MLH1, and DAPK genes in laryngeal squamous cell carcinoma and their associations with clinical profiles of the patients. Head Neck. 2014;36:1103–8. https://doi.org/10.1002/hed.23413.
75. Fan CY. Epigenetic alterations in head and neck cancer: prevalence, clinical significance, and implications. Curr Oncol Rep. 2004;6:152–61.
76. Chen K, et al. Methylation of multiple genes as diagnostic and therapeutic markers in primary head and neck squamous cell carcinoma. Arch Otolaryngol Head Neck Surg. 2007;133:1131–8. https://doi.org/10.1001/archotol.133.11.1131.
77. Demokan S, Dalay N. Role of DNA methylation in head and neck cancer. Clin Epigenetics. 2011;2:123–50. https://doi.org/10.1007/s13148-011-0045-3.
78. Ovchinnikov DA, et al. Tumor-suppressor gene promoter hypermethylation in saliva of head and neck Cancer patients. Transl Oncol. 2012;5:321–6.
79. Rettori MM, et al. Prognostic significance of TIMP3 hypermethylation in post-treatment salivary rinse from head and neck squamous cell carcinoma patients. Carcinogenesis. 2013;34:20–7. https://doi.org/10.1093/carcin/bgs311.
80. Rettori MM, et al. TIMP3 and CCNA1 hypermethylation in HNSCC is associated with an increased incidence of second primary tumors. J Transl Med. 2013;11:316. https://doi.org/10.1186/1479-5876-11-316.
81. van Kempen PM, et al. Differences in methylation profiles between HPV-positive and HPV-negative oropharynx squamous cell carcinoma: a systematic review. Epigenetics. 2014;9:194–203. https://doi.org/10.4161/epi.26881.
82. Colacino JA, et al. Comprehensive analysis of DNA methylation in head and neck squamous cell carcinoma indicates differences by survival and clinicopathologic characteristics. PLoS One. 2013;8:e54742. https://doi.org/10.1371/journal.pone.0054742.
83. Marsit CJ, et al. Epigenetic profiling reveals etiologically distinct patterns of DNA methylation in head and neck squamous cell carcinoma. Carcinogenesis. 2009;30:416–22. https://doi.org/10.1093/carcin/bgp006.
84. Marsit CJ, McClean MD, Furniss CS, Kelsey KT. Epigenetic inactivation of the SFRP genes is associated with drinking, smoking and HPV in head and neck squamous cell carcinoma. Int J Cancer. 2006;119:1761–6. https://doi.org/10.1002/ijc.22051.
85. Poage GM, et al. Global hypomethylation identifies Loci targeted for hypermethylation in head and neck cancer. Clin Cancer Res. 2011;17:3579–89. https://doi.org/10.1158/1078-0432.CCR-11-0044.
86. Richards KL, et al. Genome-wide hypomethylation in head and neck cancer is more pronounced in HPV-negative tumors and is associated with genomic instability. PLoS One. 2009;4:e4941. https://doi.org/10.1371/journal.pone.0004941.

87. Sartor MA, et al. Genome-wide methylation and expression differences in HPV(+) and HPV(−) squamous cell carcinoma cell lines are consistent with divergent mechanisms of carcinogenesis. Epigenetics. 2011;6:777–87.
88. Degli Esposti D, et al. Unique DNA methylation signature in HPV-positive head and neck squamous cell carcinomas. Genome Med. 2017;9:33. https://doi.org/10.1186/s13073-017-0419-z.
89. Toyota M, et al. CpG island methylator phenotype in colorectal cancer. Proc Natl Acad Sci U S A. 1999;96:8681–6.
90. Garcia-Manero G, et al. DNA methylation of multiple promoter-associated CpG islands in adult acute lymphocytic leukemia. Clin Cancer Res. 2002;8:2217–24.
91. Fang F, et al. Breast cancer methylomes establish an epigenomic foundation for metastasis. Sci Transl Med. 2011;3:75ra25. https://doi.org/10.1126/scitranslmed.3001875.
92. Noushmehr H, et al. Identification of a CpG island methylator phenotype that defines a distinct subgroup of glioma. Cancer Cell. 2010;17:510–22. https://doi.org/10.1016/j.ccr.2010.03.017.
93. van den Bent MJ, et al. A hypermethylated phenotype is a better predictor of survival than MGMT methylation in anaplastic oligodendroglial brain tumors: a report from EORTC study 26951. Clin Cancer Res. 2011;17:7148–55. https://doi.org/10.1158/1078-0432.CCR-11-1274.
94. Weisenberger DJ, et al. CpG island methylator phenotype underlies sporadic microsatellite instability and is tightly associated with BRAF mutation in colorectal cancer. Nat Genet. 2006;38:787–93. https://doi.org/10.1038/ng1834.
95. Figueroa ME, et al. Leukemic IDH1 and IDH2 mutations result in a hypermethylation phenotype, disrupt TET2 function, and impair hematopoietic differentiation. Cancer Cell. 2010;18:553–67. https://doi.org/10.1016/j.ccr.2010.11.015.
96. Hughes LA, et al. The CpG island methylator phenotype: what's in a name? Cancer Res. 2013;73:5858–68. https://doi.org/10.1158/0008-5472.CAN-12-4306.
97. Jithesh PV, et al. The epigenetic landscape of oral squamous cell carcinoma. Br J Cancer. 2013;108:370–9. https://doi.org/10.1038/bjc.2012.568.
98. Yoder JA, Walsh CP, Bestor TH. Cytosine methylation and the ecology of intragenomic parasites. Trends Genet. 1997;13:335–40.
99. Liu WM, Maraia RJ, Rubin CM, Schmid CW. Alu transcripts: cytoplasmic localisation and regulation by DNA methylation. Nucleic Acids Res. 1994;22:1087–95.
100. Paschos K, Allday MJ. Epigenetic reprogramming of host genes in viral and microbial pathogenesis. Trends Microbiol. 2010;18:439–47. https://doi.org/10.1016/j.tim.2010.07.003.
101. Rosl F, Arab A, Klevenz B, zur Hausen H. The effect of DNA methylation on gene regulation of human papillomaviruses. J Gen Virol. 1993;74(Pt 5):791–801. https://doi.org/10.1099/0022-1317-74-5-791.
102. Lechner M, et al. Identification and functional validation of HPV-mediated hypermethylation in head and neck squamous cell carcinoma. Genome Med. 2013;5:15. https://doi.org/10.1186/gm419.
103. Banister CE, Liu C, Pirisi L, Creek KE, Buckhaults PJ. Identification and characterization of HPV-independent cervical cancers. Oncotarget. 2017;8:13375–86. https://doi.org/10.18632/oncotarget.14533.
104. Rothbart SB, et al. An interactive database for the assessment of histone antibody specificity. Mol Cell. 2015;59:502–11. https://doi.org/10.1016/j.molcel.2015.06.022.
105. Mikkelsen TS, et al. Genome-wide maps of chromatin state in pluripotent and lineage-committed cells. Nature. 2007;448:553–60. https://doi.org/10.1038/nature06008.
106. Pauler FM, et al. H3K27me3 forms BLOCs over silent genes and intergenic regions and specifies a histone banding pattern on a mouse autosomal chromosome. Genome Res. 2009;19:221–33. https://doi.org/10.1101/gr.080861.108.
107. Volkel P, Angrand PO. The control of histone lysine methylation in epigenetic regulation. Biochimie. 2007;89:1–20. https://doi.org/10.1016/j.biochi.2006.07.009.

108. Barski A, et al. High-resolution profiling of histone methylations in the human genome. Cell. 2007;129:823–37. https://doi.org/10.1016/j.cell.2007.05.009.
109. Jones B, et al. The histone H3K79 methyltransferase Dot1L is essential for mammalian development and heterochromatin structure. PLoS Genet. 2008;4:e1000190. https://doi.org/10.1371/journal.pgen.1000190.
110. Chen C, et al. Abnormal histone acetylation and methylation levels in esophageal squamous cell carcinomas. Cancer Investig. 2011;29:548–56. https://doi.org/10.3109/07357907.2011.597810.
111. Zhang K, et al. Comparative analysis of histone H3 and H4 post-translational modifications of esophageal squamous cell carcinoma with different invasive capabilities. J Proteome. 2015;112:180–9. https://doi.org/10.1016/j.jprot.2014.09.004.
112. Esteller M Epigenetic gene silencing in cancer: the DNA hypermethylome. Hum Mol Genet. 2007;16 Spec No 1:R50–R59, doi:https://doi.org/10.1093/hmg/ddm018.
113. Lopez-Serra L, Esteller M. Proteins that bind methylated DNA and human cancer: reading the wrong words. Br J Cancer. 2008;98:1881–5. https://doi.org/10.1038/sj.bjc.6604374.
114. Thomson JP, et al. CpG islands influence chromatin structure via the CpG-binding protein Cfp1. Nature. 2010;464:1082–6. https://doi.org/10.1038/nature08924.
115. Zhang E, et al. H3K27 acetylation activated-long non-coding RNA CCAT1 affects cell proliferation and migration by regulating SPRY4 and HOXB13 expression in esophageal squamous cell carcinoma. Nucleic Acids Res. 2017;45:3086–101. https://doi.org/10.1093/nar/gkw1247.
116. Barneda-Zahonero B, Parra M. Histone deacetylases and cancer. Mol Oncol. 2012;6:579–89. https://doi.org/10.1016/j.molonc.2012.07.003.
117. Sakuma T, et al. Aberrant expression of histone deacetylase 6 in oral squamous cell carcinoma. Int J Oncol. 2006;29:117–24.
118. Feinberg AP, Koldobskiy MA, Gondor A. Epigenetic modulators, modifiers and mediators in cancer aetiology and progression. Nat Rev Genet. 2016;17:284–99. https://doi.org/10.1038/nrg.2016.13.
119. Kantidakis T, et al. Mutation of cancer driver MLL2 results in transcription stress and genome instability. Genes Dev. 2016;30:408–20. https://doi.org/10.1101/gad.275453.115.
120. Ortega-Molina A, et al. The histone lysine methyltransferase KMT2D sustains a gene expression program that represses B cell lymphoma development. Nat Med. 2015;21:1199–208. https://doi.org/10.1038/nm.3943.
121. Chen Y et al. MLL2, not MLL1, plays a major role in sustaining MLL-rearranged acute myeloid Leukemia. Cancer Cell. 2017;31:755–770 e756. doi:https://doi.org/10.1016/j.ccell.2017.05.002.
122. Rayasam GV, et al. NSD1 is essential for early post-implantation development and has a catalytically active SET domain. EMBO J. 2003;22:3153–63. https://doi.org/10.1093/emboj/cdg288.
123. Nimura K, et al. A histone H3 lysine 36 trimethyltransferase links Nkx2-5 to Wolf-Hirschhorn syndrome. Nature. 2009;460:287–91. https://doi.org/10.1038/nature08086.
124. Kuo AJ, et al. NSD2 links dimethylation of histone H3 at lysine 36 to oncogenic programming. Mol Cell. 2011;44:609–20. https://doi.org/10.1016/j.molcel.2011.08.042.
125. Lu T, et al. Regulation of NF-kappaB by NSD1/FBXL11-dependent reversible lysine methylation of p65. Proc Natl Acad Sci USA. 2010;107:46–51. https://doi.org/10.1073/pnas.0912493107.
126. Morishita M, di Luccio E. Cancers and the NSD family of histone lysine methyltransferases. Biochim Biophys Acta. 2011;1816:158–63. https://doi.org/10.1016/j.bbcan.2011.05.004.
127. Iyer NG, Ozdag H, Caldas C. p300/CBP and cancer. Oncogene. 2004;23:4225–31. https://doi.org/10.1038/sj.onc.1207118.
128. El-Naggar AK, et al. Methylation, a major mechanism of p16/CDKN2 gene inactivation in head and neck squamous carcinoma. Am J Pathol. 1997;151:1767–74.

129. Merlo A, et al. 5' CpG island methylation is associated with transcriptional silencing of the tumour suppressor p16/CDKN2/MTS1 in human cancers. Nat Med. 1995;1:686–92.
130. Reed AL, et al. High frequency of p16 (CDKN2/MTS-1/INK4A) inactivation in head and neck squamous cell carcinoma. Cancer Res. 1996;56:3630–3.
131. Diesch J, et al. A clinical-molecular update on azanucleoside-based therapy for the treatment of hematologic cancers. Clin Epigenetics. 2016;8:71. https://doi.org/10.1186/s13148-016-0237-y.
132. Heerboth S, et al. Use of epigenetic drugs in disease: an overview. Genet Epigenet. 2014;6:9–19. https://doi.org/10.4137/GEG.S12270.
133. Yoo CB, Jones PA. Epigenetic therapy of cancer: past, present and future. Nat Rev Drug Discov. 2006;5:37–50. https://doi.org/10.1038/nrd1930.
134. Greenblatt SM, Nimer SD. Chromatin modifiers and the promise of epigenetic therapy in acute leukemia. Leukemia. 2014;28:1396–406. https://doi.org/10.1038/leu.2014.94.
135. Saba HI. Decitabine in the treatment of myelodysplastic syndromes. Ther Clin Risk Manag. 2007;3:807–17.
136. Nervi C, De Marinis E, Codacci-Pisanelli G. Epigenetic treatment of solid tumours: a review of clinical trials. Clin Epigenetics. 2015;7:127. https://doi.org/10.1186/s13148-015-0157-2.
137. Mann BS, Johnson JR, Cohen MH, Justice R, Pazdur R. FDA approval summary: vorinostat for treatment of advanced primary cutaneous T-cell lymphoma. Oncologist. 2007;12:1247–52. https://doi.org/10.1634/theoncologist.12-10-1247.
138. San-Miguel JF, et al. Panobinostat plus bortezomib and dexamethasone versus placebo plus bortezomib and dexamethasone in patients with relapsed or relapsed and refractory multiple myeloma: a multicentre, randomised, double-blind phase 3 trial. Lancet Oncol. 2014;15:1195–206. https://doi.org/10.1016/S1470-2045(14)70440-1.
139. O'Connor OA, et al. Belinostat in patients with relapsed or refractory peripheral T-cell lymphoma: results of the pivotal phase II BELIEF (CLN-19) study. J Clin Oncol. 2015;33:2492–9. https://doi.org/10.1200/JCO.2014.59.2782.
140. Viet CT, et al. Decitabine rescues cisplatin resistance in head and neck squamous cell carcinoma. PLoS One. 2014;9:e112880. https://doi.org/10.1371/journal.pone.0112880.
141. Glasspool RM, et al. A randomised, phase II trial of the DNA-hypomethylating agent 5-aza-2′-deoxycytidine (decitabine) in combination with carboplatin vs carboplatin alone in patients with recurrent, partially platinum-sensitive ovarian cancer. Br J Cancer. 2014;110:1923–9. https://doi.org/10.1038/bjc.2014.116.
142. Biktasova A, et al. Demethylation therapy as a targeted treatment for human papillomavirus-associated head and neck cancer. Clin Cancer Res. 2017;23:7276–87. https://doi.org/10.1158/1078-0432.CCR-17-1438.
143. Haberland M, Montgomery RL, Olson EN. The many roles of histone deacetylases in development and physiology: implications for disease and therapy. Nat Rev Genet. 2009;10:32–42. https://doi.org/10.1038/nrg2485.
144. Ropero S, Esteller M. The role of histone deacetylases (HDACs) in human cancer. Mol Oncol. 2007;1:19–25. https://doi.org/10.1016/j.molonc.2007.01.001.
145. Marks PA, Dokmanovic M. Histone deacetylase inhibitors: discovery and development as anticancer agents. Expert Opin Investig Drugs. 2005;14:1497–511. https://doi.org/10.1517/13543784.14.12.1497.
146. Blumenschein GR Jr, et al. Phase II trial of the histone deacetylase inhibitor vorinostat (Zolinza, suberoylanilide hydroxamic acid, SAHA) in patients with recurrent and/or metastatic head and neck cancer. Investig New Drugs. 2008;26:81–7. https://doi.org/10.1007/s10637-007-9075-2.
147. Brunetto AT, et al. First-in-human, pharmacokinetic and pharmacodynamic phase I study of Resminostat, an oral histone deacetylase inhibitor, in patients with advanced solid tumors. Clin Cancer Res. 2013;19:5494–504. https://doi.org/10.1158/1078-0432.CCR-13-0735.

148. Haigentz M Jr, et al. Phase II trial of the histone deacetylase inhibitor romidepsin in patients with recurrent/metastatic head and neck cancer. Oral Oncol. 2012;48:1281–8. https://doi.org/10.1016/j.oraloncology.2012.05.024.
149. Galloway TJ, et al. A phase I study of CUDC-101, a multitarget inhibitor of HDACs, EGFR, and HER2, in combination with chemoradiation in patients with head and neck squamous cell carcinoma. Clin Cancer Res. 2015;21:1566–73. https://doi.org/10.1158/1078-0432.CCR-14-2820.
150. Bally AP, Austin JW, Boss JM. Genetic and epigenetic regulation of PD-1 expression. J Immunol. 2016;196:2431–7. https://doi.org/10.4049/jimmunol.1502643.
151. Chiappinelli KB, Zahnow CA, Ahuja N, Baylin SB. Combining epigenetic and immunotherapy to combat cancer. Cancer Res. 2016;76:1683–9. https://doi.org/10.1158/0008-5472.CAN-15-2125.
152. Di Croce L, Helin K. Transcriptional regulation by Polycomb group proteins. Nat Struct Mol Biol. 2013;20:1147–55. https://doi.org/10.1038/nsmb.2669.
153. Margueron R, Reinberg D. The Polycomb complex PRC2 and its mark in life. Nature. 2011;469:343–9. https://doi.org/10.1038/nature09784.
154. Kim KH, Roberts CW. Targeting EZH2 in cancer. Nat Med. 2016;22:128–34. https://doi.org/10.1038/nm.4036.
155. Kurmasheva RT, et al. Initial testing (stage 1) of tazemetostat (EPZ-6438), a novel EZH2 inhibitor, by the Pediatric Preclinical Testing Program. Pediatr Blood Cancer. 2017;64. https://doi.org/10.1002/pbc.26218.
156. Peri S, et al. NSD1- and NSD2-damaging mutations define a subset of laryngeal tumors with favorable prognosis. Nat Commun. 2017;8:1772. https://doi.org/10.1038/s41467-017-01877-7.
157. Rogawski DS, Grembecka J, Cierpicki T. H3K36 methyltransferases as cancer drug targets: rationale and perspectives for inhibitor development. Future Med Chem. 2016;8:1589–607. https://doi.org/10.4155/fmc-2016-0071.
158. Esteller M, et al. Inactivation of the DNA-repair gene MGMT and the clinical response of gliomas to alkylating agents. N Engl J Med. 2000;343:1350–4. https://doi.org/10.1056/NEJM200011093431901.
159. Vogelstein B, et al. Cancer genome landscapes. Science. 2013;339:1546–58. https://doi.org/10.1126/science.1235122.
160. Chung CH, et al. Molecular classification of head and neck squamous cell carcinomas using patterns of gene expression. Cancer Cell. 2004;5:489–500.
161. Walter V, et al. Molecular subtypes in head and neck cancer exhibit distinct patterns of chromosomal gain and loss of canonical cancer genes. PLoS One. 2013;8:e56823. https://doi.org/10.1371/journal.pone.0056823.
162. Chung CH, et al. Gene expression profiles identify epithelial-to-mesenchymal transition and activation of nuclear factor-kappaB signaling as characteristics of a high-risk head and neck squamous cell carcinoma. Cancer Res. 2006;66:8210–8. https://doi.org/10.1158/0008-5472.CAN-06-1213.
163. Buffa FM, Harris AL, West CM, Miller CJ. Large meta-analysis of multiple cancers reveals a common, compact and highly prognostic hypoxia metagene. Br J Cancer. 2010;102:428–35. https://doi.org/10.1038/sj.bjc.6605450.
164. Eschrich SA, et al. A gene expression model of intrinsic tumor radiosensitivity: prediction of response and prognosis after chemoradiation. Int J Radiat Oncol Biol Phys. 2009;75:489–96. https://doi.org/10.1016/j.ijrobp.2009.06.014.
165. Roepman P, et al. An expression profile for diagnosis of lymph node metastases from primary head and neck squamous cell carcinomas. Nat Genet. 2005;37:182–6. https://doi.org/10.1038/ng1502.
166. Toustrup K, et al. Development of a hypoxia gene expression classifier with predictive impact for hypoxic modification of radiotherapy in head and neck cancer. Cancer Res. 2011;71:5923–31. https://doi.org/10.1158/0008-5472.CAN-11-1182.

167. Winter SC, et al. Relation of a hypoxia metagene derived from head and neck cancer to prognosis of multiple cancers. Cancer Res. 2007;67:3441–9. https://doi.org/10.1158/0008-5472.CAN-06-3322.
168. Onken MD, et al. A surprising cross-species conservation in the genomic landscape of mouse and human oral cancer identifies a transcriptional signature predicting metastatic disease. Clin Cancer Res. 2014;20:2873–84. https://doi.org/10.1158/1078-0432.CCR-14-0205.
169. Overgaard J, et al. A randomized double-blind phase III study of nimorazole as a hypoxic radiosensitizer of primary radiotherapy in supraglottic larynx and pharynx carcinoma. Results of the Danish head and neck Cancer study (DAHANCA) protocol 5-85. Radiother Oncol. 1998;46:135–46.
170. Toustrup K, et al. Gene expression classifier predicts for hypoxic modification of radiotherapy with nimorazole in squamous cell carcinomas of the head and neck. Radiother Oncol. 2012;102:122–9. https://doi.org/10.1016/j.radonc.2011.09.010.
171. Chalise P, Koestler DC, Bimali M, Yu Q, Fridley BL. Integrative clustering methods for high-dimensional molecular data. Transl Cancer Res. 2014;3:202–16. https://doi.org/10.3978/j.issn.2218-676X.2014.06.03.
172. Mo Q, et al. A fully Bayesian latent variable model for integrative clustering analysis of multi-type omics data. Biostatistics. 2018;19:71–86. https://doi.org/10.1093/biostatistics/kxx017.
173. Mo Q, et al. Pattern discovery and cancer gene identification in integrated cancer genomic data. Proc Natl Acad Sci USA. 2013;110:4245–50. https://doi.org/10.1073/pnas.1208949110.
174. Shen R, Olshen AB, Ladanyi M. Integrative clustering of multiple genomic data types using a joint latent variable model with application to breast and lung cancer subtype analysis. Bioinformatics. 2009;25:2906–12. https://doi.org/10.1093/bioinformatics/btp543.
175. Zhang S, et al. Discovery of multi-dimensional modules by integrative analysis of cancer genomic data. Nucleic Acids Res. 2012;40:9379–91. https://doi.org/10.1093/nar/gks725.
176. Choufani S, et al. NSD1 mutations generate a genome-wide DNA methylation signature. Nat Commun. 2015;6:10207. https://doi.org/10.1038/ncomms10207.
177. Baubec T, et al. Genomic profiling of DNA methyltransferases reveals a role for DNMT3B in genic methylation. Nature. 2015;520:243–7. https://doi.org/10.1038/nature14176.
178. Li H, et al. The histone methyltransferase SETDB1 and the DNA methyltransferase DNMT3A interact directly and localize to promoters silenced in cancer cells. J Biol Chem. 2006;281:19489–500. https://doi.org/10.1074/jbc.M513249200.
179. Papillon-Cavanagh S, et al. Impaired H3K36 methylation defines a subset of head and neck squamous cell carcinomas. Nat Genet. 2017;49:180–5. https://doi.org/10.1038/ng.3757.
180. Goffin J, Eisenhauer E. DNA methyltransferase inhibitors-state of the art. Ann Oncol. 2002;13:1699–716.
181. Plummer R, et al. Phase I study of MG98, an oligonucleotide antisense inhibitor of human DNA methyltransferase 1, given as a 7-day infusion in patients with advanced solid tumors. Clin Cancer Res. 2009;15:3177–83. https://doi.org/10.1158/1078-0432.CCR-08-2859.
182. Halby L, et al. Rapid synthesis of new DNMT inhibitors derivatives of procainamide. ChemBioChem. 2012;13:157–65. https://doi.org/10.1002/cbic.201100522.
183. Stresemann C, Brueckner B, Musch T, Stopper H, Lyko F. Functional diversity of DNA methyltransferase inhibitors in human cancer cell lines. Cancer Res. 2006;66:2794–800. https://doi.org/10.1158/0008-5472.CAN-05-2821.
184. Esquela-Kerscher A, Slack FJ. Oncomirs – microRNAs with a role in cancer. Nat Rev Cancer. 2006;6:259–69. https://doi.org/10.1038/nrc1840.
185. D'Angelo B, Benedetti E, Cimini A, Giordano A. MicroRNAs: a puzzling tool in cancer diagnostics and therapy. Anticancer Res. 2016;36:5571–5. https://doi.org/10.21873/anticanres.11142.
186. Kozaki K, Imoto I, Mogi S, Omura K, Inazawa J. Exploration of tumor-suppressive microRNAs silenced by DNA hypermethylation in oral cancer. Cancer Res. 2008;68:2094–105. https://doi.org/10.1158/0008-5472.CAN-07-5194.

187. Reis PP, et al. Programmed cell death 4 loss increases tumor cell invasion and is regulated by miR-21 in oral squamous cell carcinoma. Mol Cancer. 2010;9:238. https://doi.org/10.1186/1476-4598-9-238.
188. Allegra E, Trapasso S, Pisani D, Puzzo L. The role of BMI1 as a biomarker of cancer stem cells in head and neck cancer: a review. Oncology. 2014;86:199–205. https://doi.org/10.1159/000358598.
189. An Y, Ongkeko WM. ABCG2: the key to chemoresistance in cancer stem cells? Expert Opin Drug Metab Toxicol. 2009;5:1529–42. https://doi.org/10.1517/17425250903228834.
190. Grimm M, et al. ABCB5 expression and cancer stem cell hypothesis in oral squamous cell carcinoma. Eur J Cancer. 2012;48:3186–97. https://doi.org/10.1016/j.ejca.2012.05.027.
191. Hoffmeyer K, et al. Wnt/beta-catenin signaling regulates telomerase in stem cells and cancer cells. Science. 2012;336:1549–54. https://doi.org/10.1126/science.1218370.
192. Momparler RL, Cote S. Targeting of cancer stem cells by inhibitors of DNA and histone methylation. Expert Opin Investig Drugs. 2015;24:1031–43. https://doi.org/10.1517/13543784.2015.1051220.
193. Naik PP, et al. Implications of cancer stem cells in developing therapeutic resistance in oral cancer. Oral Oncol. 2016;62:122–35. https://doi.org/10.1016/j.oraloncology.2016.10.008.
194. Shukla S, Meeran SM. Epigenetics of cancer stem cells: pathways and therapeutics. Biochim Biophys Acta. 2014;1840:3494–502. https://doi.org/10.1016/j.bbagen.2014.09.017.
195. Medina PP, Nolde M, Slack FJ. OncomiR addiction in an in vivo model of microRNA-21-induced pre-B-cell lymphoma. Nature. 2010;467:86–90. https://doi.org/10.1038/nature09284.
196. Gaykalova DA, et al. Outlier analysis defines zinc finger gene family DNA methylation in Tumors and saliva of head and neck Cancer patients. PLoS One. 2015;10:e0142148. https://doi.org/10.1371/journal.pone.0142148.

Chapter 13
Inflammation and Head and Neck Squamous Cell Carcinoma

Paul E. Clavijo, Clint T. Allen, Nicole C. Schmitt, and Carter Van Waes

Abstract Inflammation is a process that is involved in several stages of development and malignant progression of head and neck squamous cell carcinoma. Tobacco and alcohol, human papillomaviruses (HPV), or Epstein-Barr viruses (EBV) can initiate and establish chronic inflammation through a variety of mechanisms. Genomic alterations or viral oncoproteins that induce signaling via phosphatidylinositol 3-kinase (PI3K) and transcription factor nuclear factor-kappaB (NF-κB) regulate numerous genes that promote survival of cancer cells, while they induce inflammatory myeloid-derived suppressor cell (MDSC) and T regulatory (Treg) cell responses that interfere with effector T-cell immunity. Molecular therapies targeting signaling in cancer cells and these deleterious inflammatory cells are being combined with new PD-L1/PD-1 and CTLA-4 immune checkpoint inhibitors to explore better ways to harness the immune system in control of cancer.

Keywords Inflammation · Cytokines · Tumor necrosis factor · PI3K · NF-kappaB · T regulatory cells · Myeloid-derived suppressor cells

13.1 Carcinogenesis and Inflammation

Inflammation is a process that is involved in several stages of carcinogenesis and cancer development, including initiation, promotion, progression, invasion, and metastasis [1]. The hallmarks of inflammation include triggering within affected tissue the activation of genes and proteins that promote death or survival and proliferation of resident cells, and recruitment of factors and cellular mediators from the innate and adaptive immune system that normally function to combat pathogens and restore tissue homeostasis. The role of inflammation in development of head and

---

Supported by NIDCD Intramural Projects ZIA-DC-000016, 73 and 74.

P. E. Clavijo · C. T. Allen · N. C. Schmitt · C. Van Waes (✉)
Head and Neck Surgery Branch, National Institute on Deafness and Other Communication Disorders, National Institutes of Health, Bethesda, MD, USA
e-mail: vanwaesc@nidcd.nih.gov

B. Burtness, E. A. Golemis (eds.), *Molecular Determinants of Head and Neck Cancer*, Current Cancer Research, https://doi.org/10.1007/978-3-319-78762-6_13

neck squamous cell carcinomas (HNSCC) and cancers of related etiology has been the subject of considerable interest. Tobacco and alcohol, human papillomaviruses (HPV), and Epstein-Barr viruses (EBV) are the major etiologic factors involved in the development of HNSCC. These agents can initiate innate and establish chronic inflammation through a variety of mechanisms. Tobacco smoke and smokeless tobacco contain chemical carcinogens such as nitrosamine and polyaromatic hydrocarbons, and alcohol contains sulfites and formaldehyde. Their electrophilic metabolites induce reactive oxygen species (ROS) and reactive nitrogen species (RNS) that modify or damage cell membrane lipids and disrupt DNA, leading to chromosomal breaks, rearrangements, and mutations [2, 3]. Exposure of cell surface nicotinic receptors to nicotine and lipids to reactive chemical carcinogens can induce phosphatidylinositol 3-kinase (PI3K) and protein kinase A (PKA) signal activation. In parallel, DNA damage triggers activation of the nuclear kinase ataxia telangiectasia mutated (ATM), which together with PI3K induces inhibitor kappaB kinase (IKK)- and nuclear factor-kappaB (NF-κB)-dependent transcription and expression of inflammatory factors [4–7]. Additionally, metabolites of the altered oral bacterial microbiome in tobacco and alcohol users may activate Toll-like receptors (TLRs) normally involved in innate immune defenses as a separate means to activate IKK/NF-κB signaling [8, 9]. Alternatively, HPV and EBV express oncoproteins such as E6 and LMP1, respectively, that commandeer IKK signaling [10], providing another route to activate NF-κB. The role of NF-κB in summating the response to these carcinogenic stimuli is critical. This transcription factor induces numerous genes that promote survival of precancerous cells that are attempting to repair accumulating DNA damage, and cytokines that induce ongoing inflammation that further contributes to prosurvival signaling and DNA damage [10]. Unfortunately, critical damage to DNA of oncogenes encoding protein components of these signaling cascades can lead to further aberrant NF-κB activation, inflammatory factor production, and infiltration of chronic inflammatory cells producing ROS, thus perpetuating DNA damage until cumulative alterations affecting chromosomes and genes lead to cancer.

13.2 Genetic Alterations Promoting NF-κB Pathway Activation and Inflammation

The Cancer Genome Atlas (TCGA) has catalogued the major genetic alterations in tobacco, alcohol, and HPV-related HNSCC and identified genomic alterations, many of which provide constitutive signaling changes that can cause activation of NF-κB to orchestrate cell survival, inflammation, and angiogenesis [11]. The most common of these, found in ~30% of HPV(−) and ~60% of HPV(+) HNSCC, are chromosome 3q arm amplifications or mutations affecting the gene encoding PI3K catalytic subunit alpha (*PIK3CA*). PI3K can activate AKT and mTOR kinases to activate IKK/NF-κB signaling and gene expression (Fig. 13.1) [12]. This gene

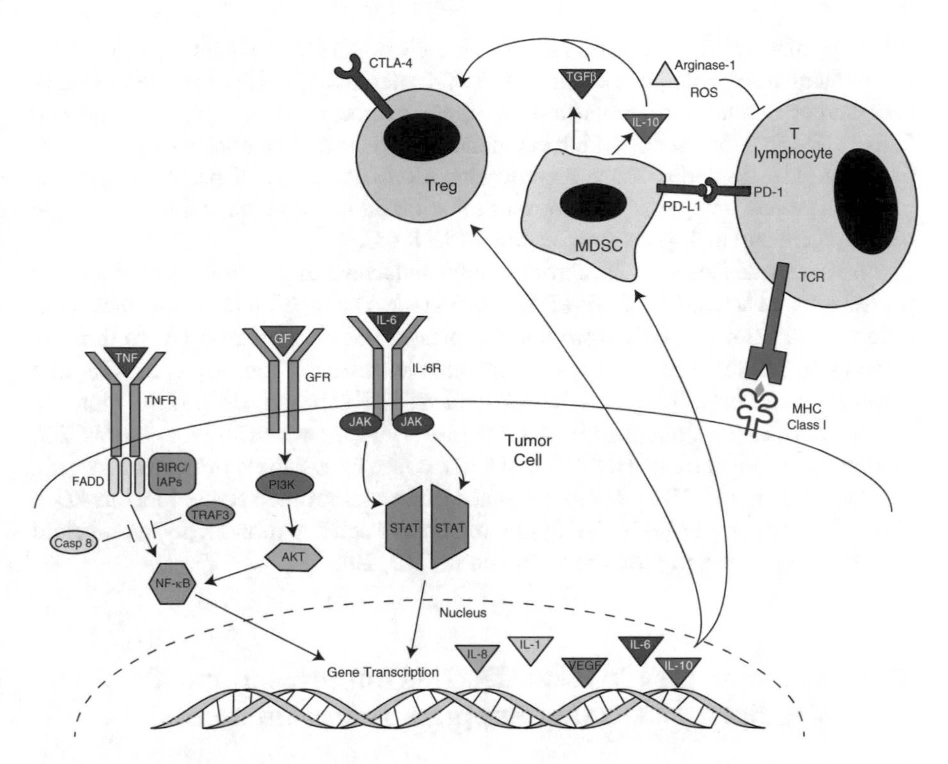

Fig. 13.1 Inflammatory signaling, cytokine gene transcription, and deleterious inflammatory cell responses suppressing T-cell immunity. Mucosal squamous cells are subject to repeated injury by chemical carcinogens or viruses that induce inflammatory factors and cytokines (TNF, tumor necrosis factor; IL-6, interleukin-6) and growth factors (e.g., EGF, epidermal growth factor) that trigger receptor-mediated signal cascades to activate transcription factors that promote inflammatory cytokine and cell responses. HNSCC tumor cells frequently acquire mutations or copy number alterations in genes that encode components of these signal pathways, commandeering them to establish aberrant transcription and expression of cytokines and chronic inflammatory responses (for details, see text). Interleukins (IL-1, IL-6, IL-8, IL-10) and vascular endothelial growth factor recruit T regulatory (Treg) and myeloid-derived suppressor cells (MDSCs) to the tumor microenvironment. These cells express immune checkpoint molecules (PDL-1, CTLA-4), immunosuppressive factors (transforming growth factor-beta, TGFβ; IL-10), and metabolites (arginase-1, reactive oxygen species (ROS)) that suppress effector T lymphocyte immunity

expression may also be enhanced by the capability of PI3K to promote expression of a shortened N-terminal isoform of the tumor suppressor TP63, ΔNp63, that enhances activation of NF-κB genes [13]. Another ~30% of HPV(−) cancers have a variety of amplifications of genes encoding growth factor receptors (EGFR, ERBB2, EPHA2, IGF1R) that can activate PI3K and NF-κB (Fig. 13.1). The remaining ~30% of HPV(−) HNSCC display chromosome 11q 13/22 amplifications of genes (*FADD*, *BIRC2*), or mutations (*CASP8*), that encode parts of the TLR and tumor necrosis factor (TNF) receptor signaling pathways that also contribute to IKK/NF-κB activation (Fig. 13.1). The elevated expression of these signaling intermediates can enhance responsiveness of NF-κB to cytokines and factors produced in an autocrine manner by neoplastic and cancer cells or paracrine signaling by infiltrating inflammatory cells [10]. Besides *PIK3CA* alterations, a subset of HPV(+) cancers harbor deletions or mutations of tumor necrosis factor receptor-associated factor (*TRAF3*), the loss of which can enhance activation of a noncanonical NF-κB pathway [11]. Together, these genetic alterations and affected pathways provide potential targets for agents to prevent or inhibit inflammation that contributes to the development and malignant progression of HNSCC.

The consequences of such chronic injury-induced and genomic alterations that promote signal activation of NF-κB are pathologic. Damage-induced and oncogene-induced activation of NF-κB promotes expression of interleukin-6 (IL-6) that can serve as an autocrine factor to coactivate another transcription factor, STAT3, that amplifies its effects. Together, NF-κB and STAT3 can activate proliferation via *Cyclin D1 (CCND1)*, and also induce the prosurvival genes *BCL-XL* and *BIRC2/3,* that promote resistance of HNSCC cells, in the face of cytotoxic inflammatory cells and factors [14, 15]. NF-κB also promotes expression of the secreted factors *IL-8*, *GRO1/CXCL1*, and *VEGF* [16–18] that recruit and activate monocytic and myeloid inflammatory cells and promote angiogenesis [19, 20].

13.3 Inflammatory Myeloid-Derived Suppressor and T Regulatory Cells that Suppress Tumor Immunity

Tumor cells with aberrant activation of NF-κB express the aforementioned cytokines IL-6, IL-8, and GRO1 and inflammatory factors such as VEGF [16] that induce inflammatory cells that can enable cancer cells to evade or suppress the action of the immune system [10, 19–21]. Such mechanisms include the recruitment of immunosuppressive cells such as immature myeloid-derived suppressor cells (MDSCs) and regulatory T cells (Tregs) (Fig. 13.1). MDSCs were first described in patients with HNSCC as a group of immunosuppressive cells expressing CD34 [20, 21]. MDSCs originate in the bone marrow from common myeloid progenitor cells in the presence of cytokines and growth factors such as granulocyte-macrophage colony-stimulating factor (GM-CSF) and granulocyte colony-stimulating factor (G-CSF) and VEGF, among others [22, 23]. MDSCs move from the bone marrow to other lymphoid organs or the tumor microenvironment (TME) in response to a chemotactic gradient. Tregs were first described by Sakaguchi et al. as a population of T cells regulating self-tolerance and autoimmunity [24]. Tregs are a subset of $CD4^+$ T cells distinguished by expression of CD25 and the transcription factor forkhead box P3 (FOXP3). Tregs have been divided into "natural" and "induced" Tregs. Natural Tregs are generated in the thymus and are involved in the process of central tolerance against self-antigens. Induced Tregs are generated mainly in the periphery and require further antigen stimulation [25]. Tregs are highly immunosuppressive, inhibiting $CD4^+CD25^-$ T cells (T helper cells), $CD8^+$ T cells (cytolytic T cells), dendritic (antigen-presenting) cells, natural killer (NK) cells, and B cells. High levels of Tregs infiltrating tumors are associated with poor disease prognosis [26, 27]. Treg suppression may be mediated by production of inhibitory cytokines IL-10 and TGF-β and by expression of inhibitory immune checkpoint receptors such as CTLA-4, PD-1, and TIM-3.

13.3.1 Myeloid-Derived Suppressor Cells

MDSCs are highly immunosuppressive and have been especially associated with suppression of T lymphocyte and natural killer (NK) cell immunity. T helper (CD4+) and T cytotoxic (CD8+) lymphocytes and NK cells recognize and promote killing of tumor cells bearing mutated protein antigens recognized by T cells, or molecular activators of NK cells. MDSCs can suppress activation of T cells by expression of a variety of mediators, including arginase 1 (ARG1), the inducible nitric oxide synthase (iNOS), reactive oxygen species (ROS), indoleamine dioxygenase (IDO), and cytokines such as IL-10 and TGF-β (Fig. 13.1) [28]. ARG1 expression depletes the amino acid arginine from the tumor environment; as arginine is critical to protein synthesis and expression of the T-cell receptor CD3 ζ chain and other proteins necessary for antigen-specific T-cell activation and response, this impairs T-cell responses [29]. iNOS inhibits the activation of T lymphocytes by inhibiting the phosphorylation and intracellular signal activation of JAK-1, JAK-3, and transcription factor STAT-5, required for T-cell activation [30]. Moreover, it has been reported that several kinases involved in phosphorylation and activation of T cells, including JAK3 and STAT5, are inactivated by NO action [30, 31]. ROS damage cell membranes and intracellular contents, decreasing viability of both T cells and NK cells. Increasing levels of IDO induce polarization of the immune response to recruitment of T helper type 1 (Th1) cells that induce antibodies rather than cytotoxic T cells needed for tumor cell killing or stimulate T-cell apoptosis [32]. Conversely, MDSCs also produce cytokines such as IL-10 and TGF-β that induce Treg cells, which suppress T helper and cytotoxic T lymphocyte responses [33].

MDSCs are a heterogeneous group of immature myeloid cells that has been divided into two main subtypes: granulocytic (gMDSCs; also known as polymorphonuclear (PMN-MDSCs)) that share features of neutrophils, and monocytic (mMDSCs) that share features of monocytes. gMDSCs are characterized by the expression of $CD11b^+$ $Ly\text{-}6G^+$ $Ly\text{-}6C^{low}$ in mice and $CD11b^+CD15^+HLA\text{-}DR^-CD33^+$ in human. gMDSCs also express the chemokine receptor CXCR2, which is activated by the ligands CXCL1, CXCL5, and CXCL8, produced by tumor and other cells in the microenvironment [28]. In contrast, mMDSCs are characterized by the expression of $CD11b^+$ $Ly\text{-}6G^-$ $Ly\text{-}6C^{high}$ in mice and $CD11b^+CD14^+HLA\text{-}DR^-CD33^+$ in humans. mMDSCs express the chemokine receptor CCR2 and are recruited to the tumor microenvironment by the chemokines CCL2 and CCL5 [28, 34]. gMDSCs are highly expressed in most cancers compared to mMDSCs and secrete many of the aforementioned mediators that suppress T and NK cells.

mMDSCs can transform into tumor-associated macrophages (TAMs) which act as immunosuppressive macrophages [28]. Youn et al. showed that in patients with HNSCC and other types of cancer, 10–20% of mononuclear cells were categorized as MDSCs, with a 5:1 ratio gMDSCs/mMDSCs [35]. Another study demonstrated that the percentage of $CD14^{+}$ $HLA\text{-}DR^{-}$ mMDSCs and Tregs are also increased in peripheral blood from HNSCC patients. mMDSCs suppressed T-cell proliferation and interferon (IFN)γ production, which are indicative of suppression of functional activation of T lymphocytes. The suppressive activity of mMDSC may be mediated by expression of TGF-β and programmed death ligand-1 (PD-L1) [36], which inhibit T lymphocyte proliferation and function. Chen et al. demonstrated high levels of $CD11b^{+}HLA\text{-}DR^{-}CD\text{-}14^{+}CD33^{+}$ MDSCs in peripheral blood from HNSCC patients compared to normal controls (28.6 ± 13.1% vs 7.5 ± 6.3%) [37]. Corzo et al. obtained tissue from patients with HNSCC and analyzed the effect of MDSCs isolated from tumor or peripheral blood on T-cell proliferation. This study found higher inhibition of T-cell proliferation by tumor rather than peripheral blood MDSCs [38].

Recent experimental studies have demonstrated the importance of MDSCs in pathogenesis of immune evasion during tumor progression, reflected in part by the fact that depletion of gMDSCs increases the efficacy of new immune checkpoint therapies for HNSCC [39]. Similar to the clinical observations made in human HNSCC [35], analysis of a mouse oral carcinoma model during tumor progression showed high infiltration of gMDSCs compared to mMDSCs [39]. Moreover, these gMDSCs potently suppress $CD4^{+}$ T helper and $CD8^{+}$ cytotoxic T-cell proliferation and effector functions. MDSCs express high levels of CXCR2, a receptor mediating their recruitment by tumor cytokines induced by NF-κB such as CXCL1/2 (also known as GRO1/2) and CXCL8 (IL-8), and therefore CXCR2 has potential for therapeutically targeting MDSCs. Supporting this, our laboratory showed that these cytokines are highly expressed by human HNSCC cell lines as well [40]. Further, infiltration of MDSCs, tumor growth, and metastasis of murine tumors overexpressing the CXCL1/GRO1 orthologue Gro1 were inhibited in CXCR2 knockout mice [19]. Targeting another inflammatory pathway, Chen et al. demonstrated that inhibitors of cyclooxygenase-2 (COX2), an enzyme involved in synthesis of prostaglandins expressed by MDSCs, decreased the numbers of CD11b+ MDSCs and decreased levels of ARG-1 and ROS produced from these cells that have deleterious effects on T-cell immunity [37].

Preclinical and clinical studies have identified another drug that unexpectedly inhibits MDSCs and Tregs, and enhances effector T-cell immunity. Tadalafil is a phosphodiesterase-5 (PDE5) inhibitor originally investigated for modulation of NOS and blood pressure, which was found to inhibit production of NOS and Arg-1 and reduce numbers of MDSCs [41]. Weed et al. performed a double-blinded, randomized, three-arm phase I study in which patients undergoing definitive surgical resection of oral and oropharyngeal HNSCC were treated with tadalafil 10 mg/day, 20 mg/day, or placebo (3:3:1 ratio) for at least 20 days preoperatively (Clinicaltrials.gov: NCT00843635) [42]. The primary endpoints were to (i) determine the effect of PDE5 inhibition on MDSC and Treg, (ii) evaluate the effect of PDE5 inhibition on tumor T-cell immunity, and (iii) evaluate if these effects were dose dependent. Compared to the placebo group, a significant decrease of both m-MDSC and Treg was observed in most of the tadalafil-treated patients. Conversely, treatment with tadalafil also significantly increased proliferation of T-cell receptor subunit CD3+ T cells in response to dendritic cells pulsed with tumor cell lysate. There was no significant difference in effect between the dose levels. In a phase II study, the Califano group further quantified these results in patients who were treated with tadalafil or placebo and undergoing definitive surgical resection of oral and oropharyngeal HNSCC (NCT00894413) [42]. ARG-1 and iNOS activity was significantly decreased in tadalafil-treated patients. ARG1 showed a mean 0.83-fold change ($P = 0.004$) in tadalafil-treated patients versus control patients; iNOS showed a mean 0.66-fold change ($P = 0.003$) compared to a slight increase (1.02-fold) in control patients. This was accompanied by a relative reduction in MDSC numbers with a mean 0.81 fold change in the treated cohort compared to a 1.28-fold increase in control patients ($P < 0.0001$). The analysis also demonstrated a statistically significant reduction in Tregs in the tadalafil-treated patients with a relative increase in the placebo group (Treg, mean placebo 1.79, tadalafil 0.84; $P = 0.0006$). Tadalafil-treated patients showed a significant increase in T-cell proliferation stimulated with anti-CD3/CD28, with a mean 2.4-fold increase compared with a 1.1-fold increase in control patients ($P = 0.003$). Tadalafil increased T-cell activation on both CD4+ T cells, mean 1.6-fold in tadalafil-treated versus 1.3-fold in control treated patients ($P = 0.042$), and CD8 T cells increased 1.4-fold versus no change ($P = 0.005$) [41].

In a recent experimental study in the murine oral cancer (MOC)-1 model, Davis et al. investigated targeting signal kinase isoforms PI3Kδ and PI3Kγ, important in activation and function of MDSCs, which can inhibit immune checkpoint antagonist facilitated T-cell responses. They examined if IPI-145, a PI3Kδ and PI3Kγ inhibitor, could inhibit MDSC function and enhance T-cell responses in combination with PD-L1 checkpoint blockade in the murine oral carcinoma (MOC) model. They demonstrated functional inhibition of MDSC with IPI-145 and combination therapy with anti-PD-L1 induced $CD8^{+}$ T lymphocyte-dependent primary tumor growth delay and prolonged survival in T-cell-inflamed MOC tumor models. However, higher doses of IPI-145 reversed the observed enhancement of anti-PD-L1 efficacy due to off-target suppression of the activity of tumor-infiltrating CD8+ cytotoxic T lymphocytes. Together, their results provide preclinical evidence supporting investigation of low-dose use, isoform-specific PI3Kδ/γ inhibitors to suppress MDSC to enhance responses to immune checkpoint blockade [43].

13.3.2 T Regulatory Cells

The link between cancer-associated inflammation and tumor-infiltrating Tregs in HNSCC is likely multifactorial. This includes chemokine-dependent recruitment of these cells into the tumor microenvironment via production of CCL5, CCL17, and CCL22, expansion of Tregs producing TGFβ and IL-10, and conversion of non-suppressive $CD25^{-}CD4+$ T cells into suppressive $FoxP3^{+}CD25^{+}CD4^{+}$Tregs [44, 45]. High levels of Tregs have been found in patients with a number of distinct types of cancer. Two groups showed an increase in the levels of Tregs in the peripheral circulation of patients with HNSCC compared to normal controls (10.1 ± 4.7 vs $5.4 \pm 2.7\%$ PMBCs) and (4.8 ± 2.3 vs $3.4 \pm 1.1\%$ p. <0.01) [46, 47]. Another group also showed an increase in the levels of Tregs in nasopharyngeal carcinoma patients (13.6 ± 6.7 vs $8.8 \pm 3.8\%$, p. 0.0001) [48]. Levels of the inhibitory receptors CTLA-4, PD-1, and TIM-3 were highly increased on tumor-infiltrating Tregs compared with Tregs in peripheral blood. In T-cell proliferation assays, tumor-infiltrating Tregs were shown to suppress T-cell proliferation to a higher degree compared to circulating Tregs, possibly due to increased upregulation of immunosuppressive markers such as CD39 and LAP in tumor-infiltrating Tregs [49]. On the contrary, Shang et al. performed a meta-analysis of 76 articles addressing the prognostic value of tumor-infiltrating Tregs across 17 types of cancer, including head and neck cancer. This study found that Treg infiltration in head and neck cancer is linked to favorable prognosis and favorable clinical outcome compared to other types of cancer (OR 0.69, 95% CI 0.50 to 0.95, $p = 0.024$), where it is associated with poor prognosis. Moreover, infiltration of Tregs in some tumors is associated with the presence of macrophages and neutrophils [50]. In other tumors, Venet et al. have shown that Tregs decrease macrophage and monocyte survival, decreasing inflammation-induced tumor progression [51]. The basis for these differences is not yet known.

The principle of therapeutic targeting of Tregs is quite similar to that of MDSCs and involves attempts to selectively target signaling or effector pathways important for Treg function or block the terminal suppressive functions themselves. Receptor tyrosine kinase signaling through PI3Kδ appears to be critical for the function of T cells, including Tregs [52]. Selective inhibition of the δ subunit of PI3K could lead to decreased Treg function, but pharmacologic inhibition could also have a narrow therapeutic window, given that effector T cells are very likely to be altered as well. Most efforts to therapeutically alter Treg presence or function have focused on the use of immune checkpoint blockade (ICB). Tregs express immune checkpoint proteins CTLA-4 and PD-L1, which both contribute to their immunosuppressive function (Fig. 13.1) [53, 54]. CTLA-4 ICB in mice leads to depletion of Tregs, supporting a role of CTLA-4 in the therapeutic effect seen with this antibody. CTLA-4 blockade of effector T cells also partially increases antitumor activity. Maximal antitumor activity is reached when CTLA-4 is blocked in both compartments, effector T cells and Tregs, showing that CTLA-4-mediated Treg depletion is essential to increase antitumor activity [55]. Whether this happens in patients with HNSCC that receive CTLA-4 ICB is yet unknown. In another study, Tregs and MDSCs were evaluated after treatment with tremelimumab (anti-CTLA-4) and interferon-α. The data showed an increase in the percentage of Tregs at day 29 and 85 that could be associated with to an increase in the total $CD4^{+}$ T-cell compartment [56]. Similarly, some preclinical evidence in melanoma suggests that receptor PD-1 ICB blocks the suppressive capacity of Tregs [57], but demonstration of this in patients with HNSCC is lacking. There is great interest in understanding how to best deplete or inhibit Treg function in HNSCC, particularly given evidence that standard-of-care cisplatin-based chemotherapy and ionizing radiation induce Treg accumulation and function. It is possible that this induction of Tregs contributes to tumor-localized immune suppression and recurrence after definitive therapy [58].

13.4 Modulating Cancer-Associated Inflammation as Therapeutic Approach to Enhancing Responses to Immune Checkpoint Blockade

Given our understanding of the underlying aberrant intracellular signaling that leads to expression of cytokines, chemokines and factors that drive inflammation in the tumor microenvironment, targeting these pathways would be an attractive means to alter the inflamed tumor microenvironment [59]. We now understand that PD-1/PD-L1-based ICB cannot induce a de novo antitumor immune response but rather unleashes existing antitumor immunity that was blocked by PD-1/PD-L1 signaling [60]. Clinically, only 15–20% of patients respond to single agent PD-based ICB even though >50% of HNSCC tumors are T-cell inflamed (tumors infiltrated with increased $CD8^+$ T lymphocytes producing type I interferon expression), indicative of an underlying antitumor immune response [61–63]. Immune suppression within the tumor microenvironment, mediated by the factors described above, likely accounts for this low level of response. As our understanding of the precise mechanisms by which dysregulated signaling within HNSCC cells ultimately leads to inflammation and the recruitment of MDSCs and Tregs into tumors increases, combining small-molecule or RNA interference-based inhibitors of these pathways may lead to additive or synergistic effects when combined with ICB.

A recent adjunct to direct targeting of inhibitory inflammatory factors and cells involves approaches to modulating the activating cytokine milieu of the tumor microenvironment. Innate immune activation, culminating in the production of type I IFN, is essential to the development of antigen-specific adaptive immunity [64]. Recent work has led to understanding the importance of signaling through STING (stimulator of interferon genes) in initiating antitumor immunity [65]. STING is an intracellular receptor that is activated by cyclic dinucleotides from DNA and induces expression of type-I interferons. Moore et al. have demonstrated potent induction of antigen-specific T-cell immunity following activation of STING signaling with a synthetic STING ligand in syngeneic murine models of oral cavity cancer [66, 67]. STING-dependent production of type I IFN further enhanced the T-cell inflamed local microenvironment, and antitumor immunity was enhanced with the addition of anti-PD-L1 ICB to reverse adaptive immune resistance in this preclinical MOC model. Such approaches to directly alter the tumor innate immune cytokine profile may be independent of genetic alterations/deregulated signaling within tumor cells and be more broadly applicable to tumors with a range of driver mutations.

References

1. Gasparoto TH, et al. Inflammatory events during murine squamous cell carcinoma development. J Inflamm (Lond). 2012;9(1):46.
2. Choudhari SK, et al. Oxidative and antioxidative mechanisms in oral cancer and precancer: a review. Oral Oncol. 2014;50(1):10–8.
3. Hecht SS. Lung carcinogenesis by tobacco smoke. Int J Cancer. 2012;131(12):2724–32.
4. West KA, et al. Tobacco carcinogen-induced cellular transformation increases Akt activation in vitro and in vivo. Chest. 2004;125(5 Suppl):101S–2S.
5. Tsurutani J, et al. Tobacco components stimulate Akt-dependent proliferation and NFkappaB-dependent survival in lung cancer cells. Carcinogenesis. 2005;26(7):1182–95.
6. Dennis PA, et al. The biology of tobacco and nicotine: bench to bedside. Cancer Epidemiol Biomark Prev. 2005;14(4):764–7.
7. Miyamoto S. Nuclear initiated NF-kappaB signaling: NEMO and ATM take center stage. Cell Res. 2011;21(1):116–30.
8. Zu Y, et al. Lipopolysaccharide-induced toll-like receptor 4 signaling in esophageal squamous cell carcinoma promotes tumor proliferation and regulates inflammatory cytokines expression. Dis Esophagus. 2017;30(2):1–8.
9. Farnebo L, et al. Targeting toll-like receptor 2 inhibits growth of head and neck squamous cell carcinoma. Oncotarget. 2015;6(12):9897–907.
10. Van Waes C. Nuclear factor-kappaB in development, prevention, and therapy of cancer. Clin Cancer Res. 2007;13(4):1076–82.
11. Cancer Genome Atlas N. Comprehensive genomic characterization of head and neck squamous cell carcinomas. Nature. 2015;517(7536):576–82.
12. Hutti JE, et al. Oncogenic PI3K mutations lead to NF-kappaB-dependent cytokine expression following growth factor deprivation. Cancer Res. 2012;72(13):3260–9.
13. Yang X, et al. DeltaNp63 versatilely regulates a broad NF-kappaB gene program and promotes squamous epithelial proliferation, migration, and inflammation. Cancer Res. 2011;71(10):3688–700.
14. Lee TL, et al. A signal network involving coactivated NF-kappaB and STAT3 and altered p53 modulates BAX/BCL-XL expression and promotes cell survival of head and neck squamous cell carcinomas. Int J Cancer. 2008;122(9):1987–98.
15. Duan J, et al. Nuclear factor-kappaB p65 small interfering RNA or proteasome inhibitor bortezomib sensitizes head and neck squamous cell carcinomas to classic histone deacetylase inhibitors and novel histone deacetylase inhibitor PXD101. Mol Cancer Ther. 2007;6(1):37–50.
16. Duffey DC, et al. Expression of a dominant-negative mutant inhibitor-kappaBalpha of nuclear factor-kappaB in human head and neck squamous cell carcinoma inhibits survival, proinflammatory cytokine expression, and tumor growth in vivo. Cancer Res. 1999;59(14):3468–74.
17. Bancroft CC, et al. Coexpression of proangiogenic factors IL-8 and VEGF by human head and neck squamous cell carcinoma involves coactivation by MEK-MAPK and IKK-NF-kappaB signal pathways. Clin Cancer Res. 2001;7(2):435–42.
18. Loukinova E, et al. Expression of proangiogenic chemokine Gro 1 in low and high metastatic variants of pam murine squamous cell carcinoma is differentially regulated by IL-1alpha, EGF and TGF-beta1 through NF-kappaB dependent and independent mechanisms. Int J Cancer. 2001;94(5):637–44.
19. Loukinova E, et al. Growth regulated oncogene-alpha expression by murine squamous cell carcinoma promotes tumor growth, metastasis, leukocyte infiltration and angiogenesis by a host CXC receptor-2 dependent mechanism. Oncogene. 2000;19(31):3477–86.
20. Young MR, et al. Human squamous cell carcinomas of the head and neck chemoattract immune suppressive CD34(+) progenitor cells. Hum Immunol. 2001;62(4):332–41.
21. Pak AS, et al. Mechanisms of immune suppression in patients with head and neck cancer: presence of CD34(+) cells which suppress immune functions within cancers that secrete granulocyte-macrophage colony-stimulating factor. Clin Cancer Res. 1995;1(1):95–103.

22. Sawanobori Y, et al. Chemokine-mediated rapid turnover of myeloid-derived suppressor cells in tumor-bearing mice. Blood. 2008;111(12):5457–66.
23. Youn JI, et al. Subsets of myeloid-derived suppressor cells in tumor-bearing mice. J Immunol. 2008;181(8):5791–802.
24. Sakaguchi S, et al. Immunologic self-tolerance maintained by activated T cells expressing IL-2 receptor alpha-chains (CD25). Breakdown of a single mechanism of self-tolerance causes various autoimmune diseases. J Immunol. 1995;155(3):1151–64.
25. Bluestone JA, Abbas AK. Natural versus adaptive regulatory T cells. Nat Rev Immunol. 2003;3(3):253–7.
26. Beyer M, Schultze JL. Regulatory T cells in cancer. Blood. 2006;108(3):804–11.
27. Tartour E, et al. Serum soluble interleukin-2 receptor concentrations as an independent prognostic marker in head and neck cancer. Lancet. 2001;357(9264):1263–4.
28. Kumar V, et al. The nature of myeloid-derived suppressor cells in the tumor microenvironment. Trends Immunol. 2016;37(3):208–20.
29. Rodriguez PC, et al. Regulation of T cell receptor CD3zeta chain expression by L-arginine. J Biol Chem. 2002;277(24):21123–9.
30. Mazzoni A, et al. Myeloid suppressor lines inhibit T cell responses by an NO-dependent mechanism. J Immunol. 2002;168(2):689–95.
31. Schindler H, Bogdan C. NO as a signaling molecule: effects on kinases. Int Immunopharmacol. 2001;1(8):1443–55.
32. Lee GK, et al. Tryptophan deprivation sensitizes activated T cells to apoptosis prior to cell division. Immunology. 2002;107(4):452–60.
33. Huang B, et al. Gr-1+CD115+ immature myeloid suppressor cells mediate the development of tumor-induced T regulatory cells and T-cell anergy in tumor-bearing host. Cancer Res. 2006;66(2):1123–31.
34. Huang B, et al. CCL2/CCR2 pathway mediates recruitment of myeloid suppressor cells to cancers. Cancer Lett. 2007;252(1):86–92.
35. Youn JI, et al. Epigenetic silencing of retinoblastoma gene regulates pathologic differentiation of myeloid cells in cancer. Nat Immunol. 2013;14(3):211–20.
36. Chikamatsu K, et al. Immunosuppressive activity of CD14+ HLA-DR- cells in squamous cell carcinoma of the head and neck. Cancer Sci. 2012;103(6):976–83.
37. Chen WC, et al. Inflammation-induced myeloid-derived suppressor cells associated with squamous cell carcinoma of the head and neck. Head Neck. 2017;39(2):347–55.
38. Corzo CA, et al. HIF-1alpha regulates function and differentiation of myeloid-derived suppressor cells in the tumor microenvironment. J Exp Med. 2010;207(11):2439–53.
39. Clavijo PE, et al. Resistance to CTLA-4 checkpoint inhibition reversed through selective elimination of granulocytic myeloid cells. Oncotarget. 2017;8(34):55804–20.
40. Chen Z, et al. Expression of proinflammatory and proangiogenic cytokines in patients with head and neck cancer. Clin Cancer Res. 1999;5(6):1369–79.
41. Califano JA, et al. Tadalafil augments tumor specific immunity in patients with head and neck squamous cell carcinoma. Clin Cancer Res. 2015;21(1):30–8.
42. Weed DT, et al. Tadalafil reduces myeloid-derived suppressor cells and regulatory T cells and promotes tumor immunity in patients with head and neck squamous cell carcinoma. Clin Cancer Res. 2015;21(1):39–48.
43. Davis RJ, et al. Anti-PD-L1 efficacy can be enhanced by inhibition of myeloid-derived suppressor cells with a selective inhibitor of PI3Kdelta/gamma. Cancer Res. 2017;77(10):2607–19.
44. Ondondo B, et al. Home sweet home: the tumor microenvironment as a haven for regulatory T cells. Front Immunol. 2013;4:197.
45. Chaudhary B, Elkord E. Regulatory T cells in the tumor microenvironment and Cancer progression: role and therapeutic targeting. Vaccines (Basel). 2016;4(3): pii, E28.
46. Chikamatsu K, et al. Relationships between regulatory T cells and CD8+ effector populations in patients with squamous cell carcinoma of the head and neck. Head Neck. 2007;29(2):120–7.
47. Schaefer C, et al. Characteristics of CD4+CD25+ regulatory T cells in the peripheral circulation of patients with head and neck cancer. Br J Cancer. 2005;92(5):913–20.

48. Lau KM, et al. Increase in circulating Foxp3+CD4+CD25(high) regulatory T cells in nasopharyngeal carcinoma patients. Br J Cancer. 2007;96(4):617–22.
49. Jie HB, et al. Intratumoral regulatory T cells upregulate immunosuppressive molecules in head and neck cancer patients. Br J Cancer. 2013;109(10):2629–35.
50. Shang B, et al. Prognostic value of tumor-infiltrating FoxP3+ regulatory T cells in cancers: a systematic review and meta-analysis. Sci Rep. 2015;5:15179.
51. Venet F, et al. Human CD4+CD25+ regulatory T lymphocytes inhibit lipopolysaccharide-induced monocyte survival through a Fas/Fas ligand-dependent mechanism. J Immunol. 2006;177(9):6540–7.
52. Ali K, et al. Inactivation of PI(3)K p110delta breaks regulatory T-cell-mediated immune tolerance to cancer. Nature. 2014;510(7505):407–11.
53. Wing K, et al. CTLA-4 control over Foxp3+ regulatory T cell function. Science. 2008;322(5899):271–5.
54. Parry RV, et al. CTLA-4 and PD-1 receptors inhibit T-cell activation by distinct mechanisms. Mol Cell Biol. 2005;25(21):9543–53.
55. Peggs KS, et al. Blockade of CTLA-4 on both effector and regulatory T cell compartments contributes to the antitumor activity of anti-CTLA-4 antibodies. J Exp Med. 2009;206(8):1717–25.
56. Tarhini AA, et al. Differing patterns of circulating regulatory T cells and myeloid-derived suppressor cells in metastatic melanoma patients receiving anti-CTLA4 antibody and interferon-alpha or TLR-9 agonist and GM-CSF with peptide vaccination. J Immunother. 2012;35(9):702–10.
57. Wang W, et al. PD1 blockade reverses the suppression of melanoma antigen-specific CTL by CD4+ CD25(hi) regulatory T cells. Int Immunol. 2009;21(9):1065–77.
58. Schuler PJ, et al. Effects of adjuvant chemoradiotherapy on the frequency and function of regulatory T cells in patients with head and neck cancer. Clin Cancer Res. 2013;19(23):6585–96.
59. Vander Broek R, et al. The PI3K/Akt/mTOR axis in head and neck cancer: functions, aberrations, cross-talk, and therapies. Oral Dis. 2015;21(7):815–25.
60. Taube JM, et al. Colocalization of inflammatory response with B7-h1 expression in human melanocytic lesions supports an adaptive resistance mechanism of immune escape. Sci Transl Med. 2012;4(127):127ra37.
61. Keck MK, et al. Integrative analysis of head and neck cancer identifies two biologically distinct HPV and three non-HPV subtypes. Clin Cancer Res. 2015;21(4):870–81.
62. Ferris RL, et al. Nivolumab for recurrent squamous-cell carcinoma of the head and neck. N Engl J Med. 2016;375(19):1856–67.
63. Seiwert TY, et al. Safety and clinical activity of pembrolizumab for treatment of recurrent or metastatic squamous cell carcinoma of the head and neck (KEYNOTE-012): an open-label, multicentre, phase 1b trial. Lancet Oncol. 2016;17(7):956–65.
64. Woo SR, Corrales L, Gajewski TF. Innate immune recognition of cancer. Annu Rev Immunol. 2015;33:445–74.
65. Corrales L, et al. The host STING pathway at the interface of cancer and immunity. J Clin Invest. 2016;126(7):2404–11.
66. Moore E, et al. Established T cell-inflamed tumors rejected after adaptive resistance was reversed by combination STING activation and PD-1 pathway blockade. Cancer Immunol Res. 2016;4(12):1061–71.
67. Gadkaree SK, et al. Induction of tumor regression by intratumoral STING agonists combined with anti-programmed death-L1 blocking antibody in a preclinical squamous cell carcinoma model. Head Neck. 2017;39(6):1086–94.

Chapter 14
Immunotherapy in Head and Neck Squamous Cell Carcinoma (HNSCC)

Jennifer Moy and Robert L. Ferris

Abstract Based on its ability to restore key signaling pathways of the host immune system and thus to counteract immune escape by malignant cells, cancer immunotherapy is now at the forefront of cancer research in the treatment of head and neck squamous cell carcinoma (HNSCC). Understanding how tumors evade immune recognition and attack, through strategies that include reducing inherent immunogenicity, dysregulating immune checkpoints, and producing an immunosuppressive tumor microenvironment, will allow the development of novel therapeutic agents to manipulate the immune response. Various forms of immunotherapy are in preclinical trials, including vaccines, oncolytic viruses, and adoptive cell transfer, with the most promising clinical results thus far associated with the use of monoclonal antibodies. This chapter will review the mechanisms of immune escape, and will describe ongoing preclinical and clinical studies, and their implications for immunotherapy in HNSCC.

Keywords NK cells · EGFR · PD-1 · PD-L1 · Cetuximab · Immune checkpoint receptors · Programmed cell death 1

J. Moy
Department of Otolaryngology, University of Pittsburgh, Pittsburgh, PA, USA

R. L. Ferris (✉)
Department of Otolaryngology, University of Pittsburgh, Pittsburgh, PA, USA

UPMC Hillman Cancer Center, Pittsburgh, PA, USA

Department of Immunology, University of Pittsburgh, Pittsburgh, PA, USA

Hillman Cancer Center Research Pavilion, Pittsburgh, PA, USA
e-mail: ferrisrl@upmc.edu

B. Burtness, E. A. Golemis (eds.), *Molecular Determinants of Head and Neck Cancer*, Current Cancer Research, https://doi.org/10.1007/978-3-319-78762-6_14

14.1 Introduction

Classically, the immune system was thought to exist to thwart invasion by foreign microbes, such as bacteria, viruses, and parasites. However, the first conception of an immune contribution to carcinogenesis was proposed in the 1900s and has recently undergone a renaissance in the study of the role of the immune system in cancer development. It is now believed that the immune system also works to prevent growth of malignantly transformed cells. Cellular immunity protects the body from foreign and mutated cells (i.e., bacteria, viruses, and malignant cells) through activating various lymphocytes aimed at destruction of these cells. Through the T-cell receptor (TCR) and the cancer cell-expressed human leukocyte antigen (HLA) signals, cytotoxic T cells are able to induce apoptosis in cells that display epitopes of foreign antigen on their surface (viral or bacterially infected cells or cancer cells displaying tumor antigens). Furthermore, cytokine release leads to the recruitment of other immune cells such as macrophages and natural killer cells (NK cells), which target these "foreign cells" for destruction. The idea of immunological control of malignant cells was first proposed by Ehrlich in 1908 [1], with the idea that premalignant cells are destroyed by the immune system before tumor formation can occur. Derangements in the immune system in conjunction with, or dependent on, alterations in transformed cancer cells may allow immune escape and tumor development.

It is now well appreciated that head and neck squamous cell carcinoma (HNSCC) is an immunosuppressive disease with an overall lower total lymphocyte number, impaired function of NK cells and effector T cells, and an induction of suppressive regulatory T cells (Tregs) that secrete suppressive cytokines and express immune inhibitory receptors, compared to healthy controls [2–8]. The goal of immunotherapy is to restore key signaling pathways of the host immune system to counteract immune escape by malignant cells. Cancer immunotherapy is now at the forefront of cancer research, as it has the potential for durable responses with less adverse effects than conventional treatments. This property provides patients with recurrent/metastatic HNSCC a viable new treatment option, with potent new forms of immunotherapeutic agent now entering trials in treatment-naïve patients.

The first anticancer immunotherapeutic agents were the multifunctional cytokines interleukin 2 (IL-2), interferon alpha (IFN-α), and tumor necrosis factor alpha (TNF-α) [9–11]. These agents were developed in the 1980s and are still widely used in various conditions, such as melanoma, renal cell carcinoma, as well as hematologic cancers such as leukemia and lymphoma. However, these agents have shown marginal efficacy, and systemic toxicities significantly limited their use [12, 13]. The real breakthrough in immunotherapy came in 2010 with the development of immune checkpoint inhibitors. The first to be approved by the Food and Drug Administration (FDA) in 2011 was ipilimumab, a cytotoxic T lymphocyte antigen-4 (CTLA4)-blocking monoclonal antibody (mAb) that was approved for treatment of melanoma. The success of ipilimumab has opened new possibilities for anticancer immunotherapy agents in other malignancies, including HNSCC.

Cetuximab, an anti-epidermal growth factor receptor (EGFR) blocking mAb, was the first biologic agent approved for HNSCC, although there remains controversy about the relative contribution of immune effects vs the effects of receptor tyrosine kinase signaling blockade within the tumor cell (please refer to the chapter on the protein/signaling effects of cetuximab, Chap. 2). This agent has a response rate of just 10–20% in the HNSCC patient population when used as a monotherapy [14]. This low response rate has led to an intense investigation into resistance mechanisms, biomarkers of response, and other potential targets for therapeutic intervention. Today, HNSCC immunotherapy encompasses a variety of diverse treatment approaches, including tumor-specific monoclonal antibodies [15, 16], immunomodulating antibodies [17], cancer vaccines [18], oncolytic viruses [19], and adoptive cell transfer [20]. Evaluations of therapeutic mAbs have led this endeavor in recurrent and advanced HNSCC, with recent FDA approval for two agents (pembrolizumab and nivolumab) targeting the immune checkpoint programmed cell death 1 (PD-1). However, many other targets show therapeutic promise and, when used in combination regimens, may have potential in improving outcomes for treatment-naïve patients.

14.2 Cancer Immune Surveillance and Immunoediting

Normal cells only divide when they receive growth stimulatory signals (i.e., growth factors) produced in an autocrine manner or in a paracrine manner from other cells. These signals are detected by growth factor receptors such as vascular endothelial growth factor (VEGF) or EGFR. When EGFR binds one of its ligands (e.g. EGF), either as a homodimer or as a heterodimer with related receptors such as ERBB2/HER2), certain proliferative or antiapoptotic signal transduction pathways are activated, such as the Ras/MAPK (rat sarcoma/mitogen-activated protein kinase), Akt/PKB (protein kinase B), or STAT3 (signal transducer and activator of transcription) pathway [21]. However, tumor cells are frequently no longer dependent on exogenous growth factors and can either divide without these signals or produce their own mitogenic factors via autocrine signaling that stimulates proliferation. These tumor cells are also able to evade inhibitory signals and apoptosis signals through different mechanisms including mutations of tumor suppressor genes (i.e., *TP53* encoding p53) [22]. It was suggested that these premalignant cells arise and begin to clonally expand but are quickly identified by immune cells as foreign through clonotype T-cell receptors recognizing neoantigens produced as a consequence of transformation. This allows their eradication by the immune system before an invasive tumor can develop, a concept termed immune surveillance [23].

Mounting evidence supports the role of multiple components of the immune system in preventing the growth of malignantly transformed cells. For example, NK cells provide innate immune protection from tumors through various mechanisms, including the release of perforin and granzyme, leading to cell lysis. Interferon-gamma (IFN-γ) has a proapoptotic effect on tumor growth through further activation of NK cells and increases antigen processing along with the expression of the MHC molecules required for antigen presentation on antigen-presenting cells (see Sect. 14.3.1) [24, 25]. There is a clear relationship between immunodeficiency and cancer development, as seen with the increased risk of cancer in HIV-infected individuals, who lack functional CD4 cells, macrophages, and dendritic cells, as well as in organ transplant patients maintained on immune suppressive drugs, such as renal and bone marrow transplant patients [26, 27]. However, most patients who develop HNSCC are immunocompetent, and it is thought that tumors evade the immune response through a process of *cancer immunoediting* [28]. This theory suggests that while the immune system can eradicate some transformed cells, other transformed cells have been altered in a way that renders them less immunogenic, providing them with a survival advantage that allows immune evasion and subsequent tumor formation [25].

Cancer immunoediting proceeds sequentially through three distinct phases termed "elimination," "equilibrium," and "escape" (Fig. 14.1).

Elimination In this phase, the innate and adaptive immune systems recognize developing tumor cells and destroy them before they become a clinically apparent tumor. While activation of the immune system at this stage is not fully understood, possibilities include "danger signals," such as IFN and other cytokines that are induced early in tumor development, or damage-associated molecular pattern molecules (DAMPs), which are released from dying tumor cells or damaged tissues from the invasive growth of the tumor. These danger signals lead to the activation of dendritic cells (DCs) and subsequent adaptive antitumor immune responses. If the tumor cells are completely eradicated, the cancer immunoediting process ends at this phase.

Equilibrium If a rare cancer cell variant survives the elimination phase, the adaptive immune system keeps these residual cells in a state of dormancy (latent tumor cells), for up to many years. In this phase, immunity is specifically responsible for the prevention of tumor cell outgrowth. The immunogenicity of the tumor cells is also sculpted during this stage as the genetically unstable and rapidly mutating tumor cells undergo Darwinian selection, providing new variants with increased resistance to immune attack. Evidence of an immunologically driven equilibrium phase came from primary tumorigenesis experiments showing that while immunocompetent mice harbored occult cancer cells for a prolonged time period without any apparent tumor development, when the immune system was ablated with T-cell and IFN-γ depleting monoclonal antibodies (mAbs), tumors rapidly developed in the original injection site in half the mice and resembled unedited sarcoma cells. Furthermore, adaptive immunity – specifically IL-12, IFN-γ, CD4$^+$, and CD8$^+$ T cells – but not innate immunity was responsible for maintaining the occult tumor cells in equilibrium. Thus, dormancy is maintained through growth inhibitory and cytocidal actions of immunity on the residual tumor cells, providing the selective pressure to promote tumor outgrowth in those cells that have acquired the most immunoevasive mutations.

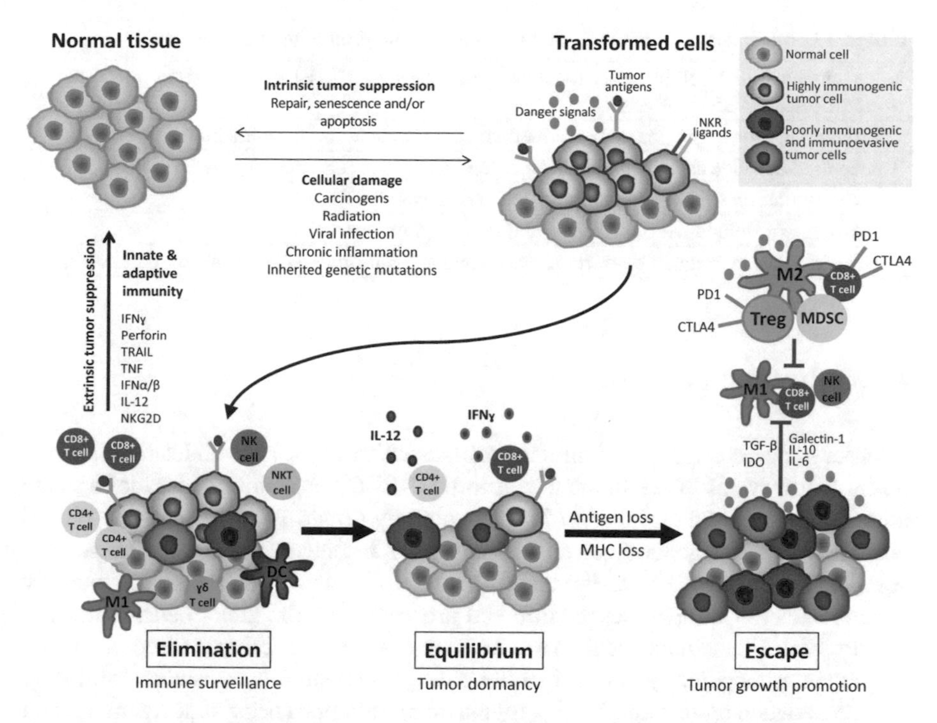

Fig. 14.1 Cancer immunoediting proceeds sequentially through three distinct phases termed "elimination," "equilibrium," and "escape". *Elimination:* The innate and adaptive immune systems recognize developing tumor cells and destroy them before they become a clinically apparent tumor. While activation of the immune system at this stage is not fully understood, possibilities include "danger signals," such as interferon (IFN) and other cytokines that are induced early in tumor development, or damage-associated molecular pattern molecules (DAMPs), which are released from dying tumor cells or damaged tissues from the invasive growth of the tumor. These danger signals lead to the activation of dendritic cells (DCs) and subsequent adaptive antitumor immune responses. If the tumor cells are completely eradicated, the cancer immunoediting process ends at this phase. *Equilibrium.* If a rare cancer cell variant survives the elimination phase, the adaptive immune system keeps these residual cells in a state of dormancy (latent tumor cells), for up to many years. The immunogenicity of the tumor cells is also sculpted during this stage as the genetically unstable and rapidly mutating tumor cells undergo Darwinian selection, providing new variants with increased resistance to immune attack. Dormancy is maintained through growth inhibitory and cytocidal actions of immunity on the residual tumor cells, providing the selective pressure to promote tumor outgrowth in those cells that have acquired the most immunoevasive mutations. *Escape.* Tumor cells acquire the ability to thwart immune recognition and destruction and are able to grow and develop into visible tumors. Progression to immune escape can occur due to immune system deterioration or changes in the tumor cell population in response to increased cancer-induced immunosuppression

Escape In the final phase, those tumor cells that have acquired the mechanisms to thwart immune recognition and destruction are able to grow and develop into visible tumors. Progression to immune escape can occur due to immune system deterioration or changes in the tumor cell population in response to increased cancer-induced immunosuppression.

Table 14.1 Mechanisms of reducing immunogenicity and promoting immune escape

1. Development of T-cell tolerance to persistent oncogenes (i.e., HPV infection) or overexpressed/mutated antigens
2. Production of low-genome copy numbers in the basal layer of the epithelium
3. Increased immune checkpoint inhibitor receptors/ligands (i.e., PD-1/PD-L1)
4. Downregulation of IFN regulatory factors and activated STAT1
5. Inhibition of inflammatory cytokines and transcription factors
6. Downregulation or mutation of HLA class I and antigen-processing machinery components

14.3 Immune Escape

In order to develop effective immunotherapies, it is necessary to understand the different pathways of tumor immune evasion. HNSCC cells can first escape immune attack by reducing their inherent immunogenicity (Table 14.1) and actively modulate four critical signals required for antitumor response (Fig. 14.2). These four signals are discussed at greater length below. In overview, they can be generally described as (1) antigen presentation and processing by the tumor cells, which are required for interactions between the TCR and the cancer cell-expressed HLA displaying immunogenic peptides; (2) the exchange of co-stimulatory and co-inhibitory signals between tumors and T cells, regulated by immune checkpoint signaling, and between Langerhans cells and antigen-presenting cells (APCs), which include dendritic cells (DCs), macrophages, and B cells; (3) the secretion of pro-inflammatory cytokines by immune cells and tumor cells; and (4) cell-extrinsic attracting chemokine signals to recruit cellular immune populations into the tumor microenvironment (TME), which can include immunosuppressive cells, such as regulatory T cells (Tregs), myeloid-derived suppressor cells (MDSCs), and tumor-associated macrophages (TAMs), dampening the immune response.

14.3.1 Signal 1. Antigen Presentation and Processing

Key signals initiating T-cell activation depend on proteins of the HLA complex, which are transmembrane proteins expressed on all somatic cells and form an HLA class I heavy chain-β2-microglobulin-antigen peptide ternary complex. Antigen peptide selection and loading as well as proper folding of the HLA complex relies on the antigen-processing machinery (APM), including transporter associated with antigen-processing (TAP) molecules, calnexin, calreticulin, and tapsin [29]. In cancer, the HLA class I proteins present captured tumor antigens (TA) to CD8 T cells [8], which results in activation of these cytotoxic T cells, leading to tumor cell death. Upon cell death or damage, released biomolecules can further perpetuate an inflammatory response. These biomolecules are termed damage-associated molecular

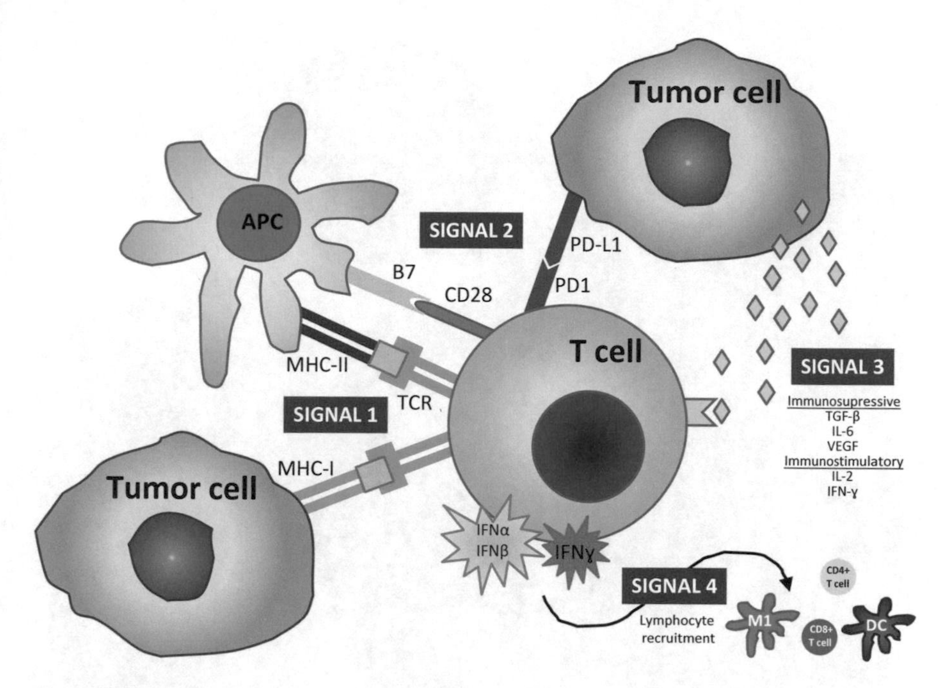

Fig. 14.2 Immune surveillance and targets for evasion. T-cell activation requires three simultaneous signals at the "immune synapse" in order to carry out its antitumor effects. Signal 1 comprises the T-cell receptor (TCR)-HLA interaction, with presentation of nonself-antigens from the tumor cell. Signal 2 is a summation of co-stimulatory and co-inhibitory signals. These signals must occur in the presence of signal 3, made up of immune activating cytokines, such as IL-2 or IFN-γ, or immunosuppressive cytokines, such as TGF-β, IL-6, or VEGF. Signal 4 is made up of chemokine signals to recruit cellular infiltrate, including antigen-presenting cells (APCs) made up of macrophages, dendritic cells (DC), and B cells, as well as cytotoxic CD8+ T cells and helper CD4+ T cells. Immune evasion can occur at any of these signals, impairing the immune system from effectively eradicating malignant cells

patterns (DAMPs) or, when associated with a pathogen (such as HPV infection), are termed pathogen-associated molecular patterns (PAMPs). However, tumor cells can reduce T-cell-mediated recognition by altering HLA class I expression. A significant proportion (20%) of HNSCC tumors show mutations in specific HLA alleles, β2-microglobulin, and genes encoding APM components, resulting in poor TA processing and presentation; these alterations are correlated with poor prognosis (32 months of disease-free survival vs 53 months in patients without APM alterations) [30–32]. While complete loss of HLA expression triggers NK activation and cell lysis, tumor cells have developed mechanisms to evade T-cell recognition while avoiding total loss of HLA expression through decreased expression or mutation of APM components [8].

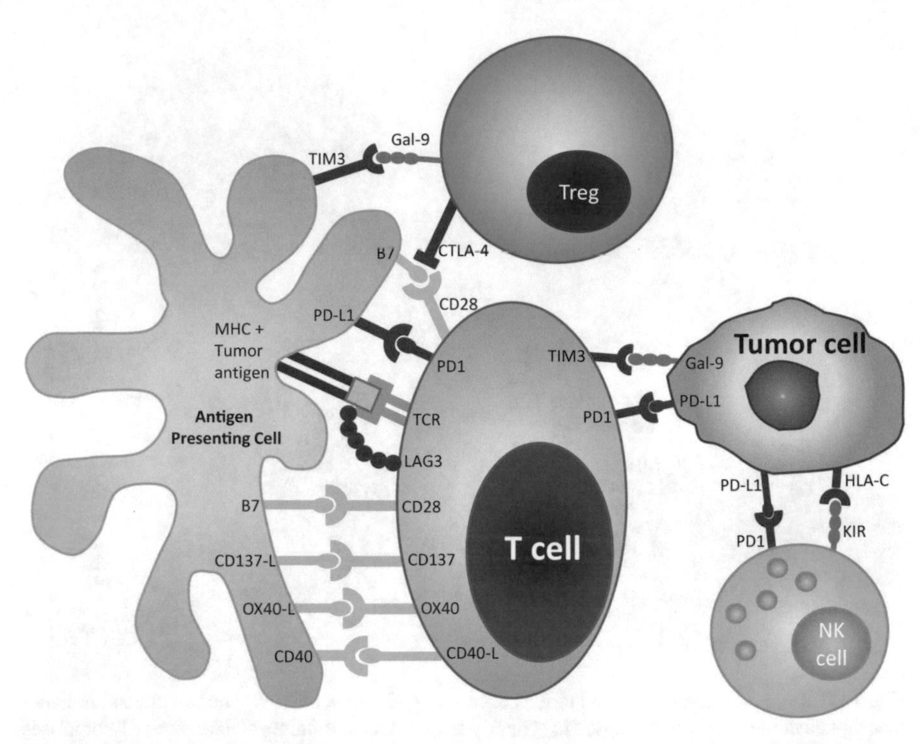

Fig. 14.3 Monoclonal antibody targets. There are multiple pathways, when summated, that can trigger either immune recognition and killing or immune tolerance and evasion. The goal of immunotherapy is to induce an immune stimulatory signal that activates T effectors and NK cells while inhibiting the immunosuppressive Treg. Red = co-inhibitory signals. Green = co-stimulatory signals. B7 = CD80 or CD86

14.3.2 Signal 2. Immune Co-Inhibitory and Co-Stimulatory Signals

In addition to antigen recognition, T-cell activation is functionally determined by the summation of many co-stimulatory and co-inhibitory signals, made up of a cell surface receptor and its corresponding ligand (Fig. 14.3). Some tumors contain cancer cells that express strong antigens and the adequate machinery to mount an immune response, yet develop T-cell tolerance to these antigens and escape immune activation during chronic antigen exposure [33]. Recent evidence suggests that T-cell exhaustion and tolerance are associated with increased expression of immune checkpoint proteins. These checkpoints are part of the adaptive immune resistance, which has evolved to thwart exaggerated immune responses and thus prevent the development of autoimmune diseases. However, these inhibitory checkpoints can be dysregulated by tumors as an important mechanism to weaken effector immune responses, contributing to tumor-promoting immunosuppression, tolerance, and even induction of T-cell apoptosis [34].

Tumors have developed mechanisms to influence adaptive immunity, expressing the ligands necessary to activate immune inhibitory signals and upregulating inhibitory checkpoint receptors. The two most well-studied checkpoint receptors are cytotoxic T lymphocyte antigen-4 (CTLA4) and programmed cell death 1 (PD-1), both of which are the targets of FDA-approved inhibitory antibodies [35]. Additional checkpoint inhibitory receptors that contribute to the immune evasion of HNSCC include lymphocyte-activation gene 3 (LAG3) [36], T-cell membrane protein 3 (TIM3) [37], and killer immunoglobulin-like receptor (KIR) [38]. In the presence of these co-inhibitory signals, the stimulatory "signal 2" will fail, leading to the induction of T-cell anergy or apoptosis [39]. Some promising therapeutic approaches that seek to target these inhibitory checkpoints are described below.

The classically described co-stimulatory pathway for immune cell activation involves interactions between CD28 expressed on T cells and a B7 ligand (CD80 or CD86), which are a family of peripheral proteins expressed on APCs that induce a series of downstream signals promoting T-cell survival and activation (Figs. 14.2 and 14.3). However, other co-stimulatory interactions between T-cell receptors and their corresponding APC ligands include CD137/CD137-L, OX40/OX40-L, and CD40/CD40-L. These co-stimulatory receptors are members of the tumor necrosis factor (TNF) receptor superfamily, and a decrease in their expression has been observed on T cells of HNSCC specimens, correlating with poor outcomes [40].

Both blocking the co-inhibitory signals and increasing the co-stimulatory signals may enhance tumor control through increased cytotoxic activity of T cells. Many mAbs targeting these signaling pathways have been developed in attempts to shift the immune system toward an antitumor function and are currently entering the therapeutic armamentarium for HNSCC. Some of the most important targets for therapeutic development are summarized below.

CTLA4 is expressed primarily on T cells in response to T-cell receptor (TCR) activation and counteracts CD28-mediated co-stimulatory signals [41, 42]. This impairs the activation of T cells, thus limiting an exaggerated immune response [43]. Although the transient expression of CTLA4 on $CD8^+$ T cells regulates effector/memory activity [44, 45, 46], the inhibitory function of CTLA4 appears to be more important on $CD4^+$ regulatory T cells (Tregs), where CTLA4 is constitutively expressed at high levels [47, 48]. CTLA4 and CD28 share the B7 family ligands CD80 and CD86, whose expression is limited to APCs [49]. Compared to CD28, CTLA4 binds these ligands with higher affinity and avidity, competitively inhibiting "signal 2" in the T-cell activation cascade [49]. Interestingly, activated $CTLA4^+$ T cells can deplete acquired B7 ligands from APCs by CTLA4-dependent endocytosis and degradation, further impairing the activation of T cells [50, 51].

PD-1 is highly expressed on the surface of activated T cells, B cells, and macrophages in the setting of chronic stimulation, with minimal expression on resting cells of the immune system [52]. PD-1 binds two distinct ligands, PD-L1 and PD-L2, additional members of the B7 family. PD-L1 is expressed constitutively on a wide range of hematopoietic cells, inhibiting T-cell cytokine production, as well as on non-hematopoietic cells, including tumor cells [53]. PD-L2 exhibits a much more restricted expression profile, largely limited to APCs, and acts to restrain effector T-cell function within inflamed sites and lymphoid organs [54, 55]. Both ligands are induced by extrinsic pro-inflammatory signals, including IFN-γ, TNF-α, granulocyte-macrophage colony-stimulating factor (GM-CSF), and IL-4. These data suggest that the inflammatory tumor microenvironment (TME) enhances tumor expression of PD-L1/2 in a feedback inhibitory loop similar to that frequently seen in chronic viral infection [56–58]. Additionally, tumor-intrinsic signaling pathways may lead to PD-L1 overexpression in tumor cells, including the PI3K-AKT pathway [59] and the EGFR-mediated JAK2/STAT1 pathway [60]. In addition to the high expression of PD-1 on $CD8^+$ and $CD4^+$ T cells in HNSCC patients, PD-L1 is expressed on 50–60% of HNSCC tumor cells [60, 61].

In the setting of chronic antigenic stimulation during viral infection, increased PD-1 expression on $CD8^+$ T cells is a marker for an exhausted phenotype, defined as a loss of effector function, coupled with loss of proliferation and cytokine production [62]. While PD-L1 expression by APCs has been shown to play a critical role in the induction of suppressive $PD\text{-}1^+$ Tregs at the tumor periphery [63], the key function of tumor-infiltrating Tregs expressing high levels of PD-1 is not clear. In malignant glioma, these cells had impaired suppressive function yet maintained high secretion of IFN-γ through increased PI3K/AKT activity [64]. Blocking the PD-1 pathway with anti-PD-1 mAb drove Treg differentiation toward a more dysfunctional, less suppressive state and increased IFN-γ production to further enhance tumor clearance [64]. PD-L1 is expressed in response to IFN-γ within the TME; therefore, even an immunogenic tumor with abundant IFN present will express PD-L1 and evade T-cell-mediated killing of tumor cells. While overexpression of PD-L1 inhibits tumor-directed T-cell cytotoxicity and permits immune evasion and tumor growth, the prognostic significance has been variable. In three studies, increased PD-L1 expression correlates with HPV positivity, which has better prognosis than HPV-negative cancers [65–67]. However, one study showed increased PD-L1 expression has a trend toward increased distant metastasis, yet no association with recurrence, overall survival, or disease-specific survival on oropharyngeal SCC [66]. More analysis is required.

LAG3 (also known as CD223) is an inhibitory checkpoint receptor that enhances the function of Tregs and inhibits $CD8^+$ effector T-cell function [68]. The only known LAG3 ligands are MHC class II molecules (also called HLA class II; these present antigens to CD4 T cells, leading to their activation), which are upregulated on some epithelial cancers in response to IFN-γ but are also expressed on dendritic cells (DCs) [69]. PD-1 and LAG3 are often co-expressed on exhausted or anergic T cells, and dual blockade of these proteins synergistically reversed the anergy of tumor-specific $CD8^+$ T cells [70]. Furthermore, $PD\text{-}1^{-/-}LAG3^{-/-}$ double knockout mice can completely reject even poorly immunogenic tumors in a T-cell-dependent manner [71].

TIM3 (also known as HAVcr2) is selectively expressed on IFN-γ producing T cells, NK cells, and Tregs and is closely involved in tumor-induced immune suppression. TIM3 marks the most exhausted or dysfunctional population of $CD8^+$ T cells and NK cells in both solid and hematologic malignancies [72, 73], where approximately 30% of patients with advanced melanoma, NSCLC, and follicular B-cell non-Hodgkin's lymphoma express TIM3 [74–76]. Importantly, T-cell exhaustion can be partially reversed with TIM3 blocking antibodies in vitro, restoring up to 30–65% of NK cell function [77]. In addition to regulating $CD8^+$ T cell and NK cell function, TIM3 is expressed on up to 60% of Tregs in the TME in HNSCC patients, compared to less than 20% expression on Tregs among peripheral blood lymphocytes [78]. This elevated expression of TIM3 is important as it leads to a more tumor-permissive environment, largely due to increased immunosuppressive cytokines and molecules (IL-10 which drives T-cell exhaustion, as well as perforin and granzymes, which are responsible for effector T-cell apoptosis) [79, 80]. Interestingly, $TIM3^+$ $CD8^+$ T cells co-express PD-1 and exhibit greater deficits in both effector cytokine production (IL-2, TNF, and IFN-γ) and cell cycle progression than are seen with expression of either receptor alone [72].

KIR proteins are expressed on NK cells and interact with HLA molecules on target cells, playing a prominent role in modulating NK cell-dependent immune surveillance and cytotoxicity [81]. While most KIRs are inhibitory, there are a limited number of activating KIRs (KIR2DL2/3-KIR2DS2, KIR2DL1-KIR2DS1, and KIR3DL1-KIR3DS1) that bind HLA molecules with less affinity [82]. Upon binding an autologous-matched HLA molecule (signaling a "self" cell), the inhibitory KIRs recruit SHP-1 and SHP-2 phosphatases, leading to subsequent suppression of activation signals and preventing NK attack on "self"-cells [83]. However, when binding a "nonself" HLA molecule (i.e., tumor cell, transplant cell, or virus-infected cell), or blockade by anti-KIR Ab, the NK cell lyses due to lack of an inhibitory signal, leading to a release of its cytolytic granules and subsequent tumor cell death. This inhibitory KIR/HLA pathway is overactive in patients with melanoma, breast cancer, and chronic leukemia [84]. Higher expression of activating KIRs is associated with better outcomes in leukemia and lymphoma [85, 86]. While there are limited studies in HNSCC, there seems to be an overexpression of KIR on the NK cells infiltrating tumors in HNSCC patients [87], suggesting this population is an excellent candidate for trials of anti-KIR Ab.

In addition to inhibitory receptors, tumors can manipulate the checkpoint system by downregulating co-stimulatory signals, including CD137, OX40, and CD40. **CD137** (also known as 4-1BB) is a co-stimulatory receptor expressed on the surface of activated T cells, DCs, and NK cells [88, 89]. When bound to its ligand (CD137-L), activated CD137 promotes antibody-dependent cell-mediated cytotoxicity (ADCC) by NK cells, differentiation of effector T cells, and inhibition of immunosuppressive Tregs [90]. **OX40** is expressed on the T-cell surface and, when activated by its ligand, OX40-L, promotes T-cell proliferation, antitumor cytokine secretion (IFN-γ), and enhanced memory T-cell function [35]. Thirty percent of activated Tregs isolated from HNSCC patients demonstrate OX40 expression in both tumors and tumor-draining lymph node samples, compared to none of the peripheral blood mononuclear cells [91]. However, OX40-L expression is low in the tumor and therefore fails to activate antitumor immune proliferation and cytokine production [92]. **CD40** is expressed on APCs, as well as non-immunogenic cells and tumor cells [93, 94]. Upon binding with its trimeric ligand (CD40-L), the downstream effects are multifaceted and depend on the type of cell expressing CD40 and the TME in which it is expressed [94]. Ligand binding on activated $CD40^+$ T cells causes the induction of adaptive immunity and cytokine release, whereas $CD40^+$ macrophages become tumoricidal, and $CD40^+$ tumor cells become apoptotic [93, 95–98]. Expression of both CD40 and CD40-L decreases with increasing HNSCC stage, and surgical resection results in increased APC expression of CD40 [99].

These immune checkpoint receptors and ligands making up *signal 2* are dysregulated on the infiltrated immune cells, favoring anergy and immunosuppression. Therefore, their manipulation to favor immune activation through mAbs can lead to improved immune-mediated killing of HNSCC cancer cells.

14.4 Signals 3 and 4: Tumor Microenvironment

Immune tolerance can be further enhanced through an immunosuppressive tumor environment, and alterations in the tumor microenvironment may prove to be a key to immune resistance and low response to therapy. Aerodigestive system tumors often arise at sites of chronic inflammation, whether from chronic toxin exposure (i.e., tobacco smoke and/or alcohol consumption), chronic infection (i.e., HPV), or inflammatory disease (i.e., Barrett's esophagus). As the uncontrolled growth of tumor cells leads to invasion through natural tissue barriers, tissue disruption leads to cytokine and chemokine release and the infiltration of inflammatory mediators. These include TGF-β, IL-6, IL-10, GM-CSF, IL-1β, IL-23, and TNF-α. Recently, evidence suggests a connection between expression of many of these cytokines and the formation of suppressive immune cells, including MDSCs, Tregs, and TAMs. These cells collaboratively contribute to an immunosuppressive, tumor-promoting environment, in part by generating genotoxic stress, supporting cellular proliferation, increasing angiogenesis, and facilitating tumor cell invasiveness and metastasis [100].

14.4.1 Signal 3. Cytokines

HNSCC cells produce many cytokines that suppress immune function [101]. TGF-β blocks the function of NK and T cells and plays a key role in the differentiation of suppressive Tregs [102]. IL-6 inhibits DC maturation and activation of NK cells,

T cells, neutrophils, and macrophages, whereas IL-10 produces T-cell anergy through the downregulation of MHC class II and the inhibition of the CD28 co-stimulatory pathway [103]. Prostaglandin E_2 (PGE2) is a pro-survival, proangiogenic molecule and is produced by many cancers, including HNSCC [104–106]. Vascular endothelial growth factor (VEGF), an inflammatory cytokine released from tumor cells, inhibits the maturation of DC, leading to T-cell dysfunction and inactivation: VEGF is overexpressed in 90% of HNSCC tumors [107, 108]. EGFR, a tyrosine kinase receptor, also facilitates immunosuppressive effects, including downregulation of HLA and APM components and the upregulation of suppressive cytokines and ligands on HNSCC cells [109]. Indoleamine-pyrrole 2,3-dioxygenase (IDO) is produced by DCs in tumors and tumor-draining lymph nodes, as well as from tumor cells themselves, inducing apoptosis in locally activated T cells through the catabolism of tryptophan [110]. T-cell proliferation is highly dependent on tryptophan, and with IDO-induced depletion, T cells are arrested in the G1 phase of the cell cycle [111]. These immunosuppressive cytokines and receptors are further driven by an imbalance of STAT proteins, leading toward increased levels of activated STAT3, which promotes immunosuppressive pathways and tumor cell proliferation, survival and invasion [112], and decreased levels of STAT1, which impairs antigen presentation and T-cell activation [113].

14.4.2 *Signal 4. Cellular Infiltrate*

The above cytokines and chemokines help to recruit a skewed cellular population of tumor-infiltrating leukocytes (TILs), favoring immunosuppression. One of the major cellular components of the TME is the immunosuppressive $CD4^+Foxp3^+CD25^+$ Treg. Tregs have emerged as central players in tolerance to tumor antigens and downregulators of the immune response [78]. Through the production of IL-10 and TGF-β and the consumption of IL-2, Tregs are responsible for taming immune responses to prevent autoimmunity and systemic inflammatory response syndrome (SIRS)-like reactions by causing anergy, apoptosis, and cell cycle arrest of activated T cells and inhibiting the action of DCs, NK cells, and B cells. Tregs are also seen in elevated numbers in both the periphery and the tumor infiltrate in HNSCC patients [78, 114], and their increased presence is correlated with poor survival [115, 116].

MDSCs are immature myeloid cells that are another cellular component of the immunosuppressive tumor environment. Studies in HNSCC show that MDSCs can suppress effector T-cell function, through influence on PD-L1 expression, TGF-β release, and arginase, and nitric oxide synthase (NOS), through depletion of key amino acids and Treg induction, as well as inhibition of the TRC-HLA interaction, signaling, and activation [117, 118]. Treatments that diminish MDSCs, such as antibody depletion and retinoic acid, gemcitabine, and STAT3 blockade, restore immune surveillance, increase T-cell activation, and improve the efficacy of immunotherapy [119].

Immature monocytes or macrophages have the ability to differentiate into various phenotypes depending on interactions with various growth factors, cytokines, and chemokines within their particular environment. Tumor-associated macrophages (TAMs) in the HNSCC tumor microenvironment are predominated by the tumor-promoting M2 phenotype, which produces tumor-permissive cytokines EGF, IL-6, and IL-10; promotes angiogenesis, cell invasiveness, and metastasis; and correlates with worse clinical outcome [120, 121].

Recent work in various solid tumors has shown a clinical correlation with two different tumor immunophenotypes: an "inflamed" phenotype characterized by a rich T-cell infiltrate, a type 1 interferon signature, and a diverse chemokine profile, and a "noninflamed" phenotype that lacks these features [122]. Tumors with an inflamed phenotype are more likely to respond to immune checkpoint therapies [123]. In HNSCC, many studies have shown improved response to various therapies, including definitive chemoradiation therapy (CRT), surgery with adjuvant therapy, and immunotherapies in patients with a higher CD8+ T cell and NK cell infiltrate. CD8+ T cells and NK cells were also shown to express higher levels of immune checkpoint receptors, including PD-1 and TIM3 [124]. Conversely, the noninflamed phonotype is dominated by CD4+ Tregs that co-express CTLA4 and PD-1 and is correlated with poor survival. These data suggest that identifying the immune phenotype may be a means of identifying patients likely to benefit from immunotherapy. Together, these immune/inflammatory cells and mediators induce an immunosuppressive, tumor-promoting environment that acts as a barrier to effective immunotherapy.

14.5 Immunotherapy Targets

14.5.1 Tumor-Targeted Monoclonal Antibodies

The discovery of aberrant expression or overexpression of specific signaling molecules in HNSCC and other cancer types has led to the development of tumor-targeted therapies. The most widely studied targeted therapy in HNC is cetuximab, a chimeric murine/human IgG1 mAb blocking epidermal growth factor receptor (EGFR) signaling. Cetuximab has both immune and nonimmune antitumor functions. The nonimmune effects of cetuximab on EGFR signaling are described in Chap. 2, and evidence supporting the effects on immune function in HNSCC is discussed here. Cetuximab activity depends significantly on the activation of Fc gamma receptor (FcγR), leading to antibody-dependent cellular cytotoxicity (ADCC), and subsequent antigen processing [125–127]. This leads to DC maturation and activation of cytotoxic T lymphocytes (CTLs), linking innate and adaptive tumor immunity [35, 128, 129]. Cetuximab could therefore overcome the immunosuppressive environment in HNSCC and increase clinical response to cancer treatment. However, patients treated with cetuximab have demonstrated an increase in CTLA4+ suppressive regulatory T cells (Tregs) that impair ADCC, which may contribute to the modest response of just 10–20% when used as a monotherapy; some data suggests it may be possible to overcome this cetuximab resistance with the addition of an anti-CTLA4 mAb [14, 128, 130].

During cetuximab therapy, patients who responded maintained stable levels of circulating and intratumoral CTLA4+ Tregs, in contrast to non-responders, who had increased levels of CTLA4+ Tregs in both peripheral blood ($p = 0.02$) and in tumor-infiltrating lymphocytes (TILs) ($p = 0.006$). However, when TILs from HNSCC patients were incubated with cetuximab, ipilimumab (anti-CTLA4 mAb), and NK cells, EGFR antagonism was able to overcome this suppression and showed enhanced ADCC by eliminating the suppressive CTLA4+ Tregs, restoring the cytolytic activity of NK cells previously suppressed by intratumoral Tregs [130]. This has stressed the need for further investigation into active immunomodulating agents, such as the inhibitory and co-stimulatory immune checkpoint receptors and their ligands. Many of these agents have been developed and are being studied clinically in HNSCC, with targets such as CTLA4, the PD-1/PD-L1 axis, LAG3, TIM3, CD137, and OX40, among others.

14.5.2 Immunomodulating Monoclonal Antibodies

In addition to targeting tumor cells to initiate an immune response, activating immune cells directly through mAbs has shown promise as anticancer therapy. Many immune mediators in the TME, such as CD8+ T cells and NK cells, have induced an upregulation of immune checkpoint inhibitory receptors, limiting their ability to elicit an effective antitumor response and leading these T cells down a path to exhaustion. However, when blocked by mAbs targeting these checkpoint receptors, the T-cell functional capacity is restored, and these effector T cells and NK cells can effectively eradicate tumor cells. The agents furthest in development for HNSCC are detailed below.

Pembrolizumab is an anti-PD-1 mAb initially tested in patients with recurrent or metastatic HNSCC in the KEYNOTE-012 trial [131]. This was an expanded phase IB trial that showed many durable responses, a 1-year overall survival of 51%, and a tolerable adverse effect profile ($n = 132$, ORR = 19%, 6-month PFS = 23%). Pembrolizumab is currently being studied as a first-line treatment of recurrent or metastatic HNSCC in the KEYNOTE-048 phase III trial (NCT02358031).

Nivolumab is another anti-PD-1 mAb, evaluated in the CheckMate 141 phase III trial in patients with platinum-refractory recurrent or metastatic HNSCC. This trial showed an improved overall survival of 7.5 months for patients treated with nivolumab ($n = 240$), compared to 5.1 months for those treated with an investigator's choice ($n = 121$) among three standard agents (methotrexate, docetaxel, and cetuximab) ($p = 0.01$), and also showed improved objective RR (13% vs 6%) and rate of PFS at 6 months (19.7% vs 9.9%), with reduced toxicity (treatment-related adverse events of grade 3 or 4 in 13% vs 35%) [24]. Correlative studies also demonstrated an improved quality of life for those treated with nivolumab [132]. Nivolumab is currently being investigated in combination with ipilimumab (anti-CTLA4 inhibitor) and as a first-line treatment in the CheckMate 651 and CheckMate 714 trials (NCT02741570 and NCT 02823574). These data have led to FDA approval of pembrolizumab and nivolumab for patients with platinum-refractory metastatic/recurrent HNSCC.

Durvalumab is an anti-PD-L1 mAb currently being investigated in a phase III trial as a monotherapy in patients with advanced solid tumors, including HNSCC. Thus far, survival results are promising, with a median overall survival of 8.9 months with grade 3–4 adverse effects in only 8% of patients [133]. Patients had higher response with PD-L1-positive tumors (18% vs 8% in PD-L1-negative tumors). There are several ongoing trials, including the HAWK trial evaluating durvalumab as a monotherapy in platinum-refractory recurrent or metastatic disease in PD-L1-positive tumors and in the CONDOR and EAGLE trials in combination with the anti-CTLA4 antibody tremelimumab in PD-L1-positive and PD-L1-negative platinum-refractory recurrent or metastatic HNSCC patients.

14.5.3 Toxicity Profiles

Immunotherapeutic agents have different mechanisms of action than conventional therapies, and therefore new specific toxicity profiles have emerged, termed "immune-related adverse events" (irAEs). Despite relatively low rates of high-grade side effects, there have been life-threatening and lethal events, warranting appropriate identification and management of these toxicities. The most common side effects include fatigue and decreased appetite. However, specific irAE profiles most commonly include gastrointestinal (GI) reactions (primarily represented by colitis, diarrhea, nausea, and vomiting, seen more often in anti-CTLA4 blockade) and skin reactions (most commonly papular rashes, vitiligo, and/or pruritus) [134]. Endocrine AEs were less frequent and include hypophysitis, hypo−/hyper-thyroiditis, and rarely adrenal insufficiency and diabetes mellitus. Pneumonitis, hepatitis, nephritis, polyarthritis, ophthalmological, and neurological irAEs were uncommon (<5%) [134]. High-grade toxicities occurred in 10–24% of patients [135] and are typically managed with systemic steroids and/or discontinuation of checkpoint inhibitor therapy.

Treatment-related deaths have occurred in up to 6% of cases in the treatment of solid tumors, most frequently occurring in the context of toxicities that include severe diarrhea/colitis, pneumonitis, neutropenic fever/sepsis, and acute hepatic toxicity [8]. Cardiac toxicity has become an emerging issue with immunotherapy, with several cases of fatal heart failure in melanoma patients treated with checkpoint inhibitors [136]. Other rare treatment-related deaths include encephalitis, anaplastic anemia, and neuromuscular disorders (myasthenia gravis, Guillain-Barre syndrome) [30, 137].

14.5.4 Vaccines

Vaccines prime the adaptive immune system against pathogens (Fig. 14.4). Vaccines have received considerable interest for both cancer prevention and cancer treatment as they can induce a robust immune response (defined as antigen-specific tumor-reactive cytotoxic T-cell expansion) coupled with a satisfactory safety profile, and

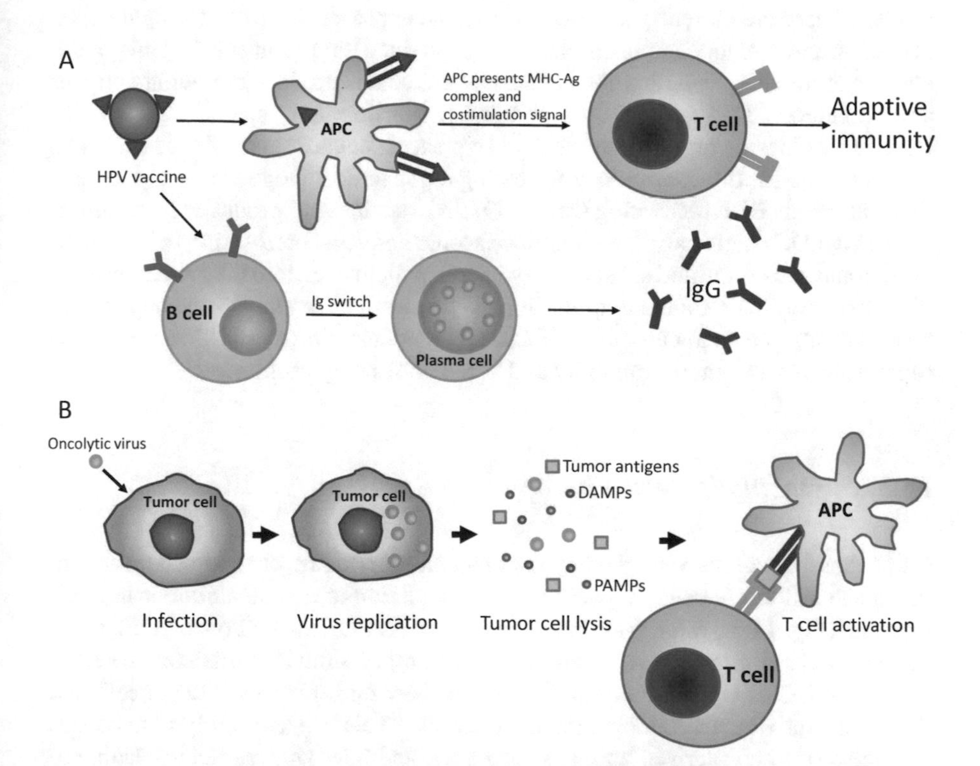

Fig. 14.4 Anticancer strategies under clinical evaluation. (**a**) Adaptive immunity activation by vaccine through priming effector T cells against specific tumor antigens and elucidating high titers of tumor-specific neutralizing antibodies. (**b**) Oncolytic viruses invade and replicate in tumor cells, leading to tumor cell lysis and release of tumor antigens, DAMPs and PAMPs. This allows recruitment of antigen-presenting cells to T cells. *HPV* human papilloma virus, *APC* antigen-presenting cell, *MHC* major histocompatibility complex, *Ag* antigen, *Ig* immunoglobulin, *DAMPs* damage-associated molecular patterns, *PAMPs* pathogen-associated molecular patterns

they can be prepared at low cost due to ease of production [138]. The viral oncogenes E6 and E7, important drivers for HPV+ cancers, are ideal targets for vaccine-mediated immune strategies (please refer to Chaps. 20 and 21 on HPV-induced HNSCC for an in-depth discussion of mechanisms of HPV+ cancer). The primary goal of prophylactic vaccination for cancers associated with infectious agents is to induce high titers of tumor-specific neutralizing antibodies capable of preventing initial infections. In contrast, therapeutic vaccines focus on the generation of CD8+ tumor-specific T-cell immune response.

Various vaccination techniques have been developed and are undergoing preclinical investigations and clinical trials for both HPV-related and HPV-unrelated cancers. CTLs are an essential component of the host immune response to tumors, and the goal of most cancer vaccine strategies is to induce a strong CTL response against tumor cells. CTLs target tumors through the recognition of HLA class I molecules and an antigenic tumor cell peptide (signal 1) in the context of co-stimulatory molecules (signal 2), usually provided by APCs, such as macrophages or DCs. There are currently a variety of vaccine approaches, aimed to trigger CTL activation and memory against tumor cells to produce long-term survival and maintenance of tumor-specific immune cells. DNA vaccines produce nonliving antigens able to induce CD4+ and CD8+ T-cell and B-cell immunity against cancer cells. Peptide vaccines incorporate amino acid sequences that are synthesized to form an immunogenic peptide molecule representing the specific epitope of a tumor antigen (TA) that binds HLA, activating CTL [139]. DC vaccines are produced from culturing ex vivo DCs and loading with tumor-specific antigens, performing in vitro maturation and activation, and subsequently injecting them back into the patient, usually via intratumoral or intranodal infiltration [39]. Several different vaccine approaches have shown promise in early clinical trials in patients with HNSCC, with evidence of induction of antitumor immunity and acceptable safety profiles.

14.5.5 Oncolytic Viral Therapy

Intratumoral oncolytic viral therapy uses naturally occurring or modified viruses to infect and kill tumor cells, in parallel with stimulation of a host antitumor immune response (Fig. 14.4). *Talimogene laherparepvec* (T-VEC), the first oncolytic virus to be approved for clinical use, is an attenuated herpes simplex virus engineered to express GM-CSF. Upon intratumoral injection, the virus replicates in tumor cells and thus alters antiviral and tumor signaling pathways. This triggers cell lysis that leads to release of tumor-derived antigens, GM-CSF, and new viral particles, leading to "infection" of surrounding tumor cells and greater immune response. Through increased IFN-γ and TNF-α production, T-VEC-expressing tumor cells can recruit effector T cells and reduce Treg-dependent immunosuppression, restoring the immunosuppressive TME [140]. In early clinical trials, viral DNA was detected in 95% of patients at the site of intralesional injection and remained through 84 days and was also present in many nontarget sites, suggesting a systemic distribution. In metastatic melanoma, responses were seen both locally in injected lesions and systemically, with >30% reduction of tumor burden seen in 62% of injected lesions, as well as in 41% of uninjected non-visceral lesions and 13% of uninjected visceral lesions [101]. Clinical trials have now confirmed the efficacy of T-VEC as a monotherapy for the treatment of advanced melanoma, with an ORR of 26%. This promising result has led to trials of T-VEC in combination with immunotherapies such as pembrolizumab in HNSCC (NCT02626000) and with ipilimumab in melanoma (NCT01740297).

14.5.6 Adoptive T-Cell Therapy (ACT)

Another promising strategy to activate the immune system is adoptive T-cell transfer. This technique involves harvesting and in vitro expansion of the patient's own TA-specific T cells and subsequent reintroduction of the expanded cell population back into the patient to enhance immunity and improve anticancer responses [141]. Early-phase trials are ongoing for ACT in HNSCC with promising results thus far. Virally induced cancers are an attractive model for ACT, as they provide virus-specific antigen targets. Epstein-Barr virus (EBV) has been strongly linked to nasopharyngeal carcinoma (NPC). Found in 95% of cases, it is believed to be important for the oncogenic process [142]. ACT can be used to expand preexisting EBV-specific CTLs to attack tumor cells presenting EBV-specific antigens. For example, a phase II clinical trial in 35 patients with stage IVC NPC who received first-line chemotherapy followed by ACT with EBV-specific CTL targeting three EBV antigens (EBNA1, latent membrane protein (LMP) 1, and LMP2) expressed by most NPC tumors showed 63% OS at 2 years and 37% at 3 years [102].

14.6 The Rationale for Combination Therapies

14.6.1 Chemotherapy- and Radiotherapy-Induced Immunogenic Cell Death and Immunogenic Modulation

The traditional dogma of radiobiology states that cytotoxicity results from DNA double-strand breaks, leading to cell death. The formative work by Zitvogel and Kroemer enhanced the understanding of radiation-induced cell death to include roles of the host immune system and the tumor microenvironment [143]. Chemotherapy and radiotherapy not only lead to tumor cell death but also induce "immunogenic cell death" (ICD) of cancer cells. After exposure to certain cytotoxic agents, dying cancer cells release endogenous "danger signals," or DAMPs, that are then recognized by APCs (such as DCs), followed by the activation of T-cell-mediated adaptive immunity [144]. Released DAMPs that are hallmarks of ICD include adenosine triphosphate (ATP), high mobility group protein B1 (HMGB1), and molecules expressed on the surface of dying cells such as calreticulin (CRT), heat shock proteins (HSP90 and HSP70), and endoplasmic reticulum (ER) sessile proteins [145]. CRT, released early in cell death, is a potent phagocytosis signal to DCs, facilitating their ability to present TAs and activate cytotoxic T cells [143]. HMGB1, released during the late stages of cell death, initiates powerful inflammatory response by binding to different surface receptors on APCs, including TLR4, and TIM3, stimulating the production of pro-inflammatory cytokines [146]. ATP binds the P2Y2 receptor on phagocytes and induces their migration into inflamed sites [147]. Furthermore, radiation triggers ATP release through the interaction with the P2X7 receptor, leading to the secretion of the pro-inflammatory IL-1β. Cytotoxic therapy thus increases antigen presentation from cancer cells as well as amplifies cytokine expression [148].

Additionally, accumulating evidence demonstrates that the surviving tumor cells recovering from conventional cytotoxic cancer therapies become more amenable to immune-mediated killing. This process, termed "immunogenic modulation," is associated with induction of a more immunostimulatory environment [149, 150]. Immunogenic modulation encompasses a variety of molecular alterations to tumor cells, including the downregulation of antiapoptotic/survival genes, modulation of APM components, and upregulating calreticulin expression on the cell surface, which acts as a phagocytosis signal to DCs [151]. These phenotypic changes translate into increased tumor sensitivity to T-cell-mediated lysis after low doses of chemotherapy or radiation [152, 153]. While these antitumor immune responses are non-curative in themselves, they can however be used in synergistic regimens with immunotherapeutic agents to improve their efficacy.

The immunogenic properties of cytotoxic therapies have been shown in several preclinical models to act synergistically with such immune checkpoint receptor inhibitors as anti-CTLA4 agents and anti-PD-1 agents [154–157]. Multiple clinical trials are now investigating immunotherapy in combination with cytotoxic treatments. In NSCLC, the KEYNOTE-021 study [158] compared the anti-PD-1 mAb pembrolizumab plus carboplatin/pemetrexed with chemotherapy alone, and the CheckMate 012 trial [159] compared the anti-PD-1 mAb nivolumab plus platinum-based chemotherapy vs chemotherapy alone. Both studies showed improved ORRs in the combination group (55% and 33–47%, respectively) compared to historical rates with PD-1 as a monotherapy (45 and 19%, respectively) [160, 161]. Studies are underway that investigate the optimal drug dosage and radiation fractionation regimens that induce adaptive immune responses. Additionally, preclinical models have shown that PD-L1 expression on tumor cells is increased following radiotherapy, highlighting a possible resistance mechanism to conventional treatment, and adding immunomodulating therapies such as anti-PD-1 mAb to conventional cytotoxic regimens may overcome this resistance [162].

While conventional cytotoxic therapies have frequent and significant acute and lifelong side effects, immunotherapy has thus far been shown to have less high-grade toxicity than conventional chemotherapy [163] (see Sect. 14.5.3). Therefore, adding immunotherapy to the clinical repertoire may be a means to not only reduce the cytotoxic load and subsequent side effects but also increase efficacy and durability compared to cytotoxic treatment alone. These findings have translated into promising clinical benefits for HNSCC patients and are the basis of many clinical trials underway as discussed below.

14.6.2 Combination Immunotherapy

Immunotherapy agents have rapidly been entering the oncologic treatment armamentarium, with the documented success of agents such as nivolumab and ipilimumab. Currently, there are over 300 active trials using immunotherapy agents for carcinoma (clinicaltrials.gov queried January 2, 2017). Despite significant promise, clinical benefit of these immunomodulatory agents as monotherapies has been limited thus far. As tumor cells gain multiple layers of compensatory immune resistance, overcoming one is often not sufficient to eradicate the tumor. However, recent evidence suggests that targeted blockade of multiple immunosuppressive pathways can induce synergistic and effective antitumor responses. In conceptualizing the rationale for combinations, T-cell exhaustion leads to upregulation of multiple immune checkpoint inhibitory receptors, including CTLA4, PD-1, TIM3, and many more. Blockade with monoclonal antibodies of just one pathway may effectively remove this inhibitory signal, only to have a separate but equally powerful inhibitory path unaffected, leaving the T cell just as dysfunctional. However, removal of multiple suppressive signals acting on unique resistant pathways may overcome the net inhibition and lead to T-cell activation and immune attack.

Many preclinical studies and clinical trials have looked into the combination of immunotherapy agents with clear differences in immunotherapeutic mechanism. The best-studied combinations are of inhibitors of CTLA4, which influences T-cell activation and proliferation, with inhibitors of the PD-1/PD-L1 axis, which is responsible for T-cell function and cytolysis [164]. Indeed, this was translated to a three-arm phase III trial in malignant melanoma, studying the combination regimen of anti-PD-1 antibody nivolumab and anti-CTLA4 antibody ipilimumab versus each agent given as monotherapy. Combined therapy had a progression-free survival (PFS) of 11.5 months, compared to 6.9 months in the nivolumab-only arm and 2.9 months in the ipilimumab-only arm, leading to FDA approval of combined nivolumab and ipilimumab for the treatment of malignant melanoma [165]. However, PD-L1-positive tumors gained no benefit with combination therapy compared to those receiving nivolumab alone, suggesting PD-L1 could be used as a biomarker to guide combination vs monotherapy. As noted above, clinical trials in HNSCC are evaluating this combination (NCT02741570) as well as that of the PD-L1-targeted antibody durvalumab with a CTLA4-directed antibody tremelimumab (NCT03212469). Other combinations with checkpoint inhibitors, such as a PD-1 inhibitor in combination with a TIM3 or LAG3 inhibitor, may also provide increased antitumor activity. Additionally, improved efficacy is seen in combinations of PD-1 blockade with agonistic antibodies for OX40 or CD137, likely due to alterations in Treg cells and changes in the ratio of effector to regulatory T cells [166, 167].

Cytokine signaling can also lead to resistance mechanisms to circumvent T-cell killing that may render monotherapy ineffective. For example, the production of IFN-γ by activated T cells in the TME upregulates PD-L1 on tumor cells [168]. The various signaling mechanisms and cells within the TME produce a variety of changes that might support the combination of immunotherapy agents with inhibitors of proteins such as VEGF and IDO. Tumor vasculature is known to exert immunosuppressive effects through many different mechanisms, including expression of immunosuppressive ligands and downregulation of adhesion molecules. Therefore, VEGF inhibition could potentially augment the activity of immunomodulatory antibodies. In a clinical trial combining ipilimumab with the VEGF-targeted antibody bevacizumab in metastatic melanoma, there were 8 partial responses and 22 instances of stable disease, with a 67% disease control rate and increased tumor lymphocyte infiltration [169]. As another example, IDO is expressed by many cancers, inhibits immune responses, promotes Treg differentiation, and may contribute to a possible resistance mechanism to anti-CTLA4 immunotherapy [170]. Preclinical studies of ipilimumab combined with IDO inhibitors show increased lymphocyte infiltration and increased response to therapy [171]. In the ECHO-202/KEYNOTE-037 phase I/II trial investigating epacadostat, an IDO inhibitor, in combination with pembrolizumab, preliminary results show ORR of 39% and disease control rate (DCR) of 65% regardless of PD-L1 expression and HPV status in recurrent or metastatic HNSCC patients [105].

While the major focus in immunology today has been on T-cell modulation and adaptive immunity, targeting the innate immune system, a population of cells that are much less suppressed in the TME, has become an attractive option for anticancer therapy. NK cells, a component of the innate immune system, are capable of spontaneously destroying cancer cells without prior sensitization in contrast to T cells of the adaptive immune system and are garnering more attention as new agents are being developed. Targeting multiple lymphocyte populations, as well as tumor-targeted agents, can be an effective approach to increase both the innate and adaptive immune response. For instance, CD137 stimulates T cells as well as augments NK cell activity and is upregulated on NK cells when exposed to cetuximab-coated, EGFR-expressing HNSCC cell lines [172]. Therefore, CD137 may be used in combination with cetuximab to target both the tumor and the host immune system to enhance NK-mediated ADCC. Additionally, blocking KIR, which is heavily upregulated on infiltrating NK cells of HNSCC patients, is another approach to improve responsiveness to anti-EGFR treatments [69, 173].

14.7 Resistance and Biomarkers

Although immunotherapy causes durable responses in recurrent and metastatic patients, a majority of patients do not respond to checkpoint inhibition, and of those that do, a minority develops acquired resistance [174]. Evolving evidence shows that alterations in several signaling pathways are involved in immediate and/or acquired resistance. For instance, cancer cell-autonomous signals may prevent tumor lymphocyte infiltration and reduce the efficacy of immune checkpoint blockade. Activation of the oncogenic Wnt/β-catenin pathway reduced T-cell and DC infiltration due to suppression of CCL4 chemokine and induced anti-PD-L1 and anti-CTLA4 mAb treatment resistance [175]. Similarly, loss of PTEN (phosphatase and tensin homolog) causing activation of PI3K pathway can also promote checkpoint inhibition resistance and increased PD-L1 expression [59, 176]. Additionally, activating mutations upstream in the MAPK signaling pathway (e.g., in BRAF) has been implicated in reduced TA antigen expression [177] and increased immunosuppressive cytokine and VEGF expression [178]. In addition to immediate resistance, long-term follow-up has now revealed a late relapse seen in 25% of patients with advanced melanoma who initially showed an objective response to PD-1 blockade [179]. Acquired resistance to PD-1 blockade in these patients was associated with loss-of-function mutations in the JAK1, JAK2, and β2-microglobulin (B2M) genes, leading to immune resistance through impaired interferon-receptor signaling and antigen presentation [174].

Recent evidence suggests the quantity and quality of the immune infiltrate into the TME dictate clinical response to treatment and outcomes. These immune signatures have been characterized in part by the Cancer Genome Atlas (TCGA), allowing the classification of tumors based on the presence of myeloid or lymphoid infiltrates and the expression of immune-related genes [180–182]. "Hot" tumors, or those with a robust infiltrate, primarily of CD8+ T cells, have the greatest implication on the clinical outcome following immunotherapy with checkpoint inhibitors. However, there are patients with an effective T-cell infiltrate who still do not respond to treatments. Possible mechanisms for immune escape in this situation include the loss of TA expression, MHC downregulation, and molecular alterations in cellular pathways (most commonly PI3K/AKT/mTOR, MAPK, Wnt/β-catenin, EGFR, Jak, and Hippo). Thus, the increased accumulation of T cells within the tumor could be a critical event, triggering tumor cell mutations and immune escape [183].

Much effort has focused on defining tumor and/or patient characteristics that predict who will and will not respond to particular therapies. In the nivolumab trials, there was a correlation of PD-L1 expression in tumors with clinical response to PD-1-blocking treatment (36%), with none of the PD-L1-negative tumors displaying response to treatment [133, 184–186]. However, since clinical response has been observed in PD-L1-negative tumors in other studies [165, 187], investigations to identify other biomarkers are ongoing. In the KEYNOTE-012 study evaluating pembrolizumab in R/M HNSCC, patients with PD-L2 expression had a trend toward higher overall response rate and longer PFS after adjusting for PD-L1 status, suggesting PD-L2 could be predictive of outcomes with pembrolizumab treatment. However, these are not perfect biomarkers, as not all PD-L1-/PD-L2-positive tumors respond to anti-PD-1 treatment, and conversely, benefit has been seen in some PD-L1-/PD-L2-deficient tumors, highlighting a need to develop additional predictive biomarkers. Currently, inquiries into IFN-γ expression, major histocompatibility complex class II (MHC II) expression, $CD8^+$ T-cell density, and PD-L1 and $CD8^+$ T-cell co-localization at the tumor margin are demonstrating potential as predictive biomarkers for PD-1/PD-L1 blockade response [188, 189].

In summary, immunotherapy is a new paradigm in the treatment of head and neck squamous cell carcinoma, harnessing and manipulating the patient's own immune system to target and destroy tumor cells, which has the potential of producing a durable immune response. Unfortunately, responses to the various immunotherapy options have been less than anticipated. Understanding the mechanisms of immediate and acquired resistance to immunotherapy will help identify the patients who are unlikely to benefit from particular treatments and help design salvage combination therapies or preventive interventions. In particular, a greater understanding of evasion mechanisms of the immune system in the development and progression of HNSCC should lead to improved treatment options and outcomes for patients. Furthermore, identification of effective biomarkers in predicting response to therapy can assist in personalized treatment.

Financial Support This work was supported by the National Institutes of Health grants R01 DE19727, P50 CA097190, and CA110249 U10 CA021115 (PITTDEVRF-00) and University of Pittsburgh Cancer Center Support Grant P30CA047904.

References

1. Ehrlich P. In: Himmelweit EI, editor. The collected papers of Paul Ehrlich. London: Pergamon Press; F1957.
2. Badoual C, et al. Better understanding tumor-host interaction in head and neck cancer to improve the design and development of immunotherapeutic strategies. Head Neck. 2010;32(7):946–58.
3. Allen CT, et al. The clinical implications of antitumor immunity in head and neck cancer. Laryngoscope. 2012;122(1):144–57.
4. Kuss I, et al. Decreased absolute counts of T lymphocyte subsets and their relation to disease in squamous cell carcinoma of the head and neck. Clin Cancer Res. 2004;10(11):3755–62.
5. Dasgupta S, et al. Inhibition of NK cell activity through TGF-beta 1 by down-regulation of NKG2D in a murine model of head and neck cancer. J Immunol. 2005;175(8):5541–50.
6. Young MR, et al. Mechanisms of immune suppression in patients with head and neck cancer: influence on the immune infiltrate of the cancer. Int J Cancer. 1996;67(3):333–8.
7. Lopez-Albaitero A, et al. Role of antigen-processing machinery in the in vitro resistance of squamous cell carcinoma of the head and neck cells to recognition by CTL. J Immunol. 2006;176(6):3402–9.
8. Ferris RL, Whiteside TL, Ferrone S. Immune escape associated with functional defects in antigen-processing machinery in head and neck cancer. Clin Cancer Res. 2006;12(13):3890–5.
9. Bindon C, et al. Clearance rates and systemic effects of intravenously administered interleukin 2 (IL-2) containing preparations in human subjects. Br J Cancer. 1983;47(1):123–33.
10. Kirkwood JM, et al. Comparison of intramuscular and intravenous recombinant alpha-2 interferon in melanoma and other cancers. Ann Intern Med. 1985;103(1):32–6.
11. Lienard D, et al. High-dose recombinant tumor necrosis factor alpha in combination with interferon gamma and melphalan in isolation perfusion of the limbs for melanoma and sarcoma. J Clin Oncol. 1992;10(1):52–60.
12. Jonasch E, Haluska FG. Interferon in oncological practice: review of interferon biology, clinical applications, and toxicities. Oncologist. 2001;6(1):34–55.
13. To, S.Q, Knower KC, Clyne CD. Origins and actions of tumor necrosis factor alpha in postmenopausal breast cancer. J Interf Cytokine Res. 2013;33(7):335–45.

14. Vermorken JB, et al. Overview of the efficacy of cetuximab in recurrent and/or metastatic squamous cell carcinoma of the head and neck in patients who previously failed platinum-based therapies. Cancer. 2008;112(12):2710–9.
15. Scott AM, Allison JP, Wolchok JD. Monoclonal antibodies in cancer therapy. Cancer Immun. 2012;12:14.
16. Rao SD, Fury MG, Pfister DG. Molecular-targeted therapies in head and neck cancer. Semin Radiat Oncol. 2012;22(3):207–13.
17. Page DB, et al. Immune modulation in cancer with antibodies. Annu Rev Med. 2014;65:185–202.
18. Whang SN, Filippova M, Duerksen-Hughes P. Recent progress in therapeutic treatments and screening strategies for the prevention and treatment of HPV-associated head and neck Cancer. Virus. 2015;7(9):5040–65.
19. Lott JB. Oncolytic viruses: a new paradigm for treatment of head and neck cancer. Oral Surg Oral Med Oral Pathol Oral Radiol. 2012;113(2):155–60.
20. Jiang P, et al. Adoptive cell transfer after chemotherapy enhances survival in patients with resectable HNSCC. Int Immunopharmacol. 2015;28(1):208–14.
21. Lynch TJ, et al. Activating mutations in the epidermal growth factor receptor underlying responsiveness of non-small-cell lung cancer to gefitinib. N Engl J Med. 2004;350(21):2129–39.
22. Lane DP. Cancer. p53, guardian of the genome. Nature. 1992;358(6381):15–6.
23. Burnet FM. The concept of immunological surveillance. Prog Exp Tumor Res. 1970;13:1–27.
24. Herberman RB, Holden HT. Natural cell-mediated immunity. Adv Cancer Res. 1978;27:305–77.
25. Dunn GP, et al. Cancer immunoediting: from immunosurveillance to tumor escape. Nat Immunol. 2002;3(11):991–8.
26. Bhatia S, et al. Solid cancers after bone marrow transplantation. J Clin Oncol. 2001;19(2):464–71.
27. Haigentz M Jr. Aerodigestive cancers in HIV infection. Curr Opin Oncol. 2005;17(5):474–8.
28. Schreiber RD, Old LJ, Smyth MJ. Cancer immunoediting: integrating immunity's roles in cancer suppression and promotion. Science. 2011;331(6024):1565–70.
29. Ferris RL, Hunt JL, Ferrone S. Human leukocyte antigen (HLA) class I defects in head and neck cancer: molecular mechanisms and clinical significance. Immunol Res. 2005;33(2):113–33.
30. Mizukami Y, et al. Downregulation of HLA class I molecules in the tumour is associated with a poor prognosis in patients with oesophageal squamous cell carcinoma. Br J Cancer. 2008;99(9):1462–7.
31. Ogino T, et al. HLA class I antigen down-regulation in primary laryngeal squamous cell carcinoma lesions as a poor prognostic marker. Cancer Res. 2006;66(18):9281–9.
32. Concha-Benavente F, et al. Immunological and clinical significance of HLA class I antigen processing machinery component defects in malignant cells. Oral Oncol. 2016;58:52–8.
33. Chatrchyan S, et al. Search for invisible decays of Higgs bosons in the vector boson fusion and associated ZH production modes. Eur Phys J C Part Fields. 2014;74(8):2980.
34. Chatrchyan S, et al. Measurement of WZ and ZZ production in pp collisions at [formula: see text] in final states with b-tagged jets. Eur Phys J C Part Fields. 2014;74(8):2973.
35. Bauman JE, Ferris RL. Integrating novel therapeutic monoclonal antibodies into the management of head and neck cancer. Cancer. 2014;120(5):624–32.
36. Andrews LP, et al. LAG3 (CD223) as a cancer immunotherapy target. Immunol Rev. 2017;276(1):80–96.
37. Anderson AC. Tim-3: an emerging target in the cancer immunotherapy landscape. Cancer Immunol Res. 2014;2(5):393–8.
38. Benson DM Jr, Caligiuri MA. Killer immunoglobulin-like receptors and tumor immunity. Cancer Immunol Res. 2014;2(2):99–104.
39. Ferris RL. Immunology and immunotherapy of head and neck Cancer. J Clin Oncol. 2015;33(29):3293–304.

40. Baruah P, et al. Decreased levels of alternative co-stimulatory receptors OX40 and 4-1BB characterise T cells from head and neck cancer patients. Immunobiology. 2012;217(7):669–75.
41. Rudd CE, Taylor A, Schneider H. CD28 and CTLA-4 coreceptor expression and signal transduction. Immunol Rev. 2009;229(1):12–26.
42. Schwartz RH. Costimulation of T lymphocytes: the role of CD28, CTLA-4, and B7/BB1 in interleukin-2 production and immunotherapy. Cell. 1992;71(7):1065–8.
43. Honeychurch J, et al. Immuno-regulatory antibodies for the treatment of cancer. Expert Opin Biol Ther. 2015;15(6):787–801.
44. Alegre ML, et al. Regulation of surface and intracellular expression of CTLA4 on mouse T cells. J Immunol. 1996;157(11):4762–70.
45. Chambers CA, Kuhns MS, Allison JP. Cytotoxic T lymphocyte antigen-4 (CTLA-4) regulates primary and secondary peptide-specific CD4(+) T cell responses. Proc Natl Acad Sci U S A. 1999;96(15):8603–8.
46. Egen JG, Allison JP. Cytotoxic T lymphocyte antigen-4 accumulation in the immunological synapse is regulated by TCR signal strength. Immunity. 2002;16(1):23–35.
47. Takahashi T, et al. Immunologic self-tolerance maintained by CD25(+)CD4(+) regulatory T cells constitutively expressing cytotoxic T lymphocyte-associated antigen 4. J Exp Med. 2000;192(2):303–10.
48. Wing K, et al. CTLA-4 control over Foxp3+ regulatory T cell function. Science. 2008;322(5899):271–5.
49. Linsley PS, et al. Human B7-1 (CD80) and B7-2 (CD86) bind with similar avidities but distinct kinetics to CD28 and CTLA-4 receptors. Immunity. 1994;1(9):793–801.
50. Qureshi OS, et al. Trans-endocytosis of CD80 and CD86: a molecular basis for the cell-extrinsic function of CTLA-4. Science. 2011;332(6029):600–3.
51. Oderup C, et al. Cytotoxic T lymphocyte antigen-4-dependent down-modulation of costimulatory molecules on dendritic cells in CD4+ CD25+ regulatory T-cell-mediated suppression. Immunology. 2006;118(2):240–9.
52. Keir ME, et al. PD-1 and its ligands in tolerance and immunity. Annu Rev Immunol. 2008;26:677–704.
53. Keir ME, et al. Tissue expression of PD-L1 mediates peripheral T cell tolerance. J Exp Med. 2006;203(4):883–95.
54. Liang SC, et al. Regulation of PD-1, PD-L1, and PD-L2 expression during normal and autoimmune responses. Eur J Immunol. 2003;33(10):2706–16.
55. Latchman Y, et al. PD-L2 is a second ligand for PD-1 and inhibits T cell activation. Nat Immunol. 2001;2(3):261–8.
56. Freeman GJ, et al. Engagement of the PD-1 immunoinhibitory receptor by a novel B7 family member leads to negative regulation of lymphocyte activation. J Exp Med. 2000;192(7):1027–34.
57. Eppihimer MJ, et al. Expression and regulation of the PD-L1 immunoinhibitory molecule on microvascular endothelial cells. Microcirculation. 2002;9(2):133–45.
58. Butte MJ, et al. Programmed death-1 ligand 1 interacts specifically with the B7-1 costimulatory molecule to inhibit T cell responses. Immunity. 2007;27(1):111–22.
59. Parsa AT, et al. Loss of tumor suppressor PTEN function increases B7-H1 expression and immunoresistance in glioma. Nat Med. 2007;13(1):84–8.
60. Concha-Benavente F, et al. Identification of the cell-intrinsic and -extrinsic pathways downstream of EGFR and IFNgamma that induce PD-L1 expression in head and neck Cancer. Cancer Res. 2016;76(5):1031–43.
61. Seiwert TY, et al. Safety and clinical activity of pembrolizumab for treatment of recurrent or metastatic squamous cell carcinoma of the head and neck (KEYNOTE-012): an open-label, multicentre, phase 1b trial. Lancet Oncol. 2016;17(7):956–65.
62. Wherry EJ. T cell exhaustion. Nat Immunol. 2011;12(6):492–9.
63. Francisco LM, et al. PD-L1 regulates the development, maintenance, and function of induced regulatory T cells. J Exp Med. 2009;206(13):3015–29.

64. Lowther DE, et al. *PD-1 marks dysfunctional regulatory T cells in malignant gliomas*. JCI Insight. 2016;1(5):e85935.
65. Badoual C, et al. PD-1-expressing tumor-infiltrating T cells are a favorable prognostic biomarker in HPV-associated head and neck cancer. Cancer Res. 2013;73(1):128–38.
66. Ukpo OC, Thorstad WL, Lewis JS Jr. B7-H1 expression model for immune evasion in human papillomavirus-related oropharyngeal squamous cell carcinoma. Head Neck Pathol. 2013;7(2):113–21.
67. Lyford-Pike S, et al. Evidence for a role of the PD-1:PD-L1 pathway in immune resistance of HPV-associated head and neck squamous cell carcinoma. Cancer Res. 2013;73(6):1733–41.
68. Goldberg MV, Drake CG. LAG-3 in Cancer immunotherapy. Curr Top Microbiol Immunol. 2011;344:269–78.
69. Camisaschi C, et al. Alternative activation of human plasmacytoid DCs in vitro and in melanoma lesions: involvement of LAG-3. J Invest Dermatol. 2014;134(7):1893–902.
70. Grosso JF, et al. Functionally distinct LAG-3 and PD-1 subsets on activated and chronically stimulated CD8 T cells. J Immunol. 2009;182(11):6659–69.
71. Woo SR, et al. Immune inhibitory molecules LAG-3 and PD-1 synergistically regulate T-cell function to promote tumoral immune escape. Cancer Res. 2012;72(4):917–27.
72. Sakuishi K, et al. Targeting Tim-3 and PD-1 pathways to reverse T cell exhaustion and restore anti-tumor immunity. J Exp Med. 2010;207(10):2187–94.
73. Zhou Q, et al. Coexpression of Tim-3 and PD-1 identifies a CD8+ T-cell exhaustion phenotype in mice with disseminated acute myelogenous leukemia. Blood. 2011;117(17):4501–10.
74. Fourcade J, et al. Upregulation of Tim-3 and PD-1 expression is associated with tumor antigen-specific CD8+ T cell dysfunction in melanoma patients. J Exp Med. 2010;207(10):2175–86.
75. Gao X, et al. TIM-3 expression characterizes regulatory T cells in tumor tissues and is associated with lung cancer progression. PLoS One. 2012;7(2):e30676.
76. Yang ZZ, et al. IL-12 upregulates TIM-3 expression and induces T cell exhaustion in patients with follicular B cell non-Hodgkin lymphoma. J Clin Invest. 2012;122(4):1271–82.
77. da Silva IP, et al. Reversal of NK-cell exhaustion in advanced melanoma by Tim-3 blockade. Cancer Immunol Res. 2014;2(5):410–22.
78. Jie HB, et al. Intratumoral regulatory T cells upregulate immunosuppressive molecules in head and neck cancer patients. Br J Cancer. 2013;109(10):2629–35.
79. Herberman RB, et al. Role of interferon in augmentation of natural and antibody-dependent cell-mediated cytotoxicity. Cancer Treat Rep. 1978;62(11):1893–6.
80. Sakuishi K, et al. TIM3+FOXP3+ regulatory T cells are tissue-specific promoters of T-cell dysfunction in cancer. Oncoimmunology. 2013;2(4):e23849.
81. Thielens A, Vivier E, Romagne F. NK cell MHC class I specific receptors (KIR): from biology to clinical intervention. Curr Opin Immunol. 2012;24(2):239–45.
82. Moesta AK, Parham P. Diverse functionality among human NK cell receptors for the C1 epitope of HLA-C: KIR2DS2, KIR2DL2, and KIR2DL3. Front Immunol. 2012;3:336.
83. Yusa S, Campbell KS. Src homology region 2-containing protein tyrosine phosphatase-2 (SHP-2) can play a direct role in the inhibitory function of killer cell Ig-like receptors in human NK cells. J Immunol. 2003;170(9):4539–47.
84. Purdy AK, Campbell KS. Natural killer cells and cancer: regulation by the killer cell Ig-like receptors (KIR). Cancer Biol Ther. 2009;8(23):2211–20.
85. Besson C, et al. Association of killer cell immunoglobulin-like receptor genes with Hodgkin's lymphoma in a familial study. PLoS One. 2007;2(5):e406.
86. Verheyden S, Bernier M, Demanet C. Identification of natural killer cell receptor phenotypes associated with leukemia. Leukemia. 2004;18(12):2002–7.
87. Mandal R, et al. The head and neck cancer immune landscape and its immunotherapeutic implications. JCI Insight. 2016;1(17):e89829.
88. Pollok KE, et al. Inducible T cell antigen 4-1BB. Analysis of expression and function. J Immunol. 1993;150(3):771–81.

89. Melero I, et al. NK1.1 cells express 4-1BB (CDw137) costimulatory molecule and are required for tumor immunity elicited by anti-4-1BB monoclonal antibodies. Cell Immunol. 1998;190(2):167–72.
90. Lynch DH. The promise of 4-1BB (CD137)-mediated immunomodulation and the immunotherapy of cancer. Immunol Rev. 2008;222:277–86.
91. Vetto JT, et al. Presence of the T-cell activation marker OX-40 on tumor infiltrating lymphocytes and draining lymph node cells from patients with melanoma and head and neck cancers. Am J Surg. 1997;174(3):258–65.
92. Bell RB, et al. OX40 signaling in head and neck squamous cell carcinoma: overcoming immunosuppression in the tumor microenvironment. Oral Oncol. 2016;52:1–10.
93. Eliopoulos AG, Young LS. The role of the CD40 pathway in the pathogenesis and treatment of cancer. Curr Opin Pharmacol. 2004;4(4):360–7.
94. van Kooten C, Banchereau J. CD40-CD40 ligand. J Leukoc Biol. 2000;67(1):2–17.
95. Cao W, et al. CD40 function in squamous cell cancer of the head and neck. Oral Oncol. 2005;41(5):462–9.
96. Posner MR, et al. Surface membrane-expressed CD40 is present on tumor cells from squamous cell cancer of the head and neck in vitro and in vivo and regulates cell growth in tumor cell lines. Clin Cancer Res. 1999;5(8):2261–70.
97. Funakoshi S, et al. Inhibition of human B-cell lymphoma growth by CD40 stimulation. Blood. 1994;83(10):2787–94.
98. von Leoprechting A, et al. Stimulation of CD40 on immunogenic human malignant melanomas augments their cytotoxic T lymphocyte-mediated lysis and induces apoptosis. Cancer Res. 1999;59(6):1287–94.
99. Sathawane D, et al. Monocyte CD40 expression in head and neck squamous cell carcinoma (HNSCC). Hum Immunol. 2013;74(1):1–5.
100. Grivennikov SI, Greten FR, Karin M. Immunity, inflammation, and cancer. Cell. 2010;140(6):883–99.
101. Jebreel A, et al. Investigation of interleukin 10, 12 and 18 levels in patients with head and neck cancer. J Laryngol Otol. 2007;121(3):246–52.
102. Moutsopoulos NM, Wen J, Wahl SM. TGF-beta and tumors--an ill-fated alliance. Curr Opin Immunol. 2008;20(2):234–40.
103. Mannino MH, et al. The paradoxical role of IL-10 in immunity and cancer. Cancer Lett. 2015;367(2):103–7.
104. Harris SG, et al. Prostaglandins as modulators of immunity. Trends Immunol. 2002;23(3):144–50.
105. Snyderman CH, et al. Prognostic significance of prostaglandin E2 production in fresh tissues of head and neck cancer patients. Head Neck. 1995;17(2):108–13.
106. Camacho M, et al. Prostaglandin E(2) pathway in head and neck squamous cell carcinoma. Head Neck. 2008;30(9):1175–81.
107. Seiwert TY, Cohen EE. Targeting angiogenesis in head and neck cancer. Semin Oncol. 2008;35(3):274–85.
108. Johnson BF, et al. Vascular endothelial growth factor and immunosuppression in cancer: current knowledge and potential for new therapy. Expert Opin Biol Ther. 2007;7(4):449–60.
109. Ciardiello F, Tortora G. Epidermal growth factor receptor (EGFR) as a target in cancer therapy: understanding the role of receptor expression and other molecular determinants that could influence the response to anti-EGFR drugs. Eur J Cancer. 2003;39(10):1348–54.
110. Uyttenhove C, et al. Evidence for a tumoral immune resistance mechanism based on tryptophan degradation by indoleamine 2,3-dioxygenase. Nat Med. 2003;9(10):1269–74.
111. Oehler JR, et al. Natural cell-mediated cytotoxicity in rats. II. In vivo augmentation of NK-cell activity. Int J Cancer. 1978;21(2):210–20.
112. Yu H, Pardoll D, Jove R. STATs in cancer inflammation and immunity: a leading role for STAT3. Nat Rev Cancer. 2009;9(11):798–809.
113. Ho HH, Ivashkiv LB. Role of STAT3 in type I interferon responses. Negative regulation of STAT1-dependent inflammatory gene activation. J Biol Chem. 2006;281(20):14111–8.

114. Schaefer C, et al. Characteristics of CD4+CD25+ regulatory T cells in the peripheral circulation of patients with head and neck cancer. Br J Cancer. 2005;92(5):913–20.
115. Wansom D, et al. Correlation of cellular immunity with human papillomavirus 16 status and outcome in patients with advanced oropharyngeal cancer. Arch Otolaryngol Head Neck Surg. 2010;136(12):1267–73.
116. Green VL, et al. Increased prevalence of tumour infiltrating immune cells in oropharyngeal tumours in comparison to other subsites: relationship to peripheral immunity. Cancer Immunol Immunother. 2013;62(5):863–73.
117. Chikamatsu K, et al. Immunosuppressive activity of CD14+ HLA-DR- cells in squamous cell carcinoma of the head and neck. Cancer Sci. 2012;103(6):976–83.
118. Ostrand-Rosenberg S, Sinha P. Myeloid-derived suppressor cells: linking inflammation and cancer. J Immunol. 2009;182(8):4499–506.
119. Grizzle WE, et al. Age-related increase of tumor susceptibility is associated with myeloid-derived suppressor cell mediated suppression of T cell cytotoxicity in recombinant inbred BXD12 mice. Mech Ageing Dev. 2007;128(11–12):672–80.
120. Costa NL, et al. Tumor-associated macrophages and the profile of inflammatory cytokines in oral squamous cell carcinoma. Oral Oncol. 2013;49(3):216–23.
121. Fujii N, et al. Cancer-associated fibroblasts and CD163-positive macrophages in oral squamous cell carcinoma: their clinicopathological and prognostic significance. J Oral Pathol Med. 2012;41(6):444–51.
122. Gajewski TF, Schreiber H, Fu YX. Innate and adaptive immune cells in the tumor microenvironment. Nat Immunol. 2013;14(10):1014–22.
123. Gajewski TF. The next hurdle in Cancer immunotherapy: overcoming the non-T-cell-inflamed tumor microenvironment. Semin Oncol. 2015;42(4):663–71.
124. Hanna GJ, et al. Defining an inflamed tumor immunophenotype in recurrent, metastatic squamous cell carcinoma of the head and neck. Oral Oncol. 2017;67:61–9.
125. Ferris RL, Jaffee EM, Ferrone S. Tumor antigen-targeted, monoclonal antibody-based immunotherapy: clinical response, cellular immunity, and immunoescape. J Clin Oncol. 2010;28(28):4390–9.
126. Lee SC, et al. Natural killer (NK): dendritic cell (DC) cross talk induced by therapeutic monoclonal antibody triggers tumor antigen-specific T cell immunity. Immunol Res. 2011;50(2–3):248–54.
127. Kondadasula SV, et al. Colocalization of the IL-12 receptor and FcgammaRIIIa to natural killer cell lipid rafts leads to activation of ERK and enhanced production of interferon-gamma. Blood. 2008;111(8):4173–83.
128. Srivastava RM, et al. Cetuximab-activated natural killer and dendritic cells collaborate to trigger tumor antigen-specific T-cell immunity in head and neck cancer patients. Clin Cancer Res. 2013;19(7):1858–72.
129. Trivedi S, Jie HB, Ferris RL. Tumor antigen-specific monoclonal antibodies and induction of T-cell immunity. Semin Oncol. 2014;41(5):678–84.
130. Jie HB, et al. CTLA-4(+) regulatory T cells increased in Cetuximab-treated head and neck Cancer patients suppress NK cell cytotoxicity and correlate with poor prognosis. Cancer Res. 2015;75(11):2200–10.
131. Chow LQ, et al. Antitumor activity of Pembrolizumab in biomarker-unselected patients with recurrent and/or metastatic head and neck squamous cell carcinoma: results from the phase Ib KEYNOTE-012 expansion cohort. J Clin Oncol. 2016;34:3838.
132. Oehler JR, et al. Natural cell-mediated cytotoxicity in rats. I. Tissue and strain distribution, and demonstration of a membrane receptor for the Fc portion of IgG. Int J Cancer. 1978;21(2):204–9.
133. Segal NH, Antonia SJ, Brahmer JR, Maio M, Blake-Haskins A, Li X, et al. Safety and efficacy of MEDI4736, an anti-PD-L1 antibody, in patients from a squamous cell carcinoma of the head and neck (SCCHN) expansion cohort. J Clin Oncol. 33(15):3011.

134. Kesselheim AS, Ferris TG, Studdert DM. Will physician-level measures of clinical performance be used in medical malpractice litigation? JAMA. 2006;295(15):1831–4.
135. Morris JC, et al. The uniform data set (UDS): clinical and cognitive variables and descriptive data from Alzheimer disease centers. Alzheimer Dis Assoc Disord. 2006;20(4):210–6.
136. Nagata Y, et al. Clinical significance of HLA class I alleles on postoperative prognosis of lung cancer patients in Japan. Lung Cancer. 2009;65(1):91–7.
137. Albers AE, et al. Alterations in the T-cell receptor variable beta gene-restricted profile of CD8+ T lymphocytes in the peripheral circulation of patients with squamous cell carcinoma of the head and neck. Clin Cancer Res. 2006;12(8):2394–403.
138. Gildener-Leapman N, Lee J, Ferris RL. Tailored immunotherapy for HPV positive head and neck squamous cell cancer. Oral Oncol. 2014;50(9):780–4.
139. DeRosier LC, et al. Combination treatment with TRA-8 anti death receptor 5 antibody and CPT-11 induces tumor regression in an orthotopic model of pancreatic cancer. Clin Cancer Res. 2007;13(18 Pt 2):5535s–43s.
140. Curran MA, et al. PD-1 and CTLA-4 combination blockade expands infiltrating T cells and reduces regulatory T and myeloid cells within B16 melanoma tumors. Proc Natl Acad Sci U S A. 2010;107(9):4275–80.
141. Costache M, et al. Ciliary body melanoma - a particularly rare type of ocular tumor. Case report and general considerations. Maedica (Buchar). 2013;8(4):360–4.
142. Chan AT, Teo PM, Johnson PJ. Nasopharyngeal carcinoma. Ann Oncol. 2002;13(7):1007–15.
143. Li L, et al. Targeting poly(ADP-ribose) polymerase and the c-Myb-regulated DNA damage response pathway in castration-resistant prostate cancer. Sci Signal. 2014;7(326):ra47.
144. Chatrchyan S, et al. Observation of the associated production of a single top quark and a W boson in pp collisions at sqrt[s] = 8 TeV. Phys Rev Lett. 2014;112(23):231802.
145. Chatrchyan S, et al. Measurement of inclusive W and Z boson production cross sections in pp collisions at sqrt[s] = 8 TeV. Phys Rev Lett. 2014;112(19):191802.
146. Chatrchyan S, et al. Search for flavor-changing neutral currents in top-quark decays t --> Zq in pp collisions at sqrt[s] = 8 TeV. Phys Rev Lett. 2014;112(17):171802.
147. Chatrchyan S, et al. Search for top squark and Higgsino production using diphoton Higgs boson decays. Phys Rev Lett. 2014;112(16):161802.
148. Thompson RF, Maity A. Radiotherapy and the tumor microenvironment: mutual influence and clinical implications. Adv Exp Med Biol. 2014;772:147–65.
149. Hodge JW, et al. The tipping point for combination therapy: cancer vaccines with radiation, chemotherapy, or targeted small molecule inhibitors. Semin Oncol. 2012;39(3):323–39.
150. Hodge JW, et al. Attacking malignant cells that survive therapy: exploiting immunogenic modulation. Oncoimmunology. 2013;2(12):e26937.
151. Martins I, et al. Surface-exposed calreticulin in the interaction between dying cells and phagocytes. Ann N Y Acad Sci. 2010;1209:77–82.
152. Hodge JW, et al. Chemotherapy-induced immunogenic modulation of tumor cells enhances killing by cytotoxic T lymphocytes and is distinct from immunogenic cell death. Int J Cancer. 2013;133(3):624–36.
153. Gameiro SR, et al. Radiation-induced immunogenic modulation of tumor enhances antigen processing and calreticulin exposure, resulting in enhanced T-cell killing. Oncotarget. 2014;5(2):403–16.
154. Folch E, et al. Adequacy of lymph node transbronchial needle aspirates using convex probe endobronchial ultrasound for multiple tumor genotyping techniques in non-small-cell lung cancer. J Thorac Oncol. 2013;8(11):1438–44.
155. Emens LA, Middleton G. The interplay of immunotherapy and chemotherapy: harnessing potential synergies. Cancer Immunol Res. 2015;3(5):436–43.
156. Formenti SC, Demaria S. Combining radiotherapy and cancer immunotherapy: a paradigm shift. J Natl Cancer Inst. 2013;105(4):256–65.
157. Demaria S, et al. Immune-mediated inhibition of metastases after treatment with local radiation and CTLA-4 blockade in a mouse model of breast cancer. Clin Cancer Res. 2005;11(2 Pt 1):728–34.

158. Langer CJ, et al. Carboplatin and pemetrexed with or without pembrolizumab for advanced, non-squamous non-small-cell lung cancer: a randomised, phase 2 cohort of the open-label KEYNOTE-021 study. Lancet Oncol. 2016;17(11):1497–508.
159. Rizvi NA, et al. Nivolumab in combination with platinum-based doublet chemotherapy for first-line treatment of advanced non-small-cell lung Cancer. J Clin Oncol. 2016;34(25):2969–79.
160. Reck M, et al. Pembrolizumab versus chemotherapy for PD-L1-positive non-small-cell lung Cancer. N Engl J Med. 2016;375(19):1823–33.
161. Borghaei H, et al. Nivolumab versus docetaxel in advanced nonsquamous non-small-cell lung Cancer. N Engl J Med. 2015;373(17):1627–39.
162. Pego-Fernandes PM, et al. Double valve replacement due to dysfunction secondary to carcinoid tumor. Arq Bras Cardiol. 2013;100(3):e32–4.
163. Ferris RL, et al. Nivolumab for recurrent squamous-cell carcinoma of the head and neck. N Engl J Med. 2016;375(19):1856–67.
164. Das R, et al. Combination therapy with anti-CTLA-4 and anti-PD-1 leads to distinct immunologic changes in vivo. J Immunol. 2015;194(3):950–9.
165. Larkin J, et al. Combined Nivolumab and Ipilimumab or monotherapy in untreated melanoma. N Engl J Med. 2015;373(1):23–34.
166. Bullhe SH. Kalāma Buḷḷe Shāha1990, Paṭiālā: Bhāshā Wibhāga, Pañjāba 12, 187 p.
167. Guo Z, et al. PD-1 blockade and OX40 triggering synergistically protects against tumor growth in a murine model of ovarian cancer. PLoS One. 2014;9(2):e89350.
168. Camirand G, et al. CD45 ligation expands Tregs by promoting interactions with DCs. J Clin Invest. 2014;124(10):4603–13.
169. Hodi FS, et al. Bevacizumab plus Ipilimumab in Patients with Metastatic Melanoma. Cancer Immunol Res. 2014;2(7):632–642.
170. Holmgaard RB, et al. Indoleamine 2,3-dioxygenase is a critical resistance mechanism in antitumor T cell immunotherapy targeting CTLA-4. J Exp Med. 2013;210(7):1389–402.
171. Cai J, et al. ATP hydrolysis catalyzed by human replication factor C requires participation of multiple subunits. Proc Natl Acad Sci U S A. 1998;95(20):11607–12.
172. Kohrt HE, et al. Targeting CD137 enhances the efficacy of cetuximab. J Clin Invest. 2014;124(6):2668–82.
173. Chatrchyan S, et al. Probing color coherence effects in pp collisions at [formula: see text]. Eur Phys J C Part Fields. 2014;74(6):2901.
174. Zaretsky JM, et al. Mutations associated with acquired resistance to PD-1 blockade in melanoma. N Engl J Med. 2016;375(9):819–29.
175. Spranger S, Bao R, Gajewski TF. Melanoma-intrinsic beta-catenin signalling prevents anti-tumour immunity. Nature. 2015;523(7559):231–5.
176. Peng W, et al. Loss of PTEN promotes resistance to T cell-mediated immunotherapy. Cancer Discov. 2016;6(2):202–16.
177. Boni A, et al. Selective BRAFV600E inhibition enhances T-cell recognition of melanoma without affecting lymphocyte function. Cancer Res. 2010;70(13):5213–9.
178. Frederick DT, et al. BRAF inhibition is associated with enhanced melanoma antigen expression and a more favorable tumor microenvironment in patients with metastatic melanoma. Clin Cancer Res. 2013;19(5):1225–31.
179. Ribas A, et al. Association of Pembrolizumab with Tumor Response and Survival among Patients with Advanced Melanoma. JAMA. 2016;315(15):1600–9.
180. Harlin H, et al. Chemokine expression in melanoma metastases associated with CD8+ T-cell recruitment. Cancer Res. 2009;69(7):3077–85.
181. Sweis RF, et al. Molecular drivers of the non-T-cell-inflamed tumor microenvironment in urothelial bladder Cancer. Cancer Immunol Res. 2016;4(7):563–8.
182. Keck MK, et al. Integrative analysis of head and neck cancer identifies two biologically distinct HPV and three non-HPV subtypes. Clin Cancer Res. 2015;21(4):870–81.
183. Sharma P, et al. Primary, adaptive, and acquired resistance to Cancer immunotherapy. Cell. 2017;168(4):707–23.

184. Topalian SL, et al. Safety, activity, and immune correlates of anti-PD-1 antibody in cancer. N Engl J Med. 2012;366(26):2443–54.
185. Gillison ML, Blumenschein G, Fayette J, Guigay J, Colevas AD, Licitra L, et al. Nivolumab (nivo) vs investigator's choice (IC) for recurrent or metastatic (R/M) head and neck squamous cell carcinoma (HNSCC): Checkmate 141. In: Proceedings of the 107th Annual Meeting of the American Association for Cancer Research. New Orleans\Philadelphia: AACR; Cancer Res; 2016. 2016;76(14 Suppl):Abstract nr CT099.
186. Seiwert TY, Burtness B, Weiss J, Gluck I, Elder JP, Pai SI, et al. A phase Ib study of MK-3475 in patients with human papillomavirus (HPV)-associated and non-HPV-associated head and neck (H/N) cancer. J Clin Oncol. 2014, May 20;32(15 Suppl):6011.
187. Aguiar PN Jr, et al. The role of PD-L1 expression as a predictive biomarker in advanced non-small-cell lung cancer: a network meta-analysis. Immunotherapy. 2016;8(4):479–88.
188. Tumeh PC, et al. PD-1 blockade induces responses by inhibiting adaptive immune resistance. Nature. 2014;515(7528):568–71.
189. Johnson DB, et al. Melanoma-specific MHC-II expression represents a tumour-autonomous phenotype and predicts response to anti-PD-1/PD-L1 therapy. Nat Commun. 2016;7:10582.

Chapter 15
The Clinical Impact of Hypoxia in Head and Neck Squamous Cell Carcinoma

Annette M. Lim, Quynh-Thu Le, and Danny Rischin

Abstract Hypoxia commonly occurs in head and neck squamous cell carcinomas and is associated with treatment resistance and poor patient outcome. The presence of tumor hypoxia can contribute to the protection of cancer cells from DNA damage induced by ionizing radiation and chemotherapy, with hypoxia also promoting alterations in tumor biology that enhance malignant progression. Significant effort has been devoted to abrogating the effects of hypoxia through approaches that include the modification of tumor oxygenation and the tumor vasculature. Recent approaches to improve therapeutic response have explored agents that can sensitize hypoxic cancer cells to chemoradiation or directly cause hypoxic cell death. However, these approaches have had limited success. There is significant clinical need to identify an appropriate predictive biomarker to select patients with tumor hypoxia that will benefit from hypoxia-modifying approaches.

Keywords Head and neck · Squamous cell carcinoma · HNSCC · Hypoxia · Radiation resistance · Oxygen enhancement ratio · Nitroimidazole · Nimorazole · Tirapazamine · Pimonidazole · FMISO · FAZA · HIF · HIF-1 · Osteopontin

15.1 Introduction

Head and neck squamous cell carcinomas (HNSCC) represent a biologically diverse group of tumors, typically referring to squamous cell carcinomas arising from the mucosa of the oral cavity, oropharynx, hypopharynx, and larynx. In 2012, more

A. M. Lim (✉)
Sir Charles Gairdner Hospital and University of Western Australia, Perth, Australia
e-mail: Annette.Lim@health.wa.gov.au

Q.-T. Le
Stanford University Medical Center, Stanford, CA, USA

D. Rischin
Peter MacCallum Cancer Centre, Melbourne, VIC, Australia

B. Burtness, E. A. Golemis (eds.), *Molecular Determinants of Head and Neck Cancer*, Current Cancer Research, https://doi.org/10.1007/978-3-319-78762-6_15

than half a million patients were diagnosed with HNSCC, with the incidence predicted to increase to more than 856,000 cases by 2035 [1–3]. As most patients present with locally advanced disease (AJCC stage III/IV), cure rates are suboptimal despite the common use of trimodality treatment, with 2–3-year overall survival (OS) rates of approximately 65% [4, 5]. For patients with recurrent or metastatic disease, the median survival is less than 5–11 months even with palliative treatment [6–8]. Recently, a clinicopathologically and biologically distinct subset of HNSCC – oropharyngeal carcinomas induced by human papillomavirus (HPV) – have been identified (see Chaps. 20 and 21), which are associated with a significantly improved survival regardless of the treatment modality used [4, 9–14]. This discovery highlights the concept that though the HNSCC in individual patients may be histologically similar and with primary tumors arising in close geographical proximity, variations in tumor biology and pathogenesis crucially drive patient outcome. As HPV-negative HNSCC still accounts for a significant proportion of HNSCC diagnosed, many patients still face a high chance of a poor outcome, and therefore devising improved treatments for this tumor class represents an area of unmet clinical need.

The identification of biological factors contributing to inferior patient outcome facilitates the study of targeted treatment approaches that can improve therapeutic efficacy. Hypoxia, the condition of a low oxygen level, is known to be a common pathophysiological characteristic of solid tumors including HNSCC, and has been found to correlate with treatment resistance and poor patient outcome [15–17]. Intra-tumoral hypoxia results from an imbalance between oxygen delivery and oxygen consumption within the tumor. There is abundant evidence that hypoxia induces resistance to radiation therapy (XRT), which is one of the most commonly used treatment modalities in HNSCC [18, 19]. Hypoxia also contributes to resistance to systemic therapy and facilitates molecular events that enhance propensity for nodal and distant tumor spread [20–22]. Therefore, significant effort has been devoted to targeting hypoxia in combination with radiotherapy to overcome the adverse effects of low tumor oxygenation.

Clinical trials have evaluated a number of distinct approaches to increase intratumoral oxygenation such as the use of hyperbaric oxygen (HBO) during radiotherapy, or carbogen and nicotinamide which aim to improve perfusion and oxygen diffusion into the tumor. More recently, hypoxic cell radiosensitizers and hypoxic cell cytotoxins have been investigated [5, 17, 19, 23–25]. Nevertheless, overall, most randomized trials of oxygenation approaches have shown limited improvement in disease control and patient survival. This is despite a favorable body of data that will be discussed within this chapter that demonstrates that tumors are vulnerable to hypoxia modification and that hypoxia-targeting agents can be exploited to enhance tumor control. However, as for many targeted therapies, the availability of an appropriate predictive marker to identify hypoxic tumors and select patients for hypoxia-modifying therapy is a key missing factor that has so far hindered the success of these approaches. The development of such biomarkers is further complicated by the evolving understanding of pathology underlying different HNSCC subsites represented by the HPV-related and HPV-unrelated subgroups, the impact of the tumor microenvironment and host factors such as the immune system on patient outcome, and perhaps most pertinently, the inherent heterogeneous nature of tumor biology.

This chapter provides an overview of the clinical relevance of hypoxia in HNSCC, focusing on the definition of hypoxia and its assessment, and the different treatment approaches that have been investigated to alter the impact of hypoxia on patient outcome.

15.2 The Definition of Hypoxia in Solid Tumors

One of the biggest challenges to understanding the role of hypoxia in HNSCC is the lack of uniform means to define and assess clinically meaningful levels of hypoxia. Simply, hypoxia refers to the imbalance between the supply and demand of oxygen. The objective definition of hypoxia in solid neoplasms is described as the presence of low oxygen tension below normoxic levels of 40–60 mmHg [26, 27], with hypoxic tumor pO_2 measurements most commonly measuring less than 2.5 mmHg. Two main states of hypoxia have been described, *perfusional* and *diffusional* hypoxia, with likely differential impact on the tumor microenvironment, expression of proteins, and response to therapy [15, 28, 29]. Perfusional hypoxia refers to a disturbance in the blood and oxygen delivery to a tumor, which usually results either from a disruption of the vascular supply or due to significant variations in the delivery of oxygen-carrying red blood cells to the tumor. Perfusional ischemia can occur in tumors in a homogeneous or heterogeneous fashion, when the increased metabolic requirements of a rapidly growing cancer and its supportive microenvironment outweigh the normal supply of oxygen or when neovascularization of the tumor does not adequately support the rate of growth. Diffusion-limited hypoxia or diffusional hypoxia can also occur, which arises due to large diffusion distances between tumor cells and blood vessels, commonly defined as greater than 70 micrometers. Indeed, both perfusional and diffusional hypoxia can occur in the same tumor [29]. Hypoxia is sometimes described according to temporality or chronology, as acute or chronic hypoxia. However the underlying causes of hypoxia remain as perfusional and diffusional changes in oxygen delivery. Additionally, given the potential for heterogeneous temporal and spatial hypoxia, it may be more pertinent to define hypoxia functionally – that is, when tumor hypoxia is sufficient to lead to stabilization of hypoxia-inducible factor (HIF) subunits [28]. However, it is difficult to assess the expression of these subunits, spatially and temporaneously, within tumors. Proteins of the HIF family are key regulators of the hypoxia response pathway, and the family is discussed in detail in Sect. 12.6.2.

15.3 Tumor Hypoxia and Poor Prognosis in HNSCC

As most patients with HNSCC present with locally advanced disease, treatment typically involves the use of combination therapy. Radiation can be administered alone or with chemotherapy, either adjuvantly after surgery or as the primary treatment modality. When chemotherapy is necessary, cisplatin-based chemotherapy is known to be the most efficacious radiosensitizer, providing an improved OS benefit [30].

Given the early conceptual understanding in the 1950s of the ability of hypoxia to confer radiation resistance [18], the clinical impact of hypoxia on patient outcome in HNSCC has been a matter of interest [23, 31, 32]. Early non-randomized studies provided indirect evidence of the negative effects of hypoxia on tumor control and also demonstrated the promise that hypoxia-modulating therapy could indeed be beneficial for patients. Commencing in 1964, the first prospective randomized controlled trial using HBO to address potential radioresistance due to hypoxia was performed in HNSCC patients treated with definitive XRT [23]. The authors observed that patients who were randomized to receive room air rather than HBO with XRT experienced a significantly worse 5-year regional control rate of 30% versus 53% ($P < 0.001$) and an increased need for salvage surgery. A similar subsequent study performed which randomized HNSCC patients to a lower XRT dose (10Gy) with HBO versus a higher XRT dose (30Gy) and room air again found that patients who received room air had both worse rate of locoregional control ($P < 0.05$) and inferior crude overall survival rate ($P < 0.02$) [24]. These early studies provided an indirect observation of the negative impact of oxygenation on tumor control.

A number of seminal studies subsequently confirmed the negative effect of hypoxia on patient outcome in HNSCC through the direct measurement of intratumoral hypoxia [2, 33–35]. Using oxygen electrodes ("polarographically") to examine the levels of oxygenation in involved lymph nodes of locally advanced HNSCC patients being treated with conventional radiation, Nordsmark and colleagues found that intra-tumoral pO_2 levels below 2.5 mmHg were an independent predictor of worse locoregional tumor control ($P = 0.018$) [36]. Brizel and colleagues similarly assessed hypoxia polarographically (described in greater detail in Sect. 12.6.1) with electrodes placed in primary tumors and pathologically enlarged lymph nodes of patients with advanced HNSCC. The authors reported a significantly lower 1-year disease-free survival (DFS) rate of 22% for patients with median tumor pO_2 less than 10 mmHg versus 78% for those with a median tumor pO_2 greater than 10 mmHg ($P = 0.009$) [33]. Rudat and colleagues pooled pre-treatment data from studies using electrode measurements of hypoxia in 194 patients with HNSCC treated with greater than 60Gy of XRT and again confirmed the poor prognostic impact of pO2 levels less than 2.5mmHg ($P = 0.004$) [34]. However, the negative and positive predictive values of pre-treatment pO2 did not adequately identify patients in need of alternative treatment approaches, with the authors recognizing the complexity of variables influencing hypoxia assessment, including intra-individual variation in hypoxia measurements with electrodes, and other factors affecting oxygenation such as anemia.

Beyond the direct intra-tumoral measurement of hypoxia, the relationship between tumor oxygenation and poor prognosis has also been confirmed with less invasive means. When tumor hypoxia is measured with exogenous detection methods such as nitroimidazoles, or endogenous hypoxia markers such as carbonic anhydrase-9 (CA-IX), or imaging-based detection methods including PET-based hypoxia tracers, in general the presence of hypoxia has been found to correlate with worse locoregional control, worse distant control, and worse overall survival regardless of treatment provided [37–43].

It should be noted that not all studies have reported a correlation between detection of tumoral hypoxia and an inferior patient outcome [42]; this highlights the heterogeneity of tumor biology and methods to detect tumor hypoxia that undoubtedly impact on correlative results. However overall, it is widely accepted that the presence of hypoxia confers a negative impact on patient outcome, regardless of how it is assessed.

15.4 Theory of Hypoxia-Induced Treatment Resistance

As ionizing radiation is a key treatment in HNSCC, one important mechanism by which hypoxia confers an adverse effect on patient survival is due to the necessity of oxygen for radiation cytotoxicity. Mechanistically, the therapeutic benefit of ionizing radiation reflects its ability to cause cell death by both directly and indirectly damaging DNA. DNA can be damaged by radiation by direct photon interaction (photo-interaction). Alternatively, indirect DNA damage can be caused by the generation of secondary radicals. Ionizing radiation-induced water lysis leads to the production of H_2O_2 and hydroxyl radicals which cause double-stranded DNA damage, single-stranded DNA damage, base breaks, and DNA-protein crosslinking. In addition, sugar radical production and chemical modification of purine and pyrimidine bases can also occur. Oxygen serves to stabilize these radiation-induced free radical species, so that DNA in close proximity to regions of high oxygen sustains increased damage [44].

In the early twentieth century, Hanh and Schwartz were the first to propose the importance of oxygenation on the efficacy of radiation therapy, based on observations of altered radiation effect with vascular compression causing relative “anemia” in tissues [29]. Later in vivo work from the 1960s onward using mouse and human xenograft tumor models treated with XRT in differing hypoxic conditions demonstrated that a fraction of tumor cells were resistant to the effects of radiation and remained viable even at high doses of greater than 3000 rad [45, 46]. Using survival models, tumor growth models, and tumor control models, a fraction of approximately 10–20% hypoxic cells were found to be radioresistant, and importantly, these hypoxic cells accounted for the radiation response observed. A proportion of radioresistant cells was consistently noted regardless of tumor size and was even identified in microscopic foci, with the hypoxic fraction of cells increasing proportionally with tumor size [45]. Furthermore, it was observed that tumors could become radioresistant in hypoxic conditions and conversely could be made more radiosensitive when oxygenation was improved [29].

In 1953, Gray observed that approximately a three times greater radiation dose was required to kill hypoxic cells compared to their well-oxygenated counterparts [18]. This phenomenon is quantified as an oxygen enhancement ratio (OER) and refers to the ratio of the radiation dose given under hypoxic conditions versus normoxic conditions that would yield the same biological effect. This ratio typically falls between 2.7 and 3.0 and can transition from 1.0 to 3.0 when tumor oxygen tension is below 5 mm Hg. The OER is also dependent on other factors such as the oxygen partial pressure of the hypoxic and normoxic conditions, cell or tissue histotype, radiation dose, and linear energy transfer of the radiation applied [47] .

In addition to modifying radiation response, hypoxia is known to contribute to chemotherapy resistance [20, 48, 49]. Oxygen levels can directly alter the efficacy of some chemotherapy agents such as the alkylating agents, which also require the presence of oxygen for maximal efficacy through the generation of free radicals [49]. Additionally in hypoxic conditions, the effects of many drugs are reduced due to the presence of substrates that directly compete with DNA for alkylation, hinder the ability of the chemotherapy to produce DNA damage, or slow cell cycling making cells less vulnerable to chemotherapy [20, 50–52]. Additionally, hypoxia itself exerts selective pressure biased for cells that can survive and proliferate in otherwise unviable conditions low in oxygen and nutrients through adaptive processes mediated by the hypoxia signaling pathways discussed further in this chapter [53].

A hypoxic environment also drives upregulation of genes that control angiogenesis, cell proliferation, survival, glucose metabolism, invasion, and metastasis [54]. One of the most extensively studied regulators of the oxygen-responsive pathways is the family of hypoxia-inducible transcription factors (HIF; Fig. 15.1) that bind to hypoxia response elements (HREs) of hundreds of target genes [55–57]. One of the key mechanisms by which hypoxia is thought to contribute to chemoresistance is via HIF-1 pathway activation [22, 58–61]. HIF-1 also regulates the transcription of members of the ABC-transporter family, which facilitate cancer survival in hypoxic conditions. These ABC transporters include members of the multidrug resistance (MDR) gene family, which promote chemoresistance to agents including taxanes and anthracyclines through overexpression of the P-gp membrane efflux pump and related pumps [22, 61–63].

Hypoxia and resultant HIF-1 activation also lead to acidification of the tumor microenvironment through facilitation of anaerobic glycolytic metabolism beyond the expected aerobic glycolytic metabolism (Warburg effect) in tumors, mainly through the induction of the GLUT-1 and GLUT-3 glucose transporters and via direct activation of carbonic anhydrases [22, 64, 65, 66]. The relative environmental acidity hinders passive diffusion of some chemotherapy agents, such as anthracyclines. HIF-1 signaling also promotes other biological mechanisms that enhance malignant potential including genomic instability and facilitation of survival of cancer stem cells and TP53-mutated tumors which inherently possess greater clonogenic and metastatic potential that can drive repopulation and relapse after therapy [60, 67, 68, 69, 70].

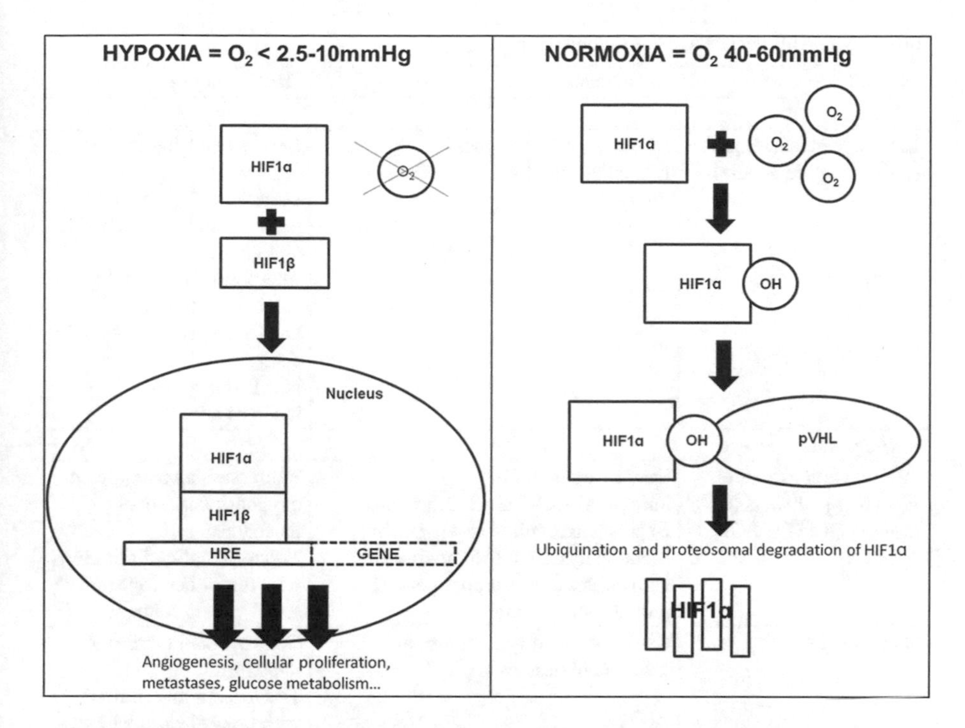

Fig. 15.1 Hypoxic regulation of HIF-1 function. In hypoxic conditions, the HIF1 alpha and beta subunits bind to the hypoxic response elements (HRE) in the promotor region of a variety of hypoxia response genes including carbonic anhydrase-9 (CA-IX), glucose transporters (GLUT), vascular endothelial growth factor (VEGF) proteins, programmed cell-death-1-ligand (PDL1), and erythropoietin receptor. This results in activation of downstream signaling pathways which facilitate angiogenesis, cellular proliferation, invasion and metastases, and glucose metabolism. In contrast, in normoxic conditions HIF1 hydroxylation can occur due to the presence of oxygen which facilitates binding to the von Hippel-Lindau protein which then instigates ubiquitination and proteasomal degradation of HIF1

15.5 Tumor Hypoxia Detection

Due in part to the many different characteristics of hypoxia and variable ways of assessing tumoral hypoxia, no global biomarker of hypoxia has yet been identified to select appropriate patients for therapeutic strategies. The ability to consistently detect and measure biologically significant hypoxia is crucial for the improvement of hypoxia-modulating approaches for the treatment of HNSCC. The well-established and extensively studied relationship between low tumor oxygenation and poor prognosis has utilized two main detection methods: (1) direct electrode measurement of hypoxia, and (2) indirect hypoxia assessment with endogenous and exogenous biomarkers or with imaging modalities (Table 15.1) [42, 43]. Importantly, the choice of methodology can impact patient comfort but also has the capacity to

Table 15.1 Approaches to hypoxia detection

Method	Advantages	Disadvantages
Direct		
Electrode measurement (polarographic needle)	Direct in situ assessment of hypoxia Immediate results	Invasive and anesthetic required Unable to distinguish between areas of necrosis and viable tumor Interobserver variability Limited by tumor heterogeneity and dynamic changes Needle placement inconsistent
Indirect: Endogenous		
HIF and downstream signaling pathways (e.g., CA-IX, GLUT, VEGF, VEGFR)	Easy to assess with immunohistochemical techniques Expression regulated by hypoxia Meta-analyses confirm relationship with expression and poor patient prognosis in general	Expression also regulated by oxygen-independent mechanisms Tissue sampling limitations with tumor heterogeneity and dynamic changes
Osteopontin	Can be isolated from serum or assessed in tissue samples	Controversial marker of hypoxia Tissue sampling limitations with tumor heterogeneity and dynamic changes
Gene signatures	Assesses the complex hypoxia signaling pathways Next-generation sequencing can be used	Gene lists do not overlap with other reported methodology No clear concordance with other endogenous markers Tissue sampling limitations with tumor heterogeneity and dynamic changes
Indirect: Exogenous		
Nitroimidazoles	Combined with imaging can facilitate simultaneous tumor hypoxia assessment, anatomical imaging, and management Can be used for immunohistochemical assessment Permit real-time assessment of hypoxia with imaging Noninvasive Permits quantification of hypoxia	Radioactive tracers can be slow to accumulate in tumor tissue with slow clearance kinetics Role as predictive marker for therapy not yet clearly established

CA-IX carbonic anhydrase-9, *GLUT* glucose transporter, *HIF* hypoxia-inducible factor, *IHC* immunohistochemistry, *VEGF* vascular endothelial growth factor

direct clinical management. For example, the merging of positron emission tomography (PET)-based hypoxia imaging with computer tomography (CT) anatomical information provides greater tumor volume resolution which can facilitate staging, XRT planning, and evaluation of tumor response to treatment.

15.5.1 *Direct Measurement of Tumor Hypoxia*

Previously considered the "gold standard" of tumor hypoxia measurement, direct measurement of tissue oxygenation was first conducted using polarographic needle electrodes (pO_2 histograph; Eppendorf, Hamburg, Germany) in cancers of the head and neck, cervix, and breast [35, 71, 72]. Several independent investigators subsequently showed that tumor oxygenation measured by these probes was associated with treatment outcome in HNSCC, as discussed previously in the chapter [16, 33, 73]. This method involves the placement of an oxygen sensor at the tip of a needle that is positioned and inserted by ultrasound or CT guidance and advanced through the tissue by a step motor. Rapid measurements can be made every 1.4 seconds to collect 50–80 measurements along five to six tracks which generate a histogram of oxygen partial pressure in the tumor or tissue of interest. Notably, this approach has been validated by comparison of electrode measurements against exogenous and endogenous hypoxia marker staining [74, 75].

Unfortunately, electrode techniques are not without limitations. Invasive by nature, probe insertion and probe migration may cause tissue damage. Measurements can only be made in superficial regions, rendering many deeply seated tumors difficult to access for direct evaluation. Significantly, electrode measurements cannot distinguish between hypoxia arising in viable tumor or necrotic areas. Variation in readings collected by different observers (interobserver variability) is unfortunately common, and hypoxia can be heterogeneous within a tumor making the decision of where to place the probe within the tumor challenging and the determination of the most clinically relevant measure of hypoxia variable (measurement of the most hypoxic region, versus the mean level of hypoxia, versus the hypoxic range) [35, 76, 77]. Finally, dynamic changes in oxygen tension throughout the tumor cannot be adequately resolved due to the inability to acquire spatially comprehensive readings over time.

15.5.2 *Indirect Measurement of Tumor Hypoxia*

Indirect measurement of hypoxia represents the most feasible approach for widespread adoption into clinical care, with the benefit that these methods are minimally or noninvasive. Indirect hypoxia assessment methods also have the ability to combine hypoxia assessment with other clinical management. Important methods include measurement of both endogenous and exogenous biomarkers.

15.5.2.1 Endogenous Biomarkers

Endogenous hypoxia-related biomarkers generally refer to proteins whose expression are increased either through upregulation or through decreased degradation in the presence of hypoxia. Based on routine pathological techniques such as immunohistochemistry, this approach is a convenient and cost-efficient method for assessing tumor oxygenation in both fresh and archival tissue samples. Virtually all archived tumor samples are amenable to endogenous marker analysis, unlike exogenous markers that must be injected prior to biopsy. Furthermore, if reliable, assessment of hypoxia using endogenous markers can feasibly be performed as part of the pathological workup of the tumor and thus guide the selection of patients for further management. However overall, regardless of which marker is used, there are issues that confound correlative analyses of which the most significant is the fact that the expression of most hypoxia-linked proteins is not specific to hypoxia alone.

Hypoxia-Inducible Factors (HIF) and Other HIF-Mediated Signaling Pathways

Of all endogenous markers, the HIF family (and particularly HIF-1) is most well studied given its crucial role in initiating and orchestrating downstream biological pathways in response to hypoxia (Fig. 15.1) [55, 78–80]. Overall, the pivotal role of HIF-1 in the context of tumor hypoxia is to facilitate the shift of metabolism to meet the available oxygen supply, while maintaining sufficient energy production through glycolysis and decrease of mitochondrial function [81]. HIF-1 was first reported in 1992 as a nuclear factor that could induce the transcription of erythropoietin in response to hypoxia [82]. HIF-1 is a basic-helix-loop-helix heterodimer that consists of two units: the 120-kDa HIF-1alpha subunit which is regulated by oxygen levels and a constitutively expressed 91- to 94-kDa HIF-1beta subunit [83]. HIF-1 function and activity are tightly regulated (Fig. 15.1). In normoxic conditions, HIF-1 degradation is initiated by oxygen-dependent hydroxylation that facilitates binding of HIF-1 to the von Hippel-Lindau protein (pVHL), which marks it for proteasomal degradation through recruitment of an E3 ubiquitin ligase complex. In hypoxic conditions, degradation is inhibited through substrate limitation, as hydroxylation is oxygen-dependent. Therefore, this forms the basis for HIF-1alpha activity being regulated by hypoxia. In low oxygen states, accumulation and dimerization of HIF-1alpha and beta subunits occur; these dimers then bind to the HREs in the promotor region of genes including CA-IX, vascular endothelial growth factor (VEGF), and the glucose transporters (GLUTs) [28, 75, 84–89].

However, HIF-1 activity is not regulated only by hypoxia but other oxygen-independent mechanisms as well, with a vast array of other triggers including nitric oxide [90], cytokines and growth factors such as TGF-beta [91], cholesterol [92], other signal transduction pathways such as the phosphatidylinositol 3-kinase/AKT/mammalian target of rapamycin (PI3K/AKT/mTOR) and mitogen-activated protein kinase (MAPK) pathways [55, 93, 94], and those induced by viruses of the HPV family [95, 96]. This immediately highlights potential confounding issues for the use of HIF-1 measurement as a reliable endogenous marker of hypoxia, given that other factors apart from hypoxia can stimulate its activity.

Unsurprisingly, the literature contains discordant correlative analyses regarding the impact of hypoxia measured by HIF assessment [97–100]. Many of the reported studies investigate different anatomical HNSCC subsites, vary in study number size and are often based on small numbers of patients, and use different criteria for reporting of the number of cells or degree of staining of immunohistochemical analyses of HIF. Given these issues, a meta-analysis of 28 studies using the collated data from 2293 HNSCC patients has been performed [97]. This meta-analysis reported a significant relationship between increased mortality risk and HIF expression (HR=2.12; 95% CI 1.52–2.94), which persists regardless of what HIF isoform is analyzed [97]. A significant correlation between poor prognosis and HIF expression has been confirmed in another meta-analysis focusing only on oral carcinomas [101]. In general, other systematic reviews comparing the utility of HIF analyses as a hypoxia biomarker compared to other endogenous markers identify HIF-1 as the most consistently correlated with poor patient outcome [42].

Other endogenous markers downstream of HIF have been investigated for their utility as biomarkers of hypoxia, including CA-IX [85–87, 102]. CA-IX is a transmembrane protein and a member of the zinc metalloenzyme family; it is responsible for the reversible hydration of carbon dioxide and water produced as a byproduct of glycolysis, to bicarbonate and hydrogen ($C0_2 + H_20 > HCO_3^- + H^+$). Therefore, CA-IX contributes to the regulation of the pH of the cellular microenvironment promoting intracellular alkalosis and extracellular acidosis, promoting tumor cell survival in hypoxic conditions. The identification of a HIF-1 HRE in the promoter region of the CA-IX gene has confirmed the direct regulation of CA-IX expression by HIF [87]. However similar to the HIF family, CA-IX expression is not specific to hypoxia and is regulated by other signaling pathways such as the PI3K pathway [103]. Furthermore, CA-IX expression in hypoxic conditions is not universal for all cancer cell lines and also may not be reflective of hypoxia given that overexpression of CA-IX can persist despite reoxygenation due to a long protein half-life [104, 105]. Again, correlative analyses between CA-IX expression and patient prognosis are discordant in the literature with studies varying in numbers, reporting methods, and cutoffs of significance and with some studies using tissue microarrays. A recent meta-analysis of the prognostic significance of CA-IX expression in all cancers examined 147 studies and confirmed a significant relationship between poor patient outcome and expression of the protein. The subgroup analysis for HNSCC studies confirmed a correlation between CA-IX expression and worse overall survival (HR 1.66, 95% CI 1.29–2.13) and worse locoregional control (HR1.54, 95% CI 1.12–2.12) [102]. An earlier meta-analysis of HNSCC-specific data from 842 specimens with dichotomized reporting of CA-IX also identified a correlation with significantly worse patient survival ($P < 0.0001$) [85].

Expression of members of the family of transmembrane glucose transporter proteins, which includes GLUT-1, is transcriptionally regulated by HIF, with a HIF1-responsive element (HRE) identified in the promoter region of the gene [106, 107]. However, not surprisingly, GLUT-1 expression is also not hypoxia-specific and can also be regulated by other factors including insulin and hormones such as estrogen [106, 108]. Given the adaption of tumors to primarily glycolytic metabolism in both aerobic and anaerobic conditions, glucose transporters play a crucial role in facilitating the energy-independent transfer of glucose across cell membranes to meet the increased metabolic requirements of cellular proliferation of malignancy [89]. In oral cancers and premalignant lesions, overexpression of GLUT-1 has been associated with the increased risk of malignant transformation [109]. A meta-analysis of data from 1301 oral cavity squamous cell carcinoma patients demonstrated a significant relationship between GLUT-1 expression and more advanced stage disease and shorter survival (HR=1.88; 95% CI 1.51–2.33, $P < 0.001$) [110].

The vascular endothelial growth factor (VEGF) family of secreted ligands, and the VEGF receptors (VEGFRs), a group of related transmembrane proteins represents another hypoxia-responsive signaling pathway triggered by HIF [88, 111, 112]. VEGF signaling is involved in angiogenesis, cell proliferation, and survival [113]. As with the other endogenous markers, literature exists confirming the relationship of VEGFR expression and poor patient outcome in HNSCC but with acknowledgment that this is not a consistent finding [114, 115] (and Chap. 16).

Osteopontin

Osteopontin (OPN), a member of the small integrin-binding ligand N-linked glycoprotein (SIBLING) family, is a putative biomarker of hypoxia that differs from the other proteins discussed above, as its expression is not primarily regulated by HIF but is rather mediated by a Ras-activated enhancer [116, 117]. SIBLINGs are secreted soluble glycoproteins that have autocrine and paracrine signal transduction function and can be detected in plasma or on a variety of cell surfaces including epithelial cells and bone and stromal tissues given their ability to bind to multiple protein partners (e.g., integrins, matrix metalloproteinase family, and complement factor H) [116]. Thus, SIBLINGs are known to contribute to invasion and metastases through facilitation of cell adhesion and migration and through modulation of the immune and inflammatory responses [118–120]. The relationship between OPN and poor prognosis in cancer patients has been confirmed in a meta-analysis of more than 200 studies [121]. However, the use of OPN as a hypoxia marker is controversial (Table 15.1). Key reports have confirmed an inverse correlation between plasma OPN levels and oxygen tension, but this is inconsistently replicated when OPN levels are assessed in tumor samples or correlated with other hypoxia markers or compared with direct measurements of oxygen tension [74, 75, 122, 123].

A seminal report by the Danish Head and Neck Cancer (DAHANCA) group, which investigated the benefit of the hypoxia radiosensitizer nimorazole with XRT in 320 HNSCC patients, found that high plasma OPN was able to select patients who benefited from the use of this hypoxia-modulating approach (DAHANCA-5) [124]. Patients with high OPN levels who received XRT alone had inferior locoregional control rates ($P = 0.01$) and disease-specific survival ($P = 0.0004$). Thus, it was with interest that the subsequent Trans-Tasman Radiation Oncology Group (TROG) 02.02 phase III international study of tirapazamine (TPZ; a hypoxic cell cytotoxin), in combination with cisplatin-based chemoXRT, also examined the utility of OPN as a hypoxia biomarker [125]. With larger patient numbers (n=578) and the use of samples from younger patients for the analysis, the prognostic significance of plasma OPN could not be confirmed and nor were levels found to be predictive of benefit for hypoxia-modulating therapy. Therefore, the utility of this marker is uncertain.

Gene Signatures

Given the complexity of signaling pathways in response to hypoxia, gene expression or proteomic analyses may represent a more realistic assessment of the network of activated pathways (Table 15.1). Through these approaches, HIF-regulated and non-HIF-regulated proteins such as connective tissue growth factor (CTGF), hypoxia-inducible gene-2 (HIG2), dihydrofolate reductase, and lysyl oxidase (LOX) have been examined histologically in HNSCC to identify markers that can predict cancer-specific survival, overall survival, and identify HNSCC patients best suited for hypoxia-targeted therapies [122]. The most promising hypoxia-related gene signature in HNSCC has been reported by the DAHANCA group to be both prognostic and predictive for benefit of hypoxia modulation with nimorazole [126, 127]. The authors first utilized in vivo xenografts to confirm the relevance of a number of reported hypoxia-related genes and then quantified gene expression in a set of 58 HNSCC samples of "more" hypoxic (~greater than 60% hypoxic fraction with pO_2 less than 2.5mm Hg) and "less" hypoxic (~less than 60% hypoxic fraction with pO_2 less than 2.5mm Hg) tumors defined by oxygen electrode measurements. A final 15 gene expression signature was identified and validated in 323 samples from patients treated on the DAHANCA-5 study. Patient tumors were classified by the gene signature as less or more hypoxic. For those with a more hypoxic signature who received nimorazole plus XRT, the 5-year locoregional failure rates were lower compared to those who received XRT alone (46% vs. 79%, P = 0.0001, respectively). The test for interaction was significant for those who received nimorazole (P = 0.003). Interestingly, the 15-gene signature (*ADM*, *ALDOA*, *ANKRD37*, *BNIP3*, *BNIP3L*, *C3orf28*, *EGLN3*, *KCTD11*, *LOX*, *NDRG1*, *P4HA1*, *P4HA2*, *PDK1*, *PFKFB3,* and *SLC2A1*) did not include any of the previously discussed, well-studied biomarkers of hypoxia [127]. A more recent report by the German Cancer Consortium Radiation Oncology Group (DKTK-ROG) investigating the utility of the same hypoxia gene signature in 158 HNSCC patients treated with cisplatin- or mitomycin-C-based chemoXRT could not confirm a significant relationship between more "hypoxic" tumors and locoregional control rates (P = 0.071) [128]. The utility of the 15-gene signature is being further investigated in an international phase III trial seeking to validate the benefit of chemotherapy plus accelerated fractionated XRT with or without nimorazole in HPV-negative tumors (EORTC 1219/TROG 14.03/NCT01880359).

Despite the convenience of the discussed endogenous markers, limitations exist for their use in the assessment of tumor hypoxia. As for any method that requires a tissue biopsy, limited sampling of a tumor will not reliably represent the known spatial and temporal heterogeneity of hypoxia within a tumor. Additionally, archival tissue assessment may not reliably reflect hypoxia of an in situ tumor with variations in vascular supply and changes that evolve during exposure to treatment. Technical issues arise with storage processes, fixation processes, staining methods with variability between antibodies and platforms used, and variability in the reporting measurement or quantification of protein expression. Furthermore, the staining patterns of certain endogenous markers can differ significantly from exogenous tracers, making it a challenge to correlate with other methods of hypoxia assessment. Additionally, as discussed above, the expression of hypoxia-induced endogenous markers is in general not hypoxia specific.

15.5.2.2 Exogenous Biomarkers

Exogenous biomarkers (Table 15.1) take advantage of the physicochemical properties of drugs and chemicals that are injected into a patient or tumor, which accumulate and only become detectable through bioreduction at levels of oxygen less than 10mm Hg [43, 99, 129]. Furthermore, an advantage of these compounds is that dead cells do not demonstrate a false signal given their inability to metabolize the bioreductive probes. Numerous injectable metabolic and bioreductive markers have been developed to measure tumor hypoxia. When radiolabeled with PET-compatible isotopes, these markers provide a means to image tumors in situ while capturing the temporal and spatial heterogeneity of hypoxia that direct probe measurement cannot achieve. The merging of PET hypoxia marker imaging with CT anatomical information provides greater tumor volume resolution. Therefore, the use of exogenous biomarkers coupled with imaging techniques represents a feasible approach to hypoxia assessment that can be easily combined with routine clinical management. However, although a variety of bioreductive complexes are available to image tumor hypoxia, further improvements are required to address the limitations of current tracers. There is need for tracers to exhibit faster, more specific localization as well as rapid clearance from well-oxygenated tissues which will enhance the signal-to-noise ratio. Efforts toward improving hypoxia tracers have yielded compounds varying in their lipophilicities and bio-distributions, several of which are still in various stages of animal and clinical testing [130–133].

The prototype of bioreductive probes, the antibiotic *2-nitroimidazole*, was identified in the 1950s as effective against bacteria which grew in hypoxic conditions and was eventually developed as a more efficient compound with increased electron affinity [134–136]. Nitroimidazoles, which were originally created using tritiated misonidazole, undergo an enzymatic reduction to a radical anion that is back-oxidized to its starting compound in well-oxygenated conditions. However, in the hypoxic environment, the radical anion is further reduced and remains bound to macromolecules, causing it to be irreversibly retained in hypoxic cells. Trapped nitroimidazole compounds are detected by specific antibodies for immunohistological analysis of biopsied tumor samples or radiolabeled for PET imaging in vivo.

Fluorine 18-fluoromisonidazole (FMISO, [1-(2-nitroimidazolyl)-2-hydroxy-3-fluoropropane] is one of the most widely used PET imaging tracers for hypoxia detection in cancer, including HNSCC [136, 137]. Regions identified as hypoxic by FMISO have been shown to correlate with other biomarkers of hypoxia, including correlation with areas staining for both exogenous pimonidazole and endogenous CA-IX hypoxia markers [138–140]. FMISO accumulation has been reported to significantly correlate with poor response to XRT, confirming its clinical utility to identify high-risk hypoxic disease [40, 141–143]. Of clinical interest, in HNSCC patients treated in a randomized trial of chemoXRT versus chemoXRT and tirapazamine (TPZ), FMISO-PET uptake was shown to identify patients at higher risk of locoregional failure and was able to identify patients with hypoxic tumors who benefited from the use of TPZ [144]. As quantification of hypoxia is also possible with FMISO imaging, a defined value of hypoxia may additionally represent an objective metric for assessment that could form a putative prognostic marker in HNSCC and other cancers [141, 142, 145]. Another application of FMISO-PET imaging in treatment of hypoxic HNSCC has been to guide dose escalations of up to 105 Gy to hypoxic areas using intensity-modulated XRT (IMRT), which has been shown to be achievable without exceeding normal tissue tolerance [146]. However, the lipophilicity of FMISO and slow specific accumulation of the tracer in target tumor tissue with slow clearance kinetics from normoxic tissues require delays of 4 h after administration for the optimal contrast between hypoxic tissue and background [147, 131]. This delay is clinically cumbersome and has hindered application of FMISO imaging. There is obvious need for agents with higher signal-to-noise ratio.

Fluoroazomycinarabinofuranoside (FAZA) is another nitroimidazole with a sugar addition that generates a better signal-to-noise ratio than FMISO but has a similar tracer distribution. It exhibits faster diffusion through cell membranes, faster accumulation in target hypoxic tumor cells, and improved clearance from normoxic tissues [148–150]. However, in animal models, FAZA-PET concentrations were shown to be lower than FMISO indicating inferior sensitivity of the agent for detection of hypoxia. Despite this, the role of FAZA imaging in the clinic is still being explored as there is suggestion that it demonstrates potential utility as a predictive marker for benefit of TPZ treatment with chemoXRT in HNSCC, and signal intensity may correlate with patient outcome [151–153].

Pimonidazole ([1-(alpha-methoxymethylethanol)-2-nitroimidazole]), which can be administered both intravenously and orally and is commonly used in animal models, has shown utility for detection of hypoxia within frozen and fixed tumor samples [154, 155]. It may be that pimonidazole and other exogenous markers detect chronic diffusion-limited hypoxia due to the observation of maximal uptake of the probe at distances greater than 100 μm from vasculature [156]. Kaanders and colleagues examined HNSCC samples from patients who were injected with pimonidazole prior to biopsy under general anesthetic, finding higher expression of this bioreductive probe correlated with significantly worse locoregional control and

DFS after treatment, when expression imaged with fluorescence microscopy was defined as a dichotomous variable [41]. Promisingly, the poor prognostic impact of the hypoxia detected by pimonidazole was abrogated in patients who were allocated treatment with accelerated XRT combined with carbogen and nicotinamide (ARCON) versus other therapy, insofar as ARCON has been reported to counteract diffusion- and perfusion-mediated hypoxia. Conversely, only weak correlation ($R = 0.36$, $P = 0.02$) was found between pimonidazole detection and the detection of the endogenous hypoxia marker CA-IX. In this study, CA-IX was not found to be a prognostic marker for patient outcome [41]. The utility of 18-F-radiolabeled pimonidazole as a radiotracer with PET imaging has been investigated but was proved inferior to other more commonly used agents [157]. There are limited further data on the usefulness of pimonidazole in HNSCC.

EF5 ([2-(2-nitro-1H-imidazol-yl)-N-(2,2,3,3,3-pentaflouropropyl) acetamine]), a fluorinated derivative of the 2-nitroimidazole etanidazole, can be 18-F labeled for PET imaging and has been shown to predict radioresistance in individual tumors in murine and rat models [158]. A simplified method of EF5 synthesis that meets the standards of purity and activity for clinic use has been described and is also amenable to automated synthesis, giving [(18)F] EF5 PET promise for clinic use [159]. Comparison can be made between EF5 PET imaging and EF5 fluorescence-based immunohistochemical assessment [160].

Finally, beyond the use of bioreductive probes as exogenous markers of hypoxia, investigation of other imaging modalities may further improve the imaging resolution of tumor hypoxia, such as intrinsic-susceptibility magnetic resonance imaging (MRI) which detects paramagnetic deoxyhemoglobin [161].

15.6 Hypoxia-Modulating Therapeutic Approaches in HNSCC

Great effort has been devoted to targeting hypoxia clinically in order to improve treatment response in solid tumors. Means to modify tumor oxygenation such as HBO [24, 31, 162, 163], ARCON [164–166], and blood transfusions [167–169] have been used to increase oxygen delivery to tumors to render them more sensitive to treatment. More recently, the additional benefit of hypoxic cell radiosensitizers that sensitize hypoxic cells to cytotoxins and chemoXRT has been tested in HNSCC. Other efforts have been devoted to identifying and targeting hypoxia-driven genes or pathways. Some of these efforts have focused on perturbing the HIF pathway, tumor metabolism, and immune responses.

15.6.1 *Tumor Oxygen Modifiers*

Hyperbaric Oxygen (HBO). The purpose of HBO is to boost oxygen delivery to tumors, thereby increasing their sensitivity to radiation. Patients are administered 100% oxygen under pressure greater than one atmosphere in order to enhance oxygen diffusion into the tumor and circumvent diffusion-limited hypoxia. Studies examining the benefit of HBO were among the earliest randomized trials conducted to test the benefits of hypoxia modulation and XRT in HNSCC [23, 163]. The earliest reported randomized study commenced in the 1960s administered 35 Gy in ten fractions over 3 weeks to patients with HNSCC, allocating patients to administration in air or HBO [22]. The study showed an improvement in the local control rates of 53% for patients administered HBO/XRT versus 30% who received XRT alone ($P < 0.001$). However, the improvement in locoregional control was primarily noted for smaller tumors and not for larger ones, and no overall survival benefit was observed [23]. Another similar study performed in the 1970s that treated HNSCC patients with 23–25.30 Gy of XRT with air or HBO to 4 atmosphere administered under general anesthetic also reported improved 5-year local control rates in the HBO arm compared to room air (29% vs. 16%) [163]. However, improvement in other outcome measures including overall survival was not observed. Furthermore, although acute toxicities were similar between groups, a trend to worse late toxicities was reported in patients who received HBO. Adding to the concerns, other groups raised queries as to the durability of benefit of HBO with XRT, observing that initial benefits were not sustained at 2 years [163, 170, 171]. Thus the role of HBO and XRT remained uncertain with other confounding issues between trials including differences in hyperbaric pressure, radiation fractionation, and total radiation dose administered.

Cochrane meta-analyses of studies investigating the benefit of HBO and XRT eventually confirmed that although a significant reduction in deaths at 5 years was found for HNSCC patients receiving HBO, this was at the increased risk of severe radiation-related tissue injury (RR 2.35, $P < 0.0001$, NNH = 8) and seizures (RR 6.76, $P = 0.03$, NNH = 22) [162, 172]. Other disadvantages to the use of HBO include poor patient tolerance of a pressurized environment due to claustrophobia and the cumbersome process of administering HBO and XRT [23, 173]. Largely due to the increased risk of adverse events secondary to HBO and the difficulties of administration, combined HBO and XRT has not been adopted in the clinic [174]. Ultimately, HBO studies have contributed more toward highlighting the significance of tumor hypoxia in radioresistance rather than improving tumor response to radiation.

Accelerated Radiotherapy with Carbogen and Nicotinamide (ARCON) Another approach tested in clinical trials to address tumor hypoxia has been the use of carbogen and nicotinamide with XRT (ARCON); this combination has been shown to enhance the OER. Carbogen hyperoxic gas breathing consisting of 98% oxygen and 2% carbon dioxide aims to combat diffusion-limited hypoxia, while the vasoactive agent, nicotinamide, counteracts perfusion-limited hypoxia. After promising early studies [165, 166], a phase III randomized trial was performed to determine the benefit of accelerated XRT with or without ARCON for patients with cT2-4 laryngeal squamous cell carcinomas [175]. Of the 345 patients treated with 64–68 Gy of XRT randomized on study, no significant difference between treatment arms was observed for the primary endpoint of locoregional control. DFS and OS also did not improve with ARCON treatment. Interestingly, a significant improvement was observed for regional control rates for patients with hypoxic tumors versus normoxic tumors defined by pimonidazole staining (100% vs. 55%, respectively; $P = 0.01$) [175]. Thus, in further attempt to identify patients who benefited from hypoxia modulation with ARCON, subsequent translational research identified that potentially a subgroup of patients with tumors with low epidermal growth factor receptor (EGFR) expression or patients with low pre-treatment hemoglobin levels may benefit from this hypoxia-modulating approach [176, 177].

Other approaches. Given the adverse prognostic factor of anemia in HNSCC patients [178], other approaches to enhance tumor oxygenation such as the use of blood transfusions have also been investigated [167, 168]. However, pooled analysis of more than 1100 patient data to assess the benefit of receipt of packed red cells prior to treatment failed to demonstrate that this was an effective method of altering patient outcome [168]. Furthermore, it may be that that receipt of blood transfusions could have an adverse effect on patient outcome [169].

Another approach investigated to combat the poor prognostic effects of anemia in HNSCC has been the use of erythropoietin-stimulating agents [179, 180]. However concern exists regarding this approach given the knowledge that many cancer cells, including head and neck cancer, have been reported to highly express the erythropoietin receptor, whether assessed by immunohistochemistry or mRNA expression [181–184]. Thus, administration of stimulating agents in this context could obviously be oncogenic (Fig. 15.1). Cochrane analyses have been performed addressing the impact of erythropoietin-stimulating agents in cancer patients [180, 185, 186] and also its effects as an adjunct therapy for head and neck cancer patients receiving adjuvant XRT or chemoXRT [179]. The most recent review of 91 trials including data from more than 20,100 cancer patients significantly demonstrated evidence of harm for those receiving stimulating agents, with increased estimated mortality rates (HR 1.17, 95% CI 1.06–1.29) and worse OS rates (HR 1.05, 95% CI 1.00–1.11) observed [180]. An increased risk of thromboembolic events and hypertension was also noted for those who received erythropoietin-stimulating agents. This Cochrane review included data of six randomized head and neck cancer studies, with over 1,449 participants' data. A separate meta-analysis of five randomized controlled trials investigated the impact of adjuvant erythropoietin in nearly 1,400 head and neck cancer patients receiving XRT with or without chemotherapy [179]. A significantly worse OS was observed for those who received erythropoietin in addition to standard adjuvant therapy (HR 0.73, 95% CI 0.58–0.91, $P = 0.005$). Therefore, the use of erythropoietin-stimulating agents is not routinely recommended in the management of head and neck cancer patients due to the evidence of harm.

Use of hyperthermia for hypoxia modulation in HNSCC has also being investigated with limited evidence for benefit beyond the potential of improved complete response rates [187, 188].

15.6.2 *Hypoxic Cell Radiosensitizers and Cytotoxins*

Since the mid-1970s, approaches to circumvent hypoxia-induced treatment resistance in HNSCC have focused on utilizing hypoxic cell radiosensitizers in combination with radiation. These sensitizers were developed as electron-affinic compounds that selectively increase radiation-induced cell kill of hypoxic cells by mimicking the effects of oxygen. Research conducted with these hypoxia-specific radiosensitizers lead to the development of hypoxic cell cytotoxins, which are directly lethal to tumor cells with low oxygen tension. Under hypoxic conditions, hypoxic cell cytotoxins are metabolized by intracellular reductases to form reactive radical species that induce cell death through formation of single-stranded DNA breaks, double-stranded DNA breaks, and chromosomal aberrations (Fig. 15.2) [189].

15.6.2.1 Nitroimidazole Studies

Nitroimidazoles have been the main agent used in the clinic to target tumor hypoxia. Multiple clinical trials conducted by the Radiation Therapy Oncology Group (RTOG), the European Organization for Research and Treatment of Cancer (EORTC), and the DAHANCA group have investigated the benefit of combining a variety of 2-nitroimidazole derivatives with radiation.

Between 1979 and 1985, the EORTC investigated a split-course accelerated fractionation regimen with or without *misonidazole* 1g/m^2/day compared to standard fractionation alone in patients with locally advanced HNSCC. Unfortunately, misonidazole did not improve 5-year locoregional control rates or OS [190]. The RTOG conducted a randomized phase III trial to compare the combination of etanidazole 2.0g/m^2 three times a week for 17 weeks plus conventional XRT (66 Gy in 33 fractions to 74 Gy in 37 fractions) versus radiation therapy alone. The addition of etanidazole to radiation therapy failed to demonstrate any benefit [191]. Similarly negative results were reported by a European *etanidazole* trial [192].

In contrast, *nimorazole* (1-(N-B-theylmorpholine)-5-nitro-imidazole) is of interest due to its reported benefit with XRT, although it is a less potent radiosensitizer and less toxic compound compared to etanidazole. Use of nimorazole was reported in a seminal randomized double-blind phase III study (DAHANCA-5) in over 400 patients with supraglottic and pharyngeal squamous cell carcinomas to improve both 5-year locoregional control rates (49% for the nimorazole group versus 33% for placebo, $P = 0.002$) and cancer-related deaths (52% for the nimorazole versus 41% for placebo, $P = 0.002$) [25]. Patients were randomized to receive conventional

XRT alone to a total dose of 62–68 Gy (2 Gy/fraction, five fractions per week) with or without nimorazole. However, receipt of nimorazole did not significantly improve OS. To date, nimorazole is the only hypoxic cell radiosensitizer that is being used in the clinic, albeit its use is limited outside of Denmark. The role of nimorazole remains controversial due to the lack of a proven survival benefit and the lack of comparative trials using the current international standard of care, being platinum-based chemoradiotherapy. Results from a phase III international trial investigating the benefit of nimorazole plus chemoXRT in HPV-unrelated tumors are eagerly awaited (EORTC 1219/TROG 14.03/ NCT01880359).

15.6.2.2 Tirapazamine Studies

Tirapazamine (TPZ) is an aromatic heterocycle di-N-oxide (3-amino-1,2,4-benzotriazine-1,4 dioxide) that was developed as a hypoxic cell cytotoxin. While inert in normoxic conditions, in hypoxic environments, TPZ is reduced to an active cytotoxic superoxide state (Fig. 15.2). Preclinical studies showed that TPZ was highly active in mammalian cells in vitro when combined with fractionated radiation at doses comparable to those used in clinical practice [193]. In other preclinical studies, Dorie and Brown reported that the combination of TPZ and various chemotherapy agents had an additive antitumor effect in an implanted fibrosarcoma mouse model [194]. 18F-FAZA-PET imaging also predicted benefit of TPZ plus cisplatin-based chemotherapy in mouse tumor models [195].

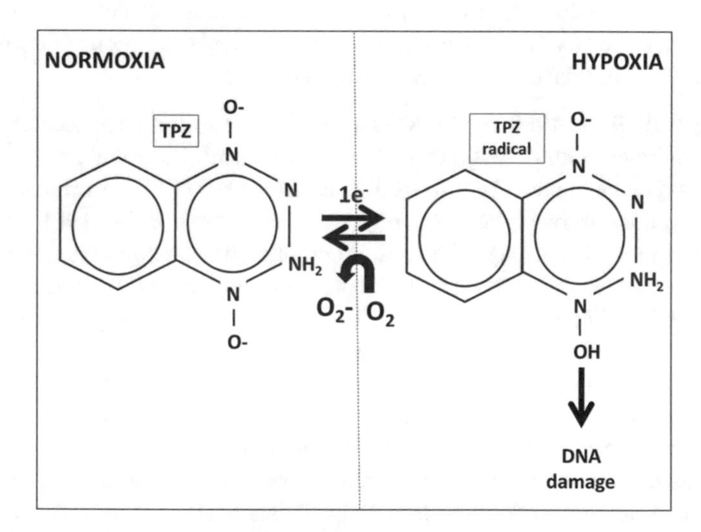

Fig. 15.2 Mechanism of action of tirapazamine (TPZ), a hypoxic cell cytotoxin, in normoxic and hypoxic conditions. In normal levels of oxygenation, TPZ exists in its inert form. However, in hypoxic conditions, TPZ undergoes one-electron reduction and forms a superoxide radical that can elicit DNA damage thus exerting its cytotoxic effect. The radical form of TPZ can be back-oxidized to its inert state in the presence of oxygen

A phase I trial in HNSCC patients was performed that confirmed the feasibility of administering TPZ with cisplatin-based chemoXRT and suggested predictive benefit of 18F-misonazole-PET imaging to select patients for hypoxia modulating therapy [196]. The subsequent randomized phase II trial (TROG 98.02) performed was designed to investigate the benefit of two chemoXRT regimens in 122 patients with stage III/IV HNSCC – one with TPZ aimed at targeting tumor hypoxia and the other cisplatin-/fluorouracil-based regimen targeting repopulation kinetics. The study found a nonsignificant difference in 3-year failure-free survival rates for cisplatin plus TPZ versus cisplatin and 5-fluorouracil (55% versus 44%, log rank $P = 0.16$) [197].

Based on these results, the TROG 02.02 (HeadSTART) randomized phase III clinical trial was designed to investigate the benefit of adding TPZ to concurrent radiation and cisplatin 100mg/m^2 in locally advanced HNSCC. The international study involved 88 centers from 13 countries and 861 patients with stage III–IV tumors arising from the oral cavity, oropharynx, hypopharynx, and larynx. Patients were randomized to receive 70 Gy of radiation over 7 weeks with either cisplatin (100mg/m^2/d on day 1 of weeks 1, 4, and 7) or cisplatin (75mg/m^2/d on day 1 of weeks 1, 4, and 7) plus TPZ (290mg/m^2/d on day 1 of weeks 1, 4, and 7 and 160mg/m^2 on day 1, 3, and 5 of weeks 2 and 3). Disappointingly, the addition of TPZ did not improve 2-year OS rates, failure-free survival, time to locoregional failure, or quality of life scores [198]. However, a seminal finding of the study was that 12% of patients had radiation deviations that were predicted to have an adverse effect on tumor control and indeed had inferior survival [199]. It is possible that poor quality radiation administration and the unrecognized emergence and inclusion of HPV-related disease may have impacted the overall results. In addition, another factor that may limit the efficacy of tirapazamine is the metabolic consumption of the prodrug during diffusion, resulting in suboptimal levels reaching the most hypoxic areas of the tumor [200]. In exploratory analyses of patients with acceptable radiation plans and in patients with p16-negative (i.e., likely HPV-negative) oropharyngeal cancer, an increased but not statistically significant locoregional control rate was observed in the TPZ arm [13, 199]. This is discussed in detail further in the chapter.

A number of new hypoxia-activated cytotoxins are under early preclinical investigation, and research is being performed to identify novel ways of drug delivery [201–204]. However, regardless of the promise of new therapies, the key issue hindering the success of any targeted hypoxia-modulating approach is in the lack of a robust method to select patients who will benefit.

15.6.3 HIF-Targeted Therapies

As HIF-1 is a key transcriptional regulator of hypoxia-responsive genes, targeting HIF-1 and its upstream and downstream pathways has become an attractive approach for circumventing hypoxia-mediated tumor aggressiveness and treatment resistance. However, despite numerous therapeutic approaches being available to potentially counteract HIF and its pathways, no agent is currently utilized in clinical practice specifically for its effect on hypoxia modulation. Given the number of oxygen and oxygen-independent mechanisms known to regulate HIF, there are many therapies available that can potentially counteract HIF activity. However, it is important to note that antitumor effects may not be directly due to hypoxia modulation or directly due to HIF modulation but are obtained through effect on other signaling pathways or mechanisms that may more significantly underpin tumor behavior.

Perhaps the most promising means to counteract HIF activity is through the use of agents currently in clinical use that target interacting signal transduction pathways. For example, as HIF activity can be precipitated via the PI3K/AKT/mTOR pathway, use of mTOR inhibitors such as temsirolimus, or PI3K inhibitors such as LY294002, may block HIF signaling [205, 206]. However, the benefit of these agents has not been proven in HNSCC [207, 208], and these drugs induce some adverse effects [209]. HIF-1 induction can also be attenuated upstream through the perturbation of EGFR signaling pathways with use of inhibitory monoclonal antibodies such as cetuximab [210–213]. However, the benefit of cetuximab is not believed to be due to HIF or hypoxia modulation, nor is resistance to cetuximab likely to arise through hypoxia-mediated mechanisms (see Chap. 15) [214, 215]. Perhaps one of the most disappointing results for therapies that counteract HIF activation pathways has been with the use of VEGF inhibitors, which target angiogenesis, cellular proliferation, and cell survival [216]. As single agents, VEGF inhibitors such as bevacizumab have limited use in HNSCC. Similarly phase I–III studies combining these agents with other therapies in HNSCC have no proven overall survival benefit but rather some evidence of harm [217–219].

Rather than interfering with upstream or downstream signaling, novel antisense oligonucleotide agents have been tested that directly target HIF-1α mRNA, albeit so far with little success [220]. Enhancement of HIF degradation is another means to limit the activity of the protein. For example, HIF-1alpha degradation can be induced through inhibition of the heat shock protein, Hsp90, with use of agents such as 17-N-allylamino-17-demethoxy geldanamycin (17-AAG) or through use of histone deacetylase inhibitors [210, 221–223]. It is unclear whether these early studies that predominantly utilize combination therapy are of benefit and, if beneficial, whether this is due to hypoxia modification. A derivative of melphalan (PX-478) has been reported to inhibit HIF-1a on multiple fronts, blocking its transcription and translation and promoting its degradation [224]. No results have as yet been reported from a completed phase 1 trial (NCT 00522652). HIF-1a transcriptional activation can be abrogated by targeting its co-activator p300 directly with chetomin or by disrupting the interaction of these proteins with the proteasome inhibitor, bortezomib [211, 225]. However, the activity of proteasome inhibitors in HNSCC is unclear.

Thus, although knowledge of HIF activation can provide therapeutic avenues for modulation, it may be that targeting HIF and its complex network of signaling pathways does not represent a means by which hypoxia can be effectively targeted.

15.7 The Future of Hypoxia Targeting in HNSCC: The Importance of Patient Selection

Clinical trials that have evaluated hypoxia and hypoxic cell cytotoxins or radiosensitizers in HNSCC have demonstrated clearly that not all tumors exhibit hypoxia detectable by current methods. Therefore, the future of hypoxia targeting in HNSCC is dependent on the evaluation of tumor oxygenation to select suitable patients suitable for hypoxia modulating therapy.

15.7.1 Predictive Tools for Patient Selection for Hypoxia-Modulating Therapy

A number of clinical studies have reported tools that were able to appropriately select patients with tumor hypoxia that benefited from hypoxia-modulating approaches. As discussed previously, pimonidazole staining from tumor biopsies to assess tumor hypoxia in larynx cancers was able to select patients with tumors that were responsive to ARCON treatment [175]. The authors reported that patients with hypoxic tumors defined by pimonidazole staining (>2.6% positive staining in the tumor) responded significantly better to ARCON compared to accelerated XRT alone. However, the utility of this approach may be limited due to the requirement for a biopsy and concerns about sampling variability within tumors.

Of all the reported approaches to pretreatment hypoxia evaluation, perhaps the most feasible is the use of exogenous marker plus imaging combinations. In the development of hypoxia cell cytotoxins, Rischin and colleagues were among the first to report the predictive capability, clinical feasibility, and prognostic ability of FMISO hypoxia imaging in HNSCC [144, 226]. A sub-study of the phase II TROG 98.02 trial examined the utility of FMISO hypoxia imaging before and during treatment with TPZ and chemoXRT. Patients with high FMISO uptake within the primary tumor were found to have a worse locoregional failure rates when treated with chemoXRT alone compared to those treated with chemoXRT and TPZ [144]. These results suggested a possible role for FMISO imaging in identifying patients who would benefit from TPZ. However, the number of patients involved in this sub-study was small (n=45). The same TROG group presented further data from 63 patients treated on the phase I trial, TROG 98.02 phase II trial, and extension cohort of the phase II trial, assessing the utility of FMISO-PET imaging for HNSCC patient selection for TPZ use [227]. Hypoxia was demonstrated in 49/63 (78%) of tumors. Importantly, patients with baseline hypoxia detected with FMISO-PET imaging who received TPZ-based therapy had superior locoregional control, failure-free survival, and control in the primary tumor site compared to those who did not ($P \leq 0.001$).

FMISO-PET imaging is also useful in detecting dynamic changes of tumoral hypoxia, which could potentially direct adaptive therapy. The optimal time point to assess tumor hypoxia is not clear, given the dynamic changes of hypoxia that occur with treatment. Studies conducted investigating the temporal changes in tumor hypoxia during XRT to define the optimal timing of assessment have shown that earlier time points during radiation (weeks 1 and 2 or receipt of 10–20 Gy) serve as stronger indicators of local progression-free survival rather than later time points during treatment [39, 145]. In addition, FMISO-PET imaging to assess spatial changes in hypoxia within HNSCC tumors evolving during treatment may help identify a clonal population at risk of treatment failure [228].

FAZA-PET imaging has also been used in the clinic to select patients who benefit from TPZ therapy, given the improved tumor to muscle ratio of the probe, with reports also suggesting that it may be prognostic in HNSCC [151, 153, 195]. In a sub-study of the TROG 02.02 trial, 41 patients received pretreatment FAZA-PET scans. Interestingly, compared to the TROG FMISO-PET imaging studies, the reported hypoxia was less common with FAZA-PET, with 21/38 (55%) patients reported with detected hypoxia. Nevertheless, FAZA-PET imaging identified patients with hypoxic tumors who benefited from the addition of TPZ relative to cisplatin alone. Further research is clearly required to identify and prospectively validate methods to select patients for hypoxia-modulating therapy.

15.7.2 Tumor HPV Status in Patient Selection

Given the recognition of a clinicopathologically distinct subset of oropharyngeal carcinomas associated with HPV, predominantly the high-risk HPV-16 subtype (see Chaps. 20 and 21), the question has been raised as to whether the predictive and prognostic role of hypoxia differs in HPV-related and HPV-unrelated tumors. In particular, given the known differences in pathogenesis coupled with the improved survival of the subgroup of HPV-related oropharyngeal tumors, it was unclear whether the presence of hypoxia further stratified patient outcome or whether hypoxia-modulating therapy affected these tumors differently [4, 9, 10, 12, 13, 229–233]. The incidence of HPV-related oropharyngeal carcinomas has been increasing and is now the most common cancer overall induced by the HPV family in the USA [234]. Epidemiologically, patients with HPV-related oropharyngeal cancers are typically younger, have less cigarette or alcohol exposure, and have better performance status with less comorbidities compared to patients with HPV-unrelated tumors [10]. The mechanisms by which HPV-related tumors are more responsive to treatment are not well understood but have been hypothesized to be related to intact p53 function, differential genomic aberrations, less overall genomic instability, and better immune surveillance to viral-specific antigens compared to HPV-negative tumors [235–240].

Preliminary questions were raised about differences between the benefit of hypoxia-modulating therapy according to HPV status due to the retrospective analyses of the seminal phase III DAHANCA-5 nimorazole trial and the TROG 02.02 TPZ trial [13, 229]. Both studies confirmed the improved overall survival of HPV-related oropharyngeal cancers but also suggested that HPV-unrelated (p16-negative) tumors may possibly derive greater benefit from hypoxia-directed intervention than HPV-related tumors. Lassen and colleagues evaluated 331/414 samples for HPV using the surrogate marker of p16 overexpression [229]. Samples with strong diffuse nuclear and cytoplasmic staining in more than 10% of cells were classified as HPV-related. The authors reported a significant improvement for tumors of any subsite that were HPV-unrelated who had received nimorazole ($P = 0.02$). A supplemental analysis restricted to the oropharynx was performed, of which 90 oropharyngeal cancers stained negative for p16, while 53 oropharyngeal cancers were positive. Five-year locoregional control rates for the 90 HPV-unrelated cases suggested a trend for benefit for the group that received nimorazole (37%) versus those who received placebo (20%; HR = 0.68, 95% CI 0.40–1.15, *P* value not reported). For the p16 positive cases, outcome was similar between treatment arms [229]. Rischin and colleagues evaluated samples from 185 oropharyngeal patients on the TROG 02.02 study and defined p16 positivity by the presence of moderate to strong (scored as 2 or 3) cytoplasmic and nuclear staining out of a scale of 0–3 (no staining to strong staining) [13]. The authors also reported a nonsignificant trend for benefit for patients who had HPV-negative oropharyngeal patients that received TPZ plus chemoradiotherapy versus those who did not receive TPZ, with a 2-year time to locoregional failure rate of 92% versus 81%, respectively (HR, 0.33; 95% CI 0.09–1.24; $P = 0.13$).

When FMISO imaging was used to assess the presence of hypoxia in patients treated on the three early-phase TROG studies examining the benefit of TPZ, and HPV status was assessed using p16 immunohistochemistry, Trinkaus and colleagues found that there were no significant differences in hypoxia between HPV-related and HPV-unrelated tumors [227]. Both subgroups of oropharyngeal tumors demonstrated a high prevalence of hypoxia with 14/19 (74%) of HPV-related and 35/44 (80%) of HPV-unrelated tumors having FMISO-detectable hypoxia. Furthermore, the pattern of distribution of hypoxia between the primary site and nodes was similar between the HPV-related and HPV-unrelated groups. Indeed, other groups have also found no difference between HPV-related and HPV-unrelated disease when hypoxia has been assessed by the presence of CA-IX [231], HIF-1α expression [241], and the 15-gene hypoxia signature [126]. These studies suggest that while hypoxia may be detectable in HPV-related oropharyngeal cancer, its biologic significance may be less than in HPV-negative HNSCC.

Lee and colleagues used FMISO-PET imaging to examine the feasibility of de-escalating treatment in HPV-related hypoxic tumors [242], finding that for 48% (16/33) HPV-positive patients, pre-treatment hypoxia detected by FMISO-PET resolved within 1 week of chemoradiotherapy. Ten eligible patients selected for nodal dose de-escalation of 10 Gy had excellent 2-year locoregional control rates and OS rates of 100%, with median follow-up of 32 months (range, 21–61 months). However, given the expected good outcome for HPV-related disease and the uncertainty about the clinical significance of hypoxia in HPV-related oropharyngeal disease [13, 153, 227, 229], further work is required to determine whether detection of resolution of hypoxia is the most appropriate test to select good responders suitable for de-escalated treatment.

15.7.3 The Role of the Immune System and Patient Selection

It has been with much excitement that new immunotherapy agents targeting immune checkpoints have been investigated in HNSCC. Recent studies demonstrated improved and durable OS in a subset of patients with metastatic cancer [243–245]. These effective immunotherapy agents have highlighted the importance of the interplay between the tumor and host environment on patient outcome [8, 246]. Therapies such as those that target programmed cell death-1 (PD-1) receptor and its ligand (PD-L1) are able to enhance the presence of an effective antitumoral response in HNSCC and are able to improve OS with less toxicity compared to traditional chemotherapy agents [8]. Furthermore, the immune system represents the only therapeutic strategy that has capacity to dynamically evolve its response as tumors undergo clonal evolution. Therefore, the effect of hypoxia on the immune response will undoubtedly be another crucial factor to take into account when selecting therapeutic strategies for HNSCC patients [247, 248].

There are two main mechanisms by which hypoxia is described to promote tumor immune escape: (1) hypoxia can directly hinder immune cell function; and (2) hypoxia can indirectly hinder immune cell function due to the generation of an acidic microenvironment arising through the metabolic reprogramming of the tumor to favor glycolytic metabolism and the downstream effects of the HIF activation (Fig. 15.3) [249, 250].

The adaptive immune system recognizes specific tumor antigens, and hypoxia can directly alter adaptive immunity. Cytotoxic T-lymphocyte development and function at levels of oxygenation of 2.5% have been observed to be impaired when compared activity at levels of oxygenation of 20%, although lytic activity of cytotoxic T cells is surprisingly greater in hypoxia [251]. Furthermore, the generation of effector cytokines and proliferative cytokines are reduced in hypoxic conditions. The acidity of the tumor microenvironment has the greatest impact on T cell function, where a progressively acidic environment suppresses activation, proliferation, and cytotoxicity and can even cause apoptosis of T-lymphocytes [250, 252]. As activated T cells are highly dependent on glycolysis for energy, the metabolic competition with highly adaptive tumor cells for the same glucose supply represents another mechanism by which hypoxia can blunt an active immune response [249]. One fascinating means by which tumor cells may achieve adaptive superiority and achieve immune escape in hypoxic conditions is through direct cannibalism of live lymphocytes by tumor cells, as a means of obtaining nutrients in a state of low nutrient supply [253]. The immunosurveillance performed by natural killer (NK) cells, which can be directly lethal to tumor cells, is also impaired in hypoxic conditions [254].

Antigen-presenting cells form an important link between the adaptive and innate immune responses. Dendritic cell maturation and activity are impaired in reduced states of oxygenation, with inhibition of expression of important co-stimulatory molecules CD80 (B7-1) and CD86 (B7-2) observed [255]. The CD80/86 proteins, expressed on dendritic cells and other activated immune cells, generate an important co-stimulatory signal required for T-cell activation. It also represents the ligand for

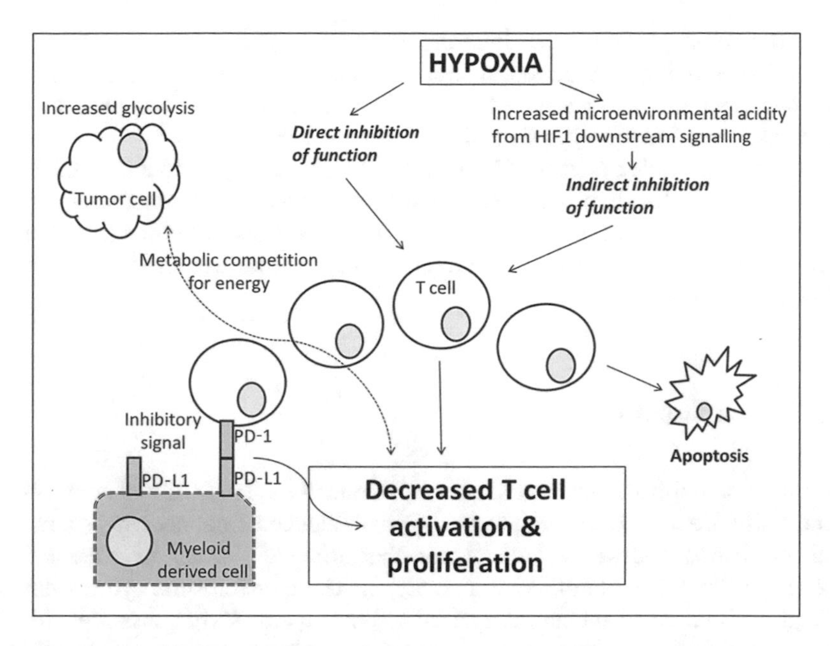

Fig. 15.3 The impact of the hypoxic on the immune system. A hypoxic microenvironment can directly hinder immune cell function or indirectly hinder immune cell function through the generation of an acidic microenvironment that arises due to the HIF1-mediated hypoxia response pathway. Tumor cells and T cells compete for glucose in the hypoxic environment which leads to substrate limitation. In hypoxic conditions, HIF1 signaling also increases expression of the inhibitory programmed cell death ligand-1 (PD-L1) on myeloid-derived stem cells, which binds to the programmed cell death-1 (PD-1) protein expressed on T cells. The interaction between PD-1 and PD-L1 results in an exhausted T-cell phenotype. Ultimately hypoxia can result in decreased T cell activation, decreased proliferation, apoptosis, and immunosuppression

cytotoxic T-lymphocyte-associated protein 4 (CTLA-4) [256]. Hypoxia also drives the differential expression of the innate immune response to favor a T-helper type 2 response, with reduced T-cell stimulation and increased dendritic cell OPN secretion which facilitates tumor cell migration [257]. Myeloid-derived cells, in particular the tumor-associated macrophages (TAMs), have a fascinating bi-faceted role directly promoting tumor progression through stimulation of angiogenesis, invasion, metastases, and tumor cell survival, while also acting as a key negative regulator of both the adaptive and innate immune system [248, 250, 258]. These immune cells are considered in general to have a pro-tumor effect. Myeloid-derived cells can also inhibit T-lymphocyte tumor infiltration and NK cell function. It appears that these effects are mediated by the myeloid-derived cells and TAMs through the HIF-1α pathway, supported by observations in animal tumor models with targeted deletion of the gene [259]. HIF-1α also regulates the maturation of TAMs, which serves to further propagate immunosuppression [260]. Furthermore, in hypoxic conditions, HIF-1α directly binds to a HRE in the PD-L1 promotor of myeloid-derived cells, upregulating the expression of PD-L1, which leads to immunosuppression [261].

Therefore, it is possible that immunotherapy combined with HIF-1 targeting approaches may improve patient outcome perhaps through simultaneous immune activation and the abrogation of microenvironmental conditions that facilitate immunosuppression via hypoxia and its HIF-1-mediated effects.

Due to the complex effects of hypoxia on the immune system and on the tumor, it is highly likely that hypoxia interferes with the effects of immunotherapy. It may be that combination approaches that concurrently address the immunosuppressive effects of hypoxia and pro-tumor effects of hypoxia are required before breakthroughs in this area are observed.

15.8 Conclusion

Perfusional and diffusional hypoxia occurs frequently in HNSCC and confers treatment resistance and poor prognosis. Hypoxia-induced treatment resistance is a well-established biological phenomenon that arises via direct interference with mechanisms that mediate radiation toxicity and chemotherapeutic cytotoxicity and through facilitation of metabolic resistance phenotypes. Hypoxia modulation has been a compelling approach to abrogate the adverse impact of hypoxia in HNSCC. Despite the knowledge that some tumors are truly amenable to hypoxia modification therapies, clinical trials of hypoxia targeting in HNSCC have been disappointing. Future investigation of hypoxia modulation will require a reliable, sensitive, and non-invasive predictive test to ensure optimal patient selection.

References

1. Shield KD, Ferlay J, Jemal A, Sankaranarayanan R, Chaturvedi AK, Bray F, et al. The global incidence of lip, oral cavity, and pharyngeal cancers by subsite in 2012. CA Cancer J Clin. 2017;67(1):51–64.
2. Nordsmark M, Overgaard M, Overgaard J. Pretreatment oxygenation predicts radiation response in advanced squamous cell carcinoma of the head and neck. Radiother Oncol. 1996;41(1):31–9.
3. Parkin DM, Bray F, Ferlay J, Pisani P. Global cancer statistics, 2002. CA Cancer J Clin. 2005;55(2):74–108.
4. Ang KK, Harris J, Wheeler R, Weber R, Rosenthal DI, Nguyen-Tan PF, et al. Human papillomavirus and survival of patients with oropharyngeal cancer. N Engl J Med. 2010;363(1):24–35.
5. Rischin D, Peters LJ, O'Sullivan B, Giralt J, Fisher R, Yuen K, et al. Tirapazamine, cisplatin, and radiation versus cisplatin and radiation for advanced squamous cell carcinoma of the head and neck (TROG 02.02, HeadSTART): a phase III trial of the Trans-Tasman Radiation Oncology Group. J Clin Oncol: official journal of the American Society of Clinical Oncology. 2010;28(18):2989–95.
6. Vermorken JB, Mesia R, Rivera F, Remenar E, Kawecki A, Rottey S, et al. Platinum-based chemotherapy plus cetuximab in head and neck ancer. N Engl J Med. 2008;359(11):1116–27.

7. Vermorken JB, Stohlmacher-Williams J, Davidenko I, Licitra L, Winquist E, Villanueva C, et al. Cisplatin and fluorouracil with or without panitumumab in patients with recurrent or metastatic squamous-cell carcinoma of the head and neck (SPECTRUM): an open-label phase 3 randomised trial. Lancet Oncol. 2013;14(8):697–710.
8. Ferris RL, Blumenschein G Jr, Fayette J, Guigay J, Colevas AD, Licitra L, et al. Nivolumab for recurrent squamous-cell carcinoma of the head and neck. N Engl J Med. 2016;375(19):1856–67.
9. Gillison ML, Koch WM, Capone RB, Spafford M, Westra WH, Wu L, et al. Evidence for a causal association between human papillomavirus and a subset of head and neck cancers. J Natl Cancer Inst. 2000;92(9):709–20.
10. Gillison ML, D'Souza G, Westra W, Sugar E, Xiao W, Begum S, et al. Distinct risk factor profiles for human papillomavirus type 16-positive and human papillomavirus type 16-negative head and neck cancers. J Natl Cancer Inst. 2008;100(6):407–20.
11. Gillison ML. HPV and prognosis for patients with oropharynx cancer. Eur J Cancer. 2009;45(Suppl 1):383–5.
12. Lassen P, Eriksen JG, Hamilton-Dutoit S, Tramm T, Alsner J, Overgaard J. Effect of HPV-associated p16INK4A expression on response to radiotherapy and survival in squamous cell carcinoma of the head and neck. J Clin Oncol: official journal of the American Society of Clinical Oncology. 2009;27(12):1992–8.
13. Rischin D, Young RJ, Fisher R, Fox SB, Le QT, Peters LJ, et al. Prognostic significance of p16INK4A and human papillomavirus in patients with oropharyngeal cancer treated on TROG 02.02 phase III trial. J Clin Oncol: official journal of the American Society of Clinical Oncology. 2010;28(27):4142–8.
14. Licitra L, Perrone F, Bossi P, Suardi S, Mariani L, Artusi R, et al. High-risk human papillomavirus affects prognosis in patients with surgically treated oropharyngeal squamous cell carcinoma. J Clin Oncol: official journal of the American Society of Clinical Oncology. 2006;24(36):5630–6.
15. Vaupel P, Kelleher DK, Hockel M. Oxygen status of malignant tumors: pathogenesis of hypoxia and significance for tumor therapy. Semin Oncol. 2001;28(2 Suppl 8):29–35.
16. Nordsmark M, Bentzen SM, Rudat V, Brizel D, Lartigau E, Stadler P, et al. Prognostic value of tumor oxygenation in 397 head and neck tumors after primary radiation therapy. An international multi-center study. Radiother Oncol. 2005;77(1):18–24.
17. Overgaard J, Horsman MR. Modification of hypoxia-induced radioresistance in tumors by the use of oxygen and sensitizers. Semin Radiat Oncol. 1996;6(1):10–21.
18. Gray LH, Conger AD, Ebert M, Hornsey S, Scott OC. Concentration of oxygen dissolved in tissues at the time of irradiation as a factor in radiotherapy. Br J Radiol. 1953;26:638–48.
19. Overgaard J. Hypoxic modification of radiotherapy in squamous cell carcinoma of the head and neck--a systematic review and meta-analysis. Radiother Oncol. 2011;100(1):22–32.
20. Shannon AM, Bouchier-Hayes DJ, Condron CM, Toomey D. Tumour hypoxia, chemotherapeutic resistance and hypoxia-related therapies. Cancer Treat Rev. 2003;29(4):297–307.
21. Becker A, Hansgen G, Bloching M, Weigel C, Lautenschlager C, Dunst J. Oxygenation of squamous cell carcinoma of the head and neck: comparison of primary tumors, neck node metastases, and normal tissue. Int J Radiat Oncol Biol Phys. 1998;42(1):35–41.
22. Monti E, Gariboldi MB. HIF-1 as a target for cancer chemotherapy, chemosensitization and chemoprevention. Curr Mol Pharmacol. 2011;4(1):62–77.
23. Henk JM, Kunkler PB, Smith CW. Radiotherapy and hyperbaric oxygen in head and neck cancer. Final report of first controlled clinical trial. Lancet. 1977;2(8029):101–3.
24. Henk JM. Late results of a trial of hyperbaric oxygen and radiotherapy in head and neck cancer: a rationale for hypoxic cell sensitizers? Int J Radiat Oncol Biol Phys. 1986;12(8):1339–41.
25. Overgaard J, Hansen HS, Overgaard M, Bastholt L, Berthelsen A, Specht L, et al. A randomized double-blind phase III study of nimorazole as a hypoxic radiosensitizer of primary radiotherapy in supraglottic larynx and pharynx carcinoma. Results of the Danish Head and Neck Cancer Study (DAHANCA) Protocol 5-85. Radiother Oncol. 1998;46(2):135–46.

26. Brown JM. Tumor microenvironment and the response to anticancer therapy. Cancer Biol Ther. 2002;1(5):453–8.
27. Rademakers SE, Span PN, Kaanders JH, Sweep FC, van der Kogel AJ, Bussink J. Molecular aspects of tumour hypoxia. Mol Oncol. 2008;2(1):41–53.
28. Bredell MG, Ernst J, El-Kochairi I, Dahlem Y, Ikenberg K, Schumann DM. Current relevance of hypoxia in head and neck cancer. Oncotarget. 2016;7(31):50781–804.
29. Rockwell S, Dobrucki IT, Kim EY, Marrison ST, Vu VT. Hypoxia and radiation therapy: past history, ongoing research, and future promise. Curr Mol Med. 2009;9(4):442–58.
30. Pignon JP, le Maitre A, Maillard E, Bourhis J. Meta-analysis of chemotherapy in head and neck cancer (MACH-NC): an update on 93 randomised trials and 17,346 patients. Radiother Oncol. 2009;92(1):4–14.
31. Henk JM, Kunkler PB, Shah NK, Smith CW, Sutherland WH, Wassif SB. Hyperbaric oxygen in radiotherapy of head and neck carcinoma. Clin Radiol. 1970;21(3):223–31.
32. Churchill-Davidson I, Sanger C, Thomlinson RH. II. Clinical application. Br J Radiol. 1957;30(356):406–22.
33. Brizel DM, Sibley GS, Prosnitz LR, Scher RL, Dewhirst MW. Tumor hypoxia adversely affects the prognosis of carcinoma of the head and neck. Int J Radiat Oncol Biol Phys. 1997;38(2):285–189.
34. Rudat V, Stadler P, Becker A, Vanselow B, Dietz A, Wannenmacher M, et al. Predictive value of the tumor oxygenation by means of pO2 histography in patients with advanced head and neck cancer. Strahlenther Onkol. 2001;177(9):462–8.
35. Nordsmark M, Bentzen SM, Overgaard J. Measurement of human tumour oxygenation status by a polarographic needle electrode. An analysis of inter- and intratumour heterogeneity. Acta Oncol. 1994;33(4):383–9.
36. Nordsmark M, Overgaard M, Overgaard J. Pretreatment oxygenation predicts radiation response in advanced squamous cell carcinoma of the head and neck. Radiother Oncol. 1996;41(1):31–9.
37. Koukourakis MI, Bentzen SM, Giatromanolaki A, Wilson GD, Daley FM, Saunders MI, et al. Endogenous markers of two separate hypoxia response pathways (hypoxia inducible factor 2 alpha and carbonic anhydrase 9) are associated with radiotherapy failure in head and neck cancer patients recruited in the CHART randomized trial. J Clin Oncol: official journal of the American Society of Clinical Oncology. 2006;24(5):727–35.
38. Ferreira MB, De Souza JA, Cohen EE. Role of molecular markers in the management of head and neck cancers. Curr Opin Oncol. 2011;23(3):259–64.
39. Zips D, Zophel K, Abolmaali N, Perrin R, Abramyuk A, Haase R, et al. Exploratory prospective trial of hypoxia-specific PET imaging during radiochemotherapy in patients with locally advanced head-and-neck cancer. Radiother Oncol. 2012;
40. Kikuchi M, Yamane T, Shinohara S, Fujiwara K, Hori SY, Tona Y, et al. 18F-fluoromisonidazole positron emission tomography before treatment is a predictor of radiotherapy outcome and survival prognosis in patients with head and neck squamous cell carcinoma. Ann Nucl Med. 2011;25(9):625–33.
41. Kaanders JH, Wijffels KI, Marres HA, Ljungkvist AS, Pop LA, van den Hoogen FJ, et al. Pimonidazole binding and tumor vascularity predict for treatment outcome in head and neck cancer. Cancer Res. 2002;62(23):7066–74.
42. Swartz JE, Pothen AJ, Stegeman I, Willems SM, Grolman W. Clinical implications of hypoxia biomarker expression in head and neck squamous cell carcinoma: a systematic review. Cancer Med. 2015;4(7):1101–16.
43. Bache M, Kappler M, Said HM, Staab A, Vordermark D. Detection and specific targeting of hypoxic regions within solid tumors: current preclinical and clinical strategies. Curr Med Chem. 2008;15(4):322–38.
44. Karam PA, Leslie SA, Anbar A. The effects of changing atmospheric oxygen concentrations and background radiation levels on radiogenic DNA damage rates. Health Phys. 2001;81(5):545–53.

45. Moulder JE, Rockwell S. Hypoxic fractions of solid tumors: experimental techniques, methods of analysis, and a survey of existing data. Int J Radiat Oncol Biol Phys. 1984;10(5):695–712.
46. Rockwell S, Moulder JE. Hypoxic fractions of human tumors xenografted into mice: a review. Int J Radiat Oncol Biol Phys. 1990;19(1):197–202.
47. Wenzl T, Wilkens JJ. Modelling of the oxygen enhancement ratio for ion beam radiation therapy. Phys Med Biol. 2011;56(11):3251–68.
48. Shannon AM, Bouchier-Hayes DJ, Condron CM, Toomey D. Tumour hypoxia, chemotherapeutic resistance and hypoxia-related therapies. Cancer Treat Rev. 29(4):297–307.
49. Teicher BA, Lazo JS, Sartorelli AC. Classification of antineoplastic agents by their selective toxicities toward oxygenated and hypoxic tumor cells. Cancer Res. 1981;41(1):73–81.
50. Wozniak AJ, Ross WE. DNA damage as a basis for 4'-demethylepipodophyllotoxin-9-(4,6-O-ethylidene-beta-D-glucopyranoside) (etoposide) cytotoxicity. Cancer Res. 1983;43(1):120–4.
51. Wozniak AJ, Glisson BS, Hande KR, Ross WE. Inhibition of etoposide-induced DNA damage and cytotoxicity in L1210 cells by dehydrogenase inhibitors and other agents. Cancer Res. 1984;44(2):626–32.
52. Walker LJ, Craig RB, Harris AL, Hickson ID. A role for the human DNA repair enzyme HAP1 in cellular protection against DNA damaging agents and hypoxic stress. Nucleic Acids Res. 1994;22(23):4884–9.
53. Graeber TG, Osmanian C, Jacks T, Housman DE, Koch CJ, Lowe SW, et al. Hypoxia-mediated selection of cells with diminished apoptotic potential in solid tumours. Nature. 1996;379(6560):88–91.
54. Le QT, Denko NC, Giaccia AJ. Hypoxic gene expression and metastasis. Cancer Metastasis Rev. 2004;23(3-4):293–310.
55. Semenza GL. Hypoxia-inducible factor 1 (HIF-1) pathway. Sci STKE: signal transduction knowledge environment. 2007;2007(407):cm8.
56. Takenaga K. Angiogenic signaling aberrantly induced by tumor hypoxia. Front Biosci: a journal and virtual library. 2011;16:31–48.
57. Harris AL. Hypoxia--a key regulatory factor in tumour growth. Nat Rev Cancer. 2002;2(1):38–47.
58. Li DW, Dong P, Wang F, Chen XW, Xu CZ, Zhou L. Hypoxia induced multidrug resistance of laryngeal cancer cells via hypoxia-inducible factor-1α. Asian Pac J Cancer Prev. 2013;14(8):4853–8.
59. Hsu DS, Lan HY, Huang CH, Tai SK, Chang SY, Tsai TL, et al. Regulation of excision repair cross-complementation group 1 by Snail contributes to cisplatin resistance in head and neck cancer. Clin Cancer Res. 2010;16(18):4561–71.
60. Gammon L, Mackenzie IC. Roles of hypoxia, stem cells and epithelial–mesenchymal transition in the spread and treatment resistance of head and neck cancer. J Oral Pathol Med. 2016;45(2):77–82.
61. Comerford KM, Wallace TJ, Karhausen J, Louis NA, Montalto MC, Colgan SP. Hypoxia-inducible factor-1-dependent regulation of the multidrug resistance (MDR1) gene. Cancer Res. 2002;62(12):3387–94.
62. Krishnamurthy P, Ross DD, Nakanishi T, Bailey-Dell K, Zhou S, Mercer KE, et al. The stem cell marker Bcrp/ABCG2 enhances hypoxic cell survival through interactions with heme. J Biol Chem. 2004;279(23):24218–25.
63. Zeng L, Kizaka-Kondoh S, Itasaka S, Xie X, Inoue M, Tanimoto K, et al. Hypoxia inducible factor-1 influences sensitivity to paclitaxel of human lung cancer cell lines under normoxic conditions. Cancer Sci. 2007;98(9):1394–401.
64. Greijer AE, de Jong MC, Scheffer GL, Shvarts A, van Diest PJ, van der Wall E. Hypoxia-induced acidification causes mitoxantrone resistance not mediated by drug transporters in human breast cancer cells. Cell Oncol: the official journal of the International Society for Cellular Oncology 2005;27(1):43-49.
65. Wykoff CC, Beasley NJ, Watson PH, Turner KJ, Pastorek J, Sibtain A, et al. Hypoxia-inducible expression of tumor-associated carbonic anhydrases. Cancer Res. 2000;60(24):7075–83.

66. Warburg O, Wind F, Negelein E. The metabolism of tumors in the body. J Gen Physiol. 1927;8(6):519–30.
67. Vaupel P. Tumor microenvironmental physiology and its implications for radiation oncology. Semin Radiat Oncol. 2004;14(3):198–206.
68. Vaupel P, Kallinowski F, Okunieff P. Blood flow, oxygen and nutrient supply, and metabolic microenvironment of human tumors: a review. Cancer Res. 1989;49(23):6449–65.
69. Moncharmont C, Levy A, Gilormini M, Bertrand G, Chargari C, Alphonse G, et al. Targeting a cornerstone of radiation resistance: cancer stem cell. Cancer Lett. 2012;322(2):139–47.
70. Sermeus A, Michiels C. Reciprocal influence of the p53 and the hypoxic pathways. Cell Death Dis. 2011;2:e164.
71. Vaupel P, Schlenger K, Knoop C, Hockel M. Oxygenation of human tumors: evaluation of tissue oxygen distribution in breast cancers by computerized O2 tension measurements. Cancer Res. 1991;51(12):3316–22.
72. Hockel M, Schlenger K, Knoop C, Vaupel P. Oxygenation of carcinomas of the uterine cervix: evaluation by computerized O2 tension measurements. Cancer Res. 1991;51(22):6098–102.
73. Nordsmark M, Overgaard J. A confirmatory prognostic study on oxygenation status and loco-regional control in advanced head and neck squamous cell carcinoma treated by radiation therapy. Radiother Oncol. 2000;57(1):39–43.
74. Le QT, Sutphin PD, Raychaudhuri S, Yu SC, Terris DJ, Lin HS, et al. Identification of osteopontin as a prognostic plasma marker for head and neck squamous cell carcinomas. Clin Cancer Res. 2003;9(1):59–67.
75. Nordsmark M, Eriksen JG, Gebski V, Alsner J, Horsman MR, Overgaard J. Differential risk assessments from five hypoxia specific assays: the basis for biologically adapted individualized radiotherapy in advanced head and neck cancer patients. Radiother Oncol. 2007;83(3):389–97.
76. Rudat V, Vanselow B, Wollensack P, Bettscheider C, Osman-Ahmet S, Eble MJ, et al. Repeatability and prognostic impact of the pretreatment pO(2) histography in patients with advanced head and neck cancer. Radiother Oncol. 2000;57(1):31–7.
77. Nozue M, Lee I, Yuan F, Teicher BA, Brizel DM, Dewhirst MW, et al. Interlaboratory variation in oxygen tension measurement by Eppendorf "Histograph" and comparison with hypoxic marker. J Surg Oncol. 1997;66(1):30–8.
78. Wiesener MS, Jürgensen JS, Rosenberger C, Scholze C, Hörstrup JH, Warnecke C, et al. Widespread, hypoxia-inducible expression of HIF-2α in distinct cell populations of different organs. FASEB J. 2002;17:271–3.
79. Unwith S, Zhao H, Hennah L, Ma D. The potential role of HIF on tumour progression and dissemination. Int J Cancer. 2015;136(11):2491–503.
80. Semenza GL. Targeting HIF-1 for cancer therapy. Nat Rev Cancer. 2003;3(10):721–32.
81. Denko NC. Hypoxia, HIF1 and glucose metabolism in the solid tumour. Nat Rev Cancer. 2008;8(9):705–13.
82. Semenza GL, Wang GL. A nuclear factor induced by hypoxia via de novo protein synthesis binds to the human erythropoietin gene enhancer at a site required for transcriptional activation. Mol Cell Biol. 1992;12(12):5447–54.
83. Wang GL, Jiang BH, Rue EA, Semenza GL. Hypoxia-inducible factor 1 is a basic-helix-loop-helix-PAS heterodimer regulated by cellular O2 tension. Proc Natl Acad Sci U S A. 1995;92(12):5510–4.
84. Nordsmark MAJ, Eriksen JG, et al. The prognostic value of serum osteopontin, HIF-1alpha and pO2 measurements in advanced head and neck tumors treated by radiotherapy. Eur J Cancer Suppl. 2003;1:S145.
85. Peridis S, Pilgrim G, Athanasopoulos I, Parpounas K. Carbonic anhydrase-9 expression in head and neck cancer: a meta-analysis. Eur Arch Oto Rhino Laryngol: official journal of the European Federation of Oto-Rhino-Laryngological Societies. 2011;268(5):661–70.
86. Li J, Zhang G, Wang X, Li X-F. Is carbonic anhydrase IX a validated target for molecular imaging of cancer and hypoxia? Future Oncol. 2015;11(10):1531–41.

87. Wykoff CC, Beasley NJP, Watson PH, Turner KJ, Pastorek J, Sibtain A, et al. Hypoxia-inducible expression of tumor-associated carbonic anhydrases. Cancer Res. 2000;60(24):7075–83.
88. Forsythe JA, Jiang BH, Iyer NV, Agani F, Leung SW, Koos RD, et al. Activation of vascular endothelial growth factor gene transcription by hypoxia-inducible factor 1. Mol Cell Biol. 1996;16(9):4604–13.
89. Macheda ML, Rogers S, Best JD. Molecular and cellular regulation of glucose transporter (GLUT) proteins in cancer. J Cell Physiol. 2005;202(3):654–62.
90. Kimura H, Weisz A, Kurashima Y, Hashimoto K, Ogura T, D'Acquisto F, et al. Hypoxia response element of the human vascular endothelial growth factor gene mediates transcriptional regulation by nitric oxide: control of hypoxia-inducible factor-1 activity by nitric oxide. Blood. 2000;95(1):189–97.
91. Shih SC, Claffey KP. Role of AP-1 and HIF-1 transcription factors in TGF-beta activation of VEGF expression. Growth Factors. 2001;19(1):19–34.
92. Anavi S, Hahn-Obercyger M, Madar Z, Tirosh O. Mechanism for HIF-1 activation by cholesterol under normoxia: a redox signaling pathway for liver damage. Free Radic Biol Med. 2014;71:61–9.
93. Fukuda R, Hirota K, Fan F, Jung YD, Ellis LM, Semenza GL. Insulin-like growth factor 1 induces hypoxia-inducible factor 1-mediated vascular endothelial growth factor expression, which is dependent on MAP kinase and phosphatidylinositol 3-kinase signaling in colon cancer cells. J Biol Chem. 2002;277(41):38205–11.
94. Richard DE, Berra E, Gothie E, Roux D, Pouyssegur J. p42/p44 mitogen-activated protein kinases phosphorylate hypoxia-inducible factor 1alpha (HIF-1alpha) and enhance the transcriptional activity of HIF-1. J Biol Chem. 1999;274(46):32631–7.
95. Nakamura M, Bodily JM, Beglin M, Kyo S, Inoue M, Laimins LA. Hypoxia-specific stabilization of HIF-1alpha by human papillomaviruses. Virology. 2009;387(2):442–8.
96. Guo Y, Meng X, Ma J, Zheng Y, Wang Q, Wang Y, et al. Human papillomavirus 16 E6 contributes HIF-1alpha induced Warburg effect by attenuating the VHL-HIF-1alpha interaction. Int J Mol Sci. 2014;15(5):7974–86.
97. Gong L, Zhang W, Zhou J, Lu J, Xiong H, Shi X, et al. Prognostic value of HIFs expression in head and neck cancer: a systematic review. PLoS One. 2013;8(9):e75094.
98. Hoogsteen IJ, Marres HA, Bussink J, van der Kogel AJ, Kaanders JH. Tumor microenvironment in head and neck squamous cell carcinomas: predictive value and clinical relevance of hypoxic markers. A review. Head Neck. 2007;29(6):591–604.
99. Bussink J, Kaanders JH, van der Kogel AJ. Tumor hypoxia at the micro-regional level: clinical relevance and predictive value of exogenous and endogenous hypoxic cell markers. Radiother Oncol. 2003;67(1):3–15.
100. PO DEL, Jorge CC, Oliveira DT, Pereira MC. Hypoxic condition and prognosis in oral squamous cell carcinoma. Anticancer Res. 2014;34(2):605–12.
101. Qian J, Wenguang X, Zhiyong W, Yuntao Z, Wei H. Hypoxia inducible factor: a potential prognostic biomarker in oral squamous cell carcinoma. Tumor Biol. 2016;37(8):10815–20.
102. van Kuijk SJA, Yaromina A, Houben R, Niemans R, Lambin P, Dubois LJ. Prognostic significance of carbonic anhydrase IX expression in cancer patients: a meta-analysis. Front Oncol. 2016;6(69)
103. Kaluz S, Kaluzova M, Chrastina A, Olive PL, Pastorekova S, Pastorek J, et al. Lowered oxygen tension induces expression of the hypoxia marker MN/carbonic anhydrase IX in the absence of hypoxia-inducible factor 1 alpha stabilization: a role for phosphatidylinositol 3'-kinase. Cancer Res. 2002;62(15):4469–77.
104. Mayer A, Hockel M, Vaupel P. Carbonic anhydrase IX expression and tumor oxygenation status do not correlate at the microregional level in locally advanced cancers of the uterine cervix. Clin Cancer Res. 2005;11(20):7220–5.
105. Li XF, Carlin S, Urano M, Russell J, Ling CC, O'Donoghue JA. Visualization of hypoxia in microscopic tumors by immunofluorescent microscopy. Cancer Res. 2007;67(16):7646–53.

106. Zelzer E, Levy Y, Kahana C, Shilo BZ, Rubinstein M, Cohen B. Insulin induces transcription of target genes through the hypoxia-inducible factor HIF-1alpha/ARNT. EMBO J. 1998;17(17):5085–94.
107. Okino ST, Chichester CH, Whitlock JP Jr. Hypoxia-inducible mammalian gene expression analyzed in vivo at a TATA-driven promoter and at an initiator-driven promoter. J Biol Chem. 1998;273(37):23837–43.
108. Rivenzon-Segal D, Boldin-Adamsky S, Seger D, Seger R, Degani H. Glycolysis and glucose transporter 1 as markers of response to hormonal therapy in breast cancer. Int J Cancer. 2003;107(2):177–82.
109. Brands RC, Köhler O, Rauthe S, Hartmann S, Ebhardt H, Seher A, et al. The prognostic value of GLUT-1 staining in the detection of malignant transformation in oral mucosa. Clin Oral Investig. 2016:1–7.
110. Li C-X, Sun J-L, Gong Z-C, Lin Z-Q, Liu H. Prognostic value of GLUT-1 expression in oral squamous cell carcinoma: a prisma-compliant meta-analysis. Medicine. 2016;95(45):e5324.
111. Vassilakopoulou M, Psyrri A, Argiris A. Targeting angiogenesis in head and neck cancer. Oral Oncol. 2015;51(5):409–15.
112. Glück AA, Aebersold DM, Zimmer Y, Medová M. Interplay between receptor tyrosine kinases and hypoxia signaling in cancer. Int J Biochem Cell Biol. 2015;62:101–14.
113. Kowanetz M, Ferrara N. Vascular endothelial growth factor signaling pathways: therapeutic perspective. Clin Cancer Res. 2006;12(17):5018–22.
114. Kyzas PA, Cunha IW, Ioannidis JP. Prognostic significance of vascular endothelial growth factor immunohistochemical expression in head and neck squamous cell carcinoma: a meta-analysis. Clin Cancer Res. 2005;11(4):1434–40.
115. Zhang LP, Chen HL. Increased vascular endothelial growth factor expression predicts a worse prognosis for laryngeal cancer patients: a meta-analysis. J Laryngol Otol. 2016;131(1):44–50.
116. Bellahcene A, Castronovo V, Ogbureke KU, Fisher LW, Fedarko NS. Small integrin-binding ligand N-linked glycoproteins (SIBLINGs): multifunctional proteins in cancer. Nat Rev Cancer. 2008;8(3):212–26.
117. Zhu Y, Denhardt DT, Cao H, Sutphin PD, Koong AC, Giaccia AJ, et al. Hypoxia upregulates osteopontin expression in NIH-3T3 cells via a Ras-activated enhancer. Oncogene. 2005;24(43):6555–63.
118. Fisher LW, Jain A, Tayback M, Fedarko NS. Small integrin binding ligand N-linked glycoprotein gene family expression in different cancers. Clin Cancer Res. 2004;10(24):8501–11.
119. Wai PY, Kuo PC. Osteopontin: regulation in tumor metastasis. Cancer Metastasis Rev. 2008;27(1):103–18.
120. Wang KX, Denhardt DT. Osteopontin: role in immune regulation and stress responses. Cytokine Growth Factor Rev. 2008;19(5-6):333–45.
121. Weber GF, Lett GS, Haubein NC. Osteopontin is a marker for cancer aggressiveness and patient survival. Br J Cancer. 2010;103(6):861–9.
122. Le QT, Kong C, Lavori PW, O'Byrne K, Erler JT, Huang X, et al. Expression and prognostic significance of a panel of tissue hypoxia markers in head-and-neck squamous cell carcinomas. Int J Radiat Oncol Biol Phys. 2007;69(1):167–75.
123. Bache M, Reddemann R, Said HM, Holzhausen HJ, Taubert H, Becker A, et al. Immunohistochemical detection of osteopontin in advanced head-and-neck cancer: prognostic role and correlation with oxygen electrode measurements, hypoxia-inducible-factor-1alpha-related markers, and hemoglobin levels. Int J Radiat Oncol Biol Phys. 2006;66(5):1481–7.
124. Overgaard J, Eriksen JG, Nordsmark M, Alsner J, Horsman MR. Plasma osteopontin, hypoxia, and response to the hypoxia sensitiser nimorazole in radiotherapy of head and neck cancer: results from the DAHANCA 5 randomised double-blind placebo-controlled trial. Lancet Oncol. 2005;6(10):757–64.
125. Lim AM, Rischin D, Fisher R, Cao H, Kwok K, Truong D, et al. Prognostic significance of plasma osteopontin in patients with locoregionally advanced head and neck squamous cell carcinoma treated on TROG 02.02 phase III trial. Clin Cancer Res. 2012;18(1):301–7.

126. Toustrup K, Sorensen BS, Nordsmark M, Busk M, Wiuf C, Alsner J, et al. Development of a hypoxia gene expression classifier with predictive impact for hypoxic modification of radiotherapy in head and neck cancer. Cancer Res. 2011;71(17):5923–31.
127. Toustrup K, Sorensen BS, Alsner J, Overgaard J. Hypoxia gene expression signatures as prognostic and predictive markers in head and neck radiotherapy. Semin Radiat Oncol. 2012;22(2):119–27.
128. Linge A, Lohaus F, Lock S, Nowak A, Gudziol V, Valentini C, et al. HPV status, cancer stem cell marker expression, hypoxia gene signatures and tumour volume identify good prognosis subgroups in patients with HNSCC after primary radiochemotherapy: a multicentre retrospective study of the German Cancer Consortium Radiation Oncology Group (DKTK-ROG). Radiother Oncol. 2016;121(3):364–73.
129. Arteel GE, Thurman RG, Yates JM, Raleigh JA. Evidence that hypoxia markers detect oxygen gradients in liver: pimonidazole and retrograde perfusion of rat liver. Br J Cancer. 1995;72(4):889–95.
130. Bourgeois M, Rajerison H, Guerard F, Mougin-Degraef M, Barbet J, Michel N, et al. Contribution of [64Cu]-ATSM PET in molecular imaging of tumour hypoxia compared to classical [18F]-MISO--a selected review. Nucl Med Rev Cent East Eur. 2011;14(2):90–5.
131. Chen L, Zhang Z, Kolb HC, Walsh JC, Zhang J, Guan Y. (1)(8)F-HX4 hypoxia imaging with PET/CT in head and neck cancer: a comparison with (1)(8)F-FMISO. Nucl Med Commun. 2012;33(10):1096–102.
132. Dubois LJ, Lieuwes NG, Janssen MH, Peeters WJ, Windhorst AD, Walsh JC, et al. Preclinical evaluation and validation of [18F]HX4, a promising hypoxia marker for PET imaging. Proc Natl Acad Sci U S A. 2011;108(35):14620–5.
133. Suh YE, Lawler K, Henley-Smith R, Pike L, Leek R, Barrington S, et al. Association between hypoxic volume and underlying hypoxia-induced gene expression in oropharyngeal squamous cell carcinoma. Br J Cancer. 2017;116:1057–64.
134. Rasey JS, Grunbaum Z, Magee S, Nelson NJ, Olive PL, Durand RE, et al. Characterization of radiolabeled fluoromisonidazole as a probe for hypoxic cells. Radiat Res. 1987;111(2):292–304.
135. Maeda K, Osato T, Umezawa H. A new antibiotic, azomycin. J Antibiot. 1953;6(4):182.
136. Hodolic M, Fettich J, Kairemo K. Hypoxia PET tracers in EBRT dose planning in head and neck cancer. Curr Radiopharm. 2015;8(1):32–7.
137. Tamaki N, Hirata K. Tumor hypoxia: a new PET imaging biomarker in clinical oncology. Int J Clin Oncol. 2016;21(4):619–25.
138. Troost EG, Laverman P, Philippens ME, Lok J, van der Kogel AJ, Oyen WJ, et al. Correlation of [18F]FMISO autoradiography and pimonidazole [corrected] immunohistochemistry in human head and neck carcinoma xenografts. Eur J Nucl Med Mol Imaging. 2008;35(10):1803–11.
139. Troost EG, Laverman P, Kaanders JH, Philippens M, Lok J, Oyen WJ, et al. Imaging hypoxia after oxygenation-modification: comparing [18F]FMISO autoradiography with pimonidazole immunohistochemistry in human xenograft tumors. Radiother Oncol. 2006;80(2):157–64.
140. Dubois L, Landuyt W, Haustermans K, Dupont P, Bormans G, Vermaelen P, et al. Evaluation of hypoxia in an experimental rat tumour model by [(18)F]fluoromisonidazole PET and immunohistochemistry. Br J Cancer. 2004;91(11):1947–54.
141. Rajendran JG, Schwartz DL, O'Sullivan J, Peterson LM, Ng P, Scharnhorst J, et al. Tumor hypoxia imaging with [F-18] fluoromisonidazole positron emission tomography in head and neck cancer. Clin Cancer Res. 2006;12(18):5435–41.
142. Eschmann SM, Paulsen F, Reimold M, Dittmann H, Welz S, Reischl G, et al. Prognostic impact of hypoxia imaging with 18F-misonidazole PET in non-small cell lung cancer and head and neck cancer before radiotherapy. J Nucl Med. 2005;46(2):253–60.
143. Dirix P, Vandecaveye V, De Keyzer F, Stroobants S, Hermans R, Nuyts S. Dose painting in radiotherapy for head and neck squamous cell carcinoma: value of repeated functional imaging with (18)F-FDG PET, (18)F-fluoromisonidazole PET, diffusion-weighted MRI, and dynamic contrast-enhanced MRI. J Nucl Med. 2009;50(7):1020–7.

144. Rischin D, Hicks RJ, Fisher R, Binns D, Corry J, Porceddu S, et al. Prognostic significance of [18F]-misonidazole positron emission tomography-detected tumor hypoxia in patients with advanced head and neck cancer randomly assigned to chemoradiation with or without tirapazamine: a substudy of Trans-Tasman Radiation Oncology Group Study 98.02. J Clin Oncol: official journal of the American Society of Clinical Oncology. 2006;24(13):2098–104.
145. Wiedenmann NE, Bucher S, Hentschel M, Mix M, Vach W, Bittner M-I, et al. Serial [18F]-fluoromisonidazole PET during radiochemotherapy for locally advanced head and neck cancer and its correlation with outcome. Radiother Oncol. 2015;117(1):113–7.
146. Lee NY, Mechalakos JG, Nehmeh S, Lin Z, Squire OD, Cai S, et al. Fluorine-18-labeled fluoromisonidazole positron emission and computed tomography-guided intensity-modulated radiotherapy for head and neck cancer: a feasibility study. Int J Radiat Oncol Biol Phys. 2008;70(1):2–13.
147. Henriques de Figueiredo B, Merlin T, de Clermont-Gallerande H, Hatt M, Vimont D, Fernandez P, et al. Potential of [18F]-Fluoromisonidazole positron-emission tomography for radiotherapy planning in head and neck squamous cell carcinomas. Strahlenther Onkol. 2013;189(12):1015–9.
148. Piert M, Machulla HJ, Picchio M, Reischl G, Ziegler S, Kumar P, et al. Hypoxia-specific tumor imaging with 18F-fluoroazomycin arabinoside. J Nucl Med. 2005;46(1):106–13.
149. Postema EJ, McEwan AJ, Riauka TA, Kumar P, Richmond DA, Abrams DN, et al. Initial results of hypoxia imaging using 1-alpha-D: -(5-deoxy-5-[18F]-fluoroarabinofuranosyl)-2-nitroimidazole (18F-FAZA). Eur J Nucl Med Mol Imaging. 2009;36(10):1565–73.
150. Kumar P, Stypinski D, Xia H, McEwan AJB, Machulla HJ, Wiebe LI. Fluoroazomycin arabinoside (FAZA): synthesis, 2H and 3H-labelling and preliminary biological evaluation of a novel 2-nitroimidazole marker of tissue hypoxia. J Label Compd Radiopharm. 1999;42(1):3–16.
151. Saga T, Inubushi M, Koizumi M, Yoshikawa K, Zhang M-R, Obata T, et al. Prognostic value of PET/CT with 18F-fluoroazomycin arabinoside for patients with head and neck squamous cell carcinomas receiving chemoradiotherapy. Ann Nucl Med. 2016;30(3):217–24.
152. Rischin D, Fisher R, Peters L, Corry J, Hicks R. Hypoxia in head and neck cancer: studies with hypoxic positron emission tomography imaging and hypoxic cytotoxins. Int J Radiat Oncol Biol Phys. 2007;69(2 Suppl):S61–3.
153. Graves EE, Hicks RJ, Binns D, Bressel M, Le Q-T, Peters L, et al. Quantitative and qualitative analysis of [18F]FDG and [18F]FAZA positron emission tomography of head and neck cancers and associations with HPV status and treatment outcome. Eur J Nucl Med Mol Imaging. 2016;43(4):617–25.
154. Graves EE, Vilalta M, Cecic IK, Erler JT, Tran PT, Felsher D, et al. Hypoxia in models of lung cancer: implications for targeted therapeutics. Clin Cancer Res. 2010;16(19):4843–52.
155. Young RJ, Moller A. Immunohistochemical detection of tumour hypoxia. Methods Mol Biol. 2010;611:151–9.
156. Rijken PF, Bernsen HJ, Peters JP, Hodgkiss RJ, Raleigh JA, van der Kogel AJ. Spatial relationship between hypoxia and the (perfused) vascular network in a human glioma xenograft: a quantitative multi-parameter analysis. Int J Radiat Oncol Biol Phys. 2000;48(2):571–82.
157. Busk M, Jakobsen S, Horsman MR, Mortensen LS, Iversen AB, Overgaard J, et al. PET imaging of tumor hypoxia using 18F-labeled pimonidazole. Acta Oncol. 2013;52(7):1300–7.
158. Evans SM, Joiner B, Jenkins WT, Laughlin KM, Lord EM, Koch CJ. Identification of hypoxia in cells and tissues of epigastric 9L rat glioma using EF5 [2-(2-nitro-1H-imidazol-1-yl)-N-(2,2,3,3,3- pentafluoropropyl) acetamide]. Br J Cancer. 1995;72(4):875–82.
159. Chitneni SK, Bida GT, Dewhirst MW, Zalutsky MR. A simplified synthesis of the hypoxia imaging agent 2-(2-Nitro-1H-imidazol-1-yl)-N-(2,2,3,3,3-[(18)F]pentafluoropropyl)-acetamide ([(18)F]EF5). Nucl Med Biol. 2012;39(7):1012–8.
160. Evans SM, Hahn S, Pook DR, Jenkins WT, Chalian AA, Zhang P, et al. Detection of hypoxia in human squamous cell carcinoma by EF5 binding. Cancer Res. 2000;60(7):2018–24.

161. Panek R, Welsh L, Baker LC, Schmidt MA, Wong KH, Riddell A, et al. Non-invasive imaging of cycling hypoxia in head and neck cancer using intrinsic susceptibility MRI. Clin Cancer Res. 2017;23(15):4233–41.
162. Bennett M, Feldmeier J, Smee R, Milross C. Hyperbaric oxygenation for tumour sensitisation to radiotherapy: a systematic review of randomised controlled trials. Cancer Treat Rev. 2008;34(7):577–91.
163. Haffty BG, Hurley R, Peters LJ. Radiation therapy with hyperbaric oxygen at 4 atmospheres pressure in the management of squamous cell carcinoma of the head and neck: results of a randomized clinical trial. Cancer J Sci Am. 1999;5(6):341–7.
164. Kaanders JH, Bussink J, van der Kogel AJ. ARCON: a novel biology-based approach in radiotherapy. Lancet Oncol. 2002;3(12):728–37.
165. Kaanders JH, Pop LA, Marres HA, Bruaset I, van den Hoogen FJ, Merkx MA, et al. ARCON: experience in 215 patients with advanced head-and-neck cancer. Int J Radiat Oncol Biol Phys. 2002;52(3):769–78.
166. Kaanders JH, Pop LA, Marres HA, Liefers J, van den Hoogen FJ, van Daal WA, et al. Accelerated radiotherapy with carbogen and nicotinamide (ARCON) for laryngeal cancer. Radiother Oncol. 1998;48(2):115–22.
167. Welsh L, Panek R, Riddell A, Wong K, Leach MO, Tavassoli M, et al. Blood transfusion during radical chemo-radiotherapy does not reduce tumour hypoxia in squamous cell cancer of the head and neck. Br J Cancer. 2017;116(1):28–35.
168. Hoff CM, Lassen P, Eriksen JG, Hansen HS, Specht L, Overgaard M, et al. Does transfusion improve the outcome for HNSCC patients treated with radiotherapy? – Results from the randomized DAHANCA 5 and 7 trials. Acta Oncol. 2011;50(7):1006–14.
169. Bhide SA, Ahmed M, Rengarajan V, Powell C, Miah A, Newbold K, et al. Anemia during sequential induction chemotherapy and chemoradiation for head and neck cancer: the impact of blood transfusion on treatment outcome. Int J Radiat Oncol Biol Phys. 2009;73(2):391–8.
170. Sealy R, Cridland S, Barry L, Norris R. Irradiation with misonidazole and hyperbaric oxygen: final report on a randomized trial in advanced head and neck cancer. Int J Radiat Oncol Biol Phys. 1986;12(8):1343–6.
171. Tobin DA, Vermund H. A randomized study of hyperbaric oxygen as an adjunct to regularly fractionated radiation therapy for clinical treatment of advanced neoplastic disease. Am J Roentgenol Radium Ther Nucl Med. 1971;111(3):613–21.
172. Bennett MH, Feldmeier J, Smee R, Milross C. Hyperbaric oxygenation for tumour sensitisation to radiotherapy. Cochrane Database Syst Rev. 2005;(4)
173. Giebfried JW, Lawson W, Biller HF. Complications of hyperbaric oxygen in the treatment of head and neck disease. Otolaryngol Head Neck Surg: official journal of American Academy of Otolaryngology-Head and Neck Surgery. 1986;94(4):508–12.
174. Bennett MH, Feldmeier J, Smee R, Milross C. Hyperbaric oxygenation for tumour sensitisation to radiotherapy. Cochrane Database Syst Rev. 2012;4:CD005007.
175. Janssens GO, Rademakers SE, Terhaard CH, Doornaert PA, Bijl HP, van den Ende P, et al. Accelerated radiotherapy with carbogen and nicotinamide for laryngeal cancer: results of a phase III randomized trial. J Clin Oncol: official journal of the American Society of Clinical Oncology. 2012;30(15):1777–83.
176. Nijkamp MM, Span PN, Terhaard CH, Doornaert PA, Langendijk JA, van den Ende PL, et al. Epidermal growth factor receptor expression in laryngeal cancer predicts the effect of hypoxia modification as an additive to accelerated radiotherapy in a randomised controlled trial. Eur J Cancer. 2013;49(15):3202–9.
177. Janssens GO, Rademakers SE, Terhaard CH, Doornaert PA, Bijl HP, van den Ende P, et al. Improved recurrence-free survival with ARCON for anemic patients with laryngeal cancer. Clin Cancer Res. 2014;20(5):1345–54.
178. Lee WR, Berkey B, Marcial V, Fu KK, Cooper JS, Vikram B, et al. Anemia is associated with decreased survival and increased locoregional failure in patients with locally advanced head and neck carcinoma: a secondary analysis of RTOG 85-27. Int J Radiat Oncol Biol Phys. 1998;42(5):1069–75.

179. Lambin P, Ramaekers BLT, van Mastrigt GAPG, Van den Ende P, de Jong J, De Ruysscher DKM, et al. Erythropoietin as an adjuvant treatment with (chemo) radiation therapy for head and neck cancer. Cochrane Database Syst Rev. 2009;3(3):CD006158. https://doi.org/10.1002/14651858.CD006158.pub2.
180. Tonia T, Mettler A, Robert N, Schwarzer G, Seidenfeld J, Weingart O, et al. Erythropoietin or darbepoetin for patients with cancer. Cochrane Database Syst Rev. 2012;12(12):CD003407. https://doi.org/10.1002/14651858.CD003407.pub5.
181. Bennett CL, Lai SY, Sartor O, et al. Consensus on the existence of functional erythropoietin receptors on cancer cells. JAMA Oncol. 2016;2(1):134–6.
182. Bennett CL, Lai SY, Henke M, Barnato SE, Armitage JO, Sartor O. Association between pharmaceutical support and basic science research on erythropoiesis-stimulating agents. Arch Intern Med. 2010;170(16):1490–8.
183. Winter SC, Shah KA, Campo L, Turley H, Leek R, Corbridge RJ, et al. Relation of erythropoietin and erythropoietin receptor expression to hypoxia and anemia in head and neck squamous cell carcinoma. Clin Cancer Res. 2005;11(21):7614–20.
184. Arcasoy MO, Amin K, Chou S-C, Haroon ZA, Varia M, Raleigh JA. Erythropoietin and erythropoietin receptor expression in head and neck cancer: relationship to tumor hypoxia. Clin Cancer Res. 2005;11(1):20–7.
185. Bohlius J, Schmidlin K, Brillant C, Schwarzer G, Trelle S, Seidenfeld J, et al. Erythropoietin or Darbepoetin for patients with cancer - meta-analysis based on individual patient data. Cochrane Database Syst Rev. 2009;8(3):CD007303. https://doi.org/10.1002/14651858.CD007303.pub2.
186. Bohlius J, Wilson J, Seidenfeld J, Piper M, Schwarzer G, Sandercock J, et al. Erythropoietin or Darbepoetin for patients with cancer. Cochrane Database Syst Rev. 2006;(3):CD003407.
187. Datta NR, Bose AK, Kapoor HK, Gupta S. Head and neck cancers: results of thermoradiotherapy versus radiotherapy. Int J Hyperthermia. 1990;6(3):479–86.
188. Datta NR, Rogers S, Ordóñez SG, Puric E, Bodis S. Hyperthermia and radiotherapy in the management of head and neck cancers: a systematic review and meta-analysis. Int J Hyperthermia. 2016;32(1):31–40.
189. Walton MI, Wolf CR, Workman P. Molecular enzymology of the reductive bioactivation of hypoxic cell cytotoxins. Int J Radiat Oncol Biol Phys. 1989;16:983–6.
190. Van den Bogaert W, van der Schueren E, Horiot JC, De Vilhena M, Schraub S, Svoboda V, et al. The EORTC randomized trial on three fractions per day and misonidazole (trial no. 22811) in advanced head and neck cancer: long-term results and side effects. Radiother Oncol. 1995;35(2):91–9.
191. Lee DJ, Cosmatos D, Marcial VA, Fu KK, Rotman M, Cooper JS, et al. Results of an RTOG phase III trial (RTOG 85-27) comparing radiotherapy plus etanidazole with radiotherapy alone for locally advanced head and neck carcinomas. Int J Radiat Oncol Biol Phys. 1995;32(3):567–76.
192. Eschwege F, Sancho-Garnier H, Chassagne D, Brisgand D, Guerra M, Malaise EP, et al. Results of a European randomized trial of Etanidazole combined with radiotherapy in head and neck carcinomas [see comments]. Int J Radiat Oncol Biol Phys. 1997;39(2):275–81.
193. Zeman EM, Brown JM, Lemmon MJ, Hirst VK, Lee WW. SR 4233: a new bioreductive agent with high selective toxicity for hypoxic mammalian cells. Int J Radiat Oncol Biol Phys. 1986;12:1239–42.
194. Dorie MJ, Brown JM. Modification of the antitumor activity of chemotherapeutic drugs by the hypoxic cytotoxic agent tirapazamine. Cancer Chemother Pharmacol. 1997;39(4):361–6.
195. Beck R, Roper B, Carlsen JM, Huisman MC, Lebschi JA, Andratschke N, et al. Pretreatment 18F-FAZA PET predicts success of hypoxia-directed radiochemotherapy using tirapazamine. J Nucl Med. 2007;48(6):973–80.
196. Rischin D, Peters L, Hicks R, Hughes P, Fisher R, Hart R, et al. Phase I trial of concurrent tirapazamine, cisplatin, and radiotherapy in patients with advanced head and neck cancer. J Clin Oncol: official journal of the American Society of Clinical Oncology. 2001;19(2):535–42.

197. Rischin D, Peters L, Fisher R, Macann A, Denham J, Poulsen M, et al. Tirapazamine, Cisplatin, and Radiation versus Fluorouracil, Cisplatin, and Radiation in patients with locally advanced head and neck cancer: a randomized phase II trial of the Trans-Tasman Radiation Oncology Group (TROG 98.02). J Clin Oncol: official journal of the American Society of Clinical Oncology. 2005;23(1):79–87.
198. Rischin D, Peters L, O'Sullivan B, Giralt J, Yuen K, Trotti A, et al. Phase III study of tirapazamine, cisplatin and radiation versus cisplatin and radiation for advanced squamous cell carcinoma of the head and neck. J Clin Oncol: official journal of the American Society of Clinical Oncology. 2008;26(May 20 suppl):abstr LBA6008.
199. Peters LJ, O'Sullivan B, Giralt J, Fitzgerald TJ, Trotti A, Bernier J, et al. Critical impact of radiotherapy protocol compliance and quality in the treatment of advanced head and neck cancer: results from TROG 02.02. J Clin Oncol: official journal of the American Society of Clinical Oncology. 2010;28(18):2996–3001.
200. Hicks KO, Pruijn FB, Secomb TW, Hay MP, Hsu R, Brown JM, et al. Use of three-dimensional tissue cultures to model extravascular transport and predict in vivo activity of hypoxia-targeted anticancer drugs. JNCI: Journal of the National Cancer Institute. 2006;98(16):1118–28.
201. Li Q, Lin Q, Yun Z. Hypoxia-activated cytotoxicity of benznidazole against clonogenic tumor cells. Cancer Biol Ther. 2016;17(12):1266–73.
202. Sun JD, Liu Q, Ahluwalia D, Ferraro DJ, Wang Y, Jung D, et al. Comparison of hypoxia-activated prodrug evofosfamide (TH-302) and ifosfamide in preclinical non-small cell lung cancer models. Cancer Biol Ther. 2016;17(4):371–80.
203. Liu Y, Liu Y, Bu W, Xiao Q, Sun Y, Zhao K, et al. Radiation-/hypoxia-induced solid tumor metastasis and regrowth inhibited by hypoxia-specific upconversion nanoradiosensitizer. Biomaterials. 2015;49:1–8.
204. Oku N, Matoba S, Yamazaki YM, Shimasaki R, Miyanaga S, Igarashi Y. Complete stereochemistry and preliminary structure–activity relationship of rakicidin A, a hypoxia-selective cytotoxin from micromonospora sp. J Nat Prod. 2014;77(11):2561–5.
205. Jiang BH, Jiang G, Zheng JZ, Lu Z, Hunter T, Vogt PK. Phosphatidylinositol 3-kinase signaling controls levels of hypoxia-inducible factor 1. Cell Growth Differ: the molecular biology journal of the American Association for Cancer Research. 2001;12(7):363–9.
206. Hudson CC, Liu M, Chiang GG, Otterness DM, Loomis DC, Kaper F, et al. Regulation of hypoxia-inducible factor 1alpha expression and function by the mammalian target of rapamycin. Mol Cell Biol. 2002;22(20):7004–14.
207. Geiger JL, Bauman JE, Gibson MK, Gooding WE, Varadarajan P, Kotsakis A, et al. Phase II trial of everolimus in patients with previously treated recurrent or metastatic head and neck squamous cell carcinoma. Head Neck. 2016;38(12):1759–64.
208. Massarelli E, Lin H, Ginsberg LE, Tran HT, Lee JJ, Canales JR, et al. Phase II trial of everolimus and erlotinib in patients with platinum-resistant recurrent and/or metastatic head and neck squamous cell carcinoma. Ann Oncol. 2015;26(7):1476–80.
209. Soulières D, Faivre S, Mesía R, Remenár É, Li S-H, Karpenko A, et al. Buparlisib and paclitaxel in patients with platinum-pretreated recurrent or metastatic squamous cell carcinoma of the head and neck (BERIL-1): a randomised, double-blind, placebo-controlled phase 2 trial. Lancet Oncol. 2017;18(3):323–35.
210. Lee NY, Le QT. New developments in radiation therapy for head and neck cancer: intensity-modulated radiation therapy and hypoxia targeting. Semin Oncol. 2008;35(3):236–50.
211. Hu Y, Liu J, Huang H. Recent agents targeting HIF-1alpha for cancer therapy. J Cell Biochem. 2012;114:498–509.
212. Lu H, Liang K, Lu Y, Fan Z. The anti-EGFR antibody cetuximab sensitizes human head and neck squamous cell carcinoma cells to radiation in part through inhibiting radiation-induced upregulation of HIF-1α. Cancer Lett. 2012;322(1):78–85.
213. Li X, Fan Z. The epidermal growth factor receptor antibody cetuximab induces autophagy in cancer cells by downregulating HIF-1alpha and Bcl-2 and activating the beclin 1/hVps34 complex. Cancer Res. 2010;70(14):5942–52.

214. Boeckx C, Baay M, Wouters A, Specenier P, Vermorken JB, Peeters M. Anti-epidermal growth factor receptor therapy in head and neck squamous cell carcinoma: focus on potential molecular mechanisms of drug resistance. Oncologist. 2013;18(7):850–64.
215. Boeckx C, Van den Bossche J, De Pauw I, Peeters M, Lardon F, Baay M, et al. The hypoxic tumor microenvironment and drug resistance against EGFR inhibitors: preclinical study in cetuximab-sensitive head and neck squamous cell carcinoma cell lines. BMC Res Notes. 2015;8(1):203.
216. Hsu HW, Wall NR, Hsueh CT, Kim S, Ferris RL, Chen CS, et al. Combination antiangiogenic therapy and radiation in head and neck cancers. Oral Oncol. 2014;50(1):19–26.
217. Argiris A, Bauman JE, Ohr J, Gooding WE, Heron DE, Duvvuri U, et al. Phase II randomized trial of radiation therapy, cetuximab, and pemetrexed with or without bevacizumab in patients with locally advanced head and neck cancer. Ann Oncol. 2016;27(8):1594–600.
218. Ahn PH, Machtay M, Anne PR, Cognetti D, Keane WM, Wuthrick E, et al. Phase I trial using induction cisplatin, docetaxel, 5-FU and erlotinib followed by cisplatin, bevacizumab and erlotinib with concurrent radiotherapy for advanced head and neck cancer. Am J Clin Oncol. 2016. [Epub ahead of print].
219. Fury MG, Xiao H, Sherman EJ, Baxi S, Smith-Marrone S, Schupak K, et al. Phase II trial of bevacizumab + cetuximab + cisplatin with concurrent intensity-modulated radiation therapy for patients with stage III/IVB head and neck squamous cell carcinoma. Head Neck. 2016;38(S1):E566–E70.
220. Jeong W, Rapisarda A, Park SR, Kinders RJ, Chen A, Melillo G, et al. Pilot trial of EZN-2968, an antisense oligonucleotide inhibitor of hypoxia-inducible factor-1 alpha (HIF-1α), in patients with refractory solid tumors. Cancer Chemother Pharmacol. 2014;73(2):343–8.
221. Giaccia A, Siim BG, Johnson RS. HIF-1 as a target for drug development. Nat Rev Drug Discov. 2003;2(10):803–11.
222. Caponigro F, Di Gennaro E, Ionna F, Longo F, Aversa C, Pavone E, et al. Phase II clinical study of valproic acid plus cisplatin and cetuximab in recurrent and/or metastatic squamous cell carcinoma of Head and Neck-V-CHANCE trial. BMC Cancer. 2016;16(1):918.
223. Galloway TJ, Wirth LJ, Colevas AD, Gilbert J, Bauman JE, Saba NF, et al. A phase I study of CUDC-101, a multitarget inhibitor of HDACs, EGFR, and HER2, in combination with chemoradiation in patients with head and neck squamous cell carcinoma. Clin Cancer Res. 2015;21(7):1566–73.
224. Welsh S, Williams R, Kirkpatrick L, Paine-Murrieta G, Powis G. Antitumor activity and pharmacodynamic properties of PX-478, an inhibitor of hypoxia-inducible factor-1alpha. Mol Cancer Ther. 2004;3(3):233–44.
225. Falchook GS, Wheler JJ, Naing A, Jackson EF, Janku F, Hong D, et al. Targeting hypoxia-inducible factor-1alpha (HIF-1alpha) in combination with antiangiogenic therapy: a phase I trial of bortezomib plus bevacizumab. Oncotarget. 2014;5(21):10280–92.
226. Hicks R, Rischin D, Fisher R, Binns D, Scott A, Peters L. Utility of FMISO PET in advanced head and neck cancer treated with chemoradiation incorporating a hypoxia-targeting chemotherapy agent. Eur J Nucl Med Mol Imaging. 2005;32(12):1384–91.
227. Trinkaus ME, Hicks RJ, Young RJ, Peters LJ, Solomon B, Bressel M, et al. Correlation of p16 status, hypoxic imaging using [18F]-misonidazole positron emission tomography and outcome in patients with loco-regionally advanced head and neck cancer. J Med Imaging Radiat Oncol. 2014;58(1):89–97.
228. Zschaeck S, Haase R, Abolmaali N, Perrin R, Stützer K, Appold S, et al. Spatial distribution of FMISO in head and neck squamous cell carcinomas during radio-chemotherapy and its correlation to pattern of failure. Acta Oncol. 2015;54(9):1355–63.
229. Lassen P, Eriksen JG, Hamilton-Dutoit S, Tramm T, Alsner J, Overgaard J. HPV-associated p16-expression and response to hypoxic modification of radiotherapy in head and neck cancer. Radiother Oncol. 2010;94(1):30–5.

230. Lassen P, Eriksen JG, Krogdahl A, Therkildsen MH, Ulhoi BP, Overgaard M, et al. The influence of HPV-associated p16-expression on accelerated fractionated radiotherapy in head and neck cancer: evaluation of the randomised DAHANCA 6&7 trial. Radiother Oncol. 2011;100(1):49–55.
231. Kong CS, Narasimhan B, Cao H, Kwok S, Erickson JP, Koong A, et al. The relationship between human papillomavirus status and other molecular prognostic markers in head and neck squamous cell carcinomas. Int J Radiat Oncol Biol Phys. 2009;74(2):553–61.
232. Fakhry C, Westra WH, Li S, Cmelak A, Ridge JA, Pinto H, et al. Improved survival of patients with human papillomavirus-positive head and neck squamous cell carcinoma in a prospective clinical trial. J Natl Cancer Inst. 2008;100(4):261–9.
233. Weinberger PM, Yu Z, Haffty BG, Kowalski D, Harigopal M, Brandsma J, et al. Molecular classification identifies a subset of human papillomavirus--associated oropharyngeal cancers with favorable prognosis. J Clin Oncol: official journal of the American Society of Clinical Oncology. 2006;24(5):736–47.
234. Jemal A, Simard EP, Dorell C, Noone A-M, Markowitz LE, Kohler B, et al. Annual report to the nation on the status of cancer, 1975–2009, featuring the burden and trends in human papillomavirus (HPV)–Associated cancers and HPV vaccination coverage levels. J Natl Cancer Inst. 2013;105(3):175–201.
235. Snow AN, Laudadio J. Human papillomavirus detection in head and neck squamous cell carcinomas. Adv Anat Pathol. 2010;17(6):394–403.
236. Fakhry C, Gillison ML. Clinical implications of human papillomavirus in head and neck cancers. J Clin Oncol: official journal of the American Society of Clinical Oncology. 2006;24(17):2606–11.
237. Chakravarthy A, Henderson S, Thirdborough SM, Ottensmeier CH, Su X, Lechner M, et al. Human papillomavirus drives tumor development throughout the head and neck: improved prognosis is associated with an immune response largely restricted to the oropharynx. J Clin Oncol. 2016;34(34):4132–41.
238. Mandal R, Şenbabaoğlu Y, Desrichard A, Havel JJ, Dalin MG, Riaz N, et al. The head and neck cancer immune landscape and its immunotherapeutic implications. JCI Insight. 2016;1(17):e89829.
239. Keck MK, Zuo Z, Khattri A, Stricker TP, Brown CD, Imanguli M, et al. Integrative analysis of head and neck cancer identifies two biologically distinct HPV and three non-HPV subtypes. Clin Cancer Res. 2015;21(4):870–81.
240. Cancer Genome Atlas Network. Comprehensive genomic characterization of head and neck squamous cell carcinomas. Nature. 2015;517(7536):576–82.
241. Hong A, Zhang M, Veillard AS, Jahanbani J, Lee CS, Jones D, et al. The prognostic significance of hypoxia inducing factor 1-alpha in oropharyngeal cancer in relation to human papillomavirus status. Oral Oncol. 2013;49(4):354–9.
242. Lee N, Schoder H, Beattie B, Lanning R, Riaz N, McBride S, et al. Strategy of using intratreatment hypoxia imaging to selectively and safely guide radiation dose de-escalation concurrent with chemotherapy for locoregionally advanced human papillomavirus-related oropharyngeal carcinoma. Int J Radiat Oncol Biol Phys. 2016;96(1):9–17.
243. Hodi FS, Chesney J, Pavlick AC, Robert C, Grossmann KF, McDermott DF, et al. Combined nivolumab and ipilimumab versus ipilimumab alone in patients with advanced melanoma: 2-year overall survival outcomes in a multicentre, randomised, controlled, phase 2 trial. Lancet Oncol. 17(11):1558–68.
244. Topalian SL, Hodi FS, Brahmer JR. Safety, activity, and immune correlates of anti-PD-1 antibody in cancer. N Engl J Med. 2012;366:2443–54
245. Hamid O, Sosman JA, Lawrence DP, Sullivan RJ, Ibrahim N, Kluger HM, et al. Clinical activity, safety, and biomarkers of MPDL3280A, an engineered PD-L1 antibody in patients with locally advanced or metastatic melanoma (mM). ASCO Meeting Abstracts. 2013;31(15_suppl):9010.

246. Seiwart T, Burtness B, Weiss J, Gluck I, Eder J, Pai S, et al. A phase Ib study of MK-3475 in patients with human papillomavirus (HPV)-associated and non-HPV–associated head and neck (H/N) cancer. J Clin Oncol. 2014;32(suppl; abstr 6011):5s.
247. Labiano S, Palazon A, Melero I. Immune response regulation in the tumor microenvironment by hypoxia. Semin Oncol. 2015;42(3):378–86.
248. Huber V, Camisaschi C, Berzi A, Ferro S, Lugini L, Triulzi T, et al. Cancer acidity: an ultimate frontier of tumor immune escape and a novel target of immunomodulation. Semin Cancer Biol. 2017;43:74–89.
249. Kareva I, Hahnfeldt P. The emerging "hallmarks" of metabolic reprogramming and immune evasion: distinct or linked? Cancer Res. 2013;73(9):2737–42.
250. Chouaib S, Noman MZ, Kosmatopoulos K, Curran MA. Hypoxic stress: obstacles and opportunities for innovative immunotherapy of cancer. Oncogene. 2017;36(4):439–45.
251. Caldwell CC, Kojima H, Lukashev D, Armstrong J, Farber M, Apasov SG, et al. Differential effects of physiologically relevant hypoxic conditions on T lymphocyte development and effector functions. J Immunol. 2001;167(11):6140–9.
252. Nakagawa Y, Negishi Y, Shimizu M, Takahashi M, Ichikawa M, Takahashi H. Effects of extracellular pH and hypoxia on the function and development of antigen-specific cytotoxic T lymphocytes. Immunol Lett. 2015;167(2):72–86.
253. Lugini L, Matarrese P, Tinari A, Lozupone F, Federici C, Iessi E, et al. Cannibalism of live lymphocytes by human metastatic but not primary melanoma cells. Cancer Res. 2006;66(7):3629–38.
254. Tittarelli A, Janji B, Van Moer K, Noman MZ, Chouaib S. The selective degradation of synaptic connexin 43 ;protein by hypoxia-induced autophagy impairs natural killer cell-mediated tumor cell killing. J Biol Chem. 2015;290(39):23670–9.
255. Mancino A, Schioppa T, Larghi P, Pasqualini F, Nebuloni M, Chen IH, et al. Divergent effects of hypoxia on dendritic cell functions. Blood. 2008;112(9):3723–34.
256. Ward RC, Kaufman HL. Targeting costimulatory pathways for tumor immunotherapy. Int Rev Immunol. 2007;26(3–4):161–96.
257. Yang M, Ma C, Liu S, Sun J, Shao Q, Gao W, et al. Hypoxia skews dendritic cells to a T helper type 2-stimulating phenotype and promotes tumour cell migration by dendritic cell-derived osteopontin. Immunology. 2009;128(1pt2):e237–e49.
258. Noy R, Pollard Jeffrey W. Tumor-associated macrophages: from mechanisms to therapy. Immunity. 2014;41(1):49–61.
259. Doedens AL, Stockmann C, Rubinstein MP, Liao D, Zhang N, DeNardo DG, et al. Macrophage expression of hypoxia-inducible factor-1 alpha suppresses T-cell function and promotes tumor progression. Cancer Res. 2010;70(19):7465–75.
260. Corzo CA, Condamine T, Lu L, Cotter MJ, Youn JI, Cheng P, et al. HIF-1alpha regulates function and differentiation of myeloid-derived suppressor cells in the tumor microenvironment. J Exp Med. 2010;207(11):2439–53.
261. Noman MZ, Desantis G, Janji B, Hasmim M, Karray S, Dessen P, et al. PD-L1 is a novel direct target of HIF-1alpha, and its blockade under hypoxia enhanced MDSC-mediated T cell activation. J Exp Med. 2014;211(5):781–90.

Chapter 16
Angiogenesis and Anti-angiogenic Therapy in Head and Neck Cancer

Lindsay Wilde, Jennifer Johnson, and Athanassios Argiris

Abstract The formation of new blood vessels, or angiogenesis, takes place through a variety of different physiologic and unique pathologic processes in tumor tissue. While the control mechanisms of some of these processes are not yet understood, as in the case of de novo vessel formation or intussusceptive angiogenesis, a closer examination of the process of sprouting angiogenesis highlights the complexity of the molecular mechanisms of angiogenesis. Through both positive regulation with proteins such as vascular endothelial growth factors (VEGFs), fibroblast growth factor (FGF), and NOTCH and negative regulation with other signals such as thrombospondin, endostatin, and angiostatin, the endothelial cells of an existing vessel can reorganize into new functional luminal architecture. As our comprehension of the regulatory machinery has improved, so has the desire to create anti-angiogenic therapies using targeted monoclonal antibodies, tyrosine kinase inhibitors, and other novel targeted small molecular inhibitors directed at interrupting these regulatory signals. Drugs directed against vascular endothelial growth factors in particular (e.g., bevacizumab or sunitinib) have been studied as antineoplastic agents either alone or in combination with cytotoxic chemotherapy or other targeted agents. A unique toxicity profile has been seen with anti-angiogenics that may include events such as bleeding, hypertension, and proteinuria. The interest in targeting angiogenesis continues, and more clinical trials are underway with new targets and evolving strategies.

Keywords Squamous cell carcinoma of the head and neck · Angiogenesis · VEGF · Anti-angiogenic therapy · Bevacizumab · Tyrosine kinase inhibitors

L. Wilde · J. Johnson · A. Argiris (✉)
Department of Medical Oncology, Thomas Jefferson University, Philadelphia, PA, USA
e-mail: Lindsay.Wilde@jefferson.edu; Jennifer.M.Johnson@jefferson.edu; Athanassios.Argiris@jefferson.edu

B. Burtness, E. A. Golemis (eds.), *Molecular Determinants of Head and Neck Cancer*, Current Cancer Research, https://doi.org/10.1007/978-3-319-78762-6_16

16.1 Introduction

Angiogenesis plays an integral role in the development and propagation of human cancers, including squamous cell carcinoma of the head and neck (HNSCC), making it an attractive therapeutic target. However, the use of anti-angiogenic agents in HNSCC has proven challenging, showing only modest benefits in clinical trials to date. There is continued interest in developing a better understanding of the process of angiogenesis in an effort to employ anti-angiogenic treatments more successfully. Further clinical trials with anti-angiogenic agents in HNSCC are ongoing, seeking to capitalize on these insights. Here, we examine the processes of normal physiologic angiogenesis and tumor angiogenesis, discuss the importance of angiogenesis in HNSCC, and review the status of clinical research on anti-angiogenic agents in HNSCC.

16.2 Physiological Angiogenesis

Blood vessels are essential for the delivery of oxygen and nutrients and the removal of waste products to and from tissues, respectively. There are several mechanisms by which blood vessels are formed. Some of these processes, such as vasculogenesis, intussusceptive angiogenesis, and sprouting angiogenesis, occur in both normal tissues and tumors. Others, such as co-option of existing vessels, vascular mimicry, and tumor cell differentiation into endothelial cells, occur only in tumors [1].

Vasculogenesis, the de novo formation of blood vessels, begins in the developing embryo and is marked by the differentiation of angioblasts into endothelial cells (ECs) [2, 3]. Through a series of complex steps, the ECs align into a network of vessels from which additional vessels can then be formed [2]. Although initially thought to be active only during the prenatal period, it is now accepted that vasculogenesis occurs throughout life, both in physiologic and pathologic states [2]. The molecular regulation of vasculogenesis, which is still being elucidated, shares some features with angiogenesis and involves many of the same pathways [4].

Intussusceptive angiogenesis (IA) is the process by which transluminal tissue pillars arise from existing vessels, which have previously been created via vasculogenesis or sprouting angiogenesis, and fuse to form new or remodeled vessels [5]. Although IA is utilized in embryonic development, in physiologic neovascularization, and in tumorigenesis, the molecular mechanisms controlling this process are yet to be elucidated [6].

Sprouting angiogenesis, the best-understood method of vessel formation (Fig. 16.1), will be discussed here in greater detail as a process relevant to HNSCC. Under normal physiologic conditions, endothelial cells are rendered quiescent by autocrine signals from vascular endothelial growth factor (VEGF), NOTCH, and angiopoietin-1 (ANG-1) [1]. These endothelial cells, which are ensheathed by, and share a basement membrane with, pericytes, possess the ability to sense hypoxic conditions and respond to angiogenic stimuli [7, 8].

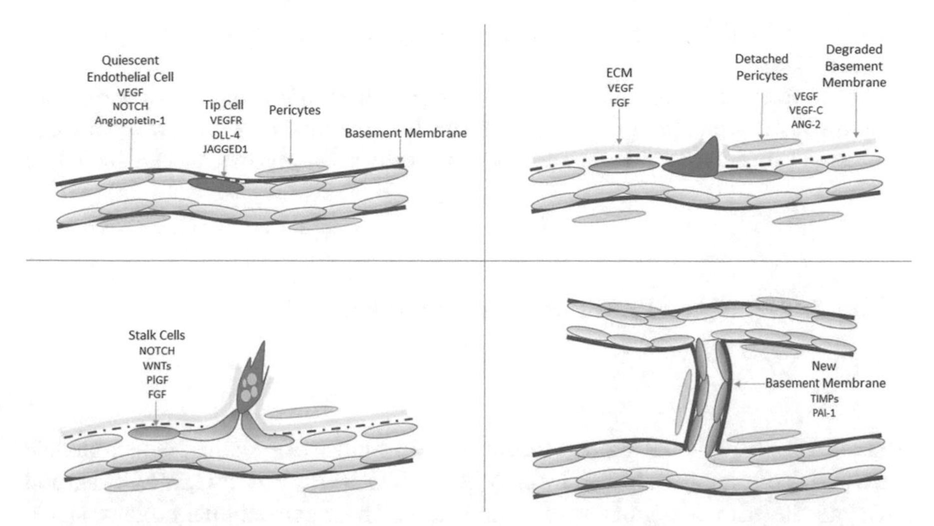

Fig. 16.1 Sprouting angiogenesis. The process of sprouting angiogenesis begins when hypoxic conditions or angiogenic stimuli are sensed, disrupting the normally quiescent cells. (1) Proteolytic degradation of the basement membrane (dark blue) and detachment of the pericytes (gold) are followed by extrusion of plasma proteins which form a sort of temporary basement membrane. (2) One lead endothelial cell, the tip cell (green), will lengthen in response to further angiogenic signaling, and adjacent cells will begin to mobilize as stalk cells. (3) Guided by VEGF gradients, the tip cell will align with a target within in the tissue environment. (4) These sprouting cells will eventually vacuolize forming a lumen, a basement membrane will again ensheath the endoluminal cells, and pericytes will join tightly to complete the new vessel

When an angiogenic signal such as VEGF, VEGF-C, or ANG-2 is sensed, proteolytic degradation of the vascular basement membrane occurs, and pericytes detach from the vessel wall [1, 9]. As the existing vessel dilates, the permeability of the endothelial cell layer is increased, and a temporary extracellular matrix (ECM) is produced from extravasated plasma proteins. This ECM acts as a platform upon which endothelial cells align themselves in response to integrin signaling. The ECM also stores signaling molecules such as VEGF and fibroblast growth factor (FGF), which are then released and serve to further stimulate the angiogenic process [10].

When these angiogenic signals are propagated, one endothelial cell, the so-called tip cell, becomes the leader for new vessel formation. The tip cell is selected in the presence of a variety of factors including VEGF receptors and the NOTCH ligands DLL-4 and JAGGED1 and can respond to environmental stimuli for guidance [10, 11]. Cells adjacent to the tip cell, called stalk cells, are under the influence of NOTCH, WNTs, placental growth factor F (PlGF), and FGFs and divide to lengthen the stalk and form a lumen. Stalk cells signal adjacent cells about their position through the release of molecules such as EGFL7 and, as such, direct elongation of the forming vessel [1, 11, 12]. The nascent vessel fuses with a mature vessel with the help of myeloid bridge cells, and blood flow begins [13]. Once the new vessel is formed, endothelial cells are again covered by pericytes and rendered quiescent. A basement membrane is formed through the action of protease inhibitors (TIMPs and

PAI-1), and stabilization of the cellular junctions occurs, concluding the process [1] (Fig. 16.1).

As expected, the molecular controls of angiogenesis are complex. However, our evolving understanding of these mechanisms has provided us with new therapeutic targets and improved outcomes for patients with a variety of cancers, including HNSCC.

16.3 Molecular Signaling in Angiogenesis

16.3.1 VEGF

The vascular endothelial growth factor (VEGF) family of proteins, which include VEGF-A (commonly referred to as VEGF), VEGF-B, VEGF-C, VEGF-D, and PlGF, are the primary regulators of angiogenesis. These extracellular proteins signal through their associated tyrosine kinase receptors, located at the plasma membrane. VEGFs exist as homodimeric polypeptides and can be spliced to generate polypeptides of varying sizes and functionalities [14]. The VEGF gene is located on chromosome 6p21.3 and consists of an ~14 kb coding region spanning eight exons [15, 16]. The VEGF isoforms are created by alternative pre-mRNA splicing of these eight exons [17]. The biologic properties of the isoforms vary, particularly with respect to their ability to bind heparin or extracellular matrix, rendering some pro-angiogenic and some anti-angiogenic [17, 18]. All VEGFs share a common core domain that consists of eight fixed cysteine residues, six of which form S-S intramolecular bonds creating three loop structures while the remaining two form intermolecular S-S bonds and influence the homodimeric structure [19, 20]. Many members of the VEGF family are upregulated by hypoxia inducible factor (HIF), specifically HIF-1α, and are typically induced in times of tissue growth (either normal or aberrant) [21, 22]. Other factors that have been shown to upregulate VEGF expression include STAT3, PI3K, RAS, and p53 [23, 24]. VEGFs are produced by most parenchymal cells and can act in both a paracrine and autocrine manner [25–27].

Three VEGF receptor (VEGFR) tyrosine kinases have been identified thus far (Fig. 16.2). Each is comprised of seven extracellular immunoglobulin-like domains for ligand binding, a transmembrane domain, and an intracytoplasmic tyrosine kinase domain [19, 28]. The VEGFRs are structurally similar to other families of tyrosine kinases such as platelet-derived growth factor receptors (PDGFRs) but differ in the amino acid sequences found in their tyrosine kinase domains, contributing to their differing downstream signaling effects [19, 29]. In general, VEGFR1 (also known as Flt-1) is expressed in monocytes and macrophages, VEGFR2 (Flk-1 and KDR) in vascular endothelial cells, and VEGFR3 (Flt-4) in lymphatic endothelial cells, although variation in this pattern does exist [30]. When a VEGF ligand binds to its associated receptor, dimerization is induced, which leads to a conformational change that exposes the intracellular ATP binding site and activates kinase function [14]. This ultimately leads to downstream signaling and regulation of a variety of

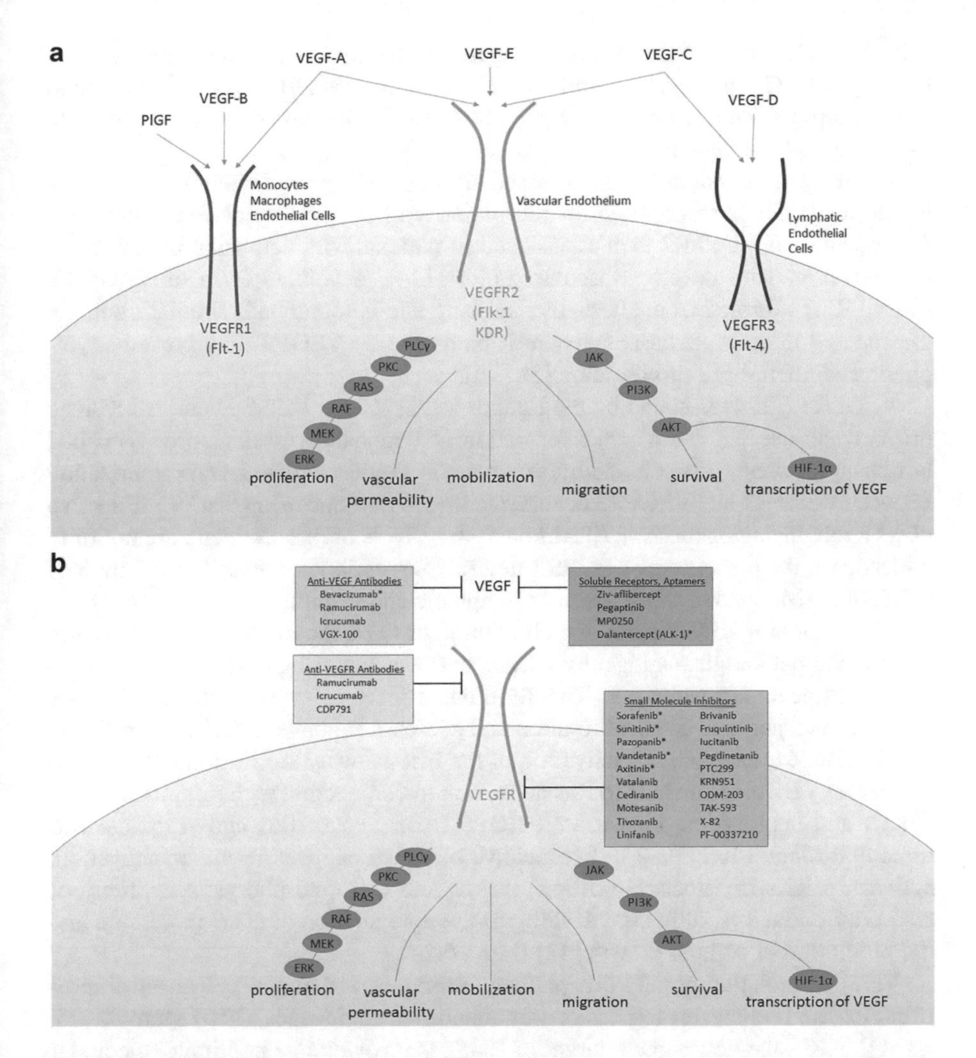

Fig. 16.2 The VEGF family. (**a**) The VEGF receptors are capable of binding multiple ligands including VEGF-A–E and PlGF, thereby activating downstream pathways that effect vasculogenesis, angiogenesis, and lymphangiogenesis. (**b**) Targeted blockade of these signals can be accomplished through monoclonal antibodies directed against either the receptor or the ligand, soluble receptor fragments that "trap" circulating ligand, or small molecule inhibitors to the pathway components. An * denotes agents that have been studied in HNSCC

endothelial cell functions including proliferation, vascular permeability, and migration [14, 30, 31].

In addition to expression on monocytes and macrophages, VEGFR1 is expressed widely on the endothelium. Ligands for VEGFR1 include VEGF, VEGF-B, and PlGF [14]. It has been shown that although VEGFR1 binds with high affinity to VEGF, its tyrosine kinase activity is less than that of VEGFR2, the major receptor involved in VEGF signaling and angiogenesis [14, 32]. The role of VEGFR1 has not fully been elucidated, but it has been suggested that this high-affinity, low-activity binding to VEGF may help to regulate the amount of free VEGF available to bind to other receptors. This is supported by the observation that loss of VEGFR1 leads to the overgrowth of vessels [33].

VEGFR2 is predominantly expressed on the vascular endothelium and, when bound to VEGF, is responsible for the majority of pro-angiogenic signaling [19]. The potency of VEGFR2 is demonstrated in patients with activating mutations of this receptor, who develop vascular tumors [34]. A soluble form of VEGFR2, sVEGFR2, is formed from alternative splicing and is found both circulating in the plasma and in multiple tissues, where it binds to free VEGF-C and prevents lymphatic endothelial cell proliferation [30, 35].

VEGFR3 (Flt-4) is bound by the ligands VEGF-C and VEGF-D and is involved primarily in the regulation of the formation of lymphatic vessels, a process called lymphangiogenesis [36]. In addition, there is evidence that heterodimerization between VEGFR2 and VEGFR3 occurs, leading to the conclusion that VEGFR3 also plays a role in angiogenesis [37]. Although the effects of this cross talk are not fully understood, the presence of VEGFR3 neutralizing antibodies leads to inhibition of heterodimerization and suppression of lymphatic and vascular formation [14, 38].

VEGF binding to VEGFR2, which is thought to be the major mediator of both normal and pathologic angiogenesis, leads to downstream signaling through a variety of intracellular pathways. The terminal effects of these pathways include increased cell survival and proliferation and decreased apoptosis. Cellular proliferation is induced via VEGFR2 activation of the Erk pathway. Activation can occur in two ways: VEGFR2 can lead to Ras activation in conjunction with protein kinase C (PKC) and sphingosine kinase, or VEGFR2 can bypass Ras and stimulate Raf directly through PKC [39–41]. Irrespective of which pathway results in activation, activated ERKs translocate to the nucleus where they phosphorylate a variety of transcription factors, ultimately leading to the upregulation of gene expression and the promotion of cellular growth [42] (Fig. 16.2a).

VEGF stimulation of the PI3K pathway promotes endothelial cell survival, proliferation, and cell cycle progression both in normal angiogenesis and in tumors [35, 43, 44]. AKT, the downstream target of PI3K, is involved in a multitude of cellular functions including cellular metabolism, protein synthesis, cell cycle activation, and inhibition of apoptosis and plays a major role in angiogenesis [45]. Additionally, activation of the PI3K pathway can lead to increased production of HIF-1α, causing increased expression of VEGF and perpetuation of vascular proliferation [35, 46] (Fig. 16.2a).

VEGF has a downstream effect on a variety of other proteins that have been identified as playing a role in angiogenesis. For example, p38MAPK is involved in modulation of the endothelial cytoskeleton and cellular migration, and Rac, a small-GTP-binding protein, has been shown to effect vascular permeability [47, 48]. The exact mechanisms of these interactions, as well as others, are still being elucidated.

16.3.2 PlGF

Placenta-derived growth factor (PlGF) is also a member of the VEGF subfamily and binds to VEGFR1. While it is produced in the placental trophoblasts and has its main role in embryogenesis, it is also expressed in normal physiologic conditions in other tissues such as the heart, lung, thyroid, adipose tissue, and skeletal muscle. It has been implicated as contributing to pathologic but not physiologic angiogenesis [49]. The 3D structure of PlGF is extremely similar to that of VEGF-A. Despite this similarity, PlGF binds exclusively to VEGFR1, unlike VEGF-A which can also directly bind to and activate VEGFR2. It may indirectly activate VEGFR2 by displacing VEGF-A from VEGFR1 and allowing it to bind VEGFR2 or potentially through transphosphorylation. PlGF/VEGFR1 is a key element of neo-angiogenesis in hypoxic conditions; however, more recently it has also been shown to recruit and activate macrophages which, in turn, secrete pro-inflammatory and pro-angiogenic cytokines such as IL-1 and TNF-α [49, 50]. Clinically, the presence of PlGF and VEGFR1 activation has been associated with the promotion of pulmonary metastases [51, 52]. Further exploration of the role of this protein in physiologic conditions and in pathologic conditions is ongoing.

16.3.3 PDGFs

Complementing the activity of the VEGF proteins, platelet-derived growth factors (PDGFs) are involved in vessel maturation and stabilization [1]. Five different PDGF isoforms exist, consisting of a combination of disulfide-linked dimers produced by four different genes. These isoforms are synthesized as precursor molecules and cleaved to form active growth factors [53]. Once active, PDGFs bind to their conjugate tyrosine kinase receptors, namely, PDGFRα and PDGFRβ [54]. The PDGFRs are similar in structure and consist of extracellular immunoglobulin-like domains and intracellular kinase domains. When a ligand is bound, the PDGFR dimerizes causing intracellular autophosphorylation and downstream signaling [53]. Physiologic signaling via PDGFRs activates many of the same pathways that are involved in VEGF signaling. Specifically, ligand binding to PDGFRβ has been shown to initiate signaling through the AKT and ERK pathways among others, leading to cell migration, proliferation, and angiogenesis [55].

16.3.4 Other Key Pathways in Angiogenesis

A variety of other molecules and pathways have been shown to play key roles in angiogenesis. For example, NOTCH and WNT signaling are involved in tip and stalk cell control, and TGF-β signaling is active in the modulation of vascular smooth muscle differentiation [1]. Activin receptor-like kinase-1 (ALK1) is a type 1 receptor in the TGF-β superfamily that is expressed on activated endothelial cells. Activation of ALK1 occurs when bound to its ligands BMP9 and BMP10. Once activated, ALK1 signaling leads to downstream phosphorylation of members of the SMAD family of proteins. SMAD proteins in turn activate cellular programs distinct from VEGF and involved in blood vessel maturation and stabilization and are engaged in cross talk with the VEGF and NOTCH pathways [56, 57]. ALK1 is present in the vasculature of multiple tumor types, including SCCHN [58]. Its upregulation in tumor angiogenesis has made it an attractive therapeutic target.

EGFR is a tyrosine kinase receptor and member of the ErbB family with some relevance to angiogenesis. It is expressed on most normal tissues, especially those of endothelial origin, and is activated by the binding of a variety of ligands including epidermal growth factor (EGF) and transforming growth factor-α (TGF-α) [59]. As with VEGFR and PDGFR, ligand binding leads to dimerization of the receptor and internal signal transduction. Once signaling occurs, the activated receptor is endocytosed in a clathrin-coated pit and either recycled to the cell surface or ubiquitinated and degraded [60]. EGFR is involved in a complex network of signaling pathways, including PI3K, Ras, and JAK/Stat, and its activation can lead to cell cycle progression, cellular proliferation, migration, and angiogenesis [59, 60]. Preclinical studies have shown that EGFR activation leads to endothelial cell migration and tube formation as well as smooth muscle cell recruitment and vessel maturation [61]. EGFR inhibition is a validated therapeutic strategy in HNSCC (see Chap. 2).

Fibroblast growth factor-1 (FGF-1), also known as acidic FGF (aFGF), is a member of the fibroblast growth factor family. It is involved in physiological processes including embryonal development, cell growth, morphogenesis, and tissue repair as well as pathological tumor growth and invasion. FGF-1 is thought to be the angiogenic factor specifically responsible for endothelial cell (EC) migration and proliferation. Stored in the vascular basement membrane, FGF-1 is upregulated during active angiogenesis. It binds to the receptor tyrosine kinases FGFR-1 and FGFR-2. Binding to FGFR-1 promotes EC migration and capillary morphogenesis and activates multiple other signaling pathways. Binding to FGFR-2 mediates proteolysis of matrix components and induction of the synthesis of collagen, fibronectin, and proteoglycans [62]. FGFRs are often overexpressed in tumors, and mutations in the FGFR genes have been noted, making this another potential therapeutic target [63].

A wide array of chemokines and cytokines including prostaglandins, COX-2, and IL-6 are also involved in the regulation of angiogenesis and may be viable therapeutic targets for control of this process [20, 64, 65]. Hepatocyte growth factor (HGF) is a mesenchymal cytokine that is a potent mitogen, pro-motility factor, and stimulator of angiogenesis [66]. It has been shown to stimulate angiogenesis both through the upregulation of VEGF expression and through the downregulation of thrombospondin, an inhibitor of angiogenesis [67].

The integrin family of adhesion receptors is also active in the regulation of angiogenesis. These molecules are heterodimeric membrane glycoproteins that are present on endothelial cells and serve as receptors for various proteins present in the extracellular matrix [68]. While many integrins exist, αvβ3 and αvβ5 have been specifically implicated in angiogenesis. Both are expressed on the endothelium of blood vessels, and their presence is necessary for angiogenesis induced by FGF, VEGF, and other cytokines [69]. Tumor cells have also been shown to express these integrins, leading to enhanced angiogenesis and tumor growth [70, 71].

A number of endogenous anti-angiogenic factors have also been identified. Some of these inhibitors, such as thrombospondins and endostatin, are matrix-derived, while others, such as angiostatin, are present in the circulation [72, 73]. These factors act to maintain the normal physiologic balance of angiogenesis; however, they also play a role in regulating tumor progression and metastasis [73]. The endogenous anti-angiogenics are under investigation in both animal and human studies as novel anticancer agents [74].

16.4 Angiogenesis and Tumorigenesis

Tumors rely on neovascularization for nutrient delivery and growth. Early studies looking at angiogenesis revealed that tumors are unable to grow larger than a few millimeters in size without inducing their own blood supply [75]. This process, termed the angiogenic switch, refers to a discrete time in tumorigenesis when pro-angiogenic factors outweigh anti-angiogenic factors and vessel growth begins to occur [76]. Initially, mouse models confirmed the theory that angiogenesis is initiated early in tumor development and is likely a rate-limiting step in the development of multiple different tumor types [75]. Subsequent in vitro studies confirmed these findings in human cancer cells, including those derived from the pancreas, cervix, breast, dermis, and epidermis [75].

Once the angiogenic switch occurs, tumor cells are able to initiate and perpetuate a cascade of events that lead to new vessel formation [77]. The bulk of neovascularization occurs through sprouting angiogenesis, although studies have shown that tumors are able to induce other physiologic mechanisms such as vasculogenesis as well [76, 77]. Tumor angiogenesis differs from physiologic angiogenesis, however, in that it does not result in the formation of stable, organized vessels. Instead, the vessels are abnormal in structure and function, lacking the ability to become quiescent and often developing leaks or disruptions in blood flow [78].

Some tumors are also able to induce angiogenesis outside of the physiologic pathways of sprouting and vasculogenesis. Vessel co-option, for example, is the process by which a tumor appropriates a pre-existing vessel for its own use [79]. Vascular mimicry, on the other hand, occurs when tumor cells undergo endothelial differentiation and organize into a vascular-like network [80, 81]. Tumors form vascular structures without the contribution of host endothelial cells. These structures are not true blood vessels bur rather patterned, ECM-rich matrices that mimic the function of blood vessels. These channels are capable of transport of both fluids and erythrocytes. This process is rare and has been identified in a variety of solid tumors, including melanoma, breast, lung, bladder, and sarcoma [82]. The signaling processes involved in vascular mimicry are complex, with contributions from VE-cadherin, PI3K, HIF1-α, NOTCH, and others [83, 84]. It has been proposed that these are potential mechanisms by which tumors evade anti-angiogenic therapy [85].

VEGF, which is expressed at normal or elevated levels by tumor cells, is the major driver of tumor angiogenesis and is induced, at least in part, by the hypoxic environment of the tumor and HIF-1α [78, 86]. Animal models have shown that VEGF can act in an autocrine manner if the tumor cells express VEGFRs or in a paracrine fashion if neighboring endothelial cells express the receptors [86]. The ultimate result of increased VEGF expression is activation of the multitude of pathways described above as integral in angiogenesis. Together with contributions from additional factors including FGF-1, PDGF-B and C, and COX-2 which are also upregulated in tumor cells, these signals contribute to neovascularization and tumor growth [76].

In HNSCC and other tumors, EGFR signaling indirectly leads to tumor angiogenesis by causing increased production of VEGF and other pro-angiogenic factors [61]. In vitro studies have demonstrated that EGFR regulates the expression of these factors through transcriptional and posttranscriptional mechanisms [87]. Evidence also suggests that EGF ligands have a direct effect on EGFR-expressing microvascular endothelial cells, which differ from endothelial cells in the microvasculature both in structure and molecular composition [87–90]. These microvascular endothelial cells are found in normal human tissue but have also been isolated from various tumors and may play a role in tumor growth and response to therapy [91, 92]. Blockade of EGFR signaling on these cells has been shown to impede angiogenesis [89].

16.5 Angiogenesis and the Immune System

The tumor microenvironment, including specifically components of the immune system, are in dialogue with factors involved in angiogenesis. For example, VEGF contributes to immune suppression in multiple ways: by binding to VEGFR1 on CD34+ hematopoietic progenitor cells and preventing their differentiation into mature immune cells, by inducing programmed death-ligand 1 (PD-L1) expression on dendritic cells which leads to decreased T cell activation, by decreasing T cell adhesion and extravasation through vessel walls, and by increasing Treg differentiation [93–96]. These discoveries have clearly identified VEGF as a link between tumor angiogenesis and immune system evasion.

Research has also shown that, in addition to a tumor's ability to induce neovascularization directly via the secretion of pro-angiogenic factors, tumors also recruit immune cells to the microenvironment and induce them to produce additional pro-angiogenic molecules [97–99]. The most prominent of these immune cells are the tumor-associated macrophages (TAMs) [100]. Now known to be a heterogeneous group of myeloid cells, TAMs are involved in a variety of processes including immune modulation and induction of the angiogenic switch [101]. Multiple studies have demonstrated that TAMs secrete pro-angiogenic cytokines including VEGF, TNF-α, basic fibroblast growth factor, thymidine phosphorylase, and urokinase-type plasminogen activator [102, 103]. TAMs also secrete anti-angiogenic factors such as thrombospondin indicating that they may also play a role in tumor remodeling [104]. Preclinical models have demonstrated that depletion of TAMs from a tumor can lead to decreased angiogenesis and slower tumor growth [105–107].

Early studies attempting to elucidate the role of TAMs in angiogenesis utilized a knockout mouse that was null for colony-stimulating factor-1 (CSF-1). It was noted that mammary tumors in CSF-1 null mice failed to undergo the angiogenic switch, leading to a marked decrease in tumor growth and progression [105]. Subsequent studies have shown that TAMs, which express the receptor for CSF-1 (CSF1R), are recruited by activation of the CSF1R signaling pathway and help to perpetuate angiogenesis in tumors [100]. In addition, TAMs express HIF1-α and are able to sense the hypoxic milieu of the tumor and respond by producing VEGF, again contributing to neovascularization [102].

16.6 Angiogenesis in Head and Neck Cancer

Angiogenesis has been extensively investigated in multiple solid tumor types, yielding in many cases therapeutic advances. The remainder of this chapter will focus on the role of angiogenesis in HNSCC development and progression and the potential of anti-angiogenic agents in this disease.

HNSCCs are a heterogeneous group both anatomically and biologically. However, the upregulation of angiogenesis is a common feature among all tumors of the head and neck, and the degree to which this occurs can be important in both prognosis and treatment. Increased VEGF expression is viewed as the most important inducer of angiogenesis and has been linked to poor prognosis in HNSCC; up to 90% of HNSCCs express VEGF [108, 109]. A meta-analysis conducted by Kyzas et al. addressed 17 studies evaluating the relationship between VEGF expression and prognosis in HNSCC [110]. They found that patients whose primary tumors overexpressed VEGF, as measured by immunohistochemistry, had a 1.88-fold higher mortality. Studies have also attempted to link VEGF overexpression to lymph node metastases in HNSCC; however, results have been mixed, and no clear conclusions can be drawn [110–113].

EGFR has also been shown to be upregulated in 70–100% of HNSCCs and correlated with poor prognosis [114, 115]. When activated, EGFR stimulates angiogenesis through a variety of pathways including PI3K and Erk. Furthermore, in a study by Wang et al., EGFR inhibition in mice led to a rapid decrease in HIF1-α, another integral mediator of vessel formation, highlighting the multitude of ways in which EGFR plays a role in angiogenesis [116]. VEGF upregulation independent of EGFR signaling may contribute to resistance to EGFR inhibition, and combination strategies have shown potential synergistic interactions between anti-EGFR and anti-VEGFR therapies [117, 118]. Plasma levels of EGFR, VEGF, and other pro-angiogenic factors have been studied as biomarkers for both the diagnosis and prognosis of HNSCC [119–121]. Results have varied, but higher levels of VEGF in the circulation do appear to be correlated to an increased risk of disease progression after up-front therapy [122].

Hypoxia is an important mediator of tumor behavior in many cancers, including HNSCC. Studies have demonstrated that both hypoxic and normoxic conditions can cause HIF1-α upregulation in HNSCC cells, and this is associated with a more aggressive tumor phenotype [123, 124]. It has also been reported that HIF1-α overexpression predicts for a decreased response to local radiation therapy in early-stage glottic cancer [125]. Conversely, there have been some studies suggesting that HIF1-α expression is actually correlated with positive outcomes in some subtypes of HNSCC [126, 127]. Further studies evaluating the prognostic and predictive significance of HIF1-α expression are ongoing.

Among the major molecular regulators of angiogenesis, the role of PDGFR in HNSCC remains most unclear. Although PDGFR is expressed on HNSCC cells, given its widespread distribution in normal tissue, the significance of this is unknown [128]. More studies are needed to better understand how PDGFR affects angiogenesis in HNSCC.

A variety of other pro-angiogenic cytokines including IL-8, IL-6, GM-CSF, and FGF are overexpressed in HNSCC [129]. A study published by Ninck et al. revealed that HNSCC cells themselves produce these cytokines in various patterns and quantities and that those tumors that secreted greater amounts had higher microvascular density and behaved more aggressively [130]. Expanding on this, Byers et al. retrospectively analyzed a number of cytokines and angiogenic factors in tumor samples from patients with HNSCC and found that eight markers, including VEGF, Gro-α (CXCL1), IL-4, IL-8, osteopontin, eotaxin, G-CSF, and SDF-1α, were expressed at higher baseline levels in patients whose disease ultimately progressed after initial therapy when compared to those patients whose disease did not progress [122]. This suggests that characterizing pro-angiogenic chemical or molecular signatures may help to refine prognosis and guide treatment decisions for HNSCC.

Cancer-associated fibroblasts (CAFs), which are fibroblasts that arise and evolve within the tumor microenvironment, have also been shown to contribute to the pro-angiogenic milieu [131]. This occurs, in part, through the upregulation of cytokines such as hepatocyte growth factor (HGF), which has been shown to work synergistically to stimulate angiogenesis with VEGF, and transforming growth factor beta (TGF-β), which contributes to angiogenesis both through its effects on endothelial cells and on immune response cells [132–134]. CAFs also produce CXCL12, which binds to its receptor, CXCR4, causing the downstream effects of increased HIF-1α expression and angiogenesis [132, 135].

Tumor immunity influences angiogenesis in HNSCC [136, 137]. As in other cancers such as the breast and lung, TAMs are present in high levels in HNSCC [132, 138, 139]. Some research has suggested that HNSCC tumors with higher numbers of TAMs are associated with more aggressive disease, invasion, and poorer prognosis [140–142]. TAMs in HNSCC have been found to produce high levels of TGF-β, VEGF, and VEGFRs which exert pro-angiogenic effects on the tumor microenvironment [143].

Although attempts to develop prognostic or predictive models based on molecular markers in HNSCC have been made, variations in sampling, laboratory techniques, and tumor heterogeneity have limited their utility up to this point [144, 145]. In the future, it is likely that markers of angiogenesis will play an increasingly important role in the diagnosis and management of HNSCC.

16.7 Methods of Targeting Angiogenesis

Preclinical models have demonstrated that inhibition of VEGF, EGFR, and other pro-angiogenic factors can lead to decreased vascular proliferation and inhibition of tumor growth [146–148]. Given the prominent role that angiogenesis plays in tumorigenesis, it is understandable that it has been an attractive therapeutic target (Fig. 16.2b). Numerous anti-angiogenic drugs have been developed and tested in clinical trials (Table 16.1). These drugs generally fall into two classes: monoclonal antibodies and tyrosine kinase inhibitors.

The development of therapeutic antibodies for the treatment of cancer and other diseases became a possibility in the mid-1970s with the creation of hybridoma technology [149]. However, it was not until the 1990s that monoclonal antibodies entered the clinic for cancer treatment. Monoclonal antibodies function by binding to a specific ligand or receptor on a cell surface and blocking signaling via the associated pathway or pathways [150]. Since their development, monoclonal antibodies targeting angiogenesis have been evaluated in combination with chemotherapy, radiation, and other targeted agents in locally advanced, recurrent, and metastatic HNSCC.

Table 16.1 Anti-angiogenic agents currently in use for the treatment of solid tumors

Agent	Drug class	Molecular targets
Bevacizumab	Monoclonal antibody	VEGF
Ramucirumab	Monoclonal antibody	VEGFR2
Ziv-aflibercept	Monoclonal antibody	VEGF, VEGF-B, PlGF
Dalantercept	Monoclonal antibody	ALK1
Sorafenib	TKI	RAF/MEK/ERK, VEGFR1–3, PDGFR- β, c-Kit, FLT3, RET
Sunitinib	TKI	VEGFR1 and VEGFR2, PDGFR-α and PDGFR-β, c-Kit, RET, CSF1R, FLT3
Vandetanib	TKI	VEGFR2 and VEGFR3, EGFR, RET
Pazopanib	TKI	VEGFR1–3, PDGFR-α and PDGFR-β, FGFR-1 and FGFR-3, c-Kit
Axitinib	TKI	VEGFR1–3, PDGFR-α and PDGFR-β, c-Kit
Regorafenib	TKI	VEGFR1–3

Abbreviations: *TKI* tyrosine kinase inhibitor

The prototypical example of an anti-angiogenic monoclonal antibody is bevacizumab, a humanized monoclonal anti-VEGF antibody that was developed for the treatment of solid tumors. Preclinical studies demonstrated it to be effective not only at decreasing tumor growth but also at decreasing tumor vascularity and increasing cellular apoptosis [151]. For example, Fujita et al. conducted an in vitro study in HNSCC that tested the effects of bevacizumab alone and in combination with paclitaxel. They found that while bevacizumab alone was able to inhibit tumor growth, bevacizumab in combination with paclitaxel caused a greater decrease in vessel density and an increase in apoptosis [152]. However, the mechanism for this is not fully understood. Somewhat paradoxically, bevacizumab and other anti-VEGF therapies have also been shown to stabilize and mature tumor vasculature, leading to lower interstitial fluid pressure and increased tumor blood flow [27, 153, 154]. It has been hypothesized that this process, termed vascular normalization, occurs because anti-angiogenic agents help to reestablish a balance between pro- and anti-angiogenic factors in the tumor [155, 156]. This may reduce tumor hypoxia and lead to improved delivery of chemotherapy to tumor tissue, providing further explanation of the synergistic effect bevacizumab has with other agents. Bevacizumab was first FDA approved in 2004 for the treatment of metastatic colorectal cancer in combination with chemotherapy (e.g., NCT00394176). It has subsequently been FDA approved in other malignancies including non-small cell lung cancer (NCT00021060), glioblastoma (NCT00345163), renal cell carcinoma (NCT00738530), cervical cancer (NCT00803062), and ovarian cancer (NCT00976911) and has been widely tested in many other cancers, including HNSCC (Table 16.2).

Ziv-aflibercept is another anti-angiogenic monoclonal antibody with a distinct mechanism of action. A recombinant protein made up of the Fc portion of a human IgG1 molecule fused to portions of the extracellular domains of human VEGFR1 and VEGFR2, it functions as a soluble receptor that binds to VEGF, VEGF-B, and PlGF and inhibits angiogenesis by blocking normal ligand-receptor interactions [157]. Currently only approved in combination with chemotherapy for the treatment of metastatic colorectal cancer, ziv-aflibercept is actively being investigated in other advanced solid tumors (e.g., NCT02298959 and NCT02159989).

Dalantercept is a soluble receptor fusion protein that consists of the extracellular domain of ALK1 (type 1 receptor in the TGF-β superfamily) linked to the Fc portion of IgG1. Similar to ziv-aflibercept, it binds and sequesters the ligands of cellular ALK1 such as BMP9 and BMP10 and thus prevent its activation [158]. Without ligands and activation, there is no downstream signaling through SMAD proteins, which consequently blocks angiogenesis in a VEGF-independent fashion.

Tyrosine kinase inhibitors (TKIs) function by competitively inhibiting ATP at the catalytic binding site of a tyrosine kinase [159]. In contrast to monoclonal antibodies whose localization is restricted to targets on the cell surface or secreted molecules, because of the large size of the antibody molecules, TKIs are small molecules that can enter a cell and act on intracellular binding domains and signaling pathways [160]. While members of this class of drugs all share a similar mechanism of action, they differ greatly in their targets, pharmacokinetics, and side effects.

Anti-angiogenic TKIs, either alone or in combination with other therapies, are widely utilized as cancer therapeutics. Many of the TKIs that block angiogenesis are multi-kinase inhibitors, targeting several signaling pathways. Examples of these are sunitinib, which targets VEGFR1 and VEGFR2, PDGFR-α and PDGFR-β, c-Kit, RET, CSF1R, and FLT3; sorafenib, which targets RAF/MEK/ERK, VEGFR1–3, PDGFR-β, c-Kit, FLT3, and RET; and pazopanib, which targets VEGFR1–3, PDGFR-α and PDGFR-β, FGFR-1 and FGFR-3, and c-Kit [160]. These so-called "dirty" TKIs can be more efficacious due to their ability to simultaneously inhibit a variety of critical signaling pathways, often at the expense of added toxicity [160]. The optimal use of anti-angiogenic TKIs both alone and in combination with other therapies is still being elucidated in HNSCC and other cancers.

16.8 Toxicities of Anti-Angiogenic Therapy

As little angiogenesis is occurring in normal adult tissues, it was contrary to expectations that anti-angiogenic therapies were shown to have a wide array of toxicities. These toxicities vary, in part, by drug class and mechanism of action and differ from those of traditional cytotoxic chemotherapy.

Bleeding complications ranging from minor epistaxis to fatal pulmonary hemorrhage have been demonstrated with all of the available anti-VEGF agents [161]. This is thought to be due to disruptions in the endothelial cell-platelet interactions that help to maintain and repair vascular integrity [161, 162]. The reported incidence of bleeding complications has varied, with bevacizumab showing the highest frequency of severe events. A meta-analysis published in 2010 found that bevacizumab use across diverse cancers was associated with a statistically significant increased risk of overall bleeding (RR 2.48, 95% CI 1.93–3.18, $I^2 = 53\%$; 20 RCTs) and fatal hemorrhage (RR 3.56, 95% CI 1.71–7.41; 14 RCTs) [163]. The FDA has issued a warning for bevacizumab indicating that severe or fatal hemoptysis, GI bleeding, CNS hemorrhage, epistaxis, and vaginal bleeding occur up to fivefold more frequently in patients receiving this agent.

Previous experience with bevacizumab in squamous cell lung cancer resulted in unacceptable rates of bleeding [164]. Given that tumoral and carotid bleeding are recognized complications of HNSCC, there was a concern for the possibility of bleeding events with the use of bevacizumab in patients with this disease. A phase II trial of the folate antimetabolite pemetrexed and bevacizumab in the first-line treatment of recurrent or metastatic HNSCC reported a rather high incidence of 15% of grades 3–5 bleeding events [165]. It was unclear whether some of these events were related to the underlying disease versus a treatment effect. On the other hand, use of the EGFR-targeted monoclonal antibody cetuximab plus bevacizumab was associated with only 4% incidence of grades 3–4 hemorrhage in patients with recurrent or metastatic HNSCC [166]. Randomized, controlled studies were required to assess the toxicity profile of bevacizumab in HNSCC. A phase III randomized trial of platinum doublets with or without bevacizumab (E1305), discussed in further detail below, demonstrated that bevacizumab increased the rates of grades 3–4 bleeding as well as neutropenia, febrile neutropenia, fatigue, diarrhea, oral mucositis, hypertension, and thromboembolic events [167]. However, all reported (including those deemed not treatment-related) grades 3–5 bleeding adverse events increased from 3.5% to 7.7%, which was not a statistically significant increase ($p = 0.08$). An increase in bleeding events by the addition of bevacizumab to a chemoradiotherapy regimen was seen in phase II randomized trial in locally advanced HNSCC [168].

Hypertension is also a common side effect of anti-angiogenic therapy. This may be related, in part, to decreased endothelial nitric oxide synthase expression, which occurs as a result of VEGF inhibition, and perturbation of vasodilation in the setting of VEGF inhibition [169–171]. Other mechanisms have been also been implicated, including increases in levels of the potent vasoconstrictor endothelin-1, and microvascular rarefaction, defined as a diminished number of perfused capillaries [170, 172]. Some studies have attempted to correlate the development of hypertension with improved treatment outcomes; however, additional studies are needed to validate this as a true biomarker of response [170, 173]. Management of hypertension often requires the initiation and titration of antihypertensive medications during the treatment period. For dalantercept, toxicities include peripheral edema, fatigue, and anemia; the side effect profile of this unique agent may reflect its ALK1-dependent, VEGF-independent mechanism of action.

The development of proteinuria frequently occurs during treatment with anti-angiogenics. VEGF is highly expressed in glomerular podocytes, and disruption of this expression can lead to a breakdown in the glomerular filtration barrier [162, 174]. The development of hypertension also influences the onset of proteinuria [161]. Patients are usually asymptomatic, and proteinuria resolves with cessation of therapy; it is rare for this to progress to permanent renal dysfunction [162].

Additional side effects from anti-angiogenic therapy include thrombosis, gastrointestinal perforation, cardiac dysfunction, hypothyroidism, and impaired wound healing [161]. Although most of these are rare, serious or fatal events can occur. The mechanisms that underlie these toxicities highlight the complex and integral role that angiogenesis plays both in tumors and in healthy tissues.

16.9 Clinical Trials of Anti-Angiogenics in HNSCC

16.9.1 Targeting VEGF

Given the prominent role that angiogenesis plays in HNSCC, it is unsurprising that anti-angiogenics have shown activity in this group of tumors. Bevacizumab, a monoclonal antibody inhibitor of VEGF-A, is the most extensively studied of these agents and has been tested in combination with chemotherapy, radiation, and other targeted therapies. Table 16.2 summarizes the major phase II and phase III studies evaluating bevacizumab in HNSCC. Bevacizumab has already been approved for use in multiple other tumor types in combination with chemotherapy (e.g., with

5-fluorouracil (5FU)-based treatment in metastatic colorectal cancer or with carboplatin and paclitaxel for metastatic non-squamous non-small cell lung cancer).

In the setting of recurrent or metastatic HNSCC, bevacizumab has been combined with the antimetabolite pemetrexed. This drug functions by inhibiting several enzymes of the folate pathway and has activity in multiple tumor types. As the antifolate methotrexate already has known activity in HNSCC, pemetrexed chemotherapy was chosen to be tested in a phase II trial in combination with bevacizumab. Forty patients with recurrent or metastatic disease were enrolled and treated with a combination of the two agents every 21 days. The time to progression for these patients was 5 months with a median overall survival of 11.3 months and an overall response rate of 30%. This was comparable to the efficacy of the reference platinum, 5-FU, and cetuximab regimen, when studied in a similar patient population [175]. Bleeding was found to be the major toxicity associated with the combination of bevacizumab and pemetrexed, with six patients (15%) experiencing grades 3–5 bleeding events, including two that were fatal [165].

E1305 was a randomized phase III trial conducted by the ECOG-ACRIN Cancer Research Group in patients with recurrent or metastatic HNSCC, testing the combination of bevacizumab with a platinum-based doublet, using either cisplatin or carboplatin plus either docetaxel or 5-fluorouracil. Patients had no prior therapy for recurrent or metastatic disease; prior chemotherapy or cetuximab was allowed if given in the setting of prior potentially curative treatment and an interval of at least 4 months had elapsed. Four hundred and three patients were accrued in this study. With a median follow-up of 23.1 months, Argiris et al. have preliminarily reported a median overall survival of 12.6 months with bevacizumab versus 11 months without bevacizumab, a difference that was not statistically significant ($p = 0.13$, HR 0.84, 95% confidence intervals, 0.67–1.05) [167]. Although the primary endpoint of a statistically significant OS improvement was not reached, there was a numerical survival advantage at 2, 3, and 4 years in the bevacizumab arm (26% vs 18% at 2 years, 16% vs 8% at 3 years, and 13% vs 6% at 4 years). Moreover, median PFS improved from 4.4 months to 6.1 months (HR 0.71; $p = 0.0012$) and response rate increased from 25% to 36% ($p = 0.013$) with bevacizumab. As mentioned in the previous section, there was an increase in the percentage of patients experiencing treatment-related toxicities, including grades 3–5 bleeding. While this study provides evidence of improved antitumor activity with the addition of an anti-angiogenic agent to chemotherapy with an expected side effect profile, randomized data proving their ability to increase overall survival are still lacking. On the other hand, based on the interesting results of E1305, further investigation of agents targeting the VEGF pathway in HNSCC is justifiable. The evaluation of molecular markers in this context will be very important.

Bevacizumab has also been studied in combination with radiation therapy. VEGF can be upregulated by different mechanisms, including by hypoxia-inducing factor (HIF-1). Radiation exposure can activate EGFR and in turn the PI3K/AKT and STAT3 pathways, upregulating VEGF. Hypothetically, the addition of an anti-angiogenic agent like bevacizumab could help to overcome radioresistance, which is associated with upregulation of multiple factors that induce angiogenesis. As

Table 16.2 Clinical trials of bevacizumab in HNSCC

Agents	Study design/ phase	# of patients	Primary efficacy endpoint	Author
Chemotherapy + bevacizumab				
Chemotherapy +/– bevacizumab (chemotherapy was investigator's choice: cisplatin + 5-FU, cisplatin + docetaxel, carboplatin + 5-FU, or carboplatin + docetaxel)	Randomized, phase III	$n = 403$	Overall survival 12.6 months in regimen with bevacizumab and 11 months without bevacizumab; $p = 0.13$	Argiris [188] and Argiris et al. [167]
Pemetrexed + bevacizumab	Single arm, phase II	$n = 47$	Median TTP 5 months	Argiris et al. [165]
Targeted agent + bevacizumab				
Cetuximab + bevacizumab	Single arm, phase II	$n = 46$	RR 16%	Argiris et al. [166]
Erlotinib + bevacizumab	Single arm, phase I/II	Phase I $n = 10$ Phase II $n = 48$	Median PFS 4.1 months	Cohen et al. [65]
RT + chemotherapy + bevacizumab				
Docetaxel + RT + bevacizumab	Single arm, phase II	$n = 30$	3-year PFS 61.7%	Yao et al. [189]
Cisplatin+ IMRT + bevacizumab	Single arm, phase II	$n = 42$	2-year PFS 75.9%	Fury et al. [177]
Chemotherapy + RT + targeted agent + bevacizumab				
Pemetrexed+ RT + cetuximab +/– bevacizumab	Randomized, phase II	$n = 78$	2-year PFS 79% (non- bevacizumab arm); 75% (bevacizumab arm)	Argiris et al. [168]
Cisplatin+ IMRT+ cetuximab + bevacizumab	Single arm, phase II	$n = 30$	2-year PFS 88.5%	Fury et al. [190]
Paclitaxel+ carboplatin+ infusional 5-FU + bevacizumab, followed by RT + paclitaxel + erlotinib + bevacizumab	Single arm, phase II	$n = 60$	3-year PFS 71%	Hainsworth et al. [178]

Abbreviations: *RT* radiotherapy, *IMRT* intensity-modulated radiation therapy, *5-FU* 5-fluorouracil, *TTP* time to progression, *PFS* progression-free survival, *RR* response rate

summarized in Table 16.2, there are multiple trials that have evaluated the combination of bevacizumab with concurrent chemoradiation. A phase I trial evaluated the combination of bevacizumab with 5-FU, hydroxyurea, and radiation for patients with recurrent or poor prognosis HNSCC with a goal of finding the maximum-tolerated dose of bevacizumab. Two thirds of the patients had been previously irradiated. Investigators determined that the side effect profile was comparable to historic re-irradiation controls. With a median overall survival for re-irradiated patients of 10.3 months, the antitumor effect was enough to spur further interest in the bevacizumab-radiation combination, and the lack of increased thrombotic or bleeding events was enough to provide reassurance that this was a viable combination [176].

Further trials looking at the combination of bevacizumab and radiation have been carried out in the locally advanced setting. In a phase II trial, bevacizumab was added to concurrent chemoradiation with high-dose cisplatin given in split doses (days 1, 2, 22, 23, 43, and 44) and 70 Gy of IMRT [177]. Forty-two patients with previously untreated disease were enrolled and treated. Their overall survival rate at 2 years was 88%, and their side effect profile did not suggest additive toxicities. Another frontline study looked at the addition of bevacizumab to both an induction chemotherapy regimen and a concurrent chemoradiotherapy regimen. Patients were first given 6 weeks of paclitaxel, carboplatin, 5-FU, and bevacizumab before continuing with concurrent chemoradiation, this time with weekly paclitaxel, bevacizumab, and erlotinib with 68.4 Gy of radiation [178]. The 2-year overall survival was a striking 90%, and no unanticipated toxicities were seen. Further work is being done to evaluate strategies of multi-target combinations with radiation.

16.9.2 Targeting VEGF and EGFR in Combination

Cetuximab is FDA approved for the treatment of HNSCC as monotherapy and in combination with radiation or chemotherapy [175, 179–181]. Given its success in these settings, and since one of the proposed mechanisms of failure of anti-VEGF therapy is escape via an alternate angiogenic pathway, cetuximab has also been studied in combination with bevacizumab [166, 182]. The combination, which was supported by preclinical models and was shown to be tolerable, appeared to be efficacious when compared to historical controls; response rate was 16% and median OS 7.5 months [166]. However, a larger randomized trial will be required to evaluate these combination treatments. Argiris et al. evaluated the combination of cetuximab and radiation with or without bevacizumab in previously untreated locally advanced HNSCC patients. Patients received radiation at 70 Gy with concurrent cetuximab and pemetrexed, with or without bevacizumab, followed by bevacizumab maintenance for 24 months. The 2-year PFS was not different at 79% without and 75% with bevacizumab, but the rate of hemorrhage was higher in the group with bevacizumab, showing some of the limitations of the combined approach [168].

16.9.3 Other Targeted Agents

Targeted anti-angiogenic agents have yielded varying degrees of success when used for the treatment of HNSCC (Table 16.3). Sorafenib, a multi-tyrosine kinase inhibitor that inhibits both VEGFR and EGFR, has been evaluated as a single agent in

Table 16.3 Selected trials of anti-angiogenic tyrosine kinase inhibitors HNSCC

Agents	Study design/ phase	# of patients	Primary efficacy endpoint	Author
Sorafenib monotherapy				
Sorafenib 400 mg twice daily	Single arm, phase II	$n = 41$	Confirmed response probability 2%	Williamson et al. [184]
Sorafenib 400 mg twice daily	Single arm, phase II	$n = 27$	RR 3.7%	Elser et al. [183]
Sunitinib monotherapy				
Sunitinib 37.5 mg daily	Single arm, phase II	$n = 38$	Rate of disease control 50%	Machiels et al. [185]
Sunitinib 50 mg daily for 4 weeks on, 2 weeks off	Single arm, phase II	$n = 17$	RR 0%	Fountzilas et al. [187]
Sunitinib 50 mg daily for 4 weeks on, 2 weeks off	Single arm, phase II	$n = 22$	RR 3–4%	Choong et al. [186]
Axitinib monotherapy				
Axitinib 5 mg twice daily (with planned dose escalation)	Single arm, phase II	$n = 30$	6-month PFS 30%	Swiecicki et al. [191]
Combination therapy				
Cetuximab +/– sorafenib	Randomized, phase II	$n = 55$	PFS 3 months (cetuximab arm); 3.2 months (combination arm) ($p = 0.87$, 95% CI 1.9–5.0 months and 1.8–4.2 months, respectively)	Gilbert et al. [192]
Docetaxel +/– vandetanib	Randomized, phase II	$n = 29$	RR 7% (single arm); 13% (combined arm)	Limaye et al. [193]

Abbreviations: *RR* response rate; *PFS* progression-free survival

recurrent and/or metastatic HNSCC. Phase II data from several studies showed a low likelihood of response to sorafenib in this setting [183, 184]. Sunitinib, a multi-kinase inhibitor with activity against PDGFRs and VEGFRs, showed more promising response results in a phase II study published in 2010, but other studies have failed to duplicate these results [185–187]. Early phase trials with other TKIs, including erlotinib, vandetanib, axitinib, and pazopanib, have resulted in mixed responses; however, studies with these agents are ongoing. In a phase II multicenter, open-label study of dalantercept in recurrent and or metastatic SCCHN, the 40 patients who were considered eligible for evaluation of response had a progression-free survival of 1.4 months and overall survival of 7.1 months. While the drug was well tolerated, these results were only modest at best.

16.9.4 Future Directions

Despite modest success in clinical trials up to this point, targeting angiogenesis in HNSCC remains an attractive strategy. A search of clinicaltrials.gov reveals more than 30 clinical trials that are ongoing. New anti-angiogenic therapies such as foretinib, ficlatuzumab, cilengitide-targeting VEGF, hepatocyte growth factor (HGF), and integrins, respectively, are being studied, as are novel therapies such as E10A, an injectable gene therapy that upregulates human endostatin, an inhibitor of VEGF, and angiogenesis.

The immune checkpoint inhibitors nivolumab and pembrolizumab have both been FDA approved for the treatment of metastatic or recurrent HNSCC. The utility of combining one of these agents with an anti-angiogenic agent such as bevacizumab is unknown. However, given the role that VEGF has been shown to play in tumor immunity, it is possible that they could have a synergistic effect, leading to greater responses. Other immunomodulatory agents, such as inhibitors of CSF1R, are also being investigated (NCT02452424).

16.10 Conclusion

The induction and sustainment of angiogenesis are some of the hallmarks of all solid tumors. Indeed, various pro-angiogenic factors have been found to be upregulated in HNSCC, correlating with more aggressive disease. Anti-angiogenic agents with differing mechanisms of action have been tested in the treatment of HNSCC with mixed success. Studies evaluating new combinations of therapies as well as novel agents are ongoing. In the future, individualized anti-angiogenic treatment based on molecular tumor characterization will likely lead to improved outcomes.

References

1. Carmeliet P, Jain RK. Molecular mechanisms and clinical applications of angiogenesis. Nature. 2011;473(7347):298–307.
2. Ribatti D, et al. Postnatal vasculogenesis. Mech Dev. 2001;100(2):157–63.
3. Tang DG, Conti CJ. Endothelial cell development, vasculogenesis, angiogenesis, and tumor neovascularization: an update. Semin Thromb Hemost. 2004;30(01):109–17.
4. Lee S-P, et al. Integrin-linked kinase, a hypoxia-responsive molecule, controls postnatal vasculogenesis by recruitment of endothelial progenitor cells to ischemic tissue. Circulation. 2006;114(2):150–9.
5. Makanya AN, Hlushchuk R, Djonov VG. Intussusceptive angiogenesis and its role in vascular morphogenesis, patterning, and remodeling. Angiogenesis. 2009;12(2):113.
6. Djonov V, Baum O, Burri PH. Vascular remodeling by intussusceptive angiogenesis. Cell Tissue Res. 2003;314(1):107–17.

7. Michiels C, Arnould T, Remacle J. Endothelial cell responses to hypoxia: initiation of a cascade of cellular interactions. Biochim Biophys Acta, Mol Cell Res. 2000;1497(1):1–10.
8. Bergers G, Song S. The role of pericytes in blood-vessel formation and maintenance. Neuro-Oncology. 2005;7(4):452–64.
9. van Hinsbergh VWM, Koolwijk P. Endothelial sprouting and angiogenesis: matrix metalloproteinases in the lead. Cardiovasc Res. 2008;78(2):203–12.
10. Potente M, Gerhardt H, Carmeliet P. Basic and therapeutic aspects of angiogenesis. Cell. 2011;146(6):873–87.
11. Siekmann AF, Affolter M, Belting H-G. The tip cell concept 10 years after: new players tune in for a common theme. Exp Cell Res. 2013;319(9):1255–63.
12. Ribatti D, Crivellato E. "Sprouting angiogenesis", a reappraisal. Dev Biol. 2012;372(2):157–65.
13. Schmidt T, Carmeliet P. Blood-vessel formation: bridges that guide and unite. Nature. 2010;465(7299):697–9.
14. Koch S, et al. Signal transduction by vascular endothelial growth factor receptors. Biochem J. 2011;437(2):169–83.
15. Robinson CJ, Stringer SE. The splice variants of vascular endothelial growth factor (VEGF) and their receptors. J Cell Sci. 2001;114(5):853–65.
16. Vincenti V, et al. Assignment of the vascular endothelial growth factor gene to human chromosome 6p21.3. Circulation. 1996;93(8):1493–5.
17. Ladomery MR, Harper SJ, Bates DO. Alternative splicing in angiogenesis: the vascular endothelial growth factor paradigm. Cancer Lett. 2007;249(2):133–42.
18. Biselli-Chicote PM, et al. VEGF gene alternative splicing: pro- and anti-angiogenic isoforms in cancer. J Cancer Res Clin Oncol. 2012;138(3):363–70.
19. Shibuya M. Vascular endothelial growth factor (VEGF) and its receptor (VEGFR) signaling in angiogenesis: a crucial target for anti- and pro-angiogenic therapies. Genes Cancer. 2011;2(12):1097–105.
20. Hoeben A, et al. Vascular endothelial growth factor and angiogenesis. Pharmacol Rev. 2004;56(4):549–80.
21. Forsythe JA, et al. Activation of vascular endothelial growth factor gene transcription by hypoxia-inducible factor 1. Mol Cell Biol. 1996;16(9):4604–13.
22. Krock BL, Skuli N, Simon MC. Hypoxia-induced angiogenesis: good and evil. Genes Cancer. 2011;2(12):1117–33.
23. Niu G, et al. Constitutive Stat3 activity up-regulates VEGF expression and tumor angiogenesis. Oncogene. 2002;21(13):2000–8.
24. McColl BK, Stacker SA, Achen MG. Molecular regulation of the VEGF family – inducers of angiogenesis and lymphangiogenesis. APMIS. 2004;112(7–8):463–80.
25. Lee S, et al. Autocrine VEGF signaling is required for vascular homeostasis. Cell. 2007;130(4):691–703.
26. Rashdan N, Lloyd P. Autocrine and paracrine effects of VEGF-A on PLGF in an in vitro model of the vessel wall. FASEB J. 2015;29(1 Supplement). Abstract number:797.7
27. Vassilakopoulou M, Psyrri A, Argiris A. Targeting angiogenesis in head and neck cancer. Oral Oncol. 2015;51(5):409–15.
28. Shibuya M. Structure and function of VEGF/VEGF-receptor system involved in angiogenesis. Cell Struct Funct. 2001;26(1):25–35.
29. Cébe-Suarez S, Zehnder-Fjällman A, Ballmer-Hofer K. The role of VEGF receptors in angiogenesis; complex partnerships. Cell Mol Life Sci. 2006;63(5):601–15.
30. Smith GA, et al. The cellular response to vascular endothelial growth factors requires co-ordinated signal transduction, trafficking and proteolysis. Biosci Rep. 2015;35(5):e00253.
31. Taimeh Z, et al. Vascular endothelial growth factor in heart failure. Nat Rev Cardiol. 2013;10(9):519–30.
32. Meyer RD, Mohammadi M, Rahimi N. A single amino acid substitution in the activation loop defines the decoy characteristic of VEGFR-1/FLT-1. J Biol Chem. 2006;281(2):867–75.

33. Fischer C, et al. FLT1 and its ligands VEGFB and PlGF: drug targets for anti-angiogenic therapy? Nat Rev Cancer. 2008;8(12):942–56.
34. Miettinen M, et al. Vascular endothelial growth factor receptor 2 (Vegfr2) as a marker for malignant vascular tumors and mesothelioma – Immunohistochemical study of 262 vascular endothelial and 1640 nonvascular tumors. Am J Surg Pathol. 2012;36(4):629–39.
35. Karar J, Maity A. PI3K/AKT/mTOR pathway in angiogenesis. Front Mol Neurosci. 2011;4:51.
36. Deng Y, Zhang X, Simons M. Molecular controls of lymphatic VEGFR3 signaling. Arterioscler Thromb Vasc Biol. 2015;35(2):421–9.
37. Alam A, et al. Heterodimerization with vascular endothelial growth factor receptor-2 (VEGFR-2) is necessary for VEGFR-3 activity. Biochem Biophys Res Commun. 2004;324(2):909–15.
38. Tvorogov D, et al. Effective suppression of vascular network formation by combination of antibodies blocking VEGFR ligand binding and receptor dimerization. Cancer Cell. 2010;18(6):630–40.
39. Meadows KN, Bryant P, Pumiglia K. Vascular endothelial growth factor induction of the Angiogenic phenotype requires Ras activation. J Biol Chem. 2001;276(52):49289–98.
40. Shu X, et al. Sphingosine kinase mediates vascular endothelial growth factor-induced activation of ras and mitogen-activated protein kinases. Mol Cell Biol. 2002;22(22):7758–68.
41. Takahashi T, Ueno H, Shibuya M. VEGF activates protein kinase C-dependent, but Ras-independent Raf-MEK-MAP kinase pathway for DNA synthesis in primary endothelial cells. Oncogene. 1999;18(13):2221–30.
42. Roberts PJ, Der CJ. Targeting the Raf-MEK-ERK mitogen-activated protein kinase cascade for the treatment of cancer. Oncogene. 2007;26(22):3291–310.
43. Abid MR, et al. Vascular endothelial growth factor activates PI3K/Akt/Forkhead signaling in endothelial cells. Arterioscler Thromb Vasc Biol. 2004;24(2):294–300.
44. Shiojima I, Walsh K. Role of Akt signaling in vascular homeostasis and angiogenesis. Circ Res. 2002;90(12):1243–50.
45. Jiang BH, Liu LZ. Chapter 2 PI3K/PTEN Signaling in Angiogenesis and Tumorigenesis. Adv Cancer Res. 2009., Academic Press;102:19–65.
46. Zhong H, et al. Modulation of hypoxia-inducible factor 1α expression by the epidermal growth factor/phosphatidylinositol 3-kinase/PTEN/AKT/FRAP pathway in human prostate Cancer cells: implications for tumor angiogenesis and therapeutics. Cancer Res. 2000;60(6):1541–5.
47. Rousseau S, et al. p38 MAP kinase activation by vascular endothelial growth factor mediates actin reorganization and cell migration in human endothelial cells. Oncogene. 1997;15(18):2169–77.
48. Eriksson A, et al. Small GTP-binding protein Rac is an essential mediator of vascular endothelial growth factor-induced endothelial fenestrations and vascular permeability. Circulation. 2003;107(11):1532–8.
49. De Falco S. The discovery of placenta growth factor and its biological activity. Exp Mol Med. 2012;44:1–9.
50. Kim K-J, Cho C-S, Kim W-U. Role of placenta growth factor in cancer and inflammation. Exp Mol Med. 2012;44(1):10–9.
51. Hiratsuka S, et al. MMP9 induction by vascular endothelial growth factor receptor-1 is involved in lung-specific metastasis. Cancer Cell. 2002;2(4):289–300.
52. Zhang W, et al. Placental growth factor promotes metastases of non-small cell lung cancer through MMP9. Cell Physiol Biochem. 2015;37(3):1210–8.
53. Heldin C-H, Lennartsson J. Structural and functional properties of platelet-derived growth factor and stem cell factor receptors. Cold Spring Harb Perspect Biol. 2013;5(8):a009100.
54. Pietras K, et al. PDGF receptors as cancer drug targets. Cancer Cell. 2003;3(5):439–43.
55. Jastrzębski K, et al. Multiple routes of endocytic internalization of PDGFRβ contribute to PDGF-induced STAT3 signaling. J Cell Sci. 2016;130(3):577–589
56. Seki T, Yun J, Oh SP. Arterial endothelium-specific Activin receptor-like kinase 1 expression suggests its role in arterialization and vascular remodeling. Circ Res. 2003;93(7):682–9.

57. de Vinuesa AG, et al. Targeting tumour vasculature by inhibiting activin receptor-like kinase (ALK)1 function. Biochem Soc Trans. 2016;44(4):1142–9.
58. Cunha SI, Pietras K. ALK1 as an emerging target for antiangiogenic therapy of cancer. Blood. 2011;117(26):6999–7006.
59. Ellis LM. Epidermal growth factor receptor in tumor angiogenesis. Hematol Oncol Clin North Am. 2004;18(5):1007–21.
60. Yarden Y. The EGFR family and its ligands in human cancer: signalling mechanisms and therapeutic opportunities. Eur J Cancer. 2001;37(Supplement 4):3–8.
61. van Cruijsen H, Giaccone G, Hoekman K. Epidermal growth factor receptor and angiogenesis: opportunities for combined anticancer strategies. Int J Cancer. 2005;117(6):883–8.
62. Ucuzian AA, et al. Molecular mediators of angiogenesis. J Burn Care Res. 2010;31(1):158.
63. Cook KM, Figg WD. Angiogenesis inhibitors – current strategies and future prospects. CA Cancer J Clin. 2010;60(4):222–43.
64. Tamura K, Sakurai T, Kogo H. Relationship between prostaglandin E2 and vascular endothelial growth factor (VEGF) in angiogenesis in human vascular endothelial cells. Vasc Pharmacol. 2006;44(6):411–6.
65. Cohen EE, et al. Erlotinib and bevacizumab in patients with recurrent or metastatic squamous-cell carcinoma of the head and neck: a phase I/II study. Lancet Oncol. 2009;10:247–57.
66. Ren Y, et al. Hepatocyte growth factor promotes cancer cell migration and angiogenic factors expression: a prognostic marker of human esophageal squamous cell carcinomas. Clin Cancer Res. 2005;11(17):6190–7.
67. Zhang Y-W, et al. Hepatocyte growth factor/scatter factor mediates angiogenesis through positive VEGF and negative thrombospondin 1 regulation. Proc Natl Acad Sci. 2003;100(22):12718–23. Abstract number: 6000
68. Avraamides CJ, Garmy-Susini B, Varner JA. Integrins in angiogenesis and lymphangiogenesis. Nat Rev Cancer. 2008;8(8):604–17.
69. Weis SM, Cheresh DA. αv Integrins in angiogenesis and cancer. Cold Spring Harb Perspect Med. 2011;1(1):a006478.
70. De S, et al. VEGF–integrin interplay controls tumor growth and vascularization. Proc Natl Acad Sci USA. 2005;102(21):7589–94.
71. Lorger M, et al. Activation of tumor cell integrin α(v)β(3) controls angiogenesis and metastatic growth in the brain. Proc Natl Acad Sci USA. 2009;106(26):10666–71.
72. Huang Z, Bao S-D. Roles of main pro- and anti-angiogenic factors in tumor angiogenesis. World J Gastroenterol. 2004;10(4):463–70.
73. Nyberg P, Xie L, Kalluri R. Endogenous inhibitors of angiogenesis. Cancer Res. 2005;65(10):3967–79.
74. Sund M, Zeisberg M, Kalluri R. Endogenous stimulators and inhibitors of angiogenesis in gastrointestinal cancers: basic science to clinical application. Gastroenterology. 2005;129(6):2076–91.
75. Hanahan D, Folkman J. Patterns and emerging mechanisms of the angiogenic switch during tumorigenesis. Cell. 1996;86(3):353–64.
76. Baeriswyl V, Christofori G. The angiogenic switch in carcinogenesis. Semin Cancer Biol. 2009;19(5):329–37.
77. Weis SM, Cheresh DA. Tumor angiogenesis: molecular pathways and therapeutic targets. Nat Med. 2011;17(11):1359–70.
78. Bergers G, Benjamin LE. Tumorigenesis and the angiogenic switch. Nat Rev Cancer. 2003;3:401–10.
79. Qian C-N, et al. Revisiting tumor angiogenesis: vessel co-option, vessel remodeling, and cancer cell-derived vasculature formation. Chin J Cancer. 2016;35(1):10.
80. Folberg R, Hendrix MJC, Maniotis AJ. Vasculogenic mimicry and tumor angiogenesis. Am J Pathol. 2000;156(2):361–81.
81. Seftor REB, et al. Tumor cell vasculogenic mimicry: from controversy to therapeutic promise. Am J Pathol. 2012;181(4):1115–25.

82. Delgado-Bellido D, et al. Vasculogenic mimicry signaling revisited: focus on non-vascular VE-cadherin. Mol Cancer. 2017;16(1):65.
83. Hendrix MJ, et al. Expression and functional significance of VE-cadherin in aggressive human melanoma cells: role in vasculogenic mimicry. Proc Natl Acad Sci USA. 2001;98:8018–23.
84. Hendrix MJ, et al. Transendothelial function of human metastatic melanoma cells: role of the microenvironment in cell-fate determination. Cancer Res. 2002;62:665–8.
85. Bergers G, Hanahan D. Modes of resistance to anti-angiogenic therapy. Nat Rev Cancer. 2008;8:592–603.
86. Kerbel RS. Tumor angiogenesis. N Engl J Med. 2008;358(19):2039–49.
87. De Luca A, et al. The role of the EGFR signaling in tumor microenvironment. J Cell Physiol. 2008;214(3):559–67.
88. Baker CH, et al. Blockade of epidermal growth factor receptor signaling on tumor cells and tumor-associated endothelial cells for therapy of human carcinomas. Am J Pathol. 2002;161(3):929–38.
89. Hirata A, et al. ZD1839 (Iressa) induces antiangiogenic effects through inhibition of epidermal growth factor receptor tyrosine kinase. Cancer Res. 2002;62(9):2554–60.
90. Nolan DJ, et al. Molecular signatures of tissue-specific microvascular endothelial cell heterogeneity in organ maintenance and regeneration. Dev Cell. 2013;26(2):204–19.
91. Nör JE, et al. Up-regulation of Bcl-2 in microvascular endothelial cells enhances intratumoral angiogenesis and accelerates tumor growth. Cancer Res. 2001;61(5):2183–8.
92. Garcia-Barros M, et al. Tumor response to radiotherapy regulated by endothelial cell apoptosis. Science. 2003;300(5622):1155–9.
93. Oyama T, et al. Vascular endothelial growth factor affects dendritic cell maturation through the inhibition of nuclear factor-κB activation in Hemopoietic progenitor cells. J Immunol. 1998;160(3):1224–32.
94. Pardoll DM. The blockade of immune checkpoints in cancer immunotherapy. Nat Rev Cancer. 2012;12(4):252–64.
95. Motz GT, Coukos G. The parallel lives of angiogenesis and immunosuppression: cancer and other tales. Nat Rev Immunol. 2011;11(10):702–11.
96. Ziogas AC, et al. VEGF directly suppresses activation of T cells from ovarian cancer patients and healthy individuals via VEGF receptor type 2. Int J Cancer. 2012;130(4):857–64.
97. Du R, et al. HIF1α induces the recruitment of bone marrow-derived vascular modulatory cells to regulate tumor angiogenesis and invasion. Cancer Cell. 2008;13(3):206–20.
98. Shojaei F, et al. Bv8 regulates myeloid-cell-dependent tumour angiogenesis. Nature. 2007;450(7171):825–31.
99. Murdoch C, et al. The role of myeloid cells in the promotion of tumour angiogenesis. Nat Rev Cancer. 2008;8(8):618–31.
100. Rivera LB, Bergers G. Intertwined regulation of angiogenesis and immunity by myeloid cells. Trends Immunol. 2015;36(4):240–9.
101. Noy R, Pollard JW. Tumor-associated macrophages: from mechanisms to therapy. Immunity. 2014;41(1):49–61.
102. Riabov V, et al. Role of tumor associated macrophages in tumor angiogenesis and lymphangiogenesis. Front Physiol. 2014;5:75.
103. Guo C, et al. The role of tumor-associated macrophages in tumor vascularization. Vasc Cell. 2013;5(1):20.
104. Ribatti D, et al. Macrophages and tumor angiogenesis. Leukemia. 2007;21(10):2085–9.
105. Lin EY, et al. Colony-stimulating factor 1 promotes progression of mammary tumors to malignancy. J Exp Med. 2001;193(6):727–40.
106. Bingle L, et al. Macrophages promote angiogenesis in human breast tumour spheroids in vivo. Br J Cancer. 2005;94(1):101–7.
107. Kobayashi N, et al. Hyaluronan deficiency in tumor stroma impairs macrophage trafficking and tumor neovascularization. Cancer Res. 2010;70(18):7073–83.
108. Mineta H, et al. Prognostic value of vascular endothelial growth factor (VEGF) in head and neck squamous cell carcinomas. Br J Cancer. 2000;83(6):775–81.

109. Singhal A, et al. Vascular endothelial growth factor expression in oral cancer and its role as a predictive marker: a prospective study. Saudi Surg J. 2016;4(2):52–6.
110. Kyzas PA, Cunha IW, Ioannidis JPA. Prognostic significance of vascular endothelial growth factor Immunohistochemical expression in head and neck squamous cell carcinoma: a meta-analysis. Clin Cancer Res. 2005;11(4):1434–40.
111. Bonhin RG, et al. Correlation between vascular endothelial growth factor expression and presence of lymph node metastasis in advanced squamous cell carcinoma of the larynx. Braz J Otorhinolaryngol. 2015;81:58–62.
112. de Sousa EA, et al. Head and neck squamous cell carcinoma lymphatic spread and survival: relevance of vascular endothelial growth factor family for tumor evaluation. Head Neck. 2015;37(10):1410–6.
113. Karatzanis AD, et al. Molecular pathways of lymphangiogenesis and lymph node metastasis in head and neck cancer. Eur Arch Otorhinolaryngol. 2012;269(3):731–7.
114. Kalyankrishna S, Grandis JR. Epidermal growth factor receptor biology in head and neck Cancer. J Clin Oncol. 2006;24(17):2666–72.
115. Ang KK, et al. Impact of epidermal growth factor receptor expression on survival and pattern of relapse in patients with advanced head and neck carcinoma. Cancer Res. 2002;62(24):7350–6.
116. Wang W-M, et al. Epidermal growth factor receptor inhibition reduces angiogenesis via hypoxia-inducible factor-1α and Notch1 in head neck squamous cell carcinoma. PLoS One. 2015;10(2):e0119723.
117. Tabernero J. The role of VEGF and EGFR inhibition: implications for combining anti–VEGF and anti–EGFR agents. Mol Cancer Res. 2007;5(3):203–20.
118. Tonra JR, et al. Synergistic antitumor effects of combined epidermal growth factor receptor and vascular endothelial growth factor receptor-2 targeted therapy. Clin Cancer Res. 2006;12(7):2197–207.
119. Polanska H, et al. Evaluation of EGFR as a prognostic and diagnostic marker for head and neck squamous cell carcinoma patients. Oncol Lett. 2016;12(3):2127–32.
120. Brøndum L, et al. Plasma proteins as prognostic biomarkers in radiotherapy treated head and neck cancer patients. ClinTransl Radiat Oncol. 2017;2:46–52.
121. Guerra ENS, et al. Diagnostic accuracy of serum biomarkers for head and neck cancer: a systematic review and meta-analysis. Crit Rev Oncog Hematol. 2016;101:93–118.
122. Byers LA, et al. Serum signature of hypoxia-regulated factors is associated with progression after induction therapy in head and neck squamous cell cancer. Mol Cancer Ther. 2010;9(6):1755–63.
123. Cohen NA, et al. Dysregulation of hypoxia inducible factor-1α in head and neck squamous cell carcinoma cell lines correlates with invasive potential. Laryngoscope. 2004;114(3):418–23.
124. Jokilehto T, et al. Overexpression and nuclear translocation of hypoxia-inducible factor prolyl hydroxylase PHD2 in head and neck squamous cell carcinoma is associated with tumor aggressiveness. Clin Cancer Res. 2006;12(4):1080–7.
125. Schrijvers ML, et al. Overexpression of intrinsic hypoxia markers HIF1α and CA-IX predict for local recurrence in stage T1-T2 Glottic laryngeal carcinoma treated with radiotherapy. Int J Radiat Oncol Biol Phys. 2008;72(1):161–9.
126. Fillies T, et al. HIF1-alpha overexpression indicates a good prognosis in early stage squamous cell carcinomas of the oral floor. BMC Cancer. 2005;5(1):84.
127. Beasley NJP, et al. Hypoxia-inducible factors HIF-1α and HIF-2α in head and neck cancer. Relationship to tumor biology and treatment outcome in surgically resected patients. Cancer Res. 2002;62(9):2493–7.
128. Ongkeko WM, et al. Expression of protein tyrosine kinases in head and neck squamous cell carcinomas. Am J Clin Pathol. 2005;124(1):71–6.
129. Chen Z, et al. Expression of proinflammatory and proangiogenic cytokines in patients with head and neck cancer. Clin Cancer Res. 1999;5(6):1369–79.

130. Ninck S, et al. Expression profiles of angiogenic growth factors in squamous cell carcinomas of the head and neck. Int J Cancer. 2003;106(1):34–44.
131. Wheeler SE, et al. Tumor associated fibroblasts enhance head and neck squamous cell carcinoma proliferation, invasion, and metastasis in preclinical models. Head Neck. 2014;36(3):385–92.
132. Curry JM, et al. Tumor microenvironment in head and neck squamous cell carcinoma. Semin Oncol. 2014;41(2):217–34.
133. Xin X, et al. Hepatocyte growth factor enhances vascular endothelial growth factor-induced angiogenesis in vitro and in vivo. Am J Pathol. 2001;158(3):1111–20.
134. Ferrari G, et al. Transforming growth factor-beta 1 (TGF-β1) induces angiogenesis through vascular endothelial growth factor (VEGF)-mediated apoptosis. J Cell Physiol. 2009;219(2):449–58.
135. Ishikawa T, et al. Hypoxia enhances CXCR4 expression by activating HIF-1 in oral squamous cell carcinoma. Oncol Rep. 2009;21(3):707–12.
136. Ohm JE, Carbone DP. VEGF as a mediator of tumor-associated immunodeficiency. Immunol Res. 2001;23:263–72.
137. Sridharan V, et al. Effects of definitive chemoradiation on circulating immunologic angiogenic cytokines in head and neck cancer patients. J Immunother Cancer. 2016;4(1):1–10.
138. Williams CB, Yeh ES, Soloff AC. Tumor-associated macrophages: unwitting accomplices in breast cancer malignancy. NPJ Breast Cancer. 2016;2:15025.
139. Quatromoni JG, Eruslanov E. Tumor-associated macrophages: function, phenotype, and link to prognosis in human lung cancer. Am J Transl Res. 2012;4(4):376–89.
140. Balermpas P, et al. Tumour-infiltrating lymphocytes predict response to definitive chemoradiotherapy in head and neck cancer. Br J Cancer. 2014;110(2):501–9.
141. Li C, et al. Infiltration of tumor-associated macrophages in human oral squamous cell carcinoma. Oncol Rep. 2002;9(6):1219–23.
142. Mori K, et al. Infiltration of M2 tumor-associated macrophages in oral squamous cell carcinoma correlates with tumor malignancy. Cancers. 2011;3(4):3726.
143. Costa NL, et al. Tumor-associated macrophages and the profile of inflammatory cytokines in oral squamous cell carcinoma. Oral Oncol. 2013;49(3):216–23.
144. Dahiya K, Dhankhar R. Updated overview of current biomarkers in head and neck carcinoma. World J Methodol. 2016;6(1):77–86.
145. Williams MD. Integration of biomarkers including molecular targeted therapies in head and neck cancer. Head Neck Pathol. 2010;4(1):62–9.
146. Kim KJ, et al. Inhibition of vascular endothelial growth factor-induced angiogenesis suppresses tumour growth in vivo. Nature. 1993;362(6423):841–4.
147. Perrotte P, et al. Anti-epidermal growth factor receptor antibody C225 inhibits angiogenesis in human transitional cell carcinoma growing Orthotopically in nude mice. Clin Cancer Res. 1999;5(2):257–64.
148. Smith BD, et al. Prognostic significance of vascular endothelial growth factor protein levels in oral and oropharyngeal squamous cell carcinoma. J Clin Oncol. 2000;18(10):2046–52.
149. Strome SE, Sausville EA, Mann D. A mechanistic perspective of monoclonal antibodies in cancer therapy beyond target-related effects. Oncologist. 2007;12(9):1084–95.
150. Suzuki M, Kato C, Kato A. Therapeutic antibodies: their mechanisms of action and the pathological findings they induce in toxicity studies. J Toxicol Pathol. 2015;28(3):133–9.
151. Gerber H-P, Ferrara N. Pharmacology and pharmacodynamics of bevacizumab as monotherapy or in combination with cytotoxic therapy in preclinical studies. Cancer Res. 2005;65(3):671–80.
152. Fujita K, Sano D, Kimura M, Yamashita Y, Kawakami M, Ishiguro Y, Nishimura G, Matsuda H, Tsukuda M. Anti-tumor effects of bevacizumab in combination with paclitaxel on head and neck squamous cell carcinoma. Oncol Rep. 2007;18(1):47–51.

153. Jain RK. Normalizing tumor microenvironment to treat cancer: bench to bedside to biomarkers. J Clin Oncol. 2013;31(17):2205–18.
154. Batchelor TT, et al. AZD2171, a pan-VEGF receptor tyrosine kinase inhibitor, normalizes tumor vasculature and alleviates edema in glioblastoma patients. Cancer Cell. 2007;11(1):83–95.
155. Goel S, et al. Normalization of the vasculature for treatment of cancer and other diseases. Physiol Rev. 2011;91(3):1071–121.
156. Weißhardt P, et al. Tumor vessel stabilization and remodeling by anti-angiogenic therapy with bevacizumab. Histochem Cell Biol. 2012;137(3):391–401.
157. Patel A, Sun W. Ziv-aflibercept in metastatic colorectal cancer. Biol: Targets Ther. 2014;8:13–25.
158. Jimeno A, et al. A phase 2 study of dalantercept, an activin receptor-like kinase-1 ligand trap, in patients with recurrent or metastatic squamous cell carcinoma of the head and neck. Cancer. 2016;122(23):3641–9.
159. Jorg Thomas H, et al. Tyrosine kinase inhibitors – a review on pharmacology, metabolism and side effects. Curr Drug Metab. 2009;10(5):470–81.
160. Gotink KJ, Verheul HMW. Anti-angiogenic tyrosine kinase inhibitors: what is their mechanism of action? Angiogenesis. 2010;13(1):1–14.
161. Verheul HMW, Pinedo HM. Possible molecular mechanisms involved in the toxicity of angiogenesis inhibition. Nat Rev Cancer. 2007;7(6):475–85.
162. Kamba T, McDonald DM. Mechanisms of adverse effects of anti-VEGF therapy for cancer. Br J Cancer. 2007;96(12):1788–95.
163. Hapani S, et al. Increased risk of serious hemorrhage with bevacizumab in cancer patients: a meta-analysis. Oncology. 2010;79(1–2):27–38.
164. Johnson DH, et al. Randomized phase II trial comparing bevacizumab plus carboplatin and paclitaxel with carboplatin and paclitaxel alone in previously untreated locally advanced or metastatic non-small-cell lung cancer. J Clin Oncol. 2004;22(11):2184–91.
165. Argiris A, et al. Phase II trial of pemetrexed and bevacizumab in patients with recurrent or metastatic head and neck cancer. J Clin Oncol. 2011;29:1140–5.
166. Argiris A, et al. Cetuximab and bevacizumab: preclinical data and phase II trial in recurrent or metastatic squamous cell carcinoma of the head and neck. Ann Oncol. 2013;24:220–5.
167. Argiris A, Li S, Savvides P, Ohr J, Gilbert J, Levine M, Haigentz M Jr, Saba NF, Chakravarti A, Ikpeazu C, Schneider C, Pinto H, Forastiere AA, Burtness B. Phase III randomized trial of chemotherapy with or without bevacizumab in patients with recurrent or metastatic squamous cell carcinoma of the head and neck: survival analysis of E1305, an ECOG-ACRIN Cancer research group trial. J Clin Oncol. 2017;35(Suppl). Abstract number :6000
168. Argiris A, et al. Phase II randomized trial of radiation therapy, cetuximab, and pemetrexed with or without bevacizumab in patients with locally advanced head and neck cancer. Ann Oncol. 2016;27(8):1594–600.
169. Horowitz JR, et al. Vascular endothelial growth factor/vascular permeability factor produces nitric oxide–dependent hypotension. Evidence for a maintenance role in quiescent adult endothelium. Arterioscler Thromb Vasc Biol. 1997;17(11):2793–800.
170. de Jesus-Gonzalez N, et al. Management of antiangiogenic therapy-induced hypertension. Hypertension. 2012;60(3):607–15.
171. PANDE A, et al. Hypertension secondary to anti-angiogenic therapy: experience with bevacizumab. Anticancer Res. 2007;27(5B):3465–70.
172. Wasserstrum Y, et al. Hypertension in cancer patients treated with anti-angiogenic based regimens. Cardio-Oncology. 2015;1(1):6.
173. Schneider BP, et al. Association of vascular endothelial growth factor and vascular endothelial growth factor receptor-2 genetic polymorphisms with outcome in a trial of paclitaxel compared with paclitaxel plus bevacizumab in advanced breast cancer: ECOG 2100. J Clin Oncol. 2008;26(28):4672–8.
174. Eremina V, et al. Glomerular-specific alterations of VEGF-A expression lead to distinct congenital and acquired renal diseases. J Clin Investig. 2003;111(5):707–16.

175. Vermorken JB, et al. Platinum-based chemotherapy plus cetuximab in head and neck cancer. N Engl J Med. 2008;359(11):1116–27.
176. Seiwert TY, et al. Phase I study of bevacizumab added to fluorouracil- and hydroxyurea-based concomitant chemoradiotherapy for poor-prognosis head and neck Cancer. J Clin Oncol. 2008;26(10):1732–41.
177. Fury MG, et al. A phase 2 study of bevacizumab with cisplatin plus intensity-modulated radiation therapy for stage III/IVB head and neck squamous cell cancer. Cancer. 2012;118(20):5008–14.
178. Hainsworth JD, et al. Combined modality treatment with chemotherapy, radiation therapy, bevacizumab, and Erlotinib in patients with locally advanced squamous carcinoma of the head and neck: a phase II trial of the Sarah Cannon oncology research consortium. Cancer J. 2011;17(5):267–72.
179. Vermorken JB, et al. Open-label, uncontrolled, multicenter phase II study to evaluate the efficacy and toxicity of cetuximab as a single agent in patients with recurrent and/or metastatic squamous cell carcinoma of the head and neck who failed to respond to platinum-based therapy. J Clin Oncol. 2007;25(16):2171–7.
180. Bonner JA, et al. Radiotherapy plus cetuximab for squamous-cell carcinoma of the head and neck. N Engl J Med. 2006;354(6):567–78.
181. Bonner JA, et al. Radiotherapy plus cetuximab for locoregionally advanced head and neck cancer: 5-year survival data from a phase 3 randomised trial, and relation between cetuximab-induced rash and survival. Lancet Oncol. 2010;11(1):21–8.
182. Dey N, De P, Brian L-J. Evading anti-angiogenic therapy: resistance to anti-angiogenic therapy in solid tumors. Am J Transl Res. 2015;7(10):1675–98.
183. Elser C, et al. Phase II trial of sorafenib in patients with recurrent or metastatic squamous cell carcinoma of the head and neck or nasopharyngeal carcinoma. J Clin Oncol. 2007;25(24):3766–73.
184. Williamson SK, et al. Phase II evaluation of sorafenib in advanced and metastatic squamous cell carcinoma of the head and neck: southwest oncology group study S0420. J Clin Oncol. 2010;28(20):3330–5.
185. Machiels J-PH, et al. Phase II study of sunitinib in recurrent or metastatic squamous cell carcinoma of the head and neck: GORTEC 2006-01. J Clin Oncol. 2010;28(1):21–8.
186. Choong NW, et al. Phase II study of sunitinib malate in head and neck squamous cell carcinoma. Investig New Drugs. 2010;28(5):677–83.
187. Fountzilas G, et al. A phase II study of sunitinib in patients with recurrent and/or metastatic non-nasopharyngeal head and neck cancer. Cancer Chemother Pharmacol. 2010;65(4):649–60.
188. Argiris A. Safety analysis of a phase III randomized trial of chemotherapy with or without bevacizumab (B) in recurrent or metastatic squamous cell carcinoma of the head and neck (R/M SCCHN). in 2015 ASCO Annual Meeting. 2015. Chicago.
189. Yao M, et al. Phase II study of bevacizumab in combination with docetaxel and radiation in locally advanced squamous cell carcinoma of the head and neck. Head Neck. 2015;37(11):1665–71.
190. Fury MG, et al. Phase II trial of bevacizumab + cetuximab + cisplatin with concurrent intensity-modulated radiation therapy for patients with stage III/IVB head and neck squamous cell carcinoma. Head Neck. 2016;38(S1):E566–70.
191. Swiecicki PL, et al. A phase II study evaluating axitinib in patients with unresectable, recurrent or metastatic head and neck cancer. Investig New Drugs. 2015;33(6):1248–56.
192. Gilbert J, et al. A randomized phase II efficacy and correlative studies of cetuximab with or without sorafenib in recurrent and/or metastatic head and neck squamous cell carcinoma. Oral Oncol. 2015;51(4):376–82.
193. Limaye S, et al. A randomized phase II study of docetaxel with or without vandetanib in recurrent or metastatic squamous cell carcinoma of head and neck (SCCHN). Oral Oncol. 2013;49(8):835–41.

Chapter 17
FAK as a Target for Therapy in Head and Neck Cancer

Nassim Khosravi, Heath Skinner, and John Heymach

Abstract Despite decades of concerted effort, treatments for head and neck squamous cell carcinoma (HNSCC) have remained largely unchanged, with tumors typically managed using a combination of surgery, radiotherapy, and cytotoxic chemotherapy. Suboptimal efficacy and often severe toxicities associated with some of these treatments have encouraged development of targeted therapies that may overcome these limitations. One promising avenue of therapeutic development in HNSCC in particular has addressed integrins and integrin-mediated signaling, which mediates interactions between the tumor and the extracellular matrix (ECM) and can potentially be targeted by inhibition of the integrin-associated focal adhesion kinase (FAK). This chapter summarizes FAK structure-function relationships and how FAK impacts multiple cellular processes relevant to HNSCC, including survival and invasion. We will discuss the development of targeted FAK inhibitors, and combinatorial strategies incorporating FAK inhibition, with comparisons between human papillomavirus (HPV)-positive and HPV-negative HNSCC.

Keywords Focal adhesion kinase · Cancer · Chemotherapy · Radiation

17.1 Introduction

Head and neck cancer (head and neck squamous cell carcinoma; HNSCC) was diagnosed in close to 60,000 patients in the USA in 2015 [1]. Despite decades of concerted effort, treatments for this malignancy have remained largely unchanged, with tumors typically managed using a combination of surgery, radiotherapy, and, in

N. Khosravi · J. Heymach (✉)
Department of Thoracic, Head and Neck Medical Oncology, Division of Cancer Medicine, The University of Texas MD Anderson Cancer Center, Houston, TX, USA
e-mail: jheymach@mdanderson.org

H. Skinner
Department of Clinical Radiation Oncology, Division of Radiation Oncology, The University of Texas MD Anderson Cancer Center, Houston, TX, USA

B. Burtness, E. A. Golemis (eds.), *Molecular Determinants of Head and Neck Cancer*, Current Cancer Research, https://doi.org/10.1007/978-3-319-78762-6_17

patients diagnosed with advanced or recurrent disease, cytotoxic chemotherapy. This disease is far less likely to metastasize to distant organs than most other solid tumors, and death due to disease occurs primarily due to local or regional failure.

To better establish prognosis for individual patients, a number of biomarkers have been investigated in this disease, with the most significant determinant of outcome currently known being the presence of human papilloma virus (HPV; discussed in depth in Chaps. 20 and 21). Over the past decade, there has been a dramatic increase in HNSCC associated with HPV, primarily in the oropharynx (tonsil and base of the tongue). HPV-related HNSCCs have dramatically better outcomes than their HPV-negative counterparts, largely due to increased sensitivity to radiation and chemotherapy [2, 3]. Indeed, in many ways, HPV-related and HPV-unrelated HNSCC represent two distinct diseases, with different mutational and gene expression profiles and distinct clinical presentations.

There has also been considerable interest in identifying novel targeted therapies that may be useful in improving outcome for HNSCC patients, regardless of tumor HPV status. One interesting avenue of therapeutic development in solid tumors in general and HNSCC in particular has been integrins and integrin-mediated signaling. Integrins are a large family of transmembrane receptors, with each integrin receptor a heterodimer composed of 1 α subunit (out of 9 total) and 1 β subunit (out of 24 total) [4]. These receptors recognize various components of the extracellular matrix (ECM), such as laminin or collagen, and act to promote cell adhesion. Following activation of integrins by their binding to extracellular ligands or by other means, they undergo clustering and structural rearrangements that lead to recruitment of a number of proteins to their cytoplasmic tails. This leads to activation of a number of downstream kinases that provide signals that increase cell survival signals and affect other processes including cytoskeletal reorganization and cell cycle. One of the most critical and potentially targetable of these downstream effectors of integrin signaling is focal adhesion kinase (FAK).

FAK (also known as protein tyrosine kinase 2, or PTK2) is a ubiquitously expressed non-receptor cytosolic protein tyrosine kinase first isolated from chicken embryo fibroblasts transformed by the viral oncogene v-src [5]. Activation of FAK can mediate diverse cellular processes, including cell adhesion, motility, metastasis, angiogenesis, lymphangiogenesis, survival, and cell cycle progression [6–8]. Although primarily associated with activation by integrins and extracellular matrix (ECM) signaling, associated with function at focal adhesions, FAK can also be activated by multiple growth factor receptors, including epidermal growth factor receptor (EGFR) [9], platelet-derived growth factor receptor (PDGFR) [10], vascular endothelial growth factor receptor (VEGFR) [11], macrophage stimulating 1 receptor/recepteur d'origine nantais (MST1R/RON) [12], and others. Activated FAK can function in multiple cellular compartments [13], with considerable recent effort focused on defining the important roles of nuclear FAK in regulating the stability of p53 [14] and promoting transcriptional patterns that support immune evasion by squamous cancer (e.g., [15]). Significant efforts have been made in generating and clinically testing inhibitors of this kinase in a variety of solid tumors.

In this chapter, we will first describe the physical structure of FAK, as well as how this structure impacts FAK activation. We will then discuss the multiple cellular processes impacted by FAK signaling. The chapter will then conclude by examining the role of FAK in cancer broadly and in HNSCC specifically, with attention given to combinatorial strategies incorporating FAK inhibition, with comparisons between HPV-positive and HPV-negative disease.

17.2 Structure and Activation of FAK

The physical structure of FAK is well characterized and has been reviewed in detail previously [6–8]. It is composed of three domains: (i) a large amino-terminal FERM (band 4.1/ezrin/radixin/moesin homology) domain with a proline-rich domain and a phosphorylation site (Y397), (ii) a central catalytic kinase domain, and (iii) a large C-terminal region containing the FAT (focal adhesion targeting) domain and two proline-rich motifs [7] (Fig. 17.1).

FAK activation is regulated by phosphorylation, as well as protein-protein binding with a variety of partners. One element of its activation is disruption of autoinhibitory interactions involving the FAK FERM domain, which binds to the kinase domain when the kinase is not activated by interactions with partner proteins. The FERM domain is composed of three subdomains (F1, F2, and F3) and a nuclear localization sequence (NLS). The F2 subdomain contains the kinase domain-binding site (KDBS), which is essential in self-regulating the kinase activity of FAK. The F3 subdomain enhances p53 turnover by promoting ubiquitination and degradation of p53 through direct binding to Mdm-2 [14]. Also located in the N-terminal domain are proline-rich regions which serve as a docking site for Src homology 3 (SH3)-containing proteins. The C-terminal domain of FAK consists of two proline-rich sequences and a FAT sequence. FAK participates in many protein-protein interactions that both contribute to its activation and engage effectors. Specifically, the proline-rich domains bind additional SH3 domain-containing proteins such p130Cas/BCAR1 [6, 16], HEF1/NEDD9 [17], Rho GTPase-activating protein 26 (ARHGAP26)/Graf [18], SH3 domain-containing GRB2 like 1/endophilin A2 (SH3GL1) [19], and others. The FAT sequence determines the subcellular localiza-

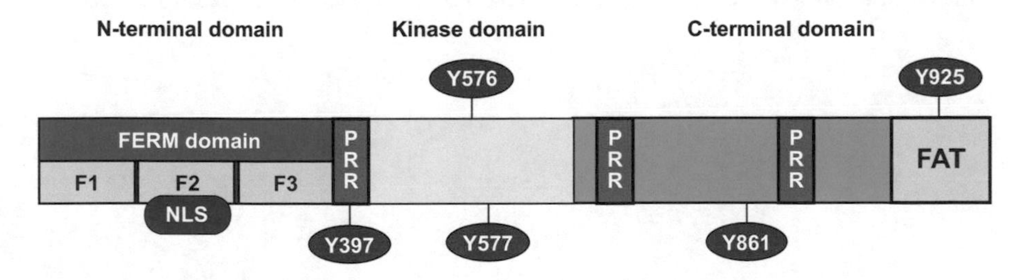

Fig. 17.1 FAK structure

tion of FAK by linking the protein to integrin cytoplasmic tails via paxillin [20] and talin [21].

FAK activation is associated with integrin binding to ECM. The heterodimeric integrin receptors recognize a variety of ECM components, depending upon the specific αβ subunit combinations. This recognition generally occurs at points of contact between the cell and the substratum termed focal adhesions. While integrins provide a mechanical linkage between the ECM and the cell, they also facilitate "inside out" and "outside in" signaling [22]. Although integrins possess no catalytic activity themselves, they can undergo conformational changes in response to external stimuli, with these changes affecting the ability of their cytoplasmic tails to provide a binding site for talin, paxillin, and other proteins [6, 23]. It is generally thought that these proteins then lead to recruitment and oligomerization of FAK [23]. However, this observation is not without controversy; indeed in some contexts it is thought that the exact opposite may be true, namely, that talin is recruited to integrins and focal adhesions by FAK [24]. Regardless, upon association with integrins, the FERM-kinase linkage seen in the "closed conformation" of FAK is released.

FAK then auto-phosphorylates at the Y397 residue, further disturbing interactions between the FERM and kinase domains. This phosphorylation event allows for FAK binding to a variety of proteins including Src family kinases, phospholipase Cγ, and the p85 subunit of PI3 kinase [25–28]. Full activation of FAK requires the binding of Src and its subsequent phosphorylation of Y576 and Y577 within the FAK kinase domain [29]. Following these phosphorylation events, FAK assumes a conformation that prevents the FERM from binding to the kinase domain, and FAK remains active. Src-dependent phosphorylation of FAK on Y861 also contributes to its activation [30, 31], while phosphorylation of FAK at Y925 in the FAT domain allows the protein to interact with GRB2 [32], connecting it to mitogenic signaling cascades [33].

In addition to its activation via integrins, FAK is also known to be activated as a consequence of the activation of numerous receptor tyrosine kinases, as noted above. This phenomenon is likely context-dependent and could be indirect via growth factor-mediated Src activation and/or directly in collaboration with integrin receptors at focal adhesions. We also note that, although beyond the scope of this chapter, humans also have a FAK paralog, known variously as PTK2B/RAFTK/PYK2 [34]. Like FAK, this protein has many similar (although not equivalent) interactions and has also been implicated in head and neck cancer, for example [35].

17.3 Cellular Processes Associated with FAK

17.3.1 Survival

FAK activity plays a significant role in pro-survival signaling. Detachment of the integrin receptors from the ECM and subsequent FAK deactivation have deleterious effects on cell survival. Indeed, in most nonmalignant cell types, this leads to a subclass of apoptosis termed "anoikis" [36], defined as cell death initiated by cell detachment from the ECM and loss of integrin signaling [37]. It has been shown both that constitutive activation of FAK rescues cells from anoikis and that this phenomenon is dependent upon FAK phosphorylation at Y397 and active FAK kinase activity (i.e., mutation of the essential K454 site in the FAK kinase domain eliminates rescue) [38]. Conversely, FAK signaling disturbance can induce cell death, after induction of cell rounding and detachment from the ECM [39].

The pro-survival function of FAK is likely due the multiple effects of FAK signaling within the cell. For example, it has been shown that FAK can directly localize to the nucleus and interact with p53 to promote cell proliferation and survival by degrading p53 [14]. Interestingly, the FAK promoter contains p53 binding sites and can be repressed by expression of p53 [40]. Thus, these two proteins appear to participate in a negative feedback loop, at least in the context of wild-type p53. As many solid tumors harbor mutations in p53, this may lead to unchecked expression of FAK. Indeed, an association between p53 mutation and FAK expression has been observed in breast cancer [41].

In more direct pro-survival signaling, hyperphosphorylated FAK complexed with Src recruits the Grb2 adaptor protein [42]. This, in turn, leads to the recruitment of son of sevenless (SOS) protein and to activation of the Ras/MEK/ERK signaling cascade. In addition, the phosphorylated moiety of FAK binds to the p85 subunit of PI3 kinase, leading to its phosphorylation and subsequent phosphorylation of the pro-survival kinase Akt [43]. Moreover, FAK can also lead to activation of NF-kB-mediated cell survival via interaction with either receptor-interacting protein (RIP) [44] or Rho/Rac GTPase [45], a phenomenon linked to clinical resistance to PI3K kinase inhibitors [46]. Following activation by FAK, NF-kB in turn leads to the induction of inhibitors of apoptosis proteins (IAPs) and cellular survival [38].

17.3.2 Motility and Invasion

FAK plays an essential role in cell migration by organizing the leading edge of migrating cells via coordinating integrin signaling and regulating focal adhesion turnover [6, 47–49]. Activated FAK binds to CAS family proteins, including p130Cas and HEF1/NEDD9, via its C-terminal proline-rich domain, which in turn can also bind to Src family kinases (Fig. 17.2). The Src kinase then phosphorylates the CAS proteins at multiple sites within their "substrate domain," which consists of

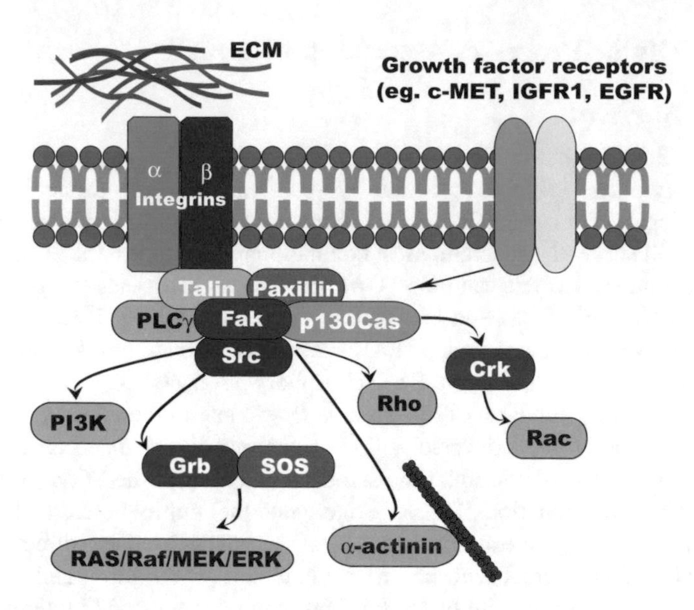

Fig. 17.2 FAK signaling

>15 tyrosines in the context of motifs that form SH2-binding motifs. This action leads to the binding of the SH2 domain-containing protein Crk to the CAS protein and the promotion of Rac GTPase activity and lamellipodia formation. In HNSCC, VEGF-dependent activation of the FAK partner NEDD9/HEF1 was shown to be essential for migratory and invasive tumor behavior, and high HEF1 expression in a primary HNSCC tumor was strongly associated with likelihood of tumor metastasis [50]. Phosphorylation of another FAK-binding partner, paxillin, can also recruit and bind the adaptor protein Crk, providing connections to control of the actin cytoskeleton [51]. FAK associations with another cytoskeletal regulator, p190RhoGEF, provides an additional input into control of focal adhesion assembly [52]. Depending on the cellular context, inhibition of any aspect of this signaling cascade can lead to decreased turnover of focal adhesions and cell motility.

Focal adhesions can form in the absence of FAK, providing some connection between cell and ECM. However, these focal adhesions lack a connection to actin stress fibers and consequently cannot mediate contractility [53]. The ability of FAK to mediate this latter process is thought to lie largely in the modulation of α-actinin (ACTN1, a spectrin-family protein that connects the actin cytoskeleton to the membrane), which is phosphorylated by FAK at Y12. Phosphorylated α-actinin does not localize to focal adhesions and has a reduced affinity for actin, ultimately allowing for turnover of focal adhesions [54]. FAK also directly binds to the actin-binding protein ARP3, and phosphorylates N-WASP (WASL) at Y256, maintaining it within the cytoplasm to activate ARP2/3 actin polymerization [55]. FAK can also mediate the activation of matrix metalloproteases (MMPs), specifically MMPs 2 and 9, at the leading edge of the cell, affecting both focal adhesion turnover and invasion [56, 57]. Two of the FAK effectors, p130Cas/BCAR1 and HEF1/NEDD9, localize to and help mature invadopodia/podosomes [50, 58], specialized structures that differentiate from focal adhesions and provide pores for the targeted secretion of MMPs.

17.3.3 Epithelial to Mesenchymal Transition (EMT)

During tumorigenesis malignant cells acquire the ability to infiltrate their surrounding stroma and travel to distant sites, a process dubbed epithelial to mesenchymal transition (EMT). This process is associated with a decrease in epithelial markers (e.g., E-cadherin, cytokeratins) and a subsequent increase in mesenchymal markers (e.g., N-cadherin), as well as the secretion of MMPs. As this process involves changes in cell polarity and migration, FAK and its effectors play key roles. Indeed, FAK-null embryonic cells remain committed to an epithelial status with subsequent increased E-cadherin and cytokeratin expression [59]. Transforming growth factor beta (TGF-β) is a major inducer of EMT for HNSCC and other tumor types. TGF-β induces SRC/FAK interactions, and FAK is important for induction of a mesenchymal transcriptional program mediated by the transcription factor Snail [60]. In a mouse model, a null FAK genotype repressed the Snail1-dependent transcription program, an effect that is reversed upon re-expression of FAK. Similar phenomena are observed in the context of cancer EMT. Inhibition of FAK (or Src) repressed of markers of EMT following TGF-β stimulation as well as alterations in E-cadherin in several tumor types [61–63]. Similar findings were observed in an IGF-1-mediated model of EMT, in which FAK expression was required for the expression of the EMT-associated biomarkers vimentin and ZEB-1 [64].

In HNSCC, the development of clinical nodal metastasis was associated with increased FAK and decreased E-cadherin [65]. Interestingly, this effect may be partially kinase independent, as at least one study using a point mutation primarily disrupting the scaffolding function of FAK has shown significant inhibition of the expression of EMT biomarkers and of metastasis in a breast cancer mouse model [66]. It is also important to note that many solid tumors are not solely dependent on EMT for invasion and metastasis. Other means of tumor escape from the primary location include collective migration [67] and mesenchymal-amoeboid transition (MAT). Given the close involvement of HEF1/NEDD9 [68, 69] and other FAK effectors in MAT, further scrutiny of this mechanism in the context of HNSCC is warranted.

17.3.4 Angiogenesis

FAK is essential for vasculogenesis and control of tumor angiogenesis. Constitutive null FAK genotype is lethal in early embryonal development [70]. However, an endothelial cell (EC)-specific knockout of FAK using a Cre-loxP mouse model showed that early vasculogenesis was unimpaired, but in late embryogenesis, FAK deletion in EC cells results in abnormalities before day E.13.5, and is lethal in mouse embryos by day E~14.5, with dramatically aberrant vessels, loss of superficial vasculature, and hemorrhage observed [71]. This phenomenon appears to be associated with significant apoptosis within the endothelial compartment [72]. Interestingly, these effects are not completely dependent upon the kinase function of FAK. A conditional EC-specific knock-in of a kinase-deficient form of FAK, although still embryonic lethal (now at day ~15.5), rescues the observed endothelial apoptosis through day 13.5. However, increased endothelial cell permeability and decreased VE-cadherin phosphorylation were observed in both FAK-null and FAK kinase-deficient embryos [72].

Endothelial FAK also has a significant role in tumor-associated angiogenesis. Work with a mouse model allowing endothelial FAK deletion in adult mice inhibited tumor growth and reduced tumor angiogenesis, following injection of syngeneic melanoma and carcinoma cell lines [73]. Indeed, this process is supported by FAK signaling both within the tumor as well as within the endothelial cell. Within the tumor, activation of FAK by integrins engaged with the tumor-associated ECM, as well as by increased growth factor receptor signaling, in turn leads to the activation of MEK/ERK signaling and VEGF secretion by the tumor [74]. Tumor-secreted VEGF then supports the survival and migration of endothelial cells, in part, via the activation of FAK within the endothelium and subsequent downstream signaling. Thus, the tumor participates in a positive feedback mechanism involving VEGF and FAK to support angiogenesis. However, this phenomenon may be more complicated than a simple stimulatory cascade, as at least one group has observed increased tumor growth and angiogenesis in mice heterozygous for FAK deletion [75]. This latter phenomenon appears to be highly dependent upon PI3 kinase activation and argues for a nonlinear relationship between FAK expression and angiogenesis. Importantly, genetic experiments have shown that FAK is not required for angiogenesis in adults, in part because loss of FAK activity is compensated for by elevated activity of its paralog PYK2 [76]. This ability of PYK2 to compensate for FAK deficiency has implications for therapeutic strategies based on inhibition of FAK.

17.3.5 Stemness

Cancer stem cells (CSCs) are a rare population of undifferentiated cancer cells with the ability to self-renew and are thought to contribute to resistance to conventional therapy. Several groups have reported a key role of FAK in maintaining CSCs in several different tumor types. In a mouse model of breast cancer, conditional knockout of FAK leads to a decreased pool of CSCs as well as a reduced functionality of those CSCs that remain [77, 78]. Interestingly, inhibition of the kinase function of FAK led to deficiency in luminal mammary SCs, while basal mammary SCs were dependent upon the scaffolding function of FAK [79]. Moreover, in this model, only the scaffolding function of FAK was required for the expression of the stem cell-associated transcription factors SNAIL, SLUG, and SOX9.

Expression and function of FAK has also been linked to CSCs in an induced mouse model of skin cancer, based on treatment with treatment with 7,12-dimethylbenz[α]anthracene/12-o-tetradecanoylphorbol-13-acetate, which generates papillomas that can progress to squamous cell tumors [80]. The ability of CSCs to self-renew and cycle in this model was dependent on maintenance of an appropriate balance between TGF-β and integrin/FAK signaling.

Finally, FAK is associated with several proteins associated with stemness, as drivers and/or biomarkers. In one study, glioblastoma cells forced to express the CSC-promoting transcription factors OCT3/4 exhibited high levels of migration and invasion, associated with increased levels of FAK expression [81]. Nanog, another marker and driver of stemness, was associated with increased FAK in a screening proteomics analysis [82]. Interestingly, the FAK promoter contains four Nanog-binding sites, to which Nanog can bind and directly increase FAK expression ([83]). Additionally, Nanog was shown to bind the N-terminal domain of FAK in at least some cancer types, leading to FAK phosphorylation of Nanog [83]. This interaction between FAK and Nanog appears to be important for FAK-mediated cellular polarization and invasion, as both were inhibited by expression of Nanog mutants incapable of binding FAK.

17.4 FAK as a Target in Cancer

FAK itself is not an oncogene and is only mutated in a very small number of tumors [84]. However, FAK is highly overexpressed in a variety of human malignancies including head and neck, ovarian, bladder, prostate, lung, pancreatic, brain, and colorectal cancer [7, 85]. Several studies have shown that FAK expression and/or activation is linked to advanced stage at presentation, poorer prognosis, or both [8]. For example, in acute myeloid leukemia (AML), elevated FAK expression was associated with resistance to daunorubicin in vitro as well as high blast count and worse survival in a cohort of 60 patients [86]. Additionally, in a large cohort of 1200 patients with breast cancer, higher FAK expression was associated poor disease-free survival [87]. However, in a separate, smaller study in lymph node-negative breast cancer patients, FAK expression was not predictive of outcome [88].

In HNSCC, FAK expression was examined in 211 tissue specimens, including 147 primary tumors (primarily with supraglottic laryngeal SCC), 56 lymph node metastases, 3 benign hyperplasias, and 5 dysplasias, and expression in primary tumor was compared with nodal metastasis [89]. In this study, high FAK expression levels was seen in 62% of the samples overall, with FAK expression higher in nodal disease. However, FAK expression was not prognostic of outcome. This finding of elevated FAK in tumors confirmed in a separate study of 95 previously untreated men with squamous cell carcinoma of the supraglottic larynx [65, 89]. In the latter study, FAK added prognostic value to primary measurement of E-cadherin but was not independently prognostic of outcome. It is important to note that the management of these patients was not discussed in the publication nor was the HPV status known, potentially confounding the results.

In a more recent analysis of a series of patients with locally advanced HPV-negative HNSCC treated uniformly with surgery and postoperative radiotherapy, FAK overexpression was associated with gene amplification. In two independent patient cohorts, FAK amplification was also highly associated with poorer disease-free survival (DFS; $P = 0.012$ and 0.034), as was overexpression of FAK mRNA ($P = 0.03$) [85]. This finding was validated in a separate cohort of HPV-negative patients from The Cancer Genome Atlas (TCGA) in the same study [85]. Moreover, in an additional retrospective study of 87 patients with hypopharyngeal cancer (who are typically HPV-negative) treated with surgery and selective postoperative radiotherapy, FAK-positive patients had worse disease-specific survival than FAK-negative patients ($p = 0.001$). Based on multivariate analyses, FAK expression was independent of extracapsular spread (ECS) as a significant adverse prognostic factor. FAK positivity also correlated with the number of metastatic lymph nodes ($p = 0.048$) and incidence of distant metastases ($p = 0.009$) [90].

17.5 Development and Evaluation of FAK Inhibitors

Based on its high expression levels in a variety of solid tumors, including HNSCC, as well as its involvement in multiple neoplastic processes, FAK may be an excellent molecule to target using small molecular inhibitors [91]. A number of small-molecule kinase inhibitors have been developed, including TAE226 (also designated NVP-TAE226) [92, 93], PF-562271 [94], PF-573228 [94], PF-04554878/VS-6063/defactinib [95, 96], VS-4718 (also designated PND-1186) [97], GSK2256098 [98, 99], NVP-TAC544 [76], 1,2,4,5-benzenetetraamine tetrahydrochloride (Y15) [100], chloropyramine hydrochloride (C4) [101, 102], OXA11 [103], INT2–31 [104, 105], M13 inhibitor [106], and Roslin 2 (R2) [107] (Table 17.1); some of these also inhibit the FAK paralog PTK2B/PYK2.

However, there are several challenges to this approach, including most notably the kinase-independent effects of FAK. As demonstrated earlier in this chapter, FAK has significant effects based largely on its scaffolding ability; thus, any drug design targeting FAK must take this function into account. Of the available FAK inhibitors in development for clinical use, several affect the protein-protein interactions of FAK (Table 17.1). Additionally, based on the general experience of clinical researchers in testing small-molecule inhibitors of unique protein targets in the treatment of cancer, inhibition of FAK as monotherapy may not be sufficient to induce significant tumor responses. However, multiple groups have generated promising data investigating the combination of FAK inhibition with cytotoxic chemotherapy, targeted therapy, immunotherapy, or radiotherapy.

Table 17.1 List of currently available FAK inhibitors

Inhibitor name	Targets	Phase	Trials identifier
TAE 226	ATP-binding site region, Pyk2	Preclinical	None
	IGF-1R		
PF-562271	ATP-binding site region, PYK2,	Phase I	NCT00666926
PF-573228	ATP-binding site region	Preclinical	None
PF-04554878	ATP-binding inhibitor	Phase I/Ib and II	NCT01951690,
(VS-6063, defactinib)			NCT00787033,
			NCT01943292,
			NCT02004028 and
			NCT01778803
VS-4718	ATP-binding site region	Phase I	NCT01849744
VS-5059	ATP-binding site region		None
GSK2256098		Phase I	NCT01938443,
			NCT01138033 and
			NCT00996671
NVP-TAC544		Preclinical	None
1,2,4,5-benzenetetraamine tetrahydrochloride (Y15)	Targets Y397 and Y418	Preclinical	None
Chloropyramine hydrochloride (C4)	Disrupts FAK-VEGFR3 interaction	Preclinical	None
OXA11	Targets Y397 and Y861	Preclinical	None
INT2-31	Targets FAK and IGFR-1	Preclinical	None
	Interaction		
M13 inhibitor	Inhibits the interaction between FAK and Mdm2	Preclinical	None
Roslin 2 (R2)	Inhibits the interaction between FAK and p53 interaction	Preclinical	None

17.5.1 FAK Inhibition as a Chemotherapeutic Sensitizer

As noted above, activated FAK can induce multiple effectors classically associated with resistance to standard chemotherapeutic agents, including PI3 Kinase and ERK. Moreover, in several solid tumor types, ECM signaling to integrin provides significant chemotherapeutic resistance, at least partially mediated by FAK [108]. Because of this, a variety of chemotherapeutic agents have been combined with FAK inhibition to achieve synergistic effect. For example, the FAK kinase inhibitor, PF-573228, was found to reverse doxorubicin resistance in lung cancer cell lines [109]. In a breast cancer model, concurrent administration of the FAK kinase inhibitors VS-6063 or VS-4718 with paclitaxel and carboplatin led to decreased enrichment of CSCs as well as delayed tumor regrowth following the cessation of treatment [110]. In ovarian cancer, the FAK kinase inhibitor TAE226 combined with docetaxel had a profound effect on several different models – much greater than either agent alone – leading to tumor regression and increased animal survival [111]. Not only did this combination lead to decreased tumor cell proliferation, but also to increased apoptosis in tumor-associated endothelial cells, arguing for the importance of FAK in maintaining tumor vasculature.

Another FAK kinase inhibitor, VS-6063, had similar synergistic effects combined with paclitaxel, also in an ovarian cancer model [95]. In a colon cancer preclinical model, the FAK kinase inhibitor 1,2,4,5-benzenetetraamine tetrahydrochloride (Y15) exhibited synergy with either 5-FU or oxaliplatin on tumor growth and apoptosis [112]. In preclinical models Y15 also demonstrated synergy with gemcitabine in pancreatic cancer and temozolomide in glioblastoma [39, 113]. An additional FAK-targeted therapy is the agent Roslin 2 (R2). This drug targets the interaction between p53 and FAK, relieving the repression of p53 transcriptional function. This agent, in combination with doxorubicin or 5-FU, led to significant tumor regression in p53 wild type but not p53 null xenografts [107].

Despite promising data from several solid tumor types for combinatorial therapy with FAK inhibition and chemotherapy, as well as several studies showing efficacy in preclinical models of FAK inhibitor alone in solid tumors, published data for combinations of FAK inhibition and cytotoxic chemotherapy in HNSCC is yet not available [114]. In our experience, VS-6063 leads to potent sensitization to cisplatin in vitro in several HNSCC cell lines based on clonogenic assay (Skinner et al., unpublished observation). We would encourage investigators to continue this avenue of exploration in HNSCC, particularly considering that the initial phase I data from PF-562271 (NCT00666926) included three patients with HNSCC, two of whom had a significant metabolic response [115].

In addition, FAK inhibitors may be promising candidates for combination with specific targeted therapies. For example, as discussed above, FAK mediates VEGFR growth factor signaling, and FAK plays a role in formation of the vasculature in embryogenesis and in tumors. In one recent study [116], combined therapy with a FAK inhibitor and an antiangiogenic agent (pazopanib or bevacizumab) reduced tumor growth and reduced negative effects after cessation of antiangiogenic therapy in a mouse model of ovarian cancer. Other combinations can be envisioned, in which a FAK inhibitor is combined with inhibition of EGFR, c-MET, or other receptor tyrosine kinases relevant to HNSCC.

17.5.2 Targeting Integrins and FAK for Radiosensitization

Similar to what has been observed with cytotoxic chemotherapy, cancer cells derived from multiple different tumor types as well as several different types of nonmalignant lineages exhibit greater radioresistance when bound to the ECM [117]. This phenomenon is dubbed cell adhesion-mediated radioresistance (CAM-RR). This process appears to be dependent upon integrin expression generally, if not necessarily the specific constituents of the ECM. The exact mechanism underlying this phenomenon is not clear; however, much research in this sphere has focused on the relationship between integrins and FAK, utilizing HNSCC as a model.

Antibody-mediated inhibition of β1-integrin led to the radiosensitization of multiple HPV-negative HNSCC cell lines maintained in 3D culture, as well as in an immunodeficient in vivo model [118, 119]. In this context, β1-integrin inhibition impaired repair of double-stranded DNA damage following radiation, primarily via effects on nonhomologous end joining (NHEJ) [120]. Moreover, the observed radiosensitization was directly dependent upon FAK, such that a similar effect was seen following FAK inhibition via siRNA, and was reversed via expression of constitutively active FAK. Interestingly, in a HNSCC cell type resistant to β1-integrin inhibition-mediated radiosensitization, no characteristic morphologic change following β1-integrin inhibition was observed. In regard to toxicity profile, the strategy of β1-integrin blockade was associated with decreased radiation-associated oral mucositis in preclinical mouse models, a toxic side effect which is usually a significant limitation to radiotherapy in head and neck cancers [121]. In independent studies, an antibody specific for αv-integrin led to increased sensitization of several tumor types, including HPV-negative HNSCC [122, 123]. Combinatorial therapy in these models increased apoptosis both within the tumor cells and in tumor-associated endothelium. Although not directly assessed in the HNSCC studies, the antibody utilized in this study, CNTO 95, is known to profoundly inhibit FAK activity [124]. Additional studies of antibodies targeting multiple α-integrins indicate that inhibition of α3-integrin led to more potent radiosensitization in HPV-negative HNSCC compared to α2-, α5-, or α6-integrin [125].

As mentioned above, direct inhibition of FAK itself has also been examined as a radiosensitizing strategy. FAK siRNA, or FAK chemical inhibition using TAE226, causes in vitro radiosensitization of a variety of HPV-negative HNSCC and several other epithelial cell lines, possibly due to downregulation or PI3 kinase and/or MEK/ERK signaling [85, 126, 127]. Moreover, inhibition of FAK via siRNA or small-molecule inhibitor leads to impaired repair of DNA damage following radiation in both HNSCC and NSCLC [85, 128]. However, in a separate study, complete knockout of FAK led to both radioresistance and improved DNA damage repair kinetics following radiotherapy in SCC cells both in vitro and following their injection as xenografts into nude mice [129]. Interestingly, this effect appeared to be mediated by functional (wild type) p53. As noted earlier, FAK and p53 participate in significant reciprocal regulation, and, as most of the cell line models used to link FAK and radiosensitization express mutant p53 with deficient functionality, this could account for the observed discrepancy. Moreover, complete knockout of FAK versus knockdown via siRNA or chemical inhibition may result in distinct phenotypes.

Given these somewhat conflicting data, the question remains as to the clinical relevance of FAK to radioresponse in HNSCC. Recently, our group examined expression of a large number of targets in an array of HPV-negative HNSCC cell lines using reverse-phase protein array (RPPA) and mRNA expression array [85].

These results were then correlated with in vitro radioresistance. One of the targets overexpressed in resistant cell lines using both assays was FAK. Similar to other groups, we found that chemical and siRNA-mediated inhibition of FAK led to radiosensitization. Moreover, we examined clinical outcomes in a cohort of HPV-negative HNSCC patients treated with surgery and postoperative radiotherapy. As discussed above, high FAK copy number and mRNA expression were both associated with significantly poorer disease-free survival in this cohort. These findings were replicated in a larger HPV-negative HNSCC cohort from TCGA. Thus, not only does targeting FAK appear to be a strategy for radiosensitization, but FAK may also play a significant role in baseline clinical radioresistance in HNSCC.

17.6 FAK and HPV in HNSCC

The most relevant biomarker in HNSCC is the presence of human papilloma virus (HPV). HPV-associated HNSCC is primarily seen in the oropharynx, specifically the tonsil and base of tongue. Its incidence is on the rise, while the incidence of HPV-negative HNSCC remains flat or is even declining [130]. While HPV-negative HNSCC is primarily related to carcinogen exposure, most notably tobacco, HPV-positive HNSCC has less of an association with environmental carcinogens. Additionally, outcomes following a diagnosis of HPV-positive HNSCC are dramatically better than those seen in HPV-negative HNSCC [2]. The question of FAK in the context of this dichotomy in HNSCCs is relatively unclear, as the majority of studies in HNSCC examining integrins and FAK have largely been performed in HPV-negative HNSCC.

It has been shown that p16 (encoded by *CDKN2A* and which inhibits CDK4/6 and contributes to RB inactivation in untransformed cells), which is highly overexpressed in HPV-associated HNSCC and is a surrogate for HPV infection, is associated with higher α4β1 integrin expression in a murine cell line-based model for HNSCC [131]. Conversely, p16 appears to bind to integrins and inhibit αvβ3-mediated cell spreading, in some contexts [132]. Given the functional differences between distinct integrin heterodimers in distinct tissue types, it is difficult to draw conclusions about the functional significance of the p16/integrin relationship without further study. Immortalization of epithelial cells via overexpression of the HPV18 E6 and E7 oncoproteins, which inhibit p53 and RB, leads to increased FAK and paxillin activity, possibly via increased production of the ECM component fibronectin [133]. Similar results were seen in several cervical SCC lines, which also harbor HPV. Moreover, increased levels of both FAK and paxillin were seen in HPV-positive HNSCC tumor samples (14 total) compared to their HPV-negative counterparts (186 total) following proteomic profiling [134].

Beside its activities in regulating p53, the E6 protein can bind to both paxillin and fibrillin 1, although the functional consequences of this binding are unclear [135, 136]. However, both E6 and E7 proteins can independently inhibit anoikis [137]. The function of E7 in this capacity appears to be related to its binding of the cytoplasmic RB-associated protein p600 [138]. Due to the integral nature of both integrins and FAK to anoikis resistance, as well as the upregulation of FAK following HPV transformation, it stands to reason that FAK may play a key role in this process; however, this has not been definitively shown.

One additional key distinction between these HPV-related and HPV-negative malignancies is p53. While in HPV-negative HNSCC, p53 is frequently mutated (in up to 80% of tumors in some series), the incidence in HPV-positive tumors is less than 10% [139]. However, p53 expression is generally repressed in HPV-positive HNSCC via binding to the E6 viral protein and its subsequent degradation. Interestingly, the activity of p53 in HPV-positive HNSCC appears to increase to some degree following radiotherapy, which is a potential mechanism that partially accounts for the enhanced radiosensitivity of these tumors [140]. As p53 participates in a reciprocal regulation with FAK, this effect may have some bearing on the potential of utilizing FAK as a therapeutic sensitizer in HPV-positive HNSCC.

17.7 Conclusions

FAK and its relationship with both upstream and downstream signaling is quite complicated, likely due to the fact that FAK participates in a broad array of cellular processes. This kinase provides a unique of opportunity for drug development and therapeutic sensitization in HNSCC due not only to its effects on cellular motility, invasion, and metastasis but also due to the observation that FAK inhibition can lead to sensitization to a wide array of conventional chemotherapeutics as well as radiotherapy. Both the question of the comparative relevance of FAK to HPV-positive versus HPV-negative HNSCC and the ideal therapeutic combination in both diseases remain open areas of investigation.

Bibliography

1. Siegel RL, Miller KD, Jemal A. Cancer statistics, 2015. CA Cancer J Clin. 2015;65(1):5–29.
2. Ang KK, Harris J, Wheeler R, Weber R, Rosenthal DI, Nguyen-Tân PF, et al. Human papillomavirus and survival of patients with oropharyngeal cancer. N Engl J Med. 2010;363(1):24–35.
3. Marur S, D'Souza G, Westra WH, Forastiere AA. HPV-associated head and neck cancer: a virus-related cancer epidemic – a review of epidemiology, biology, virus detection and issues in management. Lancet Oncol. 2010;11(8):781–9.
4. Alberts B, Johnson A, Lewis J, Raff M, Roberts K, Walter P. Integrins. 2002. Available from: https://www.ncbi.nlm.nih.gov/books/NBK26867/.
5. Kanner SB, Reynolds AB, Parsons JT. Immunoaffinity purification of tyrosine-phosphorylated cellular proteins. J Immunol Methods. 1989;120(1):115–24.
6. Mitra SK, Schlaepfer DD. Integrin-regulated FAK–Src signaling in normal and cancer cells. Curr Opin Cell Biol. 2006;18(5):516–23.
7. Yoon H, Dehart JP, Murphy JM, Lim S-TS. Understanding the roles of FAK in Cancer: inhibitors, genetic models, and new insights. J Histochem Cytochem. 2015;63(2):114–28.

8. Golubovskaya VM. Targeting FAK in human cancer: from finding to first clinical trials. Front Biosci. 2014;19:687–706.
9. Abedi H, Zachary I. Vascular endothelial growth factor stimulates tyrosine phosphorylation and recruitment to new focal adhesions of focal adhesion kinase and paxillin in endothelial cells. J Biol Chem. 1997;272(24):15442–51.
10. Laurent-Puig P, Lievre A, Blons H. Mutations and response to epidermal growth factor receptor inhibitors. Clin Cancer Res Off J Am Assoc Cancer Res. 2009;15(4):1133–9.
11. Saito Y, Mori S, Yokote K, Kanzaki T, Saito Y, Morisaki N. Phosphatidylinositol 3-kinase activity is required for the activation process of focal adhesion kinase by platelet-derived growth factor. Biochem Biophys Res Commun. 1996;224(1):23–6.
12. Danilkovitch A, Leonard EJ. Kinases involved in MSP/RON signaling. J Leukoc Biol. 1999;65(3):345–8.
13. Aoto H, Sasaki H, Ishino M, Sasaki T. Nuclear translocation of cell adhesion kinase beta/proline-rich tyrosine kinase 2. Cell Struct Funct. 2002;27(1):47–61.
14. Lim S-T, Chen XL, Lim Y, Hanson DA, Vo T-T, Howerton K, et al. Nuclear FAK promotes cell proliferation and survival through FERM-enhanced p53 degradation. Mol Cell. 2008;29(1):9–22.
15. Serrels A, Lund T, Serrels B, Byron A, McPherson RC, von Kriegsheim A, et al. Nuclear FAK controls chemokine transcription, tregs, and evasion of anti-tumor immunity. Cell. 2015;163(1):160–73.
16. Polte TR, Hanks SK. Interaction between focal adhesion kinase and Crk-associated tyrosine kinase substrate p130Cas. Proc Natl Acad Sci U S A. 1995;92(23):10678–82.
17. Law SF, Estojak J, Wang B, Mysliwiec T, Kruh G, Golemis EA. Human enhancer of filamentation 1, a novel p130cas-like docking protein, associates with focal adhesion kinase and induces pseudohyphal growth in Saccharomyces cerevisiae. Mol Cell Biol. 1996;16(7):3327–37.
18. Hildebrand JD, Taylor JM, Parsons JT. An SH3 domain-containing GTPase-activating protein for rho and Cdc42 associates with focal adhesion kinase. Mol Cell Biol. 1996;16(6):3169–78.
19. Wu X, Gan B, Yoo Y, Guan J-L. FAK-mediated src phosphorylation of endophilin A2 inhibits endocytosis of MT1-MMP and promotes ECM degradation. Dev Cell. 2005;9(2):185–96.
20. Hildebrand JD, Schaller MD, Parsons JT. Paxillin, a tyrosine phosphorylated focal adhesion-associated protein binds to the carboxyl terminal domain of focal adhesion kinase. Mol Biol Cell. 1995;6(6):637–47.
21. Chen HC, Appeddu PA, Parsons JT, Hildebrand JD, Schaller MD, Guan JL. Interaction of focal adhesion kinase with cytoskeletal protein Talin. J Biol Chem. 1995;270(28):16995–9.
22. Longhurst CM, Jennings LK. Integrin-mediated signal transduction. Cell Mol Life Sci CMLS. 1998;54(6):514–26.
23. Wang P, Ballestrem C, Streuli CH. The C terminus of Talin links integrins to cell cycle progression. J Cell Biol. 2011;195(3):499–513.
24. Lawson C, Lim S-T, Uryu S, Chen XL, Calderwood DA, Schlaepfer DD. FAK promotes recruitment of Talin to nascent adhesions to control cell motility. J Cell Biol 2012;196(3):223–232.
25. Zhang X, Chattopadhyay A, Ji Q, Owen JD, Ruest PJ, Carpenter G, et al. Focal adhesion kinase promotes phospholipase C-γ1 activity. Proc Natl Acad Sci U S A. 1999;96(16):9021–6.
26. Chen H-C, Appeddu PA, Isoda H, Guan J-L. Phosphorylation of tyrosine 397 in focal adhesion kinase is required for binding phosphatidylinositol 3-kinase. J Biol Chem. 1996;271(42):26329–34.
27. Calalb MB, Polte TR, Hanks SK. Tyrosine phosphorylation of focal adhesion kinase at sites in the catalytic domain regulates kinase activity: a role for Src family kinases. Mol Cell Biol. 1995;15(2):954–63.
28. Li L, Okura M, Imamoto A. Focal adhesions require catalytic activity of Src family kinases to mediate integrin-matrix adhesion. Mol Cell Biol. 2002;22(4):1203–17.
29. Westhoff MA, Serrels B, Fincham VJ, Frame MC, Carragher NO. Src-mediated phosphorylation of focal adhesion kinase couples actin and adhesion dynamics to survival signaling. Mol Cell Biol. 2004;24(18):8113–33.

30. Abu-Ghazaleh R, Kabir J, Jia H, Lobo M, Zachary I. Src mediates stimulation by vascular endothelial growth factor of the phosphorylation of focal adhesion kinase at tyrosine 861, and migration and anti-apoptosis in endothelial cells. Biochem J. 2001;360(Pt 1):255–64.
31. Sridhar SC, Miranti CK. Tetraspanin KAI1/CD82 suppresses invasion by inhibiting integrin-dependent crosstalk with c-met receptor and Src kinases. Oncogene. 2006;25(16):2367–78.
32. Arold ST, Hoellerer MK, Noble MEM. The structural basis of localization and signaling by the focal adhesion targeting domain. Structure. 2002; 10(3):319–27. London, England: 1993.
33. Flinder LI, Timofeeva OA, Rosseland CM, Wierød L, Huitfeldt HS, Skarpen E. EGF-induced ERK-activation downstream of FAK requires rac1-NADPH oxidase. J Cell Physiol. 2011;226(9):2267–78.
34. Sieg DJ, Ilić D, Jones KC, Damsky CH, Hunter T, Schlaepfer DD. Pyk2 and Src-family protein-tyrosine kinases compensate for the loss of FAK in fibronectin-stimulated signaling events but Pyk2 does not fully function to enhance FAK-cell migration. EMBO J. 1998;17(20):5933–47.
35. Yue Y, Li Z-N, Fang Q-G, Zhang X, Yang L-L, Sun C-F, et al. The role of Pyk2 in the CCR7-mediated regulation of metastasis and viability in squamous cell carcinoma of the head and neck cells in vivo and in vitro. Oncol Rep. 2015;34(6):3280–7.
36. Paoli P, Giannoni E, Chiarugi P. Anoikis molecular pathways and its role in cancer progression. Biochim Biophys Acta BBA – Mol Cell Biol Lipids. 2013;1833(12):3481–98.
37. Frisch SM, Francis H. Disruption of epithelial cell-matrix interactions induces apoptosis. J Cell Biol. 1994;124(4):619–26.
38. Sonoda Y, Matsumoto Y, Funakoshi M, Yamamoto D, Hanks SK, Kasahara T. Anti-apoptotic role of focal adhesion kinase (FAK) induction of inhibitor-of-apoptosis proteins and apoptosis suppression by the overexpression of fak in a human leukemic cell line, HL-60. J Biol Chem. 2000;275(21):16309–15.
39. Golubovskaya VM, Huang G, Ho B, Yemma M, Morrison CD, Lee J, Eliceiri BP, Cance WG. Pharmacologic blockade of FAK autophosphorylation decreases human glioblastoma tumor growth and synergizes with temozolomide. Mol Cancer Ther. 2013;12(2):162–72. https://doi.org/10.1158/1535-7163. MCT-12-0701. Epub 2012 Dec 12. PubMed PMID: 23243059; PubMed Central PMCID: PMC3570595.
40. Golubovskaya VM, Finch R, Kweh F, Massoll NA, Campbell-Thompson M, Wallace MR, et al. p53 regulates FAK expression in human tumor cells. Mol Carcinog. 2008;47(5):373–82.
41. Golubovskaya VM, Conway-Dorsey K, Edmiston SN, Tse C-K, Lark AA, Livasy CA, et al. FAK overexpression and p53 mutations are highly correlated in human breast cancer. Int J Cancer. 2009;125(7):1735–8.
42. Schlaepfer DD, Jones KC, Hunter T. Multiple Grb2-mediated integrin-stimulated signaling pathways to ERK2/mitogen-activated protein kinase: summation of both c-Src- and focal adhesion kinase-initiated tyrosine phosphorylation events. Mol Cell Biol. 1998;18(5):2571–85.
43. Xia H, Nho RS, Kahm J, Kleidon J, Henke CA. Focal adhesion kinase is upstream of phosphatidylinositol 3-kinase/Akt in regulating fibroblast survival in response to contraction of type I collagen matrices via a beta 1 integrin viability signaling pathway. J Biol Chem. 2004;279(31):33024–34.
44. Kamarajan P, Bunek J, Lin Y, Nunez G, Kapila YL. Receptor-interacting protein shuttles between cell death and survival signaling pathways. Mol Biol Cell. 2010;21(3):481–8.
45. Tong L, Tergaonkar V. Rho protein GTPases and their interactions with NFκB: crossroads of inflammation and matrix biology. Biosci Rep. 2014;34(3):e00115. https://doi.org/10.1042/BSR20140021. Review. PubMed PMID: 24877606; PubMed Central PMCID: PMC4069681.
46. You D, Xin J, Volk A, Wei W, Schmidt R, Scurti G, et al. FAK mediates a compensatory survival signal parallel to PI3K-AKT in PTEN-null T-ALL cells. Cell Rep. 2015;10(12):2055–68.
47. O'Neill GM, Seo S, Serebriiskii IG, Lessin SR, Golemis EA. A new central scaffold for metastasis: parsing HEF1/Cas-L/NEDD9. Cancer Res. 2007;67(19):8975–9.
48. Nikonova AS, Gaponova AV, Kudinov AE, Golemis EA. CAS proteins in health and disease: an update. IUBMB Life. 2014;66(6):387–95.

49. O'Neill GM, Fashena SJ, Golemis EA. Integrin signalling: a new cas(t) of characters enters the stage. Trends Cell Biol. 2000;10(3):111–9.
50. Lucas JT, Salimath BP, Slomiany MG, Rosenzweig SA. Regulation of invasive behavior by vascular endothelial growth factor is HEF1-dependent. Oncogene. 2010;29(31):4449–59.
51. Schaller M. Paxillin: a focal adhesion-associated adaptor protein. Oncogene. 2001;20(44):6459–72.
52. Lim Y, Lim S-T, Tomar A, Gardel M, Bernard-Trifilo JA, Chen XL, et al. PyK2 and FAK connections to p190Rho guanine nucleotide exchange factor regulate RhoA activity, focal adhesion formation, and cell motility. J Cell Biol. 2008;180(1):187–203.
53. Sieg DJ, Hauck CR, Schlaepfer DD. Required role of focal adhesion kinase (FAK) for integrin-stimulated cell migration. J Cell Sci. 1999;112(Pt 16):2677–91.
54. Tilghman R, Parsons JT. Focal adhesion kinase as a regulator of cell tension in the progression of cancer. Semin Cancer Biol. 2008;18(1):45–52.
55. Ammer A, Weed S. Cortactin branches out: roles in regulating protrusive actin dynamics. Cell Motil Cytoskelet. 2008;65(9):687–707.
56. Segarra M, Vilardell C, Matsumoto K, Esparza J, Lozano E, Serra-Pages C, et al. Dual function of focal adhesion kinase in regulating integrin-induced MMP-2 and MMP-9 release by human T lymphoid cells. FASEB J Off Publ Fed Am Soc Exp Biol. 2005;19(13):1875–7.
57. Kwiatkowska A, Kijewska M, Lipko M, Hibner U, Kaminska B. Downregulation of Akt and FAK phosphorylation reduces invasion of glioblastoma cells by impairment of MT1-MMP shuttling to lamellipodia and downregulates MMPs expression. Biochim Biophys Acta. 2011;1813(5):655–67.
58. Pan Y-R, Chen C-L, Chen H-C. FAK is required for the assembly of podosome rosettes. J Cell Biol. 2011;195(1):113–29.
59. Li X-Y, Zhou X, Rowe RG, Hu Y, Schlaepfer DD, Ilić D, et al. Snail1 controls epithelial–mesenchymal lineage commitment in focal adhesion kinase–null embryonic cells. J Cell Biol. 2011;195(5):729–38.
60. Cicchini C, Laudadio I, Citarella F, Corazzari M, Steindler C, Conigliaro A, et al. TGFbeta-induced EMT requires focal adhesion kinase (FAK) signaling. Exp Cell Res. 2008;314(1):143–52.
61. Cicchini C, Laudadio I, Citarella F, Corazzari M, Steindler C, Conigliaro A, et al. TGFβ-induced EMT requires focal adhesion kinase (FAK) signaling. Exp Cell Res. 2008;314(1):143–52.
62. Saito D, Kyakumoto S, Chosa N, Ibi M, Takahashi N, Okubo N, et al. Transforming growth factor-β1 induces epithelial–mesenchymal transition and integrin α3β1-mediated cell migration of HSC-4 human squamous cell carcinoma cells through Slug. J Biochem (Tokyo). 2013;153(3):303–15.
63. Avizienyte E, Wyke AW, Jones RJ, McLean GW, Westhoff MA, Brunton VG, et al. Src-induced de-regulation of E-cadherin in colon cancer cells requires integrin signalling. Nat Cell Biol. 2002;4(8):632–8.
64. Taliaferro-Smith L, Oberlick E, Liu T, McGlothen T, Alcaide T, Tobin R, et al. FAK activation is required for IGF1R-mediated regulation of EMT, migration, and invasion in mesenchymal triple negative breast cancer cells. Oncotarget. 2015;6(7):4757–72.
65. Rodrigo JP, Dominguez F, Suárez V, Canel M, Secades P, Chiara MD. Focal adhesion kinase and E-cadherin as markers for nodal metastasis in laryngeal cancer. Arch Otolaryngol Head Neck Surg. 2007;133(2):145–50.
66. Fan H, Zhao X, Sun S, Luo M, Guan J-L. Function of focal adhesion kinase scaffolding to mediate Endophilin A2 phosphorylation promotes epithelial-mesenchymal transition and mammary cancer stem cell activities in vivo. J Biol Chem. 2013;288(5):3322–33.
67. Te Boekhorst V, Preziosi L, Friedl P. Plasticity of cell migration in vivo and in silico. Annu Rev Cell Dev Biol. 2016;32:491–526.
68. Sanz-Moreno V, Gadea G, Ahn J, Paterson H, Marra P, Pinner S, et al. Rac activation and inactivation control plasticity of tumor cell movement. Cell. 2008;135(3):510–23.

69. Ahn J, Sanz-Moreno V, Marshall CJ. The metastasis gene NEDD9 product acts through integrin β3 and Src to promote mesenchymal motility and inhibit amoeboid motility. J Cell Sci. 2012;125(Pt 7):1814–26.
70. Ilić D, Furuta Y, Kanazawa S, Takeda N, Sobue K, Nakatsuji N, et al. Reduced cell motility and enhanced focal adhesion contact formation in cells from FAK-deficient mice. Nature. 1995;377(6549):539–44.
71. Shen T-L, Park AY-J, Alcaraz A, Peng X, Jang I, Koni P, et al. Conditional knockout of focal adhesion kinase in endothelial cells reveals its role in angiogenesis and vascular development in late embryogenesis. J Cell Biol. 2005;169(6):941–52.
72. Zhao X, Peng X, Sun S, Park AYJ, Guan J-L. Role of kinase-independent and -dependent functions of FAK in endothelial cell survival and barrier function during embryonic development. J Cell Biol. 2010;189(6):955–65.
73. Tavora B, Batista S, Reynolds LE, Jadeja S, Robinson S, Kostourou V, et al. Endothelial FAK is required for tumour angiogenesis. EMBO Mol Med. 2010;2(12):516–28.
74. Mitra SK, Mikolon D, Molina JE, Hsia DA, Hanson DA, Chi A, et al. Intrinsic FAK activity and Y925 phosphorylation facilitate an angiogenic switch in tumors. Oncogene. 2006;25(44):5969–84.
75. Kostourou V, Lechertier T, Reynolds LE, Lees DM, Baker M, Jones DT, et al. FAK-heterozygous mice display enhanced tumour angiogenesis. Nat Commun. 2013;4:2020: 1–11.
76. Weis SM, Lim S-T, Lutu-Fuga KM, Barnes LA, Chen XL, Göthert JR, et al. Compensatory role for Pyk2 during angiogenesis in adult mice lacking endothelial cell FAK. J Cell Biol. 2008;181(1):43–50.
77. Luo M, Fan H, Nagy T, Wei H, Wang C, Liu S, et al. Mammary epithelial-specific ablation of the focal adhesion kinase suppresses mammary tumorigenesis by affecting mammary cancer stem/progenitor cells. Cancer Res. 2009;69(2):466–74.
78. Fan H, Guan J-L. Compensatory function of Pyk2 protein in the promotion of focal adhesion kinase (FAK)-null mammary cancer stem cell tumorigenicity and metastatic activity. J Biol Chem. 2011;286(21):18573–82.
79. Luo M, Zhao X, Chen S, Liu S, Wicha MS, Guan J-L. Distinct FAK activities determine progenitor and mammary stem cell characteristics. Cancer Res. 2013;73(17):5591–602.
80. Schober M, Fuchs E. Tumor-initiating stem cells of squamous cell carcinomas and their control by TGF-β and integrin/focal adhesion kinase (FAK) signaling. Proc Natl Acad Sci U S A. 2011;108(26):10544–9.
81. Kobayashi K, Takahashi H, Inoue A, Harada H, Toshimori S, Kobayashi Y, et al. Oct-3/4 promotes migration and invasion of glioblastoma cells. J Cell Biochem. 2012;113(2):508–17.
82. Lin Y-L, Han Z-B, Xiong F-Y, Tian L-Y, Wu X-J, Xue S-W, et al. Malignant transformation of 293 cells induced by ectopic expression of human Nanog. Mol Cell Biochem. 2011;351(1–2):109–16.
83. Ho B, Olson G, Figel S, Gelman I, Cance WG, Golubovskaya VM. Nanog increases focal adhesion kinase (FAK) promoter activity and expression and directly binds to FAK protein to be phosphorylated. J Biol Chem. 2012;287(22):18656–73.
84. Kandoth C, McLellan MD, Vandin F, Ye K, Niu B, Lu C, et al. Mutational landscape and significance across 12 major cancer types. Nature. 2013;502(7471):333–9.
85. Skinner HD, Giri U, Yang L, Woo SH, Story MD, Pickering CR, Byers LA, Williams MD, El-Naggar A, Wang J, Diao L, Shen L, Fan YH, Molkentine DP, Beadle BM, Meyn RE, Myers JN, Heymach JV. Proteomic Profiling Identifies PTK2/FAK as a Driver of Radioresistance in HPV-negative Head and Neck Cancer. Clin Cancer Res. 2016;22(18):4643–50. https://doi.org/10.1158/1078-0432.CCR-15-2785. Epub 2016 Apr 1. PubMed PMID: 27036135; PubMed Central PMCID: PMC5061056.
86. Recher C, Ysebaert L, Beyne-Rauzy O, Mas VM-D, Ruidavets J-B, Cariven P, et al. Expression of focal adhesion kinase in acute myeloid leukemia is associated with enhanced blast migration, increased cellularity, and poor prognosis. Cancer Res. 2004;64(9):3191–7.

87. Charpin C, Secq V, Giusiano S, Carpentier S, Andrac L, Lavaut M-N, et al. A signature predictive of disease outcome in breast carcinomas, identified by quantitative immunocytochemical assays. Int J Cancer. 2009;124(9):2124–34.
88. Schmitz KJ, Grabellus F, Callies R, Otterbach F, Wohlschlaeger J, Levkau B, et al. High expression of focal adhesion kinase (p125FAK) in node-negative breast cancer is related to overexpression of HER-2/neu and activated Akt kinase but does not predict outcome. Breast Cancer Res. 2005;7:R194.
89. Canel M, Secades P, Rodrigo J-P, Cabanillas R, Herrero A, Suarez C, et al. Overexpression of focal adhesion kinase in head and neck squamous cell carcinoma is independent of fak gene copy number. Clin Cancer Res Off J Am Assoc Cancer Res. 2006;12(11 Pt 1):3272–9.
90. Omura G, Ando M, Saito Y, Kobayashi K, Yoshida M, Ebihara Y, et al. Association of the upregulated expression of focal adhesion kinase with poor prognosis and tumor dissemination in hypopharyngeal cancer. Head Neck. 2016;38(8):1164–9.
91. Lee BY, Timpson P, Horvath LG, Daly RJ. FAK signaling in human cancer as a target for therapeutics. Pharmacol Ther. 2015;146:132–49.
92. Liu W, Bloom DA, Cance WG, Kurenova EV, Golubovskaya VM, Hochwald SN. FAK and IGF-IR interact to provide survival signals in human pancreatic adenocarcinoma cells. Carcinogenesis. 2008;29(6):1096–107.
93. Golubovskaya VM, Ho B, Zheng M, Magis A, Ostrov D, Cance WG. Mitoxantrone targets the ATP-binding site of FAK, binds the FAK kinase domain and decreases FAK, Pyk-2, c-Src, and IGF-1R in vitro kinase activities. Anti Cancer Agents Med Chem. 2013;13(4):546–54.
94. Wendt MK, Schiemann WP. Therapeutic targeting of the focal adhesion complex prevents oncogenic TGF-beta signaling and metastasis. Breast Cancer Res BCR. 2009;11(5):R68.
95. Kang Y, Hu W, Ivan C, Dalton HJ, Miyake T, Pecot CV, et al. Role of focal adhesion kinase in regulating YB-1-mediated paclitaxel resistance in ovarian cancer. J Natl Cancer Inst. 2013;105(19):1485–95.
96. Hallur G, Tamizharasan N, Sulochana SP, Saini NK, Zainuddin M, Mullangi R. LC-ESI-MS/MS determination of defactinib, a novel FAK inhibitor in mice plasma and its application to a pharmacokinetic study in mice. J Pharm Biomed Anal. 2017;149:358–64.
97. Walsh C, Tanjoni I, Uryu S, Tomar A, Nam J-O, Luo H, et al. Oral delivery of PND-1186 FAK inhibitor decreases tumor growth and spontaneous breast to lung metastasis in preclinical models. Cancer Biol Ther. 2010;9(10):778–90.
98. Zhang J, He D-H, Zajac-Kaye M, Hochwald SN. A small molecule FAK kinase inhibitor, GSK2256098, inhibits growth and survival of pancreatic ductal adenocarcinoma cells. Cell Cycle Georget Tex. 2014;13(19):3143–9.
99. Soria JC, Gan HK, Blagden SP, Plummer R, Arkenau HT, Ranson M, et al. A phase I, pharmacokinetic and pharmacodynamic study of GSK2256098, a focal adhesion kinase inhibitor, in patients with advanced solid tumors. Ann Oncol Off J Eur Soc Med Oncol. 2016;27(12):2268–74.
100. O'Brien S, Golubovskaya VM, Conroy J, Liu S, Wang D, Liu B, et al. FAK inhibition with small molecule inhibitor Y15 decreases viability, clonogenicity, and cell attachment in thyroid cancer cell lines and synergizes with targeted therapeutics. Oncotarget. 2014;5(17):7945–59.
101. Kurenova EV, Hunt DL, He D, Magis AT, Ostrov DA, Cance WG. Small molecule chloropyramine hydrochloride (C4) targets the binding site of focal adhesion kinase and vascular endothelial growth factor receptor 3 and suppresses breast cancer growth in vivo. J Med Chem. 2009;52(15):4716–24.
102. Kurenova E, Ucar D, Liao J, Yemma M, Gogate P, Bshara W, et al. A FAK scaffold inhibitor disrupts FAK and VEGFR-3 signaling and blocks melanoma growth by targeting both tumor and endothelial cells. Cell Cycle Georget Tex. 2014;13(16):2542–53.
103. Moen I, Gebre M, Alonso-Camino V, Chen D, Epstein D, McDonald DM. Anti-metastatic action of FAK inhibitor OXA-11 in combination with VEGFR-2 signaling blockade in pancreatic neuroendocrine tumors. Clin Exp Metastasis. 2015;32(8):799–817.

104. Ucar DA, Cox A, He D-H, Ostrov DA, Kurenova E, Hochwald SN. A novel small molecule inhibitor of FAK and IGF-1R protein interactions decreases growth of human esophageal carcinoma. Anti Cancer Agents Med Chem. 2011;11(7):629–37.
105. Ucar DA, Kurenova E, Garrett TJ, Cance WG, Nyberg C, Cox A, et al. Disruption of the protein interaction between FAK and IGF-1R inhibits melanoma tumor growth. Cell Cycle Georget Tex. 2012;11(17):3250–9.
106. Golubovskaya VM, Palma NL, Zheng M, Ho B, Magis A, Ostrov D, et al. A small-molecule inhibitor, 5′-O-tritylthymidine, targets FAK and Mdm-2 interaction, and blocks breast and colon tumorigenesis in vivo. Anti Cancer Agents Med Chem. 2013;13(4):532–45.
107. Golubovskaya VM, Ho B, Zheng M, Magis A, Ostrov D, Morrison C, et al. Disruption of focal adhesion kinase and p53 interaction with small molecule compound R2 reactivated p53 and blocked tumor growth. BMC Cancer. 2013;13(1):342.
108. Aoudjit F, Vuori K. Integrin signaling in cancer cell survival and chemoresistance [internet]. Chemotherapy Research and Practice. 2012. Available from: https://www.hindawi.com/journals/cherp/2012/283181/.
109. Dragoj M, Milosevic Z, Bankovic J, Tanic N, Pesic M, Stankovic T. Targeting CXCR4 and FAK reverses doxorubicin resistance and suppresses invasion in non-small cell lung carcinoma. Cell Oncol Dordr. 2017;40(1):47–62.
110. Kolev VN, Tam WF, Wright QG, McDermott SP, Vidal CM, Shapiro IM, et al. Inhibition of FAK kinase activity preferentially targets cancer stem cells. Oncotarget. 2017;8(31):51733–47.
111. Halder J, Lin YG, Merritt WM, Spannuth WA, Nick AM, Honda T, et al. Therapeutic efficacy of a novel focal adhesion kinase inhibitor TAE226 in ovarian carcinoma. Cancer Res. 2007;67(22):10976–83.
112. Heffler M, Golubovskaya VM, Dunn KMB, Cance W. Focal adhesion kinase autophosphorylation inhibition decreases colon cancer cell growth and enhances the efficacy of chemotherapy. Cancer Biol Ther. 2013;14(8):761–72.
113. Hochwald SN, Nyberg C, Zheng M, Zheng D, Wood C, Massoll NA, et al. A novel small molecule inhibitor of FAK decreases growth of human pancreatic cancer. Cell Cycle. 2009;8(15):2435–43.
114. Kurio N, Shimo T, Fukazawa T, Okui T, Hassan NMM, Honami T, et al. Anti-tumor effect of a novel FAK inhibitor TAE226 against human oral squamous cell carcinoma. Oral Oncol. 2012;48(11):1159–70.
115. Infante JR, Camidge DR, Mileshkin LR, Chen EX, Hicks RJ, Rischin D, et al. Safety, pharmacokinetic, and pharmacodynamic phase I dose-escalation trial of PF-00562271, an inhibitor of focal adhesion kinase, in advanced solid tumors. J Clin Oncol. 2012;30(13):1527–33.
116. Haemmerle M, Bottsford-Miller J, Pradeep S, Taylor ML, Choi H-J, Hansen JM, et al. FAK regulates platelet extravasation and tumor growth after antiangiogenic therapy withdrawal. J Clin Invest. 2016;126(5):1885–96.
117. Sandfort V, Koch U, Cordes DN. Cell adhesion-mediated radioresistance revisited. Int J Radiat Biol. 2007;83(11–12):727–32.
118. Eke I, Deuse Y, Hehlgans S, Gurtner K, Krause M, Baumann M, et al. β_1Integrin/FAK/cortactin signaling is essential for human head and neck cancer resistance to radiotherapy. J Clin Invest. 2012;122(4):1529–40.
119. Eke I, Dickreuter E, Cordes N. Enhanced radiosensitivity of head and neck squamous cell carcinoma cells by β1 integrin inhibition. Radiother Oncol. 2012;104(2):235–42.
120. Dickreuter E, Eke I, Krause M, Borgmann K, van Vugt MA, Cordes N. Targeting of β1 integrins impairs DNA repair for radiosensitization of head and neck cancer cells. Oncogene. 2016;35(11):1353–62.
121. Albert M, Schmidt M, Cordes N, Dörr W. Modulation of radiation-induced oral mucositis (mouse) by selective inhibition of β1 integrin. Radiother Oncol. 2012;104(2):230–4.
122. Ning S, Tian J, Marshall DJ, Knox SJ. Anti-alphav integrin monoclonal antibody intetumumab enhances the efficacy of radiation therapy and reduces metastasis of human cancer xenografts in nude rats. Cancer Res. 2010;70(19):7591–9.

123. Ning S, Nemeth JA, Hanson RL, Forsythe K, Knox SJ. Anti-integrin monoclonal antibody CNTO 95 enhances the therapeutic efficacy of fractionated radiation therapy in vivo. Mol Cancer Ther. 2008;7(6):1569–78.
124. Chen Q, Manning CD, Millar H, McCabe FL, Ferrante C, Sharp C, et al. CNTO 95, a fully human anti alphav integrin antibody, inhibits cell signaling, migration, invasion, and spontaneous metastasis of human breast cancer cells. Clin Exp Metastasis. 2008;25(2):139–48.
125. Steglich A, Vehlow A, Eke I, Cordes N. α integrin targeting for radiosensitization of three-dimensionally grown human head and neck squamous cell carcinoma cells. Cancer Lett. 2015;357(2):542–8.
126. Hehlgans S, Eke I, Cordes N. Targeting FAK radiosensitizes 3-dimensional grown human HNSCC cells through reduced Akt1 and MEK1/2 signaling. Int J Radiat Oncol Biol Phys. 2012;83(5):e669–76.
127. Hehlgans S, Lange I, Eke I, Cordes N. 3D cell cultures of human head and neck squamous cell carcinoma cells are radiosensitized by the focal adhesion kinase inhibitor TAE226. Radiother Oncol. 2009;92(3):371–8.
128. Tang K-J, Constanzo JD, Venkateswaran N, Melegari M, Ilcheva M, Morales JC, et al. Focal adhesion kinase regulates the DNA damage response and its inhibition Radiosensitizes mutant KRAS lung Cancer. Clin Cancer Res Off J Am Assoc Cancer Res. 2016;22(23):5851–63.
129. Graham K, Moran-Jones K, Sansom OJ, Brunton VG, Frame MC. FAK deletion promotes p53-mediated induction of p21, DNA-damage responses and radio-resistance in advanced squamous cancer cells. PLoS One. 2011;6(12):e27806.
130. Chaturvedi AK, Engels EA, Anderson WF, Gillison ML. Incidence trends for human papillomavirus-related and -unrelated oral squamous cell carcinomas in the United States. J Clin Oncol. 2008;26(4):612–9.
131. Dok R, Glorieux M, Holacka K, Bamps M, Nuyts S. Dual role for p16 in the metastasis process of HPV positive head and neck cancers. Mol Cancer. 2017;16:113.
132. Fåhraeus R, Lane DP. The p16(INK4a) tumour suppressor protein inhibits alphavbeta3 integrin-mediated cell spreading on vitronectin by blocking PKC-dependent localization of alphavbeta3 to focal contacts. EMBO J. 1999;18(8):2106–18.
133. McCormack SJ, Brazinski SE, Moore JL, Werness BA, Goldstein DJ. Activation of the focal adhesion kinase signal transduction pathway in cervical carcinoma cell lines and human genital epithelial cells immortalized with human papillomavirus type 18. Oncogene. 1997;15(3):265–74.
134. Byers LA, Diao L, Ng PKS, Heymach C, Fan YH, El-Naggar AK, et al. Proteomic profiling of HPV-positive head and neck cancer to identify new candidates for targeted therapy. J Clin Oncol. 2014;32(15_suppl):6030.
135. Du M, Fan X, Hong E, Chen JJ. Interaction of oncogenic papillomavirus E6 proteins with fibulin-1. Biochem Biophys Res Commun. 2002;296(4):962–9.
136. Vande Pol SB, Brown MC, Turner CE. Association of Bovine Papillomavirus Type 1 E6 oncoprotein with the focal adhesion protein paxillin through a conserved protein interaction motif. Oncogene. 1998;16(1):43–52.
137. DeMasi J, Chao MC, Kumar AS, Howley PM. Bovine papillomavirus E7 Oncoprotein inhibits Anoikis. J Virol. 2007;81(17):9419–25.
138. Huh K-W, DeMasi J, Ogawa H, Nakatani Y, Howley PM, Münger K. Association of the human papillomavirus type 16 E7 oncoprotein with the 600-kDa retinoblastoma protein-associated factor, p600. Proc Natl Acad Sci U S A. 2005;102(32):11492–7.
139. Cancer Genome Atlas Network. Comprehensive genomic characterization of head and neck squamous cell carcinomas. Nature. 2015;517(7536):576–82.
140. Kimple RJ, Smith MA, Blitzer GC, Torres AD, Martin JA, Yang RZ, et al. Enhanced radiation sensitivity in HPV-positive head and neck Cancer. Cancer Res. 2013;73(15):4791–800.

Chapter 18
Diversity of Wnt/β-Catenin Signaling in Head and Neck Cancer: Cancer Stem Cells, Epithelial-to-Mesenchymal Transition, and Tumor Microenvironment

Khalid Alamoud and Maria A. Kukuruzinska

Abstract The Wnt/β-catenin signaling pathway is increasingly recognized for its roles in head and neck cancer, a devastating malignancy that presents primarily as head and neck squamous cell carcinoma (HNSCC). Wnt/β-catenin signaling impacts multiple cellular processes that endow cancer cells with the ability to maintain and expand immature stemlike phenotypes and proliferate, extend cancer cell survival, and promote aggressive characteristics resulting from loss of epithelial features and adoption of mesenchymal traits. A central component of the canonical Wnt signaling pathway is β-catenin, which balances a role as a structural component of cadherin junctions with function as a transcriptional coactivator of numerous target genes. While β-catenin is not frequently mutated in HNSCC, its activity is enhanced by some of the more common HNSCC mutations in NOTCH1, FAT1, and AJUBA. The impact of β-catenin on a wide range of epigenetic, transcriptional, and cellular processes is mediated by its interaction with numerous transcription factors, as well as with a multitude of transcriptional coactivators and corepressors, in a cell- and tissue-context-dependent manner. In addition, intrinsic β-catenin activity plays important roles in the tumor microenvironment and thus regulates extracellular matrix remodeling and immune response. Lastly, Wnt/β-catenin signaling collaborates with, and converges on, other signaling and metabolic pathways and cellular processes that modulate outputs of its activity. Unraveling the complex circuitries of Wnt/β-catenin signaling will facilitate its effective targeting for HNSCC therapy.

Keywords β-catenin · carcinoma · epigenomics · metabolism · stroma · threapeutics

---

K. Alamoud · M. A. Kukuruzinska (✉)
Department of Molecular and Cell Biology, Boston University School of Dental Medicine, Boston, MA, USA
e-mail: mkukuruz@bu.edu

B. Burtness, E. A. Golemis (eds.), *Molecular Determinants of Head and Neck Cancer*, Current Cancer Research, https://doi.org/10.1007/978-3-319-78762-6_18

18.1 Introduction

The Wnt signaling pathway is a conserved ancestral pathway with a pivotal role in organismal development and adult tissue homeostasis. Wnt signaling regulates diverse cellular processes including cell proliferation, patterning, cell fate determination, survival, and differentiation [1, 2]. Along with a small number of other pathways that include Notch, Sonic Hedgehog (Shh), Hippo, and TGF-β, Wnt signaling belongs to one of the few master pathways that organize the basic architecture of the cellular regulatory network [3]. These master pathways share a number of components through which they collaborate, regulate each other, and converge on each other's activities. Not surprisingly, co-opting these pathways is a recognized strategy by which cancer cells extend their lifespan, increase proliferation, acquire stem-like characteristics, and assume new phenotypes that eventually facilitate their exit from the primary tumor to set up residence at distant sites.

The Wnt signaling pathway was recognized more than 40 years ago for its role in *Drosophila melanogaster* embryonic patterning by identification of the segment polarity gene wingless (*Wg*, also known as DWnt1) [4]. The discovery of the *Wg* homolog, the mouse mammary tumor virus (MMTV) *int1* oncogene, led to the development of the portmanteau term Wnt and provided support for Wnt signaling in mammalian cancer [5]. Over the years, numerous genetic and biochemical studies have elucidated the Wnt signaling cascade, defining the Wnt/β-catenin "canonical" Wnt signaling pathway, the β-catenin-independent planar cell polarity (Wnt-PCP) pathway, and the Wnt-calcium signaling pathway [6, 7]. These studies have defined multiple Wnt ligands, Wnt inhibitory factors, Wnt receptors and co-receptors, as well as many nuclear partners of β-catenin [6]. Collectively, these studies identified critical roles for Wnt signaling in tissue development, homeostasis, and disease.

Cancer develops through a multistep process driven by mechanisms that promote the accumulation of epigenetic, genetic, and cytogenetic changes [8, 9]. While the three branches of the Wnt signaling pathway have been recognized for their distinct roles in cancer, the Wnt/β-catenin branch is best understood. Mounting evidence indicates that Wnt/β-catenin signaling plays important roles in head and neck cancer. During the past decade, several excellent reviews described the involvement of the Wnt/β-catenin pathway in the pathobiology of head and neck cancer [10–13]. This chapter describes seminal findings and recent advances in Wnt/β-catenin signaling and its emerging roles in head and neck cancer stem cells, in epithelial-mesenchymal transition, and in modulating the tumor microenvironment. In addition, convergence between Wnt/β-catenin and other signaling and metabolic pathways is summarized. Lastly, we discuss current strategies to inhibit Wnt/β--catenin signaling in cancer along with new promising druggable targets in this pathway.

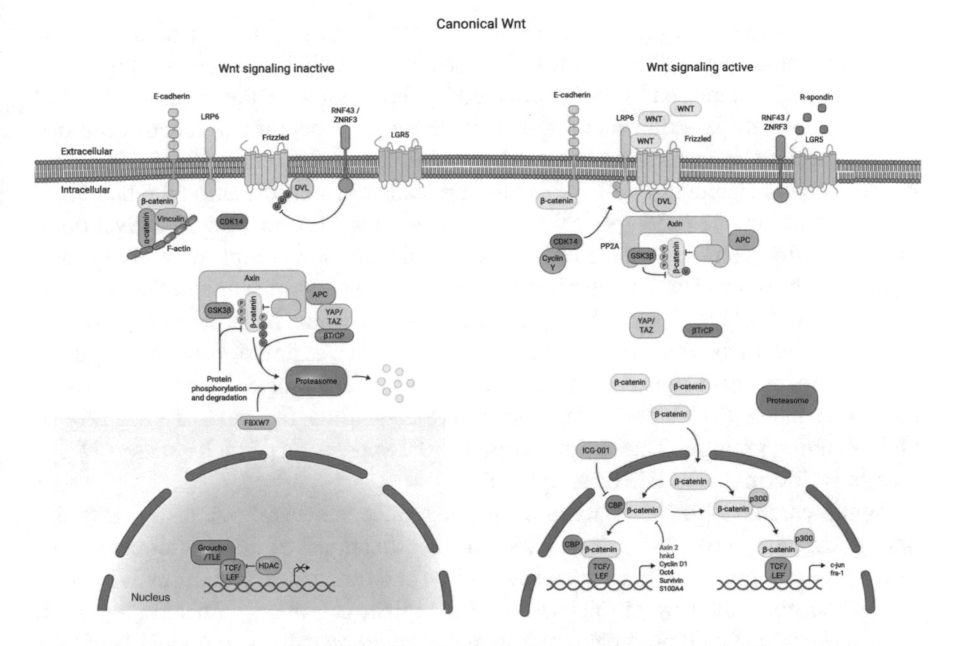

Fig. 18.1 Schematic of the Wnt/β-catenin signaling cascade. In the absence of Wnt ligand, cytoplasmic β-catenin is maintained at low levels by the destruction complex composed of the scaffold protein Axin in complex with APC, GSK3β, and CK1α that phosphorylates the N-terminal region of β-catenin and targets it for proteasomal degradation via βTrCP E3 ubiquitin ligase. The binding of Wnt ligand to the Fzd receptor-Lrp6/5 co-receptor complex inactivates the destruction complex and leads to the accumulation of cytoplasmic β-catenin and its subsequent translocation to the nucleus where it interacts with TCF-LEF transcription factor and activates target genes. Additional effectors of Wnt/β-catenin signaling include R-spondins and their receptors (LGR54/5) that potentiate Wnt signaling, and RNF43/ZRNF3 E3 ubiquitin ligases that antagonize it. In addition, E-cadherin functions as a tumor suppressor that interferes with Wnt/β-catenin activity

18.2 Wnt/β-Catenin Signaling in Cancer

The canonical Wnt/β-catenin signaling pathway is highly conserved in the animal kingdom, having arisen early in metazoan evolution and acquired increasing complexity during progression from basal metazoans such as sponges (Porifera) to mammals [14–16]. During organismal development, the Wnt/β-catenin pathway regulates cell fate determination, cell proliferation, cell survival, and cell differentiation. Under normal conditions of epithelial tissue homeostasis in adults, Wnt/β-catenin activity is maintained at low levels, and β-catenin is prominently localized to E-cadherin junctional complexes or adherens junctions (AJs) at the basolateral cell surfaces (Fig. 18.1). In the absence of Wnt signal, cytoplasmic β-catenin is kept at low levels by a complex of proteins that include the scaffold proteins adenomatous polyposis coli (APC) and Axin1, glycogen synthase kinase 3β (GSK3β), casein kinase 1α (CK1α), protein phosphatase 2A (PP2A), and additional components. These interactions sequester excess cytoplasmic β-catenin and target it for ubiquitination and proteasomal degradation via the β-TrCP, an E3 ubiquitin ligase subunit (Fig. 18.1).

The Wnt/β-catenin pathway is activated by the binding of the extracellular Wnt ligand to a seven-pass transmembrane Frizzled (Fzd) receptor and its co-receptors, the low-density lipoprotein receptor-related protein 6 (Lrp6) and/or Lrp5. The binding of the Wnt ligand to Fzd-Lrp5/Fzd-Lrp6 leads to a conformational change and phosphorylation of Lrp5/Lrp6, causing association with Axin 1 and inactivation of the destruction complex. In the absence of the destruction complex, newly synthesized, un-phosphorylated β-catenin accumulates in the cytoplasm and then translocates to the nucleus, where it displaces repressors from the promoters of target genes and interacts with transcription factors to drive expression of target genes. The most prominent transcriptional partners of nuclear β-catenin are members of the T-cell factor (TCF) and lymphoid enhancer-binding factor (LEF1) family of DNA-binding proteins. This complex induces the expression of such genes as MYC, Cyclin D1, Axin 2, and BIRC, among others.

Wnt/β-catenin signaling is dynamically regulated at several points along its signaling cascade through posttranslational modifications of its components that impact their conformation, secretion, stability, and activity. Secretion of Wnt components is controlled by the activity of Porcupine, a protein with features of an acyltransferase, which promotes their its palmitoylation. Many components of Wnt signaling are glycosylated, including the Wnt ligands and the Fzd receptors and co-receptors, imposing additional regulatory controls. Wnt secretion is also regulated by Wntless, a G-protein-coupled receptor (also known as GPR177) which regulates vesicular trafficking, and by Evenness Interrupted (Evi), which associates Wnt ligands with exosomes. The binding of Wnts to the Fzd receptor is antagonized by proteins including Dickkopf (Dkk), Wnt inhibitory factor (WIF1), and secreted Frizzled-related proteins (SFRP), among others [17, 18]. Conversely, R-spondins synergize with Wnt to activate β-catenin-mediated signaling [19, 20]. R-spondins (RSPO1–RSPO4) and their receptors (LGR4–LGR6) together form a ternary complex that promotes clearance of the E3 ligases, interaction with IQGAP1 to enhance MEK1/MEK2 phosphorylation of LRP5/LRP6 and increase in Wnt receptor levels to potentiate Wnt signaling. Other regulators of Wnt/β-catenin signaling include the RNF43/ZRNF3 E3 ubiquitin ligases, which antagonize Wnt/β-catenin signaling by ubiquitination of Wnt receptors, promoting their degradation [21, 22].

The broader Wnt signaling cascade also includes two noncanonical branches, the planar cell polarity (PCP) signaling pathway and the Wnt-calcium signaling pathway, both pathways being β-catenin-independent [6, 7] (Fig. 18.2). The PCP pathway utilizes different co-receptors including NRH1, Ryk, PTK7, or ROR1/ROR2. Binding of Wnt to the Fzd/PCP co-receptor complex activates Dvl and regulates the cytoskeleton via Rho-associated kinase (ROCK) and Rac1/JNK activity [23, 24]. In the Wnt/calcium pathway, Wnt ligand-activated Fzd receptor interacts with Dvl and also with a trimeric G-protein to promote calcium release from the ER by activating either phospholipase C (PLC) or phosphodiesterase (PDE). In addition, these interactions activate calcineurin and Ca-calmodulin protein kinase II (CAMKII) to induce nuclear factor of activated T cells (NFAT), influencing cell adhesion and migration [24].

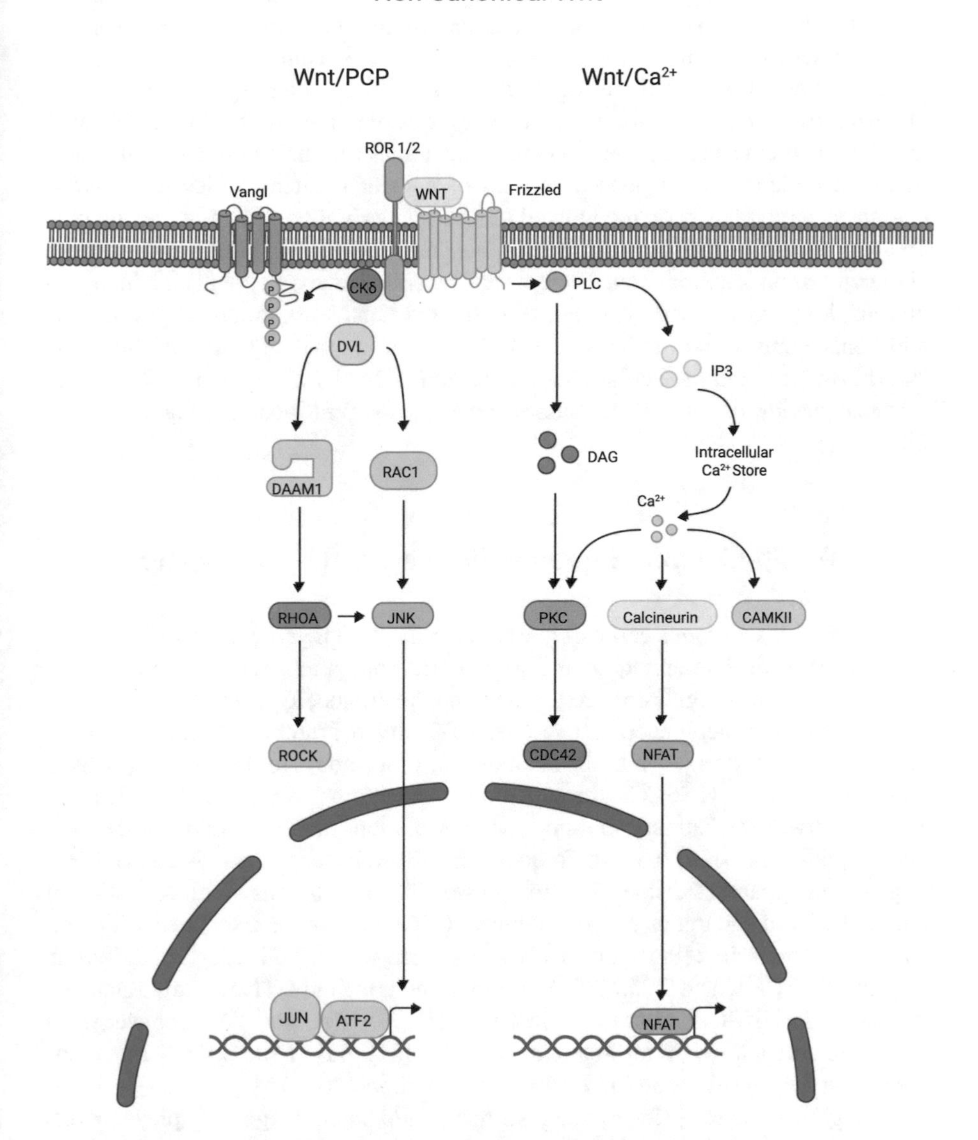

Fig. 18.2 Non-canonical Wnt signaling. The Wnt signaling pathway includes two β-catenin-independent signal transduction mechanisms, the Wnt/PCP pathway and Wnt/Ca^{2+} signaling. In the Wnt/PCP pathway, the binding of Wnt5a to the ROR1/2-Frizzled receptors complex leads to the sequestration and activation of DVL. Activated DVL, in turn, induces small GTPAses, Rac1, and Rho that activate ROCK and JNK and promote reorganization of the cytoskeleton and transcriptional activities of Jun and ATF2. The Wnt/Ca^{2+} pathway involves activation of phospholipase C which induces intracellular calcium fluxes leading to Ca^{2+}-dependent cytoskeletal rearrangements and activation of transcription factors, such as NFAT

Mutations affecting the Wnt/β-catenin pathway underlie many human cancers [25–28]. The most prominent are mutations of the adenomatous polyposis coli (APC) protein that occur in 42% of colorectal cancers, and mutations in the β-catenin gene (CTNNB1) that stabilize the β-catenin protein and consequently activate its nuclear activity in colorectal cancer. The frequency of mutations in APC or CTNNB1 is lower in other cancers, although many carcinomas are characterized by increases in the levels and transcriptional activity of nuclear β-catenin. Elevated Wnt/β--catenin signaling has been recognized to be an underlying cause of leukemia, pancreatic ductal adenocarcinoma, melanoma, gastrointestinal cancers, cholangiocarcinoma, mammary carcinoma, intestinal adenoma, hepatocellular carcinoma, lung cancer, and synovial, ovarian, and renal carcinoma, suggesting that additional control mechanisms affect the transcriptional activity of β-catenin. Recent genomic and molecular characterization of head and neck cancer has added this malignancy to the list of tumors impacted by Wnt/β-catenin signaling [11, 29–31].

18.3 Wnt/β-Catenin Signaling in Head and Neck Cancer

Head and neck squamous cell carcinoma (HNSCC) is a heterogeneous malignancy that comprises diverse epithelial and stromal cell populations whose interplay contributes to disease progression [32]. This tumor heterogeneity stems from combinations of genetic and epigenetic alterations within distinct tumor cell subpopulations, accompanied by changes in their metabolism as they adapt to the evolving cues in their environment. HNSCCs that are human papilloma virus (HPV)-driven and HPV-negative have distinct genomic profiles. Mutations in PIK3CA, loss of BRAF3, and amplification of E2F1 are frequent in HPV-related tumors, whereas HPV-negative malignancies display loss-of-function TP53 mutations and inactivation of CDKN2A. Oral squamous cell carcinomas (OSCC) are often associated with activating mutations in HRAS or PIK3CA together with inactivating mutations in CASP8, NOTCH1, and TP53 [30, 31]. While mutations in CTNNB1 are relatively infrequent in HNSCC, alterations in NOTCH1, AJUBA, and FAT1 converge on β-catenin, deregulating Wnt/β-catenin signaling [31]. AJUBA and FAT1 mutations are almost exclusively found in HPV-negative tumors [30, 31].

Emerging evidence [33–36] suggests that Wnt/β-catenin signaling plays pivotal roles in the pathobiology of HNSCC (reviewed in [11, 12, 31, 37–45]). Besides genomic alterations impacting Wnt pathway components, Wnt1 expression increases in basal cells of the head and neck mucosa [46], and the Wnt pathway is activated in OSCC by epigenetic alterations affecting the Wnt inhibitory factors, SRFP, WIF, and DKK3 [47, 48]. Accumulation of nuclear β-catenin linked to cyclin D1 overexpression has been reported in mucoepidermoid carcinoma of the head and neck [49]. Increased tissue levels of β-catenin in OSCC have been aligned with its increased transcriptional activity [33], and inappropriate stabilization of β-catenin has been also correlated with dedifferentiation and poor prognosis in HNSCC [50].

The complex nature of the Wnt signaling pathway stems, in part, from the diversity of Wnt receptors, Wnt co-receptors, and Wnt ligands. To date, ten human Wnt Frizzled (Fzd) receptors have been identified of four distinct phylogenetic lineages that display some differences in tissue-specific expression. Fzd receptors are seven transmembrane proteins of 537–706 amino acids (aa) with extracellular N-terminus and intercellular C-terminus. The N-terminal region contains a cysteine-rich domain that forms an alpha-helix of 130 aa and binds Wnts as well as other Wnt receptors. In addition, the N-terminal regions of many Fzd receptors contain potential N-linked glycosylation sites and conserved disulfide bonds. Further, the transmembrane regions of some Fzd receptors contain a leucine zipper that functions in dimer formation. Typically, Fzd receptors are involved in β-catenin-dependent and β-catenin-independent pathways, while Wnt co-receptors define these distinctions. Some Fzd, such as Fzd3, can activate Wnt/β-catenin pathway in the absence of the Wnt ligand, while others such as Fzd6 can inhibit Wnt/β-catenin signaling. In the Wnt/β-catenin pathway, the interaction of Fzd receptors with some co-receptors, such as the low-density lipoprotein receptor-related proteins 5 and 6 (LRP-5/LRP-6), promotes signaling. On the other hand, some Wnt co-receptors, like LRP-1 and Kremens 1 and 2, function as negative regulators of Wnt/β-catenin signaling. Additional inhibitors of Wnt/β-catenin signaling include ROR2, a receptor tyrosine kinase-like orphan receptor 2, and the Strabismus/Van Gogh-like protein that activate the Wnt/PCP pathway. In this process, ROR2 binds Wnt5a and Fzd, while Strabismus interacts with Dsh and competes for Fzd binding. Additional Wnt receptors involved in Wnt signaling include protein tyrosine kinase 7 (PTK7), receptor tyrosine kinase (RYK), and muscle skeletal receptor tyrosine kinase (MUSK), as well as selected members of the proteoglycan family, such as glypican and syndecan (reviewed in [51]).

The Wnt ligands belong to a family of 19 secreted proteins 350–400 aa in length with 22–24 conserved cysteine residues, and 20–85% identity within the family. Some of the Wnt ligands, Wnt1, Wnt3a, and Wnt8, display preference for the Wnt/β-catenin-dependent pathway, whereas Wnt5a and Wnt 11 are mostly involved in the Wnt/PCP pathway. However, the complexity of Fzd receptors and co-receptors renders such distinctions difficult. For instance, Wnt1, Wnt3, Wnt3a, Wnt2, and Wnt8 are known to function in the Wnt/β-catenin pathway with Fzd1, while Fzd2 interacts with Wnt5a that has been aligned with Wnt/PCP pathway. Other Fzds exhibit a range of specificities for Wnts, with Fzd3 preferentially interacting with Wnt1 and Wnt8 and Fzd5 with Wnt2, Wnt5a, Wnt7a, Wnt8, Wnt10b, and Wnt11. Furthermore, Fzd7 interacts with Wnt 5a, Wnt8, and Wnt11, Fzd8 with Wnt8, Fxd9 with Wnt2, and Fzd10 with Wnt8 (reviewed in [52]).

18.4 β-Catenin as a Central Effector of Wnt Signaling

The ability of β-catenin to regulate diverse cellular events is defined by a protein structure that provides a platform for multiple activities [53]. β-catenin encompasses 12 repeats of approximately 40 amino acids long, named armadillo (ARM) repeats. Together they form a single, rigid protein domain, the ARM domain, characterized by an elongated shape with a convex outer and a concave inner surface. Each armadillo repeat is composed of three alpha-helices. The first repeat near the N-terminus contains a helix with a kink, formed by the fusion of helices 1 and 2; the inner surface serves as a ligand-binding site for the various interaction partners [53, 54].

The N-terminal region of β-catenin is disordered. When phosphorylated at Ser33 and Ser37, it binds the β-TrCP E3 ubiquitin ligase that targets β-catenin for degradation by the proteasome . In contrast, the C-terminal region is not fully disordered but rather forms a stable helix (helix C) that interacts with additional binding partners [55, 56]. While not required for β-catenin function in cell-cell adhesion, helix C is critical for Wnt signaling, and fusion of the C-terminal segment of β-catenin to the DNA-binding domain of LEF1 transcription factor can mimic activation of the entire Wnt pathway. Cellular abundance of β-catenin may not be always indicative of its signaling potential; rather, the magnitude of change in response to external or internal signals may be an important determinant [57].

Similar to β-catenin, its homolog γ-catenin (plakoglobin) has the ARM domain structure with ligand-binding capacity and conserved N-terminal β-TrCP-binding motif. However, γ-catenin lacks motifs in the C-terminal region and thus does not have the ability to function as a strong transcriptional coactivator. The role of γ-catenin in β-catenin transcriptional activity remains unclear, but it has been suggested that it modulates the Wnt pathway target genes [58–63]. Like β-catenin, γ-catenin is also found at the promoter region of the β-catenin target gene, *DPAGT1*, in OSCC, although in vitro studies suggest that its abundance is regulated via mechanisms distinct from those controlling β-catenin [33].

Transcriptional regulation Transcriptional activity of β-catenin is determined by its interactions with components that regulate nuclear translocation, transcriptional co-activation, and epigenetic changes [64]. In addition to binding Tcf/Lef, β-catenin directly interacts with other transcription factors with significant impact on tumor cell behavior. For instance, interaction with nuclear receptors, like androgen receptor, can either enhance their transcriptional activity or interfere with it [65, 66] (reviewed in [67]). Under hypoxic conditions, hypoxia-induced factor 1α (HIF1α) is upregulated and competes with Tcf for β-catenin binding [68]. Furthermore, increased oxidative stress is accompanied by enhanced levels of reactive oxygen species (ROS) leading to upregulation of the forkhead FOXO transcription factors, which also compete with Tcf for β-catenin [69].

Among the key nuclear binding partners of β-catenin that enhance its coactivator activity is Bcl9, shown to promote early stages of intestinal cancer by inducing a subset of β-catenin target genes with roles in EMT and invasion [70]. β-catenin also recruits the homeodomain-interacting protein kinase 2 (HIPK2) that promotes dissociation of Tcf3 from target genes, leading to derepression [71]. FoxM1 and β-catenin form a complex with Tcfs at some target gene promoters. Kindlin-2 binding to β-catenin and Tcf4 promotes Axin2-Snail target gene expression critical for EMT, invasion, CSCs, and metastases [72]. One key feature of stem cells is their ability to preserve long telomeres by maintaining high *TERT* gene expression. In CSCs, β-catenin interacts with Klf4 and directly binds to the *TERT* gene promoter to enhance telomerase expression [73].

A significant aspect of β-catenin-mediated regulation of gene expression involves its central role in the epigenetic modification of chromatin. Whereas epigenetic changes in acetylation and methylation play critical roles in regulating gene expression in normal mammalian development, their deregulation is associated with cancer [64, 74, 75]. For instance, in breast cancer cells, the β-catenin-Pygo2 complex recruits histone methyltransferase MLL1/MLL2 to promote H3K4me3 tri-methylation to enhance stem cell expansion [76]. In human embryonic kidney cells, β-catenin participates in SET8-directed mono-methylation of H4K20me1 that facilitates the removal of Groucho from Tcfs [77], while methylation at H3K79me3 is driven by a Dot1-containing complex of MLL partners [78]. In HNSCC, the β-catenin-histone acetyltransferase cAMP-responsive element-binding protein (CBP) complex recruits MLL to methylate H3K4me3 and activate stem-cell-related genes [34]. In addition to CBP, another co-activating partner of β-catenin is a closely related histone acetyltransferase p300, although CBP and p300 exhibit partially distinct patterns of gene activation [79]. Moreover, nuclear β-catenin interacts with the effector of the Hippo pathway YAP1 to regulate expression of target genes associated with oncogenic features [80–83].

Lastly, recent studies have revealed that nuclear levels of β-catenin can be subject to regulation by an E3 ubiquitin ligase, SHPRH (SNF2 histone-linker PHD and RING finger domain-containing helicase), independent of the GSK3β/APC complex [84].

Cytoplasmic regulation Cytoplasmic levels of β-catenin are regulated by the APC destruction complex, which becomes inactivated when binding of Wnt ligands to the Fzd receptors and LRP co-receptors causes phosphorylation and activation of Dishevelled, with consequent Ser9 phosphorylation and inactivation of GSK3β [85]. In the absence of Wnt stimulation, phosphorylated β-catenin at Ser33 and Ser37 interacts with the E3 ubiquitin ligase β-TrCP which targets it for proteasomal degradation [86].

Additional regulation of β-catenin may be also imposed by E-cadherin. As β-catenin binds to E-cadherin during synthesis in the ER, it is thought to serve as a chaperone in effective targeting of E-cadherin to the cell membrane. Whether β-catenin can disengage from E-cadherin junctions and enter a soluble pool in the cytoplasm that then serves as a source of nuclear β-catenin remains controversial. Nonetheless, several studies have shown that the cytoplasmic pool of β-catenin may be augmented by increased recycling of E-cadherin-β-catenin complexes from the cell membrane [87, 88]. Notably, β-catenin itself regulates its own levels by impacting the expression of components of the Wnt/β-catenin pathway. For instance, β-catenin regulates expression of E3 ubiquitin ligases ZNRF3 and RNF43 that induce endocytosis of Wnt receptors in an R-spondin-dependent manner and enhance Wnt signaling [21, 22]. Further, β-catenin regulates the expression of the Wnt/β-catenin signaling inhibitor, Dkk-1 [17]. Frequently, Dkk-1 is lost in carcinoma and HNSCC, leading to unchecked β-catenin signaling [33].

18.5 Wnt/β-Catenin Signaling in Cancer Stem Cells

An acknowledged pivotal function of β-catenin is in the control of cell fates [89]. Specifically, β-catenin regulates asymmetric cell division by promoting unequal distribution of Dvl, Fzd, Axin, and APC in the cytoplasm of the mother cell [90]. This mechanism generates cells capable of renewing themselves as well as committed progenitor cells that differentiate into functionally specified cells [91]. The decision to divide or differentiate is controlled by a stem cell niche formed by extrinsic factors, including Wnt ligands, and by cells that produce them. The niche provides signals that promote stem-cell phenotypes in multiple epithelial stem cell compartments. Recent elegant work has revealed the existence of discrete stem cell niches at different tissue sites during development [92]. It is likely that similar scenarios are at play in cancer, including HNSCC.

The cancer stem cell hypothesis postulates that cancer stem cells (CSCs) give rise to tumors with hierarchical organization similar to that during development [93]. Thus, some tumor cells are now thought to hijack Wnt/β-catenin signaling to promote asymmetric cell division that leads to the expansion of daughter cells with stemlike characteristics to drive advanced disease [89, 93]. Increasing evidence supports the role of Wnt/β-catenin signaling in subpopulations of cancer cells that display stemlike characteristics [26, 89, 94]. These CSCs can efficiently seed tumors and promote aggressive disease, with Wnt/β-catenin signaling playing pivotal roles in their maintenance and expansion.

During the past decade, numerous studies have provided evidence that HNSCC arises from populations of CSCs that drive tumor development and resistance to therapy, and increasingly, the molecular mechanisms underlying the generation and expansion of CSCs in this malignancy are being unraveled [95, 96]. First characterized by the expression of the CD44 surface marker and the Bmi1 oncogene [95], HNSCC CSCs were shown to be marked by high aldehyde dehydrogenase activity and by the expression of c-Met and SOX2. Moreover, HNSCC CSCs had the ability to efflux vital dyes, to grow under non-adherent conditions as tumor spheres, to seed tumors at low numbers in nude mice, and to drive cell expansion [34, 95, 97–99]. In addition, these CSCs were characterized by an aberrant activation of Wnt/β-catenin activity [42, 99]. Significantly, HNSCC CSCs have been aligned with mesenchymal properties and resistance to treatments [100–103].

Multiple components of Wnt/β-catenin signaling have been associated with CSCs. For instance, the R-spondin receptor Lgr5 is a potential CSC marker in intestinal stem cells that can promote tumor growth when APC is deleted [104]. In colon adenoma, Lgr5-positive cells can give rise to Lgr5-positive cells and other cell types. Further, RAC1 is required for expansion of the Lgr5 population after APC loss [105]. RAC1 activation drives ROS production and activates NFκB signaling, which then enhances Wnt signaling [105].

Wnt/β-catenin signaling in CSCs is impacted by the tumor environment. For instance, myofibroblasts secrete hepatocyte growth factor (HGF) that increases Wnt activity and induces stem cell-like features in colorectal cancer cells. Additionally, breast tumor cells induce the stromal expression of the extracellular matrix protein periostin to form a metastatic niche [106]. Periostin interacts with Wnt1 and Wnt3a, thus inducing Wnt signaling and sustaining a CSC phenotype. Further, matrix metalloprotease MMP3 secreted by mammary epithelial cells stimulates canonical Wnt signaling in mammary stem cells by sequestration of Wnt5b and antagonizing the inhibitory effect of the PCP Wnt pathway. Along with local cytokines secreted from tumor-associated cells, Wnt/β-catenin signaling promotes CSCs coincident with increased metastatic capacity. β-catenin induction of TERT mRNA expression helps maintain the long telomeres needed by stem cells [73]. Lastly, emerging evidence has revealed new links between noncoding RNAs, microRNAs (miRs), and long noncoding RNAs (lncs) with Wnt/β-catenin signaling in CSCs [107, 108].

As Wnt/β-catenin signaling controls epigenetic changes that define different chromatin states, its activity has been associated with pluripotency in part via its role in epigenetic modulation of target gene expression [64, 109]. Tri-methylation at lysine 4 (H3K4me3) promotes active chromatin structure, while methylation of lysine 9 (H3K9me3) supports heterochromatin state, and methylation at lysine 27 (H3K27me3) correlates with repressed chromatin [110, 111]. As described above, β-catenin promotes H3K4me3 by recruitment of CBP [112] and histone methyltransferase, MLL, to promote active chromatin [113]. Significantly, increased interaction of β-catenin with CBP underlies HNSCC CSCs and disease progression [34, 79, 114–116]. This ability to promote aggressive properties is specific to CBP, as recruitment of the p300 histone acetyltransferase by β-catenin drives cellular differentiation [117].

Recent reports have shown that Wnt/β-catenin signaling promotes HNSCC by epigenetic regulation [34, 118]. Importantly, inappropriate activation of Wnt/β--catenin signaling and concomitant inhibition of Bmp using a genetically engineered mouse model with gain-of-function mutation in β-catenin and loss of function in BMP led to the development of aggressive salivary gland SCC-derived and HNSCC tumors [34]. These tumors were enriched in subpopulations of CD44+CD24+CD29+ cells, which had the ability to self-renew.

Further, β-catenin/CBP/MLL complexes drive tri-methylation of H3K4 and maintain immature cell phenotypes. Specifically, aggressive tumors in mouse xenografts were shown to arise from the interaction of β-catenin with the histone acetyltransferase CREB-binding protein (CBP) and a histone methyltransferase or myeloid/lymphoid or mixed-lineage leukemia 1 (MLL) known to be a positive global regulator of gene transcription. MLL belongs to the group of histone-modifying enzymes comprising transactivation domain 9aaTAD involved in the epigenetic maintenance of transcriptional memory. In HNSCC and salivary gland SCC tumors, β-catenin recruits both CBP and MLL onto regulatory elements for genes encoding factors implicated in stemness, survival, and proliferation [34]. Inhibition of either β-catenin or CBP or MLL inhibits stem cell phenotypes and promotes differentiation.

A recent elegant study using lineage tracing in 4NQO-induced oral SCC has shown that Bmi1+ cells express AP1 and have high invasive capacity and self-renewal properties. Moreover, these cells are the underlying cause of cisplatin resistance and cancer recurrence [102]. Since both Bmi1 and AP1 interact with β-catenin and depend on its activity, it is tempting to speculate that these cells also require Wnt/β-catenin signaling and its interacting signaling circuitries for the maintenance of their stem cell-like phenotypes.

The presence of CSCs has also been associated with poor disease prognosis. Perhaps best documented is the connection between intestinal CSCs and colorectal cancer [119, 120]. However, some Wnt/β-catenin target genes, such as *Axin2* and *Lgr5*, are reduced in expression in patients with poor prognosis [119, 120] due to promoter methylation of those genes. Likewise, recent studies have shown that methylation of R-spondins contributes to the etiology of gastric cancer [121]. Similar to other solid tumors, HNSCC CSCs metastasize, with invasion of cervical lymph nodes being the most common feature of HNSCC tumor spread and poor prognosis [122, 123].

18.6 Wnt/β-Catenin Signaling in Epithelial-to-Mesenchymal Transition (EMT)

EMT is a developmental process regulated by changes in intercellular adhesion that promote loss of epithelial characteristics and acquisition of a mesenchymal phenotype with important roles in cancer progression and metastasis [124–126]. In development, EMT is critical for the spatial and temporal cell sorting and migration that accompany normal tissue and organ development. Major changes that occur during EMT include the downregulation of epithelial markers, E-cadherin, TJP/ZO-1, and occludin, reorganization of the cytoskeleton, loss of intercellular adhesion and apical-basal polarity, and acquisition of a mesenchymal phenotype as indicated by increased expression of vimentin, N-cadherin, and fibronectin [126]. Many tumors co-opt these key developmental pathways and adapt them to execute changes in phenotypes that will allow them to survive, expand, and move to distant sites to establish new niches [9, 127, 128].

One of the key transcription factors responsible for EMT that is regulated by Wnt/β-catenin signaling is SNAI2 [124, 129]. Cytoplasmic SNAI2 concentration is kept in check by GSK3β phosphorylation and subsequent ubiquitination by β-TrCP. In breast cancer cells, the activation of canonical Wnt signaling stabilizes SNAI2 by inhibiting GSK3β kinase activity and initiates EMT transcriptional programs. Other key regulators of EMT are the zinc finger E-box-binding homeobox 1 (ZEB1) protein and a haploinsufficient tumor suppressor, apoptosis-stimulating protein of p53 2 (ASPP2), both affecting β-catenin activity by suppressing E-cadherin adhesion. ZEB1 is a transcriptional inhibitor of E-cadherin expression via SWI/SNF chromatin-remodeling protein, BRG1 [130], while ASPP2 binds to a β-catenin in complex with E-cadherin and inhibits N-terminal phosphorylation of β-catenin, leading to its stabilization while preventing β-catenin-mediated activation of ZEB1 [131]. Accordingly, reduced expression of ASPP2 leads to EMT and is associated with poor survival in hepatocellular and breast cancer, suggesting its important role in regulating cell plasticity [131]. In colon cancer cells with hyperactivated canonical Wnt signaling, pharmacological inhibition of PI3K-Akt signaling leads to a nuclear accumulation of β-catenin and FOXO3a that results in increased cell scattering and metastasis [132].

EMT has been associated with HNSCC progression to advanced disease [10, 100, 133, 134]. Among transcription factors associated with EMT in HNSCC is TWIST1, shown to mark loss of epithelial morphology and increased mesenchymal traits [134, 135]. β-catenin-dependent increased expression of collagen triple helix repeat containing 1 (CTHRC1) is associated with the acquisition of mesenchymal traits leading to increased cell migration through the Wnt/PCP pathway in several carcinomas [136, 137]. Wnt/β-catenin collaborates with protein N-glycosylation to induce CTHRC1 in OSCC via distinct transcriptional and posttranslational mechanisms [13]. In human OSCC tumors, CTHRC1 localizes to the tumor front in vivo and promotes cell migration by inducing the interaction between ROR2 and Fzd6 and subsequent activation of Rac1 via the Wnt/PCP pathway [13].

In addition to transcription factors and posttranslational mechanisms, microRNAs are critical regulators of the EMT process, and Wnt/β-catenin signaling influences their expression in cancer [138–141]. MicroRNAs suppress transcription factors that directly regulate EMT and downregulate other genes and pathways that indirectly impact this process. In HNSCC, the deregulation of miRNA200b and miRNA15b has been linked to EMT and advanced disease [142, 143]. Since these miRNAs have a strong association with EMT, they may represent effective therapeutic targets for HNSCC [144].

The process of metastasis reflects a succession of complex steps leading to the macroscopic outgrowth of disseminated tumor cells at the secondary site [145]. Metastasis is a hallmark of late-stage cancer and a major challenge to therapy. A main adaptive change of tumors during therapy is EMT [94, 127]. Recently, exosomes that function in intercellular communication were implicated in metastasis through their ability to transport active Wnt ligands or incorporate β-catenin [146, 147]. Another route by which distant metastasis occurs may involve circulating tumor cells (CTCs) that have been linked to Wnt signaling in prostate and pancreatic cancers [148]. In pancreatic cancer CTCs, canonical Wnt2 expression promoted anchorage-independent sphere formation and metastatic behavior. In addition, noncanonical Wnt signaling is upregulated in prostate cancer CTCs resistant to androgen receptor inhibition. Our studies also showed that CTHRC1 promotes HNSCC cell migration by activating the noncanonical Wnt/PCP pathway [13]. Collectively, increasing evidence implicates both canonical and noncanonical Wnt signaling in HNSCC metastases.

18.7 Cross Talk with E-Cadherin-Mediated Adhesion

In addition to its signaling function, β-catenin serves as a major structural component of E-cadherin-mediated AJs [149–152] (Fig. 18.1). E-cadherin is a member of classic cadherins, single-span transmembrane cell-cell adhesion receptors characterized by their extracellular repeats, ectodomains, a transmembrane region, and a short cytoplasmic tail. As a major epithelial adhesion receptor, E-cadherin organizes AJs that mediate numerous cell functions, including reorganization of the actin cytoskeleton and microtubules to promote the formation of tight junctions and the establishment of cell polarity. The major components of AJs include members of the catenin family, β-catenin, α-catenin, and p120 catenin. E-cadherin interacts with β-catenin in the ER where the latter acts as a chaperone and assists in the transit of E-cadherin to the membrane [153]. E-cadherin/β-catenin complexes recruit α-catenin through the binding to the β-catenin C-terminal region, promoting interaction with the actin cytoskeleton, which is critical for the stabilization of AJs and downregulation of Wnt/β-catenin signaling [154] (reviewed in [155]). Indeed, carcinoma progression, including HNSCC, is associated with a dramatic loss of E-cadherin-mediated adhesion and EMT [156, 157]. In addition, HNSCC is characterized by dysregulated expression of FAT1, a member of the cadherin superfamily, and mutations in FAT1 occur in 32% of HPV-negative HNSCC tumors and are associated with an unchecked activity of β-catenin [31, 158]. Lastly, retinal cadherin, CDH4, has been reported to be important for maintaining cell-cell adhesion in HNSCC epithelia, in part, through a negative regulation of the Wnt/β-catenin/CBP axis (*Kartha et al., unpublished*), and its loss correlates with the acquisition of malignant traits in carcinomas, including adenoid cystic carcinoma [159].

18.8 Interactions Between Wnt/β-Catenin and Other Signaling Pathways

As a master regulatory pathway, Wnt/β-catenin signaling receives inputs and converges on and cross talks with other signaling pathways. In epithelial cells, Wnt/β--catenin signaling has been shown to collaborate with signaling pathways with central roles in cancer, including the noncanonical branches of Wnt signaling, Notch1, TGF-β, Hippo and its effectors, YAP/TAZ, as well as EGFR and PI3K. These interactive pathways display multifaceted activities that are highly cell- and tissue-context dependent, and, as in the case of TGF-β and Notch, they can be either oncogenic or tumor suppressive.

Reciprocal interactions with non-canonical Wnt signaling The canonical Wnt/β--catenin and noncanonical Wnt pathways affect each other's activities. The Wnt/PCP pathway inhibits Wnt/β-catenin signaling [160] and can also mediate the motility of cancer cells during metastasis via exosomes secreted from fibroblasts in the tumor microenvironment. Likewise, the upregulation of calcineurin/NFAT activity via the Wnt/calcium pathway interferes with the canonical Wnt pathway [161]. On the other hand, the Wnt/Ca^{2+} and Wnt/β-catenin pathways collaborate, in that Ca^{2+} release facilitates β-catenin entry into the nucleus [162]. Further, prostaglandin E2 is a potent activator of canonical Wnt signaling through cAMP-/PKA-mediated phosphorylation and stabilization of β-catenin [163].

β-catenin interactions with TGFβ signaling Similar to Wnt/β-catenin, transforming growth factor (TGF)-β has pivotal roles in cell fate determination and in homeostasis of adult tissues [164]. TGF-β and Wnt/β-catenin collaborate on a transcriptional level via direct interaction between Smads and Tcf proteins to generate a novel oncogenic transcriptional program that promotes the development of mammary and intestinal tumors [165]. Integrity of intercellular adhesion has been shown to be critical for the regulation of TGF-β-induced EMT and to depend on β-catenin [166] and on the β-catenin/CBP axis in promoting EMT [114]. Further, Wnt/β-catenin signaling is essential for the development of fibrosis through a mechanism that involves TGF-β-mediated inhibition of Dkk1 expression [167].

NOTCH1 and β-catenin activity Notch signaling regulates stem cell maintenance, cell plasticity, proliferation, and differentiation [168]. In addition, through regulation of cell-cell communication, Notch synchronizes the behavior of closely associated cells [169]. In keratinocytes, Notch is critical for guiding the differentiation program and tumor suppression [170]. Likewise, Notch has an important role in the homeostasis of oral and esophageal epithelia [171] and functions primarily as a tumor suppressor in head and neck epithelia [30, 172–174].

Wnt/β-catenin has been shown to collaborate with Notch in different tissues during development and in tumorigenesis. Both pathways control cell fate decisions of intestinal progenitor cells where Wnt signaling plays an essential role in the proliferation of both stem cells and transit amplifying cells [175]. On the other hand, in cardiac progenitor cells, Notch antagonizes Wnt/β-catenin signaling to determine cell fate [176]. Further, Notch and Wnt/β-catenin pathways are crucial for gastric CSC development [25], and both pathways are activated in CSCs and associated with carcinogenesis, poor prognosis, and chemotherapy resistance [29–31]. Additionally, membrane-bound Notch physically associates with β-catenin in stem and colon cancer cells and negatively regulates the accumulation of β-catenin in Notch cleavage- and GSK3β-independent ways [177].

The role of Notch1-β-catenin interplay in HNSCC is likely to be complex, in part because both Notch and Wnt/β-catenin have roles in proliferation and differentiation. Additionally, substantial heterogeneity of tumors and changes in spatiotemporal oncogenic cues with HNSCC progression to advanced disease are likely to drive complicated responses. As some HNSCC tumors harbor inactivating mutations in NOTCH1 [30], they are likely to be associated with unchecked β-catenin activity [31]. Indeed, activation of Notch has been shown to inhibit Wnt/β-catenin activity in OSCC [36].

β-catenin interacts with the Hippo pathway and its downstream effectors YAP/TAZ The Hippo tumor suppressor pathway regulates adult tissue size primarily via inhibition of the oncogenic activities of its downstream effectors, the paralogous transcriptional coactivators, YAP and TAZ (YAP/TAZ), via their retention in the cytosol [178–181]. Accordingly, hyperactivation of YAP/TAZ has been shown to contribute to numerous aggressive carcinomas, including breast, lung, prostate, and HNSCC [182, 183]. YAP/TAZ respond to multiple and complex cues, including changes in the mechanical properties of the extracellular matrix, cell size, and density, as well as in cell polarity and organization of the cytoskeleton [184], and their nuclear activity drives cell proliferation, stemlike properties, migration, invasion, and drug resistance [185–187]. Recently, upregulated YAP levels were reported in premalignant oral tissues, and YAP/TAZ gene signatures were shown to track with advanced OSCC, suggesting their involvement in the pathobiology of HNSCC on multiple levels [183].

Several studies have shown that the Wnt/β-catenin and Hippo pathways interact [80, 188]. A kinase in the Hippo signaling cascade, LATS2, inhibits Wnt/β-catenin-mediated transcription via TCF by interfering in the interaction between β-catenin and BCL9, independent of LATS2 kinase activity [189]. Notably, YAP/TAZ have been shown to interact with nuclear β-catenin and to either restrain its activity or collaborate in oncogenic signaling by activating SOX2, SNAI2, BCL2L1, and BIRC5 [83, 190]. In addition to nuclear interactions, under conditions of low Wnt/β-catenin signaling, YAP and TAZ are recruited to the APC complex and inactivated, while stimulation with Wnt ligands promotes their nuclear activity [81]. An additional link between YAP/TAZ and Wnt/β-catenin signaling has been shown to involve YAP/TAZ in "alternative" Wnt signaling. This alternative Wnt activity is induced by Wnt ligands, Wnt5a/b and Wnt3a, to activate the Fzd/ROR1/2-Gα12/13-Rho axis that suppresses Lats1/2 activity and promotes nuclear YAP/TAZ-TEAD signaling. The latter drives the upregulation of genes, including *WNT5A/B, CTGF, BMP4, CYR61*, and *DKK1*, among others [191].

Convergence of epidermal growth factor receptor (EGFR) signaling with β-catenin activity While somatic mutations in EGFR are relatively infrequent in HNSCC, the activity of EGFR is aberrantly upregulated in >90% of tumors, and these increases in RTK activity are frequently not reflected in transcript levels [192]. Thus, it is likely that posttranscriptional, translational, and posttranslational changes are responsible for EGFR activation in this malignancy. Increasing evidence indicates that the Wnt/β-catenin and EGFR signaling pathways cross talk [193]. These pathways have the ability to transactivate one another in development and cancer at several convergence points: Wnt ligands can activate EGFR signaling through Fzd receptors, while EGFR can activate β-catenin via PI3K/Akt or ERK pathway activation [193, 194]. Further, EGFR can form a complex with β-catenin and increase the invasion and metastasis of cancer cells. EGFR regulates the localization and stability of β-catenin [195]. Lastly, EGFR impacts the activities of the Src family kinases that also regulate the stability of E-cadherin/β-catenin complexes by phosphorylating tyrosine residues of β-catenin and either driving destabilization of AJs and enhanced pool of cytoplasmic β-catenin or promoting collective cell migration [196], a feature common in carcinoma spread [197].

PI3K and β-catenin signaling Phosphoinositide 3-kinase (PI3K) is one of the most important effectors of RTK signaling in cancer [198], as documented by key roles in cell survival, growth, metabolism, motility, and tumor progression [199, 200]. Activating mutations in the PI3K pathway are frequent in different cancer types, with a loss of its negative regulator, phosphatase and tensin homolog (PTEN) ranking among the most common events in cancer, including HNSCC, resembling TP53 mutations. In HNSCC, PIK3CA is frequently either amplified or harbors activating mutations [30, 31, 201]. β-catenin/Tcf contributes to the transcriptional regulation of the AKT1, a key PI3K effector [202]. Furthermore, the PI3K and Wnt/β-catenin pathways converge at the level of the Akt substrate FOXO3a, where downregulation of the AKT activity induces apoptosis driven by nuclear FOXO3a. However, high nuclear β-catenin collaborates with nuclear FOXO3a to induce IQ motif-containing GTPase-activating protein (IQGAP2) to downregulate E-cadherin junctions, leading to increased cell migration and increased metastatic potential [132]. OSCC is associated with increased IQGAP1 levels, with prominently enhanced localization to AJs [203], suggesting a broader involvement of IQGAP family members in HNSCC.

18.9 Cross Talk of Wnt/β-Catenin with Cellular Metabolism

Cancer cells undergo dramatic changes in their metabolism that depend, in part, on β-catenin transcriptional activity. Alterations in cellular metabolism support extensive biosynthetic needs for rapid tumor growth, coupled with a demand for ATP and availability of oxygen. While normal cells generate ATP from oxidative phosphorylation (OXPHOS), many cancer cells use aerobic glycolysis [204, 205]. With disease progression, oxygen gradients become limiting, and β-catenin is diverted from its interaction with Tcf to enhance transcriptional activity of HIF1α to upregulate glucose transporters and glycolytic enzymes to drive glycolysis and advanced disease [53, 68, 206, 207].

Wnt/β-catenin cross-regulation with protein N-glycosylation Increased reliance on glycolysis also drives the glucose flux into the hexosamine and N-glycosylation pathways to assure adequate glycosylation of oncogenic glycoproteins, such as RTKs, as well as glycoprotein components of the Wnt pathway, including Wnt ligands, receptors, and co-receptors [208]. The N-glycosylation pathway, in particular, has been recognized for its key roles in cancer, although its complexity is only beginning to be unraveled [209].

The N-glycosylation pathway links Wnt/β-catenin signaling and E-cadherin adhesion to glucose metabolism. Cancer cells utilize more glucose by increasing aerobic glycolysis, known as the Warburg effect [163, 210]. One consequence of increased glucose levels entering into the glycolysis pathway in cancer cells is the elevation of UDP-GlcNAc production via the hexosamine pathway. The last enzyme in the pathway, the UDP-N-acetylhexosamine pyrophosphorylase, encoded by the *UAP1* gene, is aberrantly upregulated in prostate cancer and maintains N-glycosylation in the presence of its inhibitors [211]. The UAP1-dependent production of UDP-GlcNAc drives N-glycosylation as UDP-GlcNAc is a substrate for the GPT enzyme that initiates protein N-glycosylation in the ER [156]. Indeed, the hexosamine pathway and protein N-glycosylation have been implicated in mediating the effects of increased glucose levels on the induction of canonical Wnt signaling in macrophage cell lines [212]. Thus, both pathways may collaborate to drive aerobic glycolysis and contribute to the Warburg effect.

N-glycosylation regulates Wnt/β-catenin signaling as Wnts are both N-glycosylated and lipid modified in the ER [213]. In addition, N-glycosylation is positively regulated by Wnt/β-catenin signaling through inducing expression of the *DPAGT1* gene that functions at the first committed step in the lipid-linked assembly pathway in the ER, impacting the balance between proliferation and adhesion in homeostatic tissues [154, 214]. N-glycosylation also modifies E-cadherin ectodomains and thus impacts its adhesive activity as well as the assembly and paracellular permeability of tight junctions [203]. Thus, aberrant induction of DPAGT1 promotes a positive feedback network with Wnt/β-catenin that represses E-cadherin-based adhesion and drives tumorigenic phenotypes. Further, modification of receptor tyrosine kinases (RTKs) with N-glycans is known to control their surface presentation via the galectin lattice, and thus increased DPAGT1 expression likely contributes to the activation of RTKs in oral cancer [215]. These studies suggest that dysregulation of the DPAGT1/Wnt/E-cadherin network underlies the etiology and pathogenesis of HNSCC.

Further, N-glycosylation is likely to impact Wnt/β-catenin signaling by modulating the N-glycosylation status and activity of EGFR [215]. As a high-multiplicity N-glycoprotein receptor, EGFR requires proper modification with N-glycans for dimerization, as well as targeting to and retention at the membrane [216]. Modification of EGFR terminal N-glycans with galactose residues promotes the interaction between EGFR and members of the lectin family, galectins, in the stroma that forms a lattice which, in turn, enhances membrane retention of EGFR and its pro-tumorigenic signaling [215], predicted to activate β-catenin signaling as discussed above. Such diminished recycling of RTKs has been shown to promote chemoresistance [217].

18.10 Wnt/β-Catenin Signaling in the Tumor Microenvironment

The plasticity of cancer cells endows them with the ability to adapt to changes in their environment and to undergo phenotypic switches that accompany cellular discohesion, cell migration, and metastasis. Similar dynamic changes in cellular features apply to the tumor microenvironment (TME) that undergoes alterations with cancer progression to advanced disease and contributes to cancer cell plasticity. For instance, recent studies identified 28 TME cytokines and growth factors that promote EMT and acquisition of CSC properties [218]. An increasing number of studies highlight contributions of distinct components of the TME to HNSCC tumorigenesis [219, 220].

The TME includes the extracellular matrix (ECM), a noncellular structure that provides physical support and elasticity to tissues. The ECM transmits signals to epithelia by providing ligands for receptors and regulating the availability of growth factors and morphogens, such as Wnts, TGF-β, and amphiregulin. In addition, the TME includes the stroma with cellular components such as fibroblasts and cells of the innate and adaptive immune systems. Indeed, cancer-associated fibroblasts (CAFs) are major components of the stroma that acquire their phenotypes from signals derived from tumor cells. Biochemical cross talk between tumor epithelia and CAFs and mechanical remodeling of the stromal ECM by CAFs are important contributors to tumor cell migration and invasion, which are critical for cancer progression from a primary tumor to metastatic disease. In addition, changes in the mechanical properties of the ECM induced by CAFs facilitate migration and invasion of cancer cells [221]. CAFs produce different tumor components at distinct stages of cancer progression, thus remodeling the ECM and promoting metabolic and immune reprogramming of the tumor microenvironment, which impacts immune response to therapy. The pleiotropic actions of CAFs on tumor cells are probably reflective of them being a heterogeneous and plastic population with context-dependent influence on cancer [222].

β-catenin in ECM remodeling The ECM is a dynamic structure that undergoes controlled remodeling during homeostasis and whose deregulation is associated with cancer. The ECM comprises ~ 300 proteins that form a collective structure referred to as the "core matrisome" composed of collagen, proteoglycans, and glycoproteins. Many downstream targets of β-catenin are components of the ECM, such as laminin, a key protein in the basement membrane, and lysyl oxidase and fibronectin, that reside in the interstitial matrix [223, 224]. The β-catenin target genes also include invasion-associated genes such as MMP-74 and CD44 [225], suggesting a direct involvement of nuclear β-catenin in tumor spread and metastasis. CAFs organize the ECM fibronectin assembly, promoting directed tumor cell migration of human prostate and pancreatic carcinoma cells [226]. Also, β-catenin indirectly controls the abundance of lysyl oxidase. The latter has been shown to be aberrantly activated in HNSCC [227], and inhibition of its activity using genetic and pharmacological perturbation effectively reverted pro-tumorigenic effects in cellular and nude mouse models. Further, a major contributor to ECM remodeling is MMP-9, a target of Wnt/β-catenin signaling, with important roles in cell migration [228].

β-catenin in Immune Response in HNSCC Wnt/β-catenin signaling regulates both the adaptive and innate immune systems in cancer, and its activity has been inversely correlated with CD8+ T-cell infiltration of tumors [229–231]. β-catenin signaling has been aligned with non-T-cell-inflamed tumors by promoting Treg persistence, activity, and survival [231]. Tregs suppress adaptive responses by reducing CD8+ T-cell proliferation, activation, and effector functions. Further, β-catenin controls innate immunity by regulating dendritic cells [230] via activation of the ATF3 repressor activity to downregulate the CCL4 chemokine required for the recruitment of CD103+ dendritic cells to the tumor with coincident diminished activation of CD8+ T cells. Lastly, β-catenin mediates interactions between tumor cells and tumor-associated macrophages. The inhibitory effects of β-catenin signaling on adaptive and innate immune systems have been associated with immune tolerance, a major mechanism underlying failed host antitumor immune responses [232–234].

Thus, inhibiting aberrantly active β-catenin signaling represents a promising therapeutic approach likely to overcome immune evasion by cancer cells. Indeed, immune checkpoint blockade has been shown to be highly effective in the treatment of melanoma and other tumor types, although its success in the treatment of HNSCC has been limited. Recent studies reveal complexities linked to changes in β-catenin activity. For instance, increased β-catenin signaling in melanoma leads to upregulation of IL-12 and consequent impaired maturation of dendritic cells. On the other hand, autocrine inhibition of Wnt/β-catenin signaling by DKK1 in lung and breast cancers leads to the evasion of innate immune response. To date, single agent responses have been limited as multiple inhibitory checkpoint receptors play compensatory roles in the suppression of cellular immunity. Future studies focused on the interplay between intrinsic β-catenin signaling in the tumor epithelia and the microenvironment will provide more insights about the complex interactions that define cellular and innate immune responses mediated by changes in β-catenin activity.

Because of their inherent heterogeneity [30, 95, 235], HNSCC cells express a wide variety of antigens including tumor-specific or tumor-associated antigens, differentiation antigens, and lectin-binding sites. These antigens are unevenly distributed in tumor subpopulations and induce different immune responses to the same determinant. Such tumor heterogeneity has important implications for diagnosis, treatment efficacy, and the identification of potential treatment targets [236]. As tumor heterogeneity and CSCs play pivotal roles in tumor growth, tumor development and survival involve the interplay between these heterogeneous populations of cancer cells, stromal cells, and host defense mechanisms [237]. Although cells of the immune system can inhibit tumor growth and progression through the recognition and rejection of malignant cells, tumors exploit several immunological processes for survival [238]. Tumors, including HNSCC, can evade immune surveillance by interfering with cytotoxic CD8+ T-cell function via production of immune suppressive cytokines either by the cancer cells or by the stromal cells in the tumor microenvironment and by expression of inhibitory receptors [235, 239].

18.11 Wnt/β-Catenin Signaling as a Therapeutic Target for HNSCC

HNSCC remains fraught with poor survival rates largely due to the late stage of disease at diagnosis, reliance on surgery as a treatment modality, and resistance to radiation therapy (RT), chemoradiotherapy (CRT), and current targeted therapies. Despite advances in the understanding of genomic alterations in HNSCC, there has been only modest improvement in therapeutic strategies. Therefore, identifying novel molecular events that contribute to the development and progression of the disease will advance effective prevention, detection, and durable therapeutic strategies.

Due to the important roles of β-catenin as a structural component of AJs and as a regulator of CSCs in cell renewal during homeostasis, the benefits of targeting of Wnt/β-catenin signaling have been controversial [240]. To date, a number of Wnt inhibitors have been identified and tested for antitumor properties in preclinical models, but only a few moved on to clinical trials. Nonetheless, increased insights into the mechanisms of β-catenin function in the nucleus have provided substantial support for its potential to serve as a druggable target [241]. Inhibitors of Wnt signaling have been developed to target different components of the Wnt/β-catenin pathway, including the secreted factors, as well as those that function in the cytoplasm and in the nucleus.

Inhibition of Wnt/β-catenin signaling in the cytoplasm A key component of the Wnt/β-catenin cascade is the acyltransferase Porcupine, and its small molecule inhibitor, LGK974 (WNT974), was shown to inhibit lung adenocarcinoma in mice [242]. Recently, LGK974 was shown to inhibit the growth of HNSCC cell line-derived chicken chorioallantoic membrane xenografts containing CD44 + ALDH + CSCs [243]. A phase I/II trial of LGK974 for patients with metastatic colorectal cancer harboring mutations of RNF43 or R-spondin fusions is underway. In addition, a novel orally active Porcupine inhibitor, ETC-1922159 (ETC-159), has been shown to prevent the growth of R-spondin translocation-bearing colorectal cancer patient-derived xenografts [244]. Although therapeutics targeting Wnt secretion appear promising, the number and impact of potential side effects are currently unclear. In general, genetically defined cancers displaying Wnt addiction are likely to benefit from these inhibitors.

In addition to drugs targeting Wnt secretion, compounds targeting extracellular Wnt ligands and their receptors are under development (reviewed in [245]). They include OMP-54F28, a fusion Fzd8-Fc decoy receptor, shown to reduce the number of CSCs and tumor xenograft growth in preclinical studies of ovarian cancer and hepatocellular carcinoma [245]. Furthermore, an antibody against R-spondin 3, OMP131R10, is being tested in phase I clinical studies with colorectal cancer patients. Lastly, anti-Fzd7 antibody, OMP-18R5, and Wnt5A mimetic, Foxy5, are currently being evaluated in phase I/II clinical trials with breast cancer, colorectal cancer, and prostate cancer patients.

Targeting β-catenin in the nucleus In addition to approaches targeting Wnt secretion and ligands, recent efforts have focused on the discovery of small molecule inhibitors interfering with the downstream components of the Wnt pathway. The general consensus is that targeting β-catenin activity in the nucleus is likely to be less damaging to the cellular architecture while inhibiting undesired transcriptional outcomes [240]. Indeed, recently a small molecule inhibitor, axitinib, has been shown to block Wnt/β-catenin signaling by targeting E3 ubiquitin ligase SHPRH (SNF2, histone-linker, PHD, and RING finger domain-containing helicase) by promoting asymmetric cell division and destabilizing nuclear β-catenin [84].

Studies with other cancers and ongoing investigations with HNSCC have shown that a small molecule inhibitor of the interaction between the CREB-binding protein (CBP) and β-catenin in the nucleus, ICG-001, suppresses tumor growth in preclinical models [246]. Based on our functional and genomic analyses, a third-generation derivative of ICG-001, E7386, is currently in phase I clinical trials with HNSCC patients. PRI-724, a compound closely related to ICG-001 but with increased affinity, inhibits the self-renewing downstream effects of the β-catenin/CBP axis. Significantly, an active metabolite of PRI-724, C-82, interferes with fibrotic skin phenotypes of scleroderma patients [247]. Since activation of the tumor stroma accompanies HNSCC progression to advanced disease, inhibitors of the β-catenin/CBP axis are likely to target not only tumor epithelia but also to intercept the fibrosis that accompanies HNSCC [248].

Lastly, inhibiting interacting pathways in combination with β-catenin signaling offers improvement over current monotherapies or standard RT/chemotherapy for the treatment of HNSCC. Indeed, recent studies provide evidence that dual inhibition of Notch and Wnt/β-catenin pathways using γ-secretase inhibitor, GSI, represents a promising new treatment. This strategy targets CD44+ cells co-expressing Hes1+, where CD44+ Hes1+ signature tracks with poor overall patient survival [249]. Also, combination therapies linking dual targeting of homeostatic pathways in tumor epithelia with components of the tumor microenvironment increasingly present an appealing approach to intercept the cross talk between these two tumor compartments. Here, a mixture of targeted therapy for epithelial homeostatic oncogenes with immune checkpoint blockade may enhance the deletion of aggressive tumor cells along with increased anticancer immune cell function. Dual targeting of immune checkpoint receptors is likely to enhance antitumor response.

18.12 Conclusion

Although recent years have witnessed extraordinary progress in the deciphering of the molecular mechanisms underlying cancer biology and therapy, there remain many gaps in our understanding of HNSCC pathobiology, including factors driving its initiation, progression to advanced disease, recurrence, and response to treatment. HNSCC is a complex malignancy involving multiple tissue sites with distinct biology, extensive heterogeneity, and poorly understood intercellular communication, as well as dynamic changes in the cellular metabolism and the tumor microenvironment. Here, the Wnt/β-catenin signaling pathway is at the nexus of both tumor epithelial and stromal deregulation, including immune dysfunction, which collectively contribute to this malignancy. As a central homeostatic pathway, Wnt/β--catenin cross regulates other major homeostatic pathways in tumor epithelia, including Hippo-TAZ/YAP signaling, Notch, and TGF-β, and converges on key pro-tumorigenic pathways, such as EGFR and PI3K signaling. Through the interaction with different co-transcriptional and transcriptional partners, β-catenin directs both the induction of oncogenic activities and the inhibition of tumor-suppressive functions of its downstream targets, whose number in the catalog of its downstream effectors continues to expand. One pivotal activity of β-catenin includes collaboration with YAP/TAZ to regulate CSC maintenance and expansion, and to organize distinct CSC niches. At the same time, Wnt/β-catenin inhibits E-cadherin-mediated intercellular adhesion via a number of mechanisms that involve both transcriptional repression and regulation of E-cadherin junctional stability, thus promoting cellular motility and EMT. These diverse interactions support β-catenin's ability to direct the adaptation of tumor cells to metabolic changes dictated by available nutrients and physiological cues and to regulate alterations in the ECM and the stroma to further promote tumor growth and eventual spread. Lastly, deregulated β-catenin activity is associated with non-inflamed tumor stroma and T-cell exclusion, further contributing to disease progression.

Clearly, decoding such dynamic changes in cellular signaling regulated by β-catenin and designing therapeutic strategies that target oncogenic functions of β-catenin while not impacting its structural contributions to cellular architecture represent a daunting challenge. Nonetheless, rapid progress in biomedicine coupled with advances in technologies and molecular tools has highlighted key areas that are likely to elucidate HNSCC biology and translation of these findings to the clinic. These include decoding the biology of stem cells, their lineages and evolving niches, control of pluripotency and epigenetic regulation, and understanding of how cellular context and metabolic plasticity dictate the interpretation of signals in diverse Wnt/β-catenin activities. Such studies will likely rely on the application of single-cell transcriptomics and organoid-based tumor modeling coupled with biomedical engineering of tumors to simulate β-catenin functions in this malignancy. Indeed, recent single-cell transcriptomic analyses of HNSCC have defined a complex and heterogeneous ecosystem along with a subpopulation of cells likely to promote metastasis [32]. Future studies will continue to leverage computational methodologies with emerging technologies in genome and epigenome editing tools to elucidate causal relationships between chromatin features, gene expression, and cancer cell behavior and how chromatin organization and regulatory processes across the genome are deregulated by β-catenin and its colluding partners in HNSCC. Also, comprehensive characterization of gene expression changes at the protein level, including changes in posttranslational modifications (i.e., lipidation, glycosylation), and their impact on cancer cell signaling, resistance to therapy, tumor recurrence, and dynamics of epithelial-stromal cross talk in driving advanced disease and immune evasion are paramount. Lastly, insights from emerging research fields where progress currently occurs at a rapid pace will greatly benefit our understanding of how to interfere with unchecked Wnt/β-catenin transcriptional signals to improve HNSCC therapy.

References

1. Logan CY, Nusse R. The Wnt signaling pathway in development and disease. Annu Rev Cell Dev Biol. 2004;20:781–810.
2. MacDonald BT, Tamai K, He X. Wnt/beta-catenin signaling: components, mechanisms, and diseases. Dev Cell. 2009;17(1):9–26.
3. McNeill H, Woodgett JR. When pathways collide: collaboration and connivance among signalling proteins in development. Nat Rev Mol Cell Biol. 2010;11(6):404–13.
4. Nusslein-Volhard C, Wieschaus E. Mutations affecting segment number and polarity in Drosophila. Nature. 1980;287(5785):795–801.
5. Tsukamoto AS, et al. Expression of the int-1 gene in transgenic mice is associated with mammary gland hyperplasia and adenocarcinomas in male and female mice. Cell. 1988;55(4):619–25.
6. Gordon MD, Nusse R. Wnt signaling: multiple pathways, multiple receptors, and multiple transcription factors. J Biol Chem. 2006;281(32):22429–33.
7. Widelitz R. Wnt signaling through canonical and non-canonical pathways: recent progress. Growth Factors. 2005;23(2):111–6.
8. Hanahan D, Weinberg RA. Hallmarks of cancer: the next generation. Cell. 2011;144(5):646–74.
9. Valastyan S, Weinberg RA. Tumor metastasis: molecular insights and evolving paradigms. Cell. 2011;147(2):275–92.
10. Pectasides E, et al. Markers of epithelial to mesenchymal transition in association with survival in head and neck squamous cell carcinoma (HNSCC). PLoS One. 2014;9(4):e94273.
11. Zhou G. Wnt/beta-catenin signaling and oral cancer metastasis. In: Oral cancer metastasis. New York: Springer; 2010. p. 231–64.
12. Castilho R., Gutkind J. (2014) The Wnt/β-catenin Signaling Circuitry in Head and Neck Cancer. In: Burtness B., Golemis E. (eds) Molecular Determinants of Head and Neck Cancer. Current Cancer Research. Springer, New York, NY
13. Liu G, et al. N-glycosylation induces the CTHRC1 protein and drives oral cancer cell migration. J Biol Chem. 2013;288(28):20217–27.
14. Adamska M, et al. Structure and expression of conserved Wnt pathway components in the demosponge Amphimedon queenslandica. Evol Dev. 2010;12(5):494–518.
15. Lapebie P, et al. WNT/beta-catenin signalling and epithelial patterning in the homoscleromorph sponge Oscarella. PLoS One. 2009;4(6):e5823.

16. Kusserow A, et al. Unexpected complexity of the Wnt gene family in a sea anemone. Nature. 2005;433(7022):156–60.
17. Gonzalez-Sancho JM, et al. The Wnt antagonist DICKKOPF-1 gene is a downstream target of beta-catenin/TCF and is downregulated in human colon cancer. Oncogene. 2005;24(6):1098–103.
18. Katoh M. Comparative genomics on SFRP2 orthologs. Oncol Rep. 2005;14(3):783–7.
19. Carmon KS, et al. R-spondins function as ligands of the orphan receptors LGR4 and LGR5 to regulate Wnt/beta-catenin signaling. Proc Natl Acad Sci U S A. 2011;108(28):11452–7.
20. Carmon KS, et al. RSPO-LGR4 functions via IQGAP1 to potentiate Wnt signaling. Proc Natl Acad Sci U S A. 2014;111(13):E1221–9.
21. Hao HX, et al. ZNRF3 promotes Wnt receptor turnover in an R-spondin-sensitive manner. Nature. 2012;485(7397):195–200.
22. Koo BK, et al. Tumour suppressor RNF43 is a stem-cell E3 ligase that induces endocytosis of Wnt receptors. Nature. 2012;488(7413):665–9.
23. Gonzalez-Sancho JM, et al. Functional consequences of Wnt-induced dishevelled 2 phosphorylation in canonical and noncanonical Wnt signaling. J Biol Chem. 2013;288(13):9428–37.
24. Grumolato L, et al. Canonical and noncanonical Wnts use a common mechanism to activate completely unrelated coreceptors. Genes Dev. 2010;24(22):2517–30.
25. Clevers H. Wnt/beta-catenin signaling in development and disease. Cell. 2006;127(3):469–80.
26. Reya T, Clevers H. Wnt signalling in stem cells and cancer. Nature. 2005;434(7035):843–50.
27. Korinek V, et al. Constitutive transcriptional activation by a beta-catenin-Tcf complex in APC−/− colon carcinoma. Science. 1997;275(5307):1784–7.
28. Morin PJ, et al. Activation of beta-catenin-Tcf signaling in colon cancer by mutations in beta-catenin or APC. Science. 1997;275(5307):1787–90.
29. Yang F, et al. Wnt/beta-catenin signaling inhibits death receptor-mediated apoptosis and promotes invasive growth of HNSCC. Cell Signal. 2006;18(5):679–87.
30. Beck TN, Golemis EA. Genomic insights into head and neck cancer. Cancers Head Neck. 2016;1(1):1.
31. Cancer Genome Atlas, N. Comprehensive genomic characterization of head and neck squamous cell carcinomas. Nature. 2015;517(7536):576–82.
32. Puram SV, et al. Single-Cell Transcriptomic Analysis of Primary and Metastatic Tumor Ecosystems in Head and Neck Cancer. Cell. 2017;171(7):1611–24. e24.
33. Jamal B, et al. Aberrant amplification of the crosstalk between canonical Wnt signaling and N-glycosylation gene DPAGT1 promotes oral cancer. Oral Oncol. 2012;48:523–9.
34. Wend P, et al. Wnt/beta-catenin signalling induces MLL to create epigenetic changes in salivary gland tumours. EMBO J. 2013;32(14):1977–89.
35. Chang HW, et al. Knockdown of beta-catenin controls both apoptotic and autophagic cell death through LKB1/AMPK signaling in head and neck squamous cell carcinoma cell lines. Cell Signal. 2013;25(4):839–47.
36. Duan L, et al. Growth suppression induced by Notch1 activation involves Wnt-beta-catenin down-regulation in human tongue carcinoma cells. Biol Cell. 2006;98(8):479–90.
37. Fu L, et al. Wnt2 secreted by tumour fibroblasts promotes tumour progression in oesophageal cancer by activation of the Wnt/beta-catenin signalling pathway. Gut. 2011;60(12):1635–43.
38. Ge C, et al. miR-942 promotes cancer stem cell-like traits in esophageal squamous cell carcinoma through activation of Wnt/beta-catenin signalling pathway. Oncotarget. 2015;6(13):10964–77.
39. Gonzalez-Moles MA, et al. Beta-catenin in oral cancer: an update on current knowledge. Oral Oncol. 2014;50(9):818–24.
40. Goto M, et al. Rap1 stabilizes beta-catenin and enhances beta-catenin-dependent transcription and invasion in squamous cell carcinoma of the head and neck. Clin Cancer Res. 2010;16(1):65–76.
41. Iwai S, et al. Involvement of the Wnt-beta-catenin pathway in invasion and migration of oral squamous carcinoma cells. Int J Oncol. 2010;37(5):1095–103.

42. Lee SH, et al. Wnt/beta-catenin signalling maintains self-renewal and tumourigenicity of head and neck squamous cell carcinoma stem-like cells by activating Oct4. J Pathol. 2014;234(1):99–107.
43. Li L, et al. Overexpression of beta-catenin induces cisplatin resistance in oral squamous cell carcinoma. Biomed Res Int. 2016;2016:5378567.
44. Li M, et al. Aberrant expression of CDK8 regulates the malignant phenotype and associated with poor prognosis in human laryngeal squamous cell carcinoma. Eur Arch Otorhinolaryngol. 2017;274:2205–13.
45. Liang S, et al. LncRNA, TUG1 regulates the oral squamous cell carcinoma progression possibly via interacting with Wnt/beta-catenin signaling. Gene. 2017;608:49–57.
46. Lo Muzio L, et al. WNT-1 expression in basal cell carcinoma of head and neck. An immunohistochemical and confocal study with regard to the intracellular distribution of beta-catenin. Anticancer Res. 2002;22(2A):565–76.
47. Takei S, et al. Roles of beta-catenin overexpression and adenomatous polyposis coli mutation in head and neck cancer. Nihon Jibiinkoka Gakkai Kaiho. 2003;106(6):692–9.
48. Pannone G, et al. WNT pathway in oral cancer: epigenetic inactivation of WNT-inhibitors. Oncol Rep. 2010;24(4):1035–41.
49. Shiratsuchi H, et al. beta-Catenin nuclear accumulation in head and neck mucoepidermoid carcinoma: its role in cyclin D1 overexpression and tumor progression. Head Neck. 2007;29(6):577–84.
50. Padhi S, et al. Clinico-pathological correlation of beta-catenin and telomere dysfunction in head and neck squamous cell carcinoma patients. J Cancer. 2015;6(2):192–202.
51. Niehrs C. The complex world of WNT receptor signalling. Nat Rev Mol Cell Biol. 2012;13(12):767–79.
52. Mikels AJ, Nusse R. Wnts as ligands: processing, secretion and reception. Oncogene. 2006;25(57):7461–8.
53. Valenta T, Hausmann G, Basler K. The many faces and functions of beta-catenin. EMBO J. 2012;31(12):2714–36.
54. Gottardi CJ, Peifer M. Terminal regions of beta-catenin come into view. Structure. 2008;16(3):336–8.
55. Xing Y, et al. Crystal structure of a beta-catenin/axin complex suggests a mechanism for the beta-catenin destruction complex. Genes Dev. 2003;17(22):2753–64.
56. Xing Y, et al. Crystal structure of a full-length beta-catenin. Structure. 2008;16(3):478–87.
57. Lee E, et al. The roles of APC and Axin derived from experimental and theoretical analysis of the Wnt pathway. PLoS Biol. 2003;1(1):E10.
58. Maeda O, et al. Plakoglobin (gamma-catenin) has TCF/LEF family-dependent transcriptional activity in beta-catenin-deficient cell line. Oncogene. 2004;23(4):964–72.
59. Ben-Ze'ev A, Geiger B. Differential molecular interactions of beta-catenin and plakoglobin in adhesion, signalling and cancer. Curr Opin Cell Biol. 1998;10:629–39.
60. Simcha I, et al. Suppression of tumorigenicity by plakoglobin: an augmenting effect of N-cadherin. J Cell Biol. 1996;133(1):199–209.
61. Zhurinsky J, Shtutman M, Ben-Ze'ev A. Plakoglobin and beta-catenin: protein interactions, regulation and biological roles. J Cell Sci. 2000;113(Pt 18):3127–39.
62. Williams BO, Barish GD, Klymkowsky MW, Varmus HE. A comparative evaluation of beta-catenin and plakoglobin signaling activity. Oncogene. 2000;19:5720–8.
63. Narkio-Makela M, et al. Reduced gamma-catenin expression and poor survival in oral squamous cell carcinoma. Arch Otolaryngol Head Neck Surg. 2009;135(10):1035–40.
64. Mosimann C, Hausmann G, Basler K. Beta-catenin hits chromatin: regulation of Wnt target gene activation. Nat Rev Mol Cell Biol. 2009;10(4):276–86.
65. Mulholland DJ, et al. Functional localization and competition between the androgen receptor and T-cell factor for nuclear beta-catenin: a means for inhibition of the Tcf signaling axis. Oncogene. 2003;22(36):5602–13.

66. Pawlowski JE, et al. Liganded androgen receptor interaction with beta-catenin: nuclear co-localization and modulation of transcriptional activity in neuronal cells. J Biol Chem. 2002;277(23):20702–10.
67. Beildeck ME, Gelmann EP, Byers SW. Cross-regulation of signaling pathways: an example of nuclear hormone receptors and the canonical Wnt pathway. Exp Cell Res. 2010;316(11):1763–72.
68. Kaidi A, Williams AC, Paraskeva C. Interaction between beta-catenin and HIF-1 promotes cellular adaptation to hypoxia. Nat Cell Biol. 2007;9(2):210–7.
69. Essers MA, et al. Functional interaction between beta-catenin and FOXO in oxidative stress signaling. Science. 2005;308(5725):1181–4.
70. Brembeck FH, et al. BCL9-2 promotes early stages of intestinal tumor progression. Gastroenterology. 2011;141(4):1359–70, 1370 e1–3.
71. Hikasa H, et al. Regulation of TCF3 by Wnt-dependent phosphorylation during vertebrate axis specification. Dev Cell. 2010;19(4):521–32.
72. Yu Y, et al. Kindlin 2 forms a transcriptional complex with beta-catenin and TCF4 to enhance Wnt signalling. EMBO Rep. 2012;13(8):750–8.
73. Hoffmeyer K, et al. Wnt/beta-catenin signaling regulates telomerase in stem cells and cancer cells. Science. 2012;336(6088):1549–54.
74. Tam WL, Weinberg RA. The epigenetics of epithelial-mesenchymal plasticity in cancer. Nat Med. 2013;19(11):1438–49.
75. Aguilera O, et al. Epigenetic inactivation of the Wnt antagonist DICKKOPF-1 (DKK-1) gene in human colorectal cancer. Oncogene. 2006;25(29):4116–21.
76. Chen J, et al. Pygo2 associates with MLL2 histone methyltransferase and GCN5 histone acetyltransferase complexes to augment Wnt target gene expression and breast cancer stem-like cell expansion. Mol Cell Biol. 2010;30(24):5621–35.
77. Li Z, et al. Histone H4 Lys 20 monomethylation by histone methylase SET8 mediates Wnt target gene activation. Proc Natl Acad Sci U S A. 2011;108(8):3116–23.
78. Mohan M, et al. Linking H3K79 trimethylation to Wnt signaling through a novel Dot1-containing complex (DotCom). Genes Dev. 2010;24(6):574–89.
79. Ma H, et al. Differential roles for the coactivators CBP and p300 on TCF/beta-catenin-mediated survivin gene expression. Oncogene. 2005;24(22):3619–31.
80. Varelas X, et al. The Hippo pathway regulates Wnt/beta-catenin signaling. Dev Cell. 2010;18(4):579–91.
81. Azzolin L, et al. YAP/TAZ incorporation in the beta-catenin destruction complex orchestrates the Wnt response. Cell. 2014;158(1):157–70.
82. Azzolin L, et al. Role of TAZ as mediator of Wnt signaling. Cell. 2012;151(7):1443–56.
83. Rosenbluh J, et al. Beta-Catenin-driven cancers require a YAP1 transcriptional complex for survival and tumorigenesis. Cell. 2012;151(7):1457–73.
84. Qu Y, et al. Axitinib blocks Wnt/beta-catenin signaling and directs asymmetric cell division in cancer. Proc Natl Acad Sci U S A. 2016;113(33):9339–44.
85. Stamos JL, Weis WI. The beta-catenin destruction complex. Cold Spring Harb Perspect Biol. 2013;5(1):a007898.
86. Hart M, et al. The F-box protein beta-TrCP associates with phosphorylated beta-catenin and regulates its activity in the cell. Curr Biol. 1999;9(4):207–10.
87. Brembeck FH, Rosario M, Birchmeier W. Balancing cell adhesion and Wnt signaling, the key role of beta-catenin. Curr Opin Genet Dev. 2006;16(1):51–9.
88. Heuberger J, Birchmeier W. Interplay of cadherin-mediated cell adhesion and canonical Wnt signaling. Cold Spring Harb Perspect Biol. 2010;2(2):a002915.
89. Holland JD, et al. Wnt signaling in stem and cancer stem cells. Curr Opin Cell Biol. 2013;25(2):254–64.
90. Asahina M, et al. Crosstalk between a nuclear receptor and beta-catenin signaling decides cell fates in the C. elegans somatic gonad. Dev Cell. 2006;11(2):203–11.

91. Lien WH, Fuchs E. Wnt some lose some: transcriptional governance of stem cells by Wnt/beta-catenin signaling. Genes Dev. 2014;28(14):1517–32.
92. Yang H, et al. Epithelial-Mesenchymal micro-niches govern stem cell lineage choices. Cell. 2017;169(3):483–96. e13.
93. Oskarsson T, Batlle E, Massague J. Metastatic stem cells: sources, niches, and vital pathways. Cell Stem Cell. 2014;14(3):306–21.
94. Scheel C, Weinberg RA. Phenotypic plasticity and epithelial-mesenchymal transitions in cancer and normal stem cells? Int J Cancer. 2011;129(10):2310–4.
95. Prince ME, et al. Identification of a subpopulation of cells with cancer stem cell properties in head and neck squamous cell carcinoma. Proc Natl Acad Sci U S A. 2007;104(3):973–8.
96. Monroe MM, et al. Cancer stem cells in head and neck squamous cell carcinoma. J Oncol. 2011;2011:762780.
97. Krishnamurthy S, et al. Endothelial cell-initiated signaling promotes the survival and self-renewal of cancer stem cells. Cancer Res. 2010;70(23):9969–78.
98. Clay MR, et al. Single-marker identification of head and neck squamous cell carcinoma cancer stem cells with aldehyde dehydrogenase. Head Neck. 2010;32(9):1195–201.
99. Song J, et al. Characterization of side populations in HNSCC: highly invasive, chemoresistant and abnormal Wnt signaling. PLoS One. 2010;5(7):e11456.
100. Chen C, et al. Epithelial-to-mesenchymal transition and cancer stem(-like) cells in head and neck squamous cell carcinoma. Cancer Lett. 2013;338(1):47–56.
101. Nor C, et al. Cisplatin induces Bmi-1 and enhances the stem cell fraction in head and neck cancer. Neoplasia. 2014;16(2):137–46.
102. Chen D, et al. Targeting BMI1+ cancer stem cells overcomes chemoresistance and inhibits metastases in squamous cell carcinoma. Cell Stem Cell. 2017;20(5):621–34. e6.
103. Chen YW, et al. Cucurbitacin I suppressed stem-like property and enhanced radiation-induced apoptosis in head and neck squamous carcinoma--derived CD44(+)ALDH1(+) cells. Mol Cancer Ther. 2010;9(11):2879–92.
104. Schepers AG, et al. Lineage tracing reveals Lgr5+ stem cell activity in mouse intestinal adenomas. Science. 2012;337(6095):730–5.
105. Myant KB, et al. ROS production and NF-kappaB activation triggered by RAC1 facilitate WNT-driven intestinal stem cell proliferation and colorectal cancer initiation. Cell Stem Cell. 2013;12(6):761–73.
106. Malanchi I, et al. Cutaneous cancer stem cell maintenance is dependent on beta-catenin signalling. Nature. 2008;452(7187):650–3.
107. Peng Y, et al. The crosstalk between microRNAs and the Wnt/beta-catenin signaling pathway in cancer. Oncotarget. 2017;8(8):14089–106.
108. Huang K, et al. MicroRNA roles in beta-catenin pathway. Mol Cancer. 2010;9:252.
109. Behrens J, et al. Functional interaction of beta-catenin with the transcription factor LEF-1. Nature. 1996;382(6592):638–42.
110. Goldberg AD, Allis CD, Bernstein E. Epigenetics: a landscape takes shape. Cell. 2007;128(4):635–8.
111. Albert M, Peters AH. Genetic and epigenetic control of early mouse development. Curr Opin Genet Dev. 2009;19(2):113–21.
112. Parker DS, et al. Wingless signaling induces widespread chromatin remodeling of target loci. Mol Cell Biol. 2008;28(5):1815–28.
113. Sierra J, et al. The APC tumor suppressor counteracts beta-catenin activation and H3K4 methylation at Wnt target genes. Genes Dev. 2006;20(5):586–600.
114. Zhou B, et al. Interactions between beta-catenin and transforming growth factor-beta signaling pathways mediate epithelial-mesenchymal transition and are dependent on the transcriptional co-activator cAMP-response element-binding protein (CREB)-binding protein (CBP). J Biol Chem. 2012;287(10):7026–38.
115. Lenz HJ, Kahn M. Safely targeting cancer stem cells via selective catenin coactivator antagonism. Cancer Sci. 2014;105(9):1087–92.

116. Chan KC, et al. Therapeutic targeting of CBP/beta-catenin signaling reduces cancer stem-like population and synergistically suppresses growth of EBV-positive nasopharyngeal carcinoma cells with cisplatin. Sci Rep. 2015;5:9979.
117. Li J, et al. CBP/p300 are bimodal regulators of Wnt signaling. EMBO J. 2007;26(9):2284–94.
118. Wend P, et al. Wnt signaling in stem and cancer stem cells. Semin Cell Dev Biol. 2010;21(8):855–63.
119. de Sousa EM, et al. Targeting Wnt signaling in colon cancer stem cells. Clin Cancer Res. 2011;17(4):647–53.
120. de Sousa EMF, et al. Methylation of cancer-stem-cell-associated Wnt target genes predicts poor prognosis in colorectal cancer patients. Cell Stem Cell. 2011;9(5):476–85.
121. Wilhelm F, et al. Novel insights into gastric cancer: methylation of R-spondins and regulation of LGR5 by SP1. Mol Cancer Res. 2017;15(6):776–85.
122. Chinn SB, Myers JN. Oral cavity carcinoma: current management, controversies, and future directions. J Clin Oncol. 2015;33(29):3269–76.
123. Hedberg ML, et al. Genetic landscape of metastatic and recurrent head and neck squamous cell carcinoma. J Clin Invest. 2016;126(4):1606.
124. Baum B, Settleman J, Quinlan MP. Transitions between epithelial and mesenchymal states in development and disease. Semin Cell Dev Biol. 2008;19(3):294–308.
125. Thiery JP, et al. Epithelial-mesenchymal transitions in development and disease. Cell. 2009;139(5):871–90.
126. Kalluri R, Weinberg RA. The basics of epithelial-mesenchymal transition. J Clin Invest. 2009;119(6):1420–8.
127. Scheel C, Weinberg RA. Cancer stem cells and epithelial-mesenchymal transition: concepts and molecular links. Semin Cancer Biol. 2012;22(5–6):396–403.
128. Ye X, et al. Distinct EMT programs control normal mammary stem cells and tumour-initiating cells. Nature. 2015;525(7568):256–60.
129. Katoh M. Comparative genomics on SNAI1, SNAI2, and SNAI3 orthologs. Oncol Rep. 2005;14(4):1083–6.
130. Sanchez-Tillo E, et al. ZEB1 represses E-cadherin and induces an EMT by recruiting the SWI/SNF chromatin-remodeling protein BRG1. Oncogene. 2010;29(24):3490–500.
131. Wang Y, et al. ASPP2 controls epithelial plasticity and inhibits metastasis through beta-catenin-dependent regulation of ZEB1. Nat Cell Biol. 2014;16(11):1092–104.
132. Tenbaum SP, et al. Beta-catenin confers resistance to PI3K and AKT inhibitors and subverts FOXO3a to promote metastasis in colon cancer. Nat Med. 2012;18(6):892–901.
133. Nijkamp MM, et al. Expression of E-cadherin and vimentin correlates with metastasis formation in head and neck squamous cell carcinoma patients. Radiother Oncol. 2011;99(3):344–8.
134. Smith A, Teknos TN, Pan Q. Epithelial to mesenchymal transition in head and neck squamous cell carcinoma. Oral Oncol. 2013;49(4):287–92.
135. Zheng L, et al. Twist-related protein 1 enhances oral tongue squamous cell carcinoma cell invasion through beta-catenin signaling. Mol Med Rep. 2015;11(3):2255–61.
136. Ma MZ, et al. CTHRC1 acts as a prognostic factor and promotes invasiveness of gastrointestinal stromal tumors by activating Wnt/PCP-Rho signaling. Neoplasia. 2014;16(3):265–78, 278 e1–13.
137. Park EH, et al. Collagen triple helix repeat containing-1 promotes pancreatic cancer progression by regulating migration and adhesion of tumor cells. Carcinogenesis. 2013;34:694–702.
138. Zhang J, Ma L. MicroRNA control of epithelial-mesenchymal transition and metastasis. Cancer Metastasis Rev. 2012;31(3–4):653–62.
139. Park SM, et al. The miR-200 family determines the epithelial phenotype of cancer cells by targeting the E-cadherin repressors ZEB1 and ZEB2. Genes Dev. 2008;22(7):894–907.
140. Yan J, et al. Regulation of mesenchymal phenotype by MicroRNAs in cancer. Curr Cancer Drug Targets. 2013;13(9):930–4.

141. Ghahhari NM, Babashah S. Interplay between microRNAs and WNT/beta-catenin signalling pathway regulates epithelial-mesenchymal transition in cancer. Eur J Cancer. 2015;51(12):1638–49.
142. Sun L, et al. MiR-200b and miR-15b regulate chemotherapy-induced epithelial-mesenchymal transition in human tongue cancer cells by targeting BMI1. Oncogene. 2012;31(4):432–45.
143. Jung AC, et al. A poor prognosis subtype of HNSCC is consistently observed across methylome, transcriptome, and miRNome analysis. Clin Cancer Res. 2013;19(15):4174–84.
144. Coordes A, et al. Cancer stem cell phenotypes and miRNA: therapeutic targets in head and neck squamous cell carcinoma. HNO. 2014;62(12):867–72.
145. Shibue T, Weinberg RA. Metastatic colonization: settlement, adaptation and propagation of tumor cells in a foreign tissue environment. Semin Cancer Biol. 2011;21(2):99–106.
146. Luga V, et al. Exosomes mediate stromal mobilization of autocrine Wnt-PCP signaling in breast cancer cell migration. Cell. 2012;151(7):1542–56.
147. Chairoungdua A, et al. Exosome release of beta-catenin: a novel mechanism that antagonizes Wnt signaling. J Cell Biol. 2010;190(6):1079–91.
148. Chaffer CL, Weinberg RA. A perspective on cancer cell metastasis. Science. 2011;331(6024):1559–64.
149. Hinck L, Nelson WJ, Papkoff J. Wnt-1 modulates cell-cell adhesion in mammalian cells by stabilizing beta-catenin binding to the cell adhesion protein cadherin. J Cell Biol. 1994;124:729–41.
150. Hoschuetzky H, Aberle H, Kemler R. Beta-catenin mediates the interaction of the cadherin-catenin complex with epidermal growth factor receptor. J Cell Biol. 1994;127:1375–80.
151. Kemler R. From cadherins to catenins: cytoplasmic protein interactions and regulation of cell adhesion. Trends Genet. 1993;9:317–21.
152. Birchmeier W, Behrens J. Cadherin expression in carcinomas: role in the formation of cell junctions and the prevention of invasiveness. Biochim Biophys Acta. 1994;1198:11–26.
153. Chen Y-T, Stewart DB, Nelson WJ. Coupling assembly of the E-cadherin/β-catenin complex to efficient endoplasmic reticulum exit and basal-lateral membrane targeting of E-cadherin in polarized MDCK cells. J Cell Biol. 1999;144:687–99.
154. Sengupta PK, et al. Coordinate regulation of N-glycosylation gene DPAGT1, canonical Wnt signaling and E-cadherin adhesion. J Cell Sci. 2012;126:484–496.
155. Varelas X, Bouchie MP, Kukuruzinska MA. Protein N-glycosylation in oral cancer: dysregulated cellular networks among DPAGT1, E-cadherin adhesion and canonical Wnt signaling. Glycobiology. 2014;24(7):579–91.
156. Nita-Lazar M, et al. Overexpression of DPAGT1 leads to aberrant N-glycosylation of E-cadherin and cellular discohesion in oral cancer. Cancer Res. 2009;69(14):5673–80.
157. Beavon IR. The E-cadherin-catenin complex in tumour metastasis: structure, function and regulation. Eur J Cancer. 2000;36:1607–20.
158. Morris LG, et al. Recurrent somatic mutation of FAT1 in multiple human cancers leads to aberrant Wnt activation. Nat Genet. 2013;45(3):253–61.
159. Xie J, et al. CDH4 suppresses the progression of salivary adenoid cystic carcinoma via E-cadherin co-expression. Oncotarget. 2016;7(50):82961–71.
160. Yamamoto S, et al. Cthrc1 selectively activates the planar cell polarity pathway of Wnt signaling by stabilizing the Wnt-receptor complex. Dev Cell. 2008;15(1):23–36.
161. Wang Q, et al. NFAT5 represses canonical Wnt signaling via inhibition of beta-catenin acetylation and participates in regulating intestinal cell differentiation. Cell Death Dis. 2013;4:e671.
162. Thrasivoulou C, Millar M, Ahmed A. Activation of intracellular calcium by multiple Wnt ligands and translocation of beta-catenin into the nucleus: a convergent model of Wnt/Ca2+ and Wnt/beta-catenin pathways. J Biol Chem. 2013;288(50):35651–9.
163. Lecarpentier Y, et al. Thermodynamics in cancers: opposing interactions between PPAR gamma and the canonical WNT/beta-catenin pathway. Clin Transl Med. 2017;6(1):14.

164. Whitman M. Smads and early developmental signaling by the TGFbeta superfamily. Genes Dev. 1998;12(16):2445–62.
165. Labbe E, et al. Transcriptional cooperation between the transforming growth factor-beta and Wnt pathways in mammary and intestinal tumorigenesis. Cancer Res. 2007;67(1):75–84.
166. Masszi A, et al. Integrity of cell-cell contacts is a critical regulator of TGF-beta 1-induced epithelial-to-myofibroblast transition: role for beta-catenin. Am J Pathol. 2004;165(6):1955–67.
167. Akhmetshina A, et al. Activation of canonical Wnt signalling is required for TGF-beta-mediated fibrosis. Nat Commun. 2012;3:735.
168. Nowell CS, Radtke F. Notch as a tumour suppressor. Nat Rev Cancer. 2017;17(3):145–59.
169. Kopan R, Ilagan MX. The canonical Notch signaling pathway: unfolding the activation mechanism. Cell. 2009;137(2):216–33.
170. Dotto GP. Notch tumor suppressor function. Oncogene. 2008;27(38):5115–23.
171. Croagh D, et al. Esophageal stem cells and genetics/epigenetics in esophageal cancer. Ann N Y Acad Sci. 2014;1325:8–14.
172. Stransky N, et al. The mutational landscape of head and neck squamous cell carcinoma. Science. 2011;333(6046):1157–60.
173. Agrawal N, et al. Exome sequencing of head and neck squamous cell carcinoma reveals inactivating mutations in NOTCH1. Science. 2011;333(6046):1154–7.
174. Lawrence MS, et al. Comprehensive genomic characterization of head and neck squamous cell carcinomas. Nature. 2015;517(7536):576–82.
175. van Es JH, et al. Notch/gamma-secretase inhibition turns proliferative cells in intestinal crypts and adenomas into goblet cells. Nature. 2005;435(7044):959–63.
176. Munoz-Chapuli R, Perez-Pomares JM. Cardiogenesis: an embryological perspective. J Cardiovasc Transl Res. 2010;3(1):37–48.
177. Kwon C, et al. Notch post-translationally regulates beta-catenin protein in stem and progenitor cells. Nat Cell Biol. 2011;13(10):1244–51.
178. Harvey KF, Zhang X, Thomas DM. The Hippo pathway and human cancer. Nat Rev Cancer. 2013;13(4):246–57.
179. Lamar JM, et al. The Hippo pathway target, YAP, promotes metastasis through its TEAD-interaction domain. Proc Natl Acad Sci U S A. 2012;109(37):E2441–50.
180. Camargo FD, et al. YAP1 increases organ size and expands undifferentiated progenitor cells. Current biology : CB. 2007;17(23):2054–60.
181. Moroishi T, Hansen CG, Guan KL. The emerging roles of YAP and TAZ in cancer. Nat Rev Cancer. 2015;15(2):73–9.
182. Chan SW, et al. A role for TAZ in migration, invasion, and tumorigenesis of breast cancer cells. Cancer Res. 2008;68(8):2592–8.
183. Hiemer SE, et al. A YAP/TAZ-regulated molecular signature is associated with oral squamous cell carcinoma. Mol Cancer Res. 2015;13(6):957–68.
184. Low BC, et al. YAP/TAZ as mechanosensors and mechanotransducers in regulating organ size and tumor growth. FEBS Lett. 2014;588(16):2663–70.
185. Hiemer SE, Varelas X. Stem cell regulation by the Hippo pathway. Biochim Biophys Acta. 2012;18:2323–2334.
186. Mauviel A, Nallet-Staub F, Varelas X. Integrating developmental signals: a Hippo in the (path)way. Oncogene. 2012;31(14):1743–56.
187. Mo JS, Park HW, Guan KL. The Hippo signaling pathway in stem cell biology and cancer. EMBO Rep. 2014;15(6):642–56.
188. Attisano L, Wrana JL. Signal integration in TGF-beta, WNT, and Hippo pathways. F1000Prime Rep. 2013;5:17.
189. Li J, et al. LATS2 suppresses oncogenic Wnt signaling by disrupting beta-catenin/BCL9 interaction. Cell Rep. 2013;5(6):1650–63.
190. Heallen T, et al. Hippo pathway inhibits Wnt signaling to restrain cardiomyocyte proliferation and heart size. Science. 2011;332(6028):458–61.

191. Park HW, et al. Alternative Wnt signaling activates YAP/TAZ. Cell. 2015;162(4):780–94.
192. Boldrup L, et al. Expression of p63, COX-2, EGFR and beta-catenin in smokers and patients with squamous cell carcinoma of the head and neck reveal variations in non-neoplastic tissue and no obvious changes in smokers. Int J Oncol. 2005;27(6):1661–7.
193. Hu T, Li C. Convergence between Wnt-beta-catenin and EGFR signaling in cancer. Mol Cancer. 2010;9:236.
194. Kim SE, Choi KY. EGF receptor is involved in WNT3a-mediated proliferation and motility of NIH3T3 cells via ERK pathway activation. Cell Signal. 2007;19(7):1554–64.
195. Lee CH, et al. Epidermal growth factor receptor regulates beta-catenin location, stability, and transcriptional activity in oral cancer. Mol Cancer. 2010;9:64.
196. Veracini L, et al. Elevated Src family kinase activity stabilizes E-cadherin-based junctions and collective movement of head and neck squamous cell carcinomas. Oncotarget. 2015;6(10):7570–83.
197. Friedl P, Gilmour D. Collective cell migration in morphogenesis, regeneration and cancer. Nat Rev Mol Cell Biol. 2009;10(7):445–57.
198. Whitman M, et al. Type I phosphatidylinositol kinase makes a novel inositol phospholipid, phosphatidylinositol-3-phosphate. Nature. 1988;332(6165):644–6.
199. Locasale JW, Cantley LC. Altered metabolism in cancer. BMC Biol. 2010;8:88.
200. Klempner SJ, Myers AP, Cantley LC. What a tangled web we weave: emerging resistance mechanisms to inhibition of the phosphoinositide 3-kinase pathway. Cancer Discov. 2013;3(12):1345–54.
201. Lui VW, et al. Frequent mutation of the PI3K pathway in head and neck cancer defines predictive biomarkers. Cancer Discov. 2013;3(7):761–9.
202. Dihlmann S, et al. Regulation of AKT1 expression by beta-catenin/Tcf/Lef signaling in colorectal cancer cells. Carcinogenesis. 2005;26(9):1503–12.
203. Nita-Lazar M, et al. Hypoglycosylated E-cadherin promotes the assembly of tight junctions through the recruitment of PP2A to adherens junctions. Exp Cell Res. 2010;316(11):1871–84.
204. Warburg O. On respiratory impairment in cancer cells. Science. 1956;124(3215):269–70.
205. Jose C, Bellance N, Rossignol R. Choosing between glycolysis and oxidative phosphorylation: a tumor's dilemma? Biochim Biophys Acta. 2011;1807(6):552–61.
206. Cairns RA, Harris IS, Mak TW. Regulation of cancer cell metabolism. Nat Rev Cancer. 2011;11(2):85–95.
207. Mazumdar J, et al. O2 regulates stem cells through Wnt/beta-catenin signalling. Nat Cell Biol. 2010;12(10):1007–13.
208. Varki A, Kannagi R, Toole BP. Glycosylation changes in cancer. In: Varki A, et al., editors. Essentials of glycobiology. Cold Spring Harbor: Cold Spring Harbor Laboratory Press; 2009.
209. Pinho SS, Reis CA. Glycosylation in cancer: mechanisms and clinical implications. Nat Rev Cancer. 2015;15(9):540–55.
210. Gillies RJ, Gatenby RA. Adaptive landscapes and emergent phenotypes: why do cancers have high glycolysis? J Bioenerg Biomembr. 2007;39(3):251–7.
211. Itkonen HM, et al. UAP1 is overexpressed in prostate cancer and is protective against inhibitors of N-linked glycosylation. Oncogene. 2015;34(28):3744–50.
212. Anagnostou SH, Shepherd PR. Glucose induces an autocrine activation of the Wnt/beta-catenin pathway in macrophage cell lines. Biochem J. 2008;416(2):211–8.
213. Kurayoshi M, et al. Post-translational palmitoylation and glycosylation of Wnt-5a are necessary for its signalling. Biochem J. 2007;402(3):515–23.
214. Sengupta PK, Bouchie MP, Kukuruzinska MA. N-glycosylation gene DPAGT1 is a target of the Wnt/beta-catenin signaling pathway. J Biol Chem. 2010;285(41):31164–73.
215. Lau KS, et al. Complex N-glycan number and degree of branching cooperate to regulate cell proliferation and differentiation. Cell. 2007;129(1):123–34.
216. Contessa JN, et al. Inhibition of N-linked glycosylation disrupts receptor tyrosine kinase signaling in tumor cells. Cancer Res. 2008;68(10):3803–9.

217. Miller MA, et al. Reduced proteolytic shedding of receptor tyrosine kinases is a post-translational mechanism of kinase inhibitor resistance. Cancer Discov. 2016;6(4):382–99.
218. Junk DJ, et al. Oncostatin M promotes cancer cell plasticity through cooperative STAT3-SMAD3 signaling. Oncogene. 2017;36(28):4001–13.
219. Koontongkaew S. The tumor microenvironment contribution to development, growth, invasion and metastasis of head and neck squamous cell carcinomas. J Cancer. 2013;4(1):66–83.
220. Salo T, et al. Insights into the role of components of the tumor microenvironment in oral carcinoma call for new therapeutic approaches. Exp Cell Res. 2014;325(2):58–64.
221. Erdogan B, Webb DJ. Cancer-associated fibroblasts modulate growth factor signaling and extracellular matrix remodeling to regulate tumor metastasis. Biochem Soc Trans. 2017;45(1):229–36.
222. Kalluri R. The biology and function of fibroblasts in cancer. Nat Rev Cancer. 2016;16(9):582–98.
223. Bonnans C, Chou J, Werb Z. Remodelling the extracellular matrix in development and disease. Nat Rev Mol Cell Biol. 2014;15(12):786–801.
224. Gradl D, Kuhl M, Wedlich D. The Wnt/Wg signal transducer beta-catenin controls fibronectin expression. Mol Cell Biol. 1999;19(8):5576–87.
225. Wielenga VJ, et al. Expression of CD44 in Apc and Tcf mutant mice implies regulation by the WNT pathway. Am J Pathol. 1999;154(2):515–23.
226. Gopal S, et al. Fibronectin-guided migration of carcinoma collectives. Nat Commun. 2017;8:14105.
227. Bais MV, Kukuruzinska M, Trackman PC. Orthotopic non-metastatic and metastatic oral cancer mouse models. Oral Oncol. 2015;51(5):476–82.
228. Lu KW, et al. Gypenosides inhibited invasion and migration of human tongue cancer SCC4 cells through down-regulation of NFkappaB and matrix metalloproteinase-9. Anticancer Res. 2008;28(2A):1093–9.
229. van Loosdregt J, et al. Canonical Wnt signaling negatively modulates regulatory T cell function. Immunity. 2013;39(2):298–310.
230. Swafford D, Manicassamy S. Wnt signaling in dendritic cells: its role in regulation of immunity and tolerance. Discov Med. 2015;19(105):303–10.
231. Spranger S, Gajewski TF. A new paradigm for tumor immune escape: beta-catenin-driven immune exclusion. J Immunother Cancer. 2015;3:43.
232. Pai SG, et al. Wnt/beta-catenin pathway: modulating anticancer immune response. J Hematol Oncol. 2017;10(1):101.
233. Lyford-Pike S, et al. Evidence for a role of the PD-1:PD-L1 pathway in immune resistance of HPV-associated head and neck squamous cell carcinoma. Cancer Res. 2013;73(6):1733–41.
234. Spranger S, Bao R, Gajewski TF. Melanoma-intrinsic beta-catenin signalling prevents anti-tumour immunity. Nature. 2015;523(7559):231–5.
235. Marusyk A, Polyak K. Tumor heterogeneity: causes and consequences. Biochim Biophys Acta. 2010;1805(1):105–17.
236. Spranger S. Tumor heterogeneity and tumor immunity: a chicken-and-egg problem. Trends Immunol. 2016;37(6):349–51.
237. Lehuede C, et al. Metabolic plasticity as a determinant of tumor growth and metastasis. Cancer Res. 2016;76(18):5201–8.
238. Chen DS, Mellman I. Oncology meets immunology: the cancer-immunity cycle. Immunity. 2013;39(1):1–10.
239. Shayan G, et al. Adaptive resistance to anti-PD1 therapy by Tim-3 upregulation is mediated by the PI3K-Akt pathway in head and neck cancer. Oncoimmunology. 2017;6(1):e1261779.
240. Kahn M. Can we safely target the WNT pathway? Nat Rev Drug Discov. 2014;13(7):513–32.
241. Aminuddin A, Ng PY. Promising druggable target in head and neck squamous cell carcinoma: Wnt signaling. Front Pharmacol. 2016;7:244.
242. Tammela T, et al. A Wnt-producing niche drives proliferative potential and progression in lung adenocarcinoma. Nature. 2017;545(7654):355–9.

243. Rudy SF, et al. In vivo Wnt pathway inhibition of human squamous cell carcinoma growth and metastasis in the chick chorioallantoic model. J Otolaryngol Head Neck Surg. 2016;45:26.
244. Madan B, et al. Wnt addiction of genetically defined cancers reversed by PORCN inhibition. Oncogene. 2016;35(17):2197–207.
245. Zhan T, Rindtorff N, Boutros M. Wnt signaling in cancer. Oncogene. 2017;36(11):1461–73.
246. Arensman MD, et al. The CREB-binding protein inhibitor ICG-001 suppresses pancreatic cancer growth. Mol Cancer Ther. 2014;13(10):2303–14.
247. Lafyatis R, et al. Inhibition of beta-catenin signaling in the skin rescues cutaneous adipogenesis in systemic sclerosis: a randomized, double-blind, placebo-controlled trial of C-82. J Invest Dermatol. 2017;137:2473–83.
248. Kartha VK, et al. PDGFRbeta is a novel marker of stromal activation in oral squamous cell carcinomas. PLoS One. 2016;11(4):e0154645.
249. Barat S, et al. Gamma-secretase inhibitor IX (GSI) impairs concomitant activation of notch and Wnt-beta-catenin pathways in CD44+ gastric cancer stem cells. Stem Cells Transl Med. 2017;6(3):819–29.

Chapter 19
Hyaluronan-Mediated CD44 Signaling Activates Cancer Stem Cells in Head and Neck Cancer

Lilly Y. W. Bourguignon

Abstract Head and neck squamous cell carcinoma (HNSCC) is an aggressive disease associated with high morbidity and mortality. Hyaluronan (HA), a major component in the extracellular matrix (ECM) of most mammalian tissues, is accumulated in many types of tumors including HNSCC and is also highly concentrated in stem cell niches. The unique HA-enriched microenvironment appears to be involved in both the self-renewal and differentiation of human cancer stem cells (CSCs). HA binds to a ubiquitous, abundant, and functionally important family of cell surface receptors, defined by CD44. This article reviews the current evidence for the existence of a subpopulation of CD44-expressing cancer stem cells (CSCs) in HNSCC. A special emphasis is placed on HA/CD44-dependent expression of stem cell transcription factors (Nanog, OCT4, and SOX2), cell signaling, and oncogenic microRNA activation, and how the action of these factors supports CSC functions including formation of spheroid cells, self-renewal, clone formation, and chemotherapeutic drug resistance. All of these events are known to contribute to CSC-associated tumor initiation and HNSCC progression in head and neck cancer. HA/CD44-mediated CSC signaling pathways are emerging as important structural and functional tumor markers. In addition, these proteins may be valuable as drug targets in strategies to inhibit tumor cell growth, survival, and invasion/metastasis as well as to overcome chemoresistance in HNSCC.

Keywords Cancer stem cells (CSCs) · Hyaluronan (HA) · CD44 · Stemness · miRNAs · Chemoresistance · HNSCC

L. Y. W. Bourguignon (✉)
San Francisco Veterans Affairs Medical Center and Department of Medicine,
University of California at San Francisco & Endocrine Unit (111N2),
San Francisco, CA, USA
e-mail: lilly.bourguignon@ucsf.edu

B. Burtness, E. A. Golemis (eds.), *Molecular Determinants of Head and Neck Cancer*, Current Cancer Research, https://doi.org/10.1007/978-3-319-78762-6_19

19.1 Introduction

Head and neck squamous cell carcinoma (HNSCC) represents the sixth most common cancer worldwide [1]. The 3-year survival rate for patients with advanced-stage HNSCC and treated with standard therapy is only 30–50% [1–4]. Resistance to standard therapy continues to be a limiting factor in the treatment of HNSCC. Thus, there is currently a great need to clarify the key mechanisms of tumor initiation and progression underlying the clinical behavior of HNSCC. Human HNSCC tumors have been shown to contain a subpopulation of cancer stem cells (CSCs) characterized by high expression of aldehyde dehydrogenase-1 (ALDH1), and of variant isoforms of CD44, and with functional properties including capacity for self-renewal, multipotency, and efficiency in tumor initiation [5–11]. CSCs demonstrate innate chemoresistance and are associated with metastatic progression and tumor recurrence [5–11]. A better understanding of the cellular and molecular mechanisms involved in CSC-derived pathobiology of HNSCC would profoundly impact the treatment paradigms for this malignancy.

While the functional impact of CSC markers such as ALDH1, an intracellular enzyme known to convert retinol to retinoic acid [12], in supporting the CSC phenotype is not well understood, the gene CD44 encodes cell surface glycoproteins that are expressed in a variety of cells and tissues including HNSCC cells and carcinoma tissues and contribute actively to CSC properties [13–16]. Multiple CD44 isoforms (derived by alternative splicing) are variants of the standard form, CD44s [17, 18] (Fig. 19.1). The presence of high levels of CD44 variant (CD44v) isoforms is emerging as an important metastatic tumor marker in a number of cancers includ-

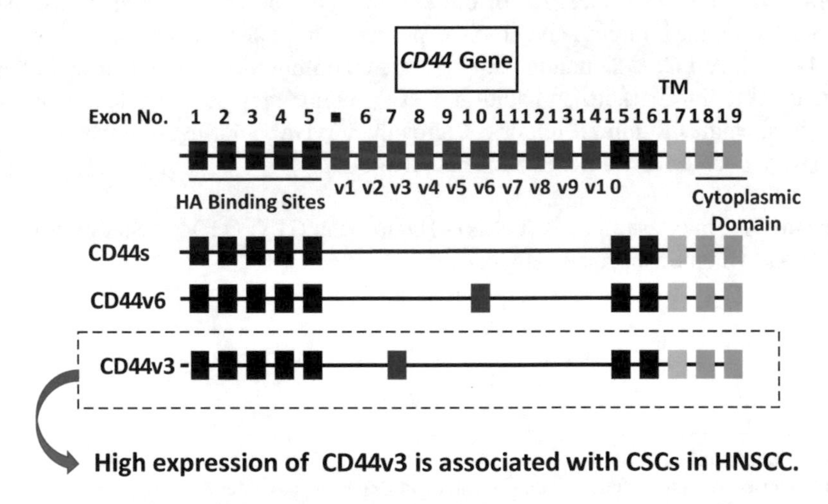

Fig. 19.1 Illustration of CD44 gene, CD44s (the standard form), and alternative spliced variants (CD44v6 and CD44v3 isoforms). High expression of CD44v3 isoform is associated with head and neck cancer stem cells (CSCs). The HA-binding domain is located at the external (in particular, N-terminal) region of all CD44 isoforms, and the signaling regulators' binding sites are located at the cytoplasmic domain of CD44 isoforms. All isoforms contain a transmembrane domain (TM)

ing HNSCC [13–16], and CD44 has been proposed to be one of the important surface markers for CSCs [19]. All stem cells are thought to exist in specialized microenvironments known as "niches" [20, 21]. Components present in these niches can regulate stem cell behavior through direct binding to stem cell surface receptors or via indirect activation of paracrine signaling [20, 21]. Extracellular matrix (ECM) components including hyaluronan (HA) are present in some of the stem cell niches [22, 23]. All CD44 isoforms contain a matrix HA-binding site in their extracellular domain and thereby serve as a major cell surface receptor for HA [24]. Both CD44v isoforms and HA are overexpressed at sites of tumor attachment to the extracellular matrix [25]. This review focuses first on matrix HA interaction with CD44 in regulating HNSCC stem cell signaling pathways and then describes downstream target functions of this signaling that contribute to tumor initiation, chemoresistance, and HNSCC progression. These data suggest that development of new therapeutic agents that effectively target HA/CD44-activated signaling events in CSCs could constitute effective HNSCC therapy.

19.2 Matrix Hyaluronan (mHA) as a Component of the CSC Niche in HNSCC

CSC interaction with their immediate microenvironment or niche is important for their survival, self-renewal, and other functions [20]. The cellular and noncellular components of the niche provide cues that regulate proliferative and self-renewal signals, thereby helping CSCs maintain their undifferentiated state [21]. Prior studies have identified specific adhesion molecule(s) expressed in HNSCCs that correlate with tumor cell invasive behavior(s). Among such candidate molecules, HA is a linear, high-molecular-weight (mega-Dalton) polymer comprised of repeating disaccharide units of (β1 → 3) D-glucuronate (β1 → 4) *N*-acetyl-D-glucosamine [26, 27], produced by both normal and tumor cells (Fig. 19.2). HA is detected in the extracellular matrix (ECM) of most mammalian tissues. Most importantly, HA accumulates at tumor cell attachment sites and appears to play an important role in promoting HNSCC tumor phenotypes [28]. HA is enriched in stem cell niches,

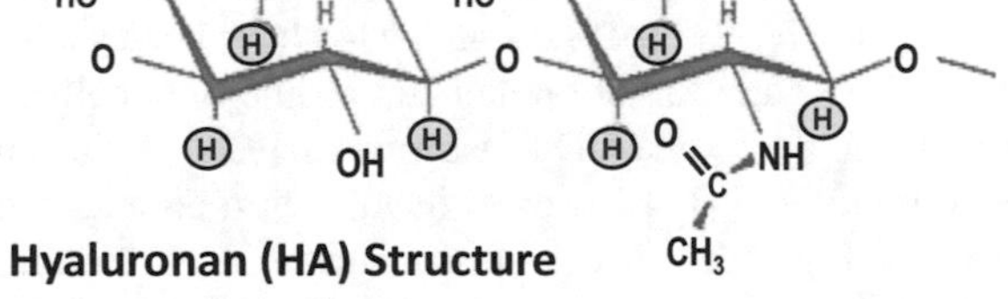

Hyaluronan (HA) Structure

Fig. 19.2 Illustration of matrix hyaluronan (HA) structure

where the unique HA-enriched microenvironment is involved in both self-renewal and differentiation of normal and cancer human stem cells [22, 23].

HA is produced by three isozymes of transmembrane HA synthase enzymes (HAS1, HAS2, and HAS3) that link the two precursor molecules in an alternating manner and extrude the growing HA strand from normal and tumor cells. While HAS1 and HAS2 produce large HA polymers (up to 10 MDa), HAS3 generates smaller-size HA with apparent molecular weight in the range of 100 KDa [29]. Dysregulation of HAS1, HAS2, and HAS3 by oncogenic signaling events often results in abnormal production of HA and directly contributes to aberrant cellular processes such as transformation and metastasis [30, 31].

In addition to functions in supporting HA production in the stem cell niche, HAS2 was shown to be expressed in at least two oral cancer cell lines, HSC-3 and SCC-4 [32]. Suppression of HAS2 expression in these cells resulted in decreased CD44-dependent migration, decreased growth, and increased cisplatin sensitivity, suggesting the importance of autocrine tumor cell HA production to promote in vitro tumor progression and treatment resistance in oral cancer cells. Increased HAS2 expression in oral cavity carcinoma clinical specimens was associated with poor clinicopathologic characteristics and worse disease-free survival [32]. These findings suggest that tumor-intrinsic HAS expression is closely associated with HNSCC progression.

During tumorigenesis, matrix HA can also be digested into a variety of biologically active smaller-sized fragments by hyaluronidases [9, 33]. Currently, at least six hyaluronidase-like proteins [e.g., Hyal-1, Hyal-2, Hyal-3, Hyal-4, *PHYAL*-1, and PH20 (or Spam1)] have been reported as expressed in both tumor and stromal cells. HNSCC patients have been shown to have elevated hyaluronidase levels in their saliva, with Hyal-1 reported as the major hyaluronidase that is expressed in HNSCC tumor tissues [34]. In another study, RT-PCR analysis indicated PH20 was also expressed in HNSCC, especially in laryngeal carcinomas [35, 36]. Since Hyal-1 is one of the major tumor-derived hyaluronidases expressed in HNSCC, the fact that it is detectable in saliva may make it useful as a specific marker for noninvasive detection of primary HNSCC and monitoring its recurrence.

19.3 CD44 Isoform Expression in HNSCC CSCs

Since CD44 is an HA receptor, it provides a physical linkage between matrix HA and various transcription factors that regulate tumor cell functions through distinct signaling pathways [11, 37, 38]. Both HA and CD44 have been shown to be involved in self-renewal and differentiation of human embryonic stem cells (hESC) [39, 40]. There is some evidence that some CD44 isoforms have specific functions in these processes. Several hundred CD44 isoforms have been reported, with those studied in HNSCC shown in Fig. 19.1. All isoforms encode an N-terminal extracellular HA-binding domain (135 amino acids, encoded by exons 1–5), an additional 83 amino acids encoded by exons 15 and 16, and a C-terminal transmembrane domain followed by a short (72 amino acids) cytoplasmic region (Fig. 19.1) [17, 18]. Between the exons encoding the HA-binding domain and exon 15 are 10 variant exons expressed in distinct variant isoforms. Previous study indicated that the expression of two CD44 isoforms (i.e., CD44s and CD44v6) is found in the majority of the cells in head and neck tissue and that this type of marker by itself was not able to distinguish normal from benign or malignant epithelial cells from the head and neck region [37]. The expression of CD44s was not significantly different in metastatic lymph nodes and primary tumors. However, overexpression of at least one CD44 variant isoform has been associated with HNSCC progression [13–16], suggesting that these CD44 isoforms may have unique signaling properties. The expression of CD44v3 appears to be closely associated with advanced T stage and regional metastasis [15]. The role of the CD44v3 isoforms in HNSCC progression is highlighted by studies showing a close correlation between the expression of v3-containing isoforms with HNSCC growth, migration, and MMP expression [14]. Transfection of a CD44v3 isoform into the FaDu HNSCC cell line, which typically expresses a low level of this variant, resulted in significantly increased tumor cell migration but not proliferation [41]. Treatment of a CD44v3 isoform-expressing HNSCC cell line with antibody specifically binding CD44v3 decreased in vitro proliferation and increased cisplatin sensitivity [14, 15]. Using the same anti-CD44v3 antibody, immunohistochemical tissue analysis revealed that CD44v3 isoforms were preferentially expressed in metastatic lymph nodes. In addition, there are several in vitro and histopathological studies suggesting that CD44 v3 isoforms are involved in HNSCC progression behaviors [14, 15].

Tumor cells with $CD44^{high}ALDH1^{high}$ (and to a much lesser extent, $CD44^{low}ALDH1^{low}$ cells) form tumors starting with a low number of tumor cells injected into animals [42, 43]. Recent studies indicate that HNSCC tumors also contain a cell subpopulation characterized by high levels of CD44v3 and ALDH1 expression ($CD44v3^{high}ALDH1^{high}$ cells) (Fig. 19.3) [8, 9]. Purified $CD44v3^{high}ALDH1^{high}$ cells injected into immunodeficient mice can generate multiple classes of phenotypically distinct cells, resulting in heterogeneous tumors [8–10]. Specifically, 5-week female NOD/SCID were injected subcutaneously with sorted $CD44v3^{high}ALDH1^{high}$ cells or $CD44v3^{low}ALDH1^{low}$ or unsorted HNSCC cells (50, 500, or 5000 cells). The results revealed that $CD44v3^{high}ALDH1^{high}$ cells (and to a lesser extent $CD44v3^{low}ALDH1^{low}$ cells or unsorted cells) were capable of forming tumors with high efficiency in mice injected with as few as 50 $CD44v3^{high}ALDH1^{high}$ cells [8–10]. These findings indicate that $CD44v3^{high}ALDH1^{high}$ cells (but not $CD44v3^{low}ALDH1^{low}$ cells or unsorted cells) are very tumorigenic. Further analysis showed that $CD44v3^{high}ALDH1^{high}$ cells display a significantly higher levels of stem cell marker (e.g., Nanog, OCT4, and SOX2) expression, self-renewal/growth, and clone formation capacity compared to $CD44v3^{low}ALDH1^{low}$ cells or unsorted cells [8–10]. These findings clearly indicate that $CD44v3^{high}ALDH1^{high}$ cells display the hallmark CSC characteristics and support the contention that HA-CD44 signaling is directly involved in the regulation of CSC-like properties.

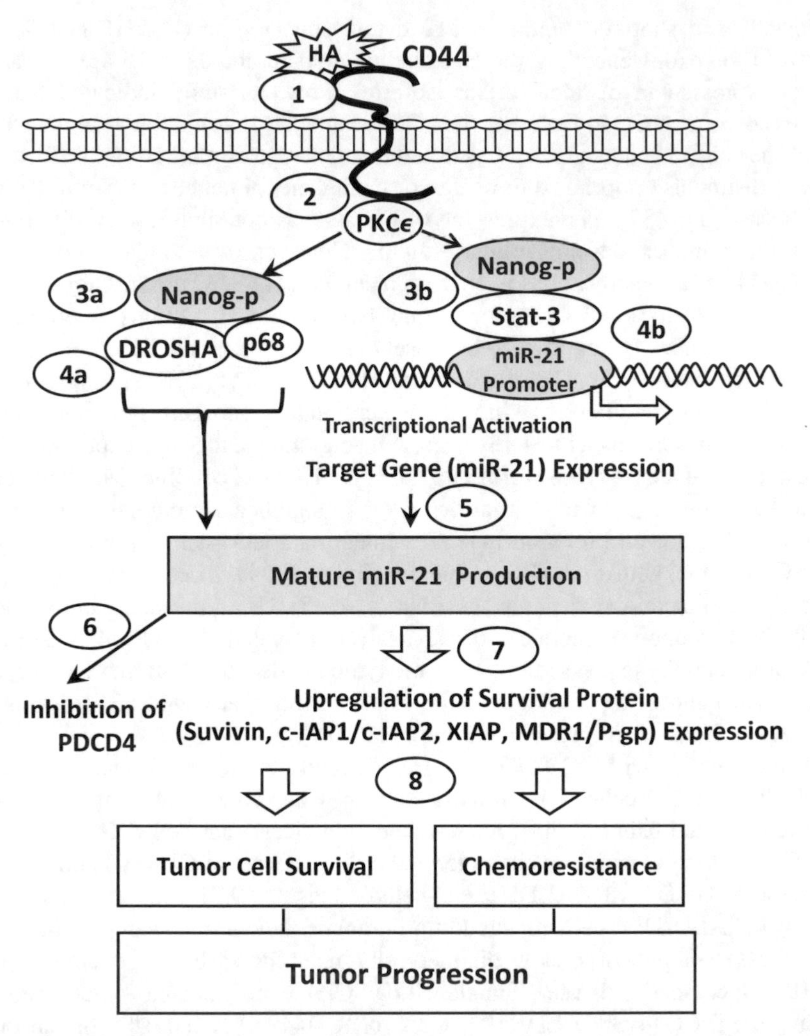

Fig. 19.3 A proposed model for HA-CD44-mediated Nanog signaling via PKCϵ activation and/or STAT3 binding in the regulation of miRNA-21 production, oncogenesis, and chemoresistance in tumor cells. The binding of HA to CD44 (*step 1*) promotes PKCϵ activity (*step 2*) which, in turn, causes phosphorylation of Nanog *(step 3a and 3b)*. Phosphorylated Nanog then translocates from the cytosol to the nucleus and interacts with the microprocessor complex containing the RNase III (DROSHA) and the RNA helicase (p68) (*step 4a*), resulting in miR-21 production (*step 5*). Nuclear translocated phospho-Nanog (step 3b) can also bind to STAT3 and the miR-21 promoter (step 4b) resulting in miR-21 gene expression and miR-21 production (*step 5*). The resultant miR-21 then functions to downregulate the tumor suppressor protein (PDCD4) (step 6) and promotes oncogenesis (step 7), leading to upregulation of IAP (survivin and XIAP)/MDR1 (p-gp) expression, tumor cell anti-apoptosis/survival, and chemoresistance (step 8)

Although most of the examples described above derive from oral cancers, accumulating evidence has suggested that CSCs also are present and hence reasonably contribute to disease aggressiveness and treatment resistance in laryngeal squamous cell carcinoma (LSCC) [44–46] and HPV-positive tumors originating in the oropharynx [47–49]. Thus, investigation of HA/CD44 signaling-induced CSC properties could help elucidate pathogenesis of all forms of head and neck cancer (e.g., oral cancer, LSCC, and HPV-positive oropharynx tumors) and the development of targeted therapies.

19.4 HA/CD44 Targeting Therapies

Because of these important roles in CSC function, a number of therapies have been devised to target CD44. In designing anti-CD44 therapies, anti-CD44v6 antibodies were used in clinical trials for patients suffering from HNSCC. Although phase I clinical trials looked promising, the studies were abruptly ended after the death of a patient [50]. Despite the termination of the trials, CD44 certainly remains a valid target for anticancer therapy. Recent studies indicate that negatively charged and biocompatible HA-based nanoparticles may be used as a therapeutic system for targeting CD44-positive cancer cells [51–53]. For example, a study indicated that HA-based nanoparticles encapsulated with epigallocatechin-3-gallate can inhibit cell cycle progression and prostate cancer growth [51]. Another study also showed that HA-based nanoparticles containing doxorubicin display therapeutic effects on CD44-positive human breast cancer by accurately delivering doxorubicin into breast tumor xenografts and greatly enhancing chemosensitivities of cancer treatment [52]. Although very little information regarding HA/CD44v3-related targeting therapies is available in HNSCC, these newly developed HA-based nanoparticles coupled with different drugs could be used to target CD44v3-expressing CSCs in HNSCC in the future.

19.5 HA/CD44-Induced Head and Neck Cancer Stem Cell Signaling Pathways

There is compelling evidence showing the master stem cell transcription factors (e.g., Nanog, OCT4, and SOX2) often form a self-organized core of transcription factors that maintain pluripotency and self-renewal of human embryonic stem cells [54, 55]. Expression of Nanog, OCT4, and SOX2 is crucial for progression of many human malignancies, including HNSCC [56, 57]. High expression of OCT4 and Nanog (and to a lesser extent, SOX2) has been shown to be associated with worse survival and strongly independent prognostic effects on HNSCC progression [56, 57]. Both mRNA and protein expression of OCT4, SOX2, and Nanog have also been shown to be significantly elevated in CSCs isolated from HNSCC cells [8]. In addition, expression of the OCT4, SOX2, and Nanog proteins can be detected in tumor tissue samples from human HNSCC patient specimens [8, 56, 57]. These findings clearly demonstrate that the master stem cell transcription factors such as Nanog, OCT4, and SOX2 are closely associated with each other during HNSCC progression. Despite intense research into the role of Nanog, OCT4, and SOX2 in normal stem cell biology in recent years, understanding of the regulatory role of these master stem cell transcription factors in tumor initiation and the pathogenesis of HNSCC is still quite limited.

19.5.1 Nanog

Several studies indicate that Nanog is an important transcription factor involved in the self-renewal and maintenance of pluripotency in the inner cell mass (ICM) of mammalian embryos and embryonic stem (ES) cells [58, 59]. Nanog signaling involves interactions with various pluripotent stem cell regulators (e.g., OCT4 and SOX2) which together control the expression of a set of target genes required for ES cell pluripotency [60, 61]. Several tumor cell types have also been shown to express Nanog and other ES markers [62–64]. The Nanog family of proteins was found to act as growth-promoting regulators in many types of tumor cell [65]. HA binding to CD44-expressing tumor cells promotes Nanog upregulation and physical association with CD44 at the plasma membrane, followed by Nanog nuclear translocation and expression of stemness biomarkers [8]. Importantly, the HA-induced CD44 interaction with Nanog plays a pivotal role in miR-21 production leading to programmed cell death 4 (PDCD4) reduction, IAP upregulation, and chemoresistance in CSC isolated from HNSCC cells [8, 62].

19.5.1.1 HA/CD44 Signaling Through PKC$_\varepsilon$ and STAT3 Activates Nanog, Inducing miRNA-21 Production During Oncogenesis

HA-CD44 signaling to promote CSC characteristics may partially result from control of microRNAs (miRNAs). These evolutionarily conserved negative regulators of gene expression work by inhibiting translation of mRNAs that contain complementary target sites, referred to as “seed regions” [60]. Accumulating evidence indicates that noncoding miRNAs are involved in both cancer development and multidrug resistance [66]. Several transcription factors, including Nanog, participate in regulation of the expression of some miRNAs during development [67]. During the production of human miRNAs, capped and polyadenylated longer transcripts are precursors to mature miRNAs [61]. In mammalian miRNA biogenesis, primary transcripts of miRNA genes (pri-mRNAs) are subsequently cleaved to produce an intermediate molecule containing a stem loop of ~70 nucleotides (pre-mRNAs) by the nuclear RNase III enzyme DROSHA and exported from the nucleus by Exportin 5 [60]. A second RNase III enzyme Dicer then generates mature miRNA, which is loaded into the RNA-induced silencing complex (RISC) in association with the argonaute protein (Ago) that induces silencing via the RNA interference pathway [68]. Although Dicer has an important role in the silencing action of miRNAs, recent studies have shown that silencing can still occur in cells that lack Dicer. In addition, the nuclear p68-RNA helicase appears to be required in the uptake of certain miRNAs into the silencing complex [69]. The p68 protein belongs to a family of proteins that are involved in RNA metabolism processes such as translation and RNA degradation [70].

HA-CD44 control of miRNA expression is thought to proceed in part through regulation of protein kinase C_{ε} (PKC_{ε}), a member of a family of serine-threonine kinases that plays a pivotal role in signal transduction and has a number of cellular functions [71]. PKC_{ε} exists as at least 11 different isoforms, some of which have distinct functionality [72]. HA-CD44 binding promotes PKC_{ε} activation and Nanog phosphorylation, leading to miRNA biogenesis as described above. Most importantly, this process leads to microRNA-21 (miR-21) production; miR-21 in turn reduces expression of the tumor suppressor protein PDCD4 and elevates expression of proteins that promote tumor cell survival and chemoresistance (survivin, IAP & MDR/Pgp-1) (Fig. 19.3).

Beyond activity in upregulating Nanog and miR-21, PKCϵ supports tumorigenesis in other ways. For example, PKCϵ activity is required for the function of the anti-apoptotic Bcl-2 family of proteins [73]. PKC also functions to prevent apoptosis in a number of cells by upregulating inhibitors of apoptosis (IAP) proteins (e.g., X-linked IAP (XIAP) and survivin) and by inhibiting caspases [73, 74]. Downregulation of PKCϵ by treating cells with PKC inhibitors sensitizes tumor necrosis factor-α (TNFα)-mediated cell death in breast tumor cells [75]. Thus, PKCϵ has multiple functional linkages to anti-apoptotic effects and survival pathways in CSCs. It seems likely that HA-mediated CD44 signaling stimulates PKCϵ activities leading to Nanog-regulated miR-21 production and upregulation of survival proteins in HNSCC cells. However, we cannot preclude the possibility that other as yet unknown signaling pathways may activate PKCϵ in HNSCC.

HA-CD44 control of miRNA expression also involves activation of signal transducer and activator of transcription protein 3 (STAT3). The transcription factor STAT3 was initially identified as APRF (acute phase response factor), an inducible DNA-binding protein that binds to the IL-6-responsive element within the promoters of hepatic acute phase genes [76, 77]. Accumulating evidence indicates that STAT3 also plays an important role in regulating cell growth, differentiation, and survival [78]. Nanog and STAT3 also appear to be functionally coupled in HA/CD44 signaling during epithelial tumor cell activation [63]. HA induces CD44 interaction with Nanog and STAT3 in HNSCC cells (HSC-3 cells) [62]. Specifically, HA binding to CD44 promotes Nanog interaction with STAT3 and tyrosine phosphorylated STAT3, followed by their nuclear translocation and transcriptional activation. Detailed analysis of the miR-21 promoter indicates the presence of STAT3 binding site(s), while chromatin immunoprecipitation (ChIP) assays demonstrate that stimulation of miR-21 expression by HA/CD44 signaling is Nanog/STAT3-dependent in HNSCC cells. The processes involved in miR-21 production by Nanog/Stat-3 signaling also result in a decrease of the tumor suppressor protein PDCD4 and an upregulation of inhibitors of the apoptosis family of proteins (IAPs) as well as chemoresistance in HNSCC cells [62].

Treatment of HNSCC cells with Nanog- and/or STAT3-specific (or PKCϵ-specific) small interfering RNAs (siRNAs) effectively blocks HA-mediated Nanog-STAT3 and PKCc-Nanog signaling events, abrogates miR-21 production, increases PDCD4 expression, downregulates IAP expression, and enhances chemosensitivity. HNSCC cells were also transfected with a specific anti-miR-21 inhibitor [62] to silence miR-21 expression. This resulted in upregulated PDCD4 expression, decreased IAP expression, and enhanced chemosensitivity in HA-treated HNSCC cells [62]. These novel Nanog/STAT3 and the PKCϵ/Nanog signaling mechanisms involved in miR-21 production may suggest future intervention strategies in the treatment of HA/CD44-activated HNSCC CSCs (Fig. 19.3).

19.5.2 HA/CD44 Activates OCT4/SOX2/Nanog Signaling and HNSCC Function in CSCs

19.5.2.1 Oct 4

Previous reports showed that Oct 4 (encoded by *POU5f*, also known as Oct3 or Oct3/4) belongs to the POU transcription factor family, in which the members regulate their target gene expression by binding to an octameric sequence motif containing the AGTCAAAT consensus sequence [79, 80]. The OCT4 protein contains three functional domains including a POU (pit-Oct-Unc) DNA-binding domain, an N-terminal transactivation domain, and a C-terminal cell type-specific transactivation domain [79, 80]. OCT4 was first identified as a stem cell-specific and germline-specific transcription factor in mice [81]. In humans, OCT4 is the product of the *OTF3* gene and has been shown to maintain “stemness” in pluripotent stem cells [82]. Functionally, OCT4 is essential for early embryonic development and functions as a master regulator of the initiation and maintenance of pluripotent cells during embryonic development. It also interacts with other embryonic regulators, such as Nanog and SOX2, to oversee a vast regulatory network that maintains pluripotency and inhibits differentiation [83, 84]. HA binding to CD44-expressing epithelial tumor cells promotes OCT4 upregulation and association with CD44, followed by nuclear translocation and transcriptional activation [8–10]. Most importantly, both HA/CD44 signaling and OCT4 function appear to be closely linked to stemness, anti-apoptosis, cell survival, and chemoresistance in HNSCC CSC [8–10].

19.5.2.2 SOX2

The Sox (short for sex-determining region Y-box2) gene family encodes a group of transcription factors that are characterized by a highly conserved high-mobility group (HMG) domain [85]. These genes play an important role in stem cell functions, organogenesis, and animal development [86, 87]. Sox proteins including SOX2 bind to specific DNA sequences (C(T/A)TTG(T/A)(T/A)) by means of their HMG domains and promote transcription of their target genes [85]. Since SOX2 by itself lacks strong affinity for DNA binding, it relies on the formation of a complex with other transcription factors to exert influence on DNA [85]. Recently, several studies have linked SOX2 overexpression with human cancers including HNSCC [88]. HA binding to CD44-expressing epithelial tumor cells promotes SOX2 protein association with CD44, followed by SOX2 activation and the expression of pluripotent stem cell regulators [8–10]. The exact nature of CD44-Sox2 binding has not yet been established. From studies in breast cancer models, it is known that matrix HA interaction with tumor cells triggers the cytoplasmic domain of CD44 to bind unique downstream effectors (which include the cytoskeletal protein ankyrin and the oncogenic signaling molecules – Tiam1, RhoA-activated ROK, c-Src kinase, and p185HER2). These binding events are important to coordinate activation of intracellular signaling pathways involving the GTPases Rho and Ras, and both receptor-linked and non-receptor-linked tyrosine kinase pathways, leading to tumor cell growth, migration and invasion, and tumor progression [89, 90]. It is possible that Sox2 association with CD44 is also regulated by the cytoskeleton and additional kinase pathways during HA signaling.

19.5.2.3 HA Binding to CSCs Activates CD44 Association with Nanog/OCT4/SOX2 Leading to miRNA-302 Production and Stemness Properties (e.g., Spheroids, Self-Renewal, and Clone Formation)

Genetic studies using mouse models revealed that OCT4, SOX2, and Nanog have distinct roles but use overlapping signaling pathways to maintain stemness functions during development [91]. Another MicroRNA, miR-302, has been identified as an important target of this signaling relevant to HNSCC. Mir-302 encodes a cluster of eight miRNAs that are expressed specifically in pluripotent embryonic stem cells (ESCs) and/or embryonic carcinoma cells (ECCs) [92, 93]. Most importantly, the miR-302 family is involved in maintaining "stemness" of human embryonal carcinoma cells [94] and reprogramming human skin cancer cells to a pluripotent ESC-like state [95]. These findings strongly suggest that miR-302 plays an important role in the pluripotency of ESCs and ECCs. Some studies have also indicated that OCT4, SOX2, and Nanog co-occupy the promoter of miR-302, which has been shown to target genes required for development and oncogenesis [8, 92–94]. At the transcriptional level, OCT4, SOX2, and Nanog form a positive autoregulatory loop that is important for the maintenance of the undifferentiated state. At the

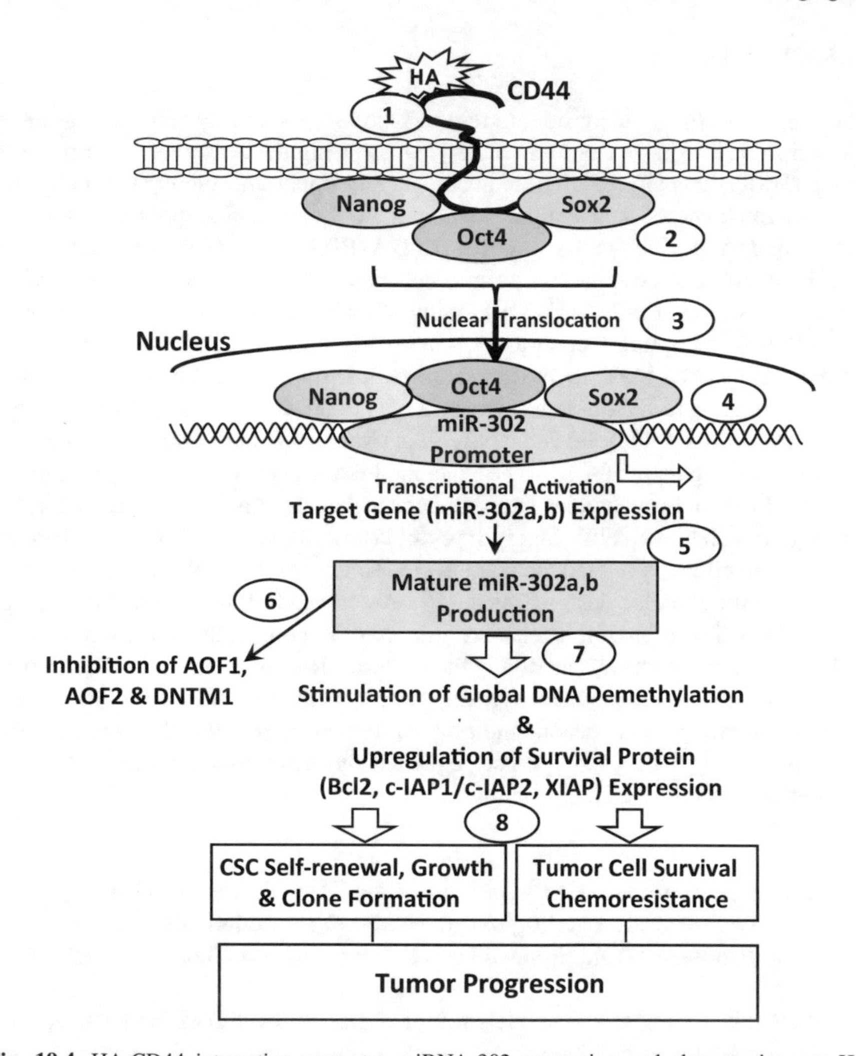

Fig. 19.4 HA-CD44 interaction promotes miRNA-302 expression and chemoresistance: HA binding to CD44 (*step 1*) promotes an association between CD44v3 and OCT4/SOX2/Nanog (*step 2*). Subsequently, OCT4/SOX2/Nanog translocates from the cytosol to the nucleus and interacts with the promoter region (containing OCT4-, SOX2-, and Nanog-binding sites) of the miR-302 cluster (*step 3*) resulting in miR-302 cluster gene expression (*step 4*) and mature miR-302a and miR-302b production (*step 5*). The resultant miR-302a/miR-302b then functions to downregulate the lysine-specific histone demethylases (namely, AOF1 and AOF2) and DNA (cytosine-5)-methyltransferase 1 (DNMT1) (*step 6*) and induce global DNA demethylation (*step 7*) leading to IAP (cIAP-1, cIAP-2, and XIAP) expression, self-renewal, clonal formation, anti-apoptosis/survival, and chemoresistance (step 8) in head and neck cancer stem cells. Taken together, these findings strongly suggest that targeting HA-CD44-mediated OCT4-SOX2-Nanog signaling pathways and miR-302 cluster function may provide important new drug targets to induce CSC apoptosis/death and overcome chemotherapy resistance in head and neck cancer cells

posttranslational level, miR-302 family is emerging as a key player in the control of cell proliferation and cell fate determination during differentiation.

HA-CD44 binding has also been shown to upregulate the expression of Nanog, OCT4, and SOX2 [8–10]. Moreover, the miR-302 cluster appears to be controlled by a promoter containing OCT4-SOX2-Nanog-binding sites in these cells, whereas ChIP assays demonstrate that stimulation of miR-302a and miR-302b production by HA is OCT4/SOX2 and Nanog complex-dependent in CSC (Fig. 19.4) [8]. Importantly, overexpression of OCT4/SOX2/Nanog and miR-302a/miR-302b occurs in both mouse tumors induced by CSC and human HNSCC patient samples [8]. These findings clearly established the existence of a close association between OCT4/SOX2/Nanog-regulated miR-302 clusters (e.g., miR-302a and miR-302b) and HNSCC development.

HA-CD44-activated miR-302 in CSCs has been shown to affect expression of epigenetic regulators such as the histone demethylase KDM1B and the DNA methyltransferase DNMT1, resulting in DNA global demethylation [8]. Activation of miR-302 in CSCs treated with HA also promotes overexpression of several survival proteins leading to self-renewal, clonal formation, and cisplatin resistance [8]. Inhibition of miR-302 expression/function (using anti-miR-302 inhibitor) not only results in KDM1B/DNMT1 upregulation and DNA demethylation downregulation but also causes epigenetic changes leading to a reduction of survival protein (cIAP-1/cIAP-2 and XIAP) expression and inhibition of CSC functions (e.g., self-renewal, clonal formation, and cisplatin resistance) (Fig. 19.4) [8]. These findings strongly suggest the existence of a novel linkage between OCT4/SOX2/Nanog signaling and miR-302 expression that regulate CSC functions in HA/CD44-activated HNSCC.

19.6 Additional Factors Affecting CSCs in HNSCC

Additional HA/CD44 receptors may be relevant to control of HNSCC CSCs. For example, HA also binds to RHAMM (receptor of HA-mediated motility), an extracellular protein that lacks a transmembrane domain but is GPI-anchored to the cell membrane. RHAMM interacts with CD44 and contributes to many of its cell functions, including cell motility, wound healing, and modification of signal transduction of the Ras signaling cascade [95–97]. The expression of RHAMM has been detected in HNSCC [98]. Downregulation by lentiviral shRNA of RHAMM but not CD44 inhibited proliferation and migration in tumor cells [99]. Since RHAMM bind to HA, targeting against RHAMM may be an additional option for the treatment of HNSCC.

In addition to matrix HA, osteopontin (OPN), a multifunctional glycosylated phosphoprotein that is secreted by tumor-associated macrophages (TAMs) [100], is also found in HNSCC [101]. Plasma OPN levels appeared to correlate with tumor hypoxia in HNSCC patients and may serve as noninvasive tests to identify patients at high risk for tumor recurrence [102]. Importantly, OPN binds to CD44v isoforms (but not CD44s) and promotes cellular signaling events leading to cell migration and stemlike properties [103, 104]. Currently, the application of RNAi, including both siRNA and shRNA targeting OPN production, appears to be a potential therapeutic strategy [105].

Although this chapter has focused on HA/CD44, the aberrant expression of other CSC markers such as the transmembrane glycoprotein CD133 and the RNA-binding protein and translational regulator Musashi 1 (MSI1) has been linked to stemness properties such as Nanog/Oct4/Sox2 expression, spheroid formation, epithelial-to-mesenchymal transition (EMT), and tumor cell migration in different tumor subpopulations during oral cancer progression [106–108]. Most importantly, anti-CD133 antibody conjugated to a genetically modified cytolethal distending toxin (Cdt) [isolated from the periodontal pathogen *Aggregatibacter actinomycetemcomitans*] has been shown to inhibit the proliferation of $CD133^+$ cells in cultures of established cell lines derived from HNSCC [109]. Thus, the discovery of different CSC markers could be very useful for designing anticancer therapies for the treatment of HNSCC patients.

19.7 Conclusion and Future Directions

The discovery of matrix hyaluronan (HA)/CD44v3-mediated signaling pathways in a subpopulation of HNSCC cells that displays cancer stem cell (CSC) properties, high tumorigenic potential, and chemoresistance could provide important novel target(s) for HNSCC therapy. It is possible to develop new cancer stem cell-based therapies targeting CD44v3 using anti-CD44v3-specific antibody and antisense strategies. An alternative approach would be to target stem cell transcription factors (using Nanog/OCT4/SOX2-specific siRNA and shRNA vector approaches) and/or miR-302/miR-21 (using anti-miR-302 inhibitor/anti-miR-21 inhibitor) to specifically block HA/CD44v3-mediated Nanog/OCT4/SOX2 signaling and miR-302/miR-21 expression and function. These approaches would be anticipated to reduce chemoresistance and HNSCC progression. If the proposed approaches are successful, these novel signaling perturbation techniques could synergistically cause apoptotic responses and indicate that chemotherapy combined with the suppression of HA/CD44v3-activated signaling, stemness, and miR-302/miR-21 is more effective than chemotherapy alone. Furthermore, HA-based nanoparticles containing therapeutic drugs (e.g., cisplatin or doxorubicin) may be used to accurately deliver therapeutic drugs into CD44v3 and RHAMM-expressing HNSCC cells to enhance chemosensitivity and downregulate CD44v3-mediated oncogenic signaling. The use of RNAi to limit OPN production may be a potential therapeutic strategy to reduce oncogenic signaling in HNSCC cells. Finally, it is also feasible to develop CSC-based therapies by targeting other tumor subpopulation expressing CD133 or MSI1 using anti-CD133-specific antibody and/or antisense strategies to downregulate Nanog/OCT4/SOX2 expression and stemness properties. The knowledge obtained from these studies on CSCs can provide the groundwork necessary for developing important tumor biomarkers and potentially yield novel drugs targeted against HNSCC progression, improving patient care.

Acknowledgment We gratefully acknowledge the assistance of Dr. Gerard J. Bourguignon in the preparation and review of this manuscript. This work was supported by Veterans Affairs (VA) Merit Review Awards (RR & D-1I01 RX000601 and BLR & D-5I01 BX000628) and United States Public Health grants (R01 CA66163). LYWB is a VA senior research career scientist.

References

1. Parkin DM, Bray F, Ferlay J, Pisani P. Global cancer statistics, 2002. CA Cancer J Clin. 2002;55:74–108.
2. Haddad RI, Shin DM. Recent advances in head and neck cancer. N Engl J Med. 2008;359:1143–54.
3. Leemans CR, Braakhuis BJ, Brakenhoff RH. The molecular biology of head and neck cancer. Nat Rev Cancer. 2011;11:9–22.
4. Pfister DG, et al. Head and neck cancers. J Natl Compr Cancer Netw. 2011;9:596–649.
5. Chen YC, Chen YW, Hsu HS, Tseng LM, Huang PI, Lu KH, et al. Aldehyde dehydrogenase 1 is a putative marker for cancer stem cells in head and neck squamous cancer. Biochem Biophys Res Commun. 2009;385:307–13.
6. Krishnamurthy S, Dong Z, Vodopyanov D, Imai A, Helman JI, Prince ME, et al. Endothelial cell-initiated signaling promotes the survival and self-renewal of cancer stem cells. Cancer Res. 2010;70:9969–78.
7. Krishnamurthy S, Nör JE. Head and neck cancer stem cells. J Dent Res. 2012;91:334–40.
8. Bourguignon LY, Wong G, Earle C, Chen L. Hyaluronan-CD44v3 interaction with OCT4-SOX2-Nanog promotes miR-302 expression leading to self-renewal, clonal formation, and cisplatin resistance in cancer stem cells from head and neck squamous cell carcinoma. J Biol Chem. 2012;287:32800–24.
9. Shiina M, Bourguignon LY. Selective activation of cancer stem cells by size-specific hyaluronan in head and neck cancer. Int J Cell Biol. 2015;989070. https://doi.org/10.1155/2015/989070.
10. Bourguignon LY, Wong G, Shiina M. Up-regulation of histone methyltransferase, DOT1L, by matrix hyaluronan promotes microRNA-10 expression leading to tumor cell invasion and chemoresistance in cancer stem cells from head and neck squamous cell carcinoma. J Biol Chem. 2016;291:10571–85.
11. Bourguignon LY, Shiina M, Li JJ. Hyaluronan-CD44 interaction promotes oncogenic signaling, microRNA functions, chemoresistance, and radiation resistance in cancer stem cells leading to tumor progression. Adv Cancer Res. 2014;123:255–75.
12. Sobreira TJ, Marlétaz F, Simões-Costa M, Schechtman D, Pereira AC, Brunet F, et al. Structural shifts of aldehyde dehydrogenase enzymes were instrumental for the early evolution of retinoid dependent axial patterning in metazoans. Proc Natl Acad Sci U S A. 2011;108:226–31.
13. Franzmann EJ, Weed DT, Civantos FJ, Goodwin WJ, Bourguignon LY. A novel CD44v3 isoform is involved in head and neck squamous cell carcinoma progression. Otolaryngol Head Neck Surg. 2001;124:426–32.
14. Wang SJ, Wreesmann VB, Bourguignon LY. Association of CD44v3-containing isoforms with tumor cell growth, migration, matrix metalloproteinase expression, and lymph node metastasis in head and neck cancer. Head Neck. 2007;29:550–8.
15. Wang SJ, Wong G, de Heer AM, Xia W, Bourguignon LY. CD44 variant isoforms in head and neck squamous cell carcinoma progression. Laryngoscope. 2009;119:1518–30.
16. Wang SJ, Bourguignon LY. Role of hyaluronan-mediated CD44 signaling in head and neck squamous cell carcinoma progression and chemoresistance. Am J Pathol. 2011;178:956–63.

17. Screaton GR, Bell MV, Jackson DG, Cornelis FB, Gerth U, Bell JI. Genomic structure of DNA coding the lymphocyte homing receptor CD44 reveals 12 alternatively spliced exons. Proc Natl Acad Sci U S A. 1992;89:12160–4.
18. Screaton GR, Bell MV, Bell JI, Jackson DG. The identification of a new alternative exon with highly restricted tissue expression in transcripts encoding the mouse Pgp-1 (CD44) homing receptor. Comparison of all 10 variable exons between mouse, human and rat. J Biol Chem. 1993;268:12235–8.
19. Al-Hajj M, Wicha MS, Benito-Hernandez A, Morrison SJ, Clarke MF. Prospective identification of tumorigenic breast cancer cells. Proc Natl Acad Sci U S A. 2003;100:3983–8.
20. Borovski T, De Souza E, Melo F, Vermeulen L, Medema JP. Cancer stem cell niche: the place to be. Cancer Res. 2011;71:634–9.
21. Kuhn NZ, Tuan RS. Regulation of stemness and stem cell niche of mesenchymal stem cells: implications in tumorigenesis and metastasis. J Cell Physiol. 2010;222:268–77.
22. Haylock DN, Nilsson SK. The role of hyaluronic acid in hemopoietic stem cell biology. Regen Med. 2006;1:437–45.
23. Astachov L, Vago R, Aviv M, Nevo Z. Hyaluronan and mesenchymal stem cells: from germ layer to cartilage and bone. Front Biosci. 2011;16:261–76.
24. Peach RJ, Hollenbaugh D, Stamenkovic I, Aruffo A. Identification of hyaluronic acid binding sites in the extracellular domain of CD44. J Cell Biol. 1993;122:257–64.
25. Yeo TK, Nagy JA, Yeo KT, Dvorak HF, Toole BP. Increased hyaluronan at sites of attachment to mesentery by CD44-positive mouse ovarian and breast tumor cells. Am J Pathol. 1996;148:1733–40.
26. Toole BP. Proteoglycans and hyaluronan in morphogenesis and differentiation. In: Hay ED, editor. Cell biology of extracellular matrix. New York: Plenum Press; 1991. p. 305–34.
27. Lee JY, Spicer AP. Hyaluronan: a multifunctional, megadalton, stealth molecule. Curr Opin Cell Biol. 2000;12:581–6.
28. Toole BP, Wight T, Tammi M. Hyaluronan-cell interactions in cancer and vascular disease. J Biol Chem. 2002;277:4593–6.
29. Weigel PH, Hascall VC, Tammi K. Hyaluronan synthases. J Biol Chem. 1997;272:13997–4000.
30. Zhang L, Underhill CB, Chen L. Hyaluronan on the surface of tumor cells is correlated with metastatic behavior. Cancer Res. 1995;55:428–33.
31. Bourguignon LY, Gilad E, Peyrollier K. Heregulin-mediated ErbB2-ERK signaling activates hyaluronan synthases leading to CD44-dependent ovarian tumor cell growth and migration. J Biol Chem. 2007;282:19426–41.
32. Wang SJ, Earle C, Wong G, Bourguignon LY. Role of hyaluronan synthase 2 to promote CD44-dependent oral cavity squamous cell carcinoma progression. Head Neck. 2013;35:511–20.
33. Stern R, Jedrzejas MJ. Hyaluronidases: their genomics, structures, and mechanisms of action. Chem Rev. 2006;106:818–39.
34. Franzmann EJ, Schroeder GL, Goodwin WJ, Weed DT, Fisher P, Lokeshwar VB. Expression of tumor markers hyaluronic acid and hyaluronidase (HYAL1) in head and neck tumors. Int J Cancer. 2003;106:438–45.
35. Christopoulos TA, Papageorgakopoulou N, Theocharis DA, Mastronikolis NS, Papadas TA, Vynios DH. Hyaluronidase and CD44 hyaluronan receptor expression in squamous cell laryngeal carcinoma. Biochim Biophys Acta. 2006;1760:1039–45.
36. Godin DA, Fitzpatrick PC, Scandurro AB, Belafsky PC, Woodworth BA, Amedee RG, et al. PH20: a novel tumor marker for laryngeal cancer. Arch Otolaryngol Head Neck Surg. 2000;126:402–4.
37. Mack B, Gires O. CD44s and CD44v6 expression in head and neck epithelia. PLoS One. 2008;3:e3360. https://doi.org/10.1371/journal.pone.0003360.
38. Bourguignon LY. Matrix hyaluronan promotes specific MicroRNA upregulation leading to drug resistance and tumor progression. Int J Mol Sci. 2016;17:517–27.

39. Stojkovic P, Hyslop L, Anyfantis G, Herbert M, Murdoch AP, Stojkovic M, Lako M. Putative role of hyaluronan and its related genes, HAS2 and RHAMM, in human early preimplantation embryogenesis and embryonic stem cell characterization. Stem Cells. 2007;25:3045–57.
40. Wheatley SC, Isacke CM. Induction of a hyaluronan receptor, CD44, during embryonal carcinoma and embryonic stem cell differentiation. Cell Adhes Commun. 1995;3:217–30.
41. Reategui EP, de Mayolo AA, Das PM, Astor FC, Singal R, Hamilton KL, Goodwin WJ, Carraway KL, Franzmann EJ. Characterization of CD44v3-containing isoforms in head and neck cancer. Cancer Biol Ther. 2006;5:1163–8.
42. Prince ME, Sivanandan R, Kaczorowski A, Wolf GT, Kaplan MJ, Dalerba P, et al. Identification of a subpopulation of cells with cancer stem cell properties in head and neck squamous cell carcinoma. Proc Natl Acad Sci U S A. 2007;104:973–8.
43. Chikamatsu K, Ishii H, Takahashi G, Okamoto A, Moriyama M, Sakakura K, et al. Resistance to apoptosis-inducing stimuli in CD44+ head and neck squamous cell carcinoma cells. Head Neck. 2012;34:336–43.
44. Suer I, Karatas OF, Yuceturk B, et al. Characterization of stem-like cells directly isolated from freshly resected laryngeal squamous cell carcinoma specimens. Curr Stem Cell Res Ther. 2014;9:347–53.
45. Wu CP, Zhou L, Xie M, et al. Identification of cancer stem-like side population cells in purified primary cultured human laryngeal squamous cell carcinoma epithelia. PLoS One. 2013;8:e65750.
46. Wang J, Wu Y, Gao W, Li F, Bo Y, Zhu M, Fu R, Liu Q, Wen S, Wang B. Identification and characterization of CD133$^+$CD44$^+$ cancer stem cells from human laryngeal squamous cell carcinoma cell lines. J Cancer. 2017;8:497–506.
47. Zhang M, Kumar B, Piao L, Xie X, Schmitt A, Arradaza N, Cippola M, Old M, Agrawal A, Ozer E, et al. Elevated intrinsic cancer stem cell population in human papillomavirus-associated head and neck squamous cell carcinoma. Cancer. 2014;120:992–1001. https://doi.org/10.1002/cncr.28538.
48. Zhang M, Kumar B, Piao L, Xie X, Schmitt A, Arradaza N, Cippola M, Old M, Agrawal A, Ozer E, Schuller D, Teknos T, Pan Q. Elevated intrinsic cancer stem cell population in human papillomavirus-associated head and neck squamous cell carcinoma. Cancer. 2014;120:992–1001.
49. Swanson MS, Kokot N, Sinha UK. The role of HPV in head and neck cancer stem cell formation and tumorigenesis. Cancers. 2016;8:24.
50. Orian-Rousseau V. CD44, a therapeutic target for metastasising tumours. Eur J Cancer. 2010;46:1271–7.
51. Huang W-Y, Lin J-N, Hsieh J-T, Chou S-C, Lai C-H, Yun E-J, Lo U-G, Pong R-C, Lin J-H, Lin Y-H. Nanoparticle targeting CD44-positive cancer cells for site-specific drug delivery in prostate cancer therapy. ACS Appl Mater Interfaces. 2016;8:30722–34.
52. Zhong Y, Zhang J, Cheng R, Deng C, Meng F, Xie F, Zhong Z. Reversibly crosslinked hyaluronic acid nanoparticles for active targeting and intelligent delivery of doxorubicin to drug resistant CD44+ human breast tumor xenografts. J Control Release. 2015;205:144–54.
53. Mattheolabakis G, Milane L, Singh A, Amiji MM. Hyaluronic acid targeting of CD44 for cancer therapy: from receptor biology to nanomedicine. J Drug Target. 2015;23:605–18.
54. Li YQ. Master stem cell transcription factors and signaling regulation. Cell Reprogram. 2010;12:3–13. https://doi.org/10.1089/cell.2009.0033.
55. Heng JC, Ng HH. Transcriptional regulation in embryonic stem cells. Adv Exp Med Biol. 2010;695:76–91.
56. Habu N, Imanishi Y, Kameyama K, Shimoda M, Tokumaru Y, Sakamoto K, Fujii R, Shigetomi S, Otsuka K, Sato Y, Watanabe Y, Ozawa H, Tomita T, Fujii M, Ogawa K. Expression of Oct3/4 and Nanog in the head and neck squamous carcinoma cells and its clinical implications for delayed neck metastasis in stage I/II oral tongue squamous cell carcinoma. BMC Cancer. 2015;15:730. https://doi.org/10.1186/s12885-015-1732-9.

57. Luo W, Li S, Peng B, Ye Y, Deng X, Yao K. Embryonic stem cells markers SOX2, OCT4 and Nanog expression and their correlations with epithelial-mesenchymal transition in nasopharyngeal carcinoma. PLoS One. 2013;8:e56324. https://doi.org/10.1371/journal.pone.0056324.
58. Chambers I, Colby D, Robertson M, Nichols J, Lee S, Tweedie S, Smith A. Functional expression cloning of Nanog, a pluripotency sustaining factor in embryonic stem cells. Cell. 2003;113:643–55.
59. Mitsui K, Tokuzawa Y, Itoh H, Segawa K, Murakami M, Takahashi K, Maruyama M, Maeda M, Yamanaka S. The homeoprotein Nanog is required for maintenance of pluripotency in mouse epiblast and ES cells. Cell. 2003;113:631–42.
60. Kuroda T, Tada M, Kubota H, Kimura H, Hatano SY, Suemori H, Nakatsuji N, Tada T. Octamer and Sox elements are required for transcriptional cis regulation of Nanog gene expression. Mol Cell Biol. 2005;25:2475–85.
61. Rodda DJ, Chew JL, Lim LH, Loh YH, Wang B, Ng HH, Robson P. Transcriptional regulation of nanog by OCT4 and SOX2. J Biol Chem. 2005;280:24731–2473.
62. Bourguignon LY, Earle C, Wong G, Spevak CC, Krueger K. Stem cell marker (Nanog) and STAT3 signaling promote MicroRNA-21 expression and chemoresistance in hyaluronan/CD44-activated head and neck squamous cell carcinoma cells. Oncogene. 2012;31:149–60.
63. Bourguignon LY, Peyrollier K, Xia W, Gilad E. Hyaluronan-CD44 interaction activates stem cell marker Nanog, STAT3-mediated MDR1 gene expression, and ankyrin-regulated multidrug efflux in breast and ovarian tumor cells. J Biol Chem. 2008;283:17635–51.
64. Bourguignon LY, Spevak CC, Wong G, Xia W, Gilad E. Hyaluronan-CD44 interaction with protein kinase C(epsilon) promotes oncogenic signaling by the stem cell marker Nanog and the production of microRNA-21, leading to down-regulation of the tumor suppressor protein PDCD4, anti-apoptosis, and chemotherapy resistance in breast tumor cells. J Biol Chem. 2009;284:26533–46.
65. Zhang J, Wang X, Li M, Han J, Chen B, Wang B, Dai J. NANOGP8 is a retrogene expressed in cancers. FEBS J. 2006;273:1723–30.
66. Chan JA, Krichevsky AM, Kosik KS. MicroRNA-21 is an antiapoptotic factor in human glioblastoma cells. Cancer Res. 2005;65:6029–33.
67. Hanahan D, Weinberg RA. The hallmarks of cancer. Cell. 2000;100:57–70.
68. Loh PG, Yang HS, Walsh MA, Wang Q, Wang X, Cheng Z, Liu D, Song H. Structural basis for translational inhibition by the tumour suppressor Pdcd4. EMBO J. 2009;28:274–85.
69. Do JT, Schöler HR. Cell-cell fusion as a means to establish pluripotency. Ernst Schering Res Found Workshop. 2006;60:35–45.
70. Hunter AM, LaCasse EC, Korneluk RG. The inhibitors of apoptosis (IAPs) as cancer targets. Apoptosis. 2007;12:1543–68.
71. Gorin MA, Pan Q. Protein kinase C epsilon: an oncogene and emerging tumor biomarker. Mol Cancer. 2009;8:9–16.
72. Stabel S, Parker PJ. Protein kinase C. Pharmacol Ther. 1991;51:71–95.
73. Steinberg R, Harari OA, Lidington EA, Boyle JJ, Nohadani M, Samarel AM, Ohba M, Haskard DO, Mason JC. A protein kinase Cepsilon-anti-apoptotic kinase signaling complex protects human vascular endothelial cells against apoptosis through induction of Bcl-2. J Biol Chem. 2007;282:32288–97.
74. Pardo OE, Wellbrock C, Khanzada UK, Aubert M, Arozarena I, Davidson S, Bowen F, Parker PJ, Filonenko VV, Gout IT, Sebire N, Marais R, Downward J, Seckl MJ. FGF-2 protects small cell lung cancer cells from apoptosis through a complex involving PKCε, B-Raf and S6K2. EMBO J. 2006;25:3078–88.
75. Basu A, Mohanty S, Sun B. Differential sensitivity of breast cancer cells to tumor necrosis factor-alpha: involvement of protein kinase C. Biochem Biophys Res Commun. 2001;280:883–91.
76. Darnell JE Jr. STATs and gene regulation. Science. 1997;277:1630–5.

77. Heinrich PC, Behrmann I, Haan S, Hermanns HM, Muller-Newen G, Schaper F. Principles of interleukin (IL)-6-type cytokine signalling and its regulation. Biochem J. 2003;374:1–20.
78. Huang S. Regulation of metastases by signal transducer and activator of transcription 3 signaling pathway: clinical implications. Clin Cancer Res. 2007;13:1362–6.
79. Pesce M, Schöler HR. Oct-4. Gatekeeper in the beginnings of mammalian development. Stem Cells. 2001;19:271–8.
80. Herr W, Cleary MA. The POU domain. Versatility in transcriptional regulation by a flexible two-in-one DNA-binding domain. Genes Dev. 1995;9:1679–93.
81. Nichols J, Zevnik B, Anastassiadis K, Niwa H, Klewe-Nebenius D, Chambers I, Schöler H, Smith A. Formation of pluripotent stem cells in the mammalian embryo depends on the POU transcription factor OCT4. Cell. 1998;95:379–91.
82. Wang X, Dai J. Concise review. Isoforms of OCT4 contribute to the confusing diversity in stem cell biology. Stem Cells. 2010;28:885–93.
83. Kashyap V, Rezende NC, Scotland KB, Shaffer SM, Persson JL, Gudas LJ, Mongan NP. Regulation of stem cell pluripotency and differentiation involves a mutual regulatory circuit of the NANOG, OCT4, and SOX2 pluripotency transcription factors with polycomb repressive complexes and stem cell microRNAs. Stem Cells Dev. 2009;18:1093–108.
84. Boyer LA, Lee TI, Cole MF, Johnstone SE, Levine SS, Zucker JP, Guenther MG, Kumar RM, Murray HL, Jenner RG, Gifford DK, Melton DA, Jaenisch R, Young RA. Core transcriptional regulatory circuitry in human embryonic stem cells. Cell. 2005;122:947–56.
85. Kamachi Y, Uchikawa M, Kondoh H. Pairing SOX off. With partners in the regulation of embryonic development. Trends Genet. 2000;16:182–7.
86. Avilion AA, Nicolis SK, Pevny LH, Perez L, Vivian N, Lovell-Badge R. Multipotent cell lineages in early mouse development depend on SOX2 function. Genes Dev. 2003;17:126–40.
87. Gontan C, de Munck A, Vermeij M, Grosveld F, Tibboel D, Rottier R. SOX2 is important for two crucial processes in lung development. Branching morphogenesis and epithelial cell differentiation. Dev Biol. 2008;317:296–309.
88. Dong C, Wilhelm D, Koopman P. Sox genes and cancer. Genome Res. 2004;105:442–7.
89. Bourguignon LY. CD44-mediated oncogenic signaling and cytoskeleton activation during mammary tumor progression. J Mammary Gland Biol Neoplasia. 2001;6:287–97.
90. Bourguignon LY. Hyaluronan-mediated CD44 activation of RhoGTPase signaling and cytoskeleton function promotes tumor progression. Semin Cancer Biol. 2008;18:251–9.
91. Card DA, Hebbar PB, Li L, Trotter KW, Komatsu Y, Mishina Y, Archer TK. OCT4/SOX2-regulated miR-302 targets cyclin D1 in human embryonic stem cells. Mol Cell Biol. 2008;28:6426–38.
92. Liu H, Deng S, Zhao Z, Zhang H, Xiao J, Song W, Gao F, Guan Y. OCT4 regulates the miR-302 cluster in P19 mouse embryonic carcinoma cells. Mol Biol Rep. 2011;38:2155–60.
93. Lin SL, Chang DC, Chang-Lin S, Lin CH, Wu DT, Chen DT, Ying SY. Mir-302 reprograms human skin cancer cells into a pluripotent ES-cell-like state. RNA. 2008;14:2115–24.
94. Lin SL, Chang DC, Lin CH, Ying SY, Leu D, Wu DT. Regulation of somatic cell reprogramming through inducible mir-302 expression. Nucleic Acids Res. 2011;39:1054–65.
95. Misra S, Hascll V, Markwald RR, Ghatak S. Interactions between hyaluronan and its receptors (CD44, RHAMM) regulate the activities of inflammation and cancer. Front Immunol. 2015. https://doi.org/10.3389/fimmu.2015.00201.
96. Entwistle J, Zhang S, Yang B, Wong C, Li Q, Hall CL, et al. Characterization of the murine gene encoding the hyaluronan receptor RHAMM. Gene. 1995;163:233–238. https://doi.org/10.1016/0378-119(95) 00398.
97. Turley EA, Noble PW, Bourguignon LY. Signaling properties of hyaluronan receptors. J Biol Chem. 2002;277:4589–92.
98. Schmitts A, Barth TFE, Beyer E, et al. The tumor antigens RHAMM and G250/CAIX are expressed in head and neck squamous cell carcinomas and elicit specific CD8+ T cell responses. Int J Oncol. 2009;34:629–39.

99. Twarock S, Tammi MI, Savani RC, Fischer JW. Hyaluronan stabilizes focal adhesions, filopodia, and the proliferative phenotype in esophageal squamous carcinoma cells. J Biol Chem. 2010;285:23276–2384.
100. Rao G, Du LCQ. Osteopontin, a possible modulator of cancer stem cells and their malignant niche. Oncoimmunology. 2013;2:e24169.
101. Chien CY, Tsai HT, Su LJ, Chuang HC, Shiu LY, Huang CC, Fang FM, Yu CC, Su HT, Chen CH. Aurora-A signaling is activated in advanced stage of squamous cell carcinoma of head and neck cancer and requires osteopontin to stimulate invasive behavior. Oncotarget. 2014;5:2243–62.
102. Le QT, Sutphin PD, Raychaudhuri S, Yu SC, Terris DJ, Lin HS, Lum B, Pinto HA, Koong AC, Giaccia AJ. Identification of osteopontin as a prognostic plasma marker for head and neck squamous cell carcinomas. Clin Cancer Res. 2003;9:59–67.
103. Katagiri YU, Sleeman J, Fujii H, Herrlich P, Hotta H, Tanaka K, et al. CD44 variants but not CD44s cooperate with beta1-containing integrins to permit cells to bind to osteopontin independently of arginine-glycine-aspartic acid, thereby stimulating cell motility and chemotaxis. Cancer Res. 1999;59:219–26.
104. Pio GM, Xia Y, Piaseczny MM, Chu JE, Allan AL. Soluble bone-derived osteopontin promotes migration and stem-like behavior of breast cancer cells. PLoS One. 2017;12:e0177640. https://doi.org/10.1371/journal.pone.0177640. eCollection.
105. Bandopadhyay M, Bulbule A, Butti R, Chakraborty G, Ghorpade P, Ghosh P, Gorain M, Kale S, Kumar D, Kumar S, Totakura KV, Roy G, Sharma P, Shetti D, Soundararajan G, Thorat D, Tomar D, Nalukurthi R, Raja R, Mishra R, Yadav AS, Kundu GC. Osteopontin as a therapeutic target for cancer. Expert Opin Ther Targets. 2014;18:883–95.
106. Ravindran G, Devaraj H. Aberrant expression of CD133 and musashi-1 in preneoplastic and neoplastic human oral squamous epithelium and their correlation with clinicopathological factors. Head Neck. 2012;34:1129–35.
107. Moon Y, Kim D, Sohn H, Lim W. Effect of CD133 overexpression on the epithelial-to-mesenchymal transition in oral cancer cell lines. Clin Exp Metastasis. 2016;33:487–96.
108. Baillie R, Tan ST, Itinteang T. Cancer stem cells in oral cavity squamous cell carcinoma. Front Oncol. 2017;7:112.
109. Damek-Poprawa M, Volgina A, Korostoff J, Sollecito TP, Brose MS, O'Malley BW Jr, Akintoye SO, DiRienzo JM. Targeted inhibition of CD133+ cells in oral cancer cell lines. J Dent Res. 2011;90:638–45.

Chapter 20
Biology and Epidemiology of Human Papillomavirus-Related Head and Neck Cancer

Alexander Y. Deneka, Jeffrey C. Liu, and Camille C. R. Ragin

Abstract Human papillomavirus (HPV) infection is now established as a major causative agent for development of the head and neck cancers. HPV-initiated tumors of the oropharynx have better survival rates than HPV-negative cancers, and this appears likely to be associated with differences in the biology underlying these two diseases. We will discuss the role of HPV-encoded proteins in host infection and carcinogenesis; will review the emerging biology of intratypic variants of HPV, with numerous variants possessing different potential for malignancy; and will suggest areas for the further study. Finally, we will highlight global trends in HPV-associated oropharyngeal head and neck cancer incidence and prevalence rates, with recent data showing a dramatic increase of infection worldwide and differing infection rates in developed and developing nations.

Keywords Human papillomavirus · Oropharyngeal cancers · Variants · Incidence · Mortality

A. Y. Deneka
Molecular Therapeutics Program, Fox Chase Cancer Center, Philadelphia, PA, USA

Kazan Federal University, Kazan, Russia

J. C. Liu
Department of Otolaryngology, Lewis Katz School of Medicine at Temple University, Philadelphia, PA, USA

C. C. R. Ragin (✉)
Department of Otolaryngology, Lewis Katz School of Medicine at Temple University, Philadelphia, PA, USA

Cancer Prevention and Control Program, Fox Chase Cancer Center, Philadelphia, PA, USA
e-mail: Camille.Ragin@fccc.edu

B. Burtness, E. A. Golemis (eds.), *Molecular Determinants of Head and Neck Cancer*, Current Cancer Research, https://doi.org/10.1007/978-3-319-78762-6_20

20.1 Introduction

20.1.1 Human Papillomavirus (HPV)

Human papillomaviruses (HPVs) are a family of 8 kb, circular DNA viruses with tropism to basal cells of the epithelial mucosa [1]. To date, over 150 discrete genotypic variants have been described among the human papillomaviridae. These fall into broad subcategories that reflect the type of epithelial cells they are able to infect, e.g., cutaneous versus mucosal cells. HPVs are also classified according to their ability to transform epithelial cells. High-risk HPV genotypes such as HPV16, 18, 31, 45 and others are capable of transforming mucosal epithelial cells and inducing malignant lesions, while low-risk HPV genotypes such as HPV6, 11 and others are associated with benign lesions such as warts or condylomata. Benign lesions in the oral cavity are common and most often involve HPV6 and 11. These HPV genotypes are also often associated with uncommon benign conditions in the larynx such as laryngeal papillomatosis and laryngeal polyps.

The involvement of HPV in head and neck carcinogenesis was first proposed in 1983 by Syrjanen et al. [2] based on morphological and immunohistochemical evaluation of oral squamous cell carcinomas, which showed features typical of HPV lesions and were positive for immunoperoxidase staining with anti-HPV serum. This evidence was further corroborated in epidemiological studies reported by Gillison et al. [3, 4], in which HPV genomes were detected in tumors from patients diagnosed with new and recurrent head and neck squamous cell carcinomas. In head and neck cancers of the oral cavity, oropharynx, larynx, and hypopharynx, high-risk HPV infections account for approximately 20–25% of these lesions. Among these sites, the tumors that arise in the oropharynx carry the largest burden of HPV infections (estimated at 36%) [5], and HPV16 is the most common genotype isolated [6–18]. Oropharyngeal cancer sites include the tonsil, base of tongue, soft palate, uvula, and vallecula, as well as the lateral and posterior walls of the oropharynx.

20.2 HPV Proteins and Functions (E1, E2, E4, E5, E6, E7, L1, L2)

The HPV genome encodes seven early (E) genes (E1, E2, E4, E5, E6, E7, E8) and two late (L) genes (L1 and L2). Together with a noncoding upstream regulatory region (URR), these play important roles in viral replication and transcription. Phylogenetic analysis of the open-reading frames of early genes from various HPV genotypes shows that proteins from viruses associated with cancer cluster into a single group, suggesting that the genetic basis for oncogenicity is dependent upon the DNA sequences in the early region of the viral genomes [19, 20]. Among high-risk HPV genotypes, only the early genes E5, E6, and E7 play important roles in viral carcinogenicity (Fig. 20.1a). The differences in oncogenic potential between high-risk and low-risk HPVs are related to differences in the biochemical activities of the early viral gene products.

During the HPV life cycle (Fig. 20.1b), the virus invades damaged areas of stratified epithelium and targets the basal cells by binding to a cellular receptor. Heparan sulfate proteoglycan, a linear polysaccharide found in human and animal tissues, has been suggested to be the primary cellular receptor for initial attachment of the HPV virus and facilitates binding to an unidentified secondary receptor prior to entry [21, 22]. Once internalized, the virus uncoats and the viral DNA is transported to the nucleus. Infections can be nonproductive (i.e., the HPV genome is maintained in episomal form within the cell at low copy numbers) or productive (i.e., viral HPV DNA is replicated and packaged into intact infectious viral particles). Since the HPV genome does not encode key proteins required for viral DNA replication (such as DNA polymerase and other enzymes), the virus must rely on the infected cell's DNA replication machinery to induce viral replication. During the nonproductive phase of infection, the HPV genome is established as an extrachromosome (or episome) and remains tethered to the host chromosome so that it is maintained and segregated into daughter cells after cell division [23].

The switch between nonproductive and productive infections is dependent upon the host cell differentiation state. During productive infection, the transcription of HPV gene products is tightly regulated by the differentiation-specific gene expression profile of the infected cell. Upon activation of HPV DNA replication, amplification, and packaging of intact viral particles [24], infectious virions egress from terminally differentiated epithelial cells that comprise the cornified (outer surface) epithelium. HPV-associated carcinogenicity is not a normal event that occurs as integral part of the productive viral life cycle, which predominantly relies on the extrachromosomal viral genome. During carcinogenesis, the circular viral genome typically linearizes and integrates itself into the host chromosome, losing coding sequences for key regulatory genes during the linearization process. However, one recent study has suggested the distinction is not absolute, reporting an alternative mechanism of HPV tumorigenesis in a subset of HNSCC that involves activation of nuclear factor κB signaling and maintenance of episomal HPV in tumors, either without viral DNA integration or with a mix of integrated and episomal viral DNA [25]. A more detailed description of the roles of specific HPV-encoded proteins in the HPV life cycle, including chromosomal integration and viral carcinogenicity, is provided below.

20.2.1 HPV L1 and L2

The L1 and L2 proteins encode the structural components of the virus and are only transcribed in productively infected cells. These structural proteins assemble to form the viral capsid, which comprises 72 "capsomeres." Each capsomere includes a pentamer of the L1 protein and an L2 protein monomer, which docks into the center of each L1 pentamer. The L1 capsid protein is capable of self-assembling into

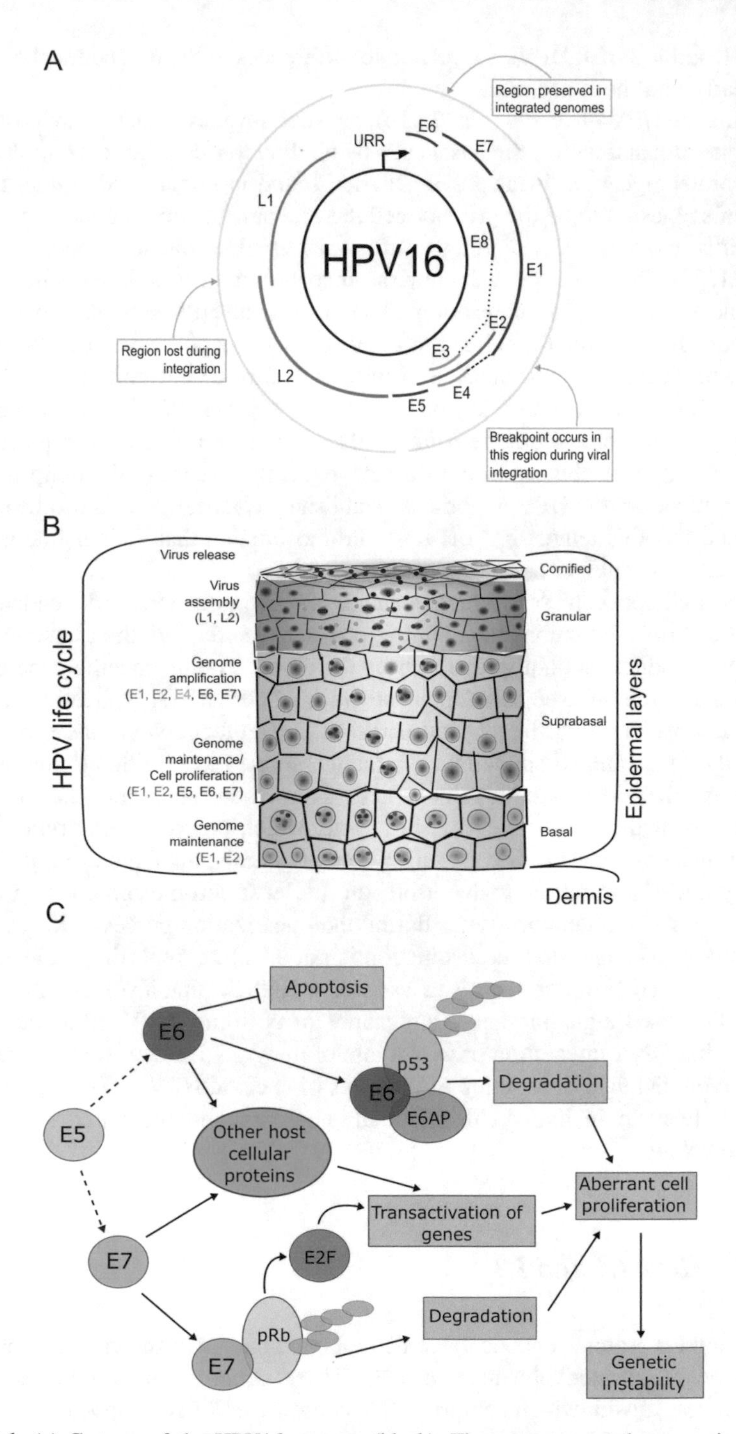

Fig. 20.1 (**a**) Cartoon of the HPV16 genome (black). The upstream regulatory region (URR) contains the enhancer elements for cell transcription factors and E1 and E2 binding sites. The open reading frames for each early (E) and late (L) gene are depicted as colored lines outside of the circular genome. The dotted lines depict the splice variants E1^E4 and E8^E2 which are derived

virus-like particles (VLPs), even *in vitro*. As the virion binds to its receptor, a conformational change occurs that exposes the N-terminus of the L2 protein, allowing it to be cleaved by the protease furin [26]. This leads to the exposure of additional regions of the L2 protein, which subsequently binds to an unidentified secondary receptor, followed by internalization of the viral particle via clathrin- or caveolin-mediated endocytosis [27]. For establishment of the infection, binding of the L2 protein to the HPV viral DNA enables its transport into the nucleus. The expression of the L2 protein later in the course of HPV infection is also important for the packaging of progeny viral genomes during viral particle assembly.

While essential structural elements of the sequence of the L1 protein are well conserved, a number of surface loops are distinct for different HPV genotypes. Exploiting these features, researchers have used HPV L1 VLPs to generate genotype-specific vaccines including a recombinant quadrivalent vaccine, targeting genotypes 6, 11, 16, and 18 (Gardasil, Merck & Co. Inc.), another recombinant HPV vaccine targeting genotypes 16 and 18 (Cervarix, Glaxo Smith Kline, provided as an adjuvanted, adsorbed form), and more recently a second-generation multivalent HPV L1 VLP vaccine that is active against the nine genotypes 6, 11, 16, 18, 31, 33, 45, 52, and 58 (Gardasil 9, Merck & Co., Inc.). These vaccines have received licenses from the FDA, and their efficacy has been demonstrated against anogenital cancers and genital warts. In addition, experimental vaccines such as chimeric L2-based VLP [28] are also evolving as promising tools for protection against cutaneous HPV infections.

20.2.2 *HPV E1, E2, E4, and E8*

The early proteins E1 and E2 are among the first viral proteins expressed in an infected cell. Both are DNA-binding proteins that regulate viral replication and gene expression [29]. E1 is a DNA helicase/ATPase that is responsible for bidirectional unwinding of viral DNA to facilitate viral genome replication. The E2 protein is a transcriptional regulator that binds at viral transcription factor binding sites within the URR to activate or repress transcription of HPV genes, with the viral oncogenes E6 and E7 particularly dependent on E2 activity. The URR is upstream of the early promoter and contains enhancer elements that are responsive to the binding of host transcription factors such as AP1, Oct-1, SP1, and YY1 and are adjacent to binding sites for the viral transcription factors E1 and E2. The activation

Fig. 20.1 (continued) from fused transcripts from the E1 and E4 or E8 and E2 open reading frames. The gray lines depict the regions of the viral genome that are preserved during integration, regions where breakpoints occur during integration, and region that is lost during integration. (**b**) Schematic of HPV life cycle. Colored captions for different viral proteins match colored dots, representing their appearance during the viral life cycle. (**c**) Interaction of high-risk HPV oncogenes E5, E6, and E7 with multiple host cellular proteins including p53 and pRb. These interactions result in a series of events leading to the inhibition of apoptosis, altered cell proliferation, and ultimately genomic instability which leads to cancer development

and repression of the early promoter are tightly regulated by E2 binding. A low level of E2 binding activates the early promoter, while a high level of E2 blocks the binding of host transcription factors and therefore represses the early promoter [30]. E2 is also responsible for maintaining the viral genome as an extrachromosomal replicon by recruiting the E1 protein to the replication origins and tethering HPV genomes to the host chromosome. E1 and E2 proteins were shown to be involved in several host intracellular signaling pathways. Thus, a recent study demonstrated that in W12 human keratinocyte cell line E2 protein indirectly maintains ERBB3 expression levels via interaction with the ubiquitin ligase neuregulin receptor degradation protein 1 (Nrdp-1) that is involved in the regulation of ERBB3 receptor, via ubiquitination and degradation. E2 loss showed no or low ERBB3 positivity in clinical samples [31]. The HPV E1 protein has been shown to bind to a number of host proteins such as cyclin-CDK2 [32, 33], Hsp40/Hsp70 [34], SW1/SNF5 [35], histone H1 [36], and Ubc9 [37, 38], but the significance of all of these interactions is not fully understood.

Heat shock protein (Hsp40) stimulates viral replication by facilitating the formation of E1:hsp40 dihexameric complexes which associate initially with the viral replication origin and remain associated with the replication elongation complex [34, 39]. The E2 gene also produces an alternatively spliced transcript that encodes E8 fused to a partial sequence of E2 (E8^E2). This splice variant is thought to negatively regulate HPV replication [40, 41]. Functional studies show that the E8^E2 proteins of high-risk HPV16, 18, and 31 inhibit the promoter that drives the expression of the HPV E6 and E7 oncoproteins [40, 42, 43]. In parallel to repression of E6 and E7 promoters, the E8^E2C protein negatively regulates host genome replication via interaction with NCoR/SMRT corepressor complex components (HDAC3, GPS2, NCoR, SMRT, TBL1, and TBLR1) as identified by proteomic analyses in 293T cell lines [43]. Therefore, the E8^E2 fusion protein exhibits long-distance transcriptional-repression activities (i.e., represses viral transcription by binding to E2 binding sites distal to the early promoter) [44] and has been shown to inhibit the growth of the cervical cancer cell line HeLa. Consistent with the role of E8^E2 in repression of the viral oncogenes E6 and E7, the expression of E8^E2 results in a rapid increase in cellular p53 and p21, which negatively regulate cell growth [45]. Among the other early proteins, E4 (expressed as an E1^E4 fusion protein) interacts with cytokeratin filaments. This E1^E4 fusion protein is cleaved by calpain, a member of calcium-dependent, non-lysosomal cysteine protease family. This cleavage results in accumulation of E1^E4 protein and amyloid fibers, leading to disruption of normal keratin networks. Disruptions to the keratin networks in differentiating epithelium are hypothesized to be involved in the release of new virus particles at late stages of the virus life cycle, but further work is needed to confirm this [46–48].

20.2.3 HPV Oncogenes E5, E6, and E7

Three oncogenes are encoded by the early open reading frames of high-risk HPV genotypes (E5, E6, and E7) (Fig. 20.1c). In general, HPV oncoproteins E5, E6, and E7 promote carcinogenicity by interfering with the regulation of cell growth by host cellular proteins, thus inducing genomic instability. The HPV16 E5 protein is the most commonly studied genotype of oncogenic E5 [49]. This protein is 90 amino acids in length and localizes in intracellular membranes such as the Golgi apparatus, endoplasmic reticulum, and nuclear membrane [50–52]. While the biochemical mechanisms related to HPV16 E5 carcinogenesis remain elusive, the oncoprotein is thought to promote carcinogenesis during early stages of the established viral infection, since the gene is often deleted when the HPV genome becomes integrated in the host chromosome [53–55]. E5 gene product leads to activation of the epidermal growth factor receptor (EGFR) to stimulate viral gene expression and cell proliferation [56]. Another molecular mechanism specific to HPV16-mediated neoplastic transformation is shown in HPV16-infected cervical cancers and cell lines, where E5-dependent reduction of miR196a expression leads to upregulation of the direct miR196a target gene, *HoxB8*. This process may also occur in head and neck cancers, but needs further validation [57]. The E5 protein co-localizes with the Bcl-2 anti-apoptotic protein on intracellular membranes, supporting survival of infected cells [58]. E5 also interferes with the recycling of major histocompatibility complex (MHC) Class I and II molecules as well as other receptors to the cell surface and reduces gap-junction-mediated intercellular communication via dephosphorylation of connexin 43 [59–61]. The inhibition of MHC Class I and II expression is a common immune evasion tactic used by many viruses, while the limitation of gap-junction-mediated intercellular communication results in a deficiency in tissue homeostatic feedback, promoting carcinogenesis. Finally, in a recent study of HPV31, A4 endoplasmic reticulum protein (a lipoprotein that is localized in the endoplasmic reticulum and reported to interact with Bap31) was identified as an E5 binding partner. This protein co-localizes with E5 in basal cells of the epithelium and mediates E5-induced cell proliferation in differentiated cells [62]. Importantly, high-risk HPV E5 proteins cooperate with E6/E7 oncogenes to promote hyper-proliferation of infected cells during the early stages of the virus replication cycle [63].

The E6 protein interacts with a number of host proteins responsible for cell proliferation, with the interaction inducing the degradation of critical cellular partners that limit cell growth. Proliferation is induced when E6 is phosphorylated by protein kinase PKN, as shown in 293T cells. At present, the role of E6-PKN interaction remains speculative; however, authors suggest that phosphorylation of E6 may play a role either in E6-induced immortalization or in the regulation of viral life cycle [64]. Mechanistically, E6 targets partner proteins for degradation via its ability to recruit E6AP (UBE3A), a ubiquitin ligase, which targets interactive partners for destruction; many proteins degraded by E6AP contain PDZ domains. A particularly critical target of E6 in head and neck cancer is the tumor suppressor protein p53 [65–67]. In addition, high-risk HPV E6 proteins promote degradation of *NFX1-91*, a transcriptional repressor that regulates telomerase expression, contributing to cell immortalization [68, 69]; target IFN regulatory factor 3 (IRF3) to abrogate host interferon response [70]; and degrade the focal adhesion protein paxillin to disrupt the actin cytoskeleton (a characteristic of transformed cells) [71, 72]. In a study performed in E6/E7-expressing human foreskin keratinocytes (HFK), both E6 and E7 proteins contributed to cytoskeleton reorganization and invasion by upregulating p63, which induced transcription factors that promoted oncogenic Src-FAK signaling [73]. Overall, E6 interactions lead to inhibition of p53-mediated apoptosis, inefficient G1/S checkpoints, and deficient repair of DNA damage and eventually to chromosomal instability.

The high-risk HPV E7 protein separately contributes to cellular proliferation based on interactions with a large number of host cellular proteins. Most importantly, its interaction with the tumor suppressor protein pRb results in hyperphosphorylation and ubiquitin-mediated degradation [74] of pRb. This leads to the release of E2F from pRb/E2F complexes, thus promoting transactivation of S-phase gene expression. The inactivation of pRb leads to histone acetylation and upregulation of *p16/CDKN2A*, which is vital for cell cycle progression and is a common surrogate biomarker for HPV carcinogenesis [75, 76]. Furthermore, E7-mediated *Rb* gene inactivation leads to upregulation of serine/threonine kinase AKT, which contributes to tumor progression by promoting survival signaling [77]. E7 cellular targets also include growth inhibitory pocket proteins related to pRb (p107 and p130), cell cycle checkpoints cyclin E, cyclin A, and cyclin-dependent kinase inhibitors p21 and p27, the transcription factor JUN, as well as the TATA box-binding proteins [78–82]. Cumulatively, these activities, in combination with those of other HPV oncoproteins, lead to genomic instability including chromosome segregation defects such as structural and numerical chromosomal abnormalities [53, 54], cell transformation, and carcinogenesis.

20.2.4 HPV Carcinogenesis

The model of HPV carcinogenicity in head and neck cancer was developed based on earlier studies of cervical cancers, which demonstrated that persistence of high-risk HPV infection increased the likelihood of viral integration into the host chromosome [83–85]. HPV DNA integration into the human genome is speculated to be an important step in oropharyngeal carcinogenesis [86]. While HPV integration occurs randomly throughout the human genome, it is thought to occur as a late event with a predilection for DNA fragile sites (regions of the genome that are late replicating and with loose chromatin structures) [87, 88]. These regions are hot spots for DNA breaks. Integration loci are located predominantly in the intergenic region, with a significant enrichment of the microhomologous sequences between the human and HPV16 genomes at the integration breakpoints. Levels of methylation within the integrated HPV genome at the integration breakpoints vary. A recent study analyzing allele-specific methylation suggests that the HPV16 integrants remain hypo- or hypermethylated accordingly to the state of flanking host genome [89]. In cervical cancer [90], the linearization of the circular HPV genome prior to integration usually occurs with a disruption in the viral E2 sequence, resulting in a defective virus. Importantly, the loss of E2 expression removes repression of the promoter for expression of the E6 and E7 oncogenes, resulting in their induction. For head and neck cancer, the relationship between physical state and integration of HPV may be more complex, since variations in HPV integration status in tumors have been reported in independent studies. HPV-positive oropharyngeal cancers have been found to carry HPV genomes primarily in episomal form, yet viral oncogenes are still expressed [91], while evidence of integrated or episomal forms only or a combination of both has also been reported in the presence of viral oncogene expression [6, 92, 93]. Tumors with episomal HPV, regardless of the presence or absence of integration sites, overexpress the HPV E2, E4, and E5 genes [94]. Subsets of these tumors present with mutations in genes *TRAF3* (tumor necrosis factor receptor-associated factor 3) and *CYLD* (cylindromatosis lysine 63 deubiquitinase). These alterations correlate not only with episomal status of the tumors but also with improved patient survival [25]. This is currently an area of active study.

20.2.5 Genetic Features of HPV-Positive Cancers

HPV-related oropharyngeal cancers have some chromosomal alterations in common with squamous cell carcinomas of the cervix, such as loss at 13q and gain at 20q [95]. In contrast, the genetic profile of HPV-initiated oropharyngeal cancers is distinct from that characterizing HPV-negative tobacco-related head and neck malignancies. HPV-initiated tumors are less likely to carry damaging mutations in *TP53* and *CDKN2A* [96–99] and have differing combinations of allelic losses and gains, such as helical domain mutations of the oncogene *PIK3CA*, loss of *TRAF3*, and amplification of the cell cycle gene *E2F1* [98–100]. HPV-related tumors typically have unique transcription profiles with downregulated expression of interferon-induced proteins such as IFIT1, IFITM1-3, IFI6-16, IFI44 L, and OAS2 and upregulated expression of transcription factors such as RPA2, TAF7 L, RFC4, and TFDP2 as well as cell division/cycle regulators such as p18, CDC7, and p16 [101, 102].

20.3 HPV Variants: Functional Differences and Implications for Risk and Treatment

20.3.1 HPV Intratypic Variants

Within the differing genotypes of HPV, intratypic variants have been reported that may affect the ability of the virus to induce cancer. The nomenclature for HPV has been established by the International Committee on Taxonomy of Viruses (ICTV)

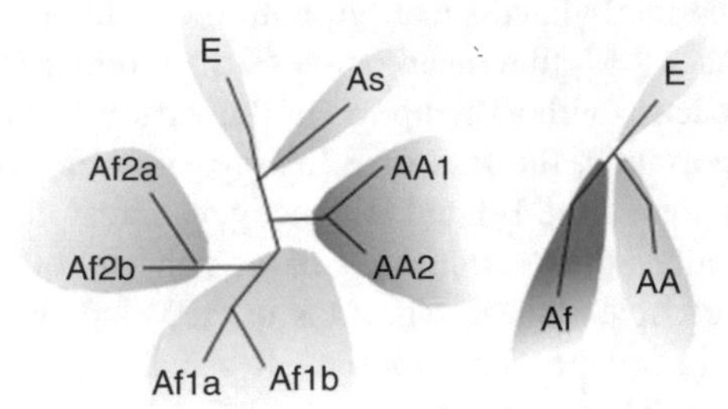

Fig. 20.2 Distinct branches of HPV16 and HPV18 evolution. Schematic representation of major variants of HPV16 (left) possessing additional sublineages: AA (Asian American), with sublineages AA1 and AA2; African Af1 with sublineages Af1a and Af1b; African Af2 with sublineages Af2a and Af2b; As (East Asian) and E (European). Three branches of HPV18 variants (right) are also described: A, African; AA, Asian Amerindian; and E, European

and is based on recommendations from the Study Group of Papillomavirus [103–105]. While definitions for genera and species have been defined, there are no set standards defined below the species level. Currently, a distinct HPV genotype is defined as having > 10% nucleotide sequence variation within the L1 gene; in contrast, the nucleotide sequences of intratypic variants typically differ between 2% and 5%. Since there are no defined standards, phylogenetic investigations of HPV16 and HPV18 variants have been limited initially to partial URR and E6 sequences, with this more recently augmented with studies of complete HPV genomes. This variability in the classification of HPV variants has resulted in limitations in the ability to compare results from independent studies, since different regions of the viral genomes have been evaluated. Nevertheless, extensive studies have been conducted on HPV16 and to a lesser extent on HPVs 18, 39, 45, 59, 68, and 70 [106–111] which have provided some insight into HPV variant lineages and sublineages [112]. For any given HPV genotype, related variants tend to cluster geographically as well as within ethnic groups. Global epidemiological trends for all major HPV genotypes are highlighted in Sect. 20.6; in the current section, we will address geographical distribution of HPV variants only.

There are five distinct phylogenetic branches of HPV16 variants possessing additional sublineages: AA with sublineages AA1 and AA2; Af1 with sublineages Af1a and Af1b; Af2 with sublineages Af2a and Af2b; As; and E. Three branches of HPV18 variants are also described: A, AA, and E [112, 113] (Fig. 20.2). All HPV16 variants are detected in different populations worldwide but at different frequencies depending on geography and ethnicity. The prototype HPV16 virus was first isolated and sequenced from a German cervical cancer patient [114] and belongs to the European (E) variant lineage. HPV16 E variants are predominantly detected in European and to a lesser degree in Western Asian, African, and American populations, including African-Americans, possibly due to extensive migration. There are two African variants (Af1 and Af2) detected primarily in African populations. In particular, Af1a, Af2a, and Af2b are predominantly described in sub-Saharan Africa and Af1b, Af2a, and to a lesser extent Af2b in North Africa. Asian American (AA) variants are primarily detected in Asian and indigenous populations in America. In contrast the East Asian (As) variants are predominant in Chinese and Japanese populations [109, 111, 113]. Similarly for HPV18, the frequencies of these variants also occur according to geography and ethnicity [106, 110, 115]. The prototype HPV18 virus belongs to the Asian Amerindian (AA) phylogenetic branch and although identified and sequenced in Germany was isolated from a cervical cancer patient from Northeast Brazil [116]. Unlike HPV16, HPV18 AA variants are found primarily in East Asians and American Indians. African populations tend to carry the A variants and Europeans the E variants.

A mixture of HPV16 and HPV18 variants is often observed in North, Central, and South American populations and is reflective of both the multiethnic groups arising from immigration and the predilection to maintain distinct ethnic groups well after immigration [109, 110]. Longitudinal studies of cervical HPV infection have shown that women of European descent are more likely to be persistently infected with E variants, while African-American women are more likely to be persistently infected with Af variants [117]. Pathogenic differences have also been noted between variants. Naturally occurring single nucleotide polymorphisms (SNPs) in the HPV16 E6 oncogene may lead to significant changes in the outcome of HPV infections and subsequent differences in viral and host transcriptomes that drive carcinogenesis [118]. A study conducted in 2016 used a Capt-HPV sequencing approach to demonstrate that AA and E variants differ by three non-synonymous SNPs in the E6 oncogene responsible for generating high-risk AA-E6 and low-risk E-E6. The high-risk variant was primarily integrated into host DNA, whereas the low-risk variant was likely to remain episomal [119]. Other studies have documented that non-European variants are more frequently persistent compared to European variants, and the risk of cancer progression in the cervix and anal mucosa is higher in non-European compared to European variants [120–125]. Specific risks of high-grade cervical intraepithelial neoplasia (CIN) were associated with different HPV variants [125–127], with the risk of developing high-grade CIN being threefold greater for women infected with non-European HPV16 variants compared to European HPV16 variants (Af2, RR = 2.7, 95% CI, 1.0–7.0, and AA, RR = 3.1, 95% CI, 1.6–6.0). Similar observations have been shown for anal carcinoma in situ [124] in HIV-positive men (HPV16 non-European HPV variants: RR = 3.2, 95% CI, 1.0–10.3). HPV18 oncogenic potential in CIN also varies among different variants due to their genomic diversity. Similarly to HPV16, non-European HPV18 variants were reported to persist more frequently and were more associated with preinvasive lesions [128].

20.3.2 Functional and Medical Implications of HPV Variants in Cancer

The genomic diversity within the viral coding sequences and regulatory regions of HPV16 and HPV18 variants may result in 1) functional differences in the viral proteins and 2) differences in the level of viral replication and transcription. Current studies have evaluated genetic variations in the HPV16, HPV18 E6, and E7 genes with primary focus on the effects on gene expression and the ability to interact with host proteins.

Variant genomes of HPV16 have been shown to be associated with different activities in degrading the p53 tumor suppressor [129]. For example, the Af2 variant (z84) shows lower activity for p53 degradation compared to the AA variant (512). It has been suggested that an R10I substitution might contribute to the lower activity observed in Af2 z84 variant and the Q14H and/or S138C substitution(s) might contribute to the higher activity of AA 512 variant. A lower ability to degrade p53 was observed among E E6, As E6, E E6 (L83 V), and As E6 (E113D) variants. A rare variation, E E6 (R10G), was reported to degrade p53 more efficiently than the other variations [130]. However, since multiple amino acid changes characterize each intratypic variant, it is not sufficient to assume that functional differences between variants might be attributed to a single amino acid change, as specific combinations of amino acid changes may be important.

Variant E6 genomes of HPV18 have also been shown to affect the degradation of p53 protein, in some cases due to differences in differential splicing in the E6 gene [131] that reduce expression of functional E6 protein. The European variant was reported to have higher functional E6 protein expression and thus less alternative splicing in the E6 gene [131]. Another study assessed the role of E6 gene variants of HPV18 on downregulation of p14(ARF), an important tumor suppressor downstream of p53 [132]. In MCF-7 cells infected with E6 gene variants AA variant induced low p53 and high p14(ARF) expression, whereas an E variant induced high p53 and low p14(ARF) expression [132]. This study suggests that each E6 variant distinctively affects p53 levels and consequently p14(ARF) expression, resulting in the differences in oncogenic potential. Other genetic variations in HPV E6 and E7 have been shown to have differential effects on cell protein interactions involving mitogen-activated protein kinase (MAPK) signaling [133] and protein kinase B/phosphatidylinositol 3-kinase (AKT/PI3K) signaling [134] which are involved in cell proliferation and survival. Genetic variations in the E2 gene as well as the URR have also been shown to affect E6 expression [135–139] and thus viral oncogenicity. In addition, sequence variations in the URR which affect the binding sites for the host-encoded transcription factors AP-1, NFI, Oct-1, TEF-1, and YY1 have been documented and shown to result in differences in HPV replication rates [140] as well as transcriptional activation of the E6/E7 promoter [135–138]. A 2015 study reported that the As variant of E6 HPV16 protein induces miR-21 in C33A cervix carcinoma cells to a greater extent than does the E variant [141]. miR-21 plays a significant role in cancer biology by promoting cell proliferation, migration, and invasion; therefore, the ability of the As HPV variant to upregulate this micro-RNA might contribute to its major role in cervical cancer development in Asian populations.

To date, there is little that is known about the clinical impact and carcinogenic relevance of HPV variants in oropharyngeal cancer. A recent study compared European HPV16 variants between 108 tonsillar squamous cell carcinoma and 52 cervical cancer samples collected from patients diagnosed between 2000 and 2008 [142]. One European HPV16 variant has an R10G change in the E6 gene, and a higher frequency of this variant was reported in tonsillar cancers compared to cervical cancer samples (19 vs. 4%) [142]. Another study of squamous cell carcinomas arising in the upper aerodigestive tract shows the predominance of the HPV16 L83 V variant (i.e., 5/8, 63% HPV-positive cases), but the prognostic significance is yet to be determined [143]. While the relevance of this difference is not currently apparent, future studies of HPV intratypic variants and their functional significance as well as their impact on the natural history and treatment for oropharyngeal cancers are warranted. In summary, HPV-positive oropharyngeal tumors represent a distinct clinicopathological profile, but there are still many unanswered questions. The mechanisms surrounding the survival advantage for patients with these tumors need further study. In addition, the significance of and mechanism related to high-risk HPV variants and disease development as well as prognosis are currently poorly understood. Further studies should address the biology of intratypic HPV variants in oropharyngeal cancers and identify additional useful clinical markers to enable appropriate risk stratification, with the goal of optimizing patient treatment.

20.4 HPV Infections in Normal Oral Mucosa

HPV infection is not a rare event, with many HPV infections not leading to cancer either because individuals are infected with non-oncogenic strains, or because an effective immune response prevents establishment of a chronic infection. Moreover, data from oral rinse studies suggest that not everyone with chronic infection develops cancer. The 2003–2004 National Health and Nutrition Examination Survey (NHANES) evaluated the serum prevalence of antibodies to HPV6, 11, 16, and 18 among 4303 persons aged 14–59 years, living in the USA. The study was conducted prior to the introduction of HPV vaccination and represents an estimate of natural HPV exposure in the USA. The overall seroprevalence of any of the four HPV genotypes was 22.4%, with significant differences between males (12.2%) and females (32.5%) [144]. The seroprevalence of HPV16 or HPV18 infection among females was also higher than males (females, HPV16 (15.6%), HPV18 (6.5%); males, HPV16 (5.1%), HPV18 (1.5%)). A much smaller percentage had serum antibodies for both HPV6 and 11 (females, 3.3%; males, 1.0%), or both HPV16 and 18 (females, 2.4%; males, 0.3%) [144].

In a subsequent NHANES study, in 2005–2006, the seroprevalence of nine HPV types was reported for 4943 persons aged 14–59 years. The overall prevalence for any of the 9 types in females was 40.5%, with seroprevalence for HPV16 and HPV18 14% and 6.7%, respectively. In males, overall seropositivity was significantly lower than in females (19.4%, $p < 0.001$) [145]. A separate study made the interesting observation that HPV16 E6 seropositivity (but not seropositivity for other HPV16 antigens—7.5% vs. 0.7%; $p = 0.005$) was significantly more common in males (87.5%) who were HIV positive (75.0%), with a median CD4 cell count of 840, than in males not infected with HIV. E6 seroprevalence was associated with reduced oral HPV16 clearance, but was not statistically significant (HR = 0.65, 95% CI, 0.16–2.70) [146].

A NHANES study conducted from 2009 to 2012 examined HPV in oral rinse samples collected from 9480 participants aged 14–69 years old. This study reported a higher prevalence of oral HPV infections among men compared to women. The overall oral HPV infection rate for any HPV was 6.8%, and the prevalence of overall oral HPV infection was higher among males (10.5%), compared to 3.1% among females. Comparing 2009–2010 and 2011–2012, the HPV prevalence for any high-risk infections (males, 6.0% vs. 7.2%, respectively, and females, 1.7% vs. 1.3%, respectively) or low-risk infections (males, 5.3% vs. 5.2%, respectively, and females, 1.8% vs. 1.9%, respectively) was similar. The most prevalent HPV type detected was HPV16, and males continued to have a higher prevalence of oral HPV16 infection (1.6%) compared to females (0.3%) [147, 148]. In this study, the potential risk factors for oral HPV infection were sexual behavior (i.e., higher number of lifetime sexual partners) and current smoking. Oral HPV infection was also more commonly detected among persons who were sexually experienced but did not report practicing oral sex, suggesting that transmission was likely to be related to sexual contact other than oral sex [147, 148].

Overall, the NHANES studies indicated differences in seroprevalence and oral HPV infection between males and females. Females tended to have higher seroprevalence rates from HPV infection and had lower prevalence of oral HPV infection, with the opposite pattern observed in males. It has been postulated that the higher HPV seroprevalence in females might be related to higher rates of genital HPV infection [144], which might result in a greater protection from subsequent oral HPV infections [149, 150]. A 2013 study assessing incidence and clearance of HPV infections in 1626 men residing in the USA, Brazil, and Mexico by Kreimer et al. reported that newly acquired oral oncogenic HPV infections in healthy men are rare, with most cleared within 1 year [151].

With respect to racial differences, the NHANES studies conducted in 2003–2004 revealed that US non-Hispanic blacks appear to have higher overall HPV seroprevalence rates compared to non-Hispanic whites and Hispanics (46.8%, 31.9%, and 22.6%, respectively) [144]. However, among males HPV16 seroprevalence appears to be similar between non-Hispanic blacks and non-Hispanic whites (7.0% vs. 5.6%), while lower rates were reported among Hispanics (1.5%) [144]. Together, these gender and racial disparities may underlie the differing frequencies in HPV-positive versus HPV-negative head and neck cancers in different populations.

20.5 Global Trends in Head and Neck Cancer Incidence and Mortality

Globally, the incidence of head and neck cancer varies by geography and gender. The overall age-adjusted incidence of head and neck cancer worldwide is 8.0 per 100,000. By geographic region, the highest rate is observed in Europe and Oceania (ASW Incidence: 10.8 per 100,000) while Western Africa has the lowest (ASW

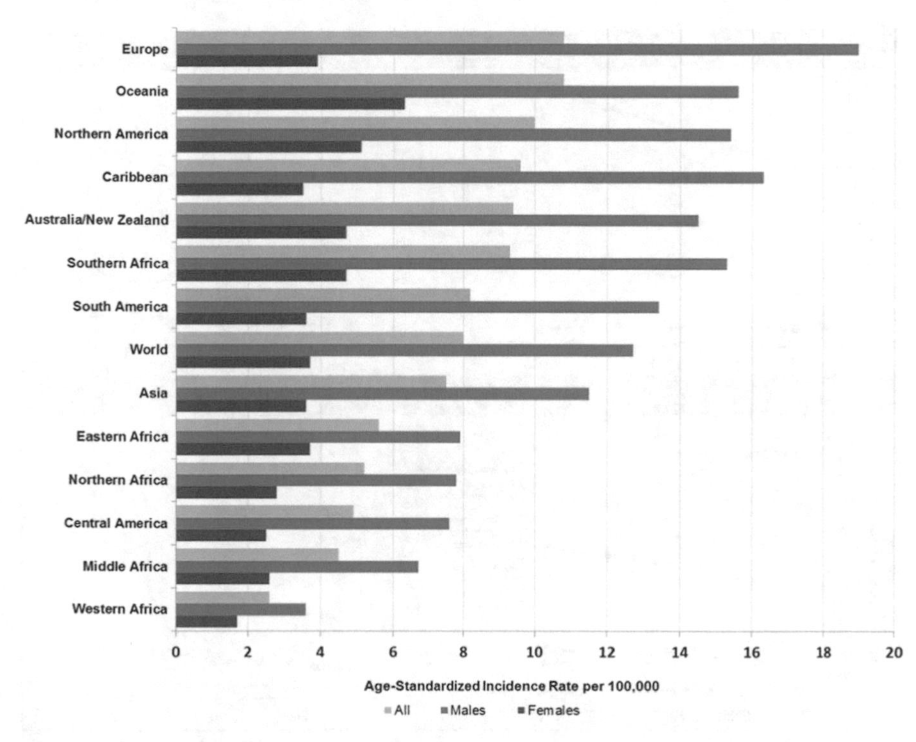

Fig. 20.3 Incidence of head and neck cancer varies by geographic regions and gender. Incidence rates in males (blue) are typically two- to fivefold higher than incidence rates in females (red). Graph reflects an age standardized rate per 100,000

Incidence: 2.6 per 100,000) [152]. The Caribbean Islands, North and South America, South Africa, and Australia and New Zealand have reported estimates that are higher than the worldwide rates, while the rates throughout other regions in Africa, Central America, and Asia are lower than the worldwide estimates (Fig. 20.3).

Head and neck cancer diagnoses worldwide are approximately threefold higher in males (12.7 per 100,000) compared to females (3.7 per 100,000), and similar trends are reported for the USA and other countries around the world (Fig. 20.3) [152, 153]. While males continue to bear the major burden of disease, for both males and females, the predicted number of new head and neck cancer cases is higher in less developed countries compared to more developed countries irrespective of age at diagnosis (Fig. 20.4). For younger persons (<65 years), it is predicted that through 2035, the number of new head and neck cancer diagnoses will increase approximately 1.4-fold for males and also 1.4-fold for females in the developing world but remain constant for males and females in developed countries. In contrast, for older persons (≥65 years), a predicted increasing trend for new cancer diagnoses is observed in both developing and developed countries.

Tremendous inequalities are seen in cancer incidence based on socioeconomic status and racial/ethnic groups. At least part of this variance is attributable to differ-

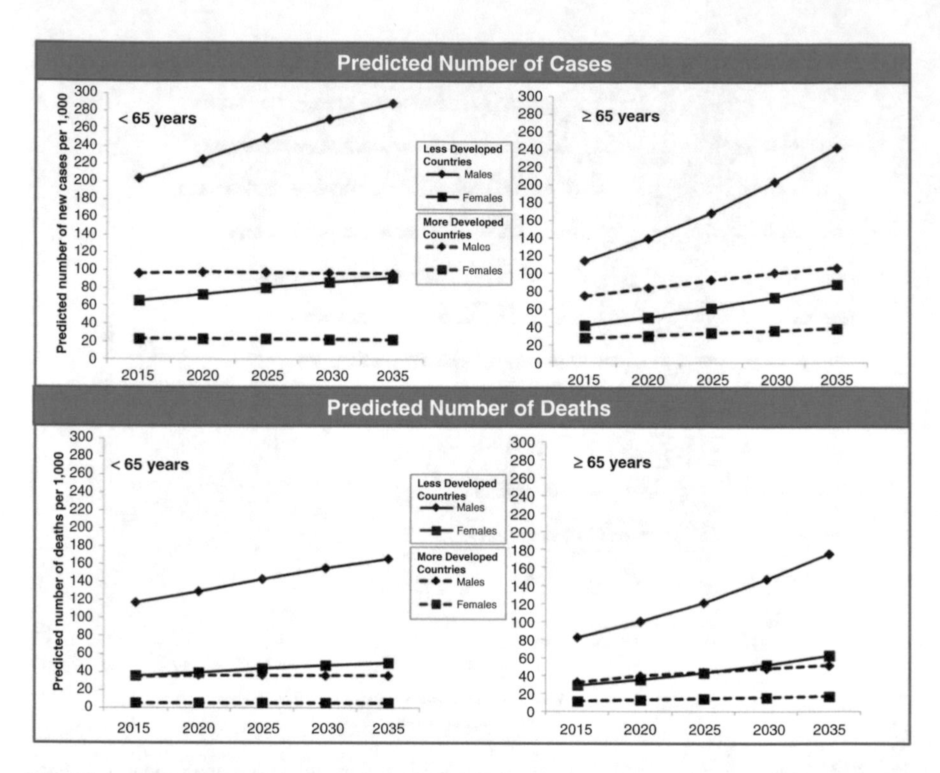

Fig. 20.4 Predicted number of head and neck cancer cases diagnosed and predicted number of deaths: 2015–2035 Globocan (IARC) 2012 are presented as stratified graphs according to age at diagnosis <65 years and ≥65 years. Frequencies for less developed countries (solid line) and for more developed countries (dotted line) are also presented separately for males (diamond) and females (square)

ences in the prevalence of tobacco and alcohol use in different populations. Globally, there are overall trends of decrease in incidence of HPV-associated cancers, except for oropharynx [154]. Increases in incidence for HPV-initiated oropharyngeal cancers, in contrast to declines in HPV-negative oral cavity cancers, were demonstrated in several studies [155]. A twofold increasing trend of new cancer diagnoses is predicted for both males and females in developing countries and 1.4-fold increase for males and females in developed countries. These trends may be explained by a number of factors including increasing tobacco use and/or poor oral health in developing countries compared to developed countries [155, 156].

The mortality rate from head and neck cancer worldwide is 4.3 per 100,000. Overall, death rates are highest in the Caribbean Islands. Asia, Europe, South America, and Southern Africa also have rates that are higher than the worldwide rate (Fig. 20.5). In contrast, North and Central America, Oceania, Australia, New Zealand, and other regions of Africa have death rates lower than worldwide rates. Mortality from head and neck cancer appears to be declining in Europe as well as the USA [153], which may be attributed to improvements in treatment and manage-

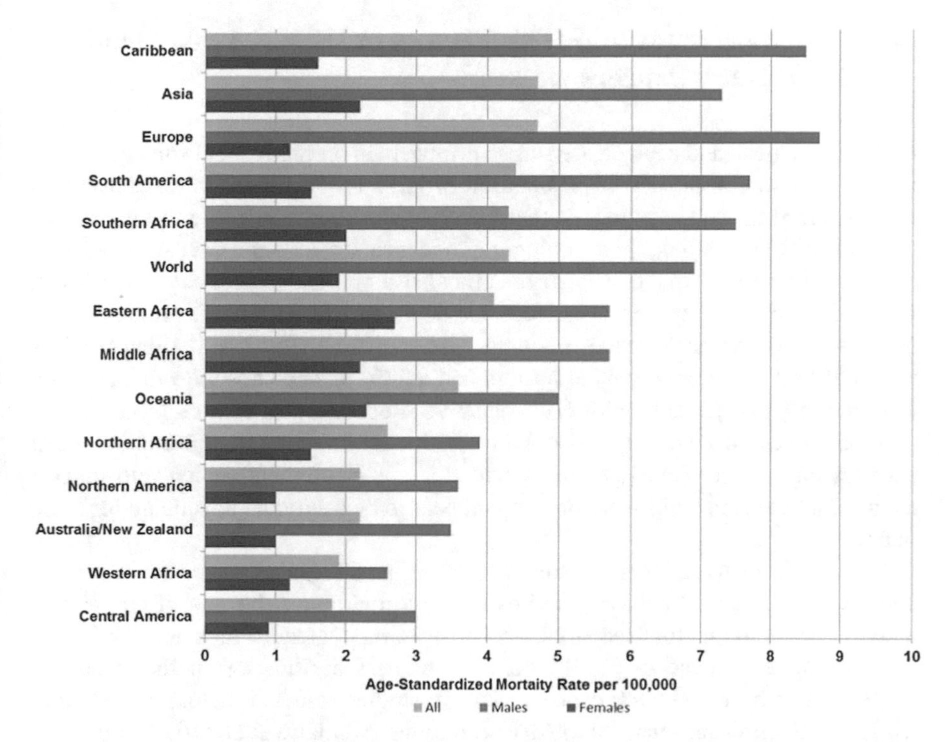

Fig. 20.5 Mortality rates of head and neck cancer according to geographic regions: age standardized rate per 100,000. Mortality rates for males and females combined (green) are highest in the Caribbean and lowest in Central America. Mortality rates for males (blue) are two to eight times greater than the mortality rates for females (red)

ment of head and neck cancer as well as the rising incidence of HPV-related cancers that are more responsive to therapy [157]. Notably, the death rate is almost fourfold lower for females compared to males (1.9 per 100,000 vs. 6.9 per 100,000) [152], and the reason for this disparity is still unclear. A recent matched analysis reported that 286 female and 286 male head and neck cancer patients with similar clinical and demographic characteristics, diagnosed and treated from a single institution, exhibited no survival disparity [158]. While this study should not be generalized to the general population, it has been speculated that the gender differences in head and neck cancer mortality might be associated with differences in incidence based on higher tobacco and alcohol exposure in males compared to females. This hypothesis needs further investigation. Males in developing countries are predicted to carry the highest burden of deaths from head and neck cancer compared to persons in developed nations, while a slight increase in deaths from this disease is also projected for females in developing countries (Fig. 20.4).

20.6 Global Trends and Risk Factors in HPV-Related Head and Neck Cancers

HPV-related head and neck cancers arise primarily in oropharyngeal subsites. There is strength and consistency of association of HPV DNA present in oropharyngeal cancers [155] as well as evidence that the HPV infection is specific to the oropharyngeal cells [3]. Strong epidemiological evidence supports the causal association of HPV with cancers in the oropharynx and also supports the idea that patients with HPV-initiated oropharyngeal cancers have a survival advantage versus HPV-negative head and neck cancer patients. Site-specific analyses show that patients with HPV-related oropharyngeal tumors had a 28% reduced absolute risk of death in comparison to patients with HPV-negative oropharyngeal tumors [159]. While there is also evidence of HPV association with viral infection in nonoropharyngeal subsites such as the oral cavity and larynx, there is insufficient evidence to suggest a causal relationship, and there does not appear to be a survival advantage for these patients.

Among the distinctive risk factors for HPV16-positive head and neck cancers are certain risks of sexual behavior and exposure to marijuana, but not of cumulative alcohol consumption, tobacco smoking, or poor oral hygiene, which are risk factors for HPV-negative cancers [160]. Racial disparities are among another risks for HPV-positive head and neck cancers. Several studies assessed factors contributing for HPV infection and survival in African settings. Mbulawa et al. [161] for the first time reported oral HPV prevalence and factors associated with oral HPV infection in South African heterosexually active couples. HPV55 and 72 were the most common oral HPV types. However, HPV16 was not reported to be common in this study despite the fact that it accounts for the majority of head and neck cancers. Infection risk decreased with age. The risk of oral infection with a specific HPV type was increased when the same type was detected in the mouth of the study partner. Women and men who whose partner had a genital HPV infection were at increased risk of oral infection of the same HPV type. Interestingly, HIV status and CD4 count did not influence oral HPV risk significantly in this study, but the genital HPV risk cohort was found to be significantly influenced by HIV status. Another study in the South African setting specifically assessed oral-genital HPV transmission and yielded similar findings [162]. For oral HPV, the most common viral subtypes detected were HPV62 (prevalence 7.4%); 72 (5.9%); 35, 52, 33, and 58 (each 2.9%); and 16, 74, and 66 (each 1.5%), with similar infection rates in women and men. HIV-positive status resulted in a nonstatistically significant increase in oral HPV prevalence among HIV-infected versus HIV-uninfected individuals.

Several studies have investigated correlations between HPV and HIV infection [163, 164]; these include the Multicenter AIDS Cohort Study (MACS; men) and Women Interagency HIV Study (WIHS; women) cohorts. HIV-positive individuals had increased odds of prevalent oral HPV infection compared with HIV-negative individuals. Risk factors also differed among these two groups: HIV-negative individuals with HPV infection had a higher number of recent oral sex partners, while in contrast, individuals positive for both HIV and HPV individuals had a lower CD4 T-cell count and higher number of lifetime sexual partners [164]. In a subsequent study [163], HIV infection (adjusted hazard ratio = 2.3, 95% CI, 1.7, 3.2), reduced CD4 cell count, and increased numbers of oral sex partners were associated with an elevated risk of oral HPV infection incidence, whereas male sex, older age, and smoking were correlated with an increased risk of oral HPV prevalence.

Increasingly, lack of vaccination might be considered an important risk factor for HPV prevalence; furthermore, it may also be implicated in the continuing increase in incidence of tonsillar cancers with shifting demographic patterns. Steinau et al. reported that 2.9% of the case patients were positive for a high-risk-type HPV infection which would not have been covered [165]. A study conducted by Gillison et al. suggested that currently available HPV vaccines targeting HPV16 and 18 may be highly effective against oropharyngeal squamous cell carcinomas [166]. A candidate 9-valent vaccine (in clinical trials as of 2017) could have the potential to prevent virtually all HPV-associated oropharyngeal cancers.

20.6.1 HPV-Induced Oropharyngeal Cancer

While the majority of head and neck cancers are caused by tobacco and alcohol use, more recently, over the past 10 years, HPV has been demonstrated to be the primary cause of most oropharyngeal cancers, with the remaining proportion of oropharyngeal tumors still attributed to habitual tobacco and/or alcohol use [167]. HPV-related oropharyngeal cancer is frequently characterized as a disease of never-smokers due to higher HPV prevalence in oropharyngeal cancers among never-smokers than ever-smokers [168]. However, a 2016 study by Chaturvedi et al. suggested that a higher HPV-attributable fraction among never-smoker oropharyngeal cancer patients is offset by lower overall incidence of oropharyngeal cancers among never-smokers. After analyzing data in SEER cancer registries, NHANES 2007/2008, NIH-AARP, and RTOG0129 studies, the authors concluded that the burden (including annual incidence rates and case counts) of HPV-positive oropharyngeal cancers during 2007/2008 in the USA was significantly higher among ever-smokers when compared to never-smokers, for both men and women [169].

The HPV virus appears to have an affinity for the epithelial cells of Waldeyer's tonsillar ring [2] and therefore is more commonly detected in tumors that arise in the tonsils and other oropharyngeal subsites. It is unclear whether persistent infection is more common in these tissues. Only a subset of HPV DNA-positive oropharyngeal cancers has been shown to actively transcribe viral genes [170]. Since HPV carcinogenicity relies on the overexpression of HPV oncogenes E6 and E7, these tumors may not be attributable to HPV [96, 97, 170, 171]. In developed countries, an increasing trend of oropharyngeal cancers versus head and neck cancers arising from other subsites has been observed for several decades, particularly in younger patients [172–176]. It has been suggested that HPV is the driver of this increased incidence since the virus (primarily HPV16) has consistently been detected in these tumors. In Europe, the prevalence of HPV in cancer affecting the base of the tongue increased from 58% in 1998–2001 to 81% in 2004–2007 [177, 178], and similar increases in HPV-positive tonsillar cancers have also been reported (from 23% in 1970–1979 to 93% in 2006–2007) [157]. In the USA, for over three decades, the prevalence of HPV-positive oropharyngeal cancers has also reportedly increased from 33% in the 1980s to 63–92% in the 2000s [179–183].

Geographic variability can be stratified as an independent risk factor in HPV-related oropharyngeal cancer prevalence, with higher prevalence in Western Europe as reported by a recent study [184]. A meta-analysis published in 2012 confirms this increasing trend of HPV-associated oropharyngeal cancers in many regions of the world (with no recent data available for Asia, Africa, the Caribbean, and South America) [185]. Various HPV detection methods have been used among these studies, from low-sensitivity assays such as Southern blotting or immunohistochemistry for HPV antigens to more sensitive methods such as in situ hybridization (ISH) or polymerase chain reaction (PCR). After adjusting for the time period during which the studies were conducted, the increasing trend did not appear to be attributed to improved sensitivity or performance of the HPV detection methods. In addition, a systematic review of published articles shows that prior to 2000, North American studies reported the highest prevalence of HPV-associated oropharyngeal cancers (North America, 50.5%; Europe, 35.3%; other countries, 32.2%). Despite the increasing trend for all geographic regions, after 2005, the pooled prevalence of HPV-positive oropharyngeal cancers was comparable in North American studies (69.7%) compared to European studies (73.1%) [185]. This suggests that the prevalence of HPV-positive oropharyngeal cancers diagnosed in Europe appears to be increasing at a faster rate compared to the USA [185]. As an example, a study conducted in Spain reported that HPV had low involvement rate, however, increasing from 1.8% in 1990–1999 to 6.1% of cases of oropharyngeal cancers in 2000–2009 [186]. It is unclear whether this may be due to competing trends such as a greater change in tobacco-related oropharynx cancers in the USA versus Europe. The increase in incidence of HPV-related oropharyngeal cancers has been observed primarily among white males. In contrast, in white women and other race groups, rates have remained unchanged or are decreasing [18, 187, 188]. The reason for the greater predilection for HPV-related oropharyngeal cancer in white males is still unclear.

While numerous studies have been conducted on the prevalence of oral HPV infection in cancer-free individuals, it is important to note that the association of oral HPV infection with disease is different from that of HPV in the oropharynx; therefore, one should be careful about correlating the prevalence of HPV infection in the oral cavity mucosa with prevalence of HPV infection in the oropharyngeal mucosa. Several studies have investigated the presence of HPV in normal oropharyngeal tissue, reporting that in normal and benign oropharyngeal tissues, the prevalence of HPV ranges from 0% to 14% [7, 189–200]. One study conducted in Belgium evaluating 80 tumor-free tonsils from cancer-free adults and children who underwent tonsillectomy due to sleep apnea or recurrent tonsillitis found that 12.5% were positive for high-risk HPV genotypes [192]. Similarly, another study reported that 14% of 50 cancer-free women from Brazil were positive for high-risk HPV in normal oropharyngeal tissues [191]. In contrast, a US study evaluating 226 archived palatine tonsils (tonsils located on the left and right sides at the back of the throat) with benign histologic findings that were surgically removed between 1975 and 2001 from adults >21 years of age did not detect HPV infections in these tissues [193]. In patients with oropharyngeal dysplasia, the prevalence of HPV was reported in one study at 9.4% [201].

Currently there is insufficient information on the prevalence of HPV in precancerous tissues in the oropharynx, and virtually no information on the prevalence rates by gender or race. Some insight has been provided by a few studies that have compared genital HPV infection with HPV status in the oropharynx and suggested the association with sexual behavior. The link between female genital HPV infection and HPV-positive oropharyngeal cancer has been demonstrated, and both male and female cancer patients appear to have increased susceptibility [202–204]. Another study in men confirmed that HPV may be transmitted through either oral-oral or oral-genital routes. Study data suggested that prevalence increased with frequency of oral sex among men whose partner had a genital infection with the same HPV type [205]. A study of 100 women with abnormal cervical cytology reported a high prevalence of high-risk HPV in the oropharynx [206]. This is not surprising since analysis of the National Cancer Institute's Surveillance, Epidemiology, and End Results (SEER) data (1973–2002) shows that women with cervical cancer diagnoses had an increased risk of developing second primary tumors in the tonsil (standardized incidence ratio (SIR), 3.1), a risk exceeding that of women diagnosed with cancers other than cervical cancer [203, 207].

A similar study of SEER data from 1973 to 1994 involving both males and females diagnosed with a HPV-associated anogenital cancer (i.e., cervical, vulvar, and anal cancers) also reported an increased risk of second primary cancers in the tonsils for female cervical cancer patients (relative risk (RR), 65.2, 95% confidence interval (CI), 2.4–10.0) as well as male anal cancer patients (RR, 6.1, 95% CI, 1.2–17.9) [202]. Furthermore, husbands of patients with invasive or in situ cervical cancers also have an excess risk of developing tonsillar cancers (SIR, 2.4, if wife had in situ cervical cancer; SIR, 2.7, if wife had invasive cervical cancer) [208]. A bidirectional association between oropharyngeal cancer and anogenital cancer has also been reported and supports the association of sexual behavior with the development of HPV-associated cancers [204]. The study shows that the risk of developing tonsillar cancer subsequent to anogenital cancer among men who were never married (SIR, 13.0) was much higher compared to men who were married (SIR, 3.8). The acquisition of genital HPV infection is associated with behaviors such as early age of sexual intercourse and multiple sexual partners; and epidemiological studies have suggested that these behaviors are associated with the transmission of HPV to the oral mucosa [209–211], as well. In addition, the association of these behaviors with the development of HPV-positive oropharyngeal cancers has also been demonstrated [212, 213]. Taking these findings altogether, it is likely that the incidence of HPV-positive oropharyngeal cancer might be related to increased intra- and/or interindividual exposures to genital high-risk HPV infections (sexual and nonsexual); nevertheless, the rising incidence of HPV-positive oropharyngeal cancer among white men still remains to be explained epidemiologically.

20.6.2 Therapeutic Responses in Patients with HPV-Positive Oropharyngeal Cancer

Patients with HPV-initiated oropharyngeal cancers have improved overall and disease-free survival compared to patients diagnosed with HPV-negative oropharyngeal cancers as well as cancers in nonoropharyngeal sites [182]. The distinct genetic features of HPV-related cancers (see Sect. 20.2.5) and tobacco-/alcohol-related malignancies are commonly correlated with the observed difference in patient response to therapy. It is possible that the genetic or epigenetic alterations associated with tobacco-associated carcinogens weigh in more heavily and negatively on the patient's response to therapy.

Importantly, in a study assessing HPV-associated *p16* expression and response to radiotherapy, with and without hypoxic modification, patients with positive expression of p16 had significantly improved outcome after radiotherapy as compared to patients with p16-negative tumors [214, 215]. A retrospective study by Bossi assessed response to therapy and smoking habits, tumor *p16* expression/HPV status, and T and N stage. The authors concluded that patients with oropharyngeal cancer from low- and intermediate-risk groups had a better survival when treated with chemotherapy compared with open surgery followed by radiation therapy [216].

The Radiation Therapy Oncology Group (RTOG) clinical trial (0129) used recursive partitioning analysis for 266 patients diagnosed with advanced stage oropharyngeal cancer who were treated with radiotherapy combined with platinum therapy. The risk of death depended primarily upon the HPV status of the tumor, followed by the number of pack-years of tobacco smoking, then tumor stage, and nodal stage [217]. The 3-year overall survival was 93% for nonsmokers with low nodal tumor stage, 70.8% for intermediate-risk patients, and 46.2% for HPV-negative smokers diagnosed with tumors at high tumor stage [217]. A clinical trial conducted in 2011 in an independent cohort reported similar results for patients treated with induction chemotherapy followed by chemoradiation or radiation only. In this study, the 3-year overall survival for low-risk patients was 100%, for intermediate-risk patients 79.6%, and for high-risk patients 70% [218]. These findings suggest that this type of risk stratification may help to define clinical decision for appropriate treatment [219, 220]. Early trials of reduced treatment intensity in HPV-related disease appear to confirm tobacco history and tumor stage as stratification factors which relate to treatment outcome [221, 222].

There is racial disparity in survival for patients diagnosed with oropharyngeal cancer. Black patients have significantly worse survival rates compared to whites. Despite the numerous studies of the impact of HPV on oropharyngeal cancer survival, investigations according to race have been limited. Nevertheless, from the few studies that have been conducted, findings suggest that blacks appear to have a lower prevalence of HPV-positive oropharyngeal cancers. In a multicenter phase III clinical trial of induction chemotherapy followed by concurrent chemoradiation, the overall and disease-free survival of patients diagnosed with stage III/IV head and neck cancer was improved as expected for patients diagnosed with HPV-positive tumors. The subset analysis of 124 oropharyngeal cancer patients (54 black, 70 white) also documented poor median overall survival for blacks compared to whites (25.2 vs. 69.4 months, $p = 0.0006$). There were no noted differences in outcome between blacks and whites diagnosed with nonoropharyngeal cancers. Further analysis of 224 head and neck tumors from 196 white and 28 black patients revealed a significant difference in the prevalence of HPV according to race. Black patients had a significantly lower prevalence of HPV-positive cancers compared to whites (4% vs. 34%). This study suggests that at least a portion of the racial disparity in oropharyngeal cancer may be attributed to differences in the prevalence of HPV-positive oropharyngeal tumors according to race [223].

In additional studies, Chernock et al. reported that the poorer disease-free survival among African-Americans may be attributed to lower HPV prevalence, the types of treatment, and higher tumor stage at diagnoses [224, 225]. Another report by Worsham et al. [226] compared the survival of 118 patients diagnosed with oropharyngeal cancer and found that 51/118 (43.2%) were HPV positive and 86 (56.8%) were HPV negative. A lower prevalence of HPV was observed for African-Americans compared to white Americans (29% vs. 71%, $p = 0.024$). The HPV-negative African-American patients had poorer survival compared to the HPV-positive African-American patients ($p = 0.0012$) and HPV-positive white Americans ($p = 0.0496$). There was no survival difference between HPV-positive African-Americans and HPV-positive white Americans. These findings were confirmed in subsequent studies [227] assessing racial disparities between African-American and white patients with oropharyngeal carcinomas. African-Americans were significantly more likely than whites to present with high T-stage disease, and silenced p16, and therefore receive nonsurgical treatment; outcome was poorer for African-American patients [228]. However, these various studies used different methods of HPV identification and small cohorts, with not all reporting both p16 status and HPV status, leading to difficulties in drawing conclusions. Some recent studies examined larger cohorts of African-American patients for both p16 and HPV status [229, 230]. These studies suggest that the majority of African-American patients with HPV-positive oropharyngeal carcinomas are p16 negative, with other hallmarks of HPV-induced cancer present, such as non-keratinizing histology, high HPV viral load, and loss of p53 and Rb expression, leading to poorer outcomes. However, more study is needed to understand the pathogenesis, epidemiology, and outcomes of p16 in HPV-positive tumors.

These findings not only support the important role that HPV plays in oropharyngeal cancer outcome but also suggest that additional factors might be contributing to the poor survival of blacks; therefore, further epidemiological studies addressing racial disparities in oropharyngeal cancer are still needed.

20.6.3 HPV-Positive Nonoropharyngeal Cancer

The role of HPV has also been extensively investigated in studies of head and neck cancers in nonoropharyngeal subsites. The pooled prevalence of HPV-positive nonoropharyngeal cancers for all regions of the world is reported to be approximately 21.8% and is higher in Europe (23.7%) than in North America (12.8%) [185]. A recent meta-analysis assessing the prevalence of HPV in HNC patients of African ancestry reported a 14.5% prevalence of HPV16-positive nonoropharyngeal cancers [227]. In contrast to the increasing prevalence of HPV-positive oropharyngeal cancers, HPV-positive nonoropharyngeal cancers do not appear to be increasing, but rather decreasing over time. For studies published prior to 2000, the prevalence of HPV-positive nonoropharyngeal cancers was 22.2% and decreased to 17.2% (2000–2004) and decreased to an even greater extent after 2004 [185, 227]. The sensitivity of the detection methods used (PCR, ISH, or both, Southern blotting or IHC) as well as the type of tissue evaluated (paraffin embedded, fresh frozen, or both) did not appear to affect the decreasing trend in HPV prevalence. However, the reason for declining HPV prevalence in nonoropharyngeal cancers is still unclear. The identification of robust fingerprints of HPV-driven carcinogenesis and the accurate exclusion of the oropharynx as the site of tumor origin are significant future challenges in the estimation of HPV-attributable nonoropharyngeal head and neck cancers [231].

Nevertheless, studies involving nonoropharyngeal subsites in the oral cavity do suggest a potential association with HPV. In oral cavity dysplasia, 12 studies conducted between 1985 and 2010 show that HPV16 and/or 18 was detected in 25.3% of dysplastic tissues, with very wide variability between studies indicating prevalence rates 0–100% [232]. The oral cavity dysplasia sites included in these studies were from the tongue, floor of the mouth/ventral tongue, buccal gingiva/vestibule, hard palate, oral commissure/lip, and other unspecified sites in the oral cavity. While DNA presence alone quantifies prevalence, it does not necessarily correlate with viral transformation. However, in 2011, Syrjanen et al. performed a systematic review of case-control studies published from 1966 to 2010 that had investigated the association of HPV with potentially malignant disorders such as oral lichen planus, leukoplakia, erythroplakia, and oral proliferative verrucous leukoplakia (956 cases and 675 controls) [233]. Case-control studies investigating HPV's association with oral squamous cell carcinoma (involving the oral cavity only) were also reviewed and summarized (1885 cases and 2248 controls). Based on this report, HPV16 appears to be associated with the development of dysplastic oral lesions (oral lichen planus, odds ratio = 5.61; 95% CI, 2.42–5.99, and leukoplakia, odds ratio = 4.47; 95% CI, 2.22–8.98) as well as oral cavity cancers (odds ratio = 3.86; 95% CI, 2.16–6.86) [233]. A similar but weaker association was also reported in an earlier study conducted in 2006 (odds ratio = 2.1; 95% CI, 1.2–3.4) [234]. In contrast, a 2016 study assessed the etiologic fraction for HPV in larynx squamous cell carcinomas and revealed that it is low (1.7%) in a geographic region with high prevalence for oropharyngeal squamous cell carcinomas [235]. Overall, this suggests that HPV may be associated with at least a subset of oral cavity tumors, but unlike HPV-positive oropharyngeal cancers, a causal association between HPV and survival advantage for patients with HPV-positive oral cavity cancers has not been clearly delineated.

Several studies documented a steady decline in the overall cumulative incidence of nonoropharyngeal head and neck cancers in the USA and worldwide for all racial/ethnic groups and genders and proposed this reflects the declining prevalence of tobacco and alcohol consumption [236–240]. In contrast, more focused studies reported an increase in the incidence of oral tongue cancers in younger adults despite the overall decline in tobacco use [241–243]. In the USA, examination of data from the NCI SEER database revealed that the incidence of oral tongue cancer was not significantly different between 1973 and 2006 among white males, but has significantly, although modestly, increased among white females and declined among African-American males and females [244]. Possible suggestions for the increasing incidence of tongue cancer among white females might be related to the recent trend of increased smoking prevalence among women in particular, or other unknown environmental factors, which may include HPV. Some evidence supports HPV involvement in at least a subset of tongue cancers, since immunostaining assays revealed a correlation between HPV16 E6 expression and p53 loss as well as HPV16 E7 expression and pRb loss [245]. However, as reported in a recent retrospective, multi-institutional study for nonoropharyngeal head and neck squamous cell carcinomas, HPV status and the p16 status are not of prognostic significance [246].

A substantial number of studies have investigated the carcinogenic role of HPV in laryngeal cancer, and recent evidence suggests that HPV16 might be associated with approximately 20% of these tumors, as suggested by results of PCR assays. Among 12 studies conducted through 2012, an association of HPV16 rather than HPV18 was observed (HPV16 pooled odds ratio, 6.07, 95% CI, 3.44–10.70, vs. HPV18 pooled odds ratio, 4.16, 95% CI, 0.87–20.04). These findings are supported by recent studies, reporting that prior infection with HPV16 may play a role in the etiology of some laryngeal cancers. A study conducted by Chen et al. reported that in 300 patients with laryngeal squamous cell carcinomas, the prevalence of HPV16 was higher in cases than healthy controls (6.7% vs. 2.7%). The risk of cancer, associated with HPV16 DNA positivity, was even higher in patients aged 55 years or younger, males, never-smokers, and never-drinkers [247].

In summary, both cutaneous and mucosal HPVs have been detected in the oral cavity and the larynx. The presence of mucosal HPVs and particularly HPV16 DNA correlated with the presence of viral transcripts and enhanced *CDKN2A (P16)* detected in some studies are compatible with a possible etiologic role in a small subset of oral cavity and laryngeal tumors [248–250]. However, unlike HPV-positive oropharyngeal cancers, there has been no documented positive impact on treatment outcomes [248, 249, 251–253].

References

1. Hausen HZ, de Villiers EM. Human papillomaviruses. Annu Rev Microbiol. 1994;48(1):427–47.
2. Syrjänen K, Syrjänen S, Lamberg M, Pyrhönen S, Nuutinen J. Morphological and immunohistochemical evidence suggesting human papillomavirus (HPV) involvement in oral squamous cell carcinogenesis. Int J Oral Surg. 1983;12(6):418–24.
3. Gillison ML, Koch WM, Capone RB, Spafford MJ, Westra W, Wu L, et al. Evidence for a causal association between human papillomavirus and a subset of head and neck cancers. J Natl Cancer Inst. 2000;92(9):709–20.
4. Gillison ML, Lowy DR. A causal role for human papillomavirus in head and neck cancer. Lancet. 2004;363(9420):1488–9.
5. Kreimer AR, Clifford GM, Boyle P, Franceschi S. Human papillomavirus types in head and neck squamous cell carcinomas worldwide: a systematic review. Cancer Epidemiol Biomark Prev. 2005;14(2):467–75.
6. Badaracco G, Venuti A, Morello R, Muller A, Marcante ML. Human papillomavirus in head and neck carcinomas: prevalence, physical status and relationship with clinical/pathological parameters. Anticancer Res. 2000;20(2B):1301–5.
7. Klussmann JP, Weissenborn SJ, Wieland U, Dries V, Kolligs J, Jungehuelsing M, et al. Prevalence, distribution, and viral load of human papillomavirus 16 DNA in tonsillar carcinomas. Cancer. 2001;92(11):2875–84.
8. Lindel K, Beer KT, Laissue J, Greiner RH, Aebersold DM. Human papillomavirus positive squamous cell carcinoma of the oropharynx. Cancer. 2001;92(4):805–13.
9. Mellin H, Friesland S, Lewensohn R, Dalianis T, Munck-Wikland E. Human papillomavirus (HPV) DNA in tonsillar cancer: clinical correlates, risk of relapse, and survival. Int J Cancer. 2000;89(3):300–4.
10. Miller CS, Johnstone BM. Human papillomavirus as a risk factor for oral squamous cell carcinoma: a meta-analysis, 1982–1997. Oral Surg Oral Med Oral Pathol Oral Radiol Endod. 2001;91(6):622–35.
11. Mork J, Lie AK, Glattre E, Clark S, Hallmans G, Jellum E, et al. Human papillomavirus infection as a risk factor for squamous-cell carcinoma of the head and neck. N Engl J Med. 2001;344(15):1125–31.
12. Sethi S, Ali-Fehmi R, Franceschi S, Struijk L, van Doorn L-J, Quint W, et al. Characteristics and survival of head and neck cancer by HPV status: a cancer registry-based study. Int J Cancer. 2011;131(5):1179–86.
13. Sisk EA, Bradford CR, Jacob A, Yian CH, Staton KM, Tang G, et al. Human papillomavirus infection in "young" versus "old" patients with squamous cell carcinoma of the head and neck. Head Neck. 2000;22(7):649–57.
14. Venuti A, Manni V, Morello R, De Marco F, Marzetti F, Marcante ML. Physical state and expression of human papillomavirus in laryngeal carcinoma and surrounding normal mucosa. J Med Virol. 2000;60(4):396–402.
15. Vietia D, Liuzzi JP, Avila M, De Guglielmo Z, Prado Y, Correnti M. Human papillomavirus detection in head and neck squamous cell carcinoma. Ecancermedicalscience 2014;8.
16. Dahlstrand HM, Dalianis T. Presence and influence of human papillomaviruses (HPV) in Tonsillar cancer. Adv in Cancer Res. 2005;93:59–89.
17. Klussmann JP, Gültekin E, Weissenborn SJ, Wieland U, Dries V, Dienes HP, et al. Expression of p16 protein identifies a distinct entity of tonsillar carcinomas associated with human papillomavirus. Am J Pathol. 2003;162(3):747–53.
18. El-Mofty SK, Lu DW. Prevalence of human papillomavirus type 16 DNA in squamous cell carcinoma of the palatine tonsil, and not the oral cavity, in young patients. Am J Surg Pathol. 2003;27(11):1463–70.
19. Burk RD, Chen Z, Van Doorslaer K. Human papillomaviruses: genetic basis of carcinogenicity. Public Health Genomics. 2009;12(5–6):281–90.

20. Van Doorslaer K, Burk RD. Evolution of human papillomavirus carcinogenicity. Adv Virus Res. 2010;77:41–62.
21. Kines RC, Thompson CD, Lowy DR, Schiller JT, Day PM. The initial steps leading to papillomavirus infection occur on the basement membrane prior to cell surface binding. Proc Natl Acad Sci. 2009;106(48):20458–63.
22. Schiller JT, Day PM, Kines RC. Current understanding of the mechanism of HPV infection. Gynecol Oncol. 2010;118(1):S12–S7.
23. McBride AA, Oliveira JG, McPhillips MG. Partitioning viral genomes in mitosis: same idea, different targets. Cell Cycle. 2006;5(14):1499–502.
24. Munger K, Baldwin A, Edwards KM, Hayakawa H, Nguyen CL, Owens M, et al. Mechanisms of human papillomavirus-induced oncogenesis. J Virol. 2004;78(21):11451–60.
25. Hajek M, Sewell A, Kaech S, Burtness B, Yarbrough WG, Issaeva N. TRAF3/CYLD mutations identify a distinct subset of human papillomavirus-associated head and neck squamous cell carcinoma. Cancer. 2017;123(10):1778–90.
26. Richards RM, Lowy DR, Schiller JT, Day PM. Cleavage of the papillomavirus minor capsid protein, L2, at a furin consensus site is necessary for infection. Proc Natl Acad Sci U S A. 2006;103(5):1522–7.
27. Letian T, Tianyu Z. Cellular receptor binding and entry of human papillomavirus. Virol J. 2010;7:2.
28. Huber B, Schellenbacher C, Shafti-Keramat S, Jindra C, Christensen N, Kirnbauer R. Chimeric L2-based Virus-Like Particle (VLP) vaccines targeting cutaneous human papillomaviruses (HPV). PLoS One. 2017;12(1):e0169533.
29. Berg M, Stenlund A. Functional interactions between papillomavirus E1 and E2 proteins. J Virol. 1997;71(5):3853–63.
30. Powell MLC, Smith JA, Sowa ME, Harper JW, Iftner T, Stubenrauch F, et al. NCoR1 mediates papillomavirus E8^E2C transcriptional repression. J Virol. 2010;84(9):4451–60.
31. Paolini F, Curzio G, Melucci E, Terrenato I, Antoniani B, Carosi M, et al. Human papillomavirus 16 E2 interacts with neuregulin receptor degradation protein 1 affecting ErbB-3 expression in vitro and in clinical samples of cervical lesions. Eur J Cancer. 2016;58:52–61.
32. Ma T, Zou N, Lin BY, Chow LT, Harper JW. Interaction between cyclin-dependent kinases and human papillomavirus replication-initiation protein E1 is required for efficient viral replication. Proc Natl Acad Sci. 1999;96(2):382–7.
33. Lin BY, Ma T, Liu J, Kuo SH, Jin G, Broker TR, et al. HeLa cells are phenotypically limiting in cyclin E/CDK2 for efficient human papillomavirus DNA replication. J Biol Chem. 2000;275(9):6167–74.
34. Liu JS, Kuo SR, Makhov AM, Cyr DM, Griffith JD, Broker TR, et al. Human Hsp70 and Hsp40 chaperone proteins facilitate human papillomavirus-11 E1 protein binding to the origin and stimulate cell-free DNA replication. J Biol Chem. 1998;273(46):30704–12.
35. Lee D, Sohn H, Kalpana GV, Choe J. Interaction of E1 and hSNF5 proteins stimulates replication of human papillomavirus DNA. Nature. 1999;399(6735):487–91.
36. Swindle CS, Engler JA. Association of the human papillomavirus type 11 E1 protein with histone H1. J Virol. 1998;72(3):1994–2001.
37. Yasugi T, Vidal M, Sakai H, Howley PM, Benson JD. Two classes of human papillomavirus type 16 E1 mutants suggest pleiotropic conformational constraints affecting E1 multimerization, E2 interaction, and interaction with cellular proteins. J Virol. 1997;71(8):5942–51.
38. Fradet-Turcotte A, Brault K, Titolo S, Howley PM, Archambault J. Characterization of papillomavirus E1 helicase mutants defective for interaction with the SUMO-conjugating enzyme Ubc9. Virology. 2009;395(2):190–201.
39. Lin BY, Makhov AM, Griffith JD, Broker TR, Chow LT. Chaperone proteins abrogate inhibition of the human papillomavirus (HPV) E1 replicative helicase by the HPV E2 protein. Mol Cell Biol. 2002;22(18):6592–604.
40. Straub E, Dreer M, Fertey J, Iftner T, Stubenrauch F. The viral E8^E2C repressor limits productive replication of human papillomavirus 16. J Virol. 2013;88(2):937–47.

41. Kurg R. The Role of E2 Proteins in Papillomavirus DNA Replication. DNA Replication–Current Advances. InTech. 2011;DOI:https://doi.org/10.5772/19609.
42. Straub E, Fertey J, Dreer M, Iftner T, Stubenrauch F. Characterization of the human papillomavirus 16 E8 promoter. J Virol. 2015;89(14):7304–13.
43. Dreer M, Fertey J, van de Poel S, Straub E, Madlung J, Macek B, et al. Interaction of NCOR/SMRT repressor complexes with papillomavirus E8^E2C proteins inhibits viral replication. PLoS Pathog. 2016;12(4):e1005556.
44. Stubenrauch F, Zobel T, Iftner T. The E8 domain confers a novel long-distance transcriptional repression activity on the E8^E2C protein of high-risk human papillomavirus type 31. J Virol. 2001;75(9):4139–49.
45. Fertey J, Hurst J, Straub E, Schenker A, Iftner T, Stubenrauch F. Growth inhibition of HeLa cells is a conserved feature of high-risk human papillomavirus E8^E2C proteins and can also be achieved by an artificial repressor protein. J Virol. 2010;85(6):2918–26.
46. Doorbar J, Ely S, Sterling J, McLean C, Crawford L. Specific interaction between HPV-16 E1–E4 and cytokeratins results in collapse of the epithelial cell intermediate filament network. Nature. 1991;352(6338):824–7.
47. Khan J, Davy CE, McIntosh PB, Jackson DJ, Hinz S, Wang Q, et al. Role of calpain in the formation of human papillomavirus type 16 E1^E4 amyloid fibers and reorganization of the Keratin network. J Virol. 2011;85(19):9984–97.
48. McIntosh PB, Laskey P, Sullivan K, Davy C, Wang Q, Jackson DJ, et al. E1--E4-mediated keratin phosphorylation and ubiquitylation: a mechanism for keratin depletion in HPV16-infected epithelium. J Cell Sci. 2010;123(Pt 16):2810–22.
49. Kumar A, Yadav IS, Hussain S, Das BC, Bharadwaj M. Identification of immunotherapeutic epitope of E5 protein of human papillomavirus-16: An in silico approach. Biologicals. 2015;43(5):344–8.
50. Conrad M, Bubb VJ, Schlegel R. The human papillomavirus type 6 and 16 E5 proteins are membrane-associated proteins which associate with the 16-kilodalton pore-forming protein. J Virol. 1993;67(10):6170–8.
51. Oetke C, Auvinen E, Pawlita M, Alonso A. Human papillomavirus type 16 E5 protein localizes to the Golgi apparatus but does not grossly affect cellular glycosylation. Arch Virol. 2000;145(10):2183–91.
52. Um SH, Mundi N, Yoo J, Palma DA, Fung K, MacNeil D, et al. Variable expression of the forgotten oncogene E5 in HPV-positive oropharyngeal cancer. J Clin Virol. 2014;61(1):94–100.
53. DiMaio D, Mattoon D. Mechanisms of cell transformation by papillomavirus E5 proteins. Oncogene. 2001;20(54):7866–73.
54. Häfner N, Driesch C, Gajda M, Jansen L, Kirchmayr R, Runnebaum IB, et al. Integration of the HPV16 genome does not invariably result in high levels of viral oncogene transcripts. Oncogene. 2007;27(11):1610–7.
55. Bouvard V, Matlashewski G, Gu Z-M, Storey A, Banks L. The human papillomavirus type 16 E5 gene cooperates with the E7 gene to stimulate proliferation of primary cells and increases viral gene expression. Virology. 1994;203(1):73–80.
56. Crusius K, Rodriguez I, Alonso A. The human papillomavirus type 16 E5 protein modulates ERK1/2 and p38 MAP kinase activation by an EGFR-independent process in stressed human keratinocytes. Virus Genes. 2000;20(1):65–9.
57. Liu C, Lin J, Li L, Zhang Y, Chen W, Cao Z, et al. HPV16 early gene E5 specifically reduces miRNA-196a in cervical cancer cells. Sci Rep. 2015;5:7653.
58. Auvinen E, Alonso A, Auvinen P. Human papillomavirus type 16 E5 protein colocalizes with the antiapoptotic Bcl-2 protein. Arch Virol. 2004;149(9):1745–59.
59. Ashrafi GH, Brown DR, Fife KH, Campo MS. Down-regulation of MHC class I is a property common to papillomavirus E5 proteins. Virus Res. 2006;120(1–2):208–11.
60. Chang JL, Tsao YP, Liu DW, Huang SJ, Lee WH, Chen SL. The expression of HPV-16 E5 protein in squamous neoplastic changes in the uterine cervix. J Biomed Sci. 2001;8(2):206–13.

61. Zhang B, Li P, Wang E, Brahmi Z, Dunn KW, Blum JS, et al. The E5 protein of human papillomavirus type 16 perturbs MHC class II antigen maturation in human foreskin keratinocytes treated with interferon-γ. Virology. 2003;310(1):100–8.
62. Kotnik Halavaty K, Regan J, Mehta K, Laimins L. Human papillomavirus E5 oncoproteins bind the A4 endoplasmic reticulum protein to regulate proliferative ability upon differentiation. Virology. 2014;452–453:223–30.
63. Paolini F, Curzio G, Cordeiro MN, Massa S, Mariani L, Pimpinelli F, et al. HPV 16 E5 oncoprotein is expressed in early stage carcinogenesis and can be a target of immunotherapy. Hum Vaccin Immunother. 2017;13(2):291–7.
64. Gao Q, Kumar A, Srinivasan S, Singh L, Mukai H, Ono Y, et al. PKN binds and phosphorylates human papillomavirus E6 Oncoprotein. J Biol Chem. 2000;275(20):14824–30.
65. Massimi P, Shai A, Lambert P, Banks L. HPV E6 degradation of p53 and PDZ containing substrates in an E6AP null background. Oncogene. 2007;27(12):1800–4.
66. Scheffner M, Huibregtse JM, Vierstra RD, Howley PM. The HPV-16 E6 and E6-AP complex functions as a ubiquitin-protein ligase in the ubiquitination of p53. Cell. 1993;75(3):495–505.
67. Werness B, Levine A, Howley P. Association of human papillomavirus types 16 and 18 E6 proteins with p53. Science. 1990;248(4951):76–9.
68. Gewin L, Myers H, Kiyono T, Galloway DA. Identification of a novel telomerase repressor that interacts with the human papillomavirus type-16 E6/E6-AP complex. Genes Dev. 2004;18(18):2269–82.
69. Katzenellenbogen RA, Egelkrout EM, Vliet-Gregg P, Gewin LC, Gafken PR, Galloway DA. NFX1-123 and poly(A) binding proteins synergistically augment activation of telomerase in human papillomavirus type 16 E6-expressing cells. J Virol. 2007;81(8):3786–96.
70. Ronco LV, Karpova AY, Vidal M, Howley PM. Human papillomavirus 16 E6 oncoprotein binds to interferon regulatory factor-3 and inhibits its transcriptional activity. Genes Dev. 1998;12(13):2061–72.
71. Tong X, Howley PM. The bovine papillomavirus E6 oncoprotein interacts with paxillin and disrupts the actin cytoskeleton. Proc Natl Acad Sci. 1997;94(9):4412–7.
72. Cavarelli J. Crystal structure of full-length Bovine Papillomavirus oncoprotein E6 in complex with LD1 motif of paxillin at 2.3A resolution. Protein Data Bank, Rutgers University;2011.
73. Srivastava K, Pickard A, McDade S, McCance DJ. p63 drives invasion in keratinocytes expressing HPV16 E6/E7 genes through regulation of Src-FAK signalling. Oncotarget. 2017;8(10):16202–19.
74. Boyer SN, Wazer DE, Band V. E7 protein of human papilloma virus-16 induces degradation of retinoblastoma protein through the ubiquitin-proteasome pathway. Cancer Res. 1996;56(20):4620–4.
75. Khleif SN, DeGregori J, Yee CL, Otterson GA, Kaye FJ, Nevins JR, et al. Inhibition of cyclin D-CDK4/CDK6 activity is associated with an E2F-mediated induction of cyclin kinase inhibitor activity. Proc Natl Acad Sci. 1996;93(9):4350–4.
76. Zhang B, Laribee RN, Klemsz MJ, Roman A. Human papillomavirus type 16 E7 protein increases acetylation of histone H3 in human foreskin keratinocytes. Virology. 2004;329(1):189–98.
77. Menges CW, Baglia LA, Lapoint R, McCance DJ. Human papillomavirus type 16 E7 Up-regulates AKT activity through the Retinoblastoma protein. Cancer Res. 2006;66(11):5555–9.
78. Arroyo M, Bagchi S, Raychaudhuri P. Association of the human papillomavirus type 16 E7 protein with the S-phase-specific E2F-cyclin A complex. Mol Cell Biol. 1993;13(10):6537–46.
79. Dyson N, Guida P, Munger K, Harlow E. Homologous sequences in adenovirus E1A and human papillomavirus E7 proteins mediate interaction with the same set of cellular proteins. J Virol. 1992;66(12):6893–902.
80. McIntyre MC, Ruesch MN, Laimins LA. Human papillomavirus E7 oncoproteins bind a single form of cyclin E in a complex with cdk2 and p107. Virology. 1996;215(1):73–82.

81. Munger K, Werness BA, Dyson N, Phelps WC, Harlow E, Howley PM. Complex formation of human papillomavirus E7 proteins with the retinoblastoma tumor suppressor gene product. EMBO J. 1989;8(13):4099–105.
82. Phelps WC, Bagchi S, Barnes JA, Raychaudhuri P, Kraus V, Munger K, et al. Analysis of trans activation by human papillomavirus type 16 E7 and adenovirus 12S E1A suggests a common mechanism. J Virol. 1991;65(12):6922–30.
83. Hausen HZ. Papillomaviruses causing cancer: evasion from host-cell control in early events in carcinogenesis. J Natl Cancer Inst. 2000;92(9):690–8.
84. Bosch FX, Lorincz A, Munoz N, Meijer CJLM, Shah KV. The causal relation between human papillomavirus and cervical cancer. J Clin Pathol. 2002;55(4):244–65.
85. Bosch FX, de Sanjose S. Chapter 1: human papillomavirus and cervical cancer–burden and assessment of causality. JNCI Monographs. 2003;2003(31):3–13.
86. Sano D, Oridate N. The molecular mechanism of human papillomavirus-induced carcinogenesis in head and neck squamous cell carcinoma. Int J Clin Oncol. 2016;21(5):819–26.
87. Ragin CCR, Reshmi SC, Gollin SM. Mapping and analysis of HPV16 integration sites in a head and neck cancer cell line. Int J Cancer. 2004;110(5):701–9.
88. Wentzensen N. Systematic review of genomic integration sites of human papillomavirus genomes in epithelial dysplasia and invasive cancer of the female lower genital tract. Cancer Res. 2004;64(11):3878–84.
89. Hatano T, Sano D, Takahashi H, Hyakusoku H, Isono Y, Shimada S, et al. Identification of human papillomavirus (HPV) 16 DNA integration and the ensuing patterns of methylation in HPV-associated head and neck squamous cell carcinoma cell lines. Int J Cancer. 2017;140(7):1571–80.
90. Arias-Pulido H, Peyton CL, Joste NE, Vargas H, Wheeler CM. Human papillomavirus type 16 integration in cervical carcinoma in situ and in invasive cervical cancer. J Clin Microbiol. 2006;44(5):1755–62.
91. Mellin H, Dahlgren L, Munck-Wikland E, Lindholm J, Rabbani H, Kalantari M, et al. Human papillomavirus type 16 is episomal and a high viral load may be correlated to better prognosis in tonsillar cancer. Int J Cancer. 2002;102(2):152–8.
92. Koskinen WJ, Chen RW, Leivo I, Mäkitie A, Bäck L, Kontio R, et al. Prevalence and physical status of human papillomavirus in squamous cell carcinomas of the head and neck. Int J Cancer. 2003;107(3):401–6.
93. Badaracco G, Rizzo C, Mafera B, Pichi B, Giannarelli D, Rahimi S, et al. Molecular analyses and prognostic relevance of HPV in head and neck tumours. Oncology Reports. 2007.
94. Parfenov M, Pedamallu CS, Gehlenborg N, Freeman SS, Danilova L, Bristow CA, et al. Characterization of HPV and host genome interactions in primary head and neck cancers. Proc Natl Acad Sci U S A. 2014;111(43):15544–9.
95. Wilting SM, Smeets SJ, Snijders PJF, van Wieringen WN, van de Wiel MA, Meijer GA, et al. Genomic profiling identifies common HPVassociated chromosomal alterations in squamous cell carcinomas of cervix and head and neck. BMC Med Genet. 2009;2(1):32.
96. van Houten VMM, Snijders PJF, van den Brekel MWM, Kummer JA, Meijer CJLM, van Leeuwen B, et al. Biological evidence that human papillomaviruses are etiologically involved in a subgroup of head and neck squamous cell carcinomas. Int J Cancer. 2001;93(2):232–5.
97. Wiest T, Schwarz E, Enders C, Flechtenmacher C, Bosch FX. Involvement of intact HPV16 E6/E7 gene expression in head and neck cancers with unaltered p53 status and perturbed pRb cell cycle control. Oncogene. 2002;21(10):1510–7.
98. Ragin CCR, Taioli E, Weissfeld JL, White JS, Rossie KM, Modugno F, et al. 11q13 amplification status and human papillomavirus in relation to p16 expression defines two distinct etiologies of head and neck tumours. Br J Cancer. 2006;95(10):1432–8.
99. The Cancer Genome Atlas N. Comprehensive genomic characterization of head and neck squamous cell carcinomas. Nature. 2015;517(7536):576–82.

100. Smeets SJ, Braakhuis BJM, Abbas S, Snijders PJF, Ylstra B, van de Wiel MA, et al. Genome-wide DNA copy number alterations in head and neck squamous cell carcinomas with or without oncogene-expressing human papillomavirus. Oncogene. 2005;25(17):2558–64.
101. Schlecht NF, Burk RD, Adrien L, Dunne A, Kawachi N, Sarta C, et al. Gene expression profiles in HPV-infected head and neck cancer. J Pathol. 2007;213(3):283–93.
102. Slebos RJC, Yi Y, Ely K, Carter JJ, Evjen A, Zhang X, et al. Gene expression differences associated with human papillomavirus status in head and neck squamous cell carcinoma. Clin Cancer Res. 2006;12(3):701–9.
103. Bernard H-U, Burk RD, Chen Z, van Doorslaer K, Hausen H, de Villiers E-M. Classification of papillomaviruses (PVs) based on 189 PV types and proposal of taxonomic amendments. Virology. 2010;401(1):70–9.
104. Fauquet CM, Mayo MA, Maniloff J, Desselberger U, Ball LA. Virus taxonomy. London: Elsevier; 2005. p. 239–55.
105. de Villiers E-M. Cross-roads in the classification of papillomaviruses. Virology. 2013;445(1–2):2–10.
106. Chen Z, DeSalle R, Schiffman M, Herrero R, Burk RD. Evolutionary dynamics of variant genomes of human papillomavirus types 18, 45, and 97. J Virol. 2008;83(3):1443–55.
107. Chen Z, Schiffman M, Herrero R, DeSalle R, Anastos K, Segondy M, et al. Evolution and taxonomic classification of human papillomavirus 16 (HPV16)-related variant genomes: HPV31, HPV33, HPV35, HPV52, HPV58 and HPV67. PLoS One. 2011;6(5):e20183.
108. Chen Z, Terai M, Fu L, Herrero R, DeSalle R, Burk RD. Diversifying selection in human papillomavirus type 16 lineages based on complete genome analyses. J Virol. 2005;79(11):7014–23.
109. Ho L, Chan SY, Burk RD, Das BC, Fujinaga K, Icenogle JP, et al. The genetic drift of human papillomavirus type 16 is a means of reconstructing prehistoric viral spread and the movement of ancient human populations. J Virol. 1993;67(11):6413–23.
110. Ong CK, Chan SY, Campo MS, Fujinaga K, Mavromara-Nazos P, Labropoulou V, et al. Evolution of human papillomavirus type 18: an ancient phylogenetic root in Africa and intratype diversity reflect coevolution with human ethnic groups. J Virol. 1993;67(11):6424–31.
111. Yamada T, Manos MM, Peto J, Greer CE, Munoz N, Bosch FX, et al. Human papillomavirus type 16 sequence variation in cervical cancers: a worldwide perspective. J Virol. 1997;71(3):2463–72.
112. Chen Z, Schiffman M, Herrero R, DeSalle R, Anastos K, Segondy M, et al. Evolution and taxonomic classification of alphapapillomavirus 7 complete genomes: HPV18, HPV39, HPV45, HPV59, HPV68 and HPV70. PLoS One. 2013;8(8):e72565.
113. Cornet I, Gheit T, Franceschi S, Vignat J, Burk RD, Sylla BS, et al. Human papillomavirus type 16 genetic variants: phylogeny and classification based on E6 and LCR. J Virol. 2012;86(12):6855–61.
114. Durst M, Gissmann L, Ikenberg H, zur Hausen H. A papillomavirus DNA from a cervical carcinoma and its prevalence in cancer biopsy samples from different geographic regions. Proc Natl Acad Sci. 1983;80(12):3812–5.
115. Chen AA, Gheit T, Franceschi S, Tommasino M, Clifford GM. Human papillomavirus 18 genetic variation and cervical cancer risk worldwide. J Virol. 2015;89(20):10680–7.
116. Boshart M, Gissmann L, Ikenberg H, Kleinheinz A, Scheurlen W, zur Hausen H. A new type of papillomavirus DNA, its presence in genital cancer biopsies and in cell lines derived from cervical cancer. EMBO J. 1984;3(5):1151–7.
117. Xi LF, Carter JJ, Galloway DA, Kuypers J, Hughes JP, Lee SK, et al. Acquisition and natural history of human papillomavirus type 16 variant infection among a cohort of female university students. Cancer Epidemiol Biomark Prev. 2002;11(4):343–51.
118. Makowsky R, Lhaki P, Wiener HW, Bhatta MP, Cullen M, Johnson DC, et al. Genomic diversity and phylogenetic relationships of human papillomavirus 16 (HPV16) in Nepal. Infect Genet Evol. 2016;46:7–11.

119. Jackson R, Rosa BA, Lameiras S, Cuninghame S, Bernard J, Floriano WB, et al. Functional variants of human papillomavirus type 16 demonstrate host genome integration and transcriptional alterations corresponding to their unique cancer epidemiology. BMC Genomics. 2016;17(1):851.
120. Berumen J, Ordonez RM, Lazcano E, Salmeron J, Galvan SC, Estrada RA, et al. Asian-American variants of human papillomavirus 16 and risk for cervical cancer: a case-control study. JNCI (Journal of the National Cancer Institute). 2001;93(17):1325–30.
121. Hildesheim A, Schiffman M, Bromley C, Wacholder S, Herrero R, Rodriguez AC, et al. Human papillomavirus type 16 variants and risk of cervical cancer. JNCI (Journal of the National Cancer Institute). 2001;93(4):315–8.
122. Matsumoto K, Yoshikawa H, Nakagava S, Tang X, Yasugi T, Kawana K, et al. Enhanced oncogenicity of human papillomavirus type 16 (HPV16) variants in Japanese population. Cancer Lett. 2000;156(2):159–65.
123. Villa LL, Caballero O, Ferenczy A, Sichero L, Rohan T, Franco EL, et al. Molecular variants of human papillomavirus types 16 and 18 preferentially associated with cervical neoplasia. J Gen Virol. 2000;81(12):2959–68.
124. Xi LF, Critchlow CW, Wheeler CM, Koutsky LA, Galloway DA, Kuypers J, et al. Risk of anal carcinoma in situ in relation to human papillomavirus type 16 variants. Cancer Res. 1998;58(17):3839–44.
125. Xi LF, Koutsky LA, Galloway DA, Kiviat NB, Kuypers J, Hughes JP, et al. Genomic variation of human papillomavirus type 16 and risk for high grade cervical intraepithelial neoplasia. JNCI (Journal of the National Cancer Institute). 1997;89(11):796–802.
126. Xi LF, Koutsky LA, Hildesheim A, Galloway DA, Wheeler CM, Winer RL, et al. Risk for high-grade cervical intraepithelial neoplasia associated with variants of human papillomavirus types 16 and 18. Cancer Epidemiol Biomark Prev. 2007;16(1):4–10.
127. Nicolas-Parraga S, Alemany L, de Sanjose S, Bosch FX, Bravo IG. Differential HPV16 variant distribution in squamous cell carcinoma, adenocarcinoma and adenosquamous cell carcinoma. Int J Cancer. 2017;140(9):2092–100.
128. Mammas IN, Spandidos DA, Sourvinos G. Genomic diversity of human papillomaviruses (HPV) and clinical implications: an overview in adulthood and childhood. Infect Genet Evol. 2014;21:220–6.
129. Stoppler MC, Ching K, Stoppler H, Clancy K, Schlegel R, Icenogle J. Natural variants of the human papillomavirus type 16 E6 protein differ in their abilities to alter keratinocyte differentiation and to induce p53 degradation. J Virol. 1996;70(10):6987–93.
130. Hang D, Gao L, Sun M, Liu Y, Ke Y. Functional effects of sequence variations in the E6 and E2 genes of human papillomavirus 16 European and Asian variants. J Med Virol. 2014;86(4):618–26.
131. De la Cruz-Hernandez E, Garcia-Carranca A, Mohar-Betancourt A, Duenas-Gonzales A, Contreras-Paredes A, Perez-Cardenas E, et al. Differential splicing of E6 within human papillomavirus type 18 variants and functional consequences. J Gen Virol. 2005;86(9):2459–68.
132. Vazquez-Vega S, Sanchez-Suarez LP, Andrade-Cruz R, Castellanos-Juarez E, Contreras-Paredes A, Lizano-Soberon M, et al. Regulation of p14 ARF expression by HPV-18 E6 variants. J Med Virol. 2013;85(7):1215–21.
133. Chakrabarti O, Veeraraghavalu K, Tergaonkar V, Liu Y, Androphy EJ, Stanley MA, et al. Human papillomavirus type 16 E6 amino acid 83 variants enhance E6-mediated MAPK signaling and differentially regulate tumorigenesis by notch signaling and oncogenic Ras. J Virol. 2004;78(11):5934–45.
134. Contreras-Paredes A, De la Cruz-Hernández E, Martínez-Ramírez I, Dueñas-González A, Lizano M. E6 variants of human papillomavirus 18 differentially modulate the protein kinase B/phosphatidylinositol 3-kinase (akt/PI3K) signaling pathway. Virology. 2009;383(1):78–85.
135. Kämmer C, Tommasino M, Syrjänen S, Delius H, Hebling U, Warthorst U, et al. Variants of the long control region and the E6 oncogene in European human papillomavirus type 16 isolates: implications for cervical disease. Br J Cancer. 2002;86(2):269–73.

136. May M, Dong XP, Beyer-Finkler E, Stubenrauch F, Fuchs PG, Pfister H. The E6/E7 promoter of extrachromosomal HPV16 DNA in cervical cancers escapes from cellular repression by mutation of target sequences for YY1. EMBO J. 1994;13(6):1460–6.
137. Park JS, Hwang ES, Lee CJ, Kim CJ, Rha JG, Kim SJ, et al. Mutational and functional analysis of HPV-16 URR derived from Korean cervical neoplasia. Gynecol Oncol. 1999;74(1):23–9.
138. Rose B, Tattersall M, Thompson C, Steger G, Pfister H, Cossart Y, et al. Point mutations in SP1 motifs in the upstream regulatory region of human papillomavirus type 18 isolates from cervical cancers increase promoter activity. J Gen Virol. 1998;79(7):1659–63.
139. Veress G, Szarka K, Gergely L, Pfister H, Dong XP. Functional significance of sequence variation in the E2 gene and the long control region of human papillomavirus type 16. J Gen Virol. 1999;80(4):1035–43.
140. Hubert WG. Variant upstream regulatory region sequences differentially regulate human papillomavirus type 16 DNA replication throughout the viral life cycle. J Virol. 2005;79(10):5914–22.
141. Chopjitt P, Pientong C, Bumrungthai S, Kongyingyoes B, Ekalaksananan T. Activities of E6 protein of human papillomavirus 16 Asian variant on miR-21 up-regulation and expression of human immune response genes. Asian Pac J Cancer Prev. 2015;16(9):3961–8.
142. Du J, Nordfors C, Näsman A, Sobkowiak M, Romanitan M, Dalianis T, et al. Human papillomavirus (HPV) 16 E6 variants in tonsillar cancer in comparison to those in cervical cancer in Stockholm, Sweden. PLoS One. 2012;7(4):e36239.
143. Boscolo-Rizzo P, Da Mosto MC, Fuson R, Frayle-Salamanca H, Trevisan R, Del Mistro A. HPV-16 E6 L83V variant in squamous cell carcinomas of the upper aerodigestive tract. J Cancer Res Clin Oncol. 2008;135(4):559–66.
144. Markowitz Lauri E, Sternberg M, Dunne Eileen F, McQuillan G, Unger Elizabeth R. Seroprevalence of human papillomavirus types 6, 11, 16, and 18 in the United States: national health and nutrition examination survey 2003–2004. J Infect Dis. 2009;200(7):1059–67.
145. Liu G, Markowitz LE, Hariri S, Panicker G, Unger ER. Seroprevalence of 9 human papillomavirus types in the United States, 2005–2006. J Infect Dis. 2016;213(2):191–8.
146. Zhang Y, Waterboer T, Pawlita M, Sugar E, Minkoff H, Cranston RD, et al. Human papillomavirus (HPV) 16 E6 seropositivity is elevated in subjects with oral HPV16 infection. Cancer Epidemiol. 2016;43:30–4.
147. Gillison ML, Broutian T, Pickard RKL, Tong Z-Y, Xiao W, Kahle L, et al. Prevalence of oral HPV infection in the United States, 2009–2010. JAMA. 2012;307(7):693.
148. Chaturvedi AK, Graubard BI, Broutian T, Pickard RKL, Tong Z, Xiao W, et al. NHANES 2009–2012 findings: association of sexual behaviors with higher prevalence of oral oncogenic human papillomavirus infections in U.S. men. Cancer Res. 2015;75(12):2468–77.
149. Safaeian M, Porras C, Schiffman M, Rodriguez AC, Wacholder S, Gonzalez P, et al. Epidemiological study of anti-HPV16/18 seropositivity and subsequent risk of HPV16 and -18 infections. JNCI (Journal of the National Cancer Institute). 2010;102(21):1653–62.
150. D'Souza G, Kluz N, Wentz A, Youngfellow R, Griffioen A, Stammer E, et al. Oral human papillomavirus (HPV) infection among unvaccinated high-risk young adults. Cancers. 2014;6(3):1691–704.
151. Kreimer AR, Pierce Campbell CM, Lin H-Y, Fulp W, Papenfuss MR, Abrahamsen M, et al. Incidence and clearance of oral human papillomavirus infection in men: the HIM cohort study. Lancet. 2013;382(9895):877–87.
152. Ferlay JSI, Ervik M, Dikshit R, Eser S, Mathers C, Rebelo M, Parkin DM, Forman D, Bray, F. GLOBOCAN 2012 v1.0, cancer incidence and mortality worldwide: IARC CancerBase No. 11 [Internet]. Lyon: International Agency for Research on Cancer. 2013.: Available from: http://globocan.iarc.fr/. Accessed on day/month/year.
153. Howlader NNA, Krapcho M, Miller D, Bishop K, Altekruse SF, Kosary CL, Yu M, Ruhl J, Tatalovich Z, Mariotto A, Lewis DR, Chen HS, Feuer EJ, Cronin KA, editors. SEER cancer statistics review, 1975–2013. National Cancer Institute Bethesda, MD, https://seer.cancer.

gov/csr/1975_2013/. based on November 2015 SEER data submission, posted to the SEER web site. 2016; April.
154. Jemal A, Simard EP, Dorell C, Noone AM, Markowitz LE, Kohler B, et al. Annual report to the nation on the status of cancer, 1975–2009, featuring the burden and trends in human papillomavirus(HPV)-associated cancers and HPV vaccination coverage levels. J Natl Cancer Inst. 2013;105(3):175–201.
155. Gillison ML, Chaturvedi AK, Anderson WF, Fakhry C. Epidemiology of human papillomavirus-positive head and neck squamous cell carcinoma. J Clin Oncol (Official Journal of the American Society of Clinical Oncology). 2015;33(29):3235–42.
156. Maritz GS, Mutemwa M. Tobacco smoking: patterns, health consequences for adults, and the long-term health of the offspring. Glob J Health Sci. 2012;4(4):62–75.
157. Näsman A, Attner P, Hammarstedt L, Du J, Eriksson M, Giraud G, et al. Incidence of human papillomavirus (HPV) positive tonsillar carcinoma in Stockholm, Sweden: an epidemic of viral-induced carcinoma? Int J Cancer. 2009;125(2):362–6.
158. Roberts JC, Li G, Reitzel LR, Wei Q, Sturgis EM. No evidence of sex-related survival disparities among head and neck cancer patients receiving similar multidisciplinary care: a matched-pair analysis. Clin Cancer Res. 2010;16(20):5019–27.
159. Ragin CCR, Taioli E. Survival of squamous cell carcinoma of the head and neck in relation to human papillomavirus infection: review and meta-analysis. Int J Cancer. 2007;121(8):1813–20.
160. Gillison ML, D'Souza G, Westra W, Sugar E, Xiao W, Begum S, et al. Distinct risk factor profiles for human papillomavirus type 16-positive and human papillomavirus type 16-negative head and neck cancers. JNCI J Natl Cancer Inst. 2008;100(6):407–20.
161. Mbulawa ZZA, Johnson LF, Marais DJ, Coetzee D, Williamson A-L. Risk factors for oral human papillomavirus in heterosexual couples in an African setting. J Infect. 2014;68(2):185–9.
162. Vogt SL, Gravitt PE, Martinson NA, Hoffmann J, D'Souza G. Concordant oral-genital HPV infection in South Africa couples: evidence for transmission. Front Oncol. 2013;3:303.
163. Beachler DC, Sugar EA, Margolick JB, Weber KM, Strickler HD, Wiley DJ, et al. Risk factors for acquisition and clearance of oral human papillomavirus infection among HIV-infected and HIV-uninfected adults. Am J Epidemiol. 2014;181(1):40–53.
164. Beachler DC, Weber KM, Margolick JB, Strickler HD, Cranston RD, Burk RD, et al. Risk factors for oral HPV infection among a high prevalence population of HIV-positive and at-risk HIV-negative adults. Cancer Epidemiol Biomark Prev. 2011;21(1):122–33.
165. Steinau M, Saraiya M, Goodman MT, Peters ES, Watson M, Cleveland JL, et al. Human papillomavirus prevalence in oropharyngeal cancer before vaccine introduction, United States. Emerg Infect Dis. 2014;20(5):822–8.
166. Gillison ML, Chaturvedi AK, Lowy DR. HPV prophylactic vaccines and the potential prevention of noncervical cancers in both men and women. Cancer. 2008;113(10 Suppl):3036–46.
167. Braakhuis BJM, Snijders PJF, Keune WJH, Meijer CJLM, Ruijter-Schippers HJ, Leemans CR, et al. Genetic patterns in head and neck cancers that contain or lack transcriptionally active human papillomavirus. JNCI (Journal of the Natl Cancer Institute). 2004;96(13):998–1006.
168. Gillison ML, Zhang Q, Jordan R, Xiao W, Westra WH, Trotti A, et al. Tobacco smoking and increased risk of death and progression for patients with p16-positive and p16-negative oropharyngeal cancer. J Clin Oncol. 2012;30(17):2102–11.
169. Chaturvedi AK, D'Souza G, Gillison ML, Katki HA. Burden of HPV-positive oropharynx cancers among ever and never smokers in the U.S. population. Oral Oncol. 2016;60:61–7.
170. Jung AC, Briolat J, Millon R, de Reynies A, Rickman D, Thomas E, et al. Biological and clinical relevance of transcriptionally active human papillomavirus (HPV) infection in oropharynx squamous cell carcinoma. Int J Cancer. 2010;126(8):1882–94.
171. Deng Z, Hasegawa M, Kiyuna A, Matayoshi S, Uehara T, Agena S, et al. Viral load, physical status, and E6/E7 mRNA expression of human papillomavirus in head and neck squamous cell carcinoma. Head Neck. 2013;35(6):800–8.

172. de Souza DLB, de Camargo Cancela M, Pérez MMB, Curado M-P. Trends in the incidence of oral cavity and oropharyngeal cancers in Spain. Head Neck. 2011;34(5):649–54.
173. Forte T, Niu J, Lockwood GA, Bryant HE. Incidence trends in head and neck cancers and human papillomavirus (HPV)-associated oropharyngeal cancer in Canada, 1992–2009. Cancer Causes Control. 2012;23(8):1343–8.
174. Golas SM. Trends in palatine tonsillar cancer incidence and mortality rates in the United States. Community Dent Oral Epidemiol. 2007;35(2):98–108.
175. McGorray SP, Guo Y, Logan H. Trends in incidence of oral and pharyngeal carcinoma in Florida: 1981–2008. J Public Health Dent. 2011;72(1):68–74.
176. Edelstein ZR, Schwartz SM, Hawes S, Hughes JP, Feng Q, Stern ME, et al. Rates and determinants of oral human papillomavirus infection in young men. Sex Transm Dis. 2012;39(11):860–7.
177. Attner P, Du J, Näsman A, Hammarstedt L. Ramqvist Tr, Lindholm J, et al. The role of human papillomavirus in the increased incidence of base of tongue cancer. Int J Cancer. 2010;126(12):2879–84.
178. Castellsagué X, Mena M, Alemany L. Epidemiology of HPV-positive tumors in Europe and in the world. In: HPV infection in head and neck cancer. Cham: Springer International Publishing; 2016. p. 27–35.
179. Chaturvedi AK, Engels EA, Pfeiffer RM, Hernandez BY, Xiao W, Kim E, et al. Human papillomavirus and rising oropharyngeal cancer incidence in the United States. J Clin Oncol. 2011;29(32):4294–301.
180. D'Souza G, Zhang HH, D'Souza WD, Meyer RR, Gillison ML. Moderate predictive value of demographic and behavioral characteristics for a diagnosis of HPV16-positive and HPV16-negative head and neck cancer. Oral Oncol. 2010;46(2):100–4.
181. Ernster JA, Sciotto CG, O'Brien MM, Finch JL, Robinson LJ, Willson T, et al. Rising incidence of oropharyngeal cancer and the role of oncogenic human papilloma virus. Laryngoscope. 2007;117(12):2115–28.
182. Fakhry C, Westra WH, Li S, Cmelak A, Ridge JA, Pinto H, et al. Improved survival of patients with human papillomavirus-positive head and neck squamous cell carcinoma in a prospective clinical trial. JNCI (Journal of the National Cancer Institute). 2008;100(4):261–9.
183. Kingma DW, Allen RA, Moore W, Caughron SK, Melby M, Gillies EM, et al. HPV genotype distribution in oral and oropharyngeal squamous cell carcinoma using seven in vitro amplification assays. Anticancer Res. 2010;30(12):5099–104.
184. Mehanna H, Franklin N, Compton N, Robinson M, Powell N, Biswas-Baldwin N, et al. Geographic variation in human papillomavirus-related oropharyngeal cancer: data from 4 multinational randomized trials. Head Neck. 2016;38(Suppl 1):E1863–9.
185. Mehanna H, Beech T, Nicholson T, El-Hariry I, McConkey C, Paleri V, et al. Prevalence of human papillomavirus in oropharyngeal and nonoropharyngeal head and neck cancer-systematic review and meta-analysis of trends by time and region. Head Neck. 2012;35(5):747–55.
186. Rodrigo JP, Heideman DA, Garcia-Pedrero JM, Fresno MF, Brakenhoff RH, Diaz Molina JP, et al. Time trends in the prevalence of HPV in oropharyngeal squamous cell carcinomas in northern Spain (1990–2009). Int J Cancer. 2014;134(2):487–92.
187. Brown LM, Check DP, Devesa SS. Oral cavity and pharynx cancer incidence trends by subsite in the United States: changing gender patterns. Journal of Oncology. 2012;2012:1–10.
188. Cole L, Polfus L, Peters ES. Examining the incidence of human papillomavirus-associated head and neck cancers by race and ethnicity in the U.S., 1995–2005. PLoS One. 2012;7(3):e32657.
189. Brandsma JL, Abramson AL. Association of papillomavirus with cancers of the head and neck. Arch Otolaryngol Head Neck Surg. 1989;115(5):621–5.
190. Chen R, Sehr P, Waterboer T, Leivo I, Pawlita M, Vaheri A, et al. Presence of DNA of human papillomavirus 16 but no other types in tumor-free tonsillar tissue. J Clin Microbiol. 2005;43(3):1408–10.

191. do Sacramento PR, Babeto E, Colombo J, Cabral Ruback MJ, Bonilha JL, Fernandes AM, et al. The prevalence of human papillomavirus in the oropharynx in healthy individuals in a Brazilian population. J Med Virol. 2006;78(5):614–8.
192. Duray A, Descamps G, Bettonville M, Sirtaine N, Ernoux-Neufcoeur P, Guenin S, et al. High prevalence of high-risk human papillomavirus in palatine tonsils from healthy children and adults. Otolaryngol Head Neck Surg. 2011;145(2):230–5.
193. Ernster JA, Sciotto CG, O'Brien MM, Robinson LJ, Willson T. Prevalence of oncogenic human papillomavirus 16 and 18 in the palatine tonsils of the general adult population. Arch Otolaryngol Head Neck Surg. 2009;135(6):554.
194. Fukushima K, Ogura H, Watanabe S, Yabe Y, Masuda Y. Human papillomavirus type 16 DNA detected by the polymerase chain reaction in non-cancer tissues of the head and neck. Eur Arch Otorhinolaryngol. 1994;251(2):109–12.
195. Mammas IN, Sourvinos G, Michael C, Spandidos DA. Human papilloma virus in hyperplastic tonsillar and adenoid tissues in children. Pediatr Infect Dis J. 2006;25(12):1158–62.
196. Sisk J, Schweinfurth JM, Wang XT, Chong K. Presence of human papillomavirus DNA in tonsillectomy specimens. Laryngoscope. 2006;116(8):1372–4.
197. Snijders PJF, Cromme FV, Van Brule AJCD, Schrijnemakers HFJ, Snow GB, Meijer CJLM, et al. Prevalence and expression of human papillomavirus in tonsillar carcinomas, indicating a possible viral etiology. Int J Cancer. 1992;51(6):845–50.
198. Strome SE, Savva A, Brissett AE, Gostout BS, Lewis J, Clayton AC, et al. Squamous cell carcinoma of the tonsils: a molecular analysis of HPV associations. Clin Cancer Res (An Official Journal of the American Association for Cancer Research). 2002;8(4):1093–100.
199. Tominaga S, Fukushima K, Nishizaki K, Watanabe S, Masuda Y, Ogura H. Presence of human papillomavirus type 6f in tonsillar condyloma acuminatum and clinically normal tonsillar mucosa. Jpn J Clin Oncol. 1996;26(6):393–7.
200. Watanabe S, Ogura H, Fukushima K, Yabe Y. Comparison of Virapap filter hybridization with polymerase chain reaction and Southern blot hybridization methods for detection of human papillomavirus in tonsillar and pharyngeal cancers. Eur Arch Otorhinolaryngol. 1993;250(2):115–9.
201. Fouret P, Martin F, Flahault A, Saint-Guily JL. Human papillomavirus infection in the malignant and premalignant head and neck epithelium. Diagn Mol Pathol. 1995;4(2):122–7.
202. Frisch M, Biggar RJ. Aetiological parallel between tonsillar and anogenital squamous-cell carcinomas. Lancet. 1999;354(9188):1442–3.
203. Rose Ragin CC, Taioli E. Second primary head and neck tumor risk in patients with cervical cancer—SEER data analysis. Head Neck. 2007;30(1):58–66.
204. Sikora AG, Morris LG, Sturgis EM. Bidirectional association of anogenital and oral cavity/pharyngeal carcinomas in men. Arch Otolaryngol Head Neck Surg. 2009;135(4):402.
205. Dahlstrom KR, Burchell AN, Ramanakumar AV, Rodrigues A, Tellier PP, Hanley J, et al. Sexual transmission of oral human papillomavirus infection among men. Cancer Epidemiol Biomark Prev. 2014;23(12):2959–64.
206. Crawford R, Grignon A-L, Kitson S, Winder DM, Ball SLR, Vaughan K, et al. High prevalence of HPV in non-cervical sites of women with abnormal cervical cytology. BMC Cancer. 2011;11(1):473.
207. Dost F, Ford PJ, Farah CS. Heightened risk of second primary carcinoma of the head and neck following cervical neoplasia. Head Neck. 2013;36(8):1132–7.
208. Hemminki K, Dong C, Frisch M. Tonsillar and other upper aerodigestive tract cancers among cervical cancer patients and their husbands. Eur J Cancer Prev. 2000;9(6):433–7.
209. D'Souza G, Kreimer AR, Viscidi R, Pawlita M, Fakhry C, Koch WM, et al. Case–control study of human papillomavirus and oropharyngeal cancer. N Engl J Med. 2007;356(19):1944–56.
210. D'Souza G, Agrawal Y, Halpern J, Bodison S, Gillison Maura L. Oral sexual behaviors associated with prevalent oral human papillomavirus infection. J Infect Dis. 2009;199(9):1263–9.
211. Pickard RKL, Xiao W, Broutian TR, He X, Gillison ML. The prevalence and incidence of oral human papillomavirus infection among young men and women, aged 18–30 years. Sex Transm Dis. 2012;39(7):559–66.

212. Dahlstrom KR, Li G, Tortolero-Luna G, Wei Q, Sturgis EM. Differences in history of sexual behavior between patients with oropharyngeal squamous cell carcinoma and patients with squamous cell carcinoma at other head and neck sites. Head Neck. 2010;33(6):847–55.
213. Heck JE, Berthiller J, Vaccarella S, Winn DM, Smith EM, Shan'gina O, et al. Sexual behaviours and the risk of head and neck cancers: a pooled analysis in the international head and neck cancer epidemiology (INHANCE) consortium. Int J Epidemiol. 2009;39(1):166–81.
214. Lassen P, Eriksen JG, Hamilton-Dutoit S, Tramm T, Alsner J, Overgaard J. Effect of HPV-associated p16INK4A expression on response to radiotherapy and survival in squamous cell carcinoma of the head and neck. J Clin Oncol. 2009;27(12):1992–8.
215. Lassen P, Eriksen JG, Hamilton-Dutoit S, Tramm T, Alsner J, Overgaard J. HPV-associated p16-expression and response to hypoxic modification of radiotherapy in head and neck cancer. Radiother Oncol. 2010;94(1):30–5.
216. Bossi P, Orlandi E, Miceli R, Perrone F, Guzzo M, Mariani L, et al. Treatment-related outcome of oropharyngeal cancer patients differentiated by HPV dictated risk profile: a tertiary cancer centre series analysis. Ann Oncol. 2014;25(3):694–9.
217. Ang KK, Harris J, Wheeler R, Weber R, Rosenthal DI, Nguyen-Tân PF, et al. Human papillomavirus and survival of patients with oropharyngeal cancer. N Engl J Med. 2010;363(1):24–35.
218. Granata R, Miceli R, Orlandi E, Perrone F, Cortelazzi B, Franceschini M, et al. Tumor stage, human papillomavirus and smoking status affect the survival of patients with oropharyngeal cancer: an Italian validation study. Ann Oncol. 2011;23(7):1832–7.
219. Urban D, Corry J, Rischin D. What is the best treatment for patients with human papillomavirus-positive and -negative oropharyngeal cancer? Cancer. 2014;120(10):1462–70.
220. Attner P, Nasman A, Du J, Hammarstedt L, Ramqvist T, Lindholm J, et al. Survival in patients with human papillomavirus positive tonsillar cancer in relation to treatment. Int J Cancer. 2012;131(5):1124–30.
221. Marur S, Li S, Cmelak AJ, Gillison ML, Zhao WJ, Ferris RL, et al. E1308: Phase II trial of induction chemotherapy followed by reduced-dose radiation and weekly Cetuximab in patients with HPV-associated resectable squamous cell carcinoma of the oropharynx- ECOG-ACRIN Cancer Research Group. J Clin Oncology Off J Am Soc Clin Oncol. 2016:JCO2016683300.
222. Chen AM, Felix C, Wang PC, Hsu S, Basehart V, Garst J, et al. Reduced-dose radiotherapy for human papillomavirus-associated squamous-cell carcinoma of the oropharynx: a single-arm, phase 2 study. Lancet Oncol. 2017;18(6):803–11.
223. Settle K, Posner MR, Schumaker LM, Tan M, Suntharalingam M, Goloubeva O, et al. Racial survival disparity in head and neck cancer results from low prevalence of human papillomavirus infection in black oropharyngeal cancer patients. Cancer Prev Res. 2009;2(9):776–81.
224. Chernock RD, Zhang Q, El-Mofty SK, Thorstad WL, Lewis JS. Human papillomavirus–related squamous cell carcinoma of the oropharynx. Arch Otolaryngol Head Neck Surg. 2011;137(2):163.
225. Worsham MJ, Stephen JK, Mahan M, Chen KM, Schweitzer V, Havard S, et al. HPV improves survival for African Americans with throat cancer. Sci Newsline Med Healthcare. 2012.
226. Worsham MJ, Stephen JK, Chen KM, Mahan M, Schweitzer V, Havard S, et al. Improved survival with HPV among African Americans with oropharyngeal cancer. Clin Cancer Res. 2013;19(9):2486–92.
227. Ragin C, Liu JC, Jones G, Shoyele O, Sowunmi B, Kennett R, et al. Prevalence of HPV infection in racial-ethnic subgroups of head and neck cancer patients. Carcinogenesis. 2016;38(2):218–29.
228. Isayeva T, Xu J, Dai Q, Whitley AC, Bonner J, Nabell L, et al. African Americans with oropharyngeal carcinoma have significantly poorer outcomes despite similar rates of human papillomavirus-mediated carcinogenesis. Hum Pathol. 2014;45(2):310–9.
229. Liu JC, Parajuli S, Blackman E, Gibbs D, Ellis A, Hull A, et al. High prevalence of discordant human papillomavirus and p16 oropharyngeal squamous cell carcinomas in an African American cohort. Head Neck. 2016;38(Suppl 1):E867–72.

230. Isayeva T, Xu J, Ragin C, Dai Q, Cooper T, Carroll W, et al. The protective effect of p16(INK4a) in oral cavity carcinomas: p16(Ink4A) dampens tumor invasion-integrated analysis of expression and kinomics pathways. Mod Pathol (An Official Journal of the United States and Canadian Academy of Pathology, Inc.). 2015;28(5):631–53.
231. Combes JD, Franceschi S. Role of human papillomavirus in non-oropharyngeal head and neck cancers. Oral Oncol. 2014;50(5):370–9.
232. Jayaprakash V, Reid M, Hatton E, Merzianu M, Rigual N, Marshall J, et al. Human papillomavirus types 16 and 18 in epithelial dysplasia of oral cavity and oropharynx: a meta-analysis, 1985–2010. Oral Oncol. 2011;47(11):1048–54.
233. Syrjänen S, Lodi G, von Bültzingslöwen I, Aliko A, Arduino P, Campisi G, et al. Human papillomaviruses in oral carcinoma and oral potentially malignant disorders: a systematic review. Oral Dis. 2011;17:58–72.
234. Hobbs CGL, Sterne JAC, Bailey M, Heyderman RS, Birchall MA, Thomas SJ. Human papillomavirus and head and neck cancer: a systematic review and meta-analysis. Clin Otolaryngol. 2006;31(4):259–66.
235. Taberna M, Resteghini C, Swanson B, Pickard RK, Jiang B, Xiao W, et al. Low etiologic fraction for human papillomavirus in larynx squamous cell carcinoma. Oral Oncol. 2016;61:55–61.
236. Brown LM. Epidemiology of alcohol-associated cancers. Alcohol. 2005;35(3):161–8.
237. LaVallee RA, Yi H-Y. Apparent per capita alcohol consumption: national, state, and regional trends, 1977–2009. Surveillance Report #92. National Institute on Alcohol Abuse and Alcoholism Division of Epidemiology and Prevention Research Alcohol Epidemiologic Data System. 2012.
238. Services. UDoHaH. Preventing tobacco use among young people: a report of the surgeon general. . Atlanta, Ga, S Dept of Health and Human Services, Public Health Service, Centers for Disease Control and Prevention, National Center for Chronic Disease Prevention and Health Promotion, Office on Smoking on Health. 1994.
239. Sturgis EM, Cinciripini PM. Trends in head and neck cancer incidence in relation to smoking prevalence. Cancer. 2007;110(7):1429–35.
240. Services ALARaP. Trends in tobacco use ref type: electronic citation. 2012.
241. Müller S, Pan Y, Li R, Chi AC. Changing trends in oral squamous cell carcinoma with particular reference to young patients: 1971–2006. The Emory university experience. Head Neck Pathol. 2008;2(2):60–6.
242. Schantz SP, Yu G-P. Head neck cancer incidence trends in young Americans, 1973–1997, with a special analysis for tongue cancer. Arch Otolaryngol Head Neck Surg. 2002;128(3):268.
243. Shiboski CH, Schmidt BL, Jordan RCK. Tongue and tonsil carcinoma. Cancer. 2005;103(9):1843–9.
244. Saba NF, Goodman M, Ward K, Flowers C, Ramalingam S, Owonikoko T, et al. Gender and ethnic disparities in incidence and survival of squamous cell carcinoma of the oral tongue, base of tongue, and tonsils: a surveillance, epidemiology and end results program-based analysis. Oncology. 2011;81(1):12–20.
245. Elango KJ, Suresh A, Erode EM, Subhadradevi L, Ravindran HK, Iyer SK, et al. Role of human papilloma virus in oral tongue squamous cell carcinoma. Asian Pac J Cancer Prev. 2011;12(4):889–96.
246. Fakhry C, Westra WH, Wang SJ, van Zante A, Zhang Y, Rettig E, et al. The prognostic role of sex, race, and human papillomavirus in oropharyngeal and nonoropharyngeal head and neck squamous cell cancer. Cancer. 2017;123(9):1566–75.
247. Chen X, Gao L, Sturgis EM, Liang Z, Zhu Y, Xia X, et al. HPV16 DNA and integration in normal and malignant epithelium: Implications for the etiology of laryngeal squamous cell carcinoma. Ann Oncol. 2017;28(5):1105–10.
248. Duray A, Descamps G, Decaestecker C, Remmelink M, Sirtaine N, Lechien J, et al. Human papillomavirus DNA strongly correlates with a poorer prognosis in oral cavity carcinoma. Laryngoscope. 2012;122(7):1558–65.

249. Isayeva T, Li Y, Maswahu D, Brandwein-Gensler M. Human papillomavirus in non-oropharyngeal head and neck cancers: a systematic literature review. Head Neck Pathol. 2012;6(S1):104–20.
250. Lingen MW, Xiao W, Schmitt A, Jiang B, Pickard R, Kreinbrink P, et al. Low etiologic fraction for high-risk human papillomavirus in oral cavity squamous cell carcinomas. Oral Oncol. 2013;49(1):1–8.
251. Li X, Gao L, Li H, Gao J, Yang Y, Zhou F, et al. Human papillomavirus infection and laryngeal cancer risk: a systematic review and meta-analysis. J Infect Dis. 2012;207(3):479–88.
252. Stephen JK, Chen KM, Shah V, Havard S, Lu M, Schweitzer VP, et al. Human papillomavirus outcomes in an access-to-care laryngeal cancer Cohort. Otolaryngol Head Neck Surg. 2012;146(5):730–8.
253. Wilson DD, Rahimi AS, Saylor DK, Stelow EB, Jameson MJ, Shonka DC, et al. p16 not a prognostic marker for hypopharyngeal squamous cell carcinoma. Arch Otolaryngol Head Neck Surg. 2012;138(6):556.

Chapter 21
Treatment Paradigms in HPV-Associated SCCHN

Christien A. Kluwe and Anthony J. Cmelak

Abstract The primary risk factor for development of HPV-related (HPV+) oropharyngeal squamous cell carcinoma (OPSCC) is oral HPV infection, most frequently by HPV type 16. The most clinically relevant features of HPV+ disease is the significantly increased incidence worldwide, improved response to chemotherapy and radiation and overall prognosis, as well as distinct patterns of failure in comparison to their HPVnegative brethren. As a result, the AJCC Cancer Staging Manual 8th edition has introduced significant changes from the seventh edition, recognizing the prognostic power of newly validated pathologic features of some primary tumors and of cervical lymph node metastases and differentiating high-risk human papilloma virus (HR-HPV)-associated oropharyngeal cancer from OPSCC with other causes. In addition, considerable thought has been directed toward reevaluating historic treatment paradigms. Recent efforts pursue treatment deintensification to reduce long-term toxicities in this relatively younger patient cohort. This also includes minimally invasive surgery techniques and the substitution or elimination of standard cytotoxic chemotherapy with biologic agents and immunotherapies. Overall, HPV-related cancers are now recognized as a distinct entity. This has resulted in a reexamination of the core principles of head and neck cancer treatment developed decades ago, with the goals of obtaining high cure rates simultaneously with meanigful reductions in long-term toxicity and improved quality of life.

Keywords Human papillomavirus · Deintensification · Transoral · De-escalation · Clinical trials · Biologics · Surveillance · Oropharynx

---

C. A. Kluwe · A. J. Cmelak (✉)
Vanderbilt-Ingram Cancer Center, Department of Radiation Oncology, Nashville, TN, USA
e-mail: anthony.cmelak@vanderbilt.edu

B. Burtness, E. A. Golemis (eds.), *Molecular Determinants of Head and Neck Cancer*, Current Cancer Research, https://doi.org/10.1007/978-3-319-78762-6_21

21.1 Introduction

The World Health Organization (WHO) has established that human papillomavirus (HPV) is an increasingly important cause of oropharyngeal squamous cell carcinoma (OPSCC) worldwide [1]. This understanding, as well as differences in prognosis, treatment response, and failure patterns, has led to a paradigm shift in our management of patients diagnosed with HPV-related (HPV+) OPSCC. The large increase in HPV+ HNSCC, associated with an improved response to treatment and overall survival (OS), has brought into question the validity of applying current staging systems to these patients, as well as historical results from clinical trials that enrolled individuals with heterogeneous tumor sites in an era when the importance of HPV was not yet understood. Thus, a number of new efforts are underway to stratify patients based on HPV status, using multiple strategies to ameliorate treatment-related toxicities for HPV+ disease.

Traditionally arising in a population with significant alcohol and tobacco exposure, squamous cell carcinoma of the head and neck (HNSCC) has been notoriously difficult to cure. Characteristically, 60% of patients have locally advanced disease at the time of diagnosis and current therapies with concurrent cisplatin yield 3-year progression-free survival (PFS) rates of 30–50% and OS of 60%, respectively [2]. Unfortunately, survival for patients with high-risk HPV-negative HNSCC has improved only marginally within the last 20 years due to the incorporation of concurrent cisplatin in curative-intent paradigms. Although locoregional control and OS are improved with concurrent cisplatin over radiation therapy (RT) alone, a meta-analysis indicated disappointing local and distant failure rates of 50% and 15%, respectively, and an absolute survival benefit of only 6.5% with the addition of chemotherapy [3]. This has led to the development of intensified treatment regimens combining cytotoxic agents, both prior to and concurrent with high-dose radiation, accelerated or hyperfractionated radiation regimens, trimodality therapies, extensive prophylactic irradiation of lymph nodes, and the addition/substitution of monoclonal antibodies (MoAbs) targeting the epidermal growth factor receptor (EGFR) [4–6]. Severe toxicities have been noted and quantified for some of these aggressive treatments, but have largely attracted less attention in treatment design than has tumor control to improve survival. Reducing such toxicities for HPV+ patients would yield a significant health benefit.

Epidemiologic data show increasing rates of OPSCC in the United States and around the world over the past several decades (see also Chap. 20) [7–9]. Simultaneously, traditional alcohol and smoking-related HNSCC has declined by 50%. Molecular analyses from patients over this period have shown a key driver to be infection with human papillomavirus (HPV) – more commonly associated with anogenital cancers but recently recognized as an important cause of OPSCC [10, 11]. A recent meta-analysis of 2099 OPSCC cases from the US literature showed an increase in the prevalence of HPV-OPSCC from 20.9% before 1990 to 51.4% between 1900 and 1999 with further increase to 65.4% for 2000 to present [12]. Other series suggest 72% or more of OPSCC cases occurring in the United States

Table 21.1 Population level incidence of HPV-associated OPSCC was shown to increase by 225% from 1988 to 2004

Prevalence of HPV in OPSCC in the United States			
Time period	Number of studies	Total number of OPSCCs analyzed	Mean HPV(+) OPSCC prevalence (95% CI)
Pre-1990	5	82	20.9 (11.8, 37.0)
1990–1999	15	684	51.4 (45.4, 58.2)
2000–present	18	1333	65.4 (60.5, 70.7)

From: Stein et al. [87]

after the year 2000 are attributable to HPV infection [13, 14]. As the vast majority of oral HPV infections are sexually acquired, sexual behavior has now been established as a risk factor for OPSCC. Men have a significantly higher prevalence of overall HPV infection (10.1% vs. 3.6%) and oral HPV type 16 infection (1.6% vs. 0.3%) than women, mimicking the gender inequality of OPSCC incidence (Table 21.1) [15]. The reasons for these differences are multifactorial; men have a higher number of sexual partners than women, have a higher per-partner increase in risk of high-risk oral HPV infection, and are less likely to seroconvert after genital HPV infection to provide protection against subsequent oral HPV infection [16]. In the United States, a bimodal distribution of HPV infection exists with peaks between 25 and 30 years and the second at 55 to 60 years [17]. Uncertainty exists regarding which of the two peaks plays the greater role in HPV-related oncogenesis.

21.2 HPV-Associated Oropharynx Cancer Biology

The primary risk factor for development of HPV+ OPSCC is oral HPV infection, most frequently by HPV type 16 [18]. Human papillomaviruses immortalize keratinocytes via the actions of two dominantly acting oncoproteins, E6 and E7, on host gene products. E6 promotes degradation of the tumor suppressor p53 in an ATP-dependent fashion operating through ubiquitination [19]. Similarly, the E7 oncoprotein segregates the tumor suppressor Rb, and efficiency of E7-Rb binding is correlated with oncogenicity of differing HPV types [20]. For transformation by the high-risk HPV type 16, the HPV type responsible for more than 85% of HPV-driven oropharynx cancers in the United States, both E6 and E7 are required [21, 22]. Head and neck tumors are associated with inflammation and markers of immune exhaustion. Profiling of the tumor immune microenvironment demonstrates significantly greater numbers of CD4-, CD8-, FOXP3-, and CD68-positive cells in HPV-associated compared with HPV-negative cancers [23]. Though the field of immuno-oncology is relatively in its infancy, this may have implications for future trials examining immunomodulatory therapy (see Chap. 14).

21.3 HPV and OPSCC: Prognostic Implications

The most clinically relevant feature of HPV+ OPSCC is the significantly improved prognosis versus HPV-negative OPSCC. Retrospective analyses indicate an improved prognosis in HPV-related tumors, a finding that was later confirmed in the prospective Eastern Cooperative Oncology Group (ECOG) 2399 and Radiation Therapy Oncology Group (RTOG) 0129 trials, even when controlling for positive prognostic indicators such as age, performance status, and smoking [24, 25]. Stenmark and colleagues reported a systematic review of multiple studies in over 8000 patients and showed advanced primary tumor stage was associated with HPV-negative disease ($P < 0.001$), whereas increasing nodal burden was associated with HPV-associated disease ($P < 0.001$). Despite less-advanced nodal disease, HPV-negative tumors were associated with a higher likelihood of metastasis at presentation ($P < 0.001$) [26].

Bonner and colleagues performed a phase III randomized study of 424 patients with locally advanced HNSCC comparing radiotherapy alone with radiotherapy plus the EGFR-targeting MoAb cetuximab, with updated 5-year results confirming improved locoregional control (LRC), PFS, and OS in the combined modality group [6, 27]. A retrospective analysis of 182 patients from the trial with OPSCC and HPV status evaluable via IHC confirmed these improved outcomes in both the HPV+ and HPV− OPSCC patient subsets suggesting no predictive ability of HPV status for response to cetuximab [28]. Additionally, improvements in LRC, PFS, and OS were noted in the HPV+ versus HPV− OPSCC patient populations, again confirming the prognostic role of HPV.

The Radiation Therapy Oncology Group (RTOG) 0129 trial included a total of 743 patients with stage III/IV HNSCC randomized to concurrent cisplatin plus accelerated fractionation radiation with a concomitant boost versus standard fractionation radiation [29]. With 8 years of follow-up, no OS difference was seen between the two radiation fractionation regimens. However, the study did confirm a survival difference between HPV+ and HPV− OPSCC regardless of treatment modality at 8 years (71% vs. 30%, respectively). A retrospective analysis performed by Ang et al. identified the dominant prognostic factors predictive of OS in the 323 OPSCC patients on trial tested for HPV which included HPV status, pack-years of tobacco smoking, tumor stage, and nodal stage [25]. Smoking history was shown to abrogate the beneficial effects of HPV status as the risks of cancer death or relapse were increased by 1% for each additional year of tobacco smoking in both HPV− and HPV+ cohorts. A cutoff point of 10 pack-years was found to be the best predictor of survival related to smoking status. Using a recursive partitioning analysis (RPA), patients were classified as having a low, intermediate, or high risk of death based on these four factors with 3-year survival rates of 93%, 71%, and 46%, respectively (Fig. 21.1).

A retrospective analysis of 505 patients with OPSCC and evaluable HPV status treated with definitive radiation (RT) or CRT between 2001 and 2009 was performed by O'Sullivan et al. in order to identify subgroups of patients suitable for

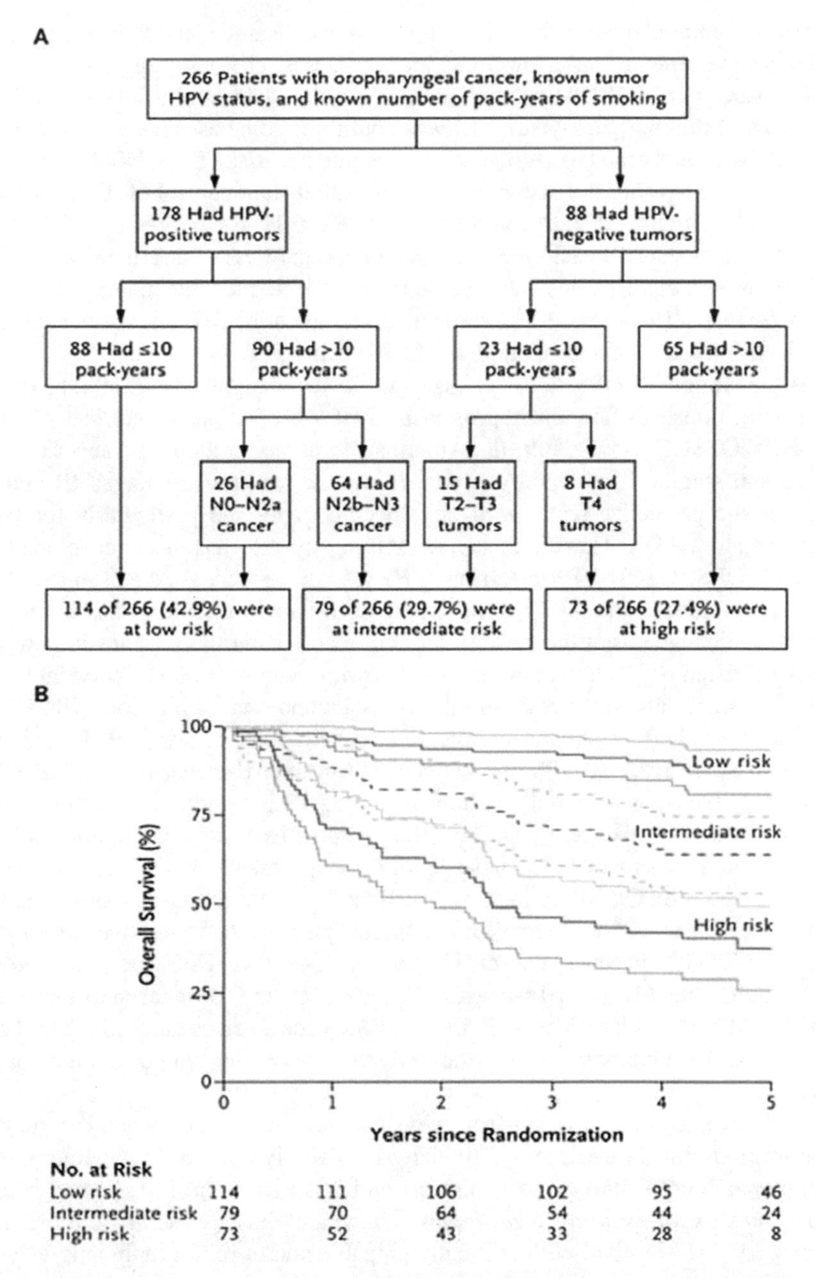

Fig. 21.1 Classification of the study patients into risk-of-death categories and Kaplan-Meier estimates of overall survival according to those categories. (From: Ang et al. [25])

treatment deintensification based on low risk of distant metastasis [30]. HPV+ patients were noted to have improved local (94% vs. 80%) and regional (95% vs. 82%) control versus HPV− patients, though similar distant control (90% vs. 86%) was found. Smoking pack-years >10 was confirmed to be associated with reduced OS. RPA was performed to segregate patients into low-risk (T1-3, N0-2c) and high-risk (T4 or N3) groups for distant metastasis with distant control of 93% and 76%, respectively. Notably, distant control rates were similar for HPV+, low-risk N0-2a patients or less than 10 pack-year N2b patients, regardless of treatment with RT or CRT. However, control rates were worse for HPV+ N2c patients managed with RT alone versus CRT (73% vs. 92%, respectively), indicating that this patient subset is not ideally suited for chemotherapy deintensification.

As the number of HPV-OPSCC has risen, it was recognized that site, histology, and staging poorly differentiated prognosis of HPV-OPSCC as it historically had for non-HPV OPSCC. As a result, the American Joint Committee on Cancer (AJCC) 7th edition staging for oropharynx cancer was skewed toward stages III and IV, reducing the predictive features of any specific stage, and particularly for HPV-related tumors [31]. Therefore, a new staging system has been developed for HR-HPV OPSCC (HR-HPV referring to HPV serotypes recognized as high risk for causing dysplasia). The AJCC Cancer Staging Manual 8th edition has introduced significant changes from the seventh edition, recognizing the prognostic power of newly validated pathologic features of some primary tumors and of cervical lymph node metastases and differentiating high-risk human papilloma virus (HR-HPV)-associated oropharyngeal cancer from OPSCC with other causes [32]. T categories in both p16-positive, HR-HPV-associated OPSCC and p16-negative, non-HR-HPV-associated OPSCC were equally valid from a prognostic standpoint and thus remain the same with several exceptions. First, the p16-positive classification now includes no carcinoma in situ (Tis) because of the nonaggressive pattern of invasive of p16-positive OPSCC and the lack of a distinct basement membrane in the epithelium of Waldeyer's ring. Second, the T4b category has been removed from p16-positive OPSCC, because the curves of the T4a and T4b categories proved indistinguishable. Third, p16-negative cancers of the oropharynx, like other non-HR-HPV-associated cancers in the head and neck, such as those of the oral cavity, larynx, hypopharynx, and paranasal sinus, will no longer include a T0 category.

cTNM employs information from physical examination and whatever imaging is performed prior to treatment. Clinically involved lymph nodes, whether one or multiple, as long as they were ipsilateral and less than 6 cm in size, had similar impact on survival (*similar hazard consistency*) and thus are included in the same N category: N1. Survival with clinically palpable and/or radiographically evident, bilateral, or contralateral lymph nodes was distinguishable with worse outcome than N1. Therefore, contralateral or bilateral lymph nodes are classified as N2. Lymph nodes greater than 6 cm foretold the worst survival from regional disease and thus warranted the highest N category: N3. This represents a significant change from the non-HR-HPV-associated (p16-negative) OPSCC N category (Tables 21.2, 21.3, and 21.4). Pathologically staged patients, likewise, have new staging (Tables 21.5 and 21.6).

Table 21.2 Clinical and Pathologic T Category for Non-Human Papillomavirus-Associated (p16-Negative) Oropharyngeal Cancer, 8th Edition Staging Manual

T category	T criteria
Tx	Primary tumor cannot be assessed
Tis	Carcinoma in situ
T1	Tumor 2 cm or smaller in greatest dimension
T2	Tumor larger than 2 cm but not larger than 4 cm in greatest dimension
T3	Tumor larger than 4 cm in greatest dimension or extension to lingual surface of epiglottis
T4	Moderately advanced or very advanced local disease
T4a	Moderately advanced local disease: tumor invades the larynx, extrinsic muscle of tongue, medial pterygoid, hard palate, or mandible
T4b	Very advanced local disease: tumor invades lateral pterygoid muscle, pterygoid plates, lateral nasopharynx, or skull base or encases carotid artery

Mucosal extension to lingual surface of epiglottis from primary tumors of the base of the tongue and vallecula does not constitute invasion of the larynx

Table 21.3 *Clinical* N Category Human Papillomavirus-Associated (p16-Positive) Oropharyngeal Cancer, 8th Edition Staging Manual

N category	N criteria
NX	Regional lymph nodes cannot be assessed
N0	No regional lymph node metastasis
N1	One or more ipsilateral lymph nodes, none larger than 6 cm
N2	Contralateral or bilateral lymph nodes, none larger than 6 cm
N3	Lymph node(s) larger than 6 cm

Table 21.4 Anatomic Stage and Prognostic Groups for *Clinical* TNM Grouping of Human Papillomavirus-Associated (p16-Positive) Oropharyngeal Cancer, 8th Edition Staging Manual

N category				
T category	N0	N1	N2	N3
Any M1 is stage IV				
T0	NA	I	II	III
T1	I	I	II	III
T2	I	I	II	III
T3	II	II	II	III
T4	III	III	III	III

Used with the permission of the American Joint Committee on Cancer (AJCC), Chicago, Illinois. The original source for this material is the AJCC Cancer Staging Manual, Eighth Edition (2017) published by Springer Science and Business Media LLC (springer.com) [86]

Table 21.5 *Pathologic* N Category Human Papillomavirus-Associated (p16-Positive) Oropharyngeal Cancer, 8th Edition Staging Manual

N category	N criteria
NX	Regional lymph nodes cannot be assessed
pN0	No regional lymph node metastasis
pN1	Metastasis in 4 or fewer lymph nodes
pN2	Metastasis in more than 4 lymph nodes

Used with the permission of the American Joint Committee on Cancer (AJCC), Chicago, Illinois. The original source for this material is the AJCC Cancer Staging Manual, Eighth Edition (2017) published by Springer Science and Business Media LLC (springer.com) [86]

Table 21.6 Anatomic Stage and Prognostic Groups for *Pathologic* TNM Grouping of Human Papillomavirus-Associated (p16-Positive) Oropharyngeal Cancer, 8th Edition Staging Manual

N category			
T category	N0	N1	N2
Any M1 is stage IV			
T0	NA	I	II
T1	I	I	II
T2	I	I	II
T3	II	II	III
T4	II	II	III

Used with the permission of the American Joint Committee on Cancer (AJCC), Chicago, Illinois. The original source for this material is the AJCC Cancer Staging Manual, Eighth Edition (2017) published by Springer Science and Business Media LLC (springer.com) [86]

21.4 Determination of HPV Status

Direct HR-HPV detection can be performed on tissue samples by in situ hybridization (ISH), but it is expensive and is not universally available, rendering ISH suboptimal for worldwide adoption. Immunohistochemistry for overexpression of the tumor suppressor protein p16 (cyclin-dependent kinase 2A) is an established, robust surrogate biomarker for HPV-mediated carcinogenesis; it is also an independent positive prognosticator in the context of OPSCC [33]. Immunohistochemical staining for p16 is inexpensive, has near universal availability, and is relatively straightforward to interpret. Hence, OPSCCs are now staged according to two distinct systems, depending on whether or not they overexpress p16. Staging by the HR-HPV-associated OPSCC system should only be assigned when p16 overexpression is determined using established criteria [34]. Specifically, the cutoff point for p16 overexpression is diffuse (≥75%) tumor expression, with at least moderate (+2/3) staining intensity. This coincides with the usual staining pattern seen in HR-HPV-associated OPSCC. Overexpression of p16 is usually localized to tumor cell nuclei and cytoplasm, and p16 staining localized only to the cytoplasm is considered nonspecific and thus not diagnostic (negative) [35, 36].

21.5 Patient Selection for Deintensification Strategies

Given the improved treatment response and outcomes of patients with HPV+ OPSCCs, considerable thought has been directed toward reevaluating historic treatment paradigms. Recent efforts pursue treatment deintensification to reduce long-term toxicities associated with chemotherapy, radiation, and radical surgery while maintaining or improving cure rates. In addition, methods of early detection and surveillance after treatment are under reexamination.

An important step in studying treatment de-escalation is identification of patients who are suitable for each strategy under evaluation. This begins with sensitive and specific determination that a particular cancer is driven by HPV, with many competing methodologies having been proposed since the first demonstration of HPV 11 and 16 in head and neck carcinoma by zur Hausen in 1985 [37, 38]. The most widely used assay in current generation cooperative group head and neck cancer trials is immunohistochemical staining for p16. This is seen as a surrogate for transcription of *E7*, because p16 is regulated by the tumor suppressor Rb and p16 levels rise when Rb levels are low [39]. Specificity for HPV-driven cancer is increased when p16 positive cancers are also tested for HPV DNA with PCR [38].

Accuracy in predicting patterns of failure, treatment responsiveness, and prognosis are required to conduct valid studies of novel treatment paradigms. For example, a group of patients in whom distant disease is the predominant mode of failure may not be appropriate for trials of radiation alone. In contrast, those with lesser treatment responsiveness may not be suitable for trials which explore the use of chemotherapy to reduce tumor burden, but may do well with minimally invasive surgery. Clinical characteristics of favorable outcome identified in a recursive partitioning analysis in R0129 – a randomized phase III trial of chemoradiation – were minimal smoking history (defined as ≤10 pack-years) and lower N stage [25]. Subsequently these data were extended in a joint analysis of the data from R0129 and an earlier trial, RTOG 9003. The risk of progression and death were both increased in a linear fashion as pack-year and years of smoking increased. In the older trial, which enrolled a higher proportion of active smokers, the risk of death doubled among those who smoked during radiotherapy after controlling for pack-years. Similar data have been reported by investigators at the University of Michigan, who observed that ever-tobacco users with HPV-related oropharynx cancer had a 35% risk of recurrence compared with a 6% risk in never smokers. Among those with HPV-related cancers, current smokers were at fivefold higher risk of disease recurrence than never smokers ($P = 0.038$) [40]. Retrospective analysis at the Princess Margaret Hospital of heterogeneously treated patients with HPV-related oropharynx cancer also identified advanced T stage (T4) as a predictor of worse outcome [30].

Using a separate cohort of 662 HPV+ OPSCC patients treated at MD Anderson between 2003 and 2012, Dhalstrom et al. were unable to validate the stage and prognostic groups proposed by Huang et al. [41]. They proposed their own staging system using nasopharyngeal (NPC) N categories rather than the traditional OPSCC regional LN categories citing the similarity between NPC and HPV-related OPSCC in regard to their viral causation and different natural history versus other HNSCC subsites. RPA stratified patients into the following stages: stage IA (T1, N0-2), stage IB (T2, N0-2), stage III (T1-3, N3), and stage IV (T4, Any N). Five-year OS for the risk groups were 94%, 87%, 76%, and 69%, respectively. Though smoking history (<10 vs. >10 pack-years) was predictive for OS for the whole cohort of patients, it was not able to stratify patient outcomes within each stage group. Though the two models differ significantly, they do share tumor volume as measured by T stage as an important prognostic variable. This finding confirms observations made by others regarding the prognostic importance of T stage versus N stage in HPV+ OPSCC [42–44].

Ideally, correlation of biomolecular signatures to treatment outcome will lead to biomarker-based characterization of tumors at diagnosis into groups suitable or unsuitable for deintensification and will identify the optimal strategy to achieve cure and minimize long-term sequelae. An understanding of the biology of HPV-related squamous cell cancer and of treatment responsiveness will be necessary in refining such patient selection strategies.

21.6 Radiation Dose De-escalation

High-dose radiation is considered the most notorious cause of both acute and long-term toxicity for patients with locally advanced head and neck cancer. Mucositis and dermatitis are extremely common and can translate into late sequelae of taste alterations, problems with xerostomia and deglutition, fibrosis, and muscle atrophy. The current treatment approach to smoking-related HNSCC is the result of a multitude of rigorous prospective trials, ultimately determining that concurrent cisplatin-based chemoradiation with a dose approaching 70 Gy demonstrates the best therapeutic ratio between locoregional control and toxicity [45, 46]. Unfortunately, this treatment paradigm is notable for its significant short- and long-term toxicities and has been advocated as a "one-size fits all" regimen for stage III–IVb disease. In the acute phase fatigue, dermatitis dysgeusia/ageusia and dysphagia (with concomitant risk of malnutrition) are common. Long-term sequelae include skin fibrosis, hypothyroidism, xerostomia, increased risk of stroke, and esophageal strictures. Radiation-induced microvascular changes can inhibit future wound healing as well, a particular concern in patients with poor dentition. In a historically elderly population with medical comorbidities driving patients' lifespans, toxicity concerns have taken a back seat. As HPV+ OPSCC is more common in younger populations, however, the morbidity and longevity of chronic toxicities are particularly burdensome and noteworthy.

The first prospective trial to evaluate the treatment responsiveness and outcome of 111 patients with HPV+ OPSCC was ECOG 2399 [47]. This large phase II study was designed as an organ-function preservation trial in resectable patients with both laryngeal and oropharyngeal primary tumors. Using induction chemotherapy with paclitaxel and carboplatin to evaluate tumor sensitivity, patients who obtained a partial clinical response (based on endoscopic exam) to two cycles then went on to concurrent 70 Gy with weekly paclitaxel. Tumor specimens were evaluated for HPV 16, 33, and 35 by in situ hybridization and PCR. Thirty-eight of 60 (63%, 95% CI = 50–75%) oropharyngeal vs. 0 of 34 (0%, 95% CI = 0–10%) laryngeal cancers were HPV positive ($P < 0.001$). Compared with patients with HPV-negative tumors, patients with HPV-associated tumors had higher response rates after induction chemotherapy (82% vs. 55%, difference = 27%, 95% CI = 9.3–44.7%, $P = 0.01$) and after chemoradiation (84% vs. 57%, difference = 27%, 95% CI = 9.7–44.3%, $P = 0.007$). After a median follow-up of 39.1 months, patients with HPV-associated tumors had improved OS (2-year OS = 95% [95% CI = 87–100%] vs. 62% [95% CI = 49–74%], difference = 33%, 95% CI = 18.6–47.4%, $P = 0.005$, log-rank test) and, after adjustment for age, tumor stage, and ECOG performance status, lower risks of progression (hazard ratio [HR] = 0.27, 95% CI = 0.10–0.75) and death from any cause (HR = 0.36, 95% CI = 0.15–0.85) than those with HPV-negative tumors (Fig. 21.2).

A follow-up prospective phase II trial of dose deintensification assessed whether a triplet induction regimen could be used to cytoreduce stage III–IVb patients with HPV-OPSCC in order to effectively treat them with reduced-dose IMRT to 54 Gy (ECOG 1308). Patients received three cycles of cisplatin 75 mg/m^2 day 1; paclitaxel

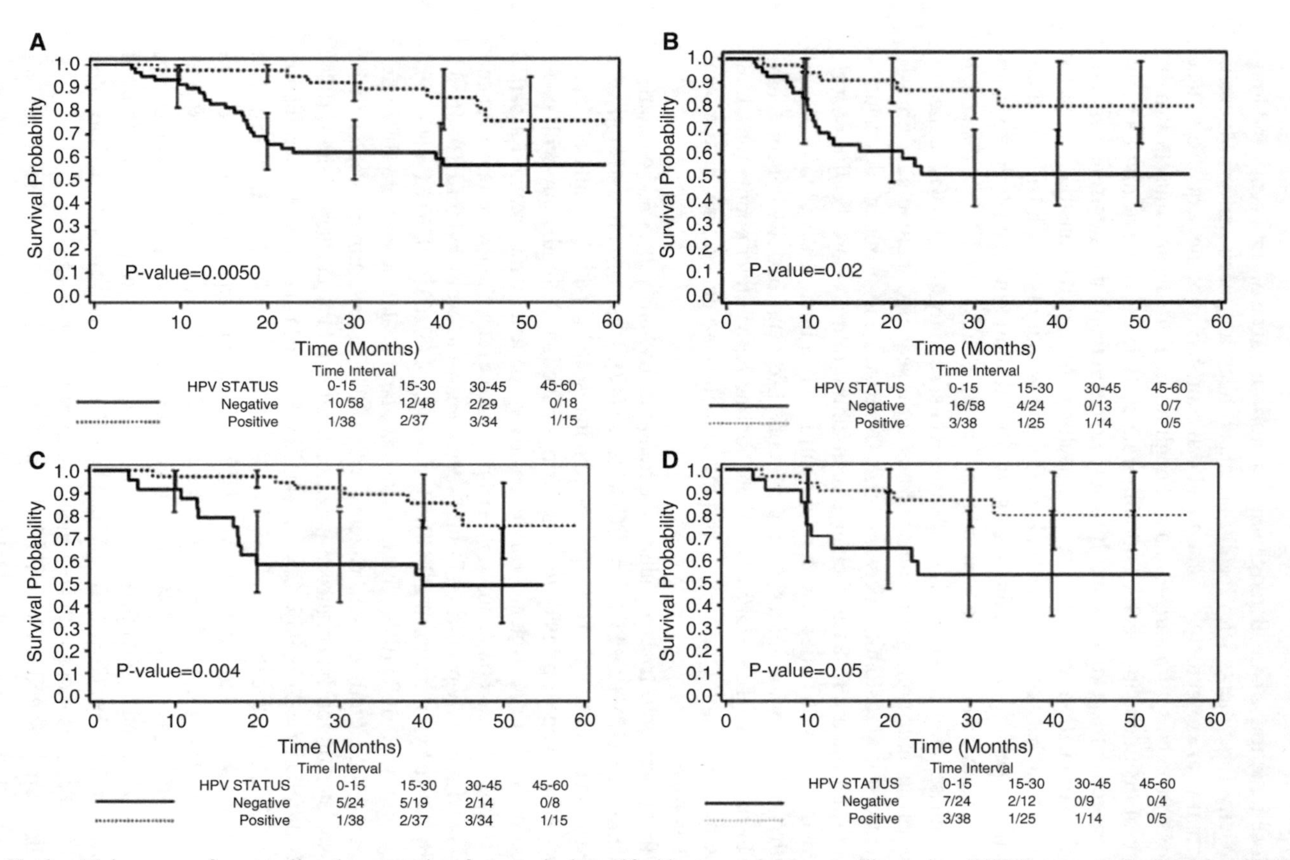

Fig. 21.2 Kaplan-Meier curves for overall and progression-free survival stratified by tumor human papillomavirus (HPV) status on the ECOG-ACRIN 2399 study. Kaplan-Meier curves for overall and progression-free survival stratified by tumor human papillomavirus (HPV) status. (**a**) Overall survival (OS) for the entire study population. (**b**) Progression-free survival (PFS) for the entire study population. (**c**) OS for patients with oropharynx cancer only. (**d**) PFS for patients with oropharynx cancer only. (From: Fakhry et al. [47] (Oxford University Press))

90 mg/m^2 days 1, 8, and 15; and cetuximab 400 mg/m^2 day 1, cycle 1, followed by cetuximab 250 mg/m^2 weekly. Cycles were repeated every 21 days for three cycles. If cisplatin was not tolerated, substitution with carboplatin AUC 5 was allowed after the first cycle. Patients were evaluated for clinical response by endoscopy and scored as a clinically complete response (cCR) or not (<cCR) before proceeding to IMRT-based radiation with concurrent weekly cetuximab 250 mg/m^2.

Of the 90 patients enrolled in the trial, 70% achieved a cCR. Of those, 78% then went on to receive reduced-dose IMRT; the incomplete responders received 69.3 Gy. After a median follow-up of 35.4 months, 2-year PFS and OS rates were 80% and 94%, respectively, for patients with primary-site cCR treated with 54 Gy of radiation ($n = 51$), and 96% and 96%, respectively, for patients with <T4, <N2c, and <10 pack-year smoking history who were treated with ≤54 Gy of radiation ($n = 27$). Patient-reported symptoms were evaluated with a battery of instruments, including the Vanderbilt Head and Neck Symptom Survey Version 2 [48]. At 12 months, significantly fewer patients treated with a radiation dose ≤54 Gy had difficulty swallowing solids (40% vs. 89%; $p = 0.011$) or had impaired nutrition (10% vs. 44%; $p = 0.025$). Only one late grade 3 toxicity was reported as an episode of hypomagnesemia, almost 3 years out from treatment [49].

Critics of this approach state that induction chemotherapy (IC) is not a necessary component in the successful treatment of this disease, particularly in good-prognosis patients with a low risk of developing metastatic disease, and treatment is prolonged by 6–10 weeks. IC might contribute to long-term toxicities, and the time of overall treatment and expense are downsides [50, 51]. To rebut these valid criticisms, study coordinators noted induction chemotherapy was not employed as a means to improve cure through control of distant metastases, as has been hypothesized for prior studies of induction chemotherapy (although it is recognized that distant metastasis is a significant cause of failure in locally advanced HPV-associated oropharynx cancer). Rather, IC was used to identify treatment-responsive patients and to reduce the tumor burden to a subclinical one comparable to that successfully treated with lower radiation doses in the postoperative setting. In addition, the extensive use of longitudinal data on acute and late toxicities was recorded prospectively with a battery of PROs at baseline, posttreatment, and every 6 months after treatment [52, 53].

NRG oncology study HN002 (NCT02254278) is a randomized trial of 296 patients asking whether accelerated radiation of 60 Gy in 5 weeks given 6 fractions each week or 60 Gy given over 6 weeks with concurrent weekly cisplatin 40 mg/m^2 in p16+ OPSCC can obtain an 85% 2-year progression-free survival. P16 status is centrally reviewed and smoking is strictly defined as ≤10 pack-years. This study also allows target volume reduction via specified elimination of distant nodal sites or even ipsilateral nodal radiation only for well lateralized tonsil primaries with low nodal burden [54].

In another deintensification phase II trial, Chera and colleagues enrolled 43 patients with favorable prognosis HPV+ OPSCC or HNSCC of unknown primary (T0-T3, N0-N2c, M0 disease with limited or remote smoking history defined as ≤10 pack-years or ≤30 pack-years with abstinence in the last 5 years). This protocol delivered weekly cisplatin and 60 Gy to the primary and gross nodal disease with 54 Gy delivered to regions at risk for subclinical disease as indicated, and post-treatment clinical and surgical investigation was performed to determine tumor response [55]. Patients were evaluated clinically by CT scan, fiber-optic laryngoscopy, and physical exam 4–8 weeks following completion of treatment, and all patients underwent a further surgical evaluation for pathologic response by resection or biopsy with neck dissection based on their primary disease state and clinical response 6–14 weeks.

Overall clinical complete tumor response rate was 64%, with 98% in the primary site and 60% in the lymph nodes of the neck. At subsequent surgical evaluation, the pathologic complete response (pCR) rate was 86%, with 98% in the primary site and 84% in the neck. For reference, the historic three-year locoregional control for HPV+ OPSCC patients with standard more intensive therapy was 87% [25]. Thorough toxicity and quality of life assessments are notable in this trial. Importantly, study patients demonstrated a low rate of progression feeding tube dependence of 39% (versus 85% in standard treatment regimens per PARADIGM trial) and none required a permanent feeding tube [56, 57]. It should be noted that the study could be critiqued, given the trimodality approach used, and it is reasonable to question whether this approach truly represents de-escalation. It does provide useful information that reduced CRT results in high pathologic complete response rates, and this could be used as a surrogate for future de-escalation studies.

21.7 Response-Adapted Volume De-escalation

In conjunction with reduction in overall radiation dose, efforts have been directed to reduce the overall radiation volume. Conventional approaches for OPSCC involve extensive coverage of the bilateral cervical levels II–IV with 50–60 Gy for clinically-negative regions, based on risk of involvement. Villaflor and colleagues incorporated a "response-adapted volume de-escalation" (RAVD) approach in a phase I/II trial in 94 patients examining the use of everolimus during induction chemotherapy (IC) [58]. Stage IVa/b HNSCC patients were enrolled to receive IC +/− everolimus. Radiographic response was determined in 89 patients completing induction therapy as "good" responder (GR, defined as ≥50% reduction in the sum of tumor diameters) or non-responder (NR). Patients with a GR to induction therapy were treated with IMRT to a single volume encompassing the gross disease (75 Gy in 1.5 Gy twice daily fractions given every other week). NR patients received additional radiation to the first uninvolved ipsilateral nodal bed (45 Gy to prophylactic volume). Both groups received concurrent multi-agent chemotherapy (Fig. 21.3).

Of the HPV+ OPSCC subset (59 total patients), 30 had GR to IC and underwent limited volume irradiation. The 2-year PFS and OS for this group were 93.1% and 92.1%, respectively. In comparison, NR patients had 2-year rates of 74.0% PFS and 95.2% OS. Though not statistically significant ($p = 0.10$ for PFS, too few deaths for

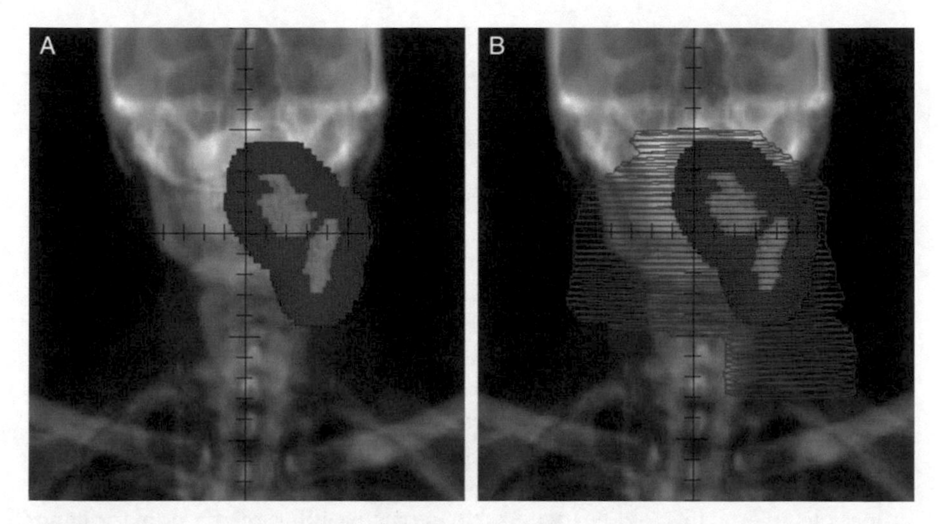

Fig. 21.3 Digitally reconstructed radiographs illustrating treatment volumes for a patient with oropharynx cancer and left level II adenopathy who experiences (**a**) good response versus (**b**) non-response to induction chemotherapy. The gross tumor volume (GTV) is shown in orange. PTV1 is shown in red and was treated to a dose of 75 Gy. PTV2 is shown in blue and was treated to a dose of 45 Gy. (From: Villaflor et al. [58]. Published by Oxford University Press on behalf of the European Society for Medical Oncology. Used by permission

OS), volume de-escalation did not appear to negatively impact disease outcomes. For all HNSCC patients in the trial, a total of 13 patients failed locally within 2 years; 12 of these 13 (92.3%) occurred infield, 11 of these in the high-dose treatment volume. The most common severe toxicities across all patients were ≥ Grade 3 mucositis and dermatitis. There was no significant difference between the treatment groups in these respects: 59.5% GR vs. 59.6% NR, 27.0% GR vs. 25.0% NR, respectively. NR patients with enlarged treatment volume however were more likely to undergo feeding tube placement during treatment (73.5% vs. 50.0% for GR) and remain dependent at 3-month (57.1% vs. 22.9%) and 6-month follow-ups (32.6% vs. 5.7%). At 1 year posttreatment, 8.6% of NR patients retained a feeding tube, whereas no GR patients remained dependent. This is in comparison to historical rates of approximately 25%. Although the addition of everolimus to IC was not beneficial, the elimination of elective nodal coverage in patients with GR to IC did not appear to compromise outcomes and resulted in significantly decreased late toxicity [29].

Additional studies evaluating the feasibility and safety of volume-reduced radiotherapy in HPV-OPSCC have been reported at the 2017 American Society of Clinical Oncology (ASCO) meeting. Riaz and colleagues from MSKCC reported a small study in 19 patients with stage III/IV p16+ OPSCC who had undergone surgical excision of the primary tumor only (the lymph nodes were not removed). They utilized dynamic 18F-FMISO (fluoromisonidazole) PET prior to chemoRT to determine tumor hypoxia in involved lymph node(s) and again after 10 Gy with one dose of cisplatin. Patients with no significant hypoxia (defined as >1.2 tumor to muscle SUV ratio in cervical lymph nodes) went on to receive one additional cycle of chemotherapy and an additional 20 Gy (30 Gy total), and those who had tumor hypoxia after 10 Gy received an additional 60–70 Gy with further high-dose cisplatin or carboplatin. Neck dissection was performed 3–5 months after completion of treatment. Of the 16 patients with known primary site, 5 had a positive margin. Of 19 pretreatment FMISO scans, 13 were positive in lymph nodes. Of 12 intra-treatment FMISO scans, 3 were positive. Ultimately, 15 patients were de-escalated (14 per protocol). With median follow-up of 10 months (range 6–18 months), 11 of 14 patients treated per protocol had a complete pathologic node response and 3 had a partial pathologic response. One patient had tumor progression before neck dissection, and 18 of 19 patients had no evidence of relapse. All patients are presently free of disease. They conclude that patient-specific imaging base treatment response if feasible in HPV-OPSCC and 30 Gy appears with short follow-up to be safe and efficacious. Correlative studies with DNA repair profiling of the primary tumor using ex vivo IR-induced RAD51 assay, and RNA sequencing and whole exome and weekly diffusion-weighted MRIs will be reported in the future [59].

Melotek and colleagues used IC with carboplatin AUC 6 with nab-paclitaxel 100 mg/m^2 d1/d18/15 for 3 cycles in 62 HPV-OPSCC patients, both low-risk (≤T3, ≤N2b, and ≤10 pack-years smoking) and high-risk, and performed radiographic assessment of response. IMRT volumes in all patients were limited to initial gross tumor and the first echelon of uninvolved nodes. Concurrent chemotherapy consisted of paclitaxel, 5-FU, hydroxyurea, and 1.5 Gy twice daily, given every other week. Primary site biopsy and neck dissection were performed after CRT. For low-risk patients with IC radiographic responses of ≥50%, patients received 50 Gy with no chemotherapy, 30–50% received 45 Gy to gross disease and 30 Gy to first echelon uninvolved nodal areas, and for <30% response to IC 75 Gy with 45 Gy to first echelon nodes. For high-risk patients, ≥50% response on IC went on to get 45 Gy to gross disease and 30 Gy to next echelon and for <50% response received 75 Gy to gross disease and 45 Gy to next echelon nodes. With short follow-up (median of 15.2 months (range 5.3–29.6 months)), pathologic compete response was noted in 94.7% (18/19) in the low-dose (no chemotherapy) patients and in 25 of 28 patients who received low-dose radiation with chemotherapy (6/6 low risk and 19/22 with high-risk disease). Only 15% of low-dose IMRT alone patients developed ≥grade 3 mucositis compared to 46.7% who received low-dose IMRT with concurrent chemotherapy and 63.6 Gy in the standard-dose (75 Gy) IMRT with concurrent chemotherapy ($p = 0.002$). PEG dependency was lower during treatment ($p = 0.001$) and at 6 and 12 months. The conclusion was that a favorable response to IC appeared to be a powerful biomarker for dose and volume de-escalation with 50 Gy IMRT or 45 Gy chemoIMRT [60].

21.8 Chemotherapy De-escalation

Randomized clinical trials and meta-analyses have shown a significant clinical benefit of combining chemotherapy with definitive radiation in locally advanced patients. An exhaustive review of OPSCC patients ($n = 5878$) from the meta-analysis of chemotherapy in head and neck cancer (MACH-NC) demonstrated an overall 8.1% improvement in OS at 5 years for concurrent chemotherapy [61]. The standard chemotherapeutic regimen is platinum-based and notorious for toxicities including emesis, nephrotoxicity, ototoxicity, mucositis, and cytopenias [62]. The substitution of a more tolerable agent such as cetuximab with radiotherapy was associated in retrospective studies with worse locoregional control, failure-free survival, and OS [63]. These reports were, as noted, retrospective and plagued by significant differences in patient selection, as cetuximab is often favored in elderly patients or those with poor performance status. A study from Memorial Sloan Kettering Cancer Center retrospectively compared 174 consecutively treated patients with HNSCC treated definitively between 2006 and 2008 with concurrent single agent cisplatin or cetuximab. Although HPV status was not reported, 76% had OPSCC. At a median follow-up of 22.5 months, patients treated with concurrent cisplatin, compared to those treated with cetuximab, had lower 2-year locoregional failure (5.7% versus 39.9%, respectively) and OS (92.8% versus 66.6%, respectively). Treatment with cisplatin, as compared to cetuximab, was associated with improved failure-free survival and OS on multivariate analysis. Late grade 3 or 4 toxicity or feeding tube dependence was similar between groups [64].

A retrospective analysis was performed of 168 HPV-associated patients with OPSCC treated both postoperatively ($n = 23$) and definitively ($n = 145$) at the Ohio State University Wexner Medical Center between 2010 and 2013. Forty-two of this group of patients received concurrent cetuximab, whereas the remainder ($n = 126$) were given concurrent platinum-based therapy. For patients receiving cetuximab compared with platinum chemotherapy, multivariable analysis revealed inferior 2-year OS (80% vs. 96%, respectively), local relapse-free survival (74% vs. 91%, respectively), and distant metastasis-free survival (74% vs. 90%, respectively) [65].

A prospective randomized phase III trial by Magrini et al. directly compared the two therapeutic options of cetuximab with cisplatin in 70 patients. One-hundred thirty patients were planned to be enrolled but slow accrual mandated stopping the trial early. All patients had stage II, IVA, or IVB HNSCC and received 70 Gy. No HPV testing was performed. No significant differences were found in LRC, OS, or metastasis-free survival in all enrolled patients; however, it should be noted that subset analysis of OPSCC ($n = 43$), LRC, and OS favored concurrent cisplatin over cetuximab [66]. Interestingly, radiation discontinuation for more than 10 days was more frequent in the cetuximab arm (13% versus 0%). Additionally, three patients were noted to have cetuximab infusion reactions requiring removal from the trial. Toxicity profiles differed; more hematologic, GI, and renal toxicity was seen with cisplatin, whereas more cutaneous toxicity and nutritional support requirements were seen with cetuximab. The discordant results of retrospective trials coupled with the conclusions of Magnani et al. would suggest that carefully selected patients may be safely treated with non-platinum-based therapies in order to reduce acute and long-term toxicities.

Rosenthal et al. retrospectively evaluated 182 patients randomized on the IMCL-9815 study to receive definitive radiation alone or combined with weekly cetuximab. Forty-one percent were HPV-related [6, 28]. HPV status was found to be strongly prognostic for patients with OPSCC. In addition, data suggested that the addition of cetuximab to RT improved clinical outcomes regardless of p16 or HPV status versus RT alone. The study was not powered for the comparison of treatments in HPV-related or negative subgroups, but of note, the hazard ratio for death with the addition of cetuximab was 0.38 among p16(+) and 0.72 among p16(−) cases. Nolan et al. also examined this relationship in a small retrospective review of HPV+ OPSCC patients at the Ohio State Wexner Medical Center. Of 168 eligible patients including those treated postoperatively and definitively, 42 (25%) received concurrent cetuximab, whereas 126 received concurrent cisplatin or carboplatin ($n = 109$, $n = 17$, respectively). The 2-year OS for cetuximab was 80% compared to 96% for platinum-based therapy; local relapse-free survival was 74% vs. 91%; distant metastasis-free survival was 74% vs. 90% [65]. A focused clinical trial comparing the two treatments in a prospective, randomized fashion has completed accrual in July 2014, with results anticipated in the near future (RTOG 1016) [29].

The Canadian Cancer Trials Group conducted a phase III randomized trial HN6 comparing standard chemoradiotherapy (70 Gy in 7 weeks/cisplatin 100 mg/m^2 every 3 weeks) versus accelerated radiotherapy (70 Gy in 6 weeks with EGFR inhibitor panitumumab 9 mg/kg ×3) in 320 locally advanced HNC patients. Quality of life was the primary endpoint with the Functional Assessment of Cancer Therapy Head and Neck (FACT-HN), MD Anderson Dysphagia Index (MDADI), and the SWAL-QOL instruments utilized. Although the study was underpowered for outcome, no clinically important differences in QOL were seen posttreatment between treatment regimens [67].

RTOG 1016 tested bolus cisplatin (100 mg/m^2 for 2 cycles) against weekly cetuximab, with accelerated radiation to 70 Gy in both arms over 6 weeks. It is the first non-inferiority trial ever conducted in head and neck cancer and as such requires a large sample size (more than 800). Accrual was completed in 2014, and data analysis is expected in 2018. Importantly, this study contains more than a dozen QOL and translational endpoints designed for this young and comparatively healthy population of patients.

Three-weekly high-dose cisplatin (100 mg/m^2) is considered the standard systemic regimen given concurrently with postoperative or definitive radiotherapy in locally advanced squamous cell carcinoma of the head and neck (LA-SCCHN). However, due to unsatisfactory patient tolerance, various weekly low-dose schedules have been increasingly used in clinical practice. A meta-analysis was conducted to compare the efficacy, safety, and compliance between these two approaches. Platinum schedules with both definitive RT and in the high-risk postoperative setting were analyzed with the primary endpoint being OS. Secondary outcomes comprised response rate, acute and late adverse events, and treatment compliance. Fifty-two studies with 4209 patients were included in two separate meta-analyses according to the two clinical settings. There was no difference in treatment efficacy as measured by overall survival or response rate between the chemoradiation settings with low-dose weekly and high-dose three-weekly cisplatin regimens. In the definitive treatment setting, the weekly regimen was more compliant and significantly less toxic with respect to severe (grade 3–4) myelosuppression (leukopenia $p = 0.0083$; neutropenia $p = 0.0024$), severe nausea and/or vomiting ($p < 0.0001$), and severe nephrotoxicity ($p = 0.0099$). Although in the postoperative setting the two approaches were more equal in compliance and with clearly less differences in the cisplatin-induced toxicities, the weekly approach induced more grade 3–4 dysphagia ($p = 0.0026$) and weight loss ($p < 0.0001$). The authors concluded that current evidence is insufficient to demonstrate a meaningful survival difference between the two dosing regimens. However, prior to its adoption into routine clinical practice, the low-dose weekly approach needs to be prospectively compared with the standard three-weekly high-dose schedule [68].

21.9 Current Practice Guidelines Based on Evidence

The American Society for Radiation Oncology convened the OPSCC Guideline Panel to perform a systematic review of the literature to investigate the following four questions: (1) When is it appropriate to add systemic therapy to definitive RT? (2) When is it appropriate to deliver postoperative RT with and without chemotherapy following surgery for OPSCC? (3) When is it appropriate to use induction chemotherapy in OPSCC? (4) What are the appropriate dose, fractionation, and volume regimens with and without systemic therapy in the treatment of OPSCC? [69].

The Panel concluded that patients with stage IV and stage T3 N0-1 OPSCC treated with definitive RT should receive concurrent high-dose intermittent cisplatin. Patients receiving adjuvant RT following surgical resection for positive surgical margins or extracapsular extension should be treated with concurrent high-dose intermittent cisplatin, and individuals with these risk factors who are intolerant of cisplatin should not routinely receive adjuvant concurrent systemic therapy. Induction chemotherapy should not be routinely delivered to patients with OPSCC. For patients with stage IV and stage T3 N0-1 OPSCC ineligible for concurrent chemoradiation therapy, altered fractionation RT should be used. The Panel also strongly recommended ipsilateral-only RT for T1/T2 tonsil with minimal soft palate or tongue involvement, given the significant volume of (mostly retrospective) data showing very rare contralateral nodal recurrences. For patients with N2a status and T1/T2 disease with limited soft palate involvement, the Panel found sufficiently encouraging data to conditionally support ipsilateral RT, given the benefits expected in QOL versus the uncertain risk of contralateral recurrence, provided patient preferences are considered. Due to the paucity of data showing low rates of contralateral node recurrence in small-volume N2b disease, the Panel did not include this group in conditional recommendation for the use of ipsilateral RT [69].

21.10 Minimally Invasive Surgery

Retrospective data reporting on locoregional control and survival rates in early stage OPSCC have shown equivalent efficacy, although no prospective randomized trials are available to confirm these results [70]. Given the assumed comparable oncologic results in both groups, complication rates and functional outcomes associated with each modality play a major role when making treatment decisions. Radiotherapy is used preferentially in many centers because a few trials have reported higher complication rates in surgical patients. However, these adverse effects were mainly due to traditional invasive open surgical approaches used for access to the oropharynx. In order to decrease the morbidity of these techniques, transoral surgical (TOS) approaches have been developed progressively. They include transoral laser microsurgery (TLM), transoral robotic surgery (TORS), and conventional transoral techniques.

A meta-analysis comparing TORS (12 studies, 772 patients) with IMRT (8 studies, 1287 patients) showed equivalent efficacy in terms of oncologic results [71]. Patients receiving definitive IMRT also received chemotherapy (43%) or neck dissections for persistent disease (30%), whereas patients receiving TORS required adjuvant radiotherapy (26%) or chemoradiotherapy (41%). Two-year overall survival estimates ranged from 84% to 96% for IMRT and from 82% to 94% for TORS. Furthermore, studies reporting on functional outcomes in patients undergoing TOS for OPSCC has shown low rates of major long-term functional impairment following treatment, including hemorrhage (2.4%), fistula (2.5%), and gastrostomy tubes at the time of surgery (1.4%) or during adjuvant treatment (30%), and tracheostomy tubes in 12% of patients at the time of surgery, but most were decannulated prior to discharge.

The Eastern Cooperative Oncology Group-American College of Radiology Imaging Network (ECOG-ACRIN) is currently conducting a multicenter randomized trial of radiation deintensification in patients amenable to transoral resection [E3311, NCT01898494]. Patients are eligible who have p16(+) T1 or T2 oropharyngeal squamous cell carcinoma and in whom radiographic and clinical assessment specifically note the absence of matted nodes. Patients undergo transoral resection of the primary and surgical neck dissection. Those with clear margins and either N0 or N1 disease are observed. Those with involved margins, evidence for extranodal extension, or involvement of five or more nodes receive postoperative radiation with weekly cisplatin. The remainder enter randomization between 50 and 60 Gy of postoperative radiation. Accrual was completed by the end of 2017.

Until further high-level evidence is available, it is generally accepted practice to treat early stage OPSCC with a single modality treatment if feasible (T1-2 and N0-1). Minimally invasive resection of the primary tumor associated with selective neck dissection, as indicated, offers the advantage of stratifying the risk of disease recurrence based on the pathological tumor features, and adjuvant treatment can be chosen for higher-risk patients. For tumors without adverse features, no adjuvant allows prevention of potential radiation-induced late complications while keeping radiotherapy as an option for any second primary lesions whenever needed. Definitive radiotherapy is generally reserved for selected patients with specific anatomical location associated with poor functional outcome following surgery, such as tumor of the soft palate, or for patients with severe comorbidities that do not allow surgical treatment.

21.11 Utilization of Biologic Agents and Immunotherapy

HPV+ OPSCC patients are typically younger and healthier than HPV- patients. With the notion that surgical and radiation effects are lifelong in this population, some have postulated that escalating chemotherapy may produce better long-term control with good tolerability.

A phase II randomized trial was recently published where patients were assigned to cetuximab-enhanced chemotherapy protocols with comparison to historical controls [72]. A total of 110 patients with stage III or IV nonmetastatic HNSCC were enrolled and received induction therapy with carboplatin, paclitaxel, and cetuximab. Patients were then treated on chemoradiation regimens including cetuximab, 5-FU, and hydroxyurea ("Cetux-FHX", $n = 57$) or cetuximab and cisplatin ("Cetux-PX", $n = 53$). All patients were assessed at the completion of induction chemotherapy with an overall response rate of 91%; no significant differences were noted between the two arms prior to chemoradiation. At follow-up, the 2-year PFS for the entire cohort was 83.6%, a significant improvement over the historical control rate of 50%. A subgroup analysis was performed for outcomes according to HPV status ($n = 47$) with 5-year PFS and OS rates of 84.4% and 91.3%, versus the entire cohort rates of 74.1% and 80.3%. Although this is a limited dataset, the evidence indicates that properly selected patient populations such as those with HPV+ tumors may benefit from combination chemotherapy and biologic agent regimens. Although no toxicity data were included, these results suggest that intensification may yield better outcomes, with properly selected patient populations such as those with HPV+ tumors receiving an even greater benefit.

Recognition of the immune system's role in contributing to cancer development was an important advancement in our original understanding of cancer immunology from the early twentieth century [73, 74]. As noted previously, HNSCC exhibit evidence of inflammation, as well as increases in markers of immune exhaustion and increased numbers of T regulatory cells, with these changes most marked in HPV-driven cancers. A substantial percentage of patients with HNSCC have underlying immunophenotypic changes that would predict a response to immune checkpoint modulation, and tumor immunophenotype has been shown to be prognostic in both HPV-associated and HPV-negative patients [75–77]. Early trials of the PD-1 directed antibodies pembrolizumab and nivolumab have demonstrated activity in metastatic/recurrent HNSCC, offering a significant benefit over traditional chemotherapy.

The initial cohort of the phase Ib study of pembrolizumab in participants with advanced solid tumors (KEYNOTE-012) trial included 60 patients with advanced HNSCC enriched for PD ligand-1 (PD-L1) expression who were administered fixed-dose biweekly pembrolizumab, an moAb targeting PD-1 [78]. Tumor RNA expression levels for interferon-γ–related genes associated with clinical outcomes in the melanoma cohort of the KEYNOTE-001 study were also collected to calculate a composite expression score. Of the 60 patients, 23 (38%) were HPV positive. Ten patients (17%) treated with pembrolizumab experienced grade 3 or 4 drug-related adverse events. No drug-related deaths were noted. The overall response rate (ORR) for the entire population was 18% per central imaging review, compared with 21% per investigator review. Importantly, HPV-associated patients had a 25% ORR, whereas the ORR was only 14% for HPV-negative patients. The median OS for the entire cohort was 13 months. Interestingly, analysis revealed that PD-L1 expression levels and presence of stromal staining were significant predictors for best overall response and PFS, as was interferon-γ–related gene composite expression score.

Subsequently, an expansion cohort of 132 patients unselected for PD-L1 expression was accrued. These patients received pembrolizumab every 3 weeks for 24 months or until disease progression or intolerable toxicity. These patients were heavily pretreated (59% had received two or more previous therapies). The ORR per Response Evaluation Criteria in Solid Tumors (RECIST) was 23.7%, with two CRs, 39 partial responses (PRs), and 25.4% of patients with stable disease. Improved ORR was seen in patients who received two or fewer prior therapies (31 of 97 patients; ORR, 32.0%; two CRs and 29 PRs) compared with patients who received more than two prior therapies (10 of 63 patients; ORR, 16%; 10 PRs). ORR was comparable between HPV-associated and HPV-negative patients (23.6% vs. 25.0%, respectively) [78].

Preliminary results from the CheckMate-141 phase III trial recently reported at the American Association for Cancer Research and ASCO annual meetings also show significant promise for PD-1 pathway blockade in HNSCC [79]. A total of 361 patients with platinum-refractory recurrent or metastatic HNSCC were randomly assigned 2:1 to nivolumab, an moAb PD-1 inhibitor, versus investigator's choice (ICh) chemotherapy with docetaxel, methotrexate, or cetuximab. Planned interim analysis after 218 patient deaths revealed a 30% reduction in risk of death (HR, 0.70) with nivolumab versus ICh. Median OS for all patients was 7.5 months with nivolumab compared with 5.1 months with ICh. At 1 year, OS was 36% in the nivolumab arm compared with 17% in the ICh arm. Importantly, the survival benefit for nivolumab was seen in both HPV-associated and HPV-negative patients (Table 21.7). The ORR for nivolumab in patients with PD-L1 expression>1%, >5%, and >10% was 18.2%, 25.9%, and 32.6%, respectively, compared with ORRs of 3.3%, 2.3%, and 2.9%, respectively, for ICh. Grade 3 or 4 treatment-related adverse effects occurred in 13.6% of patients in the nivolumab arm compared with 35.1% of patients receiving ICh.

Table 21.7 Checkmate-141: overall survival summary

	Nivolumab		Investigator's choice		Comparison of nivolumab to investigator's choice
	n	Median, months	*n*	Median, months	HR (95% CI)
All patients	240	7.5	121	5.1	0.70 (0.51–0.96)[a]
PD-L1 ≥ 1%	88	8.7	61	4.6	0.55 (0.36–0.83)
PD-L1 < 1%	73	5.7	38	5.8	0.89 (0.54–1.45)
p16-positive	63	9.1	29	4.4	0.56 (0.32–0.99)
p16-negative	50	7.5	36	5.8	0.73 (0.42–1.25)

Adapted from Gillison et al. presented at AACR 2016
[a]Hazard ratio (HR) and 97.73% CI

These demonstrations of proof of principle for immunotherapy in HNSCC have led to numerous subsequent HNSCC trials exploring inhibition of the PD-1 immunoregulatory pathway with PD-1 or PD-L1 inhibitors alone, combined with CTLA-4 antagonists, or in combination with chemotherapy in the first-line treatment of metastatic/recurrent disease, as well as in combination with definitive chemoradiation, postoperatively, or as maintenance following definitive chemoradiation. Additional early phase trials are exploring the combination of PD-1 pathway directed therapy with HPV-specific therapeutic vaccines, oncolytic tumor virus, or other immune regulators such as IDO (indoleamine 2,3-dioxygenase). These approaches are not sufficiently mature for incorporation into deintensification schemes at this time, but as biomolecular and clinical patient selection criteria are developed to identify patients with outstanding and durable responses to immunotherapy, as well as to exclude patients at high risk of serious autoimmune complications, appropriate incorporation of immunotherapy may permit substantial reductions in radiation and chemotherapy exposure.

21.12 Surveillance Following Treatment in HPV Disease

Retrospective analysis of OPSCC patients suggests that HPV+ patients have a distinct pattern of failure in comparison to their HPV-negative brethren. A review of the failure patterns of RTOG 0129 patients by Ang et al. showed a significant difference in the 3-year OS, PFS, and local-regional relapse rate (HPV+ vs. HPV-, 82.4% vs. 57.1%, 37.7% vs. 43.4%, 13.6% vs. 35.1%); distant failure at 3 years was improved but did not achieve statistical significance, 8.7% vs. 14.6%, p = 0.23, albeit the study was not powered for this endpoint [25]. Trosman et al. evaluated 291 OPSCC patients with stage III–IVB disease treated with definitive chemoradiation at the Cleveland Clinic with known HPV status: 252 HPV-related and 39

HPV-negative. The 3-year distant control rate was significantly higher in HPV+ patients (88% vs. 74%) with a longer time to distant failure as well (median 16.4 vs. 7.2 months). In addition, HPV-related patients who failed distantly had a longer overall survival (median 25.6 vs. 11.1 months). Interestingly, HPV+ patients had a more diverse set of failure sites. While OPSCC has typically been associated with lung and brain metastases, the HPV+ patients demonstrated metastases to sites including non-regional lymph nodes (axillary and intra-abdominal), kidney, skin, and skeletal muscle. This is in line with what Huang et al. have termed a "disseminating" phenotype [80].

Retrospective analysis of trials for recurrent or metastatic HNC conducted in the Eastern Cooperative Oncology Group demonstrated longer median survival for HPV-associated disease treated with systemic therapy, and a retrospective analysis of Radiation Therapy Oncology Group trials demonstrated longer survival for HPV-associated than for HPV-negative disease whether patients were candidates for local or systemic therapy [81]. These unique characteristics call into question the most appropriate form of surveillance and whether patients should be managed according to the historically accepted paradigm of planned neck dissections or CT imaging of the neck.

A retrospective review by Frakes et al. examined the posttreatment imaging of 246 patients with HPV-related OPSCC treated with definitive radiation or chemoradiation at Moffitt Cancer Center [82]. All patients underwent a PET/CT scan 3 months after completing therapy as well as close follow-up with clinical exams. At 3 years, local control was achieved in 239 of 245 patients (97.8%). All six local failures were identified on exam by direct visualization or flexible laryngoscopy. Nine patients failed regionally, with all but one identified by symptom progression or on PET/CT. Of the 21 patients who developed distant metastasis, 15 (71%) were identified by symptoms or on imaging. Overall, symptoms and/or 3-month follow-up PET/CT identified 92% of local-regional failures and 71% of distant failures. Thus, the authors suggest a 3-month PET/CT with close clinical follow-up would provide adequate surveillance.

The less invasive approach of PET/CT imaging has been directly compared with planned neck dissections as well. Mehanna et al. directed a prospective trial of 564 patients (84% OPSCC, 75% HPV-related) with N2 or N3 HNSCC in the UK [83]. Patients were randomized to undergo a planned neck dissection or PET/CT surveillance 3 months after the conclusion of chemoradiation. If the PET/CT patients showed an incomplete or equivocal response, salvage neck dissection was performed. Of the 270 patients who underwent PET/CT, 185 (69%) showed a complete response and an additional 19 showed incomplete response in the primary with a complete response in the neck. A total of 52 patients (19.3%) underwent neck dissection, with similar complication rates to the planned dissection group (42% vs. 38%). The 2-year OS rates between the two groups were similar (84.9% for PET/CT vs. 81.5% for planned dissection). The authors note that nodal disease in HPV+ tumors may take longer to involute, suggesting a 3-month scan may be too early to assess response but nonetheless indicate surveillance is a suitable approach associated with less mortality and cost than planned posttreatment neck dissection.

With the success of PSA monitoring for recurrence and response in prostate cancer, efforts have been directed toward developing a biochemical surveillance test for oropharyngeal cancers. HPV DNA is an attractive target as it can be extracted from oral exfoliated cells in rinses. A prospective trial of 124 HPV+ OPSCC patients by Rettig et al. examined the potential of HPV DNA in oral rinses as a prospective marker [84]. Prior to treatment, 67 of the 124 patients (54%) tested positive for HPV16 DNA. After treatment, only 6 participants (5%) tested positive, including one who was not in the original 67. This patient subsequently cleared HPV DNA and did not develop recurrence. Of the 5 patients with persistent HPV DNA, all recurred within 1 year, compared to a recurrence rate of 8% in the HPV DNA negative group. Though limited by small sample sizes, this trial suggests that persistent HPV DNA demonstrates high specificity and PPV for disease recurrence and could be a potentially useful tool in determining the need for adjuvant therapy or increased surveillance.

21.13 Metastatic Sites and Salvage Therapy and Prognosis after Recurrence

Prognostic factors in patients with recurrent/metastatic SCCHN such as primary tumor site, performance status, prior radiotherapy, and cell differentiation have long been established [85]. Trosman and colleagues performed a large retrospective review and found that metastatic sites differed in HPV patients compared to a non-HPV comparison group. The lung was the most common distant site involved in HPV+ and HPV- disease (HPV+ group, 23 of 28 patients [82%]; HPV- group, 7 of 9 patients [78%]). However, the HPV+ group had metastases to several subsets atypical for head and neck squamous cell carcinoma, including the brain, kidney, skin, skeletal muscle, and axillary lymph nodes (in two patients each) and in the intra-abdominal lymph nodes (in three patients). The rate of 3-year OS was higher in the HPV+ group (89.9% vs. 62.0%; $P < 0.001$), as was the median survival after the occurrence of distant metastases regardless of additional treatment (25.6 vs. 11.1 months; $P < 0.001$) [80].

Others have also shown that HPV-associated tumor status continues to confer improved prognosis at the time of disease recurrence [24, 81]. Patients treated on two clinical trials of chemotherapy for recurrent or metastatic SCCHN were evaluated retrospectively for HPV by Argiris and colleagues. They were treated on either E1395, a phase III trial of cisplatin and paclitaxel versus cisplatin and 5-fluorouracil, or E3301, a phase II trial of irinotecan and docetaxel. Sixty-four patients were shown to have HPV by ISH and 65 by p16. The objective response rate was 55% for HPV-associated versus 19% for HPV-negative ($P = 0.022$) and 50% for p16-positive versus 19% for p16-negative ($P = 0.057$), respectively. The median survival was 12.9 versus 6.7 months for HPV-associated versus HPV-negative patients ($P = 0.014$) and 11.9 versus 6.7 months for p16-positive versus p16-negative patients ($P = 0.027$), respectively. After adjusting for other covariates, hazard ratio for OS was 2.69 ($P = 0.048$) and 2.17 ($P = 0.10$), favoring HPV-associated and p16-positive patients, respectively. Other studies have corroborated these findings. It is now the general opinion that HPV is a favorable prognostic factor in recurrent or metastatic SCCHN that should be considered in the design of clinical trials in this setting.

21.14 Conclusion

HPV-associated oropharyngeal cancer has emerged as a unique subset of head and neck squamous cell carcinoma with a distinctive pathophysiology and patient population that necessitates reconsideration of our current treatment approach. While the tumor is more sensitive to treatment and has a better prognosis, these aspects present novel challenges with respect to minimizing long-term treatment-related morbidity and surveillance technique. Retrospective data indicates that HPV+ patients stratify into a lower-risk group that may benefit from deintensification of chemotherapy and/or radiation or can be treated with surgery alone in many cases without sacrificing survival. Newer studies are incorporating biologic therapies and varying degrees of surveillance based on risk of recurrence. The evolving data suggest that HPV+ OPSCC patients could be a population who would significantly benefit from limited radiation fields and reduced doses, without sacrificing survival. Clinical trials evaluating these concepts are currently underway with highly anticipated results in the years to come.

21.15 HPV-Related Treatment Strategies: The Future

Recognizing that there are high stakes in deintensification protocols and the numerous ongoing strategies, it is important to carefully define and adhere to eligibility criteria for enrollment of patients whose tumors progress, so that we can learn from our successes as well as our failures. Extensive biospecimen banking (e.g., tumor, sera, plasma, lymphocytes at baseline, and progression) from all enrolled patients would allow for the study of individuals whose tumors progress, so that we can learn from our increasing successes, but more importantly, our failures. As we reexamine the core principles of head and neck cancer care developed decades ago, we need to acknowledge the limitations of currently available assays for HPV status determination, erring on the side of caution by using the best-validated measures of tumor HPV status that are practically available with high positive predictive value, are paramount. In situ hybridization for HPV testing has a very high positive predictive value; very careful evaluation of standardized methods and interpretation in multi-institutional protocols will be crucial in achieving our goals in this distinct and special disease – that being obtaining high cure rates simultaneously with meaningful reductions in long-term toxicity and improved quality of life.

References

1. Human papillomaviruses. IARC Monogr Eval Carcinog Risks Hum. 2007;90:1–636.
2. Ang KK, Zhang Q, Rosenthal DI, Nguyen-Tan PF, Sherman EJ, Weber RS, et al. Randomized phase III trial of concurrent accelerated radiation plus cisplatin with or without cetuximab for stage III to IV head and neck carcinoma: RTOG 0522. J Clin Oncol. 2014;32(27):2940–50.
3. Pignon JP, le Maître A, Maillard E, Bourhis J, Group M-NC. Meta-analysis of chemotherapy in head and neck cancer (MACH-NC): an update on 93 randomised trials and 17,346 patients. Radiother Oncol. 2009;92(1):4–14.
4. Brizel DM, Albers ME, Fisher SR, Scher RL, Richtsmeier WJ, Hars V, et al. Hyperfractionated irradiation with or without concurrent chemotherapy for locally advanced head and neck cancer. N Engl J Med. 1998;338(25):1798–804.
5. Posner MR, Hershock DM, Blajman CR, Mickiewicz E, Winquist E, Gorbounova V, et al. Cisplatin and fluorouracil alone or with docetaxel in head and neck cancer. N Engl J Med. 2007;357(17):1705–15.
6. Bonner JA, Harari PM, Giralt J, Azarnia N, Shin DM, Cohen RB, et al. Radiotherapy plus cetuximab for squamous-cell carcinoma of the head and neck. N Engl J Med. 2006;354(6):567–78.
7. Reddy VM, Cundall-Curry D, Bridger MW. Trends in the incidence rates of tonsil and base of tongue cancer in England, 1985-2006. Ann R Coll Surg Engl. 2010;92(8):655–9.
8. Hammarstedt L, Lindquist D, Dahlstrand H, Romanitan M, Dahlgren LO, Joneberg J, et al. Human papillomavirus as a risk factor for the increase in incidence of tonsillar cancer. Int J Cancer. 2006;119(11):2620–3.
9. Jemal A, Simard EP, Dorell C, Noone AM, Markowitz LE, Kohler B, et al. Annual Report to the Nation on the Status of Cancer, 1975-2009, featuring the burden and trends in human papillomavirus(HPV)-associated cancers and HPV vaccination coverage levels. J Natl Cancer Inst. 2013;105(3):175–201.
10. Gillison ML, Koch WM, Capone RB, Spafford M, Westra WH, Wu L, et al. Evidence for a causal association between human papillomavirus and a subset of head and neck cancers. J Natl Cancer Inst. 2000;92(9):709–20.
11. Stransky N, Egloff AM, Tward AD, Kostic AD, Cibulskis K, Sivachenko A, et al. The mutational landscape of head and neck squamous cell carcinoma. Science. 2011;333(6046):1157–60.
12. Stein AP, Saha S, Kraninger JL, Swick AD, Yu M, Lambert PF, et al. Prevalence of human papillomavirus in oropharyngeal cancer: a systematic review. Cancer J. 2015;21(3):138–46.
13. Steinau M, Saraiya M, Goodman MT, Peters ES, Watson M, Cleveland JL, et al. Human papillomavirus prevalence in oropharyngeal cancer before vaccine introduction, United States. Emerg Infect Dis. 2014;20(5):822–8.
14. Chaturvedi AK, Engels EA, Pfeiffer RM, Hernandez BY, Xiao W, Kim E, et al. Human papillomavirus and rising oropharyngeal cancer incidence in the United States. J Clin Oncol. 2011;29(32):4294–301.
15. Gillison ML, Broutian T, Pickard RK, Tong ZY, Xiao W, Kahle L, et al. Prevalence of oral HPV infection in the United States, 2009-2010. JAMA. 2012;307(7):693–703.
16. Giuliano AR, Nyitray AG, Kreimer AR, Pierce Campbell CM, Goodman MT, Sudenga SL, et al. EUROGIN 2014 roadmap: differences in human papillomavirus infection natural history, transmission and human papillomavirus-related cancer incidence by gender and anatomic site of infection. Int J Cancer. 2015;136(12):2752–60.
17. Chaturvedi AK, Engels EA, Anderson WF, Gillison ML. Incidence trends for human papillomavirus-related and -unrelated oral squamous cell carcinomas in the United States. J Clin Oncol. 2008;26(4):612–9.
18. Gillison ML, D'Souza G, Westra W, Sugar E, Xiao W, Begum S, et al. Distinct risk factor profiles for human papillomavirus type 16-positive and human papillomavirus type 16-negative head and neck cancers. J Natl Cancer Inst. 2008;100(6):407–20.

19. Scheffner M, Werness BA, Huibregtse JM, Levine AJ, Howley PM. The E6 oncoprotein encoded by human papillomavirus types 16 and 18 promotes the degradation of p53. Cell. 1990;63(6):1129–36.
20. Heck DV, Yee CL, Howley PM, Münger K. Efficiency of binding the retinoblastoma protein correlates with the transforming capacity of the E7 oncoproteins of the human papillomaviruses. Proc Natl Acad Sci U S A. 1992;89(10):4442–6.
21. Saraiya M, Unger ER, Thompson TD, Lynch CF, Hernandez BY, Lyu CW, et al. US assessment of HPV types in cancers: implications for current and 9-valent HPV vaccines. J Natl Cancer Inst. 2015;107(6):djv086.
22. Münger K, Phelps WC, Bubb V, Howley PM, Schlegel R. The E6 and E7 genes of the human papillomavirus type 16 together are necessary and sufficient for transformation of primary human keratinocytes. J Virol. 1989;63(10):4417–21.
23. Nguyen N, Bellile E, Thomas D, McHugh J, Rozek L, Virani S, et al. Tumor infiltrating lymphocytes and survival in patients with head and neck squamous cell carcinoma. Head Neck. 2016;38(7):1074–84.
24. Fakhry C, Zhang Q, Nguyen-Tan PF, Rosenthal D, El-Naggar A, Garden AS, et al. Human papillomavirus and overall survival after progression of oropharyngeal squamous cell carcinoma. J Clin Oncol. 2014;32(30):3365–73.
25. Ang KK, Harris J, Wheeler R, Weber R, Rosenthal DI, Nguyen-Tân PF, et al. Human papillomavirus and survival of patients with oropharyngeal cancer. N Engl J Med. 2010;363(1):24–35.
26. Stenmark MH, Shumway D, Guo C, Vainshtein J, Mierzwa M, Jagsi R, et al. Influence of human papillomavirus on the clinical presentation of oropharyngeal carcinoma in the United States. Laryngoscope. 2017;127(10):2270–8.
27. Bonner JA, Harari PM, Giralt J, Cohen RB, Jones CU, Sur RK, et al. Radiotherapy plus cetuximab for locoregionally advanced head and neck cancer: 5-year survival data from a phase 3 randomised trial, and relation between cetuximab-induced rash and survival. Lancet Oncol. 2010;11(1):21–8.
28. Rosenthal DI, Harari PM, Giralt J, Bell D, Raben D, Liu J, et al. Association of human papillomavirus and p16 status with outcomes in the IMCL-9815 phase III registration trial for patients with locoregionally advanced oropharyngeal squamous cell carcinoma of the head and neck treated with radiotherapy with or without cetuximab. J Clin Oncol. 2016;34(12):1300–8.
29. Nguyen-Tan PF, Zhang Q, Ang KK, Weber RS, Rosenthal DI, Soulieres D, et al. Randomized phase III trial to test accelerated versus standard fractionation in combination with concurrent cisplatin for head and neck carcinomas in the Radiation Therapy Oncology Group 0129 trial: long-term report of efficacy and toxicity. J Clin Oncol. 2014;32(34):3858–66.
30. O'Sullivan B, Huang SH, Siu LL, Waldron J, Zhao H, Perez-Ordonez B, et al. Deintensification candidate subgroups in human papillomavirus-related oropharyngeal cancer according to minimal risk of distant metastasis. J Clin Oncol. 2013;31(5):543–50.
31. Lydiatt WM, Patel SG, O'Sullivan B, Brandwein MS, Ridge JA, Migliacci JC, et al. Head and neck cancers-major changes in the American Joint Committee on cancer eighth edition cancer staging manual. CA Cancer J Clin. 2017;67(2):122–37.
32. AJCC Cancer Staging Manual. 8 ed. Springer International Publishing; 2017.
33. Kreimer AR, Clifford GM, Boyle P, Franceschi S. Human papillomavirus types in head and neck squamous cell carcinomas worldwide: a systematic review. Cancer Epidemiol Biomark Prev. 2005;14(2):467–75.
34. Jordan RC, Lingen MW, Perez-Ordonez B, He X, Pickard R, Koluder M, et al. Validation of methods for oropharyngeal cancer HPV status determination in US cooperative group trials. Am J Surg Pathol. 2012;36(7):945–54.
35. El-Naggar AK, Westra WH. p16 expression as a surrogate marker for HPV-related oropharyngeal carcinoma: a guide for interpretative relevance and consistency. Head Neck. 2012;34(4):459–61.
36. Schlecht NF, Brandwein-Gensler M, Nuovo GJ, Li M, Dunne A, Kawachi N, et al. A comparison of clinically utilized human papillomavirus detection methods in head and neck cancer. Mod Pathol. 2011;24(10):1295–305.

37. Löning T, Ikenberg H, Becker J, Gissmann L, Hoepfer I, zur Hausen H. Analysis of oral papillomas, leukoplakias, and invasive carcinomas for human papillomavirus type related DNA. J Invest Dermatol. 1985;84(5):417–20.
38. Boscolo-Rizzo P, Pawlita M, Holzinger D. From HPV-positive towards HPV-driven oropharyngeal squamous cell carcinomas. Cancer Treat Rev. 2016;42:24–9.
39. Klussmann JP, Gültekin E, Weissenborn SJ, Wieland U, Dries V, Dienes HP, et al. Expression of p16 protein identifies a distinct entity of tonsillar carcinomas associated with human papillomavirus. Am J Pathol. 2003;162(3):747–53.
40. Maxwell JH, Kumar B, Feng FY, Worden FP, Lee JS, Eisbruch A, et al. Tobacco use in human papillomavirus-positive advanced oropharynx cancer patients related to increased risk of distant metastases and tumor recurrence. Clin Cancer Res. 2010;16(4):1226–35.
41. Dahlstrom KR, Garden AS, William WN, Lim MY, Sturgis EM. Proposed staging system for patients with HPV-related oropharyngeal cancer based on nasopharyngeal cancer N categories. J Clin Oncol. 2016;34(16):1848–54.
42. Klozar J, Koslabova E, Kratochvil V, Salakova M, Tachezy R. Nodal status is not a prognostic factor in patients with HPV-positive oral/oropharyngeal tumors. J Surg Oncol. 2013;107(6):625–33.
43. Huang SH, Xu W, Waldron J, Siu L, Shen X, Tong L, et al. Refining American Joint Committee on Cancer/Union for International Cancer Control TNM stage and prognostic groups for human papillomavirus-related oropharyngeal carcinomas. J Clin Oncol. 2015;33(8):836–45.
44. Ward MJ, Mellows T, Harris S, Webb A, Patel NN, Cox HJ, et al. Staging and treatment of oropharyngeal cancer in the human papillomavirus era. Head Neck. 2015;37(7):1002–13.
45. Budach W, Bölke E, Kammers K, Gerber PA, Orth K, Gripp S, et al. Induction chemotherapy followed by concurrent radio-chemotherapy versus concurrent radio-chemotherapy alone as treatment of locally advanced squamous cell carcinoma of the head and neck (HNSCC): a meta-analysis of randomized trials. Radiother Oncol. 2016;118(2):238–43.
46. Reid SR, Losek JD. Imaging before appendectomy. Pediatrics. 2003;112(6 Pt 1):1461–2. author reply –2
47. Fakhry C, Westra WH, Li S, Cmelak A, Ridge JA, Pinto H, et al. Improved survival of patients with human papillomavirus-positive head and neck squamous cell carcinoma in a prospective clinical trial. J Natl Cancer Inst. 2008;100(4):261–9.
48. Cooperstein E, Gilbert J, Epstein JB, Dietrich MS, Bond SM, Ridner SH, et al. Vanderbilt Head and Neck Symptom Survey version 2.0: report of the development and initial testing of a subscale for assessment of oral health. Head Neck. 2012;34(6):797–804.
49. Marur S, Li S, Cmelak AJ, Gillison ML, Zhao WJ, Ferris RL, et al. E1308: phase II trial of induction chemotherapy followed by reduced-dose radiation and weekly cetuximab in patients with HPV-associated resectable squamous cell carcinoma of the oropharynx- ECOG-ACRIN Cancer Research Group. J Clin Oncol. 2017;35(5):490–7.
50. Garden AS, Blanchard P. ECOG-ACRIN 1308: commentary on a negative phase II trial. J Clin Oncol. 2017;35(17):1969–70.
51. Yom SS. Is induction chemotherapy needed to select patients for deintensified treatment of human papillomavirus-associated oropharyngeal Cancer? J Clin Oncol. 2017;35(5):479–81.
52. Marur S, Li S, Cmelak A, Burtness B. Reply to A. Garden et al. J Clin Oncol. 2017;35(17):1970–1.
53. Marur S, Cmelak AJ, Burtness B. Purpose of induction chemotherapy in E1308 and importance of patient-reported outcomes in deintensification trials. J Clin Oncol. 2017;35(17):1968–9.
54. NRG-HN002: A randomized phase II trial for patients with p16 positive, non-smoking associated, locoregionally advanced oropharyngeal cancer: NRG Oncology; Available from: www.nrgoncology.org/CLinical-Trials/NRG-HN002.
55. Chera BS, Amdur RJ, Tepper J, Qaqish B, Green R, Aumer SL, et al. Phase 2 trial of de-intensified chemoradiation therapy for favorable-risk human papillomavirus-associated oropharyngeal squamous cell carcinoma. Int J Radiat Oncol Biol Phys. 2015;93(5):976–85.
56. Haddad R, O'Neill A, Rabinowits G, Tishler R, Khuri F, Adkins D, et al. Induction chemotherapy followed by concurrent chemoradiotherapy (sequential chemoradiotherapy) versus

concurrent chemoradiotherapy alone in locally advanced head and neck cancer (PARADIGM): a randomised phase 3 trial. Lancet Oncol. 2013;14(3):257–64.
57. Yom SS, Gillison ML, Trotti AM. Dose de-escalation in human papillomavirus-associated oropharyngeal cancer: first tracks on powder. Int J Radiat Oncol Biol Phys. 2015;93(5):986–8.
58. Villaflor VM, Melotek JM, Karrison TG, Brisson RJ, Blair EA, Portugal L, et al. Response-adapted volume de-escalation (RAVD) in locally advanced head and neck cancer. Ann Oncol. 2016;27(5):908–13.
59. Riaz N, Sherman EJ, Katabi N, Leeman JE, Higginson DS, Boyle J, et al. A personalized approach using hypoxia resolution to guide curative-intent radiation dose-reduction to 30 Gy: A novel de-escalation paradigm for HPV-associated oropharynx cancers (OPC). J Clin Oncol. 2017;35(15_suppl):6076.
60. Melotek J, Seiwert TY, Blair EA, Karrison TG, Agrawal N, Portugal L, et al. Optima: A phase II dose and volume de-escalation trial for high- and low-risk HPV+ oropharynx cancers. J Clin Oncol. 2017;35(15_suppl):6066.
61. Blanchard P, Baujat B, Holostenco V, Bourredjem A, Baey C, Bourhis J, et al. Meta-analysis of chemotherapy in head and neck cancer (MACH-NC): a comprehensive analysis by tumour site. Radiother Oncol. 2011;100(1):33–40.
62. Guan J, Zhang Y, Li Q, Li L, Chen M, Xiao N, et al. A meta-analysis of weekly cisplatin versus three weekly cisplatin chemotherapy plus concurrent radiotherapy (CRT) for advanced head and neck cancer (HNC). Oncotarget. 2016;7(43):70185–93.
63. Robert F, Ezekiel MP, Spencer SA, Meredith RF, Bonner JA, Khazaeli MB, et al. Phase I study of anti-epidermal growth factor receptor antibody cetuximab in combination with radiation therapy in patients with advanced head and neck cancer. J Clin Oncol. 2001;19(13):3234–43.
64. Koutcher L, Sherman E, Fury M, Wolden S, Zhang Z, Mo Q, et al. Concurrent cisplatin and radiation versus cetuximab and radiation for locally advanced head-and-neck cancer. Int J Radiat Oncol Biol Phys. 2011;81(4):915–22.
65. Nolan N, Diavolitsis VM, Blakaj D, Pan X, Grecula JC, Savvides P, et al. Intensity modulated radiation therapy with cetuximab versus platinum chemotherapy in p16-positive oropharyngeal squamous cell carcinoma. Int J Radiat Oncol Biol Phys. 93(3):E333–E4.
66. Magrini SM, Buglione M, Corvò R, Pirtoli L, Paiar F, Ponticelli P, et al. Cetuximab and radiotherapy versus cisplatin and radiotherapy for locally advanced head and neck Cancer: a randomized phase II trial. J Clin Oncol. 2016;34(5):427–35.
67. Ringash J, Waldron JN, Siu LL, Martino R, Winquist E, Wright JR, et al. Quality of life and swallowing with standard chemoradiotherapy versus accelerated radiotherapy and panitumumab in locoregionally advanced carcinoma of the head and neck: a phase III randomised trial from the Canadian Cancer Trials Group (HN.6). Eur J Cancer. 2017;72:192–9.
68. Szturz P, Wouters K, Kiyota N, Tahara M, Prabhash K, Noronha V, et al. Weekly low-dose versus three-weekly high-dose cisplatin for concurrent chemoradiation in locoregionally advanced non-nasopharyngeal head and neck cancer: a systematic review and meta-analysis of aggregate data. Oncologist. 2017;22(9):1056–66.
69. Sher DJ, Adelstein DJ, Bajaj GK, Brizel DM, Cohen EEW, Halthore A, et al. Radiation therapy for oropharyngeal squamous cell carcinoma: executive summary of an ASTRO Evidence-Based Clinical Practice Guideline. Pract Radiat Oncol. 2017;7(4):246–53.
70. Weinstein GS, Quon H, Newman HJ, Chalian JA, Malloy K, Lin A, et al. Transoral robotic surgery alone for oropharyngeal cancer: an analysis of local control. Arch Otolaryngol Head Neck Surg. 2012;138(7):628–34.
71. de Almeida JR, Byrd JK, Wu R, Stucken CL, Duvvuri U, Goldstein DP, et al. A systematic review of transoral robotic surgery and radiotherapy for early oropharynx cancer: a systematic review. Laryngoscope. 2014;124(9):2096–102.
72. Seiwert TY, Melotek JM, Blair EA, Stenson KM, Salama JK, Witt ME, et al. Final results of a randomized phase 2 trial investigating the addition of cetuximab to induction chemotherapy and accelerated or hyperfractionated chemoradiation for locoregionally advanced head and neck cancer. Int J Radiat Oncol Biol Phys. 2016;96(1):21–9.

73. Hanahan D, Weinberg RA. Hallmarks of cancer: the next generation. Cell. 2011;144(5):646–74.
74. Thomas ED, Lochte HL, Lu WC, Ferrebee JW. Intravenous infusion of bone marrow in patients receiving radiation and chemotherapy. N Engl J Med. 1957;257(11):491–6.
75. Badoual C, Hans S, Merillon N, Van Ryswick C, Ravel P, Benhamouda N, et al. PD-1-expressing tumor-infiltrating T cells are a favorable prognostic biomarker in HPV-associated head and neck cancer. Cancer Res. 2013;73(1):128–38.
76. Jie HB, Schuler PJ, Lee SC, Srivastava RM, Argiris A, Ferrone S, et al. CTLA-4^{+} regulatory T cells increased in cetuximab-treated head and neck cancer patients suppress NK cell cytotoxicity and correlate with poor prognosis. Cancer Res. 2015;75(11):2200–10.
77. Balermpas P, Rödel F, Rödel C, Krause M, Linge A, Lohaus F, et al. CD8+ tumour-infiltrating lymphocytes in relation to HPV status and clinical outcome in patients with head and neck cancer after postoperative chemoradiotherapy: a multicentre study of the German cancer consortium radiation oncology group (DKTK-ROG). Int J Cancer. 2016;138(1):171–81.
78. Seiwert TY, Burtness B, Mehra R, Weiss J, Berger R, Eder JP, et al. Safety and clinical activity of pembrolizumab for treatment of recurrent or metastatic squamous cell carcinoma of the head and neck (KEYNOTE-012): an open-label, multicentre, phase 1b trial. Lancet Oncol. 2016;17(7):956–65.
79. Ferris RL, Blumenschein G, Fayette J, Guigay J, Colevas AD, Licitra L, et al. Nivolumab for recurrent squamous-cell carcinoma of the head and neck. N Engl J Med. 2016;375(19):1856–67.
80. Trosman SJ, Koyfman SA, Ward MC, Al-Khudari S, Nwizu T, Greskovich JF, et al. Effect of human papillomavirus on patterns of distant metastatic failure in oropharyngeal squamous cell carcinoma treated with chemoradiotherapy. JAMA Otolaryngol Head Neck Surg. 2015;141(5):457–62.
81. Argiris A, Li S, Ghebremichael M, Egloff AM, Wang L, Forastiere AA, et al. Prognostic significance of human papillomavirus in recurrent or metastatic head and neck cancer: an analysis of Eastern Cooperative Oncology Group trials. Ann Oncol. 2014;25(7):1410–6.
82. Frakes JM, Naghavi AO, Demetriou SK, Strom TJ, Russell JS, Kish JA, et al. Determining optimal follow-up in the management of human papillomavirus-positive oropharyngeal cancer. Cancer. 2016;122(4):634–41.
83. Mehanna H, Wong WL, McConkey CC, Rahman JK, Robinson M, Hartley AG, et al. PET-CT surveillance versus neck dissection in advanced head and neck cancer. N Engl J Med. 2016;374(15):1444–54.
84. Rettig EM, Wentz A, Posner MR, Gross ND, Haddad RI, Gillison ML, et al. Prognostic implication of persistent human papillomavirus type 16 DNA detection in oral rinses for human papillomavirus-related oropharyngeal carcinoma. JAMA Oncol. 2015;1(7):907–15.
85. Argiris A, Li Y, Forastiere A. Prognostic factors and long-term survivorship in patients with recurrent or metastatic carcinoma of the head and neck. Cancer. 2004;101(10):2222–9.
86. Amin MB, Edge SB, Greene FL, et al., editors. AJCC cancer staging manual. 8th ed. New York: Springer; 2017.
87. Stein AP, et al. Prevalence of human papillomavirus in oropharyngeal squamous cell carcinoma in the United States across time. Chem Res Toxicol. 2014;27(4):462–9.

Index

B. Burtness, E. A. Golemis (eds.), *Molecular Determinants of Head and Neck Cancer*, Current Cancer Research, https://doi.org/10.1007/978-3-319-78762-6

D

I

Printed in the United States
By Bookmasters